Graduate Texts in Mathematics 129

Readings in Mathematics

SPRINGER STUDY EDITION

Graduate Texts in Mathematics
Readings in Mathematics

Ebbinghaus/Hermes/Hirzebruch/Koecher/Mainzer/Neukirch/Prestel/Remmert: *Numbers*
Fulton/Harris: *Representation Theory: A First Course*
Murty: *Problems in Analytic Number Theory*
Remmert: *Theory of Complex Functions*
Walter: *Ordinary Differential Equations*

Undergraduate Texts in Mathematics
Readings in Mathematics

Anglin: *Mathematics: A Concise History and Philosophy*
Anglin/Lambek: *The Heritage of Thales*
Bressoud: *Second Year Calculus*
Hairer/Wanner: *Analysis by Its History*
Hämmerlin/Hoffmann: *Numerical Mathematics*
Isaac: *The Pleasures of Probability*
Laubenbacher/Pengelley: *Mathematical Expeditions: Chronicles by the Explorers*
Samuel: *Projective Geometry*
Stillwell: *Numbers and Geometry*
Toth: *Glimpses of Algebra and Geometry*

William Fulton Joe Harris

Representation Theory

A First Course

With 144 Illustrations

Springer

William Fulton
Department of Mathematics
University of Michigan
Ann Arbor, MI 48109
USA
wfulton@math.isa.umich.edu

Joe Harris
Department of Mathematics
Harvard University
Cambridge, MA 02138
USA
harris@abel.math.harvard.edu

Fulton and Harris: Representation Theory: A First Course

First Indian Reprint 2004
Third Indian Reprint 2009

ISBN 978-81-8128-134-0

This edition is published by Springer (India) Private Limited, (a part of Springer Science+Business Media, Registered Office: 3rd Floor, Gandharva Mahavidyalaya, 212, Deen Dayal Upadhyaya Marg, New Delhi – 110 002, India.

Printed in India by Rajkamal Electric Press, Plot No. 2, Phase-IV, HSIDC, Kundli, Haryana.

Preface

The primary goal of these lectures is to introduce a beginner to the finite-dimensional representations of Lie groups and Lie algebras. Since this goal is shared by quite a few other books, we should explain in this Preface how our approach differs, although the potential reader can probably see this better by a quick browse through the book.

Representation theory is simple to define: it is the study of the ways in which a given group may act on vector spaces. It is almost certainly unique, however, among such clearly delineated subjects, in the breadth of its interest to mathematicians. This is not surprising: group actions are ubiquitous in 20th century mathematics, and where the object on which a group acts is not a vector space, we have learned to replace it by one that is (e.g., a cohomology group, tangent space, etc.). As a consequence, many mathematicians other than specialists in the field (or even those who think they might want to be) come in contact with the subject in various ways. It is for such people that this text is designed. To put it another way, we intend this as a book for beginners to learn from and not as a reference.

This idea essentially determines the choice of material covered here. As simple as is the definition of representation theory given above, it fragments considerably when we try to get more specific. For a start, what kind of group G are we dealing with—a finite group like the symmetric group $\mathfrak{S}_n$ or the general linear group over a finite field $GL_n(\mathbb{F}_q)$, an infinite discrete group like $SL_n(\mathbb{Z})$, a Lie group like $SL_n\mathbb{C}$, or possibly a Lie group over a local field? Needless to say, each of these settings requires a substantially different approach to its representation theory. Likewise, what sort of vector space is G acting on: is it over $\mathbb{C}$, $\mathbb{R}$, $\mathbb{Q}$, or possibly a field of positive characteristic? Is it finite dimensional or infinite dimensional, and if the latter, what additional structure (such as norm, or inner product) does it carry? Various combinations of answers to these questions lead to areas of intense research activity in representation theory, and it is natural for a text intended to prepare students for a career in the subject to lead up to one or more of these areas. As a corollary, such a book tends to get through the elementary material as quickly as possible: if one has a semester to get up to and through Harish–Chandra modules, there is little time to dawdle over the representations of $\mathfrak{S}_4$ and $SL_3\mathbb{C}$.

By contrast, the present book focuses exactly on the simplest cases: representations of finite groups and Lie groups on finite-dimensional real and complex vector spaces. This is in some sense the common ground of the subject, the area that is the object of most of the interest in representation theory coming from outside.

The intent of this book to serve nonspecialists likewise dictates to some degree our approach to the material we do cover. Probably the main feature of our presentation is that we concentrate on examples, developing the general theory sparingly, and then mainly as a useful and unifying language to describe phenomena already encountered in concrete cases. By the same token, we for the most part introduce theoretical notions when and where they are useful for analyzing concrete situations, postponing as long as possible those notions that are used mainly for proving general theorems.

Finally, our goal of making the book accessible to outsiders accounts in part for the style of the writing. These lectures have grown from courses of the second author in 1984 and 1987, and we have attempted to keep the informal style of these lectures. Thus there is almost no attempt at efficiency: where it seems to make sense from a didactic point of view, we work out many special cases of an idea by hand before proving the general case; and we cheerfully give several proofs of one fact if we think they are illuminating. Similarly, while it is common to develop the whole semisimple story from one point of view, say that of compact groups, or Lie algebras, or algebraic groups, we have avoided this, as efficient as it may be.

It is of course not a strikingly original notion that beginners can best learn about a subject by working through examples, with general machinery only introduced slowly and as the need arises, but it seems particularly appropriate here. In most subjects such an approach means one has a few out of an unknown infinity of examples which are useful to illuminate the general situation. When the subject is the representation theory of complex semisimple Lie groups and algebras, however, something special happens: once one has worked through all the examples readily at hand—the "classical" cases of the special linear, orthogonal, and symplectic groups—one has not just a few useful examples, one has all but five "exceptional" cases.

This is essentially what we do here. We start with a quick tour through representation theory of finite groups, with emphasis determined by what is useful for Lie groups. In this regard, we include more on the symmetric groups than is usual. Then we turn to Lie groups and Lie algebras. After some preliminaries and a look at low-dimensional examples, and one lecture with some general notions about semisimplicity, we get to the heart of the course: working out the finite-dimensional representations of the classical groups.

For each series of classical Lie algebras we prove the fundamental existence theorem for representations of given highest weight by explicit construction. Our object, however, is not just existence, but to see the representations in action, to see geometric implications of decompositions of naturally occurring representations, and to see the relations among them caused by coincidences between the Lie algebras.

The goal of the last six lectures is to make a bridge between the example-oriented approach of the earlier parts and the general theory. Here we make an attempt to interpret what has gone before in abstract terms, trying to make connections with modern terminology. We develop the general theory enough to see that we have studied all the simple complex Lie algebras with five exceptions. Since these are encountered less frequently than the classical series, it is probably not reasonable in a first course to work out their representations as explicitly, although we do carry this out for one of them. We also prove the general Weyl character formula, which can be used to verify and extend many of the results we worked out by hand earlier in the book.

Of course, the point we reach hardly touches the current state of affairs in Lie theory, but we hope it is enough to keep the reader's eyes from glazing over when confronted with a lecture that begins: "Let G be a semisimple Lie group, P a parabolic subgroup, ..." We might also hope that working through this book would prepare some readers to appreciate the elegance (and efficiency) of the abstract approach.

In spirit this book is probably closer to Weyl's classic [We1] than to others written today. Indeed, a secondary goal of our book is to present many of the results of Weyl and his predecessors in a form more accessible to modern readers. In particular, we include Weyl's constructions of the representations of the general and special linear groups by using Young's symmetrizers; and we invoke a little invariant theory to do the corresponding result for the orthogonal and symplectic groups. We also include Weyl's formulas for the characters of these representations in terms of the elementary characters of symmetric powers of the standard representations. (Interestingly, Weyl only gave the corresponding formulas in terms of the exterior powers for the general linear group. The corresponding formulas for the orthogonal and symplectic groups were only given recently by D'Hoker, and by Koike and Terada. We include a simple new proof of these determinantal formulas.)

More about individual sections can be found in the introductions to other parts of the book.

Needless to say, a price is paid for the inefficiency and restricted focus of these notes. The most obvious is a lot of omitted material: for example, we include little on the basic topological, differentiable, or analytic properties of Lie groups, as this plays a small role in our story and is well covered in dozens of other sources, including many graduate texts on manifolds. Moreover, there are no infinite-dimensional representations, no Harish–Chandra or Verma modules, no Stiefel diagrams, no Lie algebra cohomology, no analysis on symmetric spaces or groups, no arithmetic groups or automorphic forms, and nothing about representations in characteristic $p > 0$. There is no consistent attempt to indicate which of our results on Lie groups apply more generally to algebraic groups over fields other than $\mathbb{R}$ or $\mathbb{C}$ (e.g., local fields). And there is only passing mention of other standard topics, such as universal enveloping algebras or Bruhat decompositions, which have become standard tools of representation theory. (Experts who saw drafts of this book agreed that some topic we omitted must not be left out of a modern book on representation theory—but no two experts suggested the same topic.)

We have not tried to trace the history of the subjects treated, or assign credit, or to attribute ideas to original sources—this is far beyond our knowledge. When we give references, we have simply tried to send the reader to sources that are as readable as possible for one knowing what is written here. A good systematic reference for the finite-group material, including proofs of the results we leave out, is Serre [Se2]. For Lie groups and Lie algebras, Serre [Se3], Adams [Ad], Humphreys [Hu1], and Bourbaki [Bour] are recommended references, as are the classics Weyl [We1] and Littlewood [Lit1].

We would like to thank the many people who have contributed ideas and suggestions for this manuscript, among them J-F. Burnol, R. Bryant, J. Carrell, B. Conrad, P. Diaconis, D. Eisenbud, D. Goldstein, M. Green, P. Griffiths, B. Gross, M. Hildebrand, R. Howe, H. Kraft, A. Landman, B. Mazur, N. Chriss, D. Petersen, G. Schwartz, J. Towber, and L. Tu. In particular, we would like to thank David Mumford, from whom we learned much of what we know about the subject, and whose ideas are very much in evidence in this book.

Had this book been written 10 years ago, we would at this point thank the people who typed it. That being no longer applicable, perhaps we should thank instead the National Science Foundation, the University of Chicago, and Harvard University for generously providing the various Macintoshes on which this manuscript was produced. Finally, we thank Chan Fulton for making the drawings.

Bill Fulton and Joe Harris

Note to the corrected fifth printing: We are grateful to S. Billey, M. Brion, R. Coleman, B. Gross, E. D'Hoker, D. Jaffe, R. Milson, K. Rumelhart, M. Reeder, and J. Willenbring for pointing out errors in earlier printings, and to many others for telling us about misprints.

Using This Book

A few words are in order about the practical use of this book. To begin with, prerequisites are minimal: we assume only a basic knowledge of standard first-year graduate material in algebra and topology, including basic notions about manifolds. A good undergraduate background should be more than enough for most of the text; some examples and exercises, and some of the discussion in Part IV may refer to more advanced topics, but these can readily be skipped. Probably the main practical requirement is a good working knowledge of multilinear algebra, including tensor, exterior, and symmetric products of finite dimensional vector spaces, for which Appendix B may help. We have indicated, in introductory remarks to each lecture, when any background beyond this is assumed and how essential it is.

For a course, this book could be used in two ways. First, there are a number of topics that are not logically essential to the rest of the book and that can be skimmed or skipped entirely. For example, in a minimal reading one could skip §§4, 5, 6, 11.3, 13.4, 15.3–15.5, 17.3, 19.5, 20, 22.1, 22.3, 23.3–23.4, 25.3, and 26.2; this might be suitable for a basic one-semester course. On the other hand, in a year-long course it should be possible to work through as much of the material as background and/or interest suggested. Most of the material in the Appendices is relevant only to such a long course. Again, we have tried to indicate, in the introductory remarks in each lecture, which topics are inessential and may be omitted.

Another aspect of the book that readers may want to approach in different ways is the profusion of examples. These are put in largely for didactic reasons: we feel that this is the sort of material that can best be understood by gaining some direct hands-on experience with the objects involved. For the most part, however, they do not actually develop new ideas; the reader whose tastes run more to the abstract and general than the concrete and special may skip many of them without logical consequence. (Of course, such a reader will probably wind up burning this book anyway.)

We include hundreds of exercises, of wildly different purposes and difficulties. Some are the usual sorts of variations of the examples in the text or are straightforward verifications of facts needed; a student will probably want to attempt most of these. Sometimes an exercise is inserted whose solution is a special case of something we do in the text later, if we think working on it will be useful motivation (again, there is no attempt at "efficiency," and readers are encouraged to go back to old exercises from time to time). Many exercises are included that indicate some further directions or new topics (or standard topics we have omitted); a beginner may best be advised to skim these for general information, perhaps working out a few simple cases. In exercises, we tried to include topics that may be hard for nonexperts to extract from the literature, especially the older literature. In general, much of the theory is in the exercises—and most of the examples in the text.

We have resisted the idea of grading the exercises by (expected) difficulty, although a "problem" is probably harder than an "exercise." Many exercises are starred: the * is not an indication of difficulty, but means that the reader can find some information about it in the section "Hints, Answers, and References" at the back of the book. This may be a hint, a statement of the answer, a complete solution, a reference to where more can be found, or a combination of any of these. We hope these miscellaneous remarks, as haphazard and uneven as they are, will be of some use.

Contents

PART I
FINITE GROUPS

Given that over three-quarters of this book is devoted to the representation theory of Lie groups and Lie algebras, why have a discussion of the representations of finite groups at all? There are certainly valid reasons from a logical point of view: many of the ideas, concepts, and constructions we will introduce here will be applied in the study of Lie groups and algebras. The real reason for us, however, is didactic, as we will now try to explain.

Representation theory is very much a 20th-century subject, in the following sense. In the 19th century, when groups were dealt with they were generally understood to be subsets of the permutations of a set, or of the automorphisms $GL(V)$ of a vector space V, closed under composition and inverse. Only in the 20th century was the notion of an abstract group given, making it possible to make a distinction between properties of the abstract group and properties of the particular realization as a subgroup of a permutation group or $GL(V)$. To give an analogy, in the 19th century a manifold was always a subset of $\mathbb{R}^n$; only in the 20th century did the notion of an abstract Riemannian manifold become common.

In both cases, the introduction of the abstract object made a fundamental difference to the subject. In differential geometry, one could make a crucial distinction between the intrinsic and extrinsic geometry of the manifold: which properties were invariants of the metric on the manifold and which were properties of the particular embedding in $\mathbb{R}^n$. Questions of existence or non-existence, for example, could be broken up into two parts: did the abstract manifold exist, and could it be embedded. Similarly, what would have been called in the 19th century simply "group theory" is now factored into two parts. First, there is the study of the structure of abstract groups (e.g., the classification of simple groups). Second is the companion question: given a group G, how can we describe all the ways in which G may be embedded in (or mapped to) a linear group GL(V)?. This, of course, is the subject matter of representation theory.

Given this point of view, it makes sense when first introducing representation theory to do so in a context where the nature of the groups G in question is itself simple, and relatively well understood. It is largely for this reason that we are starting off with the representation theory of finite groups: for those readers who are not already familiar with the motivations and goals of representation theory, it seemed better to establish those first in a setting where the structure of the groups was not itself an issue. When we analyze, for example, the representations of the symmetric and alternating groups on 3, 4, and 5 letters, it can be expected that the reader is already familiar with the groups and can focus on the basic concepts of representation theory being introduced.

We will spend the first six lectures on the case of finite groups. Many of the techniques developed for finite groups will carry over to Lie groups; indeed, our choice of topics is in part guided by this. For example, we spend quite a bit of time on the symmetric group; this is partly for its own interest, but also partly because what we learn here gives one way to study representations of the general linear group and its subgroups. There are other topics, such as the alternating group $\mathfrak{A}_d$, and the groups $SL_2(\mathbb{F}_q)$ and $GL_2(\mathbb{F}_q)$ that are studied purely for their own interest and do not appear later. (In general, for those readers primarily concerned with Lie theory, we have tried to indicate in the introductory notes to each lecture which ideas will be useful in the succeeding parts of this book.) Nonetheless, this is by no means a comprehensive treatment of the representation theory of finite groups; many important topics, such as the Artin and Brauer theorems and the whole subject of modular representations, are omitted.

LECTURE 1

Representations of Finite Groups

In this lecture we give the basic definitions of representation theory, and prove two of the basic results, showing that every representation is a (unique) direct sum of irreducible ones. We work out as examples the case of abelian groups, and the simplest nonabelian group, the symmetric group on 3 letters. In the latter case we give an analysis that will turn out not to be useful for the study of finite groups, but whose main idea is central to the study of the representations of Lie groups.

§1.1: Definitions
§1.2: Complete reducibility; Schur's lemma
§1.3: Examples: Abelian groups; $\mathfrak{S}_3$

§1.1. Definitions

A *representation* of a finite group G on a finite-dimensional complex vector space V is a homomorphism $\rho: G \to \mathrm{GL}(V)$ of G to the group of automorphisms of V; we say that such a map *gives V the structure of a G-module*. When there is little ambiguity about the map ρ (and, we're afraid, even sometimes when there is) we sometimes call V itself a representation of G; in this vein we will often suppress the symbol ρ and write $g \cdot v$ or gv for $\rho(g)(v)$. The dimension of V is sometimes called the *degree* of ρ.

A *map* φ between two representations V and W of G is a vector space map $\varphi: V \to W$ such that

$$\begin{array}{ccc} V & \xrightarrow{\ \varphi\ } & W \\ {\scriptstyle g}\downarrow & & \downarrow{\scriptstyle g} \\ V & \xrightarrow{\ \varphi\ } & W \end{array}$$

commutes for every $g \in G$. (We will call this a *G-linear map* when we want to distinguish it from an arbitrary linear map between the vector spaces V and W.) We can then define Ker φ, Im φ, and Coker φ, which are also G-modules.

A *subrepresentation* of a representation V is a vector subspace W of V which is invariant under G. A representation V is called *irreducible* if there is no proper nonzero invariant subspace W of V.

If V and W are representations, the *direct sum* $V \oplus W$ and the *tensor product* $V \otimes W$ are also representations, the latter via

$$g(v \otimes w) = gv \otimes gw.$$

For a representation V, the nth tensor power $V^{\otimes n}$ is again a representation of G by this rule, and the *exterior powers* $\Lambda^n(V)$ and *symmetric powers* $\mathrm{Sym}^n(V)$ are subrepresentations[1] of it. The *dual* $V^* = \mathrm{Hom}(V, \mathbb{C})$ of V is also a representation, though not in the most obvious way: we want the two representations of G to respect the natural pairing (denoted $\langle\ ,\ \rangle$) between V^* and V, so that if $\rho: G \to \mathrm{GL}(V)$ is a representation and $\rho^*: G \to \mathrm{GL}(V^*)$ is the dual, we should have

$$\langle \rho^*(g)(v^*), \rho(g)(v) \rangle = \langle v^*, v \rangle$$

for all $g \in G$, $v \in V$, and $v^* \in V^*$. This in turn forces us to define the dual representation by

$$\rho^*(g) = {}^t\rho(g^{-1}): V^* \to V^*$$

for all $g \in G$.

Exercise 1.1. Verify that with this definition of ρ^*, the relation above is satisfied.

Having defined the dual of a representation and the tensor product of two representations, it is likewise the case that if V and W are representations, then $\mathrm{Hom}(V, W)$ is also a representation, via the identification $\mathrm{Hom}(V, W) = V^* \otimes W$. Unraveling this, if we view an element of $\mathrm{Hom}(V, W)$ as a linear map φ from V to W, we have

$$(g\varphi)(v) = g\varphi(g^{-1}v)$$

for all $v \in V$. In other words, the definition is such that the diagram

$$\begin{array}{ccc} V & \xrightarrow{\ \varphi\ } & W \\ \downarrow{\scriptstyle g} & & \downarrow{\scriptstyle g} \\ V & \xrightarrow{\ g\varphi\ } & W \end{array}$$

commutes. Note that the dual representation is, in turn, a special case of this:

[1] For more on exterior and symmetric powers, including descriptions as quotient spaces of tensor powers, see Appendix B.

when $W = \mathbb{C}$ is the *trivial* representation, i.e., $gw = w$ for all $w \in \mathbb{C}$, this makes V^* into a G-module, with $g\varphi(v) = \varphi(g^{-1}v)$, i.e., $g\varphi = {}^t(g^{-1})\varphi$.

Exercise 1.2. Verify that in general the vector space of G-linear maps between two representations V and W of G is just the subspace $\text{Hom}(V, W)^G$ of elements of $\text{Hom}(V, W)$ fixed under the action of G. This subspace is often denoted $\text{Hom}_G(V, W)$.

We have, in effect, taken the identification $\text{Hom}(V, W) = V^* \otimes W$ as the definition of the representation $\text{Hom}(V, W)$. More generally, the usual identities for vector spaces are also true for representations, e.g.,

$$V \otimes (U \oplus W) = (V \otimes U) \oplus (V \otimes W),$$

$$\Lambda^k(V \oplus W) = \bigoplus_{a+b=k} \Lambda^a V \otimes \Lambda^b W,$$

$$\Lambda^k(V^*) = \Lambda^k(V)^*,$$

and so on.

Exercise 1.3*. Let $\rho: G \to \text{GL}(V)$ be any representation of the finite group G on an n-dimensional vector space V and suppose that for any $g \in G$, the determinant of $\rho(g)$ is 1. Show that the spaces $\Lambda^k V$ and $\Lambda^{n-k} V^*$ are isomorphic as representations of G.

If X is any finite set and G acts on the left on X, i.e., $G \to \text{Aut}(X)$ is a homomorphism to the permutation group of X, there is an associated *permutation representation*: let V be the vector space with basis $\{e_x : x \in X\}$, and let G act on V by

$$g \cdot \sum a_x e_x = \sum a_x e_{gx}.$$

The *regular representation*, denoted R_G or R, corresponds to the left action of G on itself. Alternatively, R is the space of complex-valued functions on G, where an element $g \in G$ acts on a function α by $(g\alpha)(h) = \alpha(g^{-1}h)$.

Exercise 1.4*. (a) Verify that these two descriptions of R agree, by identifying the element e_x with the characteristic function which takes the value 1 on x, 0 on other elements of G.

(b) The space of functions on G can also be made into a G-module by the rule $(g\alpha)(h) = \alpha(hg)$. Show that this is an isomorphic representation.

§1.2. Complete Reducibility; Schur's Lemma

As in any study, before we begin our attempt to classify the representations of a finite group G in earnest we should try to simplify life by restricting our search somewhat. Specifically, we have seen that representations of G can be built up out of other representations by linear algebraic operations, most simply by taking the direct sum. We should focus, then, on representations that are "atomic" with respect to this operation, i.e., that cannot be expressed as a direct sum of others; the usual term for such a representation is *indecomposable*. Happily, the situation is as nice as it could possibly be: a representation is atomic in this sense if and only if it is irreducible (i.e., contains no proper subrepresentations); and every representation is the direct sum of irreducibles, in a suitable sense uniquely so. The key to all this is

Proposition 1.5. *If W is a subrepresentation of a representation V of a finite group G, then there is a complementary invariant subspace W' of V, so that $V = W \oplus W'$.*

PROOF. There are two ways of doing this. One can introduce a (positive definite) Hermitian inner product H on V which is preserved by each $g \in G$ (i.e., such that $H(gv, gw) = H(v, w)$ for all $v, w \in V$ and $g \in G$). Indeed, if H_0 is any Hermitian product on V, one gets such an H by averaging over G:

$$H(v, w) = \sum_{g \in G} H_0(gv, gw).$$

Then the perpendicular subspace $W^\perp$ is complementary to W in V. Alternatively (but similarly), we can simply choose an arbitrary subspace U complementary to W, let $\pi_0 \colon V \to W$ be the projection given by the direct sum decomposition $V = W \oplus U$, and average the map π_0 over G: that is, take

$$\pi(v) = \sum_{g \in G} g(\pi_0(g^{-1}v)).$$

This will then be a G-linear map from V onto W, which is multiplication by $|G|$ on W; its kernel will, therefore, be a subspace of V invariant under G and complementary to W. □

Corollary 1.6. *Any representation is a direct sum of irreducible representations.*

This property is called *complete reducibility*, or *semisimplicity*. We will see that, for continuous representations, the circle S^1, or any compact group, has this property; integration over the group (with respect to an invariant measure on the group) plays the role of averaging in the above proof. The (additive) group $\mathbb{R}$ does not have this property: the representation

$$a \mapsto \begin{pmatrix} 1 & a \\ 0 & 1 \end{pmatrix}$$

leaves the x axis fixed, but there is no complementary subspace. We will see other Lie groups such as $\mathrm{SL}_n(\mathbb{C})$ that are semisimple in this sense. Note also that this argument would fail if the vector space V was over a field of finite characteristic since it might then be the case that $\pi(v) = 0$ for $v \in W$. The failure of complete reducibility is one of the things that makes the subject of *modular representations*, or representations on vector spaces over finite fields, so tricky.

The extent to which the decomposition of an arbitrary representation into a direct sum of irreducible ones is unique is one of the consequences of the following:

Schur's Lemma 1.7. *If V and W are irreducible representations of G and $\varphi: V \to W$ is a G-module homomorphism, then*

(1) *Either φ is an isomorphism, or $\varphi = 0$.*
(2) *If $V = W$, then $\varphi = \lambda \cdot I$ for some $\lambda \in \mathbb{C}$, I the identity.*

PROOF. The first claim follows from the fact that $\operatorname{Ker} \varphi$ and $\operatorname{Im} \varphi$ are invariant subspaces. For the second, since $\mathbb{C}$ is algebraically closed, φ must have an eigenvalue λ, i.e., for some $\lambda \in \mathbb{C}$, $\varphi - \lambda I$ has a nonzero kernel. By (1), then, we must have $\varphi - \lambda I = 0$, so $\varphi = \lambda I$. □

We can summarize what we have shown so far in

Proposition 1.8. *For any representation V of a finite group G, there is a decomposition*

$$V = V_1^{\oplus a_1} \oplus \cdots \oplus V_k^{\oplus a_k},$$

where the V_i are distinct irreducible representations. The decomposition of V into a direct sum of the k factors is unique, as are the V_i that occur and their multiplicities a_i.

PROOF. It follows from Schur's lemma that if W is another representation of G, with a decomposition $W = \bigoplus W_j^{\oplus b_j}$, and $\varphi: V \to W$ is a map of representations, then φ must map the factor $V_i^{\oplus a_i}$ into that factor $W_j^{\oplus b_j}$ for which $W_j \cong V_i$; when applied to the identity map of V to V, the stated uniqueness follows. □

In the next lecture we will give a formula for the projection of V onto $V_i^{\oplus a_i}$. The decomposition of the ith summand into a direct sum of a_i copies of V_i is not unique if $a_i > 1$, however.

Occasionally the decomposition is written

$$V = a_1 V_1 \oplus \cdots \oplus a_k V_k = a_1 V_1 + \cdots + a_k V_k, \tag{1.9}$$

especially when one is concerned only about the isomorphism classes and multiplicities of the V_i.

One more fact that will be established in the following lecture is that a finite group G admits only finitely many irreducible representations V_i up to isomorphism (in fact, we will say how many). This, then, is the framework of the classification of all representations of G: by the above, once we have described the irreducible representations of G, we will be able to describe an arbitrary representation as a linear combination of these. Our first goal, in analyzing the representations of any group, will therefore be:

(i) *Describe all the irreducible representations of* G.

Once we have done this, there remains the problem of carrying out in practice the description of a given representation in these terms. Thus, our second goal will be:

(ii) *Find techniques for giving the direct sum decomposition* (1.9), *and in particular determining the multiplicities* a_i *of an arbitrary representation* V.

Finally, it is the case that the representations we will most often be concerned with are those arising from simpler ones by the sort of linear- or multilinear-algebraic operations described above. We would like, therefore, to be able to describe, in the terms above, the representation we get when we perform these operations on a known representation. This is known generally as

(iii) *Plethysm*: *Describe the decompositions, with multiplicities, of representations derived from a given representation* V, such as $V \otimes V$, V^*, $\wedge^k(V)$, $\mathrm{Sym}^k(V)$, and $\wedge^k(\wedge^l V)$. Note that if V decomposes into a sum of two representations, these representations decompose accordingly; e.g., if $V = U \oplus W$, then

$$\wedge^k V = \bigoplus_{i+j=k} \wedge^i U \otimes \wedge^j W,$$

so it is enough to work out this plethysm for irreducible representations. Similarly, if V and W are two irreducible representations, we want to decompose $V \otimes W$; this is usually known as the *Clebsch–Gordan* problem.

§1.3. Examples: Abelian Groups; $\mathfrak{S}_3$

One obvious place to look for examples is with abelian groups. It does not take long, however, to deal with this case. Basically, we may observe in general that if V is a representation of the finite group G, abelian or not, each $g \in G$ gives a map $\rho(g)\colon V \to V$; but *this map is not generally a G-module homomorphism*: for general $h \in G$ we will have

$$g(h(v)) \neq h(g(v)).$$

Indeed, $\rho(g)\colon V \to V$ *will be G-linear for every* ρ *if* (*and only if*) g *is in the center* $Z(G)$ of G. In particular if G is abelian, and V is an irreducible representation, then by Schur's lemma every element $g \in G$ acts on V by a scalar multiple of the identity. Every subspace of V is thus invariant; so that V must be one dimensional. The irreducible representations of an abelian group G are thus simply elements of the dual group, that is, homomorphisms

$$\rho\colon G \to \mathbb{C}^*.$$

We consider next the simplest nonabelian group, $G = \mathfrak{S}_3$. To begin with, we have (as with any nontrivial symmetric group) two one-dimensional representations: we have the trivial representation, which we will denote U, and the *alternating representation* U', defined by setting

$$gv = \operatorname{sgn}(g)v$$

for $g \in G$, $v \in \mathbb{C}$. Next, since G comes to us as a permutation group, we have a natural permutation representation, in which G acts on $\mathbb{C}^3$ by permuting the coordinates. Explicitly, if $\{e_1, e_2, e_3\}$ is the standard basis, then $g \cdot e_i = e_{g(i)}$, or, equivalently,

$$g \cdot (z_1, z_2, z_3) = (z_{g^{-1}(1)}, z_{g^{-1}(2)}, z_{g^{-1}(3)}).$$

This representation, like any permutation representation, is not irreducible: the line spanned by the sum (1, 1, 1) of the basis vectors is invariant, with complementary subspace

$$V = \{(z_1, z_2, z_3) \in \mathbb{C}^3 : z_1 + z_2 + z_3 = 0\}.$$

This two-dimensional representation V is easily seen to be irreducible; we call it the *standard representation* of $\mathfrak{S}_3$.

Let us now turn to the problem of describing an arbitrary representation of $\mathfrak{S}_3$. We will see in the next lecture a wonderful tool for doing this, called *character theory*; but, as inefficient as this may be, we would like here to adopt a more ad hoc approach. This has some virtues as a didactic technique in the present context (admittedly dubious ones, consisting mainly of making the point that there are other and far worse ways of doing things than character theory). The real reason we are doing it is that it will serve to introduce an idea that, while superfluous for analyzing the representations of finite groups in general, will prove to be the key to understanding representations of Lie groups.

The idea is a very simple one: since we have just seen that the representation theory of a finite abelian group is virtually trivial, we will start our analysis of an arbitrary representation W of $\mathfrak{S}_3$ by looking just at the action of the abelian subgroup $\mathfrak{A}_3 = \mathbb{Z}/3 \subset \mathfrak{S}_3$ on W. This yields a very simple decomposition: if we take τ to be any generator of $\mathfrak{A}_3$ (that is, any three-cycle), the space W is spanned by eigenvectors v_i for the action of τ, whose eigenvalues are of course all powers of a cube root of unity $\omega = e^{2\pi i/3}$. Thus,

$$W = \bigoplus V_i,$$

where

$$V_i = \mathbb{C}v_i \quad \text{and} \quad \tau v_i = \omega^{\alpha_i} v_i.$$

Next, we ask how the remaining elements of $\mathfrak{S}_3$ act on W in terms of this decomposition. To see how this goes, let σ be any transposition, so that τ and σ together generate $\mathfrak{S}_3$, with the relation $\sigma\tau\sigma = \tau^2$. We want to know where σ sends an eigenvector v for the action of τ, say with eigenvalue ω^i; to answer this, we look at how τ acts on $\sigma(v)$. We use the basic relation above to write

$$\begin{aligned}\tau(\sigma(v)) &= \sigma(\tau^2(v))\\ &= \sigma(\omega^{2i}\cdot v)\\ &= \omega^{2i}\cdot\sigma(v).\end{aligned}$$

The conclusion, then, is that *if v is an eigenvector for τ with eigenvalue ω^i, then $\sigma(v)$ is again an eigenvector for τ, with eigenvalue ω^{2i}.*

Exercise 1.10. Verify that with $\sigma = (12)$, $\tau = (123)$, the standard representation has a basis $\alpha = (\omega, 1, \omega^2)$, $\beta = (1, \omega, \omega^2)$, with

$$\tau\alpha = \omega\alpha, \qquad \tau\beta = \omega^2\beta, \qquad \sigma\alpha = \beta, \qquad \sigma\beta = \alpha.$$

Suppose now that we start with such an eigenvector v for τ. If the eigenvalue of v is $\omega^i \neq 1$, then $\sigma(v)$ is an eigenvector with eigenvalue $\omega^{2i} \neq \omega^i$, and so is independent of v; and v and $\sigma(v)$ together span a two-dimensional subspace V' of W invariant under $\mathfrak{S}_3$. In fact, V' is isomorphic to the standard representation, which follows from Exercise 1.10. If, on the other hand, the eigenvalue of v is 1, then $\sigma(v)$ may or may not be independent of v. If it is not, then v spans a one-dimensional subrepresentation of W, isomorphic to the trivial representation if $\sigma(v) = v$ and to the alternating representation if $\sigma(v) = -v$. If $\sigma(v)$ and v are independent, then $v + \sigma(v)$ and $v - \sigma(v)$ span one-dimensional representations of W isomorphic to the trivial and alternating representations, respectively.

We have thus accomplished the first two of the goals we have set for ourselves above in the case of the group $G = \mathfrak{S}_3$. First, we see from the above that *the only three irreducible representations of $\mathfrak{S}_3$ are the trivial, alternating, and standard representations U, U' and V.* Moreover, for an arbitrary representation W of $\mathfrak{S}_3$ we can write

$$W = U^{\oplus a} \oplus U'^{\oplus b} \oplus V^{\oplus c};$$

and we have a way to determine the multiplicities a, b, and c: c, for example, is the number of independent eigenvectors for τ with eigenvalue ω, whereas $a + c$ is the multiplicity of 1 as an eigenvalue of σ, and $b + c$ is the multiplicity of -1 as an eigenvalue of σ.

In fact, this approach gives us as well the answer to our third problem, finding the decomposition of the symmetric, alternating, or tensor powers of a given representation W, since if we know the eigenvalues of τ on such a representation, we know the eigenvalues of τ on the various tensor powers of W. For example, we can use this method to decompose $V \otimes V$, where V is the standard two-dimensional representation. For $V \otimes V$ is spanned by the vectors $\alpha\otimes\alpha$, $\alpha\otimes\beta$, $\beta\otimes\alpha$, and $\beta\otimes\beta$; these are eigenvectors for τ with eigenvalues ω^2, 1, 1, and ω, respectively, and σ interchanges $\alpha\otimes\alpha$ with $\beta\otimes\beta$, and $\alpha\otimes\beta$ with $\beta\otimes\alpha$. Thus $\alpha\otimes\alpha$ and $\beta\otimes\beta$ span a subrepresentation isomorphic to V, $\alpha \otimes \beta + \beta \otimes \alpha$ spans a trivial representation U, and $\alpha \otimes \beta - \beta \otimes \alpha$ spans U', so

$$V \otimes V \cong U \oplus U' \oplus V.$$

Exercise 1.11. Use this approach to find the decomposition of the representations $\mathrm{Sym}^2 V$ and $\mathrm{Sym}^3 V$.

Exercise 1.12. (a) Decompose the regular representation R of $\mathfrak{S}_3$.

(b) Show that $\mathrm{Sym}^{k+6} V$ is isomorphic to $\mathrm{Sym}^k V \oplus R$, and compute $\mathrm{Sym}^k V$ for all k.

Exercise 1.13*. Show that $\mathrm{Sym}^2(\mathrm{Sym}^3 V) \cong \mathrm{Sym}^3(\mathrm{Sym}^2 V)$. Is $\mathrm{Sym}^m(\mathrm{Sym}^n V)$ isomorphic to $\mathrm{Sym}^n(\mathrm{Sym}^m V)$?

As we have indicated, the idea of studying a representation V of a group G by first restricting the action to an abelian subgroup, getting a decomposition of V into one-dimensional invariant subspaces, and then asking how the remaining generators of the group act on these subspaces, does not work well for finite G in general; for one thing, there will not in general be a convenient abelian subgroup to use. This idea will turn out, however, to be the key to understanding the representations of Lie groups, with a torus subgroup playing the role of the cyclic subgroup in this example.

Exercise 1.14*. Let V be an irreducible representation of the finite group G. Show that, up to scalars, there is a *unique* Hermitian inner product on V preserved by G.

LECTURE 2

Characters

This lecture contains the heart of our treatment of the representation theory of finite groups: the definition in §2.1 of the character of a representation, and the main theorem (proved in two steps in §2.2 and §2.4) that the characters of the irreducible representations form an orthonormal basis for the space of class functions on G. There will be more examples and more constructions in the following lectures, but this is what you need to know.

§2.1. Characters

As we indicated in the preceding section, there is a remarkably effective tool for understanding the representations of a finite group G, called *character theory*. This is in some ways motivated by the example worked out in the last section where we saw that a representation of $\mathfrak{S}_3$ was determined by knowing the eigenvalues of the action of the elements τ and $\sigma \in \mathfrak{S}_3$. For a general group G, it is not clear what subgroups and/or elements should play the role of $\mathfrak{A}_3$, τ, and σ; but the example certainly suggests that knowing all the eigenvalues of each element of G should suffice to describe the representation.

Of course, specifying all the eigenvalues of the action of each element of G is somewhat unwieldy; but fortunately it is redundant as well. For example, if we know the eigenvalues $\{\lambda_i\}$ of an element $g \in G$, then of course we know the eigenvalues $\{\lambda_i^k\}$ of g^k for each k as well. We can thus use this redundancy to simplify the data we have to specify. The key observation here is it is enough to give, for example, just the *sum* of the eigenvalues of each element of G, since knowing the sums $\sum \lambda_i^k$ of the kth powers of the eigenvalues of a given element $g \in G$ is equivalent to knowing the eigenvalues $\{\lambda_i\}$ of g themselves. This then suggests the following:

Definition. If V is a representation of G, its *character* χ_V is the complex-valued function on the group defined by

$$\chi_V(g) = \mathrm{Tr}(g|_V),$$

the trace of g on V.

In particular, we have

$$\chi_V(hgh^{-1}) = \chi_V(g),$$

so that χ_V is constant on the conjugacy classes of G; such a function is called a *class function.* Note that $\chi_V(1) = \dim V$.

Proposition 2.1. *Let V and W be representations of G. Then*

$$\chi_{V\oplus W} = \chi_V + \chi_W, \qquad \chi_{V\otimes W} = \chi_V \cdot \chi_W,$$

$$\chi_{V^*} = \overline{\chi}_V \quad \textit{and} \quad \chi_{\wedge^2 V}(g) = \tfrac{1}{2}[\chi_V(g)^2 - \chi_V(g^2)].$$

PROOF. We compute the values of these characters on a fixed element $g \in G$. For the action of g, V has eigenvalues $\{\lambda_i\}$ and W has eigenvalues $\{\mu_j\}$. Then $\{\lambda_i\} \cup \{\mu_j\}$ and $\{\lambda_i \cdot \mu_j\}$ are eigenvalues for $V \oplus W$ and $V \otimes W$, from which the first two formulas follow. Similarly $\{\lambda_i^{-1} = \overline{\lambda}_i\}$ are the eigenvalues for g on V^*, since all eigenvalues are nth roots of unity, with n the order of g. Finally, $\{\lambda_i\lambda_j | i < j\}$ are the eigenvalues for g on $\wedge^2 V$, and

$$\sum_{i<j} \lambda_i\lambda_j = \frac{(\sum \lambda_i)^2 - \sum \lambda_i^2}{2};$$

and since g^2 has eigenvalues $\{\lambda_i^2\}$, the last formula follows. □

Exercise 2.2. For $\mathrm{Sym}^2 V$, verify that

$$\chi_{\mathrm{Sym}^2 V}(g) = \tfrac{1}{2}[\chi_V(g)^2 + \chi_V(g^2)].$$

Note that this is compatible with the decomposition

$$V \otimes V = \mathrm{Sym}^2 V \oplus \wedge^2 V.$$

Exercise 2.3*. Compute the characters of $\mathrm{Sym}^k V$ and $\wedge^k V$.

Exercise 2.4*. Show that if we know the character χ_V of a representation V, then we know the eigenvalues of each element g of G, in the sense that we know the coefficients of the characteristic polynomial of $g\colon V \to V$. Carry this out explicitly for elements $g \in G$ of orders 2, 3, and 4, and for a representation of G on a vector space of dimension 2, 3, or 4.

Exercise 2.5. (*The original fixed-point formula*). If V is the permutation representation associated to the action of a group G on a finite set X, show that $\chi_V(g)$ is the number of elements of X fixed by g.

As we have said, the character of a representation of a group G is really a function on the set of conjugacy classes in G. This suggests expressing the basic information about the irreducible representations of a group G in the form of a *character table*. This is a table with the conjugacy classes $[g]$ of G listed across the top, usually given by a representative g, with (for reasons that will become apparent later) the number of elements in each conjugacy class over it; the irreducible representations V of G listed on the left; and, in the appropriate box, the value of the character χ_V on the conjugacy class $[g]$.

Example 2.6. We compute the character table of $\mathfrak{S}_3$. This is easy: to begin with, the trivial representation takes the values (1, 1, 1) on the three conjugacy classes [1], [(12)], and [(123)], whereas the alternating representation has values (1, −1, 1). To see the character of the standard representation, note that the permutation representation decomposes: $\mathbb{C}^3 = U \oplus V$; since the character of the permutation representation has, by Exercise 2.5, the values (3, 1, 0), we have $\chi_V = \chi_{\mathbb{C}^3} - \chi_U = (3, 1, 0) - (1, 1, 1) = (2, 0, -1)$. In sum, then, the character table of $\mathfrak{S}_3$ is

	1	3	2
$\mathfrak{S}_3$	1	(12)	(123)
trivial U	1	1	1
alternating U'	1	−1	1
standard V	2	0	−1

This gives us another solution of the basic problem posed in Lecture 1: if W is any representation of $\mathfrak{S}_3$ and we decompose W into irreducible representations $W \cong U^{\oplus a} \oplus U'^{\oplus b} \oplus V^{\oplus c}$, then $\chi_W = a\chi_U + b\chi_{U'} + c\chi_V$. In particular, since the functions χ_U, $\chi_{U'}$ and χ_V are independent, we see that *W is determined up to isomorphism by its character χ_W*.

Consider, for example, $V \otimes V$. Its character is $(\chi_V)^2$, which has values 4, 0, and 1 on the three conjugacy classes. Since $V \oplus U \oplus U'$ has the same character, this implies that $V \otimes V$ decomposes into $V \oplus U \oplus U'$, as we have seen directly. Similarly, $V \otimes U'$ has values 2, 0, and −1, so $V \otimes U' \cong V$.

Exercise 2.7*. Find the decomposition of the representation $V^{\otimes n}$ using character theory.

Characters will be similarly useful for larger groups, although it is rare to find simple closed formulas for decomposing tensor products.

§2.2. The First Projection Formula and Its Consequences

In the last lecture, we asked (among other things) for a way of locating explicitly the direct sum factors in the decomposition of a representation into irreducible ones. In this section we will start by giving an explicit formula for the projection of a representation onto the direct sum of the trivial factors in this decomposition; as it will turn out, this formula alone has tremendous consequences.

To start, for any representation V of a group G, we set

$$V^G = \{v \in V: gv = v \quad \forall g \in G\}.$$

We ask for a way of finding V^G explicitly. The idea behind our solution to this is already implicit in the previous lecture. We observed there that for any representation V of G and any $g \in G$, the endomorphism $g: V \to V$ is, in general, not a G-module homomorphism. On the other hand, if we take the *average* of all these endomorphisms, that is, we set

$$\varphi = \frac{1}{|G|} \sum_{g \in G} g \in \operatorname{End}(V),$$

then the endomorphism φ will be G-linear since $\sum g = \sum hgh^{-1}$. In fact, we have

Proposition 2.8. *The map φ is a projection of V onto V^G.*

PROOF. First, suppose $v = \varphi(w) = (1/|G|)\sum gw$. Then, for any $h \in G$,

$$hv = \frac{1}{|G|} \sum hgw = \frac{1}{|G|} \sum gw,$$

so the image of φ is contained in V^G. Conversely, if $v \in V^G$, then $\varphi(v) = (1/|G|)\sum v = v$, so $V^G \subset \operatorname{Im}(\varphi)$; and $\varphi \circ \varphi = \varphi$. □

We thus have a way of finding explicitly the direct sum of the trivial subrepresentations of a given representation, although the formula can be hard to use if it does not simplify. If we just want to know the number m of copies of the trivial representation appearing in the decomposition of V, we can do this numerically, since this number will be just the trace of the projection φ. We have

$$m = \dim V^G = \operatorname{Trace}(\varphi)$$
$$= \frac{1}{|G|} \sum_{g \in G} \operatorname{Trace}(g) = \frac{1}{|G|} \sum_{g \in G} \chi_V(g). \tag{2.9}$$

In particular, we observe that for an irreducible representation V other than the trivial one, the sum over all $g \in G$ of the values of the character χ_V is zero.

We can do much more with this idea, however. The key is to use Exercise 1.2: if V and W are representations of G, then with $\operatorname{Hom}(V, W)$, the representation defined in Lecture 1, we have

$$\operatorname{Hom}(V, W)^G = \{G\text{-module homomorphisms from } V \text{ to } W\}.$$

If V is irreducible then by Schur's lemma $\dim \operatorname{Hom}(V, W)^G$ is the multiplicity of V in W; similarly, if W is irreducible, $\dim \operatorname{Hom}(V, W)^G$ is the multiplicity of W in V, and in the case where both V and W are irreducible, we have

$$\dim \operatorname{Hom}_G(V, W) = \begin{cases} 1 & \text{if } V \cong W \\ 0 & \text{if } V \not\cong W. \end{cases}$$

But now the character $\chi_{\operatorname{Hom}(V,W)}$ of the representation $\operatorname{Hom}(V, W) = V^* \otimes W$ is given by

$$\chi_{\operatorname{Hom}(V,W)}(g) = \overline{\chi_V(g)} \cdot \chi_W(g).$$

We can now apply formula (2.9) in this case to obtain the striking

$$\frac{1}{|G|} \sum_{g \in G} \overline{\chi_V(g)} \chi_W(g) = \begin{cases} 1 & \text{if } V \cong W \\ 0 & \text{if } V \not\cong W. \end{cases} \tag{2.10}$$

To express this, let

$$\mathbb{C}_{\text{class}}(G) = \{\text{class functions on } G\}$$

and define an Hermitian inner product on $\mathbb{C}_{\text{class}}(G)$ by

$$(\alpha, \beta) = \frac{1}{|G|} \sum_{g \in G} \overline{\alpha(g)} \beta(g). \tag{2.11}$$

Formula (2.10) then amounts to

Theorem 2.12. *In terms of this inner product, the characters of the irreducible representations of G are orthonormal.*

For example, the orthonormality of the three irreducible representations of $\mathfrak{S}_3$ can be read from its character table in Example 2.6. The numbers over each conjugacy class tell how many times to count entries in that column.

Corollary 2.13. *The number of irreducible representations of G is less than or equal to the number of conjugacy classes.*

We will soon show that there are no nonzero class functions orthogonal to the characters, so that equality holds in Corollary 2.13.

Corollary 2.14. *Any representation is determined by its character.*

Indeed if $V \cong V_1^{\oplus a_1} \oplus \cdots \oplus V_k^{\oplus a_k}$, with the V_i distinct irreducible representations, then $\chi_V = \sum a_i \chi_{V_i}$, and the χ_{V_i} are linearly independent.

Corollary 2.15. *A representation V is irreducible if and only if $(\chi_V, \chi_V) = 1$.*

In fact, if $V \cong V_1^{\oplus a_1} \oplus \cdots \oplus V_k^{\oplus a_k}$ as above, then $(\chi_V, \chi_V) = \sum a_i^2$. The multiplicities a_i can be calculated via

Corollary 2.16. *The multiplicity a_i of V_i in V is the inner product of χ_V with χ_{V_i}, i.e., $a_i = (\chi_V, \chi_{V_i})$.*

We obtain some further corollaries by applying all this to the regular representation R of G. First, by Exercise 2.5 we know the character of R; it is simply

$$\chi_R(g) = \begin{cases} 0 & \text{if } g \neq e \\ |G| & \text{if } g = e. \end{cases}$$

Thus, we see first of all that R is not irreducible if $G \neq \{e\}$. In fact, if we set $R = \bigoplus V_i^{\oplus a_i}$, with V_i distinct irreducibles, then

$$a_i = (\chi_{V_i}, \chi_R) = \frac{1}{|G|} \chi_{V_i}(e) \cdot |G| = \dim V_i. \tag{2.17}$$

Corollary 2.18. *Any irreducible representation V of G appears in the regular representation* dim V *times.*

In particular, this proves again that there are only finitely many irreducible representations. As a numerical consequence of this we have the formula

$$|G| = \dim(R) = \sum_i \dim(V_i)^2. \tag{2.19}$$

Also, applying this to the value of the character of the regular representation on an element $g \in G$ other than the identity, we have

$$0 = \sum (\dim V_i) \cdot \chi_{V_i}(g) \quad \text{if } g \neq e. \tag{2.20}$$

These two formulas amount to the Fourier inversion formula for finite groups, cf. Example 3.32. For example, if all but one of the characters is known, they give a formula for the unknown character.

Exercise 2.21. The orthogonality of the rows of the character table is equivalent to an orthogonality for the columns (assuming the fact that there are as many rows as columns). Written out, this says:

(i) For $g \in G$,

$$\sum_{\chi} \overline{\chi(g)}\chi(g) = \frac{|G|}{c(g)},$$

where the sum is over all irreducible characters, and $c(g)$ is the number of elements in the conjugacy class of g.

(ii) If g and h are elements of G that are not conjugate, then

$$\sum_{\chi} \overline{\chi(g)}\chi(h) = 0.$$

Note that for $g = e$ these reduce to (2.19) and (2.20).

§2.3. Examples: $\mathfrak{S}_4$ and $\mathfrak{A}_4$

To see how the analysis of the characters of a group actually goes in practice, we now work out the character table of $\mathfrak{S}_4$. To start, we list the conjugacy classes in $\mathfrak{S}_4$ and the number of elements of $\mathfrak{S}_4$ in each. As with any symmetric group $\mathfrak{S}_d$, the conjugacy classes correspond naturally to the *partitions* of d, that is, expressions of d as a sum of positive integers $a_1, a_2, \ldots, a_k$, where the correspondence associates to such a partition the conjugacy class of a permutation consisting of disjoint cycles of length $a_1, a_2, \ldots, a_k$. Thus, in $\mathfrak{S}_4$ we have the classes of the identity element 1 ($4 = 1 + 1 + 1 + 1$), a transposition such as (12), corresponding to the partition $4 = 2 + 1 + 1$; a three-cycle (123) corresponding to $4 = 3 + 1$; a four-cycle (1234) ($4 = 4$); and the product of two disjoint transpositions (12)(34) ($4 = 2 + 2$).

Exercise 2.22. Show that the number of elements in each of these conjugacy classes is, respectively, 1, 6, 8, 6, and 3.

As for the irreducible representations of $\mathfrak{S}_4$, we start with the same ones that we had in the case of $\mathfrak{S}_3$: the trivial U, the alternating U', and the standard representation V, i.e., the quotient of the permutation representation associated to the standard action of $\mathfrak{S}_4$ on a set of four elements by the trivial subrepresentation. The character of the trivial representation on the five conjugacy classes is of course (1, 1, 1, 1, 1), and that of the alternating representation is $(1, -1, 1, -1, 1)$. To find the character of the standard representation, we observe that by Exercise 2.5 the character of the permutation representation on $\mathbb{C}^4$ is $\chi_{\mathbb{C}^4} = (4, 2, 1, 0, 0)$ and, correspondingly,

$$\chi_V = \chi_{\mathbb{C}^4} - \chi_U = (3, 1, 0, -1, -1).$$

Note that $|\chi_V| = 1$, so V is irreducible. The character table so far looks like

	1	6	8	6	3
$\mathfrak{S}_4$	1	(12)	(123)	(1234)	(12)(34)
trivial U	1	1	1	1	1
alternating U'	1	-1	1	-1	1
standard V	3	1	0	-1	-1

Clearly, we are not done yet: since the sum of the squares of the dimensions of these three representations is $1 + 1 + 9 = 11$, by (2.19) there must be additional irreducible representations of $\mathfrak{S}_4$, the squares of whose dimensions add up to $24 - 11 = 13$. Since there are by Corollary 2.13 at most two of them, there must be exactly two, of dimensions 2 and 3. The latter of these is easy to locate: if we just tensor the standard representation V with the alternating one U', we arrive at a representation V' with character $\chi_{V'} = \chi_V \cdot \chi_{U'} = (3, -1, 0, 1, -1)$. We can see that this is irreducible either from its character (since $|\chi_{V'}| = 1$) or from the fact that it is the tensor product of an irreducible representation with a one-dimensional one; since its character is not equal to that of any of the first three, this must be one of the two missing ones. As for the remaining representation of degree two, we will for now simply call it W; we can determine its character from the orthogonality relations (2.10). We obtain then the complete character table for $\mathfrak{S}_4$:

	1	6	8	6	3
$\mathfrak{S}_4$	1	(12)	(123)	(1234)	(12)(34)
trivial U	1	1	1	1	1
alternating U'	1	-1	1	-1	1
standard V	3	1	0	-1	-1
$V' = V \otimes U'$	3	-1	0	1	-1
Another W	2	0	-1	0	2

Exercise 2.23. Verify the last row of this table from (2.10) or (2.20).

We now get a dividend: we can take the character of the mystery representation W, which we have obtained from general character theory alone, and use it to describe the representation W explicitly! The key is the 2 in the last column for χ_W: this says that the action of (12)(34) on the two-dimensional vector space W is an involution of trace 2, and so must be the identity. Thus, W is really a representation of the quotient group[1]

[1] If N is a normal subgroup of a group G, a representation $\rho: G \to \mathrm{GL}(V)$ is trivial on N if and only if it factors through the quotient

$$G \to G/N \to \mathrm{GL}(V).$$

Representations of G/N can be identified with representations of G that are trivial on N.

$$\mathfrak{S}_4/\{1, (12)(34), (13)(24), (14)(23)\} \cong \mathfrak{S}_3.$$

[One may see this isomorphism by letting $\mathfrak{S}_4$ act on the elements of the conjugacy class of (12)(34); equivalently, if we realize $\mathfrak{S}_4$ as the group of rigid motions of a cube (see below), by looking at the action of $\mathfrak{S}_4$ on pairs of opposite faces.] W must then be just the standard representation of $\mathfrak{S}_3$ pulled back to $\mathfrak{S}_4$ via this quotient.

Example 2.24. As we said above, the group of rigid motions of a cube is the symmetric group on four letters; $\mathfrak{S}_4$ acts on the cube via its action on the four long diagonals. It follows, of course, that $\mathfrak{S}_4$ acts as well on the set of faces, of edges, of vertices, etc.; and to each of these is associated a permutation representation of $\mathfrak{S}_4$. We may thus ask how these representations decompose; we will do here the case of the faces and leave the others as exercises.

We start, of course, by describing the character χ of the permutation representation associated to the faces of the cube. Rotation by 180° about a line joining the midpoints of two opposite edges is a transposition in $\mathfrak{S}_4$ and fixes no faces, so $\chi(12) = 0$. Rotation by 120° about a long diagonal shows $\chi(123) = 0$. Rotation by 90° about a line joining the midpoints of two opposite faces shows $\chi(1234) = 2$, and rotation by 180° gives $\chi((12)(34)) = 2$. Now $(\chi, \chi) = 3$, so χ is the sum of three distinct irreducible representations. From the table, $(\chi, \chi_U) = (\chi, \chi_{V'}) = (\chi, \chi_W) = 1$, and the inner products with the others are zero, so this representation is $U \oplus V' \oplus W$. In fact, the sums of opposite faces span a three-dimensional subrepresentation which contains U (spanned by the sum of all faces), so this representation is $U \oplus W$. The differences of opposite faces therefore span V'.

Exercise 2.25*. Decompose the permutation representation of $\mathfrak{S}_4$ on (i) the vertices and (ii) the edges of the cube.

Exercise 2.26. The alternating group $\mathfrak{A}_4$ has four conjugacy classes. Three representations U, U', and U'' come from the representations of

$$\mathfrak{A}_4/\{1, (12)(34), (13)(24), (14)(23)\} \cong \mathbb{Z}/3,$$

so there is one more irreducible representation V of dimension 3. Compute the character table, with $\omega = e^{2\pi i/3}$:

	1	4	4	3
$\mathfrak{A}_4$	1	(123)	(132)	(12)(34)
U	1	1	1	1
U'	1	ω	ω^2	1
U''	1	ω^2	ω	1
V	3	0	0	-1

Exercise 2.27. Consider the representations of $\mathfrak{S}_4$ and their restrictions to $\mathfrak{A}_4$. Which are still irreducible when restricted, and which decompose? Which pairs of nonisomorphic representations of $\mathfrak{S}_4$ become isomorphic when restricted? Which representations of $\mathfrak{A}_4$ arise as restrictions from $\mathfrak{S}_4$?

§2.4. More Projection Formulas; More Consequences

In this section, we complete the analysis of the characters of the irreducible representations of a general finite group begun in §2.2 and give a more general formula for the projection of a general representation V onto the direct sum of the factors in V isomorphic to a given irreducible representation W. The main idea for both is a generalization of the "averaging" of the endomorphisms $g: V \to V$ used in §2.2, the point being that instead of simply averaging all the g we can ask the question: what linear combinations of the endomorphisms $g: V \to V$ are G-linear endomorphisms? The answer is given by

Proposition 2.28. *Let $\alpha: G \to \mathbb{C}$ be any function on the group G, and for any representation V of G set*

$$\varphi_{\alpha,V} = \sum \alpha(g)\cdot g: V \to V.$$

Then $\varphi_{\alpha,V}$ is a homomorphism of G-modules for all V if and only if α is a class function.

Proof. We simply write out the condition that $\varphi_{\alpha,V}$ be G-linear, and the result falls out: we have

$$\varphi_{\alpha,V}(hv) = \sum \alpha(g)\cdot g(hv)$$
$$= \sum \alpha(hgh^{-1})\cdot hgh^{-1}(hv)$$

(substituting hgh^{-1} for g)

$$= h\left(\sum \alpha(hgh^{-1})\cdot g(v)\right)$$
$$= h\left(\sum \alpha(g)\cdot g(v)\right)$$

(if α is a class function)

$$= h(\varphi_{\alpha,V}(v)).$$

Exercise 2.29*. Complete this proof by showing that conversely if α is not a class function, then there exists a representation V of G for which $\varphi_{\alpha,V}$ fails to be G-linear. □

As an immediate consequence of this proposition, we have

Proposition 2.30. *The number of irreducible representations of G is equal to the number of conjugacy classes of G. Equivalently, their characters $\{\chi_V\}$ form an orthonormal basis for $\mathbb{C}_{\text{class}}(G)$.*

PROOF. Suppose $\alpha: G \to \mathbb{C}$ is a class function and $(\alpha, \chi_V) = 0$ for all irreducible representations V; we must show that $\alpha = 0$. Consider the endomorphism

$$\varphi_{\alpha,V} = \sum \alpha(g) \cdot g: V \to V$$

as defined above. By Schur's lemma, $\varphi_{\alpha,V} = \lambda \cdot \text{Id}$; and if $n = \dim V$, then

$$\begin{aligned}\lambda &= \frac{1}{n} \cdot \text{trace}(\varphi_{\alpha,V}) \\ &= \frac{1}{n} \cdot \sum \alpha(g)\chi_V(g) \\ &= \frac{|G|}{n} \overline{(\alpha, \chi_{V^*})} \\ &= 0.\end{aligned}$$

Thus, $\varphi_{\alpha,V} = 0$, or $\sum \alpha(g) \cdot g = 0$ on any representation V of G; in particular, this will be true for the regular representation $V = R$. But in R the elements $\{g \in G\}$, thought of as elements of $\text{End}(R)$, are linearly independent. For example, the elements $\{g(e)\}$ are all independent. Thus $\alpha(g) = 0$ for all g, as required. □

This proposition completes the description of the characters of a finite group in general. We will see in more examples below how we can use this information to build up the character table of a given group. For now, we mention another way of expressing this proposition, via the *representation ring* of the group G.

The representation ring $R(G)$ of a group G is easy to define. First, as a group we just take $R(G)$ to be the free abelian group generated by all (isomorphism classes of) representations of G, and mod out by the subgroup generated by elements of the form $V + W - (V \oplus W)$. Equivalently, given the statement of complete reducibility, we can just take all integral linear combinations $\sum a_i \cdot V_i$ of the irreducible representations V_i of G; elements of $R(G)$ are correspondingly called *virtual representations*. The ring structure is then given simply by tensor product, defined on the generators of $R(G)$ and extended by linearity.

We can express most of what we have learned so far about representations of a finite group G in these terms. To begin, the character defines a map

$$\chi: R(G) \to \mathbb{C}_{\text{class}}(G)$$

from $R(G)$ to the ring of complex-valued functions on G; by the basic formulas of Proposition 2.1, this map is in fact a ring homomorphism. The statement that a representation in determined by its character then says that χ is injective;

the images of χ are called *virtual characters* and correspond thereby to virtual representations. Finally, our last proposition amounts to the statement that *χ induces an isomorphism*

$$\chi_{\mathbb{C}}: R(G) \otimes \mathbb{C} \to \mathbb{C}_{\text{class}}(G).$$

The virtual characters of G form a lattice $\Lambda \cong \mathbb{Z}^c$ in $\mathbb{C}_{\text{class}}(G)$, in which the actual characters sit as a cone $\Lambda_0 \cong \mathbb{N}^c \subset \mathbb{Z}^c$. We can thus think of the problem of describing the characters of G as having two parts: first, we have to find Λ, and then the cone $\Lambda_0 \subset \Lambda$ (once we know Λ_0, the characters of the irreducible representations will be determined). In the following lecture we will state theorems of Artin and Brauer characterizing $\Lambda \otimes \mathbb{Q}$ and Λ.

The argument for Proposition 2.30 also suggests how to obtain a more general projection formula. Explicitly, if W is a fixed irreducible representation, then for any representation V, look at the weighted sum

$$\psi = \frac{1}{|G|} \sum_{g \in G} \overline{\chi_W(g)} \cdot g \in \operatorname{End}(V).$$

By Proposition 2.28, ψ is a G-module homomorphism. Hence, if V is irreducible, we have $\psi = \lambda \cdot \mathrm{Id}$, and

$$\begin{aligned} \lambda &= \frac{1}{\dim V} \operatorname{Trace} \psi \\ &= \frac{1}{\dim V} \cdot \frac{1}{|G|} \sum \overline{\chi_W(g)} \cdot \chi_V(g) \\ &= \begin{cases} \dfrac{1}{\dim V} & \text{if } V = W \\ 0 & \text{if } V \neq W. \end{cases} \end{aligned}$$

For arbitrary V,

$$\psi_V = \dim W \cdot \frac{1}{|G|} \sum_{g \in G} \overline{\chi_W(g)} \cdot g : V \to V \tag{2.31}$$

is the projection of V onto the factor consisting of the sum of all copies of W appearing in V. In other words, if $V = \bigoplus V_i^{\oplus a_i}$, then

$$\pi_i = \dim V_i \cdot \frac{1}{|G|} \sum_{g \in G} \overline{\chi_{V_i}(g)} \cdot g \tag{2.32}$$

is the projection of V onto $V_i^{\oplus a_i}$.

Exercise 2.33*. (a) In terms of representations V and W in $R(G)$, the inner product on $\mathbb{C}_{\text{class}}(G)$ takes the simple form

$$(V, W) = \dim \operatorname{Hom}_G(V, W).$$

(b) If $\chi \in \mathbb{C}_{\text{class}}(G)$ is a virtual character, and $(\chi, \chi) = 1$, then either χ or $-\chi$ is the character of an irreducible representation, the plus sign occurring when $\chi(1) > 0$. If $(\chi, \chi) = 2$, and $\chi(1) > 0$, then χ is either the sum or the difference of two irreducible characters.

(c) If U, V, and W are irreducible representations, show that U appears in $V \otimes W$ if and only if W occurs in $V^* \otimes U$. Deduce that this cannot occur unless $\dim U \geq \dim W/\dim V$.

We conclude this lecture with some exercises that use characters to work out some standard facts about representations.

Exercise 2.34*. Let V and W be irreducible representations of G, and $L_0 \colon V \to W$ any linear mapping. Define $L \colon V \to W$ by

$$L(v) = \frac{1}{|G|} \sum_{g \in G} g^{-1} \cdot L_0(g \cdot v).$$

Show that $L = 0$ if V and W are not isomorphic, and that L is multiplication by $\operatorname{trace}(L_0)/\dim(V)$ if $V = W$.

Exercise 2.35*. Show that, if the irreducible representations of G are represented by unitary matrices [cf. Exercise 1.14], the matrix entries of these representations form an orthogonal basis for the space of *all* functions on G [with inner product given by (2.11)].

Exercise 2.36*. If G_1 and G_2 are groups, and V_1 and V_2 are representations of G_1 and G_2, then the tensor product $V_1 \otimes V_2$ is a representation of $G_1 \times G_2$, by $(g_1 \times g_2) \cdot (v_1 \otimes v_2) = g_1 \cdot v_1 \otimes g_2 \cdot v_2$. To distinguish this "external" tensor product from the internal tensor product—when $G_1 = G_2$—this *external tensor product* is sometimes denoted $V_1 \boxtimes V_2$. If χ_i is the character of V_i, then the value of the character χ of $V_1 \boxtimes V_2$ is given by the product:

$$\chi(g_1 \times g_2) = \chi_1(g_1)\chi_2(g_2).$$

If V_1 and V_2 are irreducible, show that $V_1 \boxtimes V_2$ is also irreducible and show that every irreducible representation of $G_1 \times G_2$ arises this way. In terms of representation rings,

$$R(G_1 \times G_2) = R(G_1) \otimes R(G_2).$$

In these lectures we will often be given a subgroup G of a general linear group $\mathrm{GL}(V)$, and we will look for other representations inside tensor powers of V. The following problem, which is a theorem of Burnside and Molien, shows that for a finite group G, all irreducible representations can be found this way.

Problem 2.37*. Show that if V is a faithful representation of G, i.e., $\rho: G \to \mathrm{GL}(V)$ is injective, then any irreducible representation of G is contained in some tensor power $V^{\otimes n}$ of V.

Problem 2.38*. Show that the dimension of an irreducible representation of G divides the order of G.

Another challenge:

Problem 2.39*. Show that the character of any irreducible representation of dimension greater than 1 assumes the value 0 on some conjugacy class of the group.

LECTURE 3

Examples; Induced Representations; Group Algebras; Real Representations

This lecture is something of a grabbag. We start in §3.1 with examples illustrating the use of the techniques of the preceding lecture. Section 3.2 is also by way of an example. We will see quite a bit more about the representations of the symmetric groups in general later; §4 is devoted to this and will certainly subsume this discussion, but this should provide at least a sense of how we can go about analyzing representations of a class of groups, as opposed to individual groups. In §§3.3 and 3.4 we introduce two basic notions in representation theory, induced representations and the group algebra. Finally, in §3.5 we show how to classify representations of a finite group on a real vector space, given the answer to the corresponding question over $\mathbb{C}$, and say a few words about the analogous question for subfields of $\mathbb{C}$ other than $\mathbb{R}$. Everything in this lecture is elementary except Exercises 3.9 and 3.32, which involve the notions of Clifford algebras and the Fourier transform, respectively (both exercises, of course, can be skipped).

§3.1. Examples: $\mathfrak{S}_5$ and $\mathfrak{A}_5$

We have found the representations of the symmetric and alternating groups for $n \leq 4$. Before turning to a more systematic study of symmetric and alternating groups, we will work out the next couple of cases.

Representations of the Symmetric Group $\mathfrak{S}_5$

As before, we start by listing the conjugacy classes of $\mathfrak{S}_5$ and giving the number of elements of each: we have 10 transpositions, 20 three-cycles, 30 four-cycles and 24 five-cycles; in addition, we have 15 elements conjugate to (12)(34) and 20 ~~10~~ elements conjugate to (12)(345). As for the irreducible representations, we have, of course, the trivial representation U, the alternating representation U', and the standard representation V; also, as in the case of $\mathfrak{S}_4$ we can tensor the standard representation V with the alternating one to obtain another irreducible representation V' with character $\chi_{V'} = \chi_V \cdot \chi_{U'}$.

Exercise 3.1. Find the characters of the representations V and V'; deduce in particular that V and V' are distinct irreducible representations.

The first four rows of the character table are thus

	1	10	20	30	24	15	20
$\mathfrak{S}_5$	1	(12)	(123)	(1234)	(12345)	(12)(34)	(12)(345)
U	1	1	1	1	1	1	1
U'	1	-1	1	-1	1	1	-1
V	4	2	1	0	-1	0	-1
V'	4	-2	1	0	-1	0	1

Clearly, we need three more irreducible representations. Where should we look for these? On the basis of our previous experience (and Problem 2.37), a natural place would be in the tensor products/powers of the irreducible representations we have found so far, in particular in $V \otimes V$ (the other two possible products will yield nothing new: we have $V' \otimes V = V \otimes V \otimes U'$ and $V' \otimes V' = V \otimes V$). Of course, $V \otimes V$ breaks up into $\wedge^2 V$ and $\mathrm{Sym}^2 V$, so we look at these separately. To start with, by the formula

$$\chi_{\wedge^2 V}(g) = \tfrac{1}{2}(\chi_V(g)^2 - \chi_V(g^2))$$

we calculate the character of $\wedge^2 V$:

$$\chi_{\wedge^2 V} = (6, 0, 0, 0, 1, -2, 0);$$

we see from this that it is indeed a fifth irreducible representation (and that $\wedge^2 V \otimes U' = \wedge^2 V$, so we get nothing new that way).

We can now find the remaining two representations in either of two ways. First, if n_1 and n_2 are their dimensions, we have

$$5! = 120 = 1^2 + 1^2 + 4^2 + 4^2 + 6^2 + n_1^2 + n_2^2,$$

so $n_1^2 + n_2^2 = 50$. There are no more one-dimensional representations, since these are trivial on normal subgroups whose quotient group is cyclic, and $\mathfrak{A}_5$

is the only such subgroup. So the only possibility is $n_1 = n_2 = 5$. Let W denote one of these five-dimensional representations, and set $W' = W \otimes U'$. In the table, if the row giving the character of W is

$$(5 \quad \alpha_1 \quad \alpha_2 \quad \alpha_3 \quad \alpha_4 \quad \alpha_5 \quad \alpha_6),$$

that of W' is $(5 \quad -\alpha_1 \quad \alpha_2 \quad -\alpha_3 \quad \alpha_4 \quad \alpha_5 \quad -\alpha_6)$. Using the orthogonality relations or (2.20), one sees that $W' \not\cong W$; and with a little calculation, up to interchanging W and W', the last two rows are as given:

	1	10	20	30	24	15	20
$\mathfrak{S}_5$	1	(12)	(123)	(1234)	(12345)	(12)(34)	(12)(345)
U	1	1	1	1	1	1	1
U'	1	-1	1	-1	1	1	-1
V	4	2	1	0	-1	0	-1
V'	4	-2	1	0	-1	0	1
$\wedge^2 V$	6	0	0	0	1	-2	0
W	5	1	-1	-1	0	1	1
W'	5	-1	-1	1	0	1	-1

From the decomposition $V \oplus U = \mathbb{C}^5$, we have also $\wedge^4 V = \wedge^5 \mathbb{C}^5 = U'$, and $V^* = V$. The perfect pairing[1]

$$V \times \wedge^3 V \to \wedge^4 V = U',$$

taking $v \times (v_1 \wedge v_2 \wedge v_3)$ to $v \wedge v_1 \wedge v_2 \wedge v_3$ shows that $\wedge^3 V$ is isomorphic to $V^* \otimes U' = V'$.

Another way to find the representations W and W' would be to proceed with our original plan, and look at the representation $\mathrm{Sym}^2 V$. We will leave this in the form of an exercise:

Exercise 3.2. (i) Find the character of the representation $\mathrm{Sym}^2 V$.

(ii) Without using any knowledge of the character table of $\mathfrak{S}_5$, use this to show that $\mathrm{Sym}^2 V$ is the direct sum of three distinct irreducible representations.

(iii) Using our knowledge of the first five rows of the character table, show that $\mathrm{Sym}^2 V$ is the direct sum of the representations U, V, and a third irreducible representation W. Complete the character table for $\mathfrak{S}_5$.

Exercise 3.3. Find the decomposition into irreducibles of the representations $\wedge^2 W$, $\mathrm{Sym}^2 W$, and $V \otimes W$.

[1] If V and W are n-dimensional vector spaces, and U is one dimensional, a *perfect pairing* is a bilinear map $\beta: V \times W \to U$ such that no nonzero vector v in V has $\beta(v, W) = 0$. Equivalently, the map $V \to \mathrm{Hom}(W, U) = W^* \otimes U$, $v \mapsto (w \mapsto \beta(v, w))$, is an isomorphism.

Representations of the Alternating Group $\mathfrak{A}_5$

What happens to the conjugacy classes above if we replace $\mathfrak{S}_d$ by $\mathfrak{A}_d$? Obviously, all the odd conjugacy classes disappear; but at the same time, since conjugation by a transposition is now an outer, rather than inner, automorphism, some conjugacy classes may break into two.

Exercise 3.4. Show that the conjugacy class in $\mathfrak{S}_d$ of permutations consisting of products of disjoint cycles of lengths $b_1, b_2, \ldots$ will break up into the union of two conjugacy classes in $\mathfrak{A}_d$ if all the b_k are odd and distinct; if any b_k are even or repeated, it remains a single conjugacy class in $\mathfrak{A}_d$. (We consider a fixed point as a cycle of length 1.)

In the case of $\mathfrak{A}_5$, this means we have the conjugacy class of three-cycles (as before, 20 elements), and of products of two disjoint transpositions (15 elements); the conjugacy class of five-cycles, however, breaks up into the conjugacy classes of (12345) and (21345), each having 12 elements.

As for the representations, the obvious first place to look is at restrictions to $\mathfrak{A}_5$ of the irreducible representations of $\mathfrak{S}_5$ found above. An irreducible representation of $\mathfrak{S}_5$ may become reducible when restricted to $\mathfrak{A}_5$; or two distinct representations may become isomorphic, as will be the case with U and U', V and V', or W and W'. In fact, U, V, and W stay irreducible since their characters satisfy $(\chi, \chi) = 1$. But the character of $\wedge^2 V$ has values $(6, 0, -2, 1, 1)$ on the conjugacy classes listed above, so $(\chi, \chi) = 2$, and $\wedge^2 V$ is the sum of two irreducible representations, which we denote by Y and Z. Since the sums of the squares of all the dimensions is 60, $(\dim Y)^2 + (\dim Z)^2 = 18$, so each must be three dimensional.

Exercise 3.5*. Use the orthogonality relations to complete the character table of $\mathfrak{A}_5$:

	1	20	15	12	12
$\mathfrak{A}_5$	1	(123)	(12)(34)	(12345)	(21345)
U	1	1	1	1	1
V	4	1	0	-1	-1
W	5	-1	1	0	0
Y	3	0	-1	$\frac{1+\sqrt{5}}{2}$	$\frac{1-\sqrt{5}}{2}$
Z	3	0	-1	$\frac{1-\sqrt{5}}{2}$	$\frac{1+\sqrt{5}}{2}$

The representations Y and Z may in fact be familiar: $\mathfrak{A}_5$ can be realized as the group of motions of an icosahedron (or, equivalently, of a dodecahedron) and Y is the corresponding representation. Note that the two representations $\mathfrak{A}_5 \to GL_3(\mathbb{R})$ corresponding to Y and Z have the same image, but (as you can see from the fact that their characters differ only on the conjugacy classes of (12345) and (21345)) differ by an *outer* automorphism of $\mathfrak{A}_5$.

Note also that $\wedge^2 V$ does not decompose over $\mathbb{Q}$; we could see this directly from the fact that the vertices of a dodecahedron cannot all have rational coordinates, which follows from the analogous fact for a regular pentagon in the plane.

Exercise 3.6. Find the decomposition of the permutation representation of $\mathfrak{A}_5$ corresponding to the (i) vertices, (ii) faces, and (iii) edges of the icosahedron.

Exercise 3.7. Consider the dihedral group D_{2n}, defined to be the group of isometries of a regular n-gon in the plane. Let $\Gamma \cong \mathbb{Z}/n \subset D_{2n}$ be the subgroup of rotations. Use the methods of Lecture 1 (applied there to the case $\mathfrak{S}_3 \cong D_6$) to analyze the representations of D_{2n}: that is, restrict an arbitrary representation of D_{2n} to Γ, break it up into eigenspaces for the action of Γ, and ask how the remaining generator of D_{2n} acts of these eigenspaces.

Exercise 3.8. Analyze the representations of the dihedral group D_{2n} using the character theory developed in Lecture 2.

Exercise 3.9. (a) Find the character table of the group of order 8 consisting of the quaternions $\{\pm 1, \pm i, \pm j, \pm k\}$ under multiplication. This is the case $m = 3$ of a collection of groups of order 2^m, which we denote H_m. To describe them, let C_m denote the complex Clifford algebra generated by $v_1, \ldots, v_m$ with relations $v_i^2 = -1$ and $v_i \cdot v_j = -v_j \cdot v_i$, so C_m has a basis $v_I = v_{i_1} \cdot \ldots \cdot v_{i_r}$, as $I = \{i_i < \cdots < i_r\}$ varies over subsets of $\{1, \ldots, m\}$. (See §20.1 for notation and basic facts about Clifford algebras). Set

$$H_m = \{\pm v_I \colon |I| \text{ is even}\} \subset (C_m^{\text{even}})^*.$$

This group is a 2-to-1 covering of the abelian 2-group of $m \times m$ diagonal matrices with ± 1 diagonal entries and determinant 1. The center of H_m is $\{\pm 1\}$ if m is odd and is $\{\pm 1, \pm v_{\{1,\ldots,m\}}\}$ if m is even. The other conjugacy classes consist of pairs of elements $\{\pm v_I\}$. The isomorphisms of C_m^{even} with a matrix algebra or a product of two matrix algebras give a 2^n-dimensional "spin" representation S of H_{2n+1}, and two 2^{n-1}-dimensional "spin" or "half-spin" representations S^+ and S^- of H_{2n}.

(b) Compute the characters of these spin representations and verify that they are irreducible.

(c) Deduce that the spin representations, together with the 2^{m-1} one-dimensional representations coming from the abelian group $H_m/\{\pm 1\}$ give a complete set of irreducible representations, and compute the character table for H_m.

For odd m the groups H_m are examples of *extra-special* 2-*groups*, cf. [Grie], [Qu].

Exercise 3.10. Find the character table of the group $SL_2(\mathbb{Z}/3)$.

Exercise 3.11. Let $H(\mathbb{Z}/3)$ be the *Heisenberg group* of order 27:

$$H(\mathbb{Z}/3) = \left\{ \begin{pmatrix} 1 & a & b \\ 0 & 1 & c \\ 0 & 0 & 1 \end{pmatrix}, a, b, c \in \mathbb{Z}/3 \right\} \subset SL_3(\mathbb{Z}/3).$$

Analyze the representations of $H(\mathbb{Z}/3)$, first by the methods of Lecture 1 (restricting in this case to the center

$$Z = \left\{ \begin{pmatrix} 1 & 0 & b \\ 0 & 1 & 0 \\ 0 & 0 & 1 \end{pmatrix}, b \in \mathbb{Z}/3 \right\} \cong \mathbb{Z}/3$$

of $H(\mathbb{Z}/3)$), and then by character theory.

§3.2. Exterior Powers of the Standard Representation of $\mathfrak{S}_d$

How should we go about constructing representations of the symmetric groups in general? The answer to this is not immediate; it is a subject that will occupy most of the next lecture (where we will produce all the irreducible representations of $\mathfrak{S}_d$). For now, as an example of the elementary techniques developed so far we will analyze directly one of the obvious candidates:

Proposition 3.12. *Each exterior power $\wedge^k V$ of the standard representation V of $\mathfrak{S}_d$ is irreducible,* $0 \le k \le d - 1$.

PROOF. From the decomposition $\mathbb{C}^d = V \oplus U$, we see that V is irreducible if and only if $(\chi_{\mathbb{C}^d}, \chi_{\mathbb{C}^d}) = 2$. Similarly, since

$$\wedge^k \mathbb{C}^d = (\wedge^k V \otimes \wedge^0 U) \oplus (\wedge^{k-1} V \otimes \wedge^1 U) = \wedge^k V \oplus \wedge^{k-1} V,$$

it suffices to show that $(\chi, \chi) = 2$, where χ is the character of the representation $\wedge^k \mathbb{C}^d$. Let $A = \{1, 2, \ldots, d\}$. For a subset B of A with k elements, and $g \in G = \mathfrak{S}_d$, let

$$\{g\}_B = \begin{cases} 0 & \text{if } g(B) \neq B \\ 1 & \text{if } g(B) = B \text{ and } g|_B \text{ is an even permutation} \\ -1 & \text{if } g(B) = B \text{ and } g|_B \text{ is odd.} \end{cases}$$

Here, if $g(B) = B$, $g|_B$ denotes the permutation of the set B determined by g. Then $\chi(g) = \sum \{g\}_B$, and

$$(\chi, \chi) = \frac{1}{d!} \sum_{g \in G} \left(\sum_B \{g\}_B \right)^2$$

$$= \frac{1}{d!} \sum_{g \in G} \sum_B \sum_C \{g\}_B \{g\}_C$$

$$= \frac{1}{d!} \sum_B \sum_C \sum_g (\operatorname{sgn} g|_B) \cdot (\operatorname{sgn} g|_C),$$

where the sums are over subsets B and C of A with k elements, and in the last equation, the sum is over those g with $g(B) = B$ and $g(C) = C$. Such g is given by four permutations: one of $B \cap C$, one of $B \backslash B \cap C$, one of $C \backslash B \cap C$, and one of $A \backslash B \cup C$. Letting l be the cardinality of $B \cap C$, this last sum can be written

$$\frac{1}{d!} \sum_B \sum_C \sum_{a \in \mathfrak{S}_l} \sum_{b \in \mathfrak{S}_{k-l}} \sum_{c \in \mathfrak{S}_{k-l}} \sum_{h \in \mathfrak{S}_{d-2k+l}} (\operatorname{sgn} a)^2 (\operatorname{sgn} b)(\operatorname{sgn} c)$$

$$= \frac{1}{d!} \sum_B \sum_C l!(d - 2k + l)! \left(\sum_{b \in \mathfrak{S}_{k-l}} \operatorname{sgn} b \right) \left(\sum_{c \in \mathfrak{S}_{k-l}} \operatorname{sgn} c \right).$$

These last sums are zero unless $k - l = 0$ or 1. The case $k = l$ gives

$$\frac{1}{d!} \sum_B k!(d - k)! = \frac{1}{d!} \binom{d}{k} k!(d - k)! = 1.$$

Similarly, the terms with $k - l = 1$ also add up to 1, so $(\chi, \chi) = 2$, as required. □

Note by way of contrast that the symmetric powers of the standard representation of $\mathfrak{S}_d$ are almost never irreducible. For example, we already know that the representation $\operatorname{Sym}^2 V$ contains one copy of the trivial representation: this is just the statement that every irreducible real representation (such as V) admits an inner product (unique, up to scalars) invariant under the group action; nor is the quotient of $\operatorname{Sym}^2 V$ by this trivial subrepresentation necessarily irreducible, as witness the case of $\mathfrak{S}_5$.

§3.3. Induced Representations

If $H \subset G$ is a subgroup, any representation V of G restricts to a representation of H, denoted $\operatorname{Res}_H^G V$ or simple Res V. In this section, we describe an important construction which produces representations of G from representations of H. Suppose V is a representation of G, and $W \subset V$ is a subspace which is H-invariant. For any g in G, the subspace $g \cdot W = \{g \cdot w : w \in W\}$ depends only on the left coset gH of g modulo H, since $gh \cdot W = g \cdot (h \cdot W) = g \cdot W$; for a coset

σ in G/H, we write $\sigma \cdot W$ for this subspace of V. We say that V is *induced* by W if every element in V can be written uniquely as a sum of elements in such translates of W, i.e.,

$$V = \bigoplus_{\sigma \in G/H} \sigma \cdot W.$$

In this case we write $V = \operatorname{Ind}_H^G W = \operatorname{Ind} W$.

Example 3.13. The permutation representation associated to the left action of G on G/H is induced from the trivial one-dimensional representation W of H. Here V has basis $\{e_\sigma : \sigma \in G/H\}$, and $W = \mathbb{C} \cdot e_H$, with H the trivial coset.

Example 3.14. The regular representation of G is induced from the regular representation of H. Here V has basis $\{e_g : g \in G\}$, whereas W has basis $\{e_h : h \in H\}$.

We claim that, given a representation W of H, such V exists and is unique up to isomorphism. Although we will later give several fancier ways to see this, it is not hard to do it by hand. Choose a representative $g_\sigma \in G$ for each coset $\sigma \in G/H$, with e representing the trivial coset H. To see the uniqueness, note that each element of V has a unique expression $v = \sum g_\sigma w_\sigma$, for elements w_σ in W. Given g in G, write $g \cdot g_\sigma = g_\tau \cdot h$ for some $\tau \in G/H$ and $h \in H$. Then we must have

$$g \cdot (g_\sigma w_\sigma) = (g \cdot g_\sigma) w_\sigma = (g_\tau \cdot h) w_\sigma = g_\tau (h w_\sigma).$$

This proves the uniqueness and tells us how to construct $V = \operatorname{Ind}(W)$ from W. Take a copy W^σ of W for each left coset $\sigma \in G/H$; for $w \in W$, let $g_\sigma w$ denote the element of W^σ corresponding to w in W. Let $V = \bigoplus_{\sigma \in G/H} W^\sigma$, so every element of V has a unique expression $v = \sum g_\sigma w_\sigma$ for elements w_σ in W. Given $g \in G$, define

$$g \cdot (g_\sigma w_\sigma) = g_\tau (h w_\sigma) \quad \text{if } g \cdot g_\sigma = g_\tau \cdot h.$$

To show that this defines as action of G on V, we must verify that $g' \cdot (g \cdot (g_\sigma w_\sigma)) = (g' \cdot g) \cdot (g_\sigma w_\sigma)$ for another element g' in G. Now if $g' \cdot g_\tau = g_\rho \cdot h'$, then

$$g' \cdot (g \cdot (g_\sigma w_\sigma)) = g' \cdot (g_\tau (h w_\sigma)) = g_\rho (h' (h w_\sigma)).$$

Since $(g' \cdot g) \cdot g_\sigma = g' \cdot (g \cdot g_\sigma) = g' \cdot g_\tau \cdot h = g_\rho \cdot h' \cdot h$, we have

$$(g' \cdot g) \cdot (g_\sigma w_\sigma) = g_\rho ((h' \cdot h) w_\sigma) = g_\rho (h' \cdot (h w_\sigma)),$$

as required.

Example 3.15. If $W = \bigoplus W_i$, then $\operatorname{Ind} W = \bigoplus \operatorname{Ind} W_i$.

The existence of the induced representation follows from Examples 3.14 and 3.15 since any W is a direct sum of summands of the regular representation.

Exercise 3.16. (a) If U is a representation of G and W a representation of H, show that (with all tensor products over $\mathbb{C}$)

$$U \otimes \operatorname{Ind} W = \operatorname{Ind}(\operatorname{Res}(U) \otimes W).$$

In particular, $\operatorname{Ind}(\operatorname{Res}(U)) = U \otimes P$, where P is the permutation representation of G on G/H. For a formula for $\operatorname{Res}(\operatorname{Ind}(W))$, for W a representation of H, see [Se2, p. 58].

(b) Like restriction, induction is transitive: if $H \subset K \subset G$ are subgroups, show that

$$\operatorname{Ind}_H^G(W) = \operatorname{Ind}_K^G(\operatorname{Ind}_H^K W).$$

Note that Example 3.15 says that the map Ind gives a group homomorphism between the representation rings $R(H)$ and $R(G)$, in the opposite direction from the ring homomorphism Res: $R(G) \to R(H)$ given by restriction; Exercise 3.16(a) says that this map satisfies a "push–pull" formula $\alpha \cdot \operatorname{Ind}(\beta) = \operatorname{Ind}(\operatorname{Res}(\alpha) \cdot \beta)$ with respect to the restriction map.

Proposition 3.17. *Let W be a representation of H, U a representation of G, and suppose $V = \operatorname{Ind} W$. Then any H-module homomorphism $\varphi\colon W \to U$ extends uniquely to a G-module homomorphism $\tilde{\varphi}\colon V \to U$. i.e.,*

$$\operatorname{Hom}_H(W, \operatorname{Res} U) = \operatorname{Hom}_G(\operatorname{Ind} W, U).$$

In particular, this universal property determines Ind W *up to canonical isomorphism.*

PROOF. With $V = \bigoplus_{\sigma \in G/H} \sigma \cdot W$ as before, define $\tilde{\varphi}$ on $\sigma \cdot W$ by

$$\sigma \cdot W \xrightarrow{g_\sigma^{-1}} W \xrightarrow{\varphi} U \xrightarrow{g_\sigma} U,$$

which is independent of the representative g_σ for σ since φ is H-linear. □

To compute the character of $V = \operatorname{Ind} W$, note that $g \in G$ maps σW to $g\sigma W$, so the trace is calculated from those cosets σ with $g\sigma = \sigma$, i.e., $s^{-1}gs \in H$ for $s \in \sigma$. Therefore,

$$\chi_{\operatorname{Ind} W}(g) = \sum_{g\sigma=\sigma} \chi_W(s^{-1}gs) \qquad (s \in \sigma \text{ arbitrary}). \tag{3.18}$$

Exercise 3.19. (a) If C is a conjugacy class of G, and $C \cap H$ decomposes into conjugacy classes $D_1, \ldots, D_r$ of H, (3.18) can be rewritten as: the value of the character of Ind W on C is

$$\chi_{\operatorname{Ind} W}(C) = \frac{|G|}{|H|} \sum_{i=1}^{r} \frac{|D_i|}{|C|} \chi_W(D_i).$$

(b) If W is the trivial representation of H, then

$$\chi_{\operatorname{Ind} W}(C) = \frac{[G:H]}{|C|} \cdot |C \cap H|.$$

Corollary 3.20 (Frobenius Reciprocity). *If W is a representation of H, and U a representation of G, then*

$$(\chi_{\operatorname{Ind} W}, \chi_U)_G = (\chi_W, \chi_{\operatorname{Res} U})_H.$$

PROOF. It suffices by linearity to prove this when W and U are irreducible. The left-hand side is the number of times U appears in Ind W, which is the dimension of $\operatorname{Hom}_G(\operatorname{Ind} W, U)$. The right-hand side is the dimension of $\operatorname{Hom}_H(W, \operatorname{Res} U)$. These dimensions are equal by the proposition. □

If W and U are irreducible, Frobenius reciprocity says: *the number of times U appears in* Ind W *is the same as the number of times W appears in* Res U.

Frobenius reciprocity can be used to find characters of G if characters of H are known.

Example 3.21. We compute $\operatorname{Ind}_H^G W$, when $H = \mathfrak{S}_2 \subset G = \mathfrak{S}_3$, $W = V_2$ (the standard representation) $= U_2'$ (the alternating representation). We know the irreducible represenatations of $\mathfrak{S}_3$: U_3, U_3', V_3, which restrict to U_2, $U_2' = V_2$, $U_2 \oplus U_2'$, respectively. Thus, by Frobenius, Ind $V_2 = U_3' \oplus V_3$.

Example 3.22. Consider next $H = \mathfrak{S}_3 \subset G = \mathfrak{S}_4$, $W = V_3$. Again we know the irreducible representations, and Res $U_4 = U_3$, Res $U_4' = U_3'$, Res $V_4 = U_3 \oplus V_3$ [the vector

$$(1, 1, 1, -3) \in V_4 = \{(x_1, x_2, x_3, x_4): \textstyle\sum x_i = 0\}$$

is fixed by H], Res $V_4' = U_3' \oplus V_3'$, with $V_3' = V_3$, and Res $W_4 = V_3$ (as one may see directly). Hence, Ind $V_3 = V_4 \oplus V_4' \oplus W_4$. (Note that the isomorphism Res $W_4 = V_3$ actually follows, since one W_4 is all that could be added to $V_4 \oplus V_4'$ to get Ind V_3.)

Exercise 3.23. Determine the isomorphism classes of the representations of $\mathfrak{S}_4$ induced by (i) the one-dimensional representation of the group generated by (1234) in which $(1234)\cdot v = iv$, $i = \sqrt{-1}$; (ii) the one-dimensional representation of the group generated by (123) in which $(123)\cdot v = e^{2\pi i/3}v$.

Exercise 3.24. Let $H = \mathfrak{A}_5 \subset G = \mathfrak{S}_5$. Show that Ind $U = U \oplus U'$, Ind $V = V \oplus V'$, and Ind $W = W \oplus W'$, whereas Ind Y = Ind $Z = \wedge^2 V$.

Exercise 3.25*. Which irreducible representations of $\mathfrak{S}_d$ remain irreducible when restricted to $\mathfrak{A}_d$? Which are induced from $\mathfrak{A}_d$? How much does this tell you about the irreducible representations of $\mathfrak{A}_d$?

Exercise 3.26*. There is a unique nonabelian group of order 21, which can be realized as the group of affine transformations $x \mapsto \alpha x + \beta$ of the line over the field with seven elements, with α a cube root of unity in that field. Find the irreducible representations and character table for this group.

Now that we have introduced the notion of induced representation, we can state two important theorems describing the characters of representations of a finite group. In the preceding lecture we mentioned the notion of *virtual character*; this is just an element of the image Λ of the character map

$$\chi\colon R(G) \to \mathbb{C}_{\text{class}}(G)$$

from the representation ring $R(G)$ of virtual representations. The following two theorems both state that in order to generate $\Lambda \otimes \mathbb{Q}$ (resp. Λ) it is enough to consider the simplest kind of induced representations, namely, those induced from cyclic (respective elementary) subgroups of G. For the proofs of these theorems we refer to [Se2, §9, 10]. We will not need them in these lectures.

Artin's Theorem 3.27. *The characters of induced representations from cyclic subgroups of G generate a lattice of finite index in* Λ.

A subgroup H of G is *p-elementary* if $H = A \times B$, with A cyclic of order prime to p and B a p-group.

Brauer's Theorem 3.28. *The characters of induced representations from elementary subgroups of G generate the lattice* Λ.

§3.4. The Group Algebra

There is an important notion that we have already dealt with implicitly but not explicitly; this is the group algebra $\mathbb{C}G$ associated to a finite group G. This is an object that for all intents and purposes can completely replace the group G itself; any statement about the representations of G has an exact equivalent statement about the group algebra. Indeed, to a large extent the choice of language is a matter of taste.

The underlying vector space of the group algebra of G is the vector space with basis $\{e_g\}$ corresponding to elements of the group G, that is, the underlying vector space of the regular representation. We define the algebra structure on this vector space simply by

$$e_g \cdot e_h = e_{gh}.$$

By a representation of the algebra $\mathbb{C}G$ on a vector space V we mean simply an algebra homomorphism

$$\mathbb{C}G \to \text{End}(V),$$

so that a representation V of $\mathbb{C}G$ is the same thing as a left $\mathbb{C}G$-module. Note that a representation $\rho\colon G \to \text{Aut}(V)$ will extend by linearity to a map $\tilde{\rho}\colon \mathbb{C}G \to \text{End}(V)$, so that representations of $\mathbb{C}G$ correspond exactly to representations of G; the left $\mathbb{C}G$-module given by $\mathbb{C}G$ itself corresponds to the regular representation.

If $\{W_i\}$ are the irreducible representations of G, then we have seen that the regular representation R decomposes

$$R = \bigoplus (W_i)^{\oplus \dim(W_i)}.$$

We can now refine this statement in terms of the group algebra: we have

Proposition 3.29. *As algebras,*

$$\mathbb{C}G \cong \bigoplus \mathrm{End}(W_i).$$

PROOF. As we have said, for any representation W of G, the map $G \to \mathrm{Aut}(W)$ extends by linearity to a map $\mathbb{C}G \to \mathrm{End}(W)$; applying this to each of the irreducible representations W_i gives us a canonical map

$$\varphi\colon \mathbb{C}G \to \bigoplus \mathrm{End}(W_i).$$

This is injective since the representation on the regular representation is faithful. Since both have dimension $\sum (\dim W_i)^2$, the map is an isomorphism. □

A few remarks are in order about the isomorphism φ of the proposition. First, φ can be interpreted as the Fourier transform, cf. Exercise 3.32. Note also that Proposition 2.28 has a natural interpretation in terms of the group algebra: it says that the center of $\mathbb{C}G$ consists of those $\sum \alpha(g)e_g$ for which α is a class function.

Next, we can thinkof φ as the decomposition of the semisimple algebra $\mathbb{C}G$ into a product of matrix algebras. It implies that the matrix entries of the irreducible representations give a basis for the space of all functions on G, cf. Exercise 2.35.

Note in particular that any irreducible representation is isomorphic to a (minimal) left ideal in $\mathbb{C}G$. These left ideals are generated by idempotents. In fact, we can interpret the projection formulas of the last lecture in the language of the group algebra: the formulas say simply that the elements

$$\dim W \cdot \frac{1}{|G|} \sum_{g \in G} \overline{\chi_W(g)} \cdot e_g \in \mathbb{C}G$$

are the idempotents in the group algebra corresponding to the direct sum factors in the decomposition of Proposition 3.29. To locate the irreducible representations W_i of a group G [not just a direct sum of $\dim(W_i)$ copies], we want to find other idempotents of $\mathbb{C}G$. We will see this carried out for the symmetric groups in the following lecture.

The group algebra also gives us another description of induced representations: if W is a representation of a subgroup H of G, then the induced representation may be constructed simply by

$$\mathrm{Ind}\, W = \mathbb{C}G \otimes_{\mathbb{C}H} W,$$

where G acts on the first factor: $g\cdot(e_{g'}\otimes w)=e_{gg'}\otimes w$. The isomorphism of the reciprocity theorem is then a special case of a general formula for a change of rings $\mathbb{C}H\to\mathbb{C}G$:

$$\operatorname{Hom}_{\mathbb{C}H}(W,U)=\operatorname{Hom}_{\mathbb{C}G}(\mathbb{C}G\otimes_{\mathbb{C}H}W,U).$$

Exercise 3.30*. The induced representation $\operatorname{Ind}(W)$ can also be realized concretely as a space of W-valued functions on G, which can be useful to produce matrix realizations, or when trying to decompose $\operatorname{Ind}(W)$ into irreducible pieces. Show that $\operatorname{Ind}(W)$ is isomorphic to

$$\operatorname{Hom}_H(\mathbb{C}G,W)\cong\{f\colon G\to W: f(hg)=hf(g),\quad \forall h\in H, g\in G\},$$

where G acts by $(g'\cdot f)(g)=f(gg')$.

Exercise 3.31. If $\mathbb{C}G$ is identified with the space of functions on G, the function φ corresponding to $\sum_{g\in G}\varphi(g)e_g$, show that the product in $\mathbb{C}G$ corresponds to the convolution $*$ of functions:

$$(\varphi*\psi)(g)=\sum_{h\in G}\varphi(h)\psi(h^{-1}g).$$

(With integration replacing summation, this indicates how one may extend the notion of regular representation to compact groups.)

Exercise 3.32*. If $\rho\colon G\to GL(V_\rho)$ is a representation, and φ is a function on G, define the *Fourier transform* $\hat{\varphi}(\rho)$ in $\operatorname{End}(V_\rho)$ by the formula

$$\hat{\varphi}(\rho)=\sum_{g\in G}\varphi(g)\cdot\rho(g).$$

(a) Show that $\widehat{\varphi*\psi}(\rho)=\widehat{\varphi}(\rho)\cdot\widehat{\psi}(\rho)$.

(b) Prove the *Fourier inversion formula*

$$\varphi(g)=\frac{1}{|G|}\sum\dim(V_\rho)\cdot\operatorname{Trace}(\rho(g^{-1})\cdot\hat{\varphi}(\rho)),$$

the sum over the irreducible representations ρ of G. This formula is equivalent to formulas (2.19) and (2.20).

(c) Prove the *Plancherel formula* for functions φ and ψ on G:

$$\sum_{g\in G}\varphi(g^{-1})\psi(g)=\frac{1}{|G|}\sum_{\rho}\dim(V_\rho)\cdot\operatorname{Trace}(\hat{\varphi}(\rho)\hat{\psi}(\rho)).$$

Our choice of left action of a group on a space has been perfectly arbitrary, and the entire story is the same if G acts on the *right* instead. Moreover, there is a standard way to change a right action into a left action, and vice versa: Given a right action of G on V, define the left action by

$$g\cdot v=v\cdot(g^{-1}),\qquad g\in G, v\in V.$$

If $A = \mathbb{C}G$ is the group algebra, a right action of G on V makes V a right A-module. To turn right modules into left modules, we can use the anti-involution $a \mapsto \hat{a}$ of A defined by $(\sum a_g e_g)^\wedge = \sum a_g e_{g^{-1}}$. A right A-module is then turned into a left A-module by setting $a \cdot v = v \cdot \hat{a}$.

The following exercise will take you back to the origins of representation theory in the 19th century, when Frobenius found the characters by factoring this determinant.

Exercise 3.33*. Given a finite group G of order n, take a variable x_g for each element g in G, and order the elements of G arbitrarily. Let F be the determinant of the $n \times n$ matrix whose entry in the row labeled by g and column labeled by h is $x_{g \cdot h^{-1}}$. This is a form of degree n in the n variables x_g, which is independent of the ordering. Normalize the factors of F to take the value 1 when $x_e = 1$ and $x_g = 0$ for $g \neq e$. Show that the irreducible factors of F correspond to the irreducible representations of G. Moreover, if F_ρ is the factor corresponding to the representation ρ, show that the degree of F_ρ is the degree $d(\rho)$ of the representation ρ, and that each F_ρ occurs in F $d(\rho)$ times. If χ_ρ is the character of ρ, and $g \neq e$, show that $\chi_\rho(g)$ is the coefficient of $x_g \cdot x_e^{d(\rho)-1}$ in F_ρ.

§3.5. Real Representations and Representations over Subfields of $\mathbb{C}$

If a group G acts on a real vector space V_0, then we say the corresponding complex representation of $V = V_0 \otimes_{\mathbb{R}} \mathbb{C}$ is *real*. To the extent that we are interested in the action of a group G on real rather than complex vector spaces, the problem we face is to say which of the complex representations of G we have studied are in fact real.

Our first guess might be that a representation is real if and only if its character is real-valued. This turns out not to be the case: the character of a real representation is certainly real-valued, but the converse need not be true. To find an example, suppose $G \subset \mathrm{SU}(2)$ is a finite, nonabelian subgroup. Then G acts on $\mathbb{C}^2 = V$ with a real-valued character since the trace of any matrix in $\mathrm{SU}(2)$ is real. If V were a real representation, however, then G would be a subgroup of $\mathrm{SO}(2) = S^1$, which is abelian. To produce such a group, note that $\mathrm{SU}(2)$ can be identified with the unit quaternions. Set $G = \{\pm 1, \pm i, \pm j, \pm k\}$. Then $G/\{\pm 1\}$ is abelian, so has four one-dimensional representations, which give four one-dimensional representations of G. Thus, G has one irreducible two-dimensional representation, whose character is real, but which is not real.

Exercise 3.34*. Compute the character table for this quaternion group G, and compare it with the character table of the dihedral group of order 8.

A more successful approach is to note that if V is a real representation of G, coming from V_0 as above, then one can find a positive definite symmetric bilinear form on V_0 which is preserved by G. This gives a symmetric bilinear form on V which is preserved by G. Not every representation will have such a form since degeneracies may arise when one tries to construct one following the construction of Proposition 1.5. In fact,

Lemma 3.35. *An irreducible representation V of G is real if and only if there is a nondegenerate symmetric bilinear form B on V preserved by G.*

Proof. If we have such B, and an arbitrary nondegenerate Hermitian form H, also G-invariant, then

$$V \overset{B}{\to} V^* \overset{H}{\to} V$$

gives a conjugate linear isomorphism φ from V to V: given $x \in V$, there is a unique $\varphi(x) \in V$ with $B(x, y) = H(\varphi(x), y)$, and φ commutes with the action of G. Then $\varphi^2 = \varphi \circ \varphi$ is a complex linear G-module homomorphism, so $\varphi^2 = \lambda \cdot \mathrm{Id}$. Moreover,

$$H(\varphi(x), y) = B(x, y) = B(y, x) = H(\varphi(y), x) = \overline{H(x, \varphi(y))},$$

from which it follows that $H(\varphi^2(x), y) = H(x, \varphi^2(y))$, and therefore λ is a positive real number. Changing H by a scalar, we may assume $\lambda = 1$, so $\varphi^2 = \mathrm{Id}$. Thus, V is a sum of real eigenspaces V_+ and V_- for φ corresponding to eigenvalues 1 and -1. Since φ commutes with G, V_+ and V_- are G-invariant subspaces. Finally, $\varphi(ix) = -i\varphi(x)$, so $iV_+ = V_-$, and $V = V_+ \otimes \mathbb{C}$. □

Note from the proof that a real representation is also characterized by the existence of a conjugate linear endomorphism of V whose square is the identity; if $V = V_0 \otimes_{\mathbb{R}} \mathbb{C}$, it is given by conjugation: $v_0 \otimes \lambda \mapsto v_0 \otimes \bar{\lambda}$.

A warning is in order here: an irreducible representation of G on a vector space over $\mathbb{R}$ may become reducible when we extend the group field to $\mathbb{C}$. To give the simplest example, the representation of $\mathbb{Z}/n$ on $\mathbb{R}^2$ given by

$$\rho: k \mapsto \begin{pmatrix} \cos\dfrac{2\pi k}{n} & -\sin\dfrac{2\pi k}{n} \\ \sin\dfrac{2\pi k}{n} & \cos\dfrac{2\pi k}{n} \end{pmatrix}$$

is irreducible over $\mathbb{R}$ for $n > 2$ (no line in $\mathbb{R}^2$ is fixed by the action of $\mathbb{Z}/n$), but will be reducible over $\mathbb{C}$. Thus, classifying the irreducible representations of G over $\mathbb{C}$ that are real does not mean that we have classified all the irreducible real representations. However, we will see in Exercise 3.39 below how to finish the story once we have found the real representations of G that are irreducible over $\mathbb{C}$.

Suppose V is an irreducible representation of G with χ_V real. Then there is a G-equivariant isomorphism $V \cong V^*$, i.e., there is a G-equivariant (nondegenerate) bilinear form B on V; but, in general, B need not be symmetric. Regarding B in

$$V^* \otimes V^* = \mathrm{Sym}^2 V^* \oplus \wedge^2 V^*,$$

and noting the uniqueness of B up to multiplication by scalars, we see that B is either symmetric or skew-symmetric. If B is skew-symmetric, proceeding as above one can scale so $\varphi^2 = -\mathrm{Id}$. This makes V "quaternionic," with φ becoming multiplication[2] by j:

Definition 3.36. A *quaternionic* representation is a (complex) representation V which has a G-invariant homomorphism $J: V \to V$ that is conjugate linear, and satisfies $J^2 = -\mathrm{Id}$. Thus, a skew-symmetric nondegenerate G-invariant B determines a quaternionic structure on V.

Summarizing the preceding discussion we have the

Theoem 3.37. *An irreducible representation V is one and only one of the following*:

(1) Complex: χ_V *is not real-valued*; V *does not have a G-invariant nondegenerate bilinear form.*

(2) Real: $V = V_0 \otimes \mathbb{C}$, *a real representation*; V *has a G-invariant symmetric nondegenerate bilinear form.*

(3) Quaternionic: χ_V *is real, but* V *is not real*; V *has a G-invariant skew-symmetric nondegeneate bilinear form.*

Exercise 3.38. Show that for V irreducible,

$$\frac{1}{|G|}\sum_{g \in G} \chi_V(g^2) = \begin{cases} 0 & \text{if } V \text{ is complex} \\ 1 & \text{if } V \text{ is real} \\ -1 & \text{if } V \text{ is quaternionic.} \end{cases}$$

This verifies that the three cases in the theorem are mutually exclusive. It also implies that if the order of G is odd, all nontrivial representations must be complex.

Exercise 3.39. Let V_0 be a real vector space on which G acts irreducibly, $V = V_0 \otimes \mathbb{C}$ the corresponding real representation of G. Show that if V is not irreducible, then it has exactly two irreducible factors, and they are conjugate complex representations of G.

[2] See §7.2 for more on quaternions and quaternonic representations.

Exercise 3.40. Classify the real representations of $\mathfrak{A}_4$.

Exercise 3.41*. The group algebra $\mathbb{R}G$ is a product of simple $\mathbb{R}$-algebras corresponding to the irreducible representations over $\mathbb{R}$. These simple algebras are matrix algebras over $\mathbb{C}$, $\mathbb{R}$, or the quaternions $\mathbb{H}$ according as the representation is complex, real, or quaternionic.

Exercise 3.42*. (a) Show that all characters of a group are real if and only if every element is conjugate to its inverse.

(b) Show that an element σ in a split conjugacy class of $\mathfrak{A}_d$ is conjugate to its inverse if and only if the number of cycles in σ whose length is congruent to 3 modulo 4 is even.

(c) Show that the only d's for which every character of $\mathfrak{A}_d$ is real-valued are $d = 1, 2, 5, 6, 10$, and 14.

Exercise 3.43*. Show that: (i) the tensor product of two real or two quaternionic representations is real; (ii) for any V, $V^* \otimes V$ is real; (iii) if V is real, so are all $\wedge^k V$; (iv) if V is quaternionic, $\wedge^k V$ is real for k even, quaternionic for k odd.

Representations over Subfields of $\mathbb{C}$ in General

We consider next the generalization of the preceding problem to more general subfields of $\mathbb{C}$. Unfortunately, our results will not be nearly as strong in general, but we can at least express the problem neatly in terms of the representation ring of G.

To begin with, our terminology in this general setting is a little different. Let $K \subset \mathbb{C}$ be any subfield. We define a K-representation of G to be a vector space V_0 over K on which G acts; in this case we say that the complex representation $V = V_0 \otimes \mathbb{C}$ is *defined over K*.

One way to measure how many of the representations of G are defined over a field K is to introduce the *representation ring* $R_K(G)$ *of G over K*. This is defined just like the ordinary representation ring; that is, it is just the group of formal linear combinations of K-representations of G modulo relations of the form $V + W - (V \oplus W)$, with multiplication given by tensor product.

Exercise 3.44*. Describe the representation ring of G over $\mathbb{R}$ for some of the groups G whose complex representation we have analyzed above. In particular, is the rank of $R_{\mathbb{R}}(G)$ always the same as the rank of $R(G)$?

Exercise 3.45*. (a) Show that $R_K(G)$ is the subring of the ring of class functions on G generated (as an additive group) by characters of representations defined over K.

(b) Show that the characters of irreducible representations over K form an orthogonal basis for $R_K(G)$.

(c) Show that a complex representation of G can be defined over K if and only if its character belongs to $R_K(G)$.

For more on the relation between $R_K(G)$ and $R(G)$, see [Se2].

LECTURE 4

Representations of $\mathfrak{S}_d$: Young Diagrams and Frobenius's Character Formula

In this lecture we get to work. Specifically, we give in §4.1 a complete description of the irreducible representations of the symmetric group, that is, a construction of the representations (via Young symmetrizers) and a formula (Frobenius' formula) for their characters. The proof that the representations constructed in §4.1 are indeed the irreducible representations of the symmetric group is given in §4.2; the proof of Frobenius' formula, as well as a number of others, in §4.3. Apart from their intrinsic interest (and undeniable beauty), these results turn out to be of substantial interest in Lie theory: analogs of the Young symmetrizers will give a construction of the irreducible representations of $SL_n\mathbb{C}$. At the same time, while the techniques of this lecture are completely elementary (we use only a few identities about symmetric polynomials, proved in Appendix A), the level of difficulty is clearly higher than in preceding lectures. The results in the latter half of §4.3 (from Corollary 4.39 on) in particular are quite difficult, and inasmuch as they are not used later in the text may be skipped by readers who are not symmetric group enthusiasts.

§4.1. Statements of the Results

The number of irreducible representaton of $\mathfrak{S}_d$ is the number of conjugacy classes, which is the number $p(d)$ of partitions[1] of d: $d = \lambda_1 + \cdots + \lambda_k$, $\lambda_1 \geq \cdots \geq \lambda_k \geq 1$. We have

[1] It is sometimes convenient, and sometimes a nuisance, to have partitions that end in one or more zeros; if convenient, we allow some of the λ_i on the end to be zero. Two sequences define the same partition, of course, if they differ only by zeros at the end.

$$\sum_{d=0}^{\infty} p(d)t^d = \prod_{n=1}^{\infty}\left(\frac{1}{1-t^n}\right)$$
$$= (1 + t + t^2 + \cdots)(1 + t^2 + t^4 + \cdots)(1 + t^3 + \cdots)\cdots$$

which converges exactly in $|t| < 1$. This partition number is an interesting arithmetic function, whose congruences and growth behavior as a function of d have been much studied (cf. [Har], [And]). For example, $p(d)$ is asymptotically equal to $(1/\alpha d)e^{\beta\sqrt{d}}$, with $\alpha = 4\sqrt{3}$ and $\beta = \pi\sqrt{2/3}$.

To a partition $\lambda = (\lambda_1, \ldots, \lambda_k)$ is associated a *Young diagram* (sometimes called a Young frame or Ferrers diagram)

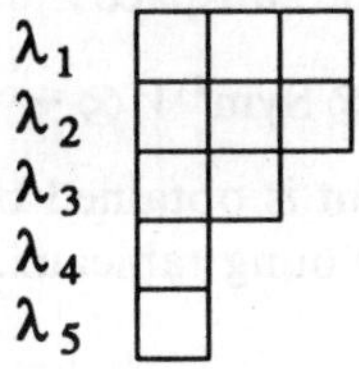

with λ_i boxes in the ith row, the rows of boxes lined up on the left. The *conjugate partition* $\lambda' = (\lambda'_1, \ldots, \lambda'_r)$ to the partition λ is defined by interchanging rows and columns in the Young diagram, i.e., reflecting the diagram in the 45° line. For example, the diagram above is that of the partition $(3, 3, 2, 1, 1)$, whose conjugate is $(5, 3, 2)$. (Without reference to the diagram, the conjugate partition to λ can be defined by saying λ'_i is the number of terms in the partition λ that are greater than or equal to i.)

Young diagrams can be used to describe projection operators for the regular representation, which will then give the irreducible representations of $\mathfrak{S}_d$. For a given Young diagram, number the boxes, say consecutively as shown:

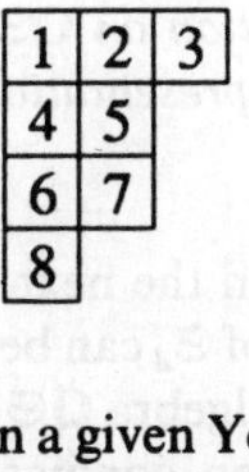

More generally, define a *tableau* on a given Young diagram to be a numbering of the boxes by the integers $1, \ldots, d$. Given a tableau, say the canonical one shown, define two subgroups[2] of the symmetric group

[2] If a tableau other than the canonical one were chosen, one would get different groups in place of P and Q, and different elements in the group ring, but the representations constructed this way will be isomorphic.

$$P = P_\lambda = \{g \in \mathfrak{S}_d : g \text{ preserves each row}\}$$

and

$$Q = Q_\lambda = \{g \in \mathfrak{S}_d : g \text{ preserves each column}\}.$$

In the group algebra $\mathbb{C}\mathfrak{S}_d$, we introduce two elements corresponding to these subgroups: we set

$$a_\lambda = \sum_{g \in P} e_g \quad \text{and} \quad b_\lambda = \sum_{g \in Q} \operatorname{sgn}(g) \cdot e_g. \tag{4.1}$$

To see what a_λ and b_λ do, observe that if V is any vector space and $\mathfrak{S}_d$ acts on the dth tensor power $V^{\otimes d}$ by permuting factors, the image of the element $a_\lambda \in \mathbb{C}\mathfrak{S}_d \to \operatorname{End}(V^{\otimes d})$ is just the subspace

$$\operatorname{Im}(a_\lambda) = \operatorname{Sym}^{\lambda_1} V \otimes \operatorname{Sym}^{\lambda_2} V \otimes \cdots \otimes \operatorname{Sym}^{\lambda_k} V \subset V^{\otimes d},$$

where the inclusion on the right is obtained by grouping the factors of $V^{\otimes d}$ according to the rows of the Young tableaux. Similarly, the image of b_λ on this tensor power is

$$\operatorname{Im}(b_\lambda) = \Lambda^{\mu_1} V \otimes \Lambda^{\mu_2} V \otimes \cdots \otimes \Lambda^{\mu_l} V \subset V^{\otimes d},$$

where μ is the conjugate partition to λ.

Finally, we set

$$c_\lambda = a_\lambda \cdot b_\lambda \in \mathbb{C}\mathfrak{S}_d; \tag{4.2}$$

this is called a *Young symmetrizer*. For example, when $\lambda = (d)$, $c_{(d)} = a_{(d)} = \sum_{g \in \mathfrak{S}_d} e_g$, and the image of $c_{(d)}$ on $V^{\otimes d}$ is $\operatorname{Sym}^d V$. When $\lambda = (1, \ldots, 1)$, $c_{(1,\ldots,1)} = b_{(1,\ldots,1)} = \sum_{g \in \mathfrak{S}_d} \operatorname{sgn}(g) e_g$, and the image of $c_{(1,\ldots,1)}$ on $V^{\otimes d}$ is $\Lambda^d V$. We will eventually see that the image of the symmetrizers c_λ in $V^{\otimes d}$ provide essentially all the finite-dimensional irreducible representations of $\operatorname{GL}(V)$. Here we state the corresponding fact for representations of $\mathfrak{S}_d$:

Theorem 4.3. *Some scalar multiple of c_λ is idempotent, i.e., $c_\lambda^2 = n_\lambda c_\lambda$, and the image of c_λ (by right multiplication on $\mathbb{C}\mathfrak{S}_d$) is an irreducible representation V_λ of $\mathfrak{S}_d$. Every irreducible representation of $\mathfrak{S}_d$ can be obtained in this way for a unique partition.*

We will prove this theorem in the next section. Note that, as a corollary, each irreducible representation of $\mathfrak{S}_d$ can be defined over the rational numbers since c_λ is in the rational group algebra $\mathbb{Q}\mathfrak{S}_d$. Note also that the theorem gives a direct correspondence between conjugacy classes in $\mathfrak{S}_d$ and irreducible representations of $\mathfrak{S}_d$, something which has never been achieved for general groups.

For example, for $\lambda = (d)$,

$$V_{(d)} = \mathbb{C}\mathfrak{S}_d \cdot \sum_{g \in \mathfrak{S}_d} e_g = \mathbb{C} \cdot \sum_{g \in \mathfrak{S}_d} e_g$$

is the trivial representation U, and when $\lambda = (1, \ldots, 1)$,

$$V_{(1,\ldots,1)} = \mathbb{C}\mathfrak{S}_d \cdot \sum_{g \in \mathfrak{S}_d} \operatorname{sgn}(g) e_g = \mathbb{C} \cdot \sum_{g \in \mathfrak{S}_d} \operatorname{sgn}(g) e_g$$

is the alternating representation U'. For $\lambda = (2, 1)$,

$$c_{(2,1)} = (e_1 + e_{(12)}) \cdot (e_1 - e_{(13)}) = 1 + e_{(12)} - e_{(13)} - e_{(132)}$$

in $\mathbb{C}\mathfrak{S}_3$, and $V_{(2,1)}$ is spanned by $c_{(2,1)}$ and $(13) \cdot c_{(2,1)}$, so $V_{(2,1)}$ is the standard representation of $\mathfrak{S}_3$.

Exercise 4.4*. Set $A = \mathbb{C}\mathfrak{S}_d$, so $V_\lambda = Ac_\lambda = Aa_\lambda b_\lambda$.

(a) Show that $V_\lambda \cong Ab_\lambda a_\lambda$.

(b) Show that V_λ is the image of the map from Aa_λ to Ab_λ given by right multiplication by b_λ. By (a), this is isomorphic to the image of $Ab_\lambda \to Aa_\lambda$ given by right multiplication by a_λ.

(c) Using (a) and the description of V_λ in the theorem show that

$$V_{\lambda'} = V_\lambda \otimes U',$$

where λ' is the conjugate partition to λ and U' is the alternating representation.

Examples 4.5. In earlier lectures we described the irreducible representations of $\mathfrak{S}_d$ for $d \leq 5$. From the construction of the representation corresponding to a Young diagram it is not hard to work out which representations come from which diagrams:

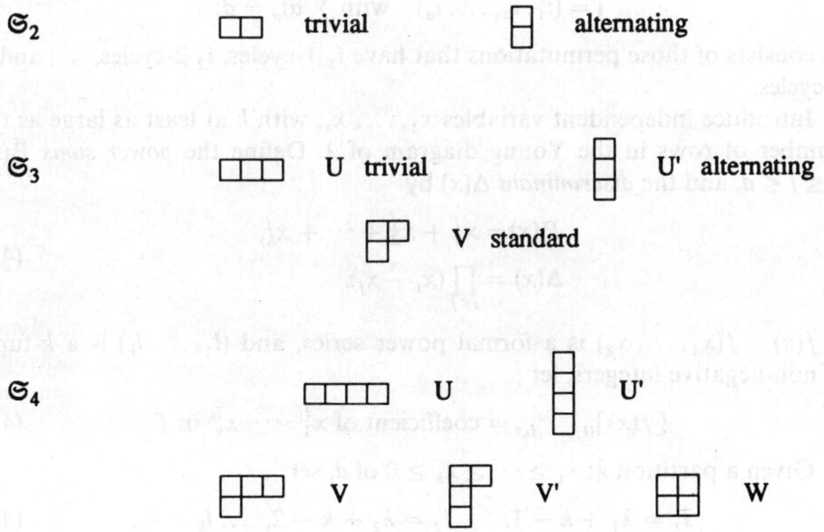

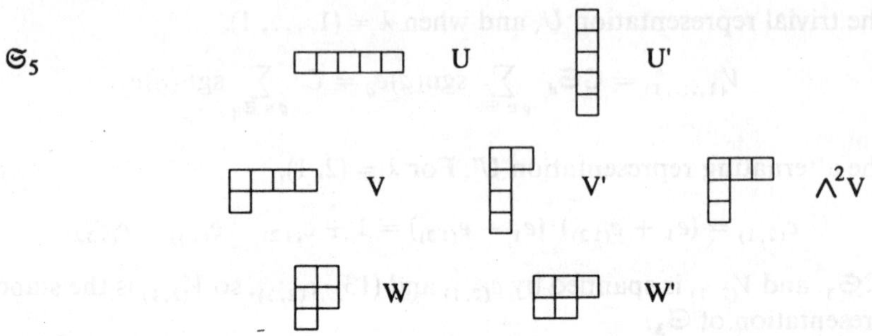

Exercise 4.6*. Show that for general d, the standard representation V corresponds to the partition $d = (d - 1) + 1$. As a challenge, you can try to prove that the exterior powers of the standard representation V are represented by a "hook":

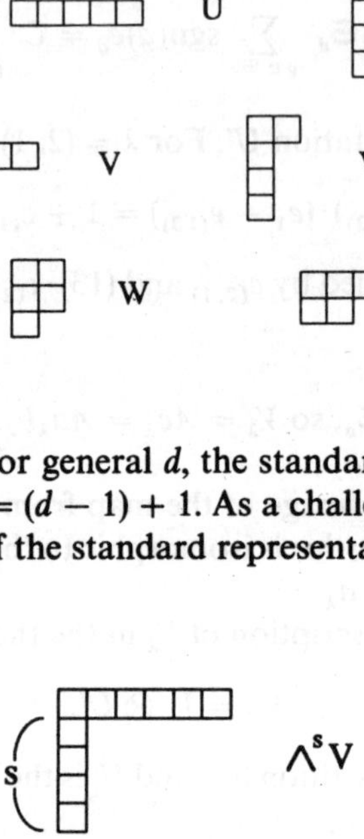

Note that this recovers our theorem that the $\Lambda^s V$ are irreducible.

Next we turn to Frobenius's formula for the character χ_λ of V_λ, which includes a formula for its dimension. Let $C_\mathbf{i}$ denote the conjugacy class in $\mathfrak{S}_d$ determined by a sequence

$$\mathbf{i} = (i_1, i_2, \ldots, i_d) \quad \text{with} \sum \alpha i_\alpha = d:$$

$C_\mathbf{i}$ consists of those permutations that have i_1 1-cycles, i_2 2-cycles, ..., and i_d d-cycles.

Introduce independent variables $x_1, \ldots, x_k$, with k at least as large as the number of rows in the Young diagram of λ. Define the *power sums* $P_j(x)$, $1 \le j \le d$, and the *discriminant* $\Delta(x)$ by

$$\begin{aligned} P_j(x) &= x_1^j + x_2^j + \cdots + x_k^j, \\ \Delta(x) &= \prod_{i<j} (x_i - x_j). \end{aligned} \tag{4.7}$$

If $f(x) = f(x_1, \ldots, x_k)$ is a formal power series, and $(l_1, \ldots, l_k)$ is a k-tuple of non-negative integers, let

$$[f(x)]_{(l_1,\ldots,l_k)} = \text{coefficient of } x_1^{l_1} \cdot \cdots \cdot x_k^{l_k} \text{ in } f. \tag{4.8}$$

Given a partition λ: $\lambda_1 \ge \cdots \ge \lambda_k \ge 0$ of d, set

$$l_1 = \lambda_1 + k - 1, \qquad l_2 = \lambda_2 + k - 2, \ldots, l_k = \lambda_k, \tag{4.9}$$

a strictly decreasing sequence of k non-negative integers. The character of V_λ evaluated on $g \in C_{\mathbf{i}}$ is given by the remarkable

Frobenius Formula 4.10

$$\chi_\lambda(C_{\mathbf{i}}) = \left[\Delta(x) \cdot \prod_j P_j(x)^{i_j} \right]_{(l_1, \ldots, l_k)}.$$

For example, if $d = 5$, $\lambda = (3, 2)$, and $C_{\mathbf{i}}$ is the conjugacy class of (12)(345), i.e., $i_1 = 0$, $i_2 = 1$, $i_3 = 1$, then

$$\chi_{(3,2)}(C_{\mathbf{i}}) = [(x_1 - x_2) \cdot (x_1^2 + x_2^2)(x_1^3 + x_2^3)]_{(4,2)} = 1.$$

Other entries in our character tables for $\mathfrak{S}_3$, $\mathfrak{S}_4$, and $\mathfrak{S}_5$ can be verified as easily, verifying the assertions of Examples 4.5.

In terms of certain symmetric functions S_λ called *Schur polynomials*, Frobenius's formula can be expressed by

$$\prod_j P_j(x)^{i_j} = \sum \chi_\lambda(C_{\mathbf{i}}) S_\lambda,$$

the sum over all partitions λ of d in at most k parts (cf. Proposition 4.37 and (A.27)). Although we do not use Schur polynomials explicitly in this lecture, they play the central role in the algebraic background developed in Appendix A.

Let us use the Frobenius formula to compute the dimension of V_λ. The conjugacy class of the identity corresponds to $\mathbf{i} = (d)$, so

$$\dim V_\lambda = \chi_\lambda(C_{(d)}) = [\Delta(x) \cdot (x_1 + \cdots + x_k)^d]_{(l_1, \ldots, l_k)}.$$

Now $\Delta(x)$ is the Vandermonde determinant:

$$\begin{vmatrix} 1 & x_k & \cdots & x_k^{k-1} \\ \vdots & \vdots & & \vdots \\ 1 & x_1 & \cdots & x_1^{k-1} \end{vmatrix} = \sum_{\sigma \in \mathfrak{S}_k} (\operatorname{sgn} \sigma) x_k^{\sigma(1)-1} \cdot \ldots \cdot x_1^{\sigma(k)-1}.$$

The other term is

$$(x_1 + \cdots + x_k)^d = \sum \frac{d!}{r_1! \cdot \ldots \cdot r_k!} x_1^{r_1} x_2^{r_2} \cdot \ldots \cdot x_k^{r_k},$$

the sum over k-tuples $(r_1, \ldots, r_k)$ that sum to d. To find the coefficient of $x_1^{l_1} \cdot \ldots \cdot x_k^{l_k}$ in the product, we pair off corresponding terms in these two sums, getting

$$\sum \operatorname{sgn}(\sigma) \cdot \frac{d!}{(l_1 - \sigma(k) + 1)! \cdots (l_k - \sigma(1) + 1)!},$$

the sum over those σ in $\mathfrak{S}_k$ such that $l_{k-i+1} - \sigma(i) + 1 \geq 0$ for all $1 \leq i \leq k$. This sum can be written as

$$\frac{d!}{l_1!\cdots l_k!}\sum_{\sigma\in\mathfrak{S}_k}\operatorname{sgn}(\sigma)\prod_{j=1}^{k} l_j(l_j-1)\cdot\ldots\cdot(l_j-\sigma(k-j+1)+2)$$

$$=\frac{d!}{l_1!\cdots l_k!}\begin{vmatrix} 1 & l_k & l_k(l_k-1) & \cdots \\ \vdots & \vdots & \vdots & \vdots \\ 1 & l_1 & l_1(l_1-1) & \cdots \end{vmatrix}.$$

By column reduction this determinant reduces to the Vandermonde determinant, so

$$\dim V_\lambda = \frac{d!}{l_1!\cdots l_k!}\prod_{i<j}(l_i-l_j), \tag{4.11}$$

with $l_i = \lambda_i + k - i$.

There is another way of expressing the dimensions of the V_λ. The *hook length* of a box in a Young diagram is the number of squares directly below or directly to the right of the box, including the box once.

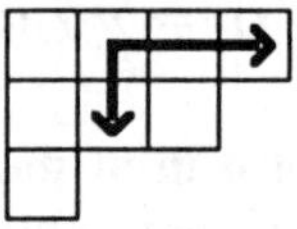

In the following diagram, each box is labeled by its hook length:

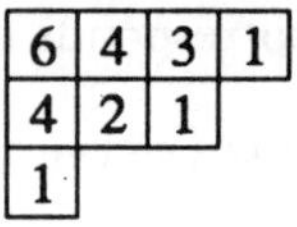

Hook Length Formula 4.12.

$$\dim V_\lambda = \frac{d!}{\prod(\text{Hook lengths})}.$$

For the above partition $4+3+1$ of 8, the dimension of the corresponding representation of $\mathfrak{S}_8$ is therefore $8!/6\cdot4\cdot4\cdot2\cdot3 = 70$.

Exercise 4.13*. Deduce the hook length formula from the Frobenius formula (4.11).

Exercise 4.14*. Use the hook length formula to show that the only irreducible representations of $\mathfrak{S}_d$ of dimension less than d are the trivial and alternating representations U and U' of dimension 1, the standard representation V and $V' = V\otimes U'$ of dimension $d-1$, and three other examples: the two-dimensional representation of $\mathfrak{S}_4$ corresponding to the partition $4 = 2+2$, and the two five-dimensional representations of $\mathfrak{S}_6$ corresponding to the partitions $6 = 3+3$ and $6 = 2+2+2$.

Exercise 4.15*. Using Frobenius's formula or otherwise, show that:

$$\chi_{(d-1,1)}(C_{\mathbf{i}}) = i_1 - 1;$$
$$\chi_{(d-2,1,1)}(C_{\mathbf{i}}) = \tfrac{1}{2}(i_1 - 1)(i_1 - 2) - i_2;$$
$$\chi_{(d-2,2)}(C_{\mathbf{i}}) = \tfrac{1}{2}(i_1 - 1)(i_1 - 2) + i_2 - 1.$$

Can you continue this list?

Exercise 4.16*. If g is a cycle of length d in $\mathfrak{S}_d$, show that $\chi_\lambda(g)$ is ± 1 if λ is a hook, and zero if λ is not a hook:

$$\chi_\lambda(g) = \begin{cases} (-1)^s & \text{if } \lambda = (d-s, 1, \ldots, 1), 0 \le s \le d-1 \\ 0 & \text{otherwise.} \end{cases}$$

Exercise 4.17. Frobenius [Fro1] used his formula to compute the value of χ_λ on a cycle of length $m \le d$.

(a) Following the procedure that led to (4.11)—which was the case $m = 1$—show that

$$\chi_\lambda((12\ldots m)) = \frac{\dim V_\lambda}{-m^2 h_m} \sum_{p=1}^{k} \frac{\psi(l_p)}{\varphi'(l_p)}, \tag{4.18}$$

where $h_m = d!/(d-m)!m$ is the number of cycles of length m (if $m > 1$), and

$$\varphi(x) = \prod_{i=1}^{k} (x - l_i), \qquad \psi(x) = \varphi(x - m) \prod_{j=1}^{m} (x - j + 1).$$

The sum in (4.18) can be realized as the coefficient of x^{-1} in the Laurent expansion of $\psi(x)/\varphi(x)$ at $x = \infty$.

Define the *rank* r of a partition to be the length of the diagonal of its Young diagram, and let a_i and b_i be the number of boxes below and to the right of the ith box of the diagonal, reading from lower right to upper left. Frobenius called $\begin{pmatrix} a_1 a_2 \ldots a_r \\ b_1 b_2 \ldots b_r \end{pmatrix}$ the *characteristics* of the partition. (Many writers now use a reverse notation for the characteristics, writing $(b_r, \ldots, b_1 | a_r, \ldots, a_r)$ instead.) For the partition (10, 9, 9, 4, 4, 4, 1):

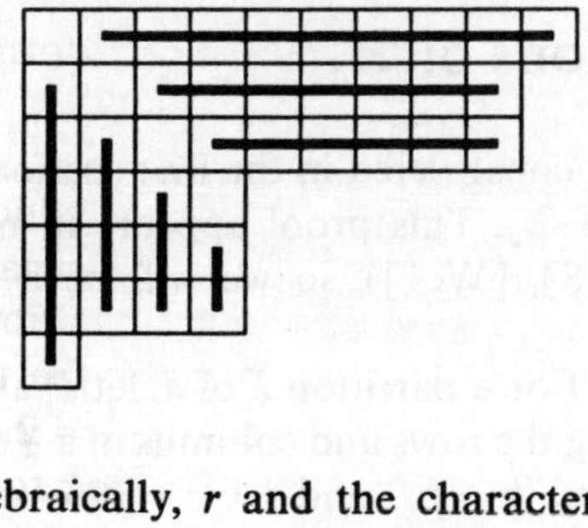

$$r = 4$$

$$\text{characteristics} = \begin{pmatrix} 2 & 3 & 4 & 6 \\ 0 & 6 & 7 & 9 \end{pmatrix}$$

Algebraically, r and the characteristics $a_1 < \cdots < a_r$ and $b_1 < \cdots < b_r$ are determined by requiring the equality of the two sets

$$\{l_1, \ldots, l_k, k-1-a_1, \ldots, k-1-a_r\} \text{ and}$$
$$\{0, 1, \ldots, k-1, k+b_1, \ldots, k+b_r\}.$$

(b) Show that $\psi(x)/\varphi(x) = g(y)/f(y)$, where $y = x - d$ and

$$f(y) = \frac{\prod_{i=1}^{r}(y - b_i)}{\prod_{i=1}^{r}(y + a_i + 1)}, \qquad g(y) = f(y-m)\prod_{j=1}^{m}(y - j + 1).$$

Deduce that the sum in (4.18) is the coefficient of x^{-1} in $g(x)/f(x)$.

(c) When $m = 2$, use this to prove the formula

$$\chi_\lambda((12)) = \frac{\dim V_\lambda}{d(d-1)}\sum_{i=1}^{r}(b_i(b_i+1) - a_i(a_i+1)).$$

Hurwitz [Hur] used this formula of Frobenius to calculate the number of ways to write a given permutation as a product of transpositions. From this he gave a formula for the number of branched coverings of the Riemann sphere with a given number of sheets and given simple branch points. Ingram [In] has given other formulas for $\chi_\lambda(g)$, when g is a somewhat more complicated conjugacy class.

Exercise 4.19*. If V is the standard representation of $\mathfrak{S}_d$, prove the decompositions into irreducible representations:

$$\mathrm{Sym}^2 V \cong U \oplus V \oplus V_{(d-2,2)},$$
$$V \otimes V = \mathrm{Sym}^2 V \oplus \wedge^2 V \cong U \oplus V \oplus V_{(d-2,2)} \oplus V_{(d-2,1,1)}.$$

Exercise 4.20*. Suppose λ is symmetric, i.e., $\lambda = \lambda'$, and let $q_1 > q_2 > \cdots > q_r > 0$ be the lengths of the symmetric hooks that form the diagram of λ; thus, $q_1 = 2\lambda_1 - 1$, $q_2 = 2\lambda_2 - 3, \ldots$. Show that if g is a product of disjoint cycles of lengths $q_1, q_2, \ldots, q_r$, then

$$\chi_\lambda(g) = (-1)^{(d-r)/2}.$$

§4.2. Irreducible Representations of $\mathfrak{S}_d$

We show next that the representations V_λ constructed in the first section are exactly the irreducible representations of $\mathfrak{S}_d$. This proof appears in many standard texts (e.g. [C-R], [Ja-Ke], [N-S], [We1]), so we will be a little concise.

Let $A = \mathbb{C}\mathfrak{S}_d$ be the group ring of $\mathfrak{S}_d$. For a partition λ of d, let P and Q be the corresponding subgroups preserving the rows and columns of a Young tableau T corresponding to λ, let $a = a_\lambda$, $b = b_\lambda$, and let $c = c_\lambda = ab$ be the corresponding Young symmetrizer, so $V_\lambda = Ac_\lambda$ is the corresponding representation. (These groups and elements should really be subscripted by T to denote dependence on the tableau chosen, but the assertions made depend only on the partition, so we usually omit reference to T.)

Note that $P \cap Q = \{1\}$, so an element of $\mathfrak{S}_d$ can be written in at most one way as a product $p \cdot q$, $p \in P$, $q \in Q$. Thus, c is the sum $\sum \pm e_g$, the sum over all g that can be written as $p \cdot q$, with coefficient ± 1 being $\operatorname{sgn}(q)$; in particular, the coefficient of e_1 in c is 1.

Lemma 4.21. (1) *For $p \in P$, $p \cdot a = a \cdot p = a$.*

(2) *For $q \in Q$, $(\operatorname{sgn}(q)q) \cdot b = b \cdot (\operatorname{sgn}(q)q) = b$.*

(3) *For all $p \in P$, $q \in Q$, $p \cdot c \cdot (\operatorname{sgn}(q)q) = c$, and, up to multiplication by a scalar, c is the only such element in A.*

Proof. Only the last assertion is not obvious. If $\sum n_g e_g$ satisfies the condition in (3), then $n_{pgq} = \operatorname{sgn}(q) n_g$ for all g, p, q; in particular, $n_{pq} = \operatorname{sgn}(q) n_1$. Thus, it suffices to verify that $n_g = 0$ if $g \notin PQ$. For such g it suffices to find a transposition t such that $p = t \in P$ and $q = g^{-1}tg \in Q$; for then $g = pgq$, so $n_g = -n_g$. If $T' = gT$ is the tableau obtained by replacing each entry i of T by $g(i)$, the claim is that there is are two distinct integers that appear in the same row of T and in the same column of T'; t is then the transposition of these two integers. We must verify that if there were no such pair of integers, then one could write $g = p \cdot q$ for some $p \in P$, $q \in Q$. To do this, first take $p_1 \in P$ and $q_1' \in Q' = gQg^{-1}$ so that $p_1 T$ and $q_1' T'$ have the same first row; repeating on the rest of the tableau, one gets $p \in P$ and $q' \in Q'$ so that $pT = q'T'$. Then $pT = q'gT$, so $p = q'g$, and therefore $g = pq$, where $q = g^{-1}(q')^{-1}g \in Q$, as required. □

We order partitions *lexicographically*:

$$\lambda > \mu \quad \text{if the first nonvanishing } \lambda_i - \mu_i \text{ is positive.} \tag{4.22}$$

Lemma 4.23. (1) *If $\lambda > \mu$, then for all $x \in A$, $a_\lambda \cdot x \cdot b_\mu = 0$. In particular, if $\lambda > \mu$, then $c_\lambda \cdot c_\mu = 0$.*

(2) *For all $x \in A$, $c_\lambda \cdot x \cdot c_\lambda$ is a scalar multiple of c_λ. In particular, $c_\lambda \cdot c_\lambda = n_\lambda c_\lambda$ for some $n_\lambda \in \mathbb{C}$.*

Proof. For (1), we may take $x = g \in \mathfrak{S}_d$. Since $g \cdot b_\mu \cdot g^{-1}$ is the element constructed from gT', where T' is the tableau used to construct b_μ, it suffices to show that $a_\lambda \cdot b_\mu = 0$. One verifies that $\lambda > \mu$ implies that there are two integers in the same row of T and the same column of T'. If t is the transposition of these integers, then $a_\lambda \cdot t = a_\lambda$, $t \cdot b_\mu = -b_\mu$, so $a_\lambda \cdot b_\mu = a_\lambda \cdot t \cdot t \cdot b_\mu = -a_\lambda \cdot b_\mu$, as required. Part (2) follows from Lemma 4.21 (3). □

Exercise 4.24*. Show that if $\lambda \neq \mu$, then $c_\lambda \cdot A \cdot c_\mu = 0$; in particular, $c_\lambda \cdot c_\mu = 0$.

Lemma 4.25. (1) *Each V_λ is an irreducible representation of $\mathfrak{S}_d$.*
(2) *If $\lambda \neq \mu$, then V_λ and V_μ are not isomorphic.*

Proof. For (1) note that $c_\lambda V_\lambda \subset \mathbb{C}c_\lambda$ by Lemma 4.23. If $W \subset V_\lambda$ is a subrepresentation, then $c_\lambda W$ is either $\mathbb{C}c_\lambda$ or 0. If the first is true, then $V_\lambda = A \cdot c_\lambda \subset W$. Otherwise $W \cdot W \subset A \cdot c_\lambda W = 0$, but this implies $W = 0$. Indeed, a projection from A onto W is given by right multiplication by an element $\varphi \in A$ with $\varphi = \varphi^2 \in W \cdot W = 0$. This argument also shows that $c_\lambda V_\lambda \neq 0$.

For (2), we may assume $\lambda > \mu$. Then $c_\lambda V_\lambda = \mathbb{C}c_\lambda \neq 0$, but $c_\lambda V_\mu = c_\lambda \cdot Ac_\mu = 0$, so they cannot be isomorphic A-modules. □

Lemma 4.26. *For any λ, $c_\lambda \cdot c_\lambda = n_\lambda c_\lambda$, with $n_\lambda = d!/\dim V_\lambda$.*

Proof. Let F be right multiplication by c_λ on A. Since F is multiplication by n_λ on V_λ, and zero on $\mathrm{Ker}(c_\lambda)$, the trace of F is n_λ times the dimension of V_λ. But the coefficient of e_g in $e_g \cdot c_\lambda$ is 1, so $\mathrm{trace}(F) = |\mathfrak{S}_d| = d!$. □

Since there are as many irreducible representations V_λ as conjugacy classes of $\mathfrak{S}_d$, these must form a complete set of isomorphism classes of irreducible representations, which completes the proof of Theorem 4.3. In the next section we will prove Frobenius's formula for the character of V_λ, and, in a series of exercises, discuss a little of what else is known about them: how to decompose tensor products or induced or restricted representations, how to find a basis for V_λ, etc.

§4.3. Proof of Frobenius's Formula

For any partition λ of d, we have a subgroup, often called a *Young subgroup*,

$$\mathfrak{S}_\lambda = \mathfrak{S}_{\lambda_1} \times \cdots \times \mathfrak{S}_{\lambda_k} \hookrightarrow \mathfrak{S}_d. \tag{4.27}$$

Let U_λ be the representation of $\mathfrak{S}_d$ induced from the trivial representation of $\mathfrak{S}_\lambda$. Equivalently, $U_\lambda = A \cdot a_\lambda$, with a_λ as in the preceding section. Let

$$\psi_\lambda = \chi_{U_\lambda} = \text{character of } U_\lambda. \tag{4.28}$$

Key to this investigation is the relation between U_λ and V_λ, i.e., between ψ_λ and the character χ_λ of V_λ. Note first that V_λ appears in U_λ, since there is a surjection

$$U_\lambda = Aa_\lambda \twoheadrightarrow V_\lambda = Aa_\lambda b_\lambda, \quad x \mapsto x \cdot b_\lambda. \tag{4.29}$$

Alternatively,

$$V_\lambda = Aa_\lambda b_\lambda \cong Ab_\lambda a_\lambda \subset Aa_\lambda = U_\lambda,$$

by Exercise 4.4. For example, we have

$$U_{(d-1,1)} \cong V_{(d-1,1)} \oplus V_{(d)}$$

which expresses the fact that the permutation representation $\mathbb{C}^d$ of $\mathfrak{S}_d$ is the sum of the standard representation and the trivial representation. Eventually we will see that every U_λ contains V_λ with multiplicity one, and contains only other V_μ for $\mu > \lambda$.

The character of U_λ is easy to compute directly since U_λ is an induced representation, and we do this next.

For $\mathbf{i} = (i_1, \ldots, i_d)$ a d-tuple of non-negative integers with $\sum \alpha i_\alpha = d$, denote by

$$C_\mathbf{i} \subset \mathfrak{S}_d$$

the conjugacy class consisting of elements made up of i_1 1-cycles, i_2 2-cycles, $\ldots$, i_d d-cycles. The number of elements in $C_\mathbf{i}$ is easily counted to be

$$|C_\mathbf{i}| = \frac{d!}{1^{i_1} i_1! \, 2^{i_2} i_2! \cdot \ldots \cdot d^{i_d} i_d!}. \tag{4.30}$$

By the formula for characters of induced representations (Exercise 3.19),

$$\begin{aligned}\psi_\lambda(C_\mathbf{i}) &= \frac{1}{|C_\mathbf{i}|}[\mathfrak{S}_d : \mathfrak{S}_\lambda] \cdot |C_\mathbf{i} \cap \mathfrak{S}_\lambda| \\ &= \frac{1^{i_1} i_1! \cdot \ldots \cdot d^{i_d} i_d!}{d!} \cdot \frac{d!}{\lambda_1! \cdot \ldots \cdot \lambda_k!} \cdot \sum \prod_{p=1}^{k} \frac{\lambda_p!}{1^{r_{p1}} r_{p1}! \cdot \ldots \cdot d^{r_{pd}} r_{pd}!},\end{aligned}$$

where the sum is over all collections $\{r_{pq}: 1 \le p \le k, 1 \le q \le d\}$ of non-negative integers satisfying

$$\begin{aligned} i_q &= r_{1q} + r_{2q} + \cdots + r_{kq}, \\ \lambda_p &= r_{p1} + 2r_{p2} + \cdots + d r_{pd}. \end{aligned}$$

(To count $C_\mathbf{i} \cap \mathfrak{S}_\lambda$, write the pth component of an element of $\mathfrak{S}_\lambda$ as a product of r_{p1} 1-cycles, r_{p2} 2-cycles, $\ldots$.) Simplifying,

$$\psi_\lambda(C_\mathbf{i}) = \sum \prod_{q=1}^{d} \frac{i_q!}{r_{1q}! r_{2q}! \cdot \ldots \cdot r_{kq}!}, \tag{4.31}$$

the sum over the same collections of integers $\{r_{pq}\}$.

This sum is exactly the coefficient of the monomial $X^\lambda = x_1^{\lambda_1} \cdot \ldots \cdot x_k^{\lambda_k}$ in the power sum symmetric polynomial

$$P^{(\mathbf{i})} = (x_1 + \cdots + x_k)^{i_1} \cdot (x_1^2 + \cdots + x_k^2)^{i_2} \cdot \ldots \cdot (x_1^d + \cdots + x_k^d)^{i_d}. \tag{4.32}$$

So we have the formula

$$\psi_\lambda(C_\mathbf{i}) = [P^{(\mathbf{i})}]_\lambda = \text{coefficient of } X^\lambda \text{ in } P^{(\mathbf{i})}. \tag{4.33}$$

To prove Frobenius's formula, we need to compare these coefficients with the coefficients $\omega_\lambda(\mathbf{i})$ defined by

$$\omega_\lambda(\mathbf{i}) = [\Delta \cdot P^{(\mathbf{i})}]_l, \quad l = (\lambda_1 + k - 1, \lambda_2 + k - 2, \ldots, \lambda_k). \tag{4.34}$$

Our goal, Frobenius's formula, is the assertion that $\chi_\lambda(C_\mathbf{i}) = \omega_\lambda(\mathbf{i})$.

There is a general identity, valid for any symmetric polynomial P, relating such coefficients:

$$[P]_\lambda = \sum_\mu K_{\mu\lambda}[\Delta \cdot P]_{(\mu_1+k-1, \mu_2+k-2, \ldots, \mu_k)},$$

where the coefficients $K_{\mu\lambda}$ are certain universally defined integers, called *Kostka numbers*. For any partitions λ and μ of d, the integer $K_{\mu\lambda}$ may be defined combinatorially as the number of ways to fill the boxes of the Young diagram for μ with λ_1 1's, λ_2 2's, up to λ_k k's, in such a way that the entries in each row are nondecreasing, and those in each column are strictly increasing; such are called *semistandard tableaux on μ of type λ*. In particular,

$$K_{\lambda\lambda} = 1, \quad \text{and } K_{\mu\lambda} = 0 \text{ for } \mu < \lambda.$$

The integer $K_{\mu\lambda}$ may be also be defined to be the coefficient of the monomial $X^\lambda = x_1^{\lambda_1} \cdot \ldots \cdot x_k^{\lambda_k}$ in the Schur polynomial S_μ corresponding to μ. For the proof that these are equivalent definitions, see (A.9) and (A.19) of Appendix A. In the present case, applying Lemma A.26 to the polynomial $P = P^{(\mathbf{i})}$, we deduce

$$\psi_\lambda(C_\mathbf{i}) = \sum_\mu K_{\mu\lambda}\omega_\mu(\mathbf{i}) = \omega_\lambda(\mathbf{i}) + \sum_{\mu > \lambda} K_{\mu\lambda}\omega_\mu(\mathbf{i}). \tag{4.35}$$

The result of Lemma A.28 can be written, using (4.30), in the form

$$\frac{1}{d!}\sum_\mathbf{i} |C_\mathbf{i}|\omega_\lambda(\mathbf{i})\omega_\mu(\mathbf{i}) = \delta_{\lambda\mu}. \tag{4.36}$$

This indicates that the functions ω_λ, regarded as functions on the conjugacy classes of $\mathfrak{S}_d$, satisfy the same orthogonality relations as the irreducible characters of $\mathfrak{S}_d$. In fact, one can deduce formally from these equations that the ω_λ must be the irreducible characters of $\mathfrak{S}_d$, which is what Frobenius proved. A little more work is needed to see that ω_λ is actually the character of the representation V_λ, that is, to prove

Proposition 4.37. *Let $\chi_\lambda = \chi_{V_\lambda}$ be the character of V_λ. Then for any conjugacy class $C_\mathbf{i}$ of $\mathfrak{S}_d$,*

$$\chi_\lambda(C_\mathbf{i}) = \omega_\lambda(\mathbf{i}).$$

PROOF. We have seen in (4.29) that the representation U_λ, whose character is ψ_λ, contains the irreducible representation V_λ. In fact, this is all that we need to know about the relation between U_λ and V_λ. It implies that we have

$$\psi_\lambda = \sum_\mu n_{\lambda\mu}\chi_\mu, \quad n_{\lambda\lambda} \geq 1, \text{ all } n_{\lambda\mu} \geq 0. \tag{4.38}$$

Consider this equation together with (4.35). We deduce first that each ω_λ is a virtual character: we can write

$$\omega_\lambda = \sum m_{\lambda\mu}\chi_\mu, \quad m_{\lambda\mu} \in \mathbb{Z}.$$

But the ω_λ, like the χ_λ, are orthonormal by (4.36), so

$$1 = (\omega_\lambda, \omega_\lambda) = \sum_\mu m_{\lambda\mu}^2,$$

and hence ω_λ is $\pm\chi$ for some irreducible character χ. (It follows from the hook length formula that the plus sign holds here, but we do not need to assume this.)

Fix λ, and assume inductively that $\chi_\mu = \omega_\mu$ for all $\mu > \lambda$, so by (4.35)

$$\psi_\lambda = \omega_\lambda + \sum_{\mu > \lambda} K_{\mu\lambda}\chi_\mu.$$

Comparing this with (4.38), and using the linear independence of characters, the only possibility is that $\omega_\lambda = \chi_\lambda$. □

Corollary 4.39 (Young's rule). *The integer $K_{\mu\lambda}$ is the multiplicity of the irreducible representation V_μ in the induced representation U_λ:*

$$U_\lambda \cong V_\lambda \oplus \bigoplus_{\mu > \lambda} K_{\mu\lambda} V_\mu, \qquad \psi_\lambda = \chi_\lambda + \sum_{\mu > \lambda} K_{\mu\lambda}\chi_\mu.$$

Note that when $\lambda = (1, \ldots, 1)$, U_λ is just the regular representation, so $K_{\mu(1,\ldots,1)} = \dim V_\mu$. This shows that *the dimension of V_λ is the number of standard tableaux on λ*, i.e., the number of ways to fill the Young diagram of λ with the numbers from 1 to d, such that all rows and columns are increasing. The hook length formula gives another combinatorial formula for this dimension. Frame, Robinson, and Thrall proved that these two numbers are equal. For a short and purely combinatorial proof, see [G-N-W]. For another proof that the dimension of V_λ is the number of standard tableaux, see [Jam]. The latter leads to a canonical decomposition of the group ring $A = \mathbb{C}\mathfrak{S}_d$ as the direct sum of left ideals Ae_T, summing over all standard tableaux, with $e_T = (\dim V_\lambda/d!)\cdot c_T$, and c_T the Young symmetrizer corresponding to T, cf. Exercises 4.47 and 4.50. This, in turn, leads to explicit calculation of matrices of the representations V_λ with integer coefficients.

For another example of Young's rule, we have a decomposition

$$U_{(d-a,a)} = \bigoplus_{i=0}^{a} V_{(d-i,i)}.$$

In fact, the only μ whose diagrams can be filled with $d - a$ 1's and a 2's, nondecreasing in rows and strictly increasing in columns, are those with at most two rows, with the second row no longer than a; and such a diagram has only one such tableau, so there are no multiplicities.

Exercise 4.40*. The characters ψ_λ of $\mathfrak{S}_d$ have been defined only when λ is a partition of d. Extend the definition to any k-tuple $a = (a_1, \ldots, a_k)$ of integers that add up to d by setting $\psi_a = 0$ if any of the a_i are negative, and otherwise $\psi_a = \psi_\lambda$, where λ is the reordering of $a_1, \ldots, a_k$ in descending order. In this case ψ_a is the character of the representation induced from the trivial representation by the inclusion of $\mathfrak{S}_{a_1} \times \cdots \times \mathfrak{S}_{a_k}$ in $\mathfrak{S}_d$. Use (A.5) and (A.9) of Appendix A to prove the *determinantal formula* for the irreducible characters χ_λ in terms of the induced characters ψ_μ:

$$\chi_\lambda = \sum_{\tau \in \mathfrak{S}_k} \operatorname{sgn}(\tau) \psi_{(\lambda_1 + \tau(1) - 1, \lambda_2 + \tau(2) - 2, \ldots, \lambda_k + \tau(k) - k)}.$$

If one writes ψ_a as a formal product $\psi_{a_1} \cdot \psi_{a_2} \cdot \ldots \cdot \psi_{a_k}$, the preceding formula can be written

$$\chi_\lambda = |\psi_{\lambda_i + j - i}| = \begin{vmatrix} \psi_{\lambda_1} & \psi_{\lambda_1 + 1} & \psi_{\lambda_1 + k - 1} \\ \psi_{\lambda_2 - 1} & \psi_{\lambda_2} \cdots & \\ \vdots & & \vdots \\ \psi_{\lambda_k - k + 1} \cdots & & \psi_{\lambda_k} \end{vmatrix}$$

The formal product of the preceding exercise is the character version of an "outer product" of representations. Given any non-negative integers $d_1, \ldots, d_k$, and representations V_i of $\mathfrak{S}_{d_i}$, denote by $V_1 \circ \cdots \circ V_k$ the (isomorphism class of the) representation of $\mathfrak{S}_d$, $d = \sum d_i$, induced from the tensor product representation $V_1 \boxtimes \cdots \boxtimes V_k$ of $\mathfrak{S}_{d_1} \times \cdots \times \mathfrak{S}_{d_k}$ by the inclusion of $\mathfrak{S}_{d_1} \times \cdots \times \mathfrak{S}_{d_k}$ in $\mathfrak{S}_d$ (see Exercise 2.36). This product is commutative and associative. It will turn out to be useful to have a procedure for decomposing such a representation into its irreducible pieces. For this it is enough to do the case of two factors, and with the individual representations V_i irreducible. In this case, one has, for V_λ the representation of $\mathfrak{S}_d$ corresponding to the partition λ of d and V_μ the representation of $\mathfrak{S}_m$ corresponding to the partition μ of m,

$$V_\lambda \circ V_\mu = \sum N_{\lambda\mu\nu} V_\nu, \tag{4.41}$$

the sum over all partitions ν of $d + m$, with $N_{\lambda\mu\nu}$ the coefficients given by the *Littlewood–Richardson rule* (A.8) of Appendix A. Indeed, by the exercise, the character of $V_\lambda \circ V_\mu$ is the product of the corresponding determinants, and, by (A.8), that is the sum of the characters $N_{\lambda\mu\nu}\chi_\nu$.

When $m = 1$ and $\mu = (m)$, V_μ is trivial; this gives

$$\operatorname{Ind}_{\mathfrak{S}_d}^{\mathfrak{S}_{d+1}} V_\lambda = \sum V_\nu, \tag{4.42}$$

the sum over all ν whose Young diagram is obtained from that of λ by adding one box. This formula uses only a simpler form of the Littlewood–Richardson rule known as Pieri's formula, which is proved in (A.7).

Exercise 4.43*. Show that the Littlewood–Richardson number $N_{\lambda\mu\nu}$ is the multiplicity of the irreducible representation $V_\lambda \boxtimes V_\mu$ in the restriction of V_ν from $\mathfrak{S}_{d+m}$ to $\mathfrak{S}_d \times \mathfrak{S}_m$. In particular, taking $m = 1$, $\mu = (1)$, Pieri's formula (A.7) gives

$$\operatorname{Res}_{\mathfrak{S}_d}^{\mathfrak{S}_{d+1}} V_\nu = \sum V_\lambda,$$

the sum over all λ obtained from ν by removing one box. This is known as the "branching theorem," and is useful for inductive proofs and constructions, particularly because the decomposition is multiplicity free. For example, you can use it to reprove the fact that the multiplicity of V_λ in U_μ is the number of semistandard tableaux on μ of type λ. It can also be used to prove the assertion made in Exercise 4.6 that the representations corresponding to hooks are exterior powers of the standard representation.

Exercise 4.44* (Pieri's rule). Regard $\mathfrak{S}_d$ as a subgroup of $\mathfrak{S}_{d+m}$ as usual. Let λ be a partition of d and ν a partition of $d + m$. Use Exercise 4.40 to show that the multiplicity of V_ν in the induced representation $\operatorname{Ind}(V_\lambda)$ is zero unless the Young diagram of λ is contained in that of ν, and then it is the number of ways to number the skew diagram lying between them with the numbers from 1 to m, increasing in both row and column. By Frobenius reciprocity, this is the same as the multiplicity of V_λ in $\operatorname{Res}(V_\nu)$.

When applied to $d = 0$ (or 1), this implies again that the dimension of V_ν is the number of standard tableaux on the Young diagram of ν.

For a sampling of the many applications of these rules, see [Dia §7, §8].

Problem 4.45*. The *Murnaghan–Nakayama rule* gives an efficient inductive method for computing character values: If λ is a partition of d, and $g \in \mathfrak{S}_d$ is written as a product of an m-cycle and a disjoint permutation $h \in \mathfrak{S}_{d-m}$, then

$$\chi_\lambda(g) = \sum (-1)^{r(\mu)} \chi_\mu(h),$$

where the sum is over all partitions μ of $d - m$ that are obtained from λ by removing a skew hook of length m, and $r(\mu)$ is the number of vertical steps in the skew hook, i.e., one less than the number of rows in the hook. A *skew hook* for λ is a connected region of boundary boxes for its Young diagram such that removing them leaves a smaller Young diagram; there is a one-to-one correspondence between skew hooks and ordinary hooks of the same size, as indicated:

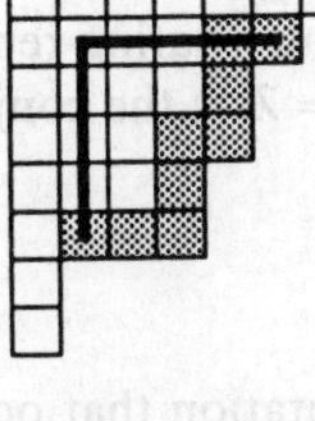

$\lambda = (7, 6, 5, 5, 4, 4, 1, 1)$

$\mu = (7, 4, 4, 3, 3, 1, 1, 1)$

hook length $= 9$, $r = 4$

For example, if λ has no hooks of length m, then $\chi_\lambda(g) = 0$.

The Murnaghan–Nakayama rule may be written inductively as follows: If g is a written as a product of disjoint cycles of lengths $m_1, m_2, \ldots, m_p$, with the lengths m_i taken in any order, then $\chi_\lambda(g)$ is the sum $\sum (-1)^{r(s)}$, where the sum is over all ways s to decompose the Young diagram of λ by successively removing p skew hooks of lengths $m_1, \ldots, m_p$, and $r(s)$ is the total number of vertical steps in the hooks of s.

(a) Deduce the Murnaghan–Nakayama rule from (4.41) and Exercise 4.16, using the Littlewood–Richardson rule. Or:

(b) With the notation of Exercise 4.40, show that

$$\psi_{a_1}\psi_{a_2}\cdot\ldots\cdot\psi_{a_k}(g) = \sum_{i=1}^{k} \psi_{a_1}\psi_{a_2}\cdot\ldots\cdot\psi_{a_i-m}\psi_{a_{i+1}}\cdot\ldots\cdot\psi_{a_k}(h).$$

Exercise 4.46*. Show that Corollary 4.39 implies the "Snapper conjecture": the irreducible representation V_μ occurs in the induced representation U_λ if and only if

$$\sum_{i=1}^{j} \lambda_i \le \sum_{i=1}^{j} \mu_i \quad \text{for all } j \ge 1.$$

Problem 4.47*. There is a more intrinsic construction of the irreducible representation V_λ, called a *Specht module*, which does not involve the choice of a tableau; it is also useful for studying representations of $\mathfrak{S}_d$ in positive characteristic. Define a *tabloid* $\{T\}$ to be an equivalence class of tableaux (numberings by the integers 1 to d) on λ, two being equivalent if the rows are the same up to order. Then $\mathfrak{S}_d$ acts by permutations on the tabloids, and the corresponding representation, with basis the tabloids, is isomorphic to U_λ. For each tableau T, define an element E_T in this representation space, by

$$E_T = b_T\{T\} = \sum \operatorname{sgn}(q)\{qT\},$$

the sum over the q that preserve the columns of T. The span of all E_T's is isomorphic to V_λ, and the E_T's, where T varies over the standard tableaux, form a basis.

Another construction of V_λ is to take the subspace of the polynomial ring $\mathbb{C}[x_1, \ldots, x_d]$ spanned by all polynomials F_T, where $F_T = \prod (x_i - x_j)$, the product over all pairs $i < j$ which occur in the same column in the tableau T.

Exercise 4.48*. Let U'_λ be the representation $A \cdot b_\lambda$, which is the representation of $\mathfrak{S}_d$ induced from the tensor product of the alternating representations on the subgroup $\mathfrak{S}_\mu = \mathfrak{S}_{\mu_1} \times \cdots \times \mathfrak{S}_{\mu_r}$, where $\mu = \lambda'$ is the conjugate partition. Show that the decomposition of U'_λ is

$$U'_\lambda = \sum_\mu K_{\mu'\lambda'} V_\mu.$$

Deduce that V_λ is the only irreducible representation that occurs in both U_λ and U'_λ, and it occurs in each with multiplicity one.

Note, however, that in general $A \cdot c_\lambda \ne A \cdot a_\lambda \cap A \cdot b_\lambda$ since $A \cdot c_\lambda$ may not be contained in $A \cdot a_\lambda$.

Exercise 4.49*. With notation as in (4.41), if $U' = V_{(1,\ldots,1)}$ is the alternating representation of $\mathfrak{S}_m$, show that $V_\lambda \circ V_{(1,\ldots,1)}$ decomposes into a direct sum $\bigoplus V_\pi$, the sum over all π whose Young diagram can be obtained from that of λ by adding m boxes, with no two in the same row.

Exercise 4.50. We have seen that $A = \mathbb{C}\mathfrak{S}_d$ is isomorphic to a direct sum of m_λ copies of $V_\lambda = Ac_\lambda$, where $m_\lambda = \dim V_\lambda$ is the number of standard tableaux on λ. This can be seen explicitly as follows. For each standard tableau T on each λ, let c_T be the element of $\mathbb{C}\mathfrak{S}_d$ constructed from T. Then $A = \bigoplus A \cdot c_T$. Indeed, an argument like that in Lemma 4.23 shows that $c_T \cdot c_{T'} = 0$ whenever T and T' are tableaux on the same diagram and $T > T'$, i.e., the first entry (reading from left to right, then top to bottom) where the tableaux differ has the entry of T larger than that of T'. From this it follows that the sum $\Sigma A \cdot c_T$ is direct. A dimension count concludes the proof. (This also gives another proof that the dimension of V_λ is the number of standard tableaux on λ, provided one verifies that the sum of the squares of the latter numbers is $d!$, cf. [Boe] or [Ke].)

Exercise 4.51*. There are several methods for decomposing a tensor product of two representations of $\mathfrak{S}_d$, which amounts to finding the coefficients $C_{\lambda\mu\nu}$ in the decomposition

$$V_\lambda \otimes V_\mu \cong \Sigma_\nu C_{\lambda\mu\nu} V_\nu,$$

for λ, μ, and ν partitions of d. Since one knows how to express V_μ in terms of the induced representations U_ν, it suffices to compute $V_\lambda \otimes U_\nu$, which is isomorphic to $\mathrm{Ind}(\mathrm{Res}(V_\lambda))$, restricting and inducing from the subgroup $\mathfrak{S}_\nu = \mathfrak{S}_{\nu_1} \times \mathfrak{S}_{\nu_2} \times \cdots$; this restriction and induction can be computed by the Littlewood–Richardson rule. For $d \leq 5$, you can work out these coefficients using only restriction to $\mathfrak{S}_{d-1}$ and Pieri's formula.

(a) Prove the following closed-form formula for the coefficients, which shows in particular that they are independent of the ordering of the subscripts λ, μ, and ν:

$$C_{\lambda\mu\nu} = \sum_{\mathbf{i}} \frac{1}{z(\mathbf{i})} \omega_\lambda(\mathbf{i}) \omega_\mu(\mathbf{i}) \omega_\nu(\mathbf{i}),$$

the sum over all $\mathbf{i} = (i_1, \ldots, i_d)$ with $\Sigma \alpha i_\alpha = d$, and with $\omega_\lambda(\mathbf{i}) = \chi_\lambda(C_{\mathbf{i}})$ and $z(\mathbf{i}) = i_1! 1^{i_1} \cdot i_2! 2^{i_2} \cdot \ldots \cdot i_d! d^{i_d}$.

(b) Show that

$$C_{\lambda\mu(d)} = \begin{cases} 1 & \text{if } \mu = \lambda \\ 0 & \text{otherwise,} \end{cases} \qquad C_{\lambda\mu(1,\ldots,1)} = \begin{cases} 1 & \text{if } \mu = \lambda' \\ 0 & \text{otherwise.} \end{cases}$$

Exercise 4.52*. Let $R_d = R(\mathfrak{S}_d)$ denote the representation ring, and set $R = \bigoplus_{d=0}^{\infty} R_d$. The outer product of (4.41) determines maps

$$R_n \otimes R_m \to R_{n+m},$$

which makes R into a commutative, graded $\mathbb{Z}$-algebra. Restriction determines maps

$$R_{n+m} = R(\mathfrak{S}_{n+m}) \to R(\mathfrak{S}_n \times \mathfrak{S}_m) = R_n \otimes R_m,$$

which defines a *co-product* $\delta: R \to R \otimes R$. Together, these make R into a (graded) Hopf algebra. (This assertion implies many of the formulas we have proved in this lecture, as well as some we have not.)

(a) Show that, as an algebra,

$$R \cong \mathbb{Z}[H_1, \ldots, H_d, \ldots],$$

where H_d is an indeterminate of degree d; H_d corresponds to the trivial representation of $\mathfrak{S}_d$. Show that the co-product δ is determined by

$$\delta(H_n) = H_n \otimes 1 + H_{n-1} \otimes H_1 + \cdots + 1 \otimes H_n.$$

If we set $\Lambda = \mathbb{Z}[H_1, \ldots, H_d, \ldots] = \bigoplus \Lambda_d$, we can identify Λ_d with the symmetric polynomials of degree d in $k \geq d$ variables. The basic symmetric polynomials in Λ_d defined in Appendix A therefore correspond to virtual representations of $\mathfrak{S}_d$.

(b) Show that E_d corresponds to the alternating representation U', and

$$H_\lambda \leftrightarrow U_\lambda, \qquad S_\lambda \leftrightarrow V_\lambda, \qquad E_\lambda \leftrightarrow U'_{\lambda'}.$$

(c) Show that the scalar product $\langle \ , \ \rangle$ defined on Λ_d in (A.16) corresponds to the scalar product defined on class functions in (2.11).

(d) Show that the involution ϑ of Exercise A.32 corresponds to tensoring a representation with the alternating representation U'.

(e) Show that the inverse map from R_d to Λ_d takes a representation W to

$$\sum_{\mathbf{i}} \frac{1}{z(\mathbf{i})} \chi_W(C_{(\mathbf{i})}) P^{(\mathbf{i})},$$

where $z(\mathbf{i}) = i_1! 1^{i_1} \cdot i_2! 2^{i_2} \cdot \ldots \cdot i_d! d^{i_d}$.

The (inner) tensor product of representations of $\mathfrak{S}_d$ gives a map $R_d \otimes R_d \to R_d$ which corresponds to an "inner product" on symmetric functions, sometimes denoted $*$.

(f) Show that

$$P^{(\mathbf{i})} * P^{(\mathbf{j})} = \begin{cases} 0 & \text{for } \mathbf{j} \neq \mathbf{i} \\ z(\mathbf{i}) P^{(\mathbf{i})} & \text{if } \mathbf{j} = \mathbf{i}. \end{cases}$$

Since these $P^{(\mathbf{i})}$ form a basis for $\Lambda_d \otimes \mathbb{Q}$, this formula determines the inner product.

LECTURE 5

Representations of $\mathfrak{A}_d$ and $GL_2(\mathbb{F}_q)$

In this lecture we analyze the representation of two more types of groups: the alternating groups $\mathfrak{A}_d$ and the linear groups $GL_2(\mathbb{F}_q)$ and $SL_2(\mathbb{F}_q)$ over finite fields. In the former case, we prove some general results relating the representations of a group to the representations of a subgroup of index two, and use what we know about the symmetric group; this should be completely straightforward given just the basic ideas of the preceding lecture. In the latter case we start essentially from scratch. The two sections can be read (or not) independently; neither is logically necessary for the remainder of the book.

§5.1. Representations of $\mathfrak{A}_d$

The alternating groups $\mathfrak{A}_d$, $d \geq 5$, form one of the infinite families of simple groups. In this section, continuing the discussion of §3.1, we describe their irreducible representations. The basic method for analyzing representations of $\mathfrak{A}_d$ is by restricting the representations we know from $\mathfrak{S}_d$.

In general when H is a subgroup of index two in a group G, there is a close relationship between their representations. We will see this phenomenon again in Lie theory for the subgroups SO_n of the orthogonal groups O_n.

Let U and U' denote the trivial and nontrivial representation of G obtained from the two representations of G/H. For any representation V of G, let $V' = V \otimes U'$; the character of V' is the same as the character of V on elements of H, but takes opposite values on elements not in H. In particular, $\mathrm{Res}_H^G V' = \mathrm{Res}_H^G V$.

If W is any representation of H, there is a *conjugate* representation defined by conjugating by any element t of G that is not in H; if ψ is the character of W, the character of the conjugate is $h \mapsto \psi(tht^{-1})$. Since t is unique up to multiplication by an element of H, the conjugate representation is unique up to isomorphism.

Proposition 5.1. *Let V be an irreducible representation of G, and let $W = \mathrm{Res}_H^G V$ be the restriction of V to H. Then exactly one of the following holds:*

(1) V is not isomorphic to V'; W is irreducible and isomorphic to its conjugate; $\mathrm{Ind}_H^G W \cong V \oplus V'$.

(2) $V \cong V'$; $W = W' \oplus W''$, where W' and W'' are irreducible and conjugate but not isomorphic; $\mathrm{Ind}_H^G W' \cong \mathrm{Ind}_H^G W'' \cong V$.

Each irreducible representation of H arises uniquely in this way, noting that in case (1) V' and V determine the same representation.

PROOF. Let χ be the character of V. We have

$$|G| = 2|H| = \sum_{h \in H} |\chi(h)|^2 + \sum_{t \notin H} |\chi(t)|^2.$$

Since the first sum is an integral multiple of $|H|$, this multiple must be 1 or 2, which are the two cases of the proposition. This shows that W is either irreducible or the sum of two distinct irreducible representations W' and W''. Note that the second case happens when $\chi(t) = 0$ for all $t \notin H$, which is the case when V' is isomorphic to V. In the second case, W' and W'' must be conjugate since W is self-conjugate, and if W' and W'' were self-conjugate V would not be irreducible. The other assertions in (1) and (2) follow from the isomorphism $\mathrm{Ind}(\mathrm{Res}\, V) = V \otimes (U \oplus U')$ of Exercise 3.16. Similarly, for any representation W of H, $\mathrm{Res}(\mathrm{Ind}\, W)$ is the direct sum of W and its conjugate—as follows say from Exercise 3.19—from which the last statement follows readily. □

Most of this discussion extends with little change to the case where H is a normal subgroup of arbitrary prime index in G, cf. [B-tD, pp. 293–296]. Clifford has extended much of this proposition to arbitrary normal subgroups of finite index, cf. [Dor, §14].

There are two types of conjugacy classes c in H: those that are also conjugacy classes in G, and those such that $c \cup c'$ is a conjugacy class in G, where $c' = tct^{-1}$, $t \notin H$; the latter are called *split*. When W is irreducible, its character assumes the same values—those of the character of the representation V of G that restricts to W—on pairs of split conjugacy classes, whereas in the other case the characters of W' and W'' agree on nonsplit classes, but they must disagree on some split classes. If $\chi_{W'}(c) = \chi_{W''}(c') = x$, and $\chi_{W'}(c') = \chi_{W''}(c) = y$, we know the sum $x + y$, since it is the value of the character of the representation V that gives rise to W' and W'' on $c \cup c'$. Often the exact values of x and y can be determined from orthogonality considerations.

Exercise 5.2*. Show that the number of split conjugacy classes is equal to the number of irreducible representations V of G that are isomorphic to V', or to the number of irreducible representations of H that are not isomorphic to their conjugates. Equivalently, the number of nonsplit classes in H is same as the number of conjugacy classes of G that are not in H.

We apply these considerations to the alternating subgroup of the symmetric group. Consider restrictions of the representations V_λ from $\mathfrak{S}_d$ to $\mathfrak{A}_d$. Recall that if λ' is the conjugate partition to λ, then

$$V_{\lambda'} = V_\lambda \otimes U',$$

with U' the alternating representation. The two cases of the proposition correspond to the cases (1) $\lambda' \neq \lambda$ and (2) $\lambda' = \lambda$. If $\lambda' \neq \lambda$, let W_λ be the restriction of V_λ to $\mathfrak{A}_d$. If $\lambda' = \lambda$, let W'_λ and W''_λ be the two representations whose sum is the restriction of V_λ. We have

$$\operatorname{Ind} W_\lambda = V_\lambda \oplus V_{\lambda'}, \qquad \operatorname{Res} V_\lambda = \operatorname{Res} V_{\lambda'} = W_\lambda \quad \text{when } \lambda' \neq \lambda,$$

$$\operatorname{Ind} W'_\lambda = \operatorname{Ind} W''_\lambda = V_\lambda, \qquad \operatorname{Res} V_\lambda = W'_\lambda \oplus W''_\lambda \quad \text{when } \lambda' = \lambda.$$

Note that

$$\begin{aligned} &\#\{\text{self-conjugate representations of } \mathfrak{S}_d\} \\ &\quad = \#\{\text{symmetric Young diagrams}\} \\ &\quad = \#\{\text{split pairs of conjugacy classes in } \mathfrak{A}_d\} \\ &\quad = \#\{\text{conjugacy classes in } \mathfrak{S}_d \text{ breaking into two classes in } \mathfrak{A}_d\}. \end{aligned}$$

Now a conjugacy class of an element written as a product of disjoint cycles is split if and only if there is no odd permutation commuting with it, which is equivalent to all the cycles having odd length, and no two cycles having the same length. So the number of self-conjugate representations is the number of partitions of d as a sum of distinct odd numbers. In fact, there is a natural correspondence between these two sets: any such partition corresponds to a symmetric Young diagram, assembling hooks as indicated:

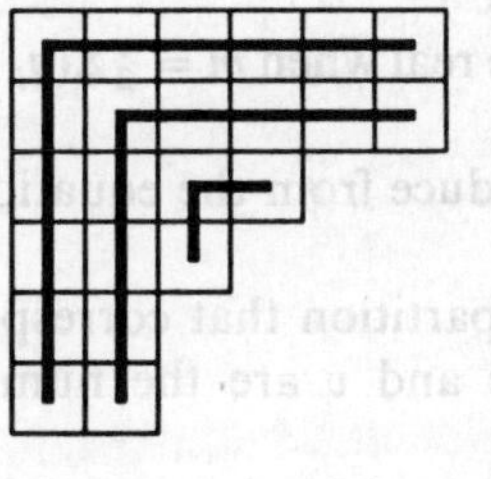

If λ is the partition, the lengths of the cycles in the corresponding split conjugacy classes are $q_1 = 2\lambda_1 - 1$, $q_2 = 2\lambda_2 - 3$, $q_3 = 2\lambda_3 - 5, \ldots$.

For a self-conjugate partition λ, let χ'_λ and χ''_λ denote the characters of W'_λ and W''_λ, and let c and c' be a pair of split conjugacy classes, consisting of cycles of odd lengths $q_1 > q_2 > \cdots > q_r$. The following proposition of Frobenius completes the description of the character table of $\mathfrak{A}_d$.

Proposition 5.3. (1) *If c and c' do not correspond to the partition λ, then*

$$\chi'_\lambda(c) = \chi'_\lambda(c') = \chi''_\lambda(c) = \chi''_\lambda(c') = \tfrac{1}{2}\chi_\lambda(c \cup c').$$

(2) *If c and c' correspond to λ, then*

$$\chi'_\lambda(c) = \chi''_\lambda(c') = x, \qquad \chi'_\lambda(c') = \chi''_\lambda(c) = y,$$

with x and y the two numbers

$$\tfrac{1}{2}\left((-1)^m \pm \sqrt{(-1)^m q_1 \cdot \ldots \cdot q_r}\right),$$

and $m = \frac{1}{2}(d - r) = \frac{1}{2}\sum (q_i - 1) \equiv \frac{1}{2}(\prod q_i - 1) \pmod 2$.

For example, if $d = 4$ and $\lambda = (2, 2)$, we have $r = 2$, $q_1 = 3$, $q_2 = 1$, and x and y are the cube roots of unity; the representations W'_λ and W''_λ are the representations labeled U' and U'' in the table in §2.3. For $d = 5$, $\lambda = (3, 1, 1)$, $r = 1$, $q_1 = 5$, and we find the representations called Y and Z in §3.1. For $d \leq 7$, there is at most one split pair, so the character table can be derived from orthogonality alone.

Note that since only one pair of character values is not taken care of by the first case of Frobenius's formula, the choice of which representation is W'_λ and which W''_λ is equivalent to choosing the plus and minus sign in (2). Note also that the integer m occurring in (2) is the number of squares above the diagonal in the Young diagram of λ.

We outline a proof of the proposition as an exercise:

Exercise 5.4*. *Step 1.* Let $q = (q_1 > \cdots > q_r)$ be a sequence of positive odd integers adding to d, and let $c' = c'(q)$ and $c'' = c''(q)$ be the corresponding conjugacy classes in $\mathfrak{A}_d$. Let λ be a self-conjugate partition of d, and let χ'_λ and χ''_λ be the corresponding characters of $\mathfrak{A}_d$. Assume that χ'_λ and χ''_λ take on the same values on each element of $\mathfrak{A}_d$ that is not in c' or c''. Let $u = \chi'_\lambda(c') = \chi''_\lambda(c'')$ and $v = \chi'_\lambda(c'') = \chi''_\lambda(c')$.

(i) Show that u and v are real when $m = \frac{1}{2}\Sigma(q_i - 1)$ is even, and $\bar{u} = v$ when m is odd.

(ii) Let $\vartheta = \chi'_\lambda - \chi''_\lambda$. Deduce from the equation $(\vartheta, \vartheta) = 2$ that $|u - v|^2 = q_1 \cdot \ldots \cdot q_r$.

(iii) Show that λ is the partition that corresponds to q and that $u + v = (-1)^m$, and deduce that u and v are the numbers specified in (2) of the proposition.

Step 2. Prove the proposition by induction on d, and for fixed d, look at that q which has smallest q_1, and for which some character has values on the classes $c'(q)$ and $c''(q)$ other than those prescribed by the proposition.

(i) If $r = 1$, so $q_1 = d = 2m + 1$, the corresponding self-conjugate partition is $\lambda = (m + 1, 1, \ldots, 1)$. By induction, Step 1 applies to χ'_λ and χ''_λ.

(ii) If $r > 1$, consider the imbedding $H = \mathfrak{A}_{q_1} \times \mathfrak{A}_{d-q_1} \subset G = \mathfrak{A}_d$, and let X' and X'' be the representations of G induced from the representations $W'_1 \boxtimes W'_2$ and $W''_1 \boxtimes W'_2$, where W'_1 and W''_1 are the representations of $\mathfrak{A}_{q_1}$ corresponding to q_1, i.e., to the self-conjugate partition $(\frac{1}{2}(q_1 - 1), 1, \ldots, 1)$ of q_1; W'_2 is one of the representations of $\mathfrak{A}_{d-q_1}$ corresponding to $(q_2, \ldots, q_r)$; and $\boxtimes$ denotes the external tensor product (see Exercise 2.36). Show that X' and X'' are conjugate representations of $\mathfrak{A}_d$, and their characters χ' and χ'' take equal values on each pair of split conjugacy classes, with the exception of $c'(q)$ and $c''(q)$, and compute the values of these characters on $c'(q)$ and $c''(q)$.

(iii) Let $\vartheta = \chi' - \chi''$, and show that $(\vartheta, \vartheta) = 2$. Decomposing X' and X'' into their irreducible pieces, deduce that $X' = Y \oplus W'_\lambda$ and $X'' = Y \oplus W''_\lambda$ for some self-conjugate representation Y and some self-conjugate partition λ of d.

(iv) Apply Step 1 to the characters χ'_λ and χ''_λ, and conclude the proof.

Exercise 5.5*. Show that if $d > 6$, the only irreducible representations of $\mathfrak{A}_d$ of dimension less than d are the trivial representation and the $(n - 1)$-dimensional restriction of the standard representation of $\mathfrak{S}_d$. Find the exceptions for $d \leq 6$.

We have worked out the character tables for all $\mathfrak{S}_d$ and $\mathfrak{A}_d$ for $d \leq 5$. With the formulas of Frobenius, an interested reader can construct the tables for a few more d—until the number of partitions of d becomes large.

§5.2. Representations of $GL_2(\mathbb{F}_q)$ and $SL_2(\mathbb{F}_q)$

The groups $GL_2(\mathbb{F}_q)$ of invertible 2×2 matrices with entries in the finite field $\mathbb{F}_q$ with q elements, where q is a prime power, form another important series of finite groups, as do their subgroups $SL_2(\mathbb{F}_q)$ consisting of matrices of determinant one. The quotient $PGL_2(\mathbb{F}_q) = GL_2(\mathbb{F}_q)/\mathbb{F}_q^*$ is the automorphism group of the finite projective line $\mathbb{P}^1(\mathbb{F}_q)$. The quotients $PSL_2(\mathbb{F}_q) = SL_2(\mathbb{F}_q)/\{\pm 1\}$ are simple groups if $q \neq 2, 3$ (Exercise 5.9). In this section we sketch the character theory of these groups.

We begin with $G = GL_2(\mathbb{F}_q)$. There are several key subgroups:

$$G \supset B = \left\{\begin{pmatrix} a & b \\ 0 & d \end{pmatrix}\right\} \supset N = \left\{\begin{pmatrix} 1 & b \\ 0 & 1 \end{pmatrix}\right\}.$$

(This "Borel subgroup" B and the group of upper triangular unipotent matrices N will reappear when we look at Lie groups.) Since G acts transitively on the projective line $\mathbb{P}^1(\mathbb{F}_q)$, with B the isotropy group of the point (1:0), we have

$$|G| = |B| \cdot |\mathbb{P}^1(\mathbb{F}_q)| = (q - 1)^2 q(q + 1).$$

We will also need the diagonal subgroup

$$D = \left\{\begin{pmatrix} a & 0 \\ 0 & d \end{pmatrix}\right\} = \mathbb{F}^* \times \mathbb{F}^*,$$

where we write $\mathbb{F}$ for $\mathbb{F}_q$. Let $\mathbb{F}' = \mathbb{F}_{q^2}$ be the extension of $\mathbb{F}$ of degree two, unique up to isomorphism. We can identify $GL_2(\mathbb{F}_q)$ as the group of all $\mathbb{F}$-linear invertible endomorphisms of $\mathbb{F}'$. This makes evident a large cyclic subgroup $K = (\mathbb{F}')^*$ of G. At least if q is odd, we may make this isomorphism explicit by choosing a generator ε for the cyclic group $\mathbb{F}^*$ and choosing a square root $\sqrt{\varepsilon}$ in $\mathbb{F}'$. Then 1 and $\sqrt{\varepsilon}$ form a basis for $\mathbb{F}'$ as a vector space over $\mathbb{F}$, so we can make the identification:

$$K = \left\{\begin{pmatrix} x & \varepsilon y \\ y & x \end{pmatrix}\right\} \cong (\mathbb{F}')^*, \qquad \begin{pmatrix} x & \varepsilon y \\ y & x \end{pmatrix} \leftrightarrow \zeta = x + y\sqrt{\varepsilon};$$

K is a cyclic subgroup of G of order $q^2 - 1$. We often make this identification, leaving it as an exercise to make the necessary modifications in case q is even.

The conjugacy classes in G are easily found:

Representative	No. Elements in Class	No. Classes
$a_x = \begin{pmatrix} x & 0 \\ 0 & x \end{pmatrix}$	1	$q-1$
$b_x = \begin{pmatrix} x & 1 \\ 0 & x \end{pmatrix}$	q^2-1	$q-1$
$c_{x,y} = \begin{pmatrix} x & 0 \\ 0 & y \end{pmatrix}, x \neq y$	q^2+q	$\dfrac{(q-1)(q-2)}{2}$
$d_{x,y} = \begin{pmatrix} x & \varepsilon y \\ y & x \end{pmatrix}, y \neq 0$	q^2-q	$\dfrac{q(q-1)}{2}$

Here $c_{x,y}$ and $c_{y,x}$ are conjugate by $\begin{pmatrix} 0 & 1 \\ -1 & 0 \end{pmatrix}$, and $d_{x,y}$ and $d_{x,-y}$ are conjugate by any $\begin{pmatrix} a & -\varepsilon c \\ c & -a \end{pmatrix}$. To count the number of elements in the conjugacy class of b_x, look at the action of G on this class by conjugation; the isotropy group is $\left\{\begin{pmatrix} a & b \\ 0 & a \end{pmatrix}\right\}$, so the number of elements in the class is the index of this group in G, which is $q^2 - 1$. Similarly the isotropy group for $c_{x,y}$ is D, and the isotropy group for $d_{x,y}$ is K. To see that the classes are disjoint, consider the eigenvalues and the Jordan canonical forms. Since they account for $|G|$ elements, the list is complete.

There are $q^2 - 1$ conjugacy classes, so we must find the same number of irreducible representations. Consider first the permutation representation of G on $\mathbb{P}^1(\mathbb{F})$, which has dimension $q + 1$. It contains the trivial representation;

let V be the complementary q-dimensional representation. The values of the character χ of V on the four types of conjugacy classes are $\chi(a_x) = q$, $\chi(b_x) = 0$, $\chi(c_{x,y}) = 1$, $\chi(d_{x,y}) = -1$, which we display as the table:

$$V: \quad q \quad 0 \quad 1 \quad -1$$

Since $(\chi, \chi) = 1$, V is irreducible.

For each of the $q-1$ characters $\alpha\colon \mathbb{F}^* \to \mathbb{C}^*$ of $\mathbb{F}^*$, we have a one-dimensional representation U_α of G defined by $U_\alpha(g) = \alpha(\det(g))$. We also have the representations $V_\alpha = V \otimes U_\alpha$. The values of the characters of these representations are

U_α:	$\alpha(x)^2$	$\alpha(x)^2$	$\alpha(x)\alpha(y)$	$\alpha(x^2 - \varepsilon y^2)$
V_α:	$q\alpha(x)^2$	0	$\alpha(x)\alpha(y)$	$-\alpha(x^2 - \varepsilon y^2)$

Note that if we identify $\begin{pmatrix} x & \varepsilon y \\ y & x \end{pmatrix}$ with $\zeta = x + y\sqrt{\varepsilon}$ in $\mathbb{F}'$, then

$$x^2 - \varepsilon y^2 = \det\begin{pmatrix} x & \varepsilon y \\ y & x \end{pmatrix} = \mathrm{Norm}_{\mathbb{F}'/\mathbb{F}}(\zeta) = \zeta \cdot \zeta^q = \zeta^{q+1}.$$

The next place to look for representations is at those that are induced from large subgroups. For each pair α, β of characters of $\mathbb{F}^*$, there is a character of the subgroup B:

$$B \to B/N = D = \mathbb{F}^* \times \mathbb{F}^* \to \mathbb{C}^* \times \mathbb{C}^* \xrightarrow{\cdot} \mathbb{C}^*,$$

which takes $\begin{pmatrix} a & b \\ 0 & d \end{pmatrix}$ to $\alpha(a)\beta(d)$. Let $W_{\alpha,\beta}$ be the representation induced from B to G by this representation; this is a representation of dimension $[G:B] = q+1$. By Exercise 3.19 its character values are found to be:

$$W_{\alpha,\beta}: \quad (q+1)\alpha(x)\beta(x) \quad \alpha(x)\beta(x) \quad \alpha(x)\beta(y) + \alpha(y)\beta(x) \quad 0$$

We see from this that $W_{\alpha,\beta} \cong W_{\beta,\alpha}$, that $W_{\alpha,\alpha} \cong U_\alpha \oplus V_\alpha$, and that for $\alpha \neq \beta$ the representation is irreducible. This gives $\frac{1}{2}(q-1)(q-2)$ more irreducible representations, of dimension $q+1$.

Comparing with the list of conjugacy classes, we see that there are $\frac{1}{2}q(q-1)$ irreducible characters left to be found. A natural way to find new characters is to induce characters from the cyclic subgroup K. For a representation

$$\varphi\colon K = (\mathbb{F}')^* \to \mathbb{C}^*,$$

the character values of the induced representation of dimension $[G:K] = q^2 - 1$ are

$$\mathrm{Ind}(\varphi): \quad q(q-1)\varphi(x) \quad 0 \quad 0 \quad \varphi(\zeta) + \varphi(\zeta)^q$$

Here again $\zeta = x + y\sqrt{\varepsilon} \in K = (\mathbb{F}')^*$. Note that $\mathrm{Ind}(\varphi^q) \cong \mathrm{Ind}(\varphi)$, so the representations $\mathrm{Ind}(\varphi)$ for $\varphi^q \neq \varphi$ give $\frac{1}{2}q(q-1)$ different representations.

However, these represenations are not irreducible: the character χ of $\mathrm{Ind}(\varphi)$ satisfies $(\chi, \chi) = q - 1$ if $\varphi^q \neq \varphi$, and otherwise $(\chi, \chi) = q$. We will have to work a little harder to get irreducible representations from these $\mathrm{Ind}(\varphi)$.

Another attempt to find more representations is to look inside tensor products of representations we know. We have $V_\alpha \otimes U_\gamma = V_{\alpha\gamma}$, and $W_{\alpha,\beta} \otimes U_\gamma \cong W_{\alpha\gamma,\beta\gamma}$, so there are no new ones to be found this way. But tensor products of the V_α's and $W_{\alpha,\beta}$'s are more promising. For example, $V \otimes W_{\alpha,1}$ has character values:

$$V \otimes W_{\alpha,1}: \qquad q(q+1)\alpha(x) \qquad 0 \qquad \alpha(x) + \alpha(y) \qquad 0$$

We can calculate some inner products of these characters with each other to estimate how many irreducible representations each contains, and how many they have in common. For example,

$$(\chi_{V\otimes W_{\alpha,1}}, \chi_{W_{\alpha,1}}) = 2,$$
$$(\chi_{\mathrm{Ind}(\varphi)}, \chi_{W_{\alpha,1}}) = 1 \quad \text{if } \varphi|_{\mathbb{F}^*} = \alpha,$$
$$(\chi_{V\otimes W_{\alpha,1}}, \chi_{V\otimes W_{\alpha,1}}) = q + 3,$$
$$(\chi_{V\otimes W_{\alpha,1}}, \chi_{\mathrm{Ind}(\varphi)}) = q \quad \text{if } \varphi|_{\mathbb{F}^*} = \alpha,$$

Comparing with the formula $(\chi_{\mathrm{Ind}(\varphi)}, \chi_{\mathrm{Ind}(\varphi)}) = q - 1$, one deduces that $V \otimes W_{\alpha,1}$ and $\mathrm{Ind}(\varphi)$ contain many of the same representations. With any luck, $\mathrm{Ind}(\varphi)$ and $W_{\alpha,1}$ should both be contained in $V \otimes W_{\alpha,1}$. This guess is easily confirmed; the virtual character

$$\chi_\varphi = \chi_{V\otimes W_{\alpha,1}} - \chi_{W_{\alpha,1}} - \chi_{\mathrm{Ind}(\varphi)}$$

takes values $(q-1)\alpha(x)$, $-\alpha(x)$, 0, and $-(\varphi(\zeta) + \varphi(\zeta)^q)$ on the four types of conjugacy classes. Therefore, $(\chi_\varphi, \chi_\varphi) = 1$, and $\chi_\varphi(1) = q - 1 > 0$, so χ_φ is, in fact, the character of an irreducible subrepresentation of $V \otimes W_{\alpha,1}$ of dimension $q - 1$. We denote this representation by X_φ. These $\frac{1}{2}q(q-1)$ representations, for $\varphi \neq \varphi^q$, and with $X_\varphi = X_{\varphi^q}$, therefore complete the list of irreducible representations for $\mathrm{GL}_2(\mathbb{F})$. The character table is

	1	$q^2 - 1$	$q^2 + q$	$q^2 - q$
$\mathrm{GL}_2(\mathbb{F}_q)$	$a_x = \begin{pmatrix} x & 0 \\ 0 & x \end{pmatrix}$	$b_x = \begin{pmatrix} x & 1 \\ 0 & x \end{pmatrix}$	$c_{x,y} = \begin{pmatrix} x & 0 \\ 0 & y \end{pmatrix}$	$d_{x,y} = \begin{pmatrix} x & \varepsilon y \\ y & x \end{pmatrix} = \zeta$
U_α	$\alpha(x^2)$	$\alpha(x^2)$	$\alpha(xy)$	$\alpha(\zeta^q)$
V_α	$q\alpha(x^2)$	0	$\alpha(xy)$	$-\alpha(\zeta^q)$
$W_{\alpha,\beta}$	$(q+1)\alpha(x)\beta(x)$	$\alpha(x)\beta(x)$	$\alpha(x)\beta(y) + \alpha(y)\beta(x)$	0
X_φ	$(q-1)\varphi(x)$	$-\varphi(x)$	0	$-(\varphi(\zeta) + \varphi(\zeta^q))$

Exercise 5.6. Find the multiplicity of each irreducible representation in the representations $V \otimes W_{\alpha,1}$ and $\mathrm{Ind}(\varphi)$.

Exercise 5.7. Find the character table of $PGL_2(\mathbb{F}) = GL_2(\mathbb{F})/\mathbb{F}^*$. Note that its characters are just the characters of $GL_2(\mathbb{F})$ that take the same values on elements equivalent mod $\mathbb{F}^*$.

We turn next to the subgroup $SL_2(\mathbb{F}_q)$ of 2×2 matrices of determinant one, with q odd. The conjugacy classes, together with the number of elements in each conjugacy class, and the number of conjugacy classes of each type, are

	Representative	No. Elements in Class	No. Classes
(1)	$e = \begin{pmatrix} 1 & 0 \\ 0 & 1 \end{pmatrix}$	1	1
(2)	$-e = \begin{pmatrix} -1 & 0 \\ 0 & -1 \end{pmatrix}$	1	1
(3)	$\begin{pmatrix} 1 & 1 \\ 0 & 1 \end{pmatrix}$	$\frac{q^2 - 1}{2}$	1
(4)	$\begin{pmatrix} 1 & \varepsilon \\ 0 & 1 \end{pmatrix}$	$\frac{q^2 - 1}{2}$	1
(5)	$\begin{pmatrix} -1 & 1 \\ 0 & -1 \end{pmatrix}$	$\frac{q^2 - 1}{2}$	1
(6)	$\begin{pmatrix} -1 & \varepsilon \\ 0 & -1 \end{pmatrix}$	$\frac{q^2 - 1}{2}$	1
(7)	$\begin{pmatrix} x & 0 \\ 0 & x^{-1} \end{pmatrix}, x \neq \pm 1$	$q(q + 1)$	$\frac{q - 3}{2}$
(8)	$\begin{pmatrix} x & y \\ \varepsilon y & x \end{pmatrix}, x \neq \pm 1$	$q(q - 1)$	$\frac{q - 1}{2}$

The verifications are very much as we did for $GL_2(\mathbb{F}_q)$. In (7), the classes of $\begin{pmatrix} x & 0 \\ 0 & x^{-1} \end{pmatrix}$ and $\begin{pmatrix} x^{-1} & 0 \\ 0 & x \end{pmatrix}$ are the same. In (8), the classes for (x, y) and $(x, -y)$ are the same; as before, a better labeling is by the element ζ in the cyclic group

$$C = \{\zeta \in (\mathbb{F}')^*: \zeta^{q+1} = 1\};$$

the elements ± 1 are not used, and the classes of ζ and ζ^{-1} are the same.

The total number of conjugacy classes is $q + 4$, so we turn to the task of finding $q + 4$ irreducible representations. We first see what we get by restricting representations from $GL_2(\mathbb{F}_q)$. Since we know the characters, there is no problem working this out, and we simply state the results:

(1) The U_α all restrict to the trivial representation U. Hence, if we restrict any representation, we will get the same for all tensor products by U_α's.

(2) The restriction V of the V_α's is irreducible.

(3) The restriction W_α of $W_{\alpha,1}$ is irreducible if $\alpha^2 \neq 1$, and $W_\alpha \cong W_\beta$ when $\beta = \alpha$ or $\beta = \alpha^{-1}$. These give $\frac{1}{2}(q-3)$ irreducible representations of dimension $q+1$.

(3′) Let τ denote the character of $\mathbb{F}^*$ with $\tau^2 = 1, \tau \neq 1$. The restriction of $W_{\tau,1}$ is the sum of two distinct irreducible representations, which we denote W' and W''.

(4) The restriction of X_φ depends only on the restriction of φ to the subgroup C, and φ and φ^{-1} determine the same representation. The representation is irreducible if $\varphi^2 \neq 1$. This gives $\frac{1}{2}(q-1)$ irreducible representations of dimension $q-1$.

(4′) If ψ denotes the character of C with $\psi^2 = 1, \psi \neq 1$, the restriction of X_ψ is the sum of two distinct irreducible representations, which we denote X' and X''.

Altogether this list gives $q+4$ distinct irreducible representations, and it is therefore the complete list. To finish the character table, the problem is to describe the four representations W', W'', X', and X''. Since we know the sum of the squares of the dimensions of all representations, we can deduce that the sum of the squares of these four representations is q^2+1, which is only possible if the first two have dimension $\frac{1}{2}(q+1)$ and the other two $\frac{1}{2}(q-1)$. This is similar to what we saw happens for restrictions of representations to subgroups of index two. Although the index here is larger, we can use what we know about index two subgroups by finding a subgroup H of index two in $GL_2(\mathbb{F}_q)$ that contains $SL_2(\mathbb{F}_q)$, and analyzing the restrictions of these four representations to H.

For H we take the matrices in $GL_2(\mathbb{F}_q)$ whose determinant is a square. The representatives of the conjugacy classes are the same as those for $GL_2(\mathbb{F}_q)$, including, of course, only those representatives whose determinant is a square, but we must add classes represented by the elements $\begin{pmatrix} x & \varepsilon \\ 0 & x \end{pmatrix}$, $x \in \mathbb{F}^*$. These are conjugate to the elements $\begin{pmatrix} x & 1 \\ 0 & x \end{pmatrix}$ in $GL_2(\mathbb{F}_q)$, but not in H. These are the $q-1$ split conjugacy classes. The procedure of the preceding section can be used to work out all the representations of H, but we need only a little of this.

Note that the sign representation U' from G/H is U_τ, so that $W_{\tau,1} \cong W_{\tau,1} \otimes U'$ and $X_\psi \cong X_\psi \otimes U'$; their restrictions to H split into sums of conjugate irreducible representations of half their dimensions. This shows these representations stay irreducible on restriction from H to $SL_2(\mathbb{F}_q)$, so that W' and W'' are conjugate representations of dimension $\frac{1}{2}(q+1)$, and X' and X'' are conjugate representations of dimension $\frac{1}{2}(q-1)$. In addition, we know that their character values on all nonsplit conjugacy classes are the same as half the characters of the representations $W_{\tau,1}$ and X_ψ, respectively. This is all the information we need to finish the character table. Indeed, the only values not covered by this discussion are

	$\begin{pmatrix}1&1\\0&1\end{pmatrix}$	$\begin{pmatrix}1&\varepsilon\\0&1\end{pmatrix}$	$\begin{pmatrix}-1&1\\0&-1\end{pmatrix}$	$\begin{pmatrix}-1&\varepsilon\\0&-1\end{pmatrix}$
W'	s	t	s'	t'
W''	t	s	t'	s'
X'	u	v	u'	v'
X''	v	u	v'	u'

The first two rows are determined as follows. We know that $s + t = \chi_{W_{\tau,1}}\left(\begin{pmatrix}1&1\\0&1\end{pmatrix}\right) = 1$. In addition, since $\begin{pmatrix}1&1\\0&1\end{pmatrix}^{-1} = \begin{pmatrix}1&-1\\0&1\end{pmatrix}$ is conjugate to $\begin{pmatrix}1&1\\0&1\end{pmatrix}$ if q is congruent to 1 modulo 4, and to $\begin{pmatrix}1&\varepsilon\\0&1\end{pmatrix}$ otherwise, and since $\chi(g^{-1}) = \overline{\chi(g)}$ for any character, we conclude that s and t are real if $q \equiv 1 \bmod(4)$, and $s = \bar{t}$ if $q \equiv 3 \bmod(4)$. In addition, since $-e$ acts as the identity or minus the identity for any irreducible representation (Schur's lemma),

$$\chi(-g) = \chi(g)\cdot\chi(1)/\chi(-e)$$

for any irreducible character χ. This gives the relations $s' = \tau(-1)s$ and $t' = \tau(-1)t$. Finally, applying the equation $(\chi, \chi) = 1$ to the character of W' gives a formula for $s\bar{t} + t\bar{s}$. Solving these equations gives $s, t = \frac{1}{2} \pm \frac{1}{2}\sqrt{\omega q}$, where $\omega = \tau(-1)$ is 1 or -1 according as $q \equiv 1$ or $3 \bmod(4)$. Similarly one computes that u and v are $-\frac{1}{2} \pm \frac{1}{2}\sqrt{\omega q}$. This concludes the computations needed to write out the character table.

Exercise 5.8. By considering the action of $SL_2(\mathbb{F}_q)$ on the set $\mathbb{P}^1(\mathbb{F}_q)$, show that $SL_2(\mathbb{F}_2) \cong \mathfrak{S}_3$, $PSL_2(\mathbb{F}_3) \cong \mathfrak{A}_4$, and $SL_2(\mathbb{F}_4) \cong \mathfrak{A}_5$.

Exercise 5.9*. Use the character table for $SL_2(\mathbb{F}_q)$ to show that $PSL_2(\mathbb{F}_q)$ is a simple group if q is odd and greater than 3.

Exercise 5.10. Compute the character table of $PSL_2(\mathbb{F}_q)$, either by regarding it as a quotient of $SL_2(\mathbb{F}_q)$, or as a subgroup of index two in $PGL_2(\mathbb{F}_q)$.

Exercise 5.11*. Find the conjugacy classes of $GL_3(\mathbb{F}_q)$, and compute the characters of the permutation representations obtained by the action of $GL_3(\mathbb{F}_q)$ on (i) the projective plane $\mathbb{P}^2(\mathbb{F}_q)$ and (ii) the "flag variety" consisting of a point on a line in $\mathbb{P}^2(\mathbb{F}_q)$. Show that the first is irreducible and that the second is a sum of the trivial representation, two copies of the first representation, and an irreducible representation.

Although the characters of the above groups were found by the early pioneers in representation theory, actually producing the representations in a natural way is more difficult. There has been a great deal of work extending this story to $GL_n(\mathbb{F}_q)$ and $SL_n(\mathbb{F}_q)$ for $n > 2$ (cf. [Gr]), and for corresponding groups, called finite Chevalley groups, related to other Lie groups. For some hints in this direction see [Hu3], as well as [Ti2]. Since all but a finite number of finite simple groups are now known to arise this way (or are cyclic or alternating groups, whose characters we already know), such representations play a fundamental role in group theory. In recent work their Lie-theoretic origins have been exploited to produce their representations, but to tell this story would go far beyond the scope of these lecture(r)s.

LECTURE 6

Weyl's Construction

In this lecture we introduce and study an important collection of functors generalizing the symmetric powers and exterior powers. These are defined simply in terms of the Young symmetrizers c_λ introduced in §4: given a representation V of an arbitrary group G, we consider the dth tensor power of V, on which both G and the symmetric group on d letters act. We then take the image of the action of c_λ on $V^{\otimes d}$; this is again a representation of G, denoted $\mathbb{S}_\lambda(V)$. This gives us a way of generating new representations, whose main application will be to Lie groups: for example, we will generate all representations of $\mathrm{SL}_n\mathbb{C}$ by applying these to the standard representation $\mathbb{C}^n$ of $\mathrm{SL}_n\mathbb{C}$. While it may be easiest to read this material while the definitions of the Young symmetrizers are still fresh in the mind, the construction will not be used again until §15, so that this lecture can be deferred until then.

§6.1. Schur Functors and Their Characters

For any finite-dimensional complex vector space V, we have the canonical decomposition

$$V \otimes V = \mathrm{Sym}^2 V \oplus \wedge^2 V.$$

The group $\mathrm{GL}(V)$ acts on $V \otimes V$, and this is, as we shall soon see, the decomposition of $V \otimes V$ into a direct sum of irreducible $\mathrm{GL}(V)$-representations. For the next tensor power,

$$V \otimes V \otimes V = \mathrm{Sym}^3 V \oplus \wedge^3 V \oplus \text{ another space.}$$

We shall see that this other space is a sum of two copies of an irreducible

GL(V)-representation. Just as $\mathrm{Sym}^d V$ and $\wedge^d V$ are images of symmetrizing operators from $V^{\otimes d} = V \otimes V \otimes \cdots \otimes V$ to itself, so are the other factors. The symmetric group $\mathfrak{S}_d$ acts on $V^{\otimes d}$, say on the right, by permuting the factors

$$(v_1 \otimes \cdots \otimes v_d) \cdot \sigma = v_{\sigma(1)} \otimes \cdots \otimes v_{\sigma(d)}.$$

This action commutes with the left action of GL(V). For any partition λ of d we have from the last lecture a Young symmetrizer c_λ in $\mathbb{C}\mathfrak{S}_d$. We denote the image of c_λ on $V^{\otimes d}$ by $\mathbb{S}_\lambda V$:

$$\mathbb{S}_\lambda V = \mathrm{Im}(c_\lambda|_{V^{\otimes d}})$$

which is again a representation of GL(V). We call the functor[1] $V \rightsquigarrow \mathbb{S}_\lambda V$ the *Schur functor* or *Weyl module*, or simply *Weyl's construction*, corresponding to λ. It was Schur who made the correspondence between representations of symmetric groups and representations of general linear groups, and Weyl who made the construction we give here.[2] We will give other descriptions later, cf. Exercise 6.14 and §15.5.

For example, the partition $d = d$ corresponds to the functor $V \rightsquigarrow \mathrm{Sym}^d V$, and the partition $d = 1 + \cdots + 1$ to the functor $V \rightsquigarrow \wedge^d V$.

We find something new for the partition $3 = 2 + 1$. The corresponding symmetrizer c_λ is

$$c_{(2,1)} = 1 + e_{(12)} - e_{(13)} - e_{(132)},$$

so the image of c_λ is the subspace of $V^{\otimes 3}$ spanned by all vectors

$$v_1 \otimes v_2 \otimes v_3 + v_2 \otimes v_1 \otimes v_3 - v_3 \otimes v_2 \otimes v_1 - v_3 \otimes v_1 \otimes v_2.$$

If $\wedge^2 V \otimes V$ is embedded in $V^{\otimes 3}$ by mapping

$$(v_1 \wedge v_3) \otimes v_2 \mapsto v_1 \otimes v_2 \otimes v_3 - v_3 \otimes v_2 \otimes v_1,$$

then the image of c_λ is the subspace of $\wedge^2 V \otimes V$ spanned by all vectors

$$(v_1 \wedge v_3) \otimes v_2 + (v_2 \wedge v_3) \otimes v_1.$$

It is not hard to verify that these vectors span the kernel of the canonical map from $\wedge^2 V \otimes V$ to $\wedge^3 V$, so we have

$$\mathbb{S}_{(2,1)} V = \mathrm{Ker}(\wedge^2 V \otimes V \to \wedge^3 V).$$

(This gives the missing factor in the decomposition of $V^{\otimes 3}$.)

Note that some of the $\mathbb{S}_\lambda V$ can be zero if V has small dimension. We will see that this is the case precisely when the number of rows in the Young diagram of λ is greater than the dimension of V.

[1] The functoriality means simply that a linear map $\varphi: V \to W$ of vector spaces determines a linear map $\mathbb{S}_\lambda(\varphi): \mathbb{S}_\lambda V \to \mathbb{S}_\lambda W$, with $\mathbb{S}_\lambda(\varphi \circ \psi) = \mathbb{S}_\lambda(\varphi) \circ \mathbb{S}_\lambda(\psi)$ and $\mathbb{S}_\lambda(\mathrm{Id}_V) = \mathrm{Id}_{\mathbb{S}_\lambda V}$.

[2] The notion goes by a variety of names and notations in the literature, depending on the context. Constructions differ markedly when not over a field of characteristic zero; and many authors now parametrize them by the conjugate partitions. Our choice of notation is guided by the correspondence between these functors and Schur polynomials, which we will see are their characters.

When $G = \mathrm{GL}(V)$, and for important subgroups $G \subset \mathrm{GL}(V)$, these $\mathbb{S}_\lambda V$ give many of the irreducible representations of G; we will come back to this later in the book. For now we can use our knowledge of symmetric group representations to prove a few facts about them—in particular, we show that they decompose the tensor powers $V^{\otimes d}$, and that they are irreducible representations of $\mathrm{GL}(V)$. We will also compute their characters; this will eventually be seen to be a special case of the Weyl character formula.

Any endomorphism g of V gives rise to an endomorphism of $\mathbb{S}_\lambda V$. In order to tell what representations we get, we will need to compute the trace of this endomorphism on $\mathbb{S}_\lambda V$; we denote this trace by $\chi_{\mathbb{S}_\lambda V}(g)$. For the computation, let $x_1, \ldots, x_k$ be the eigenvalues of g on V, $k = \dim V$. Two cases are easy. For $\lambda = (d)$,

$$\mathbb{S}_{(d)} V = \mathrm{Sym}^d V, \qquad \chi_{\mathbb{S}_{(d)} V}(g) = H_d(x_1, \ldots, x_k), \tag{6.1}$$

where $H_d(x_1, \ldots, x_k)$ is the complete symmetric polynomial of degree d. The definition of these symmetric polynomials is given in (A.1) of Appendix A. The truth of (6.1) is evident when g is a diagonal matrix, and its truth for the dense set of diagonalizable endomorphisms implies it for all endomorphisms; or one can see it directly by using the Jordan canonical form of g. For $\lambda = (1, \ldots, 1)$, we have similarly

$$\mathbb{S}_{(1,\ldots,1)} V = \textstyle\bigwedge^d V, \qquad \chi_{\mathbb{S}_{(1,\ldots,1)} V}(g) = E_d(x_1, \ldots, x_k), \tag{6.2}$$

with $E_d(x_1, \ldots, x_k)$ the elementary symmetric polynomial [see (A.3)]. The polynomials H_d and E_d are special cases of the *Schur polynomials*, which we denote by $S_\lambda = S_\lambda(x_1, \ldots, x_k)$. As λ varies over the partitions of d into at most k parts, these polynomials S_λ form a basis for the symmetric polynomials of degree d in these k variables. Schur polynomials are defined and discussed in Appendix A, especially (A.4)–(A.6). The above two formulas can be written

$$\chi_{\mathbb{S}_\lambda V}(g) = S_\lambda(x_1, \ldots, x_k) \quad \text{for } \lambda = (d) \text{ and } \lambda = (1, \ldots, 1).$$

We will show that this equation is valid for all λ:

Theorem 6.3. (1) *Let $k = \dim V$. Then $\mathbb{S}_\lambda V$ is zero if $\lambda_{k+1} \neq 0$. If $\lambda = (\lambda_1 \geq \cdots \geq \lambda_k \geq 0)$, then*

$$\dim \mathbb{S}_\lambda V = S_\lambda(1, \ldots, 1) = \prod_{1 \leq i < j \leq k} \frac{\lambda_i - \lambda_j + j - i}{j - i}.$$

(2) *Let m_λ be the dimension of the irreducible representation V_λ of $\mathfrak{S}_d$ corresponding to λ. Then*

$$V^{\otimes d} \cong \bigoplus_\lambda \mathbb{S}_\lambda V^{\oplus m_\lambda}.$$

(3) *For any $g \in \mathrm{GL}(V)$, the trace of g on $\mathbb{S}_\lambda V$ is the value of the Schur polynomial on the eigenvalues $x_1, \ldots, x_k$ of g on V:*

$$\chi_{\mathbb{S}_\lambda V}(g) = S_\lambda(x_1, \ldots, x_k).$$

(4) *Each* $\mathbb{S}_\lambda V$ *is an irreducible representation of* GL(V).

This theorem will be proved in the next section. Other formulas for the dimension of $\mathbb{S}_\lambda V$ are given in Exercises A.30 and A.31. The following is another:

Exercise 6.4*. Show that

$$\dim \mathbb{S}_\lambda V = \frac{m_\lambda}{d!}\prod (k-i+j) = \prod \frac{(k-i+j)}{h_{ij}},$$

where the products are over the d pairs (i, j) that number the row and column of boxes for λ, and h_{ij} is the hook number of the corresponding box.

Exercise 6.5. Show that $V^{\otimes 3} \cong \mathrm{Sym}^3 V \oplus \Lambda^3 V \oplus (\mathbb{S}_{(2,1)} V)^{\oplus 2}$, and
$V^{\otimes 4} \cong \mathrm{Sym}^4 V \oplus \Lambda^4 V \oplus (\mathbb{S}_{(3,1)} V)^{\oplus 3} \oplus (\mathbb{S}_{(2,2)} V)^{\oplus 2} \oplus (\mathbb{S}_{(2,1,1)} V)^{\oplus 3}$.
Compute the dimensions of each of the irreducible factors.

The proof of the theorem actually gives the following corollary:

Corollary 6.6. *If* $c \in \mathbb{C}\mathfrak{S}_d$, *and* $(\mathbb{C}\mathfrak{S}_d)\cdot c = \bigoplus_\lambda V_\lambda^{\oplus r_\lambda}$ *as representations of* $\mathfrak{S}_d$, *then there is a corresponding decomposition of* GL(V)*-spaces*:

$$V^{\otimes d}\cdot c = \bigoplus_\lambda \mathbb{S}_\lambda V^{\oplus r_\lambda}.$$

If $x_1, \ldots, x_k$ *are the eigenvalues of an endomorphism of* V, *the trace of the induced endomorphism of* $V^{\otimes d}\cdot c$ *is* $\sum r_\lambda S_\lambda(x_1, \ldots, x_k)$.

If λ and μ are different partitions, each with at most $k = \dim V$ parts, the irreducible GL(V)-spaces $\mathbb{S}_\lambda V$ and $\mathbb{S}_\mu V$ are not isomorphic. Indeed, their characters are the Schur polynomials S_λ and S_μ, which are different. More generally, at least for those representations of GL(V) which can be decomposed into a direct sum of copies of the represenations $\mathbb{S}_\lambda V$'s, *the representations are completely determined by their characters.* This follows immediately from the fact that the Schur polynomials are linearly independent.

Note, however, that we cannot hope to get *all* finite-dimensional irreducible representations of GL(V) this way, since the duals of these representations are not included. We will see in Lecture 15 that this is essentially the only omission. Note also that although the operation that takes representations of $\mathfrak{S}_d$ to representations of GL(V) preserves direct sums, the situation with respect to other linear algebra constructions such as tensor products is more complicated.

One important application of Corollary 6.6 is to the decomposition of a tensor product $\mathbb{S}_\lambda V \otimes \mathbb{S}_\mu V$ of two Weyl modules, with, say, λ a partition of

d and μ a partition of m. The result is

$$\mathbb{S}_\lambda V \otimes \mathbb{S}_\mu V \cong \bigoplus_\nu N_{\lambda\mu\nu} \mathbb{S}_\nu V; \tag{6.7}$$

here the sum is over partitions ν of $d + m$, and $N_{\lambda\mu\nu}$ are numbers determined by the *Littlewood–Richardson rule*. This is a rule that gives $N_{\lambda\mu\nu}$ as the number of ways to expand the Young diagram of λ, using μ in an appropriate way, to achieve the Young diagram for ν; see (A.8) for the precise formula. Two important special cases are easier to use and prove since they involve only the simpler Pieri formula (A.7). For $\mu = (m)$, we have

$$\mathbb{S}_\lambda V \otimes \mathrm{Sym}^m V \cong \bigoplus_\nu \mathbb{S}_\nu V, \tag{6.8}$$

the sum over all ν whose Young diagram is obtained by adding m boxes to the Young diagram of λ, with no two in the same column. Similarly for $\mu = (1, \ldots, 1)$,

$$\mathbb{S}_\lambda V \otimes \Lambda^m V = \bigoplus \mathbb{S}_\pi V, \tag{6.9}$$

the sum over all partitions π whose Young diagram is obtained from that of λ by adding m boxes, with no two in the same row.

To prove these formulas, we need only observe that

$$\begin{aligned}\mathbb{S}_\lambda V \otimes \mathbb{S}_\mu V &= V^{\otimes n} \cdot c_\lambda \otimes V^{\otimes m} \cdot c_\mu \\ &= V^{\otimes n} \otimes V^{\otimes m} \cdot (c_\lambda \otimes c_\mu) = V^{\otimes (n+m)} \cdot c,\end{aligned}$$

with $c = c_\lambda \otimes c_\mu \in \mathbb{C}\mathfrak{S}_d \otimes \mathbb{C}\mathfrak{S}_m = \mathbb{C}(\mathfrak{S}_d \times \mathfrak{S}_m) \subset \mathbb{C}\mathfrak{S}_{d+m}$. This proves that $\mathbb{S}_\lambda V \otimes \mathbb{S}_\mu V$ has a decomposition as in Corollary 6.6, and the coefficients are given by knowing the decomposition of the corresponding character. The character of a tensor product is the product of the characters of the factors; so this amounts to writing the product $S_\lambda S_\mu$ of Schur polynomials as a linear combination of Schur polynomials. This is done in Appendix A, and formulas (6.7), (6.8), and (6.9) follow from (A.8), (A.7), and Exercise A.32 (v), respectively.

For example, from $\mathrm{Sym}^d V \otimes V = \mathrm{Sym}^{d+1} V \oplus \mathbb{S}_{(d,1)} V$, it follows that

$$\mathbb{S}_{(d,1)} V = \mathrm{Ker}(\mathrm{Sym}^d V \otimes V \to \mathrm{Sym}^{d+1} V),$$

and similarly for the conjugate partition,

$$\mathbb{S}_{(2,1,\ldots,1)} V = \mathrm{Ker}(\Lambda^d V \otimes V \to \Lambda^{d+1} V).$$

Exercise 6.10*. One can also derive the preceding decompositions of tensor products directly from corresponding decompositions of representations of symmetric groups. Show that, in fact, $\mathbb{S}_\lambda V \otimes \mathbb{S}_\mu V$ corresponds to the "inner product" representation $V_\lambda \circ V_\mu$ of $\mathfrak{S}_{d+m}$ described in (4.41).

Exercise 6.11*. (a) The Littlewood–Richardson rule also comes into the decomposition of a Schur functor of a direct sum of vector spaces V and W. This generalizes the well-known identities

$$\mathrm{Sym}^n(V \oplus W) = \bigoplus_{a+b=n} (\mathrm{Sym}^a V \otimes \mathrm{Sym}^b W),$$

$$\wedge^n(V \oplus W) = \bigoplus_{a+b=n} (\wedge^a V \otimes \wedge^b W).$$

Prove the general decomposition over GL(V) × GL(W):

$$\mathbb{S}_\nu(V \oplus W) = \bigoplus N_{\lambda\mu\nu}(\mathbb{S}_\lambda V \otimes \mathbb{S}_\mu W),$$

the sum over all partitions λ, μ such that the sum of the numbers partitioned by λ and μ is the number partitioned by ν. (To be consistent with Exercise 2.36 one should use the notation ⊠ for these "external" tensor products.)

(b) Similarly prove the formula for the Schur functor of a tensor product:

$$\mathbb{S}_\nu(V \otimes W) = \bigoplus C_{\lambda\mu\nu}(\mathbb{S}_\lambda V \otimes \mathbb{S}_\mu W),$$

where the coefficients $C_{\lambda\mu\nu}$ are defined in Exercise 4.51. In particular show that

$$\mathrm{Sym}^d(V \otimes W) = \bigoplus \mathbb{S}_\lambda V \otimes \mathbb{S}_\lambda W,$$

the sum over all partitions λ of d with at most dim V or dim W rows. Replacing W by W^*, this gives the decomposition for the space of polynomial functions of degree d on the space Hom(V, W) over GL(V) × GL(W). For variations on this theme, see [Ho3]. Similarly,

$$\wedge^d(V \otimes W) = \bigoplus \mathbb{S}_\lambda V \otimes \mathbb{S}_{\lambda'} W,$$

the sum over partitions λ of d with at most dim V rows and at most dim W columns.

Exercise 6.12. Regarding

$$\mathrm{GL}_n\,\mathbb{C} = \mathrm{GL}_n\,\mathbb{C} \times \{1\} \subset \mathrm{GL}_n\,\mathbb{C} \times \mathrm{GL}_m\,\mathbb{C} \subset \mathrm{GL}_{n+m}\,\mathbb{C},$$

the preceding exercise shows how the restriction of a representation decomposes:

$$\mathrm{Res}(\mathbb{S}_\nu(\mathbb{C}^{n+m})) = \sum (N_{\lambda\mu\nu} \dim \mathbb{S}_\mu(\mathbb{C}^m))\mathbb{S}_\lambda(\mathbb{C}^n).$$

In particular, for $m = 1$, Pieri's formula gives

$$\mathrm{Res}(\mathbb{S}_\nu(\mathbb{C}^{n+1})) = \bigoplus \mathbb{S}_\lambda(\mathbb{C}^n),$$

the sum over all λ obtained from ν by removing any number of boxes from its Young diagram, with no two in any column.

Exercise 6.13*. Show that for any partition $\mu = (\mu_1, \ldots, \mu_r)$ of d,

$$\wedge^{\mu_1} V \otimes \wedge^{\mu_2} V \otimes \cdots \otimes \wedge^{\mu_r} V \cong \bigoplus_{\lambda} K_{\lambda\mu} \mathbb{S}_{\lambda'} V,$$

where $K_{\lambda\mu}$ is the Kostka number and λ' the conjugate of λ.

Exercise 6.14*. Let $\mu = \lambda'$ be the conjugate partition. Put the factors of the dth tensor power $V^{\otimes d}$ in one-to-one correspondence with the squares of the Young diagram of λ. Show that $\mathbb{S}_\lambda V$ is the image of this composite map:

$$\bigotimes_i (\Lambda^{\mu_i} V) \to \bigotimes_i (\otimes^{\mu_i} V) \to V^{\otimes d} \to \bigotimes_j (\otimes^{\lambda_j} V) \to \bigotimes_j (\mathrm{Sym}^{\lambda_j} V),$$

the first map being the tensor product of the obvious inclusions, the second grouping the factors of $V^{\otimes d}$ according to the columns of the Young diagram, the third grouping the factors according to the rows of the Young diagram, and the fourth the obvious quotient map. Alternatively, $\mathbb{S}_\lambda V$ is the image of a composite map

$$\bigotimes_i (\mathrm{Sym}^{\lambda_i} V) \to \bigotimes_i (\otimes^{\lambda_i} V) \to V^{\otimes d} \to \bigotimes_j (\otimes^{\mu_j} V) \to \bigotimes_j (\Lambda^{\mu_j} V).$$

In particular, $\mathbb{S}_\lambda V$ can be realized as a subspace of tensors in $V^{\otimes d}$ that are invariant by automorphisms that preserve the rows of a Young tableau of λ, *or* a subspace that is anti-invariant under those that preserve the columns, but not both, cf. Exercise 4.48.

Problem 6.15*. The preceding exercise can be used to describe a basis for the space $\mathbb{S}_\lambda V$. Let $v_1, \ldots, v_k$ be a basis for V. For each semistandard tableau T on λ, one can use it to write down an element v_T in $\bigotimes_i (\Lambda^{\mu_i} V)$; v_T is a tensor product of wedge products of basis elements, the ith factor in $\Lambda^{\mu_i} V$ being the wedge product (in order) of those basis vectors whose indices occur in the ith column of T. The fact to be proved is that the images of these elements v_T under the first composite map of the preceding exercise form a basis for $\mathbb{S}_\lambda V$.

At the end of Lecture 15, using more representation theory than we have at the moment, we will work out a simple variation of the construction of $\mathbb{S}_\lambda V$ which will give quick proofs of refinements of the preceding exercise and problem.

Exercise 6.16*. The Pieri formula gives a decomposition

$$\mathrm{Sym}^d V \otimes \mathrm{Sym}^d V = \bigoplus \mathbb{S}_{(d+a, d-a)} V,$$

the sum over $0 \le a \le d$. The left-hand side decomposes into a direct sum of $\mathrm{Sym}^2(\mathrm{Sym}^d V)$ and $\Lambda^2(\mathrm{Sym}^d V)$. Show that, in fact,

$$\mathrm{Sym}^2(\mathrm{Sym}^d V) = \mathbb{S}_{(2d,0)} V \oplus \mathbb{S}_{(2d-2,2)} V \oplus \mathbb{S}_{(2d-4,4)} V \oplus \cdots,$$

$$\Lambda^2(\mathrm{Sym}^d V) = \mathbb{S}_{(2d-1,1)} V \oplus \mathbb{S}_{(2d-3,3)} V \oplus \mathbb{S}_{(2d-5,5)} V \oplus \cdots.$$

Similarly using the dual form of Pieri to decompose $\Lambda^d V \otimes \Lambda^d V$ into the sum $\bigoplus \mathbb{S}_\lambda V$, the sum over all $\lambda = (2, \ldots, 2, 1, \ldots, 1)$ consisting of $d - a$ 2's and 2a 1's, $0 \le a \le d$, show that $\mathrm{Sym}^2(\Lambda^d V)$ is the sum of those factors with a even, and $\Lambda^2(\Lambda^d V)$ is the sum of those with a odd.

Exercise 6.17*. If λ and μ are any partitions, we can form the composite functor $\mathbb{S}_\mu(\mathbb{S}_\lambda V)$. The original "plethysm" problem—which remains very difficult in general—is to decompose these composites:

$$\mathbb{S}_\mu(\mathbb{S}_\lambda V) = \bigoplus_\nu M_{\lambda\mu\nu}\mathbb{S}_\nu V,$$

the sum over all partitions ν of dm, where λ is a partition of d and μ is a partition of m. The preceding exercise carried out four special cases of this.

(a) Show that there always *exists* such a decomposition for some non-negative integers $M_{\lambda\mu\nu}$ by constructing an element c in $\mathbb{C}\mathfrak{S}_{dm}$, depending on λ and μ, such that $\mathbb{S}_\mu(\mathbb{S}_\lambda V)$ is $V^{\otimes dm}\cdot c$.

(b) Compute $\mathrm{Sym}^2(\mathbb{S}_{(2,2)}V)$ and $\wedge^2(\mathbb{S}_{(2,2)}V)$.

Exercise 6.18* "Hermite reciprocity." Show that if $\dim V = 2$ there are isomorphisms

$$\mathrm{Sym}^p(\mathrm{Sym}^q V) \cong \mathrm{Sym}^q(\mathrm{Sym}^p V)$$

of $GL(V)$-representations, for all p and q.

Exercise 6.19*. Much of the story about Young diagrams and representations of symmetric and general linear groups can be generalized to *skew Young diagrams*, which are the differences of two Young diagrams. If λ and μ are partitions with $\mu_i \le \lambda_i$ for all i, λ/μ denotes the complement of the Young diagram for μ in that of λ. For example, if $\lambda = (3, 3, 1)$ and $\mu = (2, 1)$, λ/μ is the numbered part of

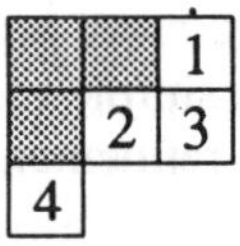

To each λ/μ we have a *skew Schur function* $S_{\lambda/\mu}$, which can be defined by any of several generalizations of constructions of ordinary Schur functions. Using the notation of Appendix A, the following definitions are equivalent:

(i) $$S_{\lambda/\mu} = |H_{\lambda_i-\mu_j-i+j}|,$$

(ii) $$S_{\lambda/\mu} = |E_{\lambda'_i-\mu'_j-i+j}|,$$

(iii) $$S_{\lambda/\mu} = \sum m_a x_1^{a_1}\cdot\ldots\cdot x_k^{a_k},$$

where m_a is the number of ways to number the boxes of λ/μ with a_1 1's, a_2 2's, ..., a_k k's, with nondecreasing rows and strictly increasing columns.

In terms of ordinary Schur polynomials, we have

(iv) $$S_{\lambda/\mu} = \sum N_{\mu\nu\lambda}S_\nu,$$

where $N_{\mu\nu\lambda}$ is the Littlewood–Richardson number.

Each λ/μ determines elements $a_{\lambda/\mu}$, $b_{\lambda/\mu}$, and Young symmetrizers $c_{\lambda/\mu} = a_{\lambda/\mu}b_{\lambda/\mu}$ in $A = \mathbb{C}\mathfrak{S}_d$, $d = \sum \lambda_i - \mu_i$, exactly as in §4.1, and hence a representation denoted $V_{\lambda/\mu} = Ac_{\lambda/\mu}$ of $\mathfrak{S}_d$. Equivalently, $V_{\lambda/\mu}$ is the image of the map $Ab_{\lambda/\mu} \to Aa_{\lambda/\mu}$ given by right multiplication by $a_{\lambda/\mu}$, or the image of the map $Aa_{\lambda/\mu} \to Ab_{\lambda/\mu}$ given by right multiplication by $b_{\lambda/\mu}$. The decomposition of $V_{\lambda/\mu}$ into irreducible representations is

(v) $$V_{\lambda/\mu} = \sum N_{\mu\nu\lambda} V_\nu.$$

Similarly there are *skew Schur functors* $\mathbb{S}_{\lambda/\mu}$, which take a vector space V to the image of $c_{\lambda/\mu}$ on $V^{\otimes d}$; equivalently, $\mathbb{S}_{\lambda/\mu}V$ is the image of a natural map (generalizing that in the Exercise 6.14)

(vi) $$\bigotimes_i (\Lambda^{\lambda_i - \mu_i} V) \to V^{\otimes d} \to \bigotimes_j (\mathrm{Sym}^{\lambda'_j - \mu'_j} V),$$

or

(vii) $$\bigotimes_i (\mathrm{Sym}^{\lambda_i - \mu_i} V) \to V^{\otimes d} \to \bigotimes_j (\Lambda^{\lambda'_j - \mu'_j} V).$$

Given a basis $v_1, \ldots, v_k$ for V and a standard tableau T on λ/μ, one can write down an element v_T in $\bigotimes_i (\Lambda^{\lambda'_j - \mu'_j} V)$; for example, corresponding to the displayed tableau, $v_T = v_4 \otimes v_2 \otimes (v_1 \wedge v_3)$. A key fact, generalizing the result of Exercise 6.15, is that the images of these elements under the map (vi) form a basis for $\mathbb{S}_{\lambda/\mu}V$.

The character of $\mathbb{S}_{\lambda/\mu}V$ is given by the Schur function $S_{\lambda/\mu}$: if g is an endomorphism of V with eigenvalues $x_1, \ldots, x_k$, then

(viii) $$\chi_{\mathbb{S}_{\lambda/\mu}V}(g) = S_{\lambda/\mu}(x_1, \ldots, x_k).$$

In terms of basic Schur functors,

(ix) $$\mathbb{S}_{\lambda/\mu}V \cong \sum N_{\mu\nu\lambda} \mathbb{S}_\nu V.$$

Exercise 6.20*. (a) Show that if $\lambda = (p, q)$, $\mathbb{S}_{(p,q)}V$ is the kernel of the contraction map

$$c_{p,q}\colon \mathrm{Sym}^p V \otimes \mathrm{Sym}^q V \to \mathrm{Sym}^{p+1} V \otimes \mathrm{Sym}^{q-1} V.$$

(b) If $\lambda = (p, q, r)$, show that $\mathbb{S}_{(p,q,r)}V$ is the intersection of the kernels of two contraction maps $c_{p,q} \otimes 1_r$ and $1_p \otimes c_{p,r}$, where 1_i denotes the identity map on $\mathrm{Sym}^i V$.

In general, for $\lambda = (\lambda_1, \ldots, \lambda_k)$, $\mathbb{S}_\lambda V \subset \mathrm{Sym}^{\lambda_1} V \otimes \cdots \otimes \mathrm{Sym}^{\lambda_k} V$ is the intersection of the kernels of the $k-1$ maps

$$\psi_i = 1_{\lambda_1} \otimes \cdots \otimes 1_{\lambda_{i-1}} \otimes c_{\lambda_i, \lambda_{i+1}} \otimes 1_{\lambda_{i+2}} \otimes \cdots \otimes 1_{\lambda_k}, \quad 1 \le i \le k-1.$$

(c) For $\lambda = (p, 1, \ldots, 1)$, show that $\mathbb{S}_\lambda V$ is the kernel of the contraction map:

$$\mathbb{S}_{(p,1,\ldots,1)}V = \mathrm{Ker}(\mathrm{Sym}^p V \otimes \Lambda^{d-p} V \to \mathrm{Sym}^{p+1} V \otimes \Lambda^{d-p-1} V).$$

In general, for any choice of a between 1 and $k-1$, the intersection of the kernels of all ψ_i except ψ_a is $\mathbb{S}_\sigma V \otimes \mathbb{S}_\tau V$, where $\sigma = (\lambda_1, \ldots, \lambda_a)$ and $\tau = (\lambda_{a+1}, \ldots, \lambda_k)$; so $\mathbb{S}_\lambda V$ is the kernel of a contraction map defined on $\mathbb{S}_\sigma V \otimes \mathbb{S}_\tau V$. For example, if a is $k - 1$, and we set $r = \lambda_k$, Pieri's formula writes $\mathbb{S}_\sigma V \otimes \mathrm{Sym}^r V$ as a direct sum of $\mathbb{S}_\lambda V$ and other factors $\mathbb{S}_\nu V$; the general assertion in (b) is equivalent to the claim that $\mathbb{S}_\lambda V$ is the only factor that is in the kernel of the contraction, ie.,

$$\mathbb{S}_\lambda V = \mathrm{Ker}(\mathbb{S}_{(\lambda_1, \ldots, \lambda_{k-1})} V \otimes \mathrm{Sym}^r V \to V^{\otimes(d-r+1)} \otimes \mathrm{Sym}^{r-1} V).$$

These results correspond to writing the representations $V_\lambda \subset U_\lambda$ of the symmetric group as the intersection of kernels of maps to $U_{\lambda_1, \ldots, \lambda_i+1, \lambda_{i+1}-1, \ldots, \lambda_k}$.

Exercise 6.21. The functorial nature of Weyl's construction has many consequences, which are not explored in this book. For example, if E_* is a complex of vector spaces, the tensor product $E_*^{\otimes d}$ is also a complex, and the symmetric group $\mathfrak{S}_d$ acts on it; when factors in E_p and E_q are transposed past each other, the usual sign $(-1)^{pq}$ is inserted. The image of the Young symmetrizer c_λ is a complex $\mathbb{S}_\lambda(E_*)$, sometimes called a *Schur complex*. Show that if E_* is the complex $E_{-1} = V \to E_0 = V$, with the boundary map the identity map, and $\lambda = (d)$, then $\mathbb{S}_\lambda(E_*)$ is the Koszul complex

$$0 \to \wedge^d \to \wedge^{d-1} \otimes S^1 \to \wedge^{d-2} \otimes S^2 \to \cdots \to \wedge^1 \otimes S^{d-1} \to S^d \to 0,$$

where $\wedge^i = \wedge^i V$, and $S^j = \mathrm{Sym}^j V$.

§6.2. The Proofs

We need first a small piece of the general story about semisimple algebras, which we work out by hand. For the moment G can be any finite group, although our application is for the symmetric group. If U is a right module over $A = \mathbb{C}G$, let

$$B = \mathrm{Hom}_G(U, U) = \{\varphi\colon U \to U\colon \varphi(v \cdot g) = \varphi(v) \cdot g, \forall v \in U, g \in G\}.$$

Note that B acts on U on the left, commuting with the right action of A; B is called the *commutator* algebra. If $U = \bigoplus U_i^{\oplus n_i}$ is an irreducible decomposition with U_i nonisomorphic irreducible right A-modules, then by Schur's Lemma 1.7

$$B = \bigoplus_i \mathrm{Hom}_G(U_i^{\oplus n_i}, U_i^{\oplus n_i}) = \bigoplus_i M_{n_i}(\mathbb{C}),$$

where $M_{n_i}(\mathbb{C})$ is the ring of $n_i \times n_i$ complex matrices.

If W is any left A-module, the tensor product

$$U \otimes_A W = U \otimes_{\mathbb{C}} W/\text{subspace generated by } \{va \otimes w - v \otimes aw\}$$

is a left B-module by acting on the first factor: $b \cdot (v \otimes w) = (b \cdot v) \otimes w$.

Lemma 6.22. *Let U be a finite-dimensional right A-module.*

(i) *For any $c \in A$, the canonical map $U \otimes_A Ac \to Uc$ is an isomorphism of left B-modules.*

(ii) *If $W = Ac$ is an irreducible left A-module, then $U \otimes_A W = Uc$ is an irreducible left B-module.*

(iii) *If $W_i = Ac_i$ are the distinct irreducible left A-modules, with m_i the dimension of W_i, then*

$$U \cong \bigoplus_i (U \otimes_A W_i)^{\oplus m_i} \cong \bigoplus_i (Uc_i)^{\oplus m_i}$$

is the decomposition of U into irreducible left B-modules.

Proof. Note first that Ac is a direct summand of A as a left A-module; this is a consequence of the semisimplicity of all representations of G (Proposition 1.5). To prove (i), consider the commutative diagram

$$\begin{array}{ccccc} U \otimes_A A & \xrightarrow{\cdot c} & U \otimes_A Ac & \hookrightarrow & U \otimes_A A \\ \downarrow & & \downarrow & & \downarrow \\ U & \xrightarrow{\cdot c} & U \cdot c & \hookrightarrow & U \end{array},$$

where the vertical maps are the maps $v \otimes a \mapsto v \cdot a$; since the left horizontal maps are surjective, the right ones injective, and the outside vertical maps are isomorphisms, the middle vertical map must be an isomorphism.

For (ii), consider first the case where U is an irreducible A-module, so $B = \mathbb{C}$. It suffices to show that $\dim U \otimes_A W \leq 1$. For this we use Proposition 3.29 to identify A with a direct sum $\bigoplus_{i=1}^{r} M_{m_i}\mathbb{C}$ of r matrix algebras. We can identify W with a minimal left ideal of A. Any minimal ideal in the sum of matrix algebras is isomorphic to one which consists of r-tuples of matrices which are zero except in one factor, and in this factor are all zero except for one column. Similarly, U can be identified with the right ideal of r-tuples which are zero except in one factor, and in that factor all are zero except in one row. Then $U \otimes_A W$ will be zero unless the factor is the same for U and W, in which case $U \otimes_A W$ can be identified with the matrices which are zero except in one row and column of that factor. This completes the proof when U is irreducible. For the general case of (ii), decompose $U = \bigoplus_i U_i^{\oplus n_i}$ into a sum of irreducible right A-modules, so $U \otimes_A W = \bigoplus_i (U_i \otimes_A W)^{\oplus n_i} = \mathbb{C}^{\oplus n_k}$ for some k, which is visibly irreducible over $B = \bigoplus M_{n_j}(\mathbb{C})$.

Part (iii) follows, since the isomorphism $A \cong \bigoplus W_i^{\oplus m_i}$ determines an isomorphism

$$U \cong U \otimes_A A \cong U \otimes_A \left(\bigoplus_i W_i^{\oplus m_i}\right) \cong \bigoplus_i (U \otimes_A W_i)^{\oplus m_i}. \qquad \square$$

To prove Theorem 6.3, we will apply Lemma 6.22 to the right $\mathbb{C}\mathfrak{S}_d$-module $U = V^{\otimes d}$. That lemma shows how to decompose U as a B-module, where B

is the algebra of all endomorphisms of U that commute with all permutations of the factors. The endomorphisms of U induced by endomorphisms of V are certainly in this algebra B. Although B is generally much larger than End(V), we have

Lemma 6.23. *The algebra B is spanned as a linear subspace of* $\mathrm{End}(V^{\otimes d})$ *by* End(V). *A subspace of $V^{\otimes d}$ is a sub-B-module if and only if it is invariant by* GL(V).

PROOF. Note that if W is any finite-dimensional vector space, then $\mathrm{Sym}^d W$ is the subspace of $W^{\otimes d}$ spanned by all $w^d = d!w \otimes \cdots \otimes w$ as w runs through W. Applying this to $W = \mathrm{End}(V) = V^* \otimes V$ proves the first statement, since $\mathrm{End}(V^{\otimes d}) = (V^*)^{\otimes d} \otimes V^{\otimes d} = W^{\otimes d}$, with compatible actions of $\mathfrak{S}_d$. The second follows from the fact that GL(V) is dense in End(V). □

We turn now to the proof of Theorem 6.3. Note that $\mathbb{S}_\lambda V$ is Uc_λ, so parts (2) and (4) follow from Lemmas 6.22 and 6.23. We use the same methods to give a rather indirect but short proof of part (3); for a direct approach see Exercise 6.28. From Lemma 6.22 we have an isomorphism of GL(V)-modules:

$$\mathbb{S}_\lambda V \cong V^{\otimes d} \otimes_A V_\lambda \tag{6.24}$$

with $V_\lambda = A \cdot c_\lambda$. Similarly for $U_\lambda = A \cdot a_\lambda$, and since the image of right multiplication by a_λ on $V^{\otimes d}$ is the tensor product of symmetric powers, we have

$$\mathrm{Sym}^{\lambda_1} V \otimes \mathrm{Sym}^{\lambda_2} V \otimes \cdots \otimes \mathrm{Sym}^{\lambda_k} V \cong V^{\otimes d} \otimes_A U_\lambda. \tag{6.25}$$

But we have an isomorphism $U_\lambda \cong \bigoplus_\mu K_{\mu\lambda} V_\mu$ of A-modules by Young's rule (4.39), so we deduce an isomorphism of GL(V)-modules

$$\mathrm{Sym}^{\lambda_1} V \otimes \mathrm{Sym}^{\lambda_2} V \otimes \cdots \otimes \mathrm{Sym}^{\lambda_k} V \cong \bigoplus_\mu K_{\mu\lambda} \mathbb{S}_\mu V. \tag{6.26}$$

By what we saw before the statement of the theorem, the trace of g on the left-hand side of (6.26) is the product $H_\lambda(x_1, \ldots, x_k)$ of the complete symmetric polynomials $H_{\lambda_i}(x_1, \ldots, x_k)$. Let $\mathbb{S}_\lambda(g)$ denote the endomorphism of $\mathbb{S}_\lambda V$ determined by an endomorphism g of V. We therefore have

$$H_\lambda(x_1, \ldots, x_k) = \Sigma_\mu K_{\mu\lambda} \mathrm{Trace}(\mathbb{S}_\mu(g)).$$

But these are precisely the relations between the functions H_λ and the Schur polynomials S_μ [see formula (A.9)], and these relations are invertible, since the matrix $(K_{\mu\lambda})$ of coefficients is triangular with 1's on the diagonal. It follows that $\mathrm{Trace}(\mathbb{S}_\lambda(g)) = S_\lambda(x_1, \ldots, x_k)$, which proves part (3).

Note that if $\lambda = (\lambda_1, \ldots, \lambda_d)$ with $d > k$ and $\lambda_{k+1} \neq 0$, this same argument shows that the trace is $S_\lambda(x_1, \ldots, x_k, 0, \ldots, 0)$, which is zero, for example by (A.6). For g the identity, this shows that $\mathbb{S}_\lambda V = 0$ in this case. From part (3) we also get

$$\dim \mathbb{S}_\lambda V = S_\lambda(1, \ldots, 1), \tag{6.27}$$

and computing $S_\lambda(1, \ldots, 1)$ via Exercise A.30(ii) yields part (1). □

Exercise 6.28. If you have given an independent proof of Problem 6.15, part (3) of Theorem 6.3 can be seen directly. The basis elements v_T for $\mathbb{S}_\lambda V$ specified in Problem 6.15 are eigenvectors for a diagonal matrix with entries $x_1, \ldots, x_k$, with eigenvalue $X^a = x_1^{a_1} \cdot \ldots \cdot x_k^{a_k}$, where the tableau T has a_1 1's, a_2 2's, $\ldots$, a_k k's. The trace is therefore $\sum K_{\lambda a} X^a$, where $K_{\lambda a}$ is the number of ways to number the boxes of the Young diagram of λ with a_1 1's, a_2 2's, $\ldots$, a_k k's. This is just the expression for S_λ obtained in Exercise A.31(a).

We conclude this lecture with a few of the standard elaborations of these ideas, in exercise form; they are not needed in these lectures.

Exercise 6.29*. Show that, in the context of Lemma 6.22, if U is a faithful A-module, then A is the commutator of its commutator B:

$$A = \{\psi \colon U \to U \colon \psi(bv) = b\psi(v), \forall v \in U, b \in B\}.$$

If U is not faithful, the canonical map from A to its bicommutator is surjective. Conclude that, in Theorem 6.3, the algebra of endomorphisms of $V^{\otimes d}$ that commute with $\mathrm{GL}(V)$ is spanned by the permutations in $\mathfrak{S}_d$.

Exercise 6.30. Show that, in Lemma 6.22, there is a natural one-to-one correspondence between the irreducible right A-modules U_i that occur in U and the irreducible left B-modules V_i. Show that there is a canonical decomposition

$$U = \bigoplus_i (V_i \otimes_{\mathbb{C}} U_i)$$

as a left B-module and as a right A-module. This shows again that the number of times V_i occurs in U is the dimension of U_i, and dually that the number of times U_i occurs is the dimension of V_i. Deduce the canonical decomposition

$$V^{\otimes d} = \bigoplus \mathbb{S}_\lambda V \otimes_{\mathbb{C}} V_\lambda,$$

the sum over partitions λ of d into at most $k = \dim V$ parts; this decomposition is compatible with the actions of $\mathrm{GL}(V)$ and $\mathfrak{S}_d$. In particular, the number of times V_λ occurs in the representation $V^{\otimes d}$ of $\mathfrak{S}_d$ is the dimension of $\mathbb{S}_\lambda V$.

Exercise 6.31. Let e be an idempotent in the group algebra $A = \mathbb{C}G$, and let $U = eA$ be the corresponding right A-module. Let $E = eAe$, a subalgebra of A. The algebra structure in A makes eA a left E-module. Show that this defines an isomorphism of $\mathbb{C}$-algebras

$$E = eAe \cong \mathrm{Hom}_A(eA, eA) = \mathrm{Hom}_G(U, U) = B.$$

Exercise 6.32. If H is a subgroup of G, and $e \in \mathbb{C}H$ is an idempotent, corresponding to a representation $W = \mathbb{C}H \cdot e$ of H, show that $\mathbb{C}G \cdot e$ is the induced representation $\operatorname{Ind}_H^G(W)$. For example, if $\vartheta: H \to \mathbb{C}^*$ is a one-dimensional representation, then

$$\operatorname{Ind}_H^G(\vartheta) = \mathbb{C}G \cdot e_\vartheta, \quad \text{where } e_\vartheta = \frac{1}{|G|} \sum_{g \in G} \overline{\vartheta(g)} e_g.$$

PART II
LIE GROUPS AND LIE ALGEBRAS

From a naive point of view, Lie groups seem to stand at the opposite end of the spectrum of groups from finite ones.[1] On the one hand, as abstract groups they seem enormously complicated: for example, being of uncountable order, there is no question of giving generators and relations. On the other hand, they do come with the additional data of a topology and a manifold structure; this makes it possible—and, given the apparent hopelessness of approaching them purely as algebraic objects, necessary—to use geometric concepts to study them.

Lie groups thus represent a confluence of algebra, topology, and geometry, which perhaps accounts in part for their ubiquity in modern mathematics. It also makes the subject a potentially intimidating one: to have to understand, both individually and collectively, all these aspects of a single object may be somewhat daunting.

Happily, just because the algebra and the geometry/topology of a Lie group are so closely entwined, there is an object we can use to approach the study of Lie groups that extracts much of the structure of a Lie group (primarily its algebraic structure) while seemingly getting rid of the topological complexity. This is, of course, the *Lie algebra*. The Lie algebra is, at least according to its definition, a purely algebraic object, consisting simply of a vector space with bilinear operation; and so it might appear that in associating to a Lie group its Lie algebra we are necessarily giving up a lot of information about the group. This is, in fact, not the case: as we shall see in many cases (and perhaps all of the most important ones), encoded in the algebraic structure of a Lie algebra is almost all of the geometry of the group. In particular, we will

[1] In spite of this there are deep, if only partially understood, relations between finite and Lie groups, extending even to their simple group classifications.

see by the end of Lecture 8 that there is a very close relationship between representations of the Lie group we start with and representations of the Lie algebra we associate to it; and by the end of the book we will make that correspondence exact.

We said that passing from the Lie group to its Lie algebra represents a simplification because it eliminates whatever nontrivial topological structure the group may have had; it "flattens out," or "linearizes," the group. This, in turn, allows for a further simplification: since a Lie algebra is just a vector space with bilinear operation, it makes perfect sense, if we are asked to study a real Lie algebra (or one over any subfield of $\mathbb{C}$) to tensor with the complex numbers. Thus, we may investigate first the structure and representations of *complex Lie algebras*, and then go back to apply this knowledge to the study of real ones. In fact, this turns out to be a feasible approach, in every respect: the structure of complex Lie algebras tends to be substantially simpler than that of real Lie algebras; and knowing the representations of the complex Lie algebra will solve the problem of classifying the representations of the real one.

There is one further reduction to be made: some very elementary Lie algebra theory allows us to narrow our focus further to the study of *semisimple Lie algebras*. This is a subset of Lie algebras analogous to simple groups in that they are in some sense atomic objects, but better behaved in a number of ways: a semisimple Lie algebra is a direct sum of simple ones; there are easy criteria for the semisimplicity of a given Lie algebra; and, most of all, their representation theory can be approached in a completely uniform manner. Moreover, as in the case of finite groups, there is a complete classification theorem for simple Lie algebras.

We may thus describe our approach to the representation theory of Lie groups by the sequence of objects

Lie group

$\rightsquigarrow$ Lie algebra

$\rightsquigarrow$ complex Lie algebra

$\rightsquigarrow$ semisimple complex Lie algebra.

We describe this progression in Lectures 7–9. In Lectures 7 and 8 we introduce the definitions of and some basic facts about Lie groups and Lie algebras. Lecture 8 ends with a description of the exponential map, which allows us to establish the close connection between the first two objects above. We then do, in Lecture 9, the very elementary classification theory of Lie algebras that motivates our focus on semisimple complex Lie algebras, and at least state the classification theorem for these. This establishes the fact that the second, third, and fourth objects above have essentially the same irreducible representations. (This lecture may also serve to give a brief taste of some general theory, which is mostly postponed to later lectures or appendices.) In Lecture 10 we discuss examples of Lie algebras in low dimensions.

From that point on we will proceed to devote ourselves almost exclusively to the study of semisimple complex Lie algebras and their representations. We do this, we have to say, in an extremely inefficient manner: we start with a couple of very special cases, which occupy us for three lectures (11–13); enunciate the general paradigm in Lecture 14; carry this out for the classical Lie algebras in Lectures 15–20; and (finally) finish off the general theory in Lectures 21–26. Thus, it will not be until the end that we go back and use the knowledge we have gained to say something about the original problem. In view of this long interlude, it is perhaps a good idea to enunciate one more time our basic

Point of View: The primary objects of interest are Lie groups and their representations; these are what actually occur in real life and these are what we want to understand. The notion of a complex Lie algebra is introduced primarily as a tool in this study; it is an essential tool[2] and we should consider ourselves incredibly lucky to have such a wonderfully effective one; but in the end it is for us a means to an end.

The special cases worked out in Lectures 11–13 are the Lie algebras of SL_2 and SL_3. Remarkably, most of the structure shared by all semisimple Lie algebras can be seen in these examples. We should probably point out that much of what we do by hand in these cases could be deduced from the Weyl construction we saw in Lecture 6 (as we will do generally in Lecture 15), but we mainly ignore this, in order to work from a "Lie algebra" point of view and motivate the general story.

[2] Perhaps not logically so; cf. Adams' book [Ad].

LECTURE 7

Lie Groups

In this lecture we introduce the definitions and basic examples of Lie groups and Lie algebras. We assume here familiarity with the definition of differentiable manifolds and maps between them, but no more; in particular, we do not mention vector fields, differential forms, Riemannian metrics, or any other tensors. Section 7.3, which discusses maps of Lie groups that are covering space maps of the underlying manifolds, may be skimmed and referred back to as needed, though working through it will help promote familiarity with basic examples of Lie groups.

§7.1. Lie Groups: Definitions

You probably already know what a Lie group is; it is just a set endowed simultaneously with the compatible structures of a group and a $\mathscr{C}^\infty$ manifold. "Compatible" here means that the multiplication and inverse operations in the group structure

$$\times: G \times G \to G$$

and

$$\iota: G \to G$$

are actually differentiable maps (logically, this is equivalent to the single requirement that the map $G \times G \to G$ sending (x, y) to $x \cdot y^{-1}$ is $\mathscr{C}^\infty$).

A *map*, or *morphism*, between two Lie groups G and H is just a map $\rho: G \to H$ that is both differentiable and a group homomorphism. In general, qualifiers applied to Lie groups refer to one or another of the two structures, usually without much ambiguity; thus, *abelian* refers to the group structure, *n-dimensional* or *connected* refers to the manifold structure. Sometimes a condition on one structure turns out to be equivalent to a condition on the other; for example, we will see below that to say that a map of connected Lie groups $\varphi: G \to H$ is a surjective map of groups is equivalent to saying that the differential $d\varphi$ is surjective at every point.

One area where there is some potential confusion is in the definition of a Lie subgroup. This is essentially a difficulty inherited directly from manifold theory, where we have to make a distinction between a *closed submanifold* of a manifold M, by which we mean a subset $X \subset M$ that inherits a manifold structure from M (i.e., that may be given, locally in M, by setting a subset of the local coordinates equal to zero), and an *immersed submanifold*, by which we mean the image of a manifold X under a one-to-one map with injective differential everywhere—that is, a map that is an embedding *locally in X*. The distinction is necessary simply because the underlying topological space structure of an immersed submanifold may not agree with the topological structure induced by the inclusion of X in M. For example, the map from X to M could be the immersion of an open interval in $\mathbb{R}$ into the plane $\mathbb{R}^2$ as a figure "6":

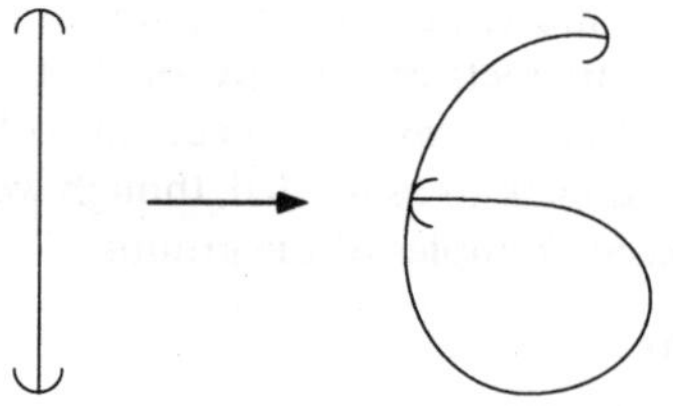

Another standard example of this, which is also an example in the category of groups, would be to take M to be the two-dimensional real torus $\mathbb{R}^2/\mathbb{Z}^2 = S^1 \times S^1$, and X the image in M of a line $V \subset \mathbb{R}^2$ having irrational slope:

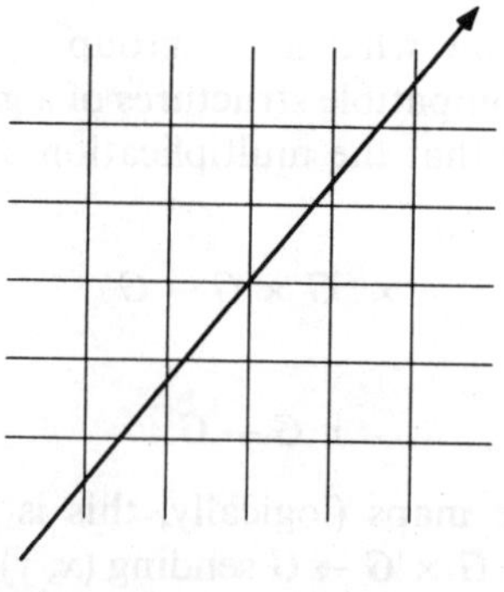

The upshot of this is that we define a *Lie subgroup* (or *closed Lie subgroup*, if we want to emphasize the point) of a Lie group G to be a subset that is simultaneously a subgroup and a *closed* submanifold; and we define an *immersed subgroup* to be the image of a Lie group H under an injective morphism to G. (That a one-to-one morphism of Lie groups has everywhere injective differential will follow from discussions later in this lecture.)

The definition of a *complex Lie group* is exactly analogous, the words "differentiable manifold" being replaced by "complex manifold" and all related notions revised accordingly. Similarly, to define an *algebraic group* one replaces "differentiable manifold" by "algebraic variety" and "differentiable map" by "regular morphism." As we will see, the category of complex Lie groups is in many ways markedly different from that of real Lie groups (for example, there are many fewer complex Lie groups than real ones). Of course, the study of algebraic groups in general is quite different from either of these since an algebraic group comes with a field of definition that may or may not be a subfield of $\mathbb{C}$ (it may, for that matter, have positive characteristic). In practice, though, while the two are not the same (we will see examples of this in Lecture 10, for example), the category of algebraic groups over $\mathbb{C}$ behaves very much like the category of complex Lie groups.

§7.2. Examples of Lie Groups

The basic example of a Lie group is of course the *general linear group* $\mathrm{GL}_n\mathbb{R}$ of invertible $n \times n$ real matrices; this is an open subset of the vector space of all $n \times n$ matrices, and gets its manifold structure accordingly (so that the entries of the matrix are coordinates on $\mathrm{GL}_n\mathbb{R}$). That the multiplication map $\mathrm{GL}_n\mathbb{R} \times \mathrm{GL}_n\mathbb{R} \to \mathrm{GL}_n\mathbb{R}$ is differentiable is clear; that the inverse map $\mathrm{GL}_n\mathbb{R} \to \mathrm{GL}_n\mathbb{R}$ is follows from Cramer's formula for the inverse. Occasionally $\mathrm{GL}_n\mathbb{R}$ will come to us as the group of automorphisms of an n-dimensional real vector space V; when we want to think of $\mathrm{GL}_n\mathbb{R}$ in this way (e.g., without choosing a basis for V and thereby identifying G with the group of matrices), we will write it as $\mathrm{GL}(V)$ or $\mathrm{Aut}(V)$. A *representation* of a Lie group G, of course, is a morphism from G to $\mathrm{GL}(V)$.

Most other Lie groups are defined initially as subgroups of GL_n (though they may appear in other contexts as subgroups of other general linear groups, which is, of course, the subject matter of these lectures). For the most part, such subgroups may be described either by equations on the entries of an $n \times n$ matrix, or as the subgroup of automorphisms of $V \cong \mathbb{R}^n$ preserving some structure on $\mathbb{R}^n$. For example, we have:

the *special linear group* $\mathrm{SL}_n\mathbb{R}$ of automorphisms of $\mathbb{R}^n$ preserving the volume element; equivalently, $n \times n$ matrices A with determinant 1.

the group B_n of *upper-triangular* matrices; equivalently, the subgroup of automorphisms of $\mathbb{R}^n$ preserving the flag[1]

[1] In general, a *flag* is a sequence of subspaces of a fixed vector space, each properly contained in the next; it is a *complete* flag if each has one dimension larger than the preceding, and *partial* otherwise.

$$0 = V_0 \subset V_1 \subset V_2 \subset \cdots V_{n-1} \subset V_n = \mathbb{R}^n,$$

where V_i is the span of the standard basis vectors $e_1, \ldots, e_i$. Note that choosing a different basis and correspondingly a different flag yields a different subgroup of $\mathrm{GL}_n\mathbb{R}$, but one isomorphic to (indeed, conjugate to) B_n. Somewhat more generally, for any sequence of positive integers $a_1, \ldots, a_k$ with sum n we can look at the group of block-upper-triangular matrices; this is the subgroup of automorphisms of $\mathbb{R}^n$ preserving a partial flag

$$0 = V_0 \subset V_1 \subset V_2 \subset \cdots \subset V_{k-1} \subset V_k = \mathbb{R}^n,$$

where the dimension of V_i is $a_1 + \cdots + a_i$. If the subspace V_i is spanned by the first $a_1 + \cdots + a_i$ basis vectors, the group will be the set of matrices of the form

$$\left(\begin{array}{c|c|c|c} * & * & * & * \\ \hline 0 & * & * & * \\ \hline 0 & 0 & * & * \\ \hline 0 & 0 & 0 & * \end{array}\right) \begin{array}{l} \}\, a_1 \\ \}\, a_2 \\ \\ \}\, a_k \end{array}$$

The group N_n of *upper-triangular unipotent* matrices (that is, upper triangular with 1's on the diagonal); equivalently, the subgroup of automorphisms of $\mathbb{R}^n$ preserving the complete flag $\{V_i\}$ where V_i is the span of the standard basis vectors $e_1, \ldots, e_i$, and acting as the identity on the successive quotients V_{i+1}/V_i. As before, we can, for any sequence of positive integers $a_1, \ldots, a_k$ with sum n, look at the group of block-upper-triangular unipotent matrices; this is the subgroup of automorphisms of $\mathbb{R}^n$ preserving a partial flag and acting as the identity on successive quotients, i.e., matrices of the form

$$\left(\begin{array}{c|c|c|c} I & * & * & * \\ \hline 0 & I & * & * \\ \hline 0 & 0 & I & * \\ \hline 0 & 0 & 0 & I \end{array}\right) \begin{array}{l} \}\, a_1 \\ \}\, a_2 \\ \\ \}\, a_k \end{array}$$

Next, there are the subgroups of $\mathrm{GL}_n\mathbb{R}$ defined as the group of transformations of $V = \mathbb{R}^n$ of determinant 1 preserving some bilinear form $Q\colon V \times V \to \mathbb{R}$. If the bilinear form Q is symmetric and positive definite, the group we get is called the (*special*) *orthogonal group* $\mathrm{SO}_n\mathbb{R}$ (sometimes written $\mathrm{SO}(n)$; see p. 100). If Q is symmetric and nondegenerate but not definite—e.g., if it has k positive eigenvalues and l negative—the group is denoted $\mathrm{SO}_{k,l}\mathbb{R}$ or $\mathrm{SO}(k, l)$; note that $\mathrm{SO}(k, l) \cong \mathrm{SO}(l, k)$. If Q is skew-symmetric and nondegenerate, the group is called the *symplectic group* and denoted $\mathrm{Sp}_n\mathbb{R}$; note that in this case n must be even.

The equations that define the subgroup of $\mathrm{GL}_n\mathbb{R}$ preserving a bilinear form Q are easy to write down. If we represent Q by a matrix M—that is, we write

$$Q(v, w) = {}^t v \cdot M \cdot w$$

for all $v, w \in \mathbb{R}^n$—then the condition

$$Q(Av, Aw) = Q(v, w)$$

translates into the condition that

$${}^t v \cdot {}^t A \cdot M \cdot A \cdot w = {}^t v \cdot M \cdot w$$

for all v and w; this is equivalent to saying that

$${}^t A \cdot M \cdot A = M.$$

Thus, for example, if Q is the symmetric form $Q(v, w) = {}^t v \cdot w$ given by the identity matrix $M = I_n$, the group $\mathrm{SO}_n\mathbb{R}$ is just the group of $n \times n$ real matrices A of determinant 1 such that ${}^t A = A^{-1}$.

Exercise 7.1*. Show that in the case of $\mathrm{Sp}_{2n}\mathbb{R}$ the requirement that the transformations have determinant 1 is redundant; whereas in the case of $\mathrm{SO}_n\mathbb{R}$ if we do not require the transformations to have determinant 1 the group we get (denoted $\mathrm{O}_n\mathbb{R}$, or sometimes $\mathrm{O}(n)$) is disconnected.

Exercise 7.2*. Show that $\mathrm{SO}(k, l)$ has two connected components if k and l are both positive. The connected component containing the identity is often denoted $\mathrm{SO}^+(k, l)$. (Composing with a projection onto $\mathbb{R}^k$ or $\mathbb{R}^l$, we may associate to an automorphism $A \in \mathrm{SO}(k, l)$ automorphisms of $\mathbb{R}^k$ and $\mathbb{R}^l$; $\mathrm{SO}^+(k, l)$ will consist of those $A \in \mathrm{SO}(k, l)$ whose associated automorphisms preserve the orientations of $\mathbb{R}^k$ and $\mathbb{R}^l$.)

Note that if the form Q is degenerate, a transformation preserving Q will carry its kernel

$$\mathrm{Ker}(Q) = \{v \in V: Q(v, w) = 0\ \forall w \in V\}$$

into itself; so that the group we get is simply the group of matrices preserving the subspace $\mathrm{Ker}(Q)$ and preserving the induced nondegenerate form $\tilde{Q}$ on the quotient $V/\mathrm{Ker}(Q)$. Likewise, if Q is a general bilinear form, that is, neither symmetric nor skew-symmetric, a linear transformation preserving Q will preserve the symmetric and skew-symmetric parts of Q individually, so we just get an intersection of the subgroups encountered already. At any rate, we usually limit our attention to nondegenerate forms that are either symmetric or skew-symmetric.

Of course, the group $\mathrm{GL}_n\mathbb{C}$ of complex linear automorphisms of a complex vector space $V = \mathbb{C}^n$ can be viewed as subgroup of the general linear group $\mathrm{GL}_{2n}\mathbb{R}$; it is, thus, a real Lie group as well, as is the subgroup $\mathrm{SL}_n\mathbb{C}$ of $n \times n$ complex matrices of determinant 1. Similarly, the subgroups $\mathrm{SO}_n\mathbb{C} \subset \mathrm{SL}_n\mathbb{C}$ and $\mathrm{Sp}_{2n}\mathbb{C} \subset \mathrm{SL}_{2n}\mathbb{C}$ of transformations of a complex vector space preserving a symmetric and skew-symmetric nondegenerate *bilinear* form, respectively, are real as well as complex Lie subgroups. Note that since all nondegenerate bilinear symmetric forms on a complex vector space are isomorphic (in particular, there is no such thing as a signature), there is only one complex orthogonal subgroup $SO_n\mathbb{C} \subset SL_n\mathbb{C}$ up to conjugation; there are no analogs of the groups $SO_{k,l}\mathbb{R}$.

Another example we can come up with here is the *unitary group* U_n or $U(n)$, defined to be the group of complex linear automorphisms of an n-dimensional complex vector space V preserving a positive definite *Hermitian* inner product H on V. (A *Hermitian* form H is required to be conjugate linear in the first[2] factor, and linear in the second: $H(\lambda v, \mu w) = \bar{\lambda} H(v, w)\mu$, and $H(w, v) = \overline{H(v, w)}$; it is *positive definite* if $H(v, v) > 0$ for $v \neq 0$.)

Just as in the case of the subgroups SO and Sp, it is easy to write down the equations for $U(n)$: for some $n \times n$ matrix M we can write the form H as

$$H(v, w) = {}^t\bar{v} \cdot M \cdot w, \quad \forall v, w \in \mathbb{C}^n$$

(note that for H to be conjugate symmetric, M must be conjugate symmetric, i.e., ${}^tM = \bar{M}$); then the group $U(n)$ is just the group of $n \times n$ complex matrices A satisfying

$${}^t\bar{A} \cdot M \cdot A = M.$$

In particular, if H is the "standard" Hermitian inner product $H(v, w) = {}^t\bar{v} \cdot w$ given by the identity matrix, $U(n)$ will be the group of $n \times n$ complex matrices A such that ${}^t\bar{A} = A^{-1}$.

Exercise 7.3. Show that if H is a Hermitian form on a complex vector space V, then the real part $R = \mathrm{Re}(H)$ of H is a symmetric form on the underlying real space, and the imaginary part $C = \mathrm{Im}(H)$ is a skew-symmetric real form; these are related by $C(v, w) = R(iv, w)$. Both R and C are invariant by multiplication by i: $R(iv, iw) = R(v, w)$. Show conversely that any such real symmetric R is the real part of a unique Hermitian H. Show that if H is standard, so is R, and C corresponds to the matrix $J = \begin{pmatrix} 0 & I_n \\ -I_n & 0 \end{pmatrix}$. Deduce that

$$U(n) = O(2n) \cap Sp_{2n}\mathbb{R}.$$

Note that the determinant of a unitary matrix can be any complex number of modulus 1; the *special unitary group*, $SU(n)$, is the subgroup of $U(n)$ of automorphisms with determinant 1. The subgroup of $GL_n\mathbb{C}$ preserving an indefinite Hermitian inner product with k positive eigenvalues and l negative ones is denoted $U_{k,l}$ or $U(k, l)$; the subgroup of those of determinant 1 is denoted $SU_{k,l}$ or $SU(k, l)$.

In a similar vein, the group $GL_n\mathbb{H}$ of quaternionic linear automorphisms of an n-dimensional vector space V over the ring $\mathbb{H}$ of quaternions is a real

[2] This choice of which factor is linear and which conjugate linear is less common than the other. It makes little difference in what follows, but it does have the small advantage of being compatible with the natural choice for quaternions.

Lie subgroup of the group $GL_{4n}\mathbb{R}$, as are the further subgroups of $\mathbb{H}$-linear transformations of V preserving a bilinear form. Since $\mathbb{H}$ is not commutative, care must be taken with the conventions here, and it may be worth a little digression to go through this now. We take the vector spaces V to be right $\mathbb{H}$-modules; $\mathbb{H}^n$ is the space of column vectors with *right* multiplication by scalars. In this way the $n \times n$ matrices with entries in $\mathbb{H}$ act in the usual way on $\mathbb{H}^n$ on the left. Scalar multiplication on the left (only) is $\mathbb{H}$-linear.

View $\mathbb{H} = \mathbb{C} \oplus j\mathbb{C} \cong \mathbb{C}^2$. Then left multiplication by elements of $\mathbb{H}$ give $\mathbb{C}$-linear endomorphisms of $\mathbb{C}^2$, which determines a mapping $\mathbb{H} \to M_2\mathbb{C}$ to the 2×2 complex matrices. In particular, $\mathbb{H}^* = GL_1\mathbb{H} \hookrightarrow GL_2\mathbb{C}$. Similarly $\mathbb{H}^n = \mathbb{C}^n \oplus j\mathbb{C}^n = \mathbb{C}^{2n}$, so we have an embedding $GL_n\mathbb{H} \hookrightarrow GL_{2n}\mathbb{C}$. Note that a $\mathbb{C}$-linear mapping $\varphi\colon \mathbb{H}^n \to \mathbb{H}^n$ is $\mathbb{H}$-linear exactly when it commutes with j: $\varphi(vj) = \varphi(v)j$. If $v = v_1 + jv_2$, then $v \cdot j = -\bar{v}_2 + j\bar{v}_1$, so multiplication by j takes $\begin{pmatrix} v_1 \\ v_2 \end{pmatrix}$ to $\begin{pmatrix} 0 & -I \\ I & 0 \end{pmatrix}\begin{pmatrix} \bar{v}_1 \\ \bar{v}_2 \end{pmatrix}$. It follows that if J is the matrix of the preceding exercise, then

$$GL_n\mathbb{H} = \{A \in GL_{2n}\mathbb{C}\colon AJ = J\bar{A}\}.$$

Those matrices with real determinant 1 form a subgroup $SL_n\mathbb{H}$.

A *Hermitian form* (or "symplectic scalar product") on a quaternionic vector space V is an $\mathbb{R}$-bilinear form $K\colon V \times V \to \mathbb{H}$ that is conjugate $\mathbb{H}$-linear in the first factor and $\mathbb{H}$-linear in the second: $K(v\lambda, w\mu) = \bar{\lambda}K(v, w)\mu$, and satisfies $K(w, v) = \overline{K(v, w)}$. It is *positive definite* if $K(v, v) > 0$ for $v \neq 0$. (The conjugate $\bar{\lambda}$ of a quaternion $\lambda = a + bi + cj + dk$ is defined to be $a - bi - cj - dk$.) The *standard* Hermitian form on $\mathbb{H}^n$ is $\Sigma\bar{v}_i w_i$. The group of automorphisms of an n-dimensional quaternionic space preserving such a form is called the *compact symplectic* group and denoted $Sp(n)$ or $U_{\mathbb{H}}(n)$. The *standard* Hermitian form on $\mathbb{H}^n$ is $\Sigma\bar{v}_i w_i$.

Exercise 7.4. Regarding V as a complex vector space, show that every quaternionic Hermitian form K has the form

$$K(v, w) = H(v, w) + jQ(v, w),$$

where H is a complex Hermitian form and Q is a skew-symmetric complex linear form on V, with H and Q related by $Q(v, w) = H(vj, w)$, and H satisfying the condition $H(vj, wj) = \overline{H(v, w)}$. Conversely, any such Hermitian H is the complex part of a unique K. If K is standard, so is H, and Q is given by the same matrix as in Exercise 7.3. Deduce that

$$Sp(n) = U(2n) \cap Sp_{2n}\mathbb{C}.$$

This shows that the two notions of "symplectic" are compatible.

More generally, if K is not positive definite, but has signature (p, q), say the standard $\sum_{i=1}^{p} \bar{v}_i w_i - \sum_{i=p+1}^{p+q} \bar{v}_i w_i$, the automorphisms preserving it form a group $U_{p,q}\mathbb{H}$. Or if the form is a skew Hermitian form (satisfying the same linearity conditions, but with $K(w, v) = -\overline{K(v, w)}$), the group is denoted $U_n^*\mathbb{H}$.

Exercise 7.5. Identify, among all the real Lie groups described above, which ones are compact.

Complex Lie Groups

So far, all of our examples have been examples of real Lie groups. As for *complex* Lie groups, these are fewer in number. The general linear group $GL_n\mathbb{C}$ is one, and again, all the elementary examples come to us as subgroups of the general linear group $GL_n\mathbb{C}$. There is, for example, the subgroup $SO_n\mathbb{C}$ of automorphisms of an n-dimensional complex vector space V having determinant 1 and preserving a nondegenerate symmetric bilinear form Q (note that Q no longer has a signature); and likewise the subgroup $Sp_n\mathbb{C}$ of transformations of determinant 1 preserving a skew-symmetric bilinear form.

Exercise 7.6. Show that the subgroup $SU(n) \subset SL_n\mathbb{C}$ is *not* a complex Lie subgroup. (It is not enough to observe that the defining equations given above are not holomorphic.)

Exercise 7.7. Show that none of the complex Lie groups described above is compact.

We should remark here that both of these exercises are immediate consequences of the general fact that *any compact complex Lie group is abelian*; we will prove this in the next lecture. A *representation* of a complex Lie group G is a map of complex Lie groups from G to $GL(V) = GL_n\mathbb{C}$ for an n-dimensional complex vector space V; note that such a map is required to be complex analytic.

Remarks on Notation

A common convention is to use a notation without subscripts or mention of ground field to denote the *real groups*:

$$O(n),\quad SO(n),\quad SO(p, q),\quad U(n),\quad SU(n),\quad SU(p, q),\quad Sp(n)$$

and to use subscripts for the algebraic groups GL_n, SL_n, SO_n, and Sp_n. This, of course, introduces some anomalies: for example, $SO_n\mathbb{R}$ is $SO(n)$, but $Sp_n\mathbb{R}$ is not $Sp(n)$; but some violation of symmetry seems inevitable in any notation. The notations $GL(n, \mathbb{R})$ or $GL(n, \mathbb{C})$ are often used in place of our $GL_n\mathbb{R}$ or $GL_n\mathbb{C}$, and similarly for SL, SO, and Sp.

Also, where we have written Sp_{2n}, some write Sp_n. In practice, it seems that those most interested in algebraic groups or Lie algebras use the former notation, and those interested in compact groups the latter. Other common notations are $U^*(2n)$ in place of our $GL_n\mathbb{H}$, $Sp(p, q)$ for our $U_{p,q}\mathbb{H}$, and $O^*(2n)$ for our $U_n^*\mathbb{H}$.

Exercise 7.8. Find the dimensions of the various real Lie groups $GL_n\mathbb{R}$, $SL_n\mathbb{R}$, B_n, N_n, $SO_n\mathbb{R}$, $SO_{k,l}\mathbb{R}$, $Sp_{2n}\mathbb{R}$, $U(n)$, $SU(n)$, $GL_n\mathbb{C}$, $SL_n\mathbb{C}$, $GL_n\mathbb{H}$, and $Sp(n)$ introduced above.

§7.3. Two Constructions

There are two constructions, in some sense inverse to one another, that arise frequently in dealing with Lie groups (and that also provide us with further examples of Lie groups). They are expressed in the following two statements.

Proposition 7.9. *Let G be a Lie group, H a connected manifold, and $\varphi\colon H \to G$ a covering space map.*[3] *Let e' be an element lying over the identity e of G. Then there is a unique Lie group structure on H such that e' is the identity and φ is a map of Lie groups; and the kernel of φ is in the center of H.*

Proposition 7.10. *Let H be a Lie group, and $\Gamma \subset Z(H)$ a discrete subgroup of its center. Then there is a unique Lie group structure on the quotient group $G = H/\Gamma$ such that the quotient map $H \to G$ is a Lie group map.*

The proof of the second proposition is straightforward. To prove the first, one shows that the multiplication on G lifts uniquely to a map $H \times H \to H$ which takes (e', e') to e', and verifies that this product satisfies the group axioms. In fact, it suffices to do this when H is the universal covering of G, for one can then apply the second proposition to intermediate coverings. □

Exercise 7.11*. (a) Show that any discrete normal subgroup of a connected Lie group G is in the center $Z(G)$.
(b) If $Z(G)$ is discrete, show that $G/Z(G)$ has trivial center.

These two propositions motivate a definition: we say that a Lie group map between two Lie groups G and H is an *isogeny* if it is a covering space map of the underlying manifolds; and we say two Lie groups G and H are *isogenous* if there is an isogeny between them (in either direction). Isogeny is not an equivalence relation, but generates one; observe that every isogeny equivalence class has an initial member (that is, one that maps to every other one by an isogeny)—that is, just the universal covering space $\tilde{G}$ of any one—and, if the center of this universal cover is discrete, as will be the case for all our semisimple groups, a final object $\tilde{G}/Z(\tilde{G})$ as well. For any group G in such an equivalence class, we will call $\tilde{G}$ the *simply connected form* of the group G, and $\tilde{G}/Z(\tilde{G})$ (if it exists) the *adjoint form* (we will see later a more general definition of adjoint form).

[3] This means that φ is a continuous map with the property that every point of G has a neighborhood U such that $\varphi^{-1}(U)$ is a disjoint union of open sets each mapping homeomorphically to U.

Exercise 7.12. If $H \to G$ is a covering of connected Lie groups, show that $Z(G)$ is discrete if and only if $Z(H)$ is discrete, and then $H/Z(H) = G/Z(G)$. Therefore, if $Z(G)$ is discrete, the adjoint form of G exists and is $G/Z(G)$.

To apply these ideas to some of the examples discussed, note that the center of SL_n (over $\mathbb{R}$ or $\mathbb{C}$) is just the subgroup of multiples of the identity by an nth root of unity; the quotient may be denoted $\mathrm{PSL}_n\mathbb{R}$ or $\mathrm{PSL}_n\mathbb{C}$. In the complex case, $\mathrm{PSL}_n\mathbb{C}$ is isomorphic to the quotient of $\mathrm{GL}_n\mathbb{C}$ by its center $\mathbb{C}^*$ of scalar matrices, and so one often writes $\mathrm{PGL}_n\mathbb{C}$ instead of $\mathrm{PSL}_n\mathbb{C}$. The center of the group SO_n is the subgroup $\{\pm I\}$ when n is even, and trivial when n is odd; in the former case the quotient will be denoted $\mathrm{PSO}_n\mathbb{R}$ or $\mathrm{PSO}_n\mathbb{C}$. Finally the center of the group Sp_{2n} is similarly the subgroup $\{\pm I\}$, and the quotient is denoted $\mathrm{PSp}_{2n}\mathbb{R}$ or $\mathrm{PSp}_{2n}\mathbb{C}$.

Exercise 7.13*. Realize $\mathrm{PGL}_n\mathbb{C}$ as a matrix group, i.e., find an embedding (faithful representation) $\mathrm{PGL}_n\mathbb{C} \hookrightarrow \mathrm{GL}_N\mathbb{C}$ for some N. Do the same for the other quotients above.

In the other direction, whenever we have a Lie group that is not simply connected, we can ask what its universal covering space is. This is, for example, how the famous *spin groups* arise: as we will see, the orthogonal groups $\mathrm{SO}_n\mathbb{R}$ and $\mathrm{SO}_n\mathbb{C}$ have fundamental group $\mathbb{Z}/2$, and so by the above there exist connected, two-sheeted covers of these groups. These are denoted $\mathrm{Spin}_n\mathbb{R}$ and $\mathrm{Spin}_n\mathbb{C}$, and will be discussed in Lecture 20; for the time being, the reader may find it worthwhile (if frustrating) to try to realize these as matrix groups. The last exercises of this section sketch a few steps in this direction which can be done now by hand.

Exercise 7.14. Show that the universal covering of $U(n)$ can be identified with the subgroup of the product $U(n) \times \mathbb{R}$ consisting of pairs (g, t) with $\det(g) = e^{\pi i t}$.

Exercise 7.15. We have seen in Exercise 7.4 that

$$\mathrm{SU}(2) = \mathrm{Sp}(2) = \{q \in \mathbb{H}: q\bar{q} = 1\}.$$

Identifying $\mathbb{R}^3$ with the imaginary quaternions (with basis i, j, k), show that, for $q\bar{q} = 1$, the map $v \mapsto qv\bar{q}$ maps $\mathbb{R}^3$ to itself, and is an isometry. Verify that the resulting map

$$\mathrm{SU}(2) = \mathrm{Sp}(2) \to \mathrm{SO}(3)$$

is a $2:1$ covering map. Since the equation $q\bar{q} = 1$ describes a 3-sphere, $\mathrm{SU}(2)$ is the universal covering of $\mathrm{SO}(3)$; and $\mathrm{SO}(3)$ is the adjoint form of $\mathrm{SU}(2)$.

Exercise 7.16. Let $M_2\mathbb{C} = \mathbb{C}^4$ be the space of 2×2 matrices, with symmetric form $Q(A, B) = \frac{1}{2}\,\mathrm{Trace}(AB^{\natural})$, where $B^{\natural}$ is the adjoint of the matrix B; the quadratic form associated to Q is the determinant. For g and h in $SL_2\mathbb{C}$, the mapping $A \mapsto gAh^{-1}$ is in $SO_4\mathbb{C}$. Show that this gives a 2 : 1 covering

$$SL_2\mathbb{C} \times SL_2\mathbb{C} \to SO_4\mathbb{C},$$

which, since $SL_2\mathbb{C}$ is simply connected, realizes the universal covering of $SO_4\mathbb{C}$.

Exercise 7.17. Identify $\mathbb{C}^3$ with the space of traceless matrices in $M_2\mathbb{C}$, so $g \in SL_2\mathbb{C}$ acts by $A \mapsto gAg^{-1}$. Show that this gives a 2 : 1 covering

$$SL_2\mathbb{C} \to SO_3\mathbb{C},$$

which realizes the universal covering of $SO_3\mathbb{C}$.

LECTURE 8

Lie Algebras and Lie Groups

In this crucial lecture we introduce the definition of the Lie algebra associated to a Lie group and its relation to that group. All three sections are logically necessary for what follows; §8.1 is essential. We use here a little more manifold theory: specifically, the differential of a map of manifolds is used in a fundamental way in §8.1, the notion of the tangent vector to an arc in a manifold is used in §8.2 and §8.3, and the notion of a vector field is introduced in an auxiliary capacity in §8.3. The Campbell-Hausdorff formula is introduced only to establish the First and Second Principles of §8.1 below; if you are willing to take those on faith the formula (and exercises dealing with it) can be skimmed. Exercises 8.27–8.29 give alternative descriptions of the Lie algebra associated to a Lie group, but can be skipped for now.

§8.1. Lie Algebras: Motivation and Definition

Given that we want to study the representations of a Lie group, how do we go about it? As we have said, the notions of generators and relations is hardly relevant here. The answer, of course, is that we have to use the continuous structure of the group. The first step in doing this is

Exercise 8.1. Let G be a connected Lie group, and $U \subset G$ any neighborhood of the identity. Show that U generates G.

This statement implies that any map $\rho: G \to H$ between connected Lie groups will be determined by what it does on any open set containing the identity in G, i.e., ρ is determined by its germ at $e \in G$. In fact, we can extend this idea a good bit further: later in this lecture we will establish the

First Principle: Let G and H be Lie groups, with G connected. A map $\rho: G \to H$ is uniquely determined by its differential $d\rho_e: T_eG \to T_eH$ at the identity.

This is, of course, great news: we can completely describe a homomorphism of Lie groups by giving a linear map between two vector spaces. It is not really worth that much, however, unless we can give at least some answer to the next, obvious question: *which maps between these two vector spaces actually arise as differentials of group homomorphisms*? The answer to this is expressed in the *Second Principle* below, but it will take us a few pages to get there. To start, we have to ask ourselves what it means for a map to be a homomorphism, and in what ways this may be reflected in the differential.

To begin with, the definition of a homomorphism is simply a $\mathscr{C}^\infty$ map ρ such that

$$\rho(gh) = \rho(g) \cdot \rho(h)$$

for all g and h in G. To express this in a more confusing way, we can say that *a homomorphism respects the action of a group on itself by left or right multiplication*: that is, for any $g \in G$ we denote by $m_g: G \to G$ the differentiable map given by multiplication by g, and observe that a $\mathscr{C}^\infty$ map $\rho: G \to H$ of Lie groups will be a homomorphism if it carries m_g to $m_{\rho(g)}$ in the sense that the diagram

$$\begin{array}{ccc} G & \xrightarrow{\rho} & H \\ {\scriptstyle m_g}\downarrow & & \downarrow{\scriptstyle m_{\rho(g)}} \\ G & \xrightarrow[\rho]{} & H \end{array}$$

commutes.

The problem with this characterization is that, since the maps m_g have no fixed points, it is hard to associate to them any operation on the tangent space to G at one point. This suggests looking, not at the diffeomorphisms m_g, but at the automorphisms of G given by conjugation. Explicitly, for any $g \in G$ we define the map

$$\Psi_g: G \to G$$

by

$$\Psi_g(h) = g \cdot h \cdot g^{-1}.$$

(Ψ_g is actually a Lie group map, but that is not relevant for our present purposes.) It is now equally the case that *a homomorphism ρ respects the action of a group G on itself by conjugation*: that is, it will carry Ψ_g into $\Psi_{\rho(g)}$ in the sense that the diagram

$$\begin{array}{ccc} G & \xrightarrow{\rho} & H \\ {\scriptstyle \Psi_g}\downarrow & & \downarrow{\scriptstyle \Psi_{\rho(g)}} \\ G & \xrightarrow[\rho]{} & H \end{array}$$

commutes. We have, in other words, a natural map

$$\Psi: G \to \mathrm{Aut}(G).$$

The advantage of working with Ψ_g is that it fixes the identity element $e \in G$; we can therefore extract some of its structure by looking at its differential at e: we set

$$\mathrm{Ad}(g) = (d\Psi_g)_e: T_eG \to T_eG. \tag{8.2}$$

This is a representation

$$\mathrm{Ad}: G \to \mathrm{Aut}(T_eG) \tag{8.3}$$

of the group G on its own tangent space, called the *adjoint representation* of the group. This gives a third characterization[1]: *a homomorphism ρ respects the adjoint action of a group G on its tangent space T_eG at the identity.* In other words, for any $g \in G$ the actions of $\mathrm{Ad}(g)$ on T_eG and $\mathrm{Ad}(\rho(g))$ on T_eH must commute with the differential $(d\rho)_e: T_eG \to T_eH$, i.e., the diagram

$$\begin{array}{ccc} T_eG & \xrightarrow{(d\rho)_e} & T_eH \\ {\scriptstyle \mathrm{Ad}(g)}\downarrow & & \downarrow{\scriptstyle \mathrm{Ad}(\rho(g))} \\ T_eG & \xrightarrow[(d\rho)_e]{} & T_eH \end{array}$$

commutes; equivalently, for any tangent vector $v \in T_eG$,

$$d\rho(\mathrm{Ad}(g)(v)) = \mathrm{Ad}(\rho(g))(d\rho(v)). \tag{8.4}$$

This is nice, but does not yet answer our question, for preservation of the adjoint representation $\mathrm{Ad}: G \to \mathrm{Aut}(T_eG)$ still involves the map ρ on the group G itself, and so is not purely a condition on the differential $(d\rho)_e$. We have instead to go one step further, and *take the differential of the map* Ad. The group $\mathrm{Aut}(T_eG)$ being just an open subset of the vector space of endomorphisms of T_eG, its tangent space at the identity is naturally identified with $\mathrm{End}(T_eG)$; taking the differential of the map Ad we arrive at a map

$$\mathrm{ad}: T_eG \to \mathrm{End}(T_eG). \tag{8.5}$$

This is essentially a trilinear gadget on the tangent space T_eG; that is, we can view the image $\mathrm{ad}(X)(Y)$ of a tangent vector Y under the map $\mathrm{ad}(X)$ as a

[1] "Characterization" is not the right word here (or in the preceding case), since we do not mean an equivalent condition, but rather something implied by the condition that ρ be a homomorphism.

function of the two variables X and Y, so that we get a bilinear map

$$T_eG \times T_eG \to T_eG.$$

We use the notation [,] for this bilinear map; that is, for a pair of tangent vectors X and Y to G at e, we write

$$[X, Y] \stackrel{\text{def}}{=} \text{ad}(X)(Y). \tag{8.6}$$

As desired, the map ad involves only the tangent space to the group G at e, and so gives us our final characterization: *the differential $(d\rho)_e$ of a homomorphism ρ on a Lie group G respects the adjoint action of the tangent space to G on itself*. Explicitly, the fact that ρ and $d\rho_e$ respect the adjoint representation implies in turn that the diagram

$$\begin{array}{ccc} T_eG & \xrightarrow{(d\rho)_e} & T_eH \\ \Big\downarrow{\scriptstyle \text{ad}(v)} & & \Big\downarrow{\scriptstyle \text{ad}(d\rho(v))} \\ T_eG & \xrightarrow[(d\rho)_e]{} & T_eH \end{array}$$

commutes; i.e., for any pair of tangent vectors X and Y to G at e,

$$d\rho_e(\text{ad}(X)(Y)) = \text{ad}(d\rho_e(X))(d\rho_e(Y)). \tag{8.7}$$

or, equivalently,

$$d\rho_e([X, Y]) = [d\rho_e(X), d\rho_e(Y)]. \tag{8.8}$$

All this may be fairly confusing (if it is not, you probably do not need to be reading this book). Two things, however, should be borne in mind. They are:

(i) *It is not so bad*, in the sense that we can make the bracket operation, as defined above, reasonably explicit. We do this first for the general linear group $G = \text{GL}_n\mathbb{R}$. Note that in this case conjugation extends to the ambient linear space $E = \text{End}(\mathbb{R}^n) = M_n\mathbb{R}$ of $\text{GL}_n\mathbb{R}$ by the same formula: $\text{Ad}(g)(M) = gMg^{-1}$, and this ambient space is identified with the tangent space T_eG; differentiation in E is usual differentiation of matrices. For any pair of tangent vectors X and Y to $\text{GL}_n\mathbb{R}$ at e, let $\gamma\colon I \to G$ be an arc with $\gamma(0) = e$ and tangent vector $\gamma'(0) = X$. Then our definition of $[X, Y]$ is that

$$[X, Y] = \text{ad}(X)(Y) = \frac{d}{dt}\bigg|_{t=0} (\text{Ad}(\gamma(t))(Y)).$$

Applying the product rule to $\text{Ad}(\gamma(t))(Y) = \gamma(t)Y\gamma(t)^{-1}$, this is

$$\begin{aligned} &= \gamma'(0)\cdot Y\cdot\gamma(0) + \gamma(0)\cdot Y(-\gamma(0)^{-1}\cdot\gamma'(0)\cdot\gamma(0)^{-1}) \\ &= X\cdot Y - Y\cdot X, \end{aligned}$$

which, of course, explains the bracket notation. In general, any time a Lie group is given as a subgroup of a general linear group $\text{GL}_n\mathbb{R}$, we can view its tangent space T_eG at the identity as a subspace of the space of endomorphisms of $\mathbb{R}^n$; and since bracket is preserved by (differentials of) maps of Lie groups, the bracket operation on T_eG will coincide with the commutator.

(ii) *Even if it were that bad, it would be worth it.* This is because it turns out that the bracket operation is exactly the answer to the question we raised before. Precisely, later in this lecture we will prove the

Second Principle: Let G and H be Lie groups, with G connected and simply connected. A linear map $T_eG \to T_eH$ is the differential of a homomorphism $\rho: G \to H$ if and only if it preserves the bracket operation, in the sense of (8.8) above.

We are now almost done: maps between Lie groups are classified by maps between vector spaces preserving the structure of a bilinear map from the vector space to itself. We have only one more question to answer: when does a vector space with this additional structure actually arise as the tangent space at the identity to a Lie group, with the adjoint or bracket product? Happily, we have the answer to this as well. First, though it is far from clear from our initial definition, it follows from our description of the bracket as a commutator that *the bracket is skew-symmetric*, i.e, $[X, Y] = -[Y, X]$. Second, it likewise follows from the description of $[X, Y]$ as a commutator that it satisfies the *Jacobi identity*: for any three tangent vectors X, Y, and Z,

$$[X, [Y, Z]] + [Y, [Z, X]] + [Z, [X, Y]] = 0.$$

We thus make the

Definition 8.9. A *Lie algebra* $\mathfrak{g}$ is a vector space together with a skew-symmetric bilinear map

$$[\ ,\ ]: \mathfrak{g} \times \mathfrak{g} \to \mathfrak{g}$$

satisfying the Jacobi identity.

We should take a moment out here to make one important point. Why, you might ask, do we define the bracket operation in terms of the relatively difficult operations Ad and ad, instead of just defining $[X, Y]$ to be the commutator $X \cdot Y - Y \cdot X$? The answer is that the "composition" $X \cdot Y$ of elements of a Lie algebra is not well defined. Specifically, any time we embed a Lie group G in a general linear group $\mathrm{GL}(V)$, we get a corresponding embedding of its Lie algebra $\mathfrak{g}$ in the space $\mathrm{End}(V)$, and can talk about the composition $X \cdot Y \in \mathrm{End}(V)$ of elements of $\mathfrak{g}$ in this context; but it must be borne in mind that this composition $X \cdot Y$ will depend on the embedding of $\mathfrak{g}$, and for that matter need not even be an element of $\mathfrak{g}$. *Only* the commutator $X \cdot Y - Y \cdot X$ is always an element of $\mathfrak{g}$, independent of the representation. The terminology sometimes heightens the confusion: for example, when we speak of embedding a Lie algebra in the algebra $\mathrm{End}(V)$ of endomorphisms of V, the word *algebra* may mean two very different things. In general, when we want to refer to the endomorphisms of a vector space V (resp. $\mathbb{R}^n$) as a Lie algebra, we will write $\mathfrak{gl}(V)$ (resp. $\mathfrak{gl}_n\mathbb{R}$) instead of $\mathrm{End}(V)$ (resp. $M_n\mathbb{R}$).

To return to our discussion of Lie algebras, a *map* of Lie algebras is a linear map of vector spaces preserving the bracket, in the sense of (8.8); notions like *Lie subalgebra* are defined accordingly. We note in passing one thing that will turn out to be significant: the definition of Lie algebra does not specify the field. Thus, we have real Lie algebras, complex Lie algebras, etc., all defined in the same way; and in addition, given a real Lie algebra $\mathfrak{g}$ we may associate to it a complex Lie algebra, whose underlying vector space is $\mathfrak{g} \otimes \mathbb{C}$ and whose bracket operation is just the bracket on $\mathfrak{g}$ extended by linearity.

Exercise 8.10*. The skew-commutativity and Jacobi identity also follow from the naturality of the bracket (8.8), without using an embedding in $\mathfrak{gl}(V)$:

(a) Deduce the skew-commutativity $[X, X] = 0$ from that fact that any X can be written the image of a vector by $d\rho_e$ for some homomorphism $\rho: \mathbb{R} \to G$. (See §8.3 for the existence of ρ.)

(b) Given that the bracket is skew-commutative, verify that the Jacobi identity is equivalent to the assertion that

$$\mathrm{ad} = d(\mathrm{Ad})_e: \mathfrak{g} \to \mathrm{End}(\mathfrak{g})$$

preserves the bracket. In particular, ad is a map of Lie algebras.

To sum up our progress so far: taking for the moment on faith the statements made, we have seen that

(i) the tangent space $\mathfrak{g}$ at the identity to a Lie group G is naturally endowed with the structure of a Lie algebra;

(ii) if G and H are Lie groups with G connected and simply connected, the maps from G to H are in one-to-one correspondence with maps of the associated Lie algebras, by associating to $\rho: G \to H$ its differential $(d\rho)_e: \mathfrak{g} \to \mathfrak{h}$.

Of course, we make the

Definition 8.11. A *representation* of a Lie algebra $\mathfrak{g}$ on a vector space V is simply a map of Lie algebras

$$\rho: \mathfrak{g} \to \mathfrak{gl}(V) = \mathrm{End}(V),$$

i.e., a linear map that preserves brackets, or an action of $\mathfrak{g}$ on V such that

$$[X, Y](v) = X(Y(v)) - Y(X(v)).$$

Statement (ii) above implies in particular that *representations of a connected and simply connected Lie group are in one-to-one correspondence with repre-*

sentations of its Lie algebra. This is, then, the first step of the series of reductions outlined in the introduction to Part II.

At this point, a few words are in order about the relation between representations of a Lie group and the corresponding representations of its Lie algebra. The first remark to make is about tensors. Recall that if V and W are representations of a Lie group G, then we define the representation $V \otimes W$ to be the vector space $V \otimes W$ with the action of G described by

$$g(v \otimes w) = g(v) \otimes g(w).$$

The definition for representations of a Lie algebra, however, is quite different. For one thing, if $\mathfrak{g}$ is the Lie algebra of G, so that the representation of G on the vector spaces V and W induces representations of $\mathfrak{g}$ on these spaces, we want the tensor product of the representations V and W of $\mathfrak{g}$ to be the representation induced by the action of G on $V \otimes W$ above. But now suppose that $\{\gamma_t\}$ is an arc in G with $\gamma_0 = e$ and tangent vector $\gamma_0' = X \in \mathfrak{g}$. Then by definition the action of X on V is given by

$$X(v) = \frac{d}{dt}\bigg|_{t=0} \gamma_t(v)$$

and similarly for $w \in W$; it follows that the action of X on the tensor product $v \otimes w$ is

$$\begin{aligned} X(v \otimes w) &= \frac{d}{dt}\bigg|_{t=0} (\gamma_t(v) \otimes \gamma_t(w)) \\ &= \left(\frac{d}{dt}\bigg|_{t=0} \gamma_t(v)\right) \otimes w + v \otimes \left(\frac{d}{dt}\bigg|_{t=0} \gamma_t(w)\right), \end{aligned}$$

so

$$X(v \otimes w) = X(v) \otimes w + v \otimes X(w). \tag{8.12}$$

This, then, is how we *define* the action of a Lie algebra $\mathfrak{g}$ on the tensor product of two representations of $\mathfrak{g}$. This describes as well other tensors: for example, if V is a representation of the group G, $v \in V$ is any vector and $v^2 \in \mathrm{Sym}^2 V$ its square, then for any $g \in G$,

$$g(v^2) = g(v)^2.$$

On the other hand, if V is a representation of the Lie algebra $\mathfrak{g}$ and $X \in \mathfrak{g}$ is any element, we have

$$X(v^2) = 2 \cdot v \cdot X(v). \tag{8.13}$$

One further example: if $\rho: G \to \mathrm{GL}(V)$ is a representation of the group G, the dual representation $\rho': G \to \mathrm{GL}(V^*)$ is defined by setting

$$\rho'(g) = {}^t\rho(g^{-1}): V^* \to V^*.$$

Differentiating this, we find that if $\rho: \mathfrak{g} \to \mathfrak{gl}(V)$ is a representation of a Lie algebra $\mathfrak{g}$, the dual representation of $\mathfrak{g}$ on V^* will be given by

$$\rho'(X) = {}^t\rho(-X) = -{}^t\rho(X): V^* \to V^*. \tag{8.14}$$

A second and related point to be made concerns terminology. Obviously, when we speak of the action of a group G on a vector space V preserving some extra structure on V, we mean that literally: for example, if we have a quadratic form Q on V, to say that G preserves Q means just that

$$Q(g(v), g(w)) = Q(v, w), \quad \forall g \in G \text{ and } v, w \in V.$$

Equivalently, we mean that the associated action of G on the vector space $\mathrm{Sym}^2 V^*$ fixes the element $Q \in \mathrm{Sym}^2 V^*$. But by the above calculation, the action of the associated Lie algebra $\mathfrak{g}$ on V satisfies

$$Q(v, X(w)) + Q(X(v), w) = 0, \quad \forall X \in \mathfrak{g} \text{ and } v, w \in V \tag{8.15}$$

or, equivalently, $Q(v, X(v)) = 0$ for all $X \in \mathfrak{g}$ and $v \in V$; in other words, *the induced action on* $\mathrm{Sym}^2 V^*$ *kills the element* Q. By way of terminology, then, *we will in general say that the action of a Lie algebra on a vector space preserves some structure when a corresponding Lie group action does.*

The next section will be spent in giving examples. In §8.3 we will establish the basic relations between Lie groups and their Lie algebras, to the point where we can prove the First and Second Principles above. The further statement that any Lie algebra is the Lie algebra of some Lie group will follow from the statement (see Appendix E) that every Lie algebra may be embedded in $\mathfrak{gl}_n\mathbb{R}$.

Exercise 8.16*. Show that if G is connected the image of $\mathrm{Ad}: G \to GL(\mathfrak{g})$ is the adjoint form of the group G when that exists.

Exercise 8.17*. Let V be a representation of a connected Lie group G and $\rho: \mathfrak{g} \to \mathrm{End}(V)$ the corresponding map of Lie algebras. Show that a subspace W of V is invariant by G if and only if it is carried into itself under the action of the Lie algebra $\mathfrak{g}$, i.e., $\rho(X)(W) \subset W$ for all X in $\mathfrak{g}$. Hence, V is irreducible over G if and only if it is irreducible over $\mathfrak{g}$.

§8.2. Examples of Lie Algebras

We start with the Lie algebras associated to each of the groups mentioned in Lecture 7. Each of these groups is given as a subgroup of $GL(V) = GL_n\mathbb{R}$, so their Lie algebras will be subspaces of $\mathrm{End}(V) = \mathfrak{gl}_n\mathbb{R}$.

Consider first the special linear group $SL_n\mathbb{R}$. If $\{A_t\}$ is an arc in $SL_n\mathbb{R}$ with $A_0 = I$ and tangent vector $A'_0 = X$ at $t = 0$, then by definition we have for any basis $e_1, \ldots, e_n$ of $V = \mathbb{R}^n$,

$$A_t(e_1) \wedge \cdots \wedge A_t(e_n) \equiv e_1 \wedge \cdots \wedge e_n.$$

Taking the derivative and evaluating at $t = 0$ we have by the product rule

$$0 = \frac{d}{dt}\bigg|_{t=0} (A_t(e_1) \wedge \cdots \wedge A_t(e_n))$$

$$= \sum e_1 \wedge \cdots \wedge X(e_i) \wedge \cdots \wedge e_n$$

$$= \mathrm{Trace}(X) \cdot (e_1 \wedge \cdots \wedge e_n).$$

The tangent vectors to $\mathrm{SL}_n\mathbb{R}$ are thus all endomorphisms of trace 0; comparing dimensions we can see that the Lie algebra $\mathfrak{sl}_n\mathbb{R}$ is exactly the vector space of traceless $n \times n$ matrices.

The orthogonal and symplectic cases are somewhat simpler. For example, the orthogonal group $\mathrm{O}_n\mathbb{R}$ is defined to be the automorphisms A of an n-dimensional vector space V preserving a quadratic form Q, so that if $\{A_t\}$ is an arc in $\mathrm{O}_n\mathbb{R}$ with $A_0 = I$ and $A'_0 = X$ we have for every pair of vectors v, $w \in V$

$$Q(A_t(v), A_t(w)) \equiv Q(v, w).$$

Taking derivatives, we see that

$$Q(X(v), w) + Q(v, X(w)) = 0 \tag{8.18}$$

for all $v, w \in V$; this is exactly the condition that describes the orthogonal Lie algebra $\mathfrak{so}_n\mathbb{R} = \mathfrak{o}_n\mathbb{R}$. In coordinates, if the quadratic form Q is given on $V = \mathbb{R}^n$ as

$$Q(v, w) = {}^tv \cdot M \cdot w \tag{8.19}$$

for some symmetric $n \times n$ matrix M, then as we have seen the condition on $A \in \mathrm{GL}_n\mathbb{R}$ to be in $\mathrm{O}_n\mathbb{R}$ is that

$${}^tA \cdot M \cdot A = M. \tag{8.20}$$

Differentiating, the condition on an $n \times n$ matrix X to be in the Lie algebra $\mathfrak{so}_n\mathbb{R}$ of the orthogonal group is that

$${}^tX \cdot M + M \cdot X = 0. \tag{8.21}$$

Note that if M is the identity matrix—i.e., Q is the "standard" quadratic form $Q(v, w) = {}^tv \cdot w$ on $\mathbb{R}^n$—then this says that $\mathfrak{so}_n\mathbb{R}$ *is the subspace of skew-symmetric* $n \times n$ matrices. To put it intrinsically, in terms of the identification of V with V^* given by the quadratic form Q, and the consequent identification $\mathrm{End}(V) = V \otimes V^* = V \otimes V$, the Lie algebra $\mathfrak{so}_n\mathbb{R} \subset \mathrm{End}(V)$ is just the subspace $\bigwedge^2 V \subset V \otimes V$ of skew-symmetric tensors:

$$\mathfrak{so}_n\mathbb{R} = \textstyle\bigwedge^2 V \subset \mathrm{End}(V) = V \otimes V. \tag{8.22}$$

All of the above, with the exception of the last paragraph, works equally well to describe the Lie algebra $\mathfrak{sp}_{2n}\mathbb{R}$ of the Lie group $\mathrm{Sp}_{2n}\mathbb{R}$ of transformations preserving a skew-symmetric bilinear form Q; that is, $\mathfrak{sp}_{2n}\mathbb{R}$ is the subspace of endomorphisms of V satisfying (8.18) for every pair of vectors v, $w \in V$, or, if Q is given by a skew-symmetric $2n \times 2n$ matrix M as in (8.19), the space of matrices satisfying (8.21). The one statement that has to be substantially modified is the last one of the last paragraph: because Q is skew-symmetric, condition (8.18) is equivalent to saying that

$$Q(X(v), w) = Q(X(w), v)$$

for all $v, w \in V$; thus, in terms of the identification of V with V^* given by Q, the Lie algebra $\mathfrak{sp}_{2n}\mathbb{R} \subset \mathrm{End}(V) = V \otimes V^* = V \otimes V$ is the subspace $\mathrm{Sym}^2 V \subset V \otimes V$:

$$\mathfrak{sp}_{2n}\mathbb{R} = \mathrm{Sym}^2 V \subset \mathrm{End}(V) = V \otimes V. \tag{8.23}$$

Exercise 8.24*. With Q a standard skew form, say of Exercise 7.3, describe $\mathrm{Sp}_{2n}\mathbb{R}$ and its Lie algebra $\mathfrak{sp}_{2n}\mathbb{R}$ (as subgroup of $\mathrm{GL}_{2n}\mathbb{R}$ and subalgebra of $\mathfrak{gl}_{2n}\mathbb{R}$). Do a corresponding calculation for $\mathrm{SO}_{k,l}\mathbb{R}$.

One more similar example is that of the Lie algebra $\mathfrak{u}_n$ of the unitary group $\mathrm{U}(n)$; by a similar calculation we find that the Lie algebra of complex linear endomorphisms of $\mathbb{C}^n$ preserving a Hermitian inner product H is just the space of matrices X satisfying

$$H(X(v), w) + H(v, X(w)) = 0, \quad \forall v, w \in V;$$

if H is given by $H(v, w) = {}^t\bar{v} \cdot w$, this amounts to saying that X is conjugate skew-symmetric, i.e., that ${}^t\bar{X} = -X$.

Exercise 8.25. Find the Lie algebras of the real Lie groups $\mathrm{SL}_n\mathbb{C}$ and $\mathrm{SL}_n\mathbb{H}$—the elements in $\mathrm{GL}_n\mathbb{H}$ whose real determinant is 1.

Exercise 8.26. Show that the Lie algebras of the Lie groups B_n and N_n introduced in §7.2 are the algebra $\mathfrak{b}_n\mathbb{R}$ of upper triangular $n \times n$ matrices and the algebra $\mathfrak{n}_n\mathbb{R}$ of strictly upper triangular $n \times n$ matrices, respectively.

If G is a complex Lie group, its Lie algebra is a complex Lie algebra. Just as in the real case, we have the complex Lie algebras $\mathfrak{gl}_n\mathbb{C}$, $\mathfrak{sl}_n\mathbb{C}$, $\mathfrak{so}_m\mathbb{C}$, and $\mathfrak{sp}_{2n}\mathbb{C}$ of the Lie groups $\mathrm{GL}_n\mathbb{C}$, $\mathrm{SL}_n\mathbb{C}$, $\mathrm{SO}_m\mathbb{C}$, and $\mathrm{Sp}_{2n}\mathbb{C}$.

Exercise 8.27. Let A be any (real or complex) algebra, not necessarily finite dimensional, or even associative. A *derivation* is a linear map $D: A \to A$ satisfying the Leibniz rule $D(ab) = aD(b) + D(a)b$.

(a) Show that the derivations $\mathrm{Der}(A)$ form a Lie algebra under the bracket $[D, E] = D \circ E - E \circ D$. If A is finite dimensional, so is $\mathrm{Der}(A)$.
(b) The group of automorphisms of A is a closed subgroup G of the group $\mathrm{GL}(A)$ of linear automorphisms of A. Show that the Lie algebra of G is $\mathrm{Der}(A)$.
(c) If the algebra A is a Lie algebra, the map $A \to \mathrm{Der}(A)$, $X \mapsto D_X$, where $D_X(Y) = [X, Y]$, is a map of Lie algebras.

Exercise 8.28*. If $\mathfrak{g}$ is a Lie algebra, the Lie algebra automorphisms of $\mathfrak{g}$ form a Lie subgroup $\mathrm{Aut}(\mathfrak{g})$ of the general linear group $GL(\mathfrak{g})$.

(a) Show that the Lie algebra of $\mathrm{Aut}(\mathfrak{g})$ is $\mathrm{Der}(\mathfrak{g})$. If G is a simply connected Lie group with Lie algebra $\mathfrak{g}$, the map $\mathrm{Aut}(G) \to \mathrm{Aut}(\mathfrak{g})$ by $\varphi \mapsto d\varphi$ is one-to-one and onto, giving $\mathrm{Aut}(G)$ the structure of a Lie group with Lie algebra $\mathrm{Der}(\mathfrak{g})$.

(b) Show that the automorphism group of any connected Lie group is a Lie subgroup of the automorphism group of its Lie algebra.

Exercise 8.29*. For any manifold M, the C^∞ vector fields on M form a Lie algebra $\mathfrak{v}(M)$, as follows: a vector field v can be identified with a derivation of the ring A of C^∞ functions on M, with $v(f)$ the function whose value at a point x of M is the value of the tangent vector v_x on f at x. Show that the vector fields on M form a Lie algebra, in fact a Lie subalgebra of the Lie algebra $\mathrm{Der}(A)$. If a Lie group G acts on M, the G-invariant vector fields form a Lie subalgebra $\mathfrak{v}_G M$ of $\mathfrak{v}(M)$. If the action is transitive, the invariant vector fields form a finite-dimensional Lie algebra.

If G is a Lie group, $\mathfrak{v}_G(G) = T_e G$ becomes a Lie algebra by the above process. Show that this bracket agrees with that defined using the adjoint map (8.6). This gives another proof that the bracket is skew-symmetric and satisfies Jacobi's identity.

§8.3. The Exponential Map

The essential ingredient in studying the relationship between a Lie group G and its Lie algebra $\mathfrak{g}$ is the exponential map. This may be defined in very straightforward fashion, using the notion of *one-parameter subgroups*, which we study next. Suppose that $X \in \mathfrak{g}$ is any element, viewed simply as a tangent vector to G at the identity. For any element $g \in G$, denote by $m_g\colon G \to G$ the map of manifolds given by multiplication on the left by g. Then we can define a vector field v_X on all of G simply by setting

$$v_X(g) = (m_g)_*(X).$$

This vector field is clearly invariant under left translation (i.e., it is carried into itself under the diffeomorphism m_g for all g); and it is not hard to see that this gives an identification of $\mathfrak{g}$ with the space of all left-invariant vector fields on G. Under this identification, the bracket operation on the Lie algebra $\mathfrak{g}$ corresponds to Lie bracket of vector fields; indeed, this may be adopted as the definition of the Lie algebra associated to a Lie group (cf. Exercise 8.29). For our present purposes, however, all we need to know is that v_X exists and is left-invariant.

Given any vector field v on a manifold M and a point $p \in M$, a basic theorem from differential equations allows us to integrate the vector field. This gives a differentiable map $\varphi: I \to M$, defined on some open interval I containing 0, with $\varphi(0) = p$, whose tangent vector at any point is the vector assigned to that point by v, i.e., such that

$$\varphi'(t) = v(\varphi(t))$$

for all t in I. The map φ is uniquely characterized by these properties. Now suppose the manifold in question is a Lie group G, the vector field the field v_X associated to an element $X \in \mathfrak{g}$, and p the identity. We arrive then at a map $\varphi: I \to G$; we claim that, at least where φ is defined, it is a *homomorphism*, i.e., $\varphi(s + t) = \varphi(s)\varphi(t)$ whenever s, t, and $s + t$ are in I. To prove this, fix s and let t vary; that is, consider the two arcs α and β given by $\alpha(t) = \varphi(s) \cdot \varphi(t)$ and $\beta(t) = \varphi(s + t)$. Of course, $\alpha(0) = \beta(0)$; and by the invariance of the vector field v_X, we see that the tangent vectors satisfy $\alpha'(t) = v_X(\alpha(t))$ and $\beta'(t) = v_X(\beta(t))$ for all t. By the uniqueness of the integral curve of a vector field on a manifold, we deduce that $\alpha(t) = \beta(t)$ for all t.

From the fact that $\varphi(s + t) = \varphi(s)\varphi(t)$ for all s and t near 0, it follows that φ extends uniquely to all of $\mathbb{R}$, defining a homomorphism

$$\varphi_X: \mathbb{R} \to G$$

with $\varphi_X'(t) = v_X(\varphi)(t) = (m_{\varphi(t)})_*(X)$ for all t.

Exercise 8.30. Establish the *product rule* for derivatives of arcs in a Lie group G: if α and β are arcs in G and $\gamma(t) = \alpha(t) \cdot \beta(t)$, then

$$\gamma'(t) = dm_{\alpha(t)}(\beta'(t)) + dn_{\beta(t)}(\alpha'(t)),$$

where for any $g \in G$, the map m_g (resp. n_g): $G \to G$ is given by left (resp. right) multiplication by g. Use this to give another proof that φ is a homomorphism.

Exercise 8.31. Show that φ_X is uniquely determined by the fact that it is a homomorphism of $\mathbb{R}$ to G with tangent vector $\varphi_X'(0)$ at the identity equal to X. Deduce that if $\psi: G \to H$ is a map of Lie groups, then $\varphi_{\psi_* X} = \psi \circ \varphi_X$.

The Lie group map $\varphi_X: \mathbb{R} \to G$.is called the *one-parameter subgroup of G with tangent vector X at the identity*. The construction of these one-parameter subgroups for each X amounts to the verification of the Second Principle of §8.1 for homomorphisms from $\mathbb{R}$ to G. The fact that there exists such a one-parameter subgroup of G with any given tangent vector at the identity is crucial. For example, it is not hard to see (we will do this in a moment) that these one-parameter subgroups fill up a neighborhood of the identity in G, which immediately implies the First Principle of §8.1. To carry this out, we define the *exponential map*

$$\exp: \mathfrak{g} \to G$$

by

$$\exp(X) = \varphi_X(1). \tag{8.32}$$

Note that by the uniqueness of φ_X, we have

$$\varphi_{(\lambda X)}(t) = \varphi_X(\lambda t);$$

so that the exponential map restricted to the lines through the origin in $\mathfrak{g}$ gives the one-parameter subgroups of G. Indeed, Exercise 8.31 implies the characterization:

Proposition 8.33. *The exponential map is the unique map from* $\mathfrak{g}$ *to* G *taking* 0 *to* e *whose differential at the origin*

$$(\exp_*)_0 \colon T_0\mathfrak{g} = \mathfrak{g} \to T_eG = \mathfrak{g}$$

is the identity, and whose restrictions to the lines through the origin in $\mathfrak{g}$ *are one-parameter subgroups of* G.

This in particular implies (cf. Exercise 8.31) that the exponential map is natural, in the sense that for any map $\psi\colon G \to H$ of Lie groups the diagram

$$\begin{array}{ccc} \mathfrak{g} & \xrightarrow{\psi_*} & \mathfrak{h} \\ {\scriptstyle \exp}\downarrow & & \downarrow{\scriptstyle \exp} \\ G & \xrightarrow[\psi]{} & H \end{array}$$

commutes.

Now, since the differential of the exponential map at the origin in $\mathfrak{g}$ is an isomorphism, the image of exp will contain a neighborhood of the identity in G. If G is connected, this will generate all of G; from this follows the First Principle: *if G is connected, then the map ψ is determined by its differential $(d\psi)_e$ at the identity.*

Using (8.32), we can write down the exponential map very explicitly in the case of $\mathrm{GL}_n\mathbb{R}$, and hence for any subgroup of $\mathrm{GL}_n\mathbb{R}$. We just use the standard power series for the function e^x, and set, for $X \in \mathrm{End}(V)$,

$$\exp(X) = 1 + X + \frac{X^2}{2} + \frac{X^3}{6} + \cdots. \tag{8.34}$$

Observe that this converges and is invertible, with inverse $\exp(-X)$. Clearly, the differential of this map from $\mathfrak{g}$ to G at the origin is the identity; and by the standard power series computation, the restriction of the map to any line through the origin in $\mathfrak{g}$ is a one-parameter subgroup of G. Thus, the map coincides with the exponential as defined originally; and by naturality the same is true for any subgroup of G. (Note that, as we have pointed out, the individual terms in the expression on the right of (8.34) are very much dependent of the particular embedding of G in a general linear group $\mathrm{GL}(V)$ and correspondingly of $\mathfrak{g}$ in $\mathrm{End}(V)$, even though the *sum* on the right in (8.34) is not.)

This explicit form of the exponential map allows us to give substance to the assertion that "the group structure of G is encoded in the Lie algebra." Explicitly, we claim that not only do the exponentials $\exp(X)$ generate G, but for X and Y in a sufficiently small neighborhood of the origin in $\mathfrak{g}$, we can write down the product $\exp(X)\cdot\exp(Y)$ as an exponential. To do this, we introduce first the "inverse" of the exponential map: for $g \in G \subset \mathrm{GL}_n\mathbb{R}$, we set

$$\log(g) = (g - I) - \frac{(g-I)^2}{2} + \frac{(g-I)^3}{3} - \cdots \in \mathfrak{gl}_n\mathbb{R}.$$

Of course, this will be defined only for g sufficiently close to the identity in G; but where it is defined it will be an inverse to the exponential map.

Now, we define a new bilinear operation on $\mathfrak{gl}_n\mathbb{R}$: we set

$$X * Y = \log(\exp(X)\cdot\exp(Y)).$$

We have to be careful what we mean by this, of course; we substitute for g in the expression above for $\log(g)$ the quantity

$$\begin{aligned}\exp(X)\cdot\exp(Y) &= \left(I + X + \frac{X^2}{2} + \cdots\right)\cdot\left(I + Y + \frac{Y^2}{2} + \cdots\right)\\ &= I + (X + Y) + \left(\frac{X^2}{2} + X\cdot Y + \frac{Y^2}{2}\right) + \cdots,\end{aligned}$$

being careful, of course, to preserve the order of the factors in each product. Doing this, we arrive at

$$\begin{aligned}X * Y &= (X + Y) + \left(-\frac{(X+Y)^2}{2} + \left(\frac{X^2}{2} + X\cdot Y + \frac{Y^2}{2}\right)\right) + \cdots\\ &= X + Y + \tfrac{1}{2}[X, Y] + \cdots.\end{aligned}$$

Observe in particular that the terms of degree 2 in X and Y do not involve the squares of X and Y or the product $X\cdot Y$ alone, but only the commutator. In fact, this is true of each term in the formula, i.e., the quantity $\log(\exp(X)\cdot\exp(Y))$ can be expressed purely in terms of X, Y, and the bracket operation; the resulting formula is called the *Campbell–Hausdorff formula* (although the actual formula in closed form was given by Dynkin). To degree three, it is

$$X * Y = X + Y + \tfrac{1}{2}[X, Y] \pm \tfrac{1}{12}[X, [X, Y]] \pm \tfrac{1}{12}[Y, [Y, X]] + \cdots.$$

Exercise 8.35*. Verify (and find the correct signs in) the cubic term of the Campbell–Hausdorff formula.

Exercise 8.36. Prove the assertion of the last paragraph that the power series $\log(\exp(X)\cdot\exp(Y))$ can be expressed purely in terms of X, Y, and the bracket operation.

Exercise 8.37. Show that for X and Y sufficiently small, the power series $\log(\exp(X)\cdot\exp(Y))$ converges.

Exercise 8.38*. (a) Show that there is a constant C such that for $X, Y \in \mathfrak{gl}_n$, $X * Y = X + Y + [X, Y] + E$, where $\|E\| \leq C(\|X\| + \|Y\|)^3$.
(b) Show that $\exp(X + Y) = \lim_{n\to\infty} (\exp(X/n)\cdot\exp(Y/n))^n$.
(c) Show that

$$\exp([X, Y]) = \lim_{n\to\infty} \left(\exp\left(\frac{X}{n}\right)\cdot\exp\left(\frac{Y}{n}\right)\cdot\exp\left(-\frac{X}{n}\right)\cdot\exp\left(-\frac{Y}{n}\right)\right)^{n^2}.$$

Exercise 8.39. Show that if G is a subgroup of $GL_n\mathbb{R}$, the elements of its Lie algebra are the "infinitesimal transformations" of G in the sense of von Neumann, i.e., they are the matrices in $\mathfrak{gl}_n\mathbb{R}$ which can be realized as limits

$$\lim_{t\to 0} \frac{A_t - I}{\varepsilon_t}, \quad A_t \in G, \varepsilon_t > 0, \varepsilon_t \to 0.$$

Exercise 8.40. Show that exp is surjective for $G = GL_n\mathbb{C}$ but not for $G = GL_n^+\mathbb{R}$ if $n > 1$, or for $G = SL_2\mathbb{C}$.

By the Campbell–Hausdorff formula, we can not only identify all the elements of G in a neighborhood of the identity, but we can also say what their pairwise products are, thus making precise the sense in which $\mathfrak{g}$ and its bracket operation determines G and its group law locally. Of course, we have not written a closed-form expression for the Campbell–Hausdorff formula; but, as we will see shortly, its very existence is significant. (For such a closed form, see [Se1, I§4.8].)

We now consider another very natural question, namely, when a vector subspace $\mathfrak{h} \subset \mathfrak{g}$ is the Lie algebra of (i.e., tangent space at the identity to) an immersed subgroup of G. Obviously, a necessary condition is that $\mathfrak{h}$ is closed under the bracket operation; we claim here that this is sufficient as well:

Proposition 8.41. *Let G be a Lie group, $\mathfrak{g}$ its Lie algebra, and $\mathfrak{h} \subset \mathfrak{g}$ a Lie subalgebra. Then the subgroup of the group G generated by* $\exp(\mathfrak{h})$ *is an immersed subgroup H with tangent space $T_eH = \mathfrak{h}$.*

PROOF. Note that the subgroup generated by $\exp(\mathfrak{h})$ is the same as the subgroup generated by $\exp(U)$, where U is any neighborhood of the origin in $\mathfrak{h}$. It will suffice, then (see Exercise 8.42), to show that the image of $\mathfrak{h}$ under the exponential map is "locally" closed under multiplication, i.e., that for a sufficiently small disc $\Delta \subset \mathfrak{h}$, the product $\exp(\Delta)\cdot\exp(\Delta)$ (that is, the set of pairwise products $\exp(X)\cdot\exp(Y)$ for $X, Y \in \Delta$) is contained in the image of $\mathfrak{h}$ under the exponential map.

We will do this under the hypothesis that G may be realized as a subgroup of a general linear group $GL_n\mathbb{R}$, so that we can use the formula (8.34) for the exponential map. This is a harmless assumption, given the statement (to be proved in Appendix E) that any finite-dimensional Lie algebra may be embedded in the Lie algebra $\mathfrak{gl}_n\mathbb{R}$: the subgroup of $\mathrm{GL}_n\mathbb{R}$ generated by $\exp(g)$ will be a group isogenous to G, and, as the reader can easily check, proving the proposition for a group isogenous to G is equivalent to proving it for G.

It thus suffices to prove the assertion in case the group G is $\mathrm{GL}_n\mathbb{R}$. But this is exactly the content of the Campbell–Hausdorff formula. □

When applied to an embedding of a Lie algebra $\mathfrak{g}$ into $\mathfrak{gl}_n$, we see, in particular, that *every finite-dimensional Lie algebra is the Lie algebra of a Lie group*. From what we have seen, this Lie group is unique if we require it to be simply connected, and then all others are obtained by dividing this simply connected model by a discrete subgroup of its center.

Exercise 8.42*. Let $G_0 = \exp(\Delta)$, where Δ is a disk centered at the origin in $\mathfrak{g}$, and let $H_0 = \exp(\Delta \cap \mathfrak{h})$. Show that $G_0^{-1} = G_0$, $H_0^{-1} = H_0$, and $H_0 \cdot H_0 \cap G_0 = H_0$. Use this to show that the subgroup H of G generated by H_0 is an immersed Lie subgroup of G.

As a fairly easy consequence of this proposition, we can finally give a proof of the Second Principle stated in §8.1, which we may restate as

Second Principle. *Let G and H be Lie groups with G simply connected, and let $\mathfrak{g}$ and $\mathfrak{h}$ be their Lie algebras. A linear map $\alpha\colon \mathfrak{g} \to \mathfrak{h}$ is the differential of a map $A\colon G \to H$ of Lie groups if and only if α is a map of Lie algebras.*

PROOF. To see this, consider the product $G \times H$. Its Lie algebra is just $\mathfrak{g} \oplus \mathfrak{h}$. Let $\mathfrak{j} \subset \mathfrak{g} \oplus \mathfrak{h}$ be the graph of the map α. Then the hypothesis that α is a map of Lie algebras is equivalent to the statement that $\mathfrak{j}$ is a Lie subalgebra of $\mathfrak{g} \oplus \mathfrak{h}$; and given this, by the proposition there exists an immersed Lie subgroup $J \subset G \times H$ with tangent space $T_eJ = \mathfrak{j}$.

Look now at the map $\pi\colon J \to G$ given by projection on the first factor. By hypothesis, the differential of this map $d\pi_e\colon \mathfrak{j} \to \mathfrak{g}$ is an isomorphism, so that the map $J \to G$ is an isogeny; but since G is simply connected *it follows that π is an isomorphism.* The projection $\eta\colon G \cong J \to H$ on the second factor is then a Lie group map whose differential at the identity is α. □

Exercise 8.43*. If $\mathfrak{g} \to \mathfrak{g}'$ is a homomorphism of Lie algebras with kernel $\mathfrak{h}$, show that the kernel H of the corresponding map of simply connected Lie groups $G \to G'$ is a closed subgroup of G with Lie group $\mathfrak{h}$. This does not extend to non-normal subgroups, i.e., to the situation when $\mathfrak{h}$ is not the kernel of a homomorphism: give an example of an immersed subgroup of a simply connected Lie group G whose image in G is not closed.

Exercise 8.44. Use the ideas of this lecture to prove the assertion that a compact complex connected Lie group G must be abelian:

(a) Verify that the map $\mathrm{Ad}\colon G \to \mathrm{Aut}(T_eG) \subset \mathrm{End}(T_eG)$ is holomorphic, and, therefore (by the maximum principle), constant.
(b) Deduce that if Ψ_g is conjugation by g, then $d\Psi_g$ is the identity, so $\Psi_g(\exp(X)) = \exp(d\Psi_g(X)) = \exp(X)$ for all $X \in T_eG$, which implies that G is abelian.
(c) Show that the exponential map from T_eG to G is surjective, with the kernel a lattice Λ, so $G = T_eG/\Lambda$ is a complex torus.

LECTURE 9

Initial Classification of Lie Algebras

In this lecture we define various subclasses of Lie algebras: nilpotent, solvable, semisimple, etc., and prove basic facts about their representations. The discussion is entirely elementary (largely because the hard theorems are stated without proof for now); there are no prerequisites beyond linear algebra. Apart from giving these basic definitions, the purpose of the lecture is largely to motivate the narrowing of our focus to semisimple algebras that will take place in the sequel. In particular, the first part of §9.3 is logically the most important for what follows.

§9.1. Rough Classification of Lie Algebras

We will give, in this section, a preliminary sort of classification of Lie algebras, reflecting the degree to which a given Lie algebra $\mathfrak{g}$ fails to be abelian. As we have indicated, the goal ultimately is to narrow our focus onto *semisimple* Lie algebras.

Before we begin, two definitions, both completely straightforward: First, we define the *center* $Z(\mathfrak{g})$ of a Lie algebra $\mathfrak{g}$ to be the subspace of $\mathfrak{g}$ of elements $X \in \mathfrak{g}$ such that $[X, Y] = 0$ for all $Y \in \mathfrak{g}$. Of course, we say $\mathfrak{g}$ is *abelian* if all brackets are zero.

Exercise 9.1. Let G be a Lie group, $\mathfrak{g}$ its Lie algebra. Show that the subgroup of G generated by exponentiating the Lie subalgebra $Z(\mathfrak{g})$ is the connected component of the identity in the center $Z(G)$ of G.

Next, we say that a Lie subalgebra $\mathfrak{h} \subset \mathfrak{g}$ of a Lie algebra $\mathfrak{g}$ is an *ideal* if it satisfies the condition

$$[X, Y] \in \mathfrak{h} \quad \text{for all } X \in \mathfrak{h}, Y \in \mathfrak{g}.$$

Just as connected subgroups of a Lie group correspond to subalgebras of its Lie algebra, the notion of ideal in a Lie algebra corresponds to the notion of normal subgroup, in the following sense:

Exercise 9.2. Let G be a connected Lie group, $H \subset G$ a connected subgroup and $\mathfrak{g}$ and $\mathfrak{h}$ their Lie algebras. Show that H is a normal subgroup of G if and only if $\mathfrak{h}$ is an ideal of $\mathfrak{g}$.

Observe also that the bracket operation on $\mathfrak{g}$ induces a bracket on the quotient space $\mathfrak{g}/\mathfrak{h}$ if and only if $\mathfrak{h}$ is an ideal in $\mathfrak{g}$.

This, in turns, motivates the next bit of terminology: we say that a Lie algebra $\mathfrak{g}$ is *simple* if $\dim \mathfrak{g} > 1$ and it contains no nontrivial ideals. By the last exercise, this is equivalent to saying that the adjoint form G of the Lie algebra $\mathfrak{g}$ has no nontrivial normal Lie subgroups.

Now, to attempt to classify Lie algebras, we introduce two descending chains of subalgebras. The first is the *lower central series* of subalgebras $\mathscr{D}_k\mathfrak{g}$, defined inductively by

$$\mathscr{D}_1\mathfrak{g} = [\mathfrak{g}, \mathfrak{g}]$$

and

$$\mathscr{D}_k\mathfrak{g} = [\mathfrak{g}, \mathscr{D}_{k-1}\mathfrak{g}].$$

Note that the subalgebras $\mathscr{D}_k\mathfrak{g}$ are in fact ideals in $\mathfrak{g}$. The other series is called the *derived series* $\{\mathscr{D}^k\mathfrak{g}\}$; it is defined by

$$\mathscr{D}^1\mathfrak{g} = [\mathfrak{g}, \mathfrak{g}]$$

and

$$\mathscr{D}^k\mathfrak{g} = [\mathscr{D}^{k-1}\mathfrak{g}, \mathscr{D}^{k-1}\mathfrak{g}].$$

Exercise 9.3. Use the Jacobi identity to show that $\mathscr{D}^k\mathfrak{g}$ is also an ideal in $\mathfrak{g}$. More generally, if $\mathfrak{h}$ is an ideal in a Lie algebra $\mathfrak{g}$, show that $[\mathfrak{h}, \mathfrak{h}]$ is also an ideal in $\mathfrak{g}$; hence all $\mathscr{D}^k\mathfrak{h}$ are ideals in $\mathfrak{g}$.

Observe that we have $\mathscr{D}^k\mathfrak{g} \subset \mathscr{D}_k\mathfrak{g}$ for all k, with equality when $k = 1$; we often write simply $\mathscr{D}\mathfrak{g}$ for $\mathscr{D}_1\mathfrak{g} = \mathscr{D}^1\mathfrak{g}$ and call this the *commutator subalgebra*. We now make the

Definitions

(i) We say that $\mathfrak{g}$ is *nilpotent* if $\mathscr{D}_k\mathfrak{g} = 0$ for some k.
(ii) We say that $\mathfrak{g}$ is *solvable* if $\mathscr{D}^k\mathfrak{g} = 0$ for some k.

(iii) We say that $\mathfrak{g}$ is *perfect* if $\mathscr{D}\mathfrak{g} = \mathfrak{g}$ (this is not a concept we will use much).
(iv) We say that $\mathfrak{g}$ is *semisimple* if $\mathfrak{g}$ has no nonzero solvable ideals.

The standard example of a nilpotent Lie algebra is the algebra $\mathfrak{n}_n\mathbb{R}$ of strictly upper-triangular $n \times n$ matrices; in this case the kth subalgebra $\mathscr{D}_k\mathfrak{g}$ in the lower central series will be the subspace $\mathfrak{n}_{k+1,n}\mathbb{R}$ of matrices $A = (a_{i,j})$ such that $a_{i,j} = 0$ whenever $j \leq i + k$, i.e., that are zero below the diagonal and within a distance k of it in each column or row. (In terms of a complete flag $\{V_i\}$ as in §7.2, these are just the endomorphisms that carry V_i into V_{i-k-1}.) It follows also that any subalgebra of the Lie algebra $\mathfrak{n}_n\mathbb{R}$ is likewise nilpotent; we will show later that any nilpotent Lie algebra is isomorphic to such a subalgebra. We will also see that if a Lie algebra $\mathfrak{g}$ is represented on a vector space V, such that each element acts as a nilpotent endomorphism, there is a basis for V such that, identifying $\mathfrak{gl}(V)$ with $\mathfrak{gl}_n\mathbb{R}$, $\mathfrak{g}$ maps to the subalgebra $\mathfrak{n}_n\mathbb{R} \subset \mathfrak{gl}_n\mathbb{R}$.

Similarly, a standard example of a solvable Lie algebra is the space $\mathfrak{b}_n\mathbb{R}$ of upper-triangular $n \times n$ matrices; in this Lie algebra the commutator $\mathscr{D}\mathfrak{b}_n\mathbb{R}$ is the algebra $\mathfrak{n}_n\mathbb{R}$ and the derived series is, thus, $\mathscr{D}^k\mathfrak{b}_n\mathbb{R} = \mathfrak{n}_{2^{k-1},n}\mathbb{R}$. Again, it follows that any subalgebra of the algebra $\mathfrak{b}_n\mathbb{R}$ is likewise solvable; and we will prove later that, conversely, *any* representation of a solvable Lie algebra on a vector space V consists, in terms of a suitable basis, entirely of upper-triangular matrices (i.e., given a solvable Lie subalgebra $\mathfrak{g}$ of $\mathfrak{gl}(V)$, there exists a basis for V such that under the corresponding identification of $\mathfrak{gl}(V)$ with $\mathfrak{gl}_n\mathbb{R}$, the subalgebra $\mathfrak{g}$ is contained in $\mathfrak{b}_n\mathbb{R} \subset \mathfrak{gl}_n\mathbb{R}$).

It is clear from the definitions that the properties of being nilpotent or solvable are inherited by subalgebras or homomorphic images. We will see that the same is true for semisimplicity in the case of homomorphic images, though not for subalgebras.

Note that $\mathfrak{g}$ is solvable if and only if $\mathfrak{g}$ has a sequence of Lie *subalgebras* $\mathfrak{g} = \mathfrak{g}_0 \supset \mathfrak{g}_1 \supset \cdots \supset \mathfrak{g}_k = 0$, such that $\mathfrak{g}_{i+1}$ is an ideal in $\mathfrak{g}_i$ and $\mathfrak{g}_i/\mathfrak{g}_{i+1}$ is abelian. Indeed, if this is the case, one sees by induction that $\mathscr{D}^i\mathfrak{g} \subset \mathfrak{g}_i$ for all i. (One may also refine such a sequence to one where each quotient $\mathfrak{g}_i/\mathfrak{g}_{i+1}$ is one dimensional.) It follows from this description that if $\mathfrak{h}$ is an ideal in a Lie algebra $\mathfrak{g}$, then $\mathfrak{g}$ *is solvable if and only if* $\mathfrak{h}$ *and* $\mathfrak{g}/\mathfrak{h}$ *are solvable Lie algebras.* (The analogous assertion for nilpotent Lie algebras is false: the ideal $\mathfrak{n}_n$ is nilpotent in the Lie algebra $\mathfrak{b}_n$ of upper-triangular matrices, and the quotient is the nilpotent algebra $\mathfrak{d}_n$ of diagonal matrices, but ~~$\mathfrak{d}_n$~~ is not nilpotent.) If $\mathfrak{g}$ is the Lie algebra of a connected Lie group G, then $\mathfrak{g}$ is solvable if and only if there is a sequence of connected subgroups, each normal in G (or in the next in the sequence), such that the quotients are abelian.

In particular, the sum of two solvable ideals in a Lie algebra $\mathfrak{g}$ is again solvable [note that $(\mathfrak{a} + \mathfrak{b})/\mathfrak{b} \cong \mathfrak{a}/(\mathfrak{a} \cap \mathfrak{b})$]. It follows that the sum of all solvable ideals in $\mathfrak{g}$ is a maximal solvable ideal, called the *radical* of $\mathfrak{g}$ and denoted $\mathrm{Rad}(\mathfrak{g})$. The quotient $\mathfrak{g}/\mathrm{Rad}(\mathfrak{g})$ is semisimple. Any Lie algebra $\mathfrak{g}$ thus fits into an exact sequence

$$0 \to \text{Rad}(\mathfrak{g}) \to \mathfrak{g} \to \mathfrak{g}/\text{Rad}(\mathfrak{g}) \to 0 \tag{9.4}$$

where the first algebra is solvable and the last is semisimple. With this somewhat shaky justification (but see Proposition 9.17), we may say that to study the representation theory of an arbitrary Lie algebra, we have to understand individually the representation theories of solvable and semisimple Lie algebras. Of these, the former is relatively easy, at least as regards irreducible representations. The basic fact about them—that any irreducible representation of a solvable Lie algebra is one dimensional—will be proved later in this lecture. The representation theory of semisimple Lie algebras, on the other hand, is extraordinarily rich, and it is this subject that will occupy us for most of the remainder of the book.

Another easy consequence of the definitions is the fact that *a Lie algebra is semisimple if and only if it has no nonzero abelian ideals.* Indeed, the last nonzero term in the derived sequence of ideals $\mathscr{D}^k\text{Rad}(\mathfrak{g})$ would be an abelian ideal in $\mathfrak{g}$ (cf. Exercise 9.3). A semisimple Lie algebra can have no center, so *the adjoint representation of a semisimple Lie algebra is faithful.*

It is a fact that the sequence (9.4) splits, in the sense that there are subalgebras of $\mathfrak{g}$ that map isomorphically onto $\mathfrak{g}/\text{Rad}(\mathfrak{g})$. The existence of such a *Levi decomposition* is part of the general theory we are postponing. To show that an arbitrary Lie algebra has a faithful representation (*Ado's theorem*), one starts with a faithful representation of the center, and then builds a representation of the radical step by step, inserting a string of ideals between the center and the radical. Then one uses a splitting to get from a faithful representation on the radical to some representation on all of $\mathfrak{g}$; the sum of this representation and the adjoint representation is then a faithful representation. See Appendix E for details.

One reason for the terminology simple/semisimple will become clear later in this lecture, when we show that a semisimple Lie algebra is a direct sum of simple ones.

Exercise 9.5. Every semisimple Lie algebra is perfect. Show that the Lie group of Euclidean motions of $\mathbb{R}^3$ has a Lie algebra $\mathfrak{g}$ which is perfect, i.e., $\mathscr{D}\mathfrak{g} = \mathfrak{g}$, but $\mathfrak{g}$ is not semisimple. More generally, if $\mathfrak{h}$ is semisimple, and V is an irreducible representation of $\mathfrak{h}$, the twisted product

$$\mathfrak{g} = \{(v, X) \mid v \in V, X \in \mathfrak{h}\} \quad \text{with } [(v, X), (w, Y)] = (Xw - Yv, [X, Y])$$

is a Lie algebra with $\mathscr{D}\mathfrak{g} = \mathfrak{g}$, $\text{Rad}(\mathfrak{g}) = V$ abelian, and $\mathfrak{g}/\text{Rad}(\mathfrak{g}) = \mathfrak{h}$.

Exercise 9.6. (a) Show that the following are equivalent for a Lie algebra $\mathfrak{g}$: (i) $\mathfrak{g}$ is nilpotent. (ii) There is a chain of ideals $\mathfrak{g} = \mathfrak{g}_0 \supset \mathfrak{g}_1 \supset \cdots \supset \mathfrak{g}_n = 0$ with $\mathfrak{g}_i/\mathfrak{g}_{i+1}$ contained in the center of $\mathfrak{g}/\mathfrak{g}_{i+1}$. (iii) There is an integer n such that

$$\text{ad}(X_1) \circ \text{ad}(X_2) \circ \cdots \circ \text{ad}(X_n)(Y) = [X_1, [X_2, \ldots, [X_n, Y] \ldots]] = 0$$

for all $X_1, \ldots, X_n$, Y in $\mathfrak{g}$.

(b) Conclude that a connected Lie group G is nilpotent if and only if it can be realized as a succession of central extensions of abelian Lie groups.

Exercise 9.7*. If G is connected and nilpotent, show that the exponential map $\exp: \mathfrak{g} \to G$ is surjective, making $\mathfrak{g}$ the universal covering space of G.

Exercise 9.8. Show that the following are equivalent for a Lie algebra $\mathfrak{g}$: (i) $\mathfrak{g}$ is solvable. (ii) There is a chain of ideals $\mathfrak{g} = \mathfrak{g}_0 \supset \mathfrak{g}_1 \supset \cdots \supset \mathfrak{g}_n = 0$ with $\mathfrak{g}_i/\mathfrak{g}_{i+1}$ abelian. (iii) There is a chain of subalgebras $\mathfrak{g} = \mathfrak{g}_0 \supset \mathfrak{g}_1 \supset \cdots \supset \mathfrak{g}_n = 0$ such that $\mathfrak{g}_{i+1}$ is an ideal in $\mathfrak{g}_i$, and $\mathfrak{g}_i/\mathfrak{g}_{i+1}$ is abelian.

§9.2. Engel's Theorem and Lie's Theorem

We will now prove the statement made above about representations of solvable Lie algebras always being upper triangular. This may give the reader an idea of how the general theory proceeds, before we go back to the concrete examples that are our main concern. The starting point is

Theorem 9.9 (Engel's Theorem). *Let $\mathfrak{g} \subset \mathfrak{gl}(V)$ be any Lie subalgebra such that every $X \in \mathfrak{g}$ is a nilpotent endomorphism of V. Then there exists a nonzero vector $v \in V$ such that $X(v) = 0$ for all $X \in \mathfrak{g}$.*

Note this implies that there exists a basis for V in terms of which the matrix representative of each $X \in \mathfrak{g}$ is strictly upper triangular: since $\mathfrak{g}$ kills v, it will act on the quotient $\bar{V}$ of V by the span of v, and by induction we can find a basis $\bar{v}_2, \ldots, \bar{v}_n$ for $\bar{V}$ in terms of which this action is strictly upper triangular. Lifting $\bar{v}_i$ to any $v_i \in V$ and setting $v_1 = v$ then gives a basis for V as desired.

Proof of Theorem 9.9. One observation before we start is that if $X \in \mathfrak{gl}(V)$ is any nilpotent element, then the adjoint action $\mathrm{ad}(X): \mathfrak{gl}(V) \to \mathfrak{gl}(V)$ is nilpotent. This is straightforward: to say that X is nilpotent is to say that there exists a flag of subspaces $0 \subset V_1 \subset V_2 \subset \cdots \subset V_k \subset V_{k+1} = V$ such that $X(V_i) \subset V_{i-1}$; we can then check that for any endomorphism Y of V the endomorphism $\mathrm{ad}(X)^m(Y)$ carries V_i into V_{i+k-m}.

We now proceed by induction on the dimension of $\mathfrak{g}$. The first step is to show that, under the hypotheses of the problem, $\mathfrak{g}$ *contains an ideal* $\mathfrak{h}$ *of codimension one*. In fact, let $\mathfrak{h} \subset \mathfrak{g}$ be any maximal proper subalgebra; we claim that $\mathfrak{h}$ has codimension one and is an ideal. To see this, we look at the adjoint representation of $\mathfrak{g}$; since $\mathfrak{h}$ is a subalgebra the adjoint action $\mathrm{ad}(\mathfrak{h})$ of $\mathfrak{h}$ on $\mathfrak{g}$ preserves the subspace $\mathfrak{h} \subset \mathfrak{g}$ and so acts on $\mathfrak{g}/\mathfrak{h}$. Moreover, by our observation above, for any $X \in \mathfrak{h}$ $\mathrm{ad}(X)$ acts nilpotently on $\mathfrak{gl}(V)$, hence on $\mathfrak{g}$, hence on $\mathfrak{g}/\mathfrak{h}$. Thus, by induction, there exists a nonzero element $\bar{Y} \in \mathfrak{g}/\mathfrak{h}$ killed by $\mathrm{ad}(X)$ for all $X \in \mathfrak{h}$; equivalently, there exists an element $Y \in \mathfrak{g}$ not in $\mathfrak{h}$ such that $\mathrm{ad}(X)(Y) \in \mathfrak{h}$ for all $X \in \mathfrak{h}$. But this is to say that the subspace $\mathfrak{h}'$ of $\mathfrak{g}$ spanned by $\mathfrak{h}$ and Y is a Lie subalgebra of $\mathfrak{g}$, in which $\mathfrak{h}$ sits as an ideal of codimension one; by the maximality of $\mathfrak{h}$ we have $\mathfrak{h}' = \mathfrak{g}$ and we are done.

We return now to the representation of $\mathfrak{g}$ on V. We may apply the induction hypothesis to the subalgebra $\mathfrak{h}$ of $\mathfrak{g}$ found in the preceding paragraph to conclude that there exists a nonzero vector $v \in V$ such that $X(v) = 0$ for all $X \in \mathfrak{h}$; let $W \subset V$ be the subspace of all such vectors $v \in V$. Let Y be any element of $\mathfrak{g}$ not in $\mathfrak{h}$; since $\mathfrak{h}$ and Y span $\mathfrak{g}$, it will suffice to show that there exists a (nonzero) vector $v \in W$ such that $Y(v) = 0$. Now for any vector $w \in W$ and any $X \in \mathfrak{h}$, we have

$$X(Y(w)) = Y(X(w)) + [X, Y](w).$$

The first term on the right is zero because by hypothesis $w \in W$, $X \in \mathfrak{h}$ and so $X(w) = 0$; likewise, the second term is zero because $[X, Y] = \mathrm{ad}(X)(Y) \in \mathfrak{h}$. Thus, $X(Y(w)) = 0$ for all $X \in \mathfrak{h}$; we deduce that $Y(w) \in W$. But this means that the action of Y on V carries the subspace W into itself; since Y acts nilpotently on V, it follows that there exists a vector $v \in W$ such that $Y(v) = 0$. □

Exercise 9.10*. Show that a Lie algebra $\mathfrak{g}$ is nilpotent if and only if $\mathrm{ad}(X)$ is a nilpotent endomorphism of $\mathfrak{g}$ for every $X \in \mathfrak{g}$.

Engel's theorem, in turn, allows us to prove the basic statement made above that every representation of a solvable Lie group can be put in upper-triangular form. This is implied by

Theorem 9.11 (Lie's Theorem). *Let $\mathfrak{g} \subset \mathfrak{gl}(V)$ be a complex solvable Lie algebra. Then there exists a nonzero vector $v \in V$ that is an eigenvector of X for all $X \in \mathfrak{g}$.*

Exercise 9.12. Show that this implies the existence of a basis for V in terms of which the matrix representative of each $X \in \mathfrak{g}$ is upper triangular.

Proof of Theorem 9.11. Once more, the first step in the argument is to assert that $\mathfrak{g}$ contains an ideal $\mathfrak{h}$ of codimension one. This time, since $\mathfrak{g}$ is solvable we know that $\mathscr{D}\mathfrak{g} \neq \mathfrak{g}$, so that the quotient $\mathfrak{a} = \mathfrak{g}/\mathscr{D}\mathfrak{g}$ is a nonzero abelian Lie algebra; the inverse image in $\mathfrak{g}$ of any codimension one subspace of $\mathfrak{a}$ will then be a codimension one ideal in $\mathfrak{g}$.

Still following the lines of the previous argument, we may by induction assume that there is a vector $v_0 \in V$ that is an eigenvector for all $X \in \mathfrak{h}$. Denote the eigenvalue of X corresponding to v_0 by $\lambda(X)$. We then consider the subspace $W \subset V$ of all vectors satisfying the same relation, i.e., we set

$$W = \{v \in V \colon X(v) = \lambda(X) \cdot v \ \forall X \in \mathfrak{h}\}.$$

Let Y now be any element of $\mathfrak{g}$ not in $\mathfrak{h}$. As before, it will suffice to show that Y carries some vector $v \in W$ into a multiple of itself, and for this it is enough to show that Y carries W into itself. We prove this in a general context in the following lemma.

Lemma 9.13. *Let* $\mathfrak{h}$ *be an ideal in a Lie algebra* $\mathfrak{g}$. *Let* V *be a representation of* $\mathfrak{g}$, *and* $\lambda: \mathfrak{h} \to \mathbb{C}$ *a linear function. Set*

$$W = \{v \in V: X(v) = \lambda(X) \cdot v \ \forall X \in \mathfrak{h}\}.$$

Then $Y(W) \subset W$ *for all* $Y \in \mathfrak{g}$.

PROOF. Let w be any nonzero element of W; to test whether $Y(w) \in W$ we let X be any element of $\mathfrak{h}$ and write

$$\begin{aligned} X(Y(w)) &= Y(X(w)) + [X, Y](w) \\ &= \lambda(X) \cdot Y(w) + \lambda([X, Y]) \cdot w \end{aligned} \tag{9.14}$$

since $[X, Y] \in \mathfrak{h}$. This differs from our previous calculation in that the second term on the right is not immediately seen to be zero; indeed, $Y(w)$ will lie in W if and only if $\lambda([X, Y]) = 0$ for all $X \in \mathfrak{h}$.

To verify this, we introduce another subspace of V, namely, the span U of the images w, $Y(w)$, $Y^2(w)$, ... of w under successive applications of Y. This subspace is clearly preserved by Y; we claim that any $X \in \mathfrak{h}$ carries U into itself as well. It is certainly the case that $\mathfrak{h}$ carries w into a multiple of itself, and hence into U, and (9.14) says that $\mathfrak{h}$ carries $Y(w)$ into a linear combination of $Y(w)$ and w, and so into U. In general, we can see that $\mathfrak{h}$ carries $Y^k(w)$ into U by induction: for any $X \in \mathfrak{h}$ we write

$$X(Y^k(w)) = Y(X(Y^{k-1}(w))) + [X, Y](Y^{k-1}(w)). \tag{9.15}$$

Since $X(Y^{k-1}(w)) \in U$ by induction the first term on the right is in U, and since $[X, Y] \in \mathfrak{h}$ the second term is in U as well.

In fact, we see something more from (9.14) and (9.15): it follows that, in terms of the basis w, $Y(w)$, $Y^2(w)$, ... for U, the action of any $X \in \mathfrak{h}$ is upper triangular, with diagonal entries all equal to $\lambda(X)$. In particular, for any $X \in \mathfrak{h}$ the trace of the restriction of X to U is just the dimension of U times $\lambda(X)$. On the other hand, for any element $X \in \mathfrak{h}$ the commutator $[X, Y]$ acts on U, and being the commutator of two endomorphisms of U the trace of this action is zero. It follows then that $\lambda([X, Y]) = 0$, and we are done. □

Exercise 9.16. Show that any irreducible representation of a solvable Lie algebra $\mathfrak{g}$ is one dimensional, and $\mathscr{D}\mathfrak{g}$ acts trivially.

At least for *irreducible* representations, Lie's theorem implies they will all be known for an arbitrary Lie algebra when they are known for the semisimple case. In fact, we have:

Proposition 9.17. *Let* $\mathfrak{g}$ *be a complex Lie algebra,* $\mathfrak{g}_{ss} = \mathfrak{g}/\mathrm{Rad}(\mathfrak{g})$. *Every irreducible representation of* $\mathfrak{g}$ *is of the form* $V = V_0 \otimes L$, *where* V_0 *is an irreducible* *representation of* $\mathfrak{g}_{ss}$ [*i.e., a representation of* $\mathfrak{g}$ *that is trivial on* Rad($\mathfrak{g}$)], *and* L *is a one-dimensional representation.*

PROOF. By Lie's theorem there is a $\lambda \in (\mathrm{Rad}(\mathfrak{g}))^*$ such that

$$W = \{v \in V: X(v) = \lambda(X)\cdot v \ \forall X \in \mathrm{Rad}(\mathfrak{g})\}$$

is not zero. Apply the preceding lemma, with $\mathfrak{h} = \mathrm{Rad}(\mathfrak{g})$. Since V is irreducible, we must have $W = V$. In particular, $\mathrm{Tr}(X) = \dim(V)\cdot\lambda(X)$ for $X \in \mathrm{Rad}(\mathfrak{g})$, so λ vanishes on $\mathrm{Rad}(\mathfrak{g}) \cap [\mathfrak{g}, \mathfrak{g}]$. Extend λ to a linear function on $\mathfrak{g}$ that vanishes on $[\mathfrak{g}, \mathfrak{g}]$, and let L be the one-dimensional representation of $\mathfrak{g}$ determined by λ; in other words, $Y(z) = \lambda(Y)\cdot z$ for all $Y \in \mathfrak{g}$ and $z \in L$. Then $V \otimes L^*$ is a representation that is trivial on $\mathrm{Rad}(\mathfrak{g})$, so it comes from a representation of $\mathfrak{g}_{ss}$, as required. □

Exercise 9.18. Show that if $\mathfrak{g}'$ is a subalgebra of $\mathfrak{g}$ that maps isomorphically onto $\mathfrak{g}/\mathrm{Rad}(\mathfrak{g})$, then any irreducible representation of $\mathfrak{g}$ restricts to an irreducible representation of $\mathfrak{g}'$, and any irreducible representation of $\mathfrak{g}'$ extends to a representation of $\mathfrak{g}$.

§9.3. Semisimple Lie Algebras

As is clear from the above, many of the aspects of the representation theory of finite groups that were essential to our approach are no longer valid in the context of general Lie algebras and Lie groups. Most obvious of these is complete reducibility, which we have seen fails for Lie groups; another is the fact that not only can the action of elements of a Lie group or algebra on a vector space be nondiagonalizable, the action of some element of a Lie algebra may be diagonalizable under one representation and not under another.

That is the bad news. The good news is that, if we just restrict ourselves to semisimple Lie algebras, everything is once more as well behaved as possible. For one thing, we have complete reducibility again:

Theorem 9.19 (Complete Reducibility). *Let* V *be a representation of the semisimple Lie algebra* $\mathfrak{g}$ *and* $W \subset V$ *a subspace invariant under the action of* $\mathfrak{g}$. *Then there exists a subspace* $W' \subset V$ *complementary to* W *and invariant under* $\mathfrak{g}$.

The proof of this basic result will be deferred to Appendix C.

The other question, the diagonalizability of elements of a Lie algebra under a representation, requires a little more discussion. Recall first the statement of *Jordan decomposition*: any endomorphism X of a complex vector space V can be uniquely written in the form

$$X = X_s + X_n$$

where X_s is diagonalizable, X_n is nilpotent, and the two commute. Moreover, X_s and X_n may be expressed as polynomials in X.

Now, suppose that $\mathfrak{g}$ is an arbitrary Lie algebra, $X \in \mathfrak{g}$ any element, and $\rho: \mathfrak{g} \to \mathfrak{gl}_n\mathbb{C}$ any representation. We have seen that the image $\rho(X)$ need not be diagonalizable; we may still ask how $\rho(X)$ behaves with respect to the Jordan decomposition. The answer is that, in general, absolutely nothing need be true. For example, just taking $\mathfrak{g} = \mathbb{C}$, we see that under the representation

$$\rho_1: t \mapsto (t)$$

every element is diagonalizable, i.e., $\rho(X)_s = \rho(X)$; under the representation

$$\rho_2: t \mapsto \begin{pmatrix} 0 & t \\ 0 & 0 \end{pmatrix}$$

every element is nilpotent [i.e., $\rho(X)_s = 0$]; whereas under the representation

$$\rho_3: t \mapsto \begin{pmatrix} t & t \\ 0 & 0 \end{pmatrix}$$

not only are the images $\rho(X)$ neither diagonalizable nor nilpotent, the diagonalizable and nilpotent parts of $\rho(X)$ are not even in the image $\rho(\mathfrak{g})$ of the representation.

If we assume the Lie algebra $\mathfrak{g}$ is semisimple, however, the situation is radically different. Specifically, we have

Theorem 9.20 (Preservation of Jordan Decomposition). *Let $\mathfrak{g}$ be a semisimple Lie algebra. For any element $X \in \mathfrak{g}$, there exist X_s and $X_n \in \mathfrak{g}$ such that for any representation $\rho: \mathfrak{g} \to \mathfrak{gl}(V)$ we have*

$$\rho(X)_s = \rho(X_s) \quad \text{and} \quad \rho(X)_n = \rho(X_n).$$

In other words, if we think of ρ as injective and $\mathfrak{g}$ accordingly as a Lie subalgebra of $\mathfrak{gl}(V)$, *the diagonalizable and nilpotent parts of any element X of $\mathfrak{g}$ are again in $\mathfrak{g}$ and are independent of the particular representation ρ.*

The proofs we will give of the last two theorems both involve introducing objects that are not essential for the rest of this book, and we therefore relegate them to Appendix C. It is worth remarking, however, that another approach was used classically by Hermann Weyl; this is the famous *unitary trick*, which we will describe briefly.

A Digression on "The Unitary Trick"

Basically, the idea is that the statements above (complete reducibility, preservation of Jordan decomposition) can be proved readily for the representations of a compact Lie group. To prove complete reducibility, for example, we can proceed more or less just as in the case of a finite group: if the compact group G acts on a vector space, we see that there is a Hermitian metric on V invariant under the action of G by taking an arbitrary metric on V and averaging its images under the action of G. If G fixes a subspace $W \subset V$, it will then fix as well its orthogonal complement $W^{\perp}$ with respect to this metric. (Alternatively, we can choose an arbitrary complement W' to W, not necessarily fixed by G, and average over G the projection map to $g(W')$ with kernel W; this average will have image invariant under G.)

How does this help us analyze the representation of a semisimple Lie algebra? The key fact here (to be proved in Lecture 26) is that *if* $\mathfrak{g}$ *is any complex semisimple Lie algebra, there exists a (unique) real Lie algebra* $\mathfrak{g}_0$ *with complexification* $\mathfrak{g}_0 \otimes \mathbb{C} = \mathfrak{g}$, *such that the simply connected form of the Lie algebra* $\mathfrak{g}_0$ *is a compact Lie group* G. Thus, restricting a given representation of $\mathfrak{g}$ to $\mathfrak{g}_0$, we can exponentiate to obtain a representation of G, for which complete reducibility holds; and we can deduce from this the complete reducibility of the original representation. For example, while it is certainly not true that any representation ρ of the Lie group $SL_n\mathbb{R}$ on a vector space V admits an invariant Hermitian metric (in fact, it cannot, unless it is the trivial representation), we can

(i) let ρ' be the corresponding (complex) representation of the Lie algebra $\mathfrak{sl}_n\mathbb{R}$;
(ii) by linearity extend the representation ρ' of $\mathfrak{sl}_n\mathbb{R}$ to a representation ρ'' of $\mathfrak{sl}_n\mathbb{C}$;
(iii) restrict to a representation ρ''' of the subalgebra $\mathfrak{su}_n \subset \mathfrak{sl}_n\mathbb{C}$;
(iv) exponentiate to obtain a representation ρ'''' of the unitary group SU_n.

We can now argue that

If a subspace $W \subset V$ is invariant under the action of $SL_n\mathbb{R}$,

it must be invariant under $\mathfrak{sl}_n\mathbb{R}$; and since $\mathfrak{sl}_n\mathbb{C} = \mathfrak{sl}_n\mathbb{R} \otimes \mathbb{C}$, it follows that
it will be invariant under $\mathfrak{sl}_n\mathbb{C}$; so of course
it will be invariant under $\mathfrak{su}_n$; and hence
it will be invariant under SU_n.

Now, since SU_n is compact, there will exist a complementary subspace W' preserved by SU_n; we argue that

W' will then be invariant under $\mathfrak{su}_n$; and since $\mathfrak{sl}_n\mathbb{C} = \mathfrak{su}_n \otimes \mathbb{C}$, it follows that

it will be invariant under $\mathfrak{sl}_n\mathbb{C}$. Restricting, we see that
it will be invariant under $\mathfrak{sl}_n\mathbb{R}$, and exponentiating,
it will be invariant under $SL_n\mathbb{R}$.

Similarly, if one wants to know that the diagonal elements of $SL_n\mathbb{R}$ act semisimply in any representation, or equivalently that the diagonal elements of $\mathfrak{sl}_n\mathbb{R}$ act semisimply, one goes through the same reasoning, coming down to the fact that the group of diagonal elements in $\mathfrak{su}_n$ is abelian and *compact*.

In general, most of the theorems about the finite-dimensional representation of semisimple Lie algebras admit proofs along two different lines: either algebraically, using just the structure of the Lie algebra; or by the unitary trick, that is, by associating to a representation of such a Lie algebra a representation of a compact Lie group and working with that. Which is preferable depends very much on taste and context; in this book we will for the most part go with the algebraic proofs, though in the case of the Weyl character formula in Part IV the proof via compact groups is so much more appealing it has to be mentioned.

The following exercises include a few applications of these two theorems.

Exercise 9.21*. Show that a Lie algebra $\mathfrak{g}$ is semisimple if and only if every finite-dimensional representation is semisimple, i.e., every invariant subspace has a complement.

Exercise 9.22. Use Weyl's unitary trick to show that, for $n > 2$, all representations of $SO_n\mathbb{C}$ are semisimple, so that, in particular, the Lie algebras $\mathfrak{so}_n\mathbb{C}$ are semisimple. Do the same for $Sp_{2s}\mathbb{C}$ and $\mathfrak{sp}_{2n}\mathbb{C}$, $n \geq 1$. Where does the argument break down for $SO_2\mathbb{C}$?

Exercise 9.23. Show that a real Lie algebra $\mathfrak{g}$ is solvable if and only if the complex Lie algebra $\mathfrak{g} \otimes_{\mathbb{R}} \mathbb{C}$ is solvable. Similarly for nilpotent and semisimple.

Exercise 9.24*. If $\mathfrak{h}$ is an ideal in a Lie algebra $\mathfrak{g}$, show that $\mathfrak{g}$ is semisimple if and only if $\mathfrak{h}$ and $\mathfrak{g}/\mathfrak{h}$ are semisimple. Deduce that every semisimple Lie algebra is a direct sum of simple Lie algebras.

Exercise 9.25*. A Lie algebra is called *reductive* if its radical is equal to its center. A Lie group is reductive if its Lie algebra is reductive. For example, $GL_n\mathbb{C}$ is reductive. Show that the following are true for a reductive Lie algebra $\mathfrak{g}$: (i) $\mathscr{D}\mathfrak{g}$ is semisimple; (ii) the adjoint representation of $\mathfrak{g}$ is semisimple; (iii) $\mathfrak{g}$ is a product of a semisimple and an abelian Lie algebra; (iv) $\mathfrak{g}$ has a finite-dimensional faithful semisimple representation. In fact, each of these conditions is equivalent to $\mathfrak{g}$ being reductive.

§9.4. Simple Lie Algebras

There is one more basic fact about Lie algebras to be stated here; though its proof will have to be considerably deferred, it informs our whole approach to the subject. This is the complete classification of simple Lie algebras:

Theorem 9.26. *With five exceptions, every simple complex Lie algebra is isomorphic to either $\mathfrak{sl}_n\mathbb{C}$, $\mathfrak{so}_n\mathbb{C}$, or $\mathfrak{sp}_{2n}\mathbb{C}$ for some n.*

The five exceptions can all be explicitly described, though none is particularly simple except in name; they are denoted $\mathfrak{g}_2$, $\mathfrak{f}_4$, $\mathfrak{e}_6$, $\mathfrak{e}_7$, and $\mathfrak{e}_8$. We will give a construction of each later in the book (§22.3). The algebras $\mathfrak{sl}_n\mathbb{C}$ (for $n > 1$), $\mathfrak{so}_n\mathbb{C}$ (for $n > 2$), and $\mathfrak{sp}_{2n}\mathbb{C}$ are commonly called the *classical Lie algebras* (and the corresponding groups the *classical Lie groups*); the other five algebras are called, naturally enough, the *exceptional Lie algebras*.

The nature of the classification theorem for simple Lie algebras creates a dilemma as to how we approach the subject: many of the theorems about simple Lie algebras can be proved either in the abstract, or by verifying them in turn for each of the particular algebras listed in the classification theorem. Another alternative is to declare that we are concerned with understanding only the representations of the classical algebras $\mathfrak{sl}_n\mathbb{C}$, $\mathfrak{so}_n\mathbb{C}$, and $\mathfrak{sp}_{2n}\mathbb{C}$, and verify any relevant theorems just in these cases.

Of these three approaches, the last is in many ways the least satisfactory; it is, however, the one that we shall for the most part take. Specifically, what we will do, starting in Lecture 11, is the following:

(i) Analyze in Lectures 11–13 a couple of examples, namely, $\mathfrak{sl}_2\mathbb{C}$ and $\mathfrak{sl}_3\mathbb{C}$, on what may appear to be an ad hoc basis.
(ii) On the basis of these examples, propose in Lecture 14 a general paradigm for the study of representations of a simple (or semisimple) Lie algebra.
(iii) Proceed in Lectures 15–20 to carry out this analysis for the classical algebras $\mathfrak{sl}_n\mathbb{C}$, $\mathfrak{so}_n\mathbb{C}$, and $\mathfrak{sp}_{2n}\mathbb{C}$.
(iv) Give in Part IV and the appendices proofs for general simple Lie algebras of the facts discovered in the preceding sections for the classical ones (as well as one further important result, the *Weyl character formula*).

We can at least partially justify this seemingly inefficient approach by saying that even if one makes a beeline for the general theorems about the structure and representation theory of a simple Lie algebra, to apply these results in practice we would still need to carry out the sort of explicit analysis of the individual algebras done in Lectures 11–20. This is, however, a fairly bald rationalization: the fact is, the reason we are doing it this way is that this is the only way we have ever been able to understand any of the general results.

LECTURE 10
Lie Algebras in Dimensions One, Two, and Three

Just to get a sense of what a Lie algebra is and what groups might be associated to it, we will classify here all Lie algebras of dimension three or less. We will work primarily with complex Lie algebras and Lie groups, but will mention the real case as well. Needless to say, this lecture is logically superfluous; but it is easy, fun, and serves a didactic purpose, so why not read it anyway. The analyses of both the Lie algebras and the Lie groups are completely elementary, with one exception: the classification of the complex Lie groups associated to abelian Lie algebras involves the theory of complex tori, and should probably be skipped by anyone not familiar with this subject.

§10.1. Dimensions One and Two

To begin with, any one-dimensional Lie algebra $\mathfrak{g}$ is clearly abelian, that is, $\mathbb{C}$ with all brackets zero.

The simply connected Lie group with this Lie algebra is just the group $\mathbb{C}$ under addition; and other connected Lie groups that have $\mathfrak{g}$ as their Lie algebra must all be quotients of $\mathbb{C}$ by discrete subgroups $\Lambda \subset \mathbb{C}$. If Λ has rank one, then the quotient is just $\mathbb{C}^*$ under multiplication. If Λ has rank two, however, G may be any one of a continuously varying family of *complex tori of dimension one* (or *Riemann surfaces of genus one*, or *elliptic curves over* $\mathbb{C}$). The set of isomorphism classes of such tori is parametrized by the complex plane with coordinate j, where the function j on the set of lattices $\Lambda \subset \mathbb{C}$ is as described in, e.g., [Ahl].

Over the real numbers, the situation is completely straightforward: the only real Lie algebra of dimension one is again $\mathbb{R}$ with trivial bracket; the simply connected Lie group associated to it is $\mathbb{R}$ under addition; and the only other connected real Lie group with this Lie algebra is $\mathbb{R}/\mathbb{Z} \cong S^1$.

Dimension Two

Here we have to consider two cases, depending on whether $\mathfrak{g}$ is abelian or not.

Case 1: $\mathfrak{g}$ *abelian.* This is very much like the previous case; the simply connected two-dimensional abelian complex Lie group is just $\mathbb{C}^2$ under addition, and the remaining connected Lie groups with Lie algebra $\mathfrak{g}$ are just quotients of $\mathbb{C}^2$ by discrete subgroups. Such a subgroup $\Lambda \subset \mathbb{C}^2$ can have rank 1, 2, 3, or 4, and we analyze these possibilities in turn (the reader who has seen enough complex tori in the preceding example may wish to skip directly to Case 2 at this point).

If the rank of Λ is 1, we can complete the generator of Λ to a basis for $\mathbb{C}^2$, so that $\Lambda = \mathbb{Z}e_1 \subset \mathbb{C}e_1 \oplus \mathbb{C}e_2$ and $G \cong \mathbb{C}^* \times \mathbb{C}$. If the rank of Λ is 2, there are two possibilities: either Λ lies in a one-dimensional complex subspace of $\mathbb{C}^2$ or it does not. If it does not, a pair of generators for Λ will also be a basis for $\mathbb{C}^2$ over $\mathbb{C}$, so that $\Lambda = \mathbb{Z}e_1 \oplus \mathbb{Z}e_2$, $\mathbb{C}^2 = \mathbb{C}e_1 \oplus \mathbb{C}e_2$, and $G \cong \mathbb{C}^* \times \mathbb{C}^*$. If on the other hand Λ does lie in a complex line in $\mathbb{C}^2$, so that we have $\Lambda = \mathbb{Z}e_1 \oplus \mathbb{Z}\tau e_1$ for some $\tau \in \mathbb{C}\setminus\mathbb{R}$, then $G = E \times \mathbb{C}$ will be the product of the torus $\mathbb{C}/(\mathbb{Z} \oplus \mathbb{Z}\tau)$ and $\mathbb{C}$; the remarks above apply to the classification of these (see Exercise 10.1).

The cases where Λ has rank 3 or 4 are a little less clear. To begin with, if the rank of Λ is 3, the main question to ask is whether any rank 2 sublattice Λ' of Λ lies in a complex line. If it does, then we can assume this sublattice is saturated (i.e., a pair of generators for Λ' can be completed to a set of generators for Λ) and write $\Lambda = \mathbb{Z}e_1 \oplus \mathbb{Z}\tau e_1 \oplus \mathbb{Z}e_2$, so that we will have $G = E \times C^*$, where E is a torus as above.

Exercise 10.1*. For two one-dimensional complex tori E and E', show that the complex Lie groups $G = E \times \mathbb{C}$ and $G' = E' \times \mathbb{C}$ are isomorphic if and only if $E \cong E'$. Similarly for $E \times \mathbb{C}^*$ and $E' \times \mathbb{C}^*$.

If, on the other hand, no such sublattice of Λ exists, the situation is much more mysterious. One way we can try to represent G is by choosing a generator for Λ and considering the projection of $\mathbb{C}^2$ onto the quotient of $\mathbb{C}^2$ by the line spanned by this generator; thus, if we write $\Lambda = \mathbb{Z}e_1 \oplus \mathbb{Z}e_2 \oplus \mathbb{Z}(\alpha e_1 + \beta e_2)$ then (assuming β is not real) we have maps

$$\begin{array}{ccc} \mathbb{C}^2 & \longrightarrow & \mathbb{C}^2/\mathbb{C}e_1 = \mathbb{C} \\ \downarrow & & \downarrow \\ G = \mathbb{C}^2/\mathbb{Z}e_1 \oplus \mathbb{Z}e_2 \oplus \mathbb{Z}(\alpha e_1 + \beta e_2) & \longrightarrow & \mathbb{C}/(\mathbb{Z} \oplus \mathbb{Z}\beta) \end{array}$$

expressing G as a bundle over a torus $E = \mathbb{C}/(\mathbb{Z} \oplus \mathbb{Z}\beta)$, with fibers isomorphic to $\mathbb{C}^*$. This expression of G does not, however, help us very much to describe the family of all such groups. For one thing, the elliptic curve E is surely not determined by the data of G: if we just exchange e_1 and e_2, for example, we replace E by $\mathbb{C}/(\mathbb{Z} \oplus \mathbb{Z}\alpha)$, which, of course, need not even be isogenous to E. Indeed, this yields an example of different algebraic groups isomorphic as complex Lie groups: expressing G as a $\mathbb{C}^*$ bundle in this way gives it the structure of an algebraic variety, which, in turn, determines the elliptic curve E (for example, the field of rational functions on G will be the field of rational functions on E with one variable adjoined). Thus, different expressions of the complex Lie group G as a $\mathbb{C}^*$ bundle yield nonisomorphic algebraic groups.

Finally, the case where Λ has rank 4 remains completely mysterious. Among such two-dimensional complex tori are the *abelian varieties*; these are just the tori that may be embedded in complex projective space (and hence may be realized as algebraic varieties). For polarized abelian varieties (that is, abelian varieties with equivalence class of embedding in projective space) there exists a reasonable moduli theory; but the set of abelian varieties forms only a countable dense union in the set of all complex tori (indeed, the general complex torus possesses no nonconstant meromorphic functions whatsoever). No satisfactory theory of moduli is known for these objects.

Needless to say, the foregoing discussion of the various abelian complex Lie groups in dimension two is completely orthogonal to our present purposes. We hope to make the point, however, that even in this seemingly trivial case there lurk some fairly mysterious phenomena. Of course, none of this occurs in the real case, where the two-dimensional abelian simply connected real Lie group is just $\mathbb{R} \times \mathbb{R}$ and any other connected two-dimensional abelian real Lie group is the quotient of this by a sublattice $\Lambda \subset \mathbb{R} \times \mathbb{R}$ of rank 1 or 2, which is to say either $\mathbb{R} \times S^1$ or $S^1 \times S^1$.

Case 2: $\mathfrak{g}$ *not abelian*. Viewing the Lie bracket as a linear map $[\ ,\ ]: \wedge^2\mathfrak{g} \to \mathfrak{g}$, we see that if it is not zero, it must have one-dimensional image. We can thus choose a basis $\{X, Y\}$ for $\mathfrak{g}$ as vector space with X spanning the image of $[\ ,\ ]$; after multiplying Y by an appropriate scalar we will have $[X, Y] = X$, which of course determines $\mathfrak{g}$ completely. There is thus a unique nonabelian two-dimensional Lie algebra $\mathfrak{g}$ over either $\mathbb{R}$ or $\mathbb{C}$.

What are the complex Lie groups with Lie algebra $\mathfrak{g}$? To find one, we start with the adjoint representation of $\mathfrak{g}$, which is faithful: we have

$$\begin{aligned} \operatorname{ad}(X)\colon X &\mapsto 0, & \operatorname{ad}(Y)\colon X &\mapsto -X,\\ Y &\mapsto X, & Y &\mapsto 0 \end{aligned}$$

or in matrix notation, in terms of the basis $\{X, Y\}$ for $\mathfrak{g}$,

$$\operatorname{ad}(X) = \begin{pmatrix} 0 & 1 \\ 0 & 0 \end{pmatrix}, \qquad \operatorname{ad}(Y) = \begin{pmatrix} -1 & 0 \\ 0 & 0 \end{pmatrix}.$$

These generate the algebra $\mathfrak{g} = \begin{pmatrix} * & * \\ 0 & 0 \end{pmatrix} \subset \mathfrak{gl}_2\mathbb{C}$; we may exponentiate to arrive at the adjoint form

$$G_0 = \left\{\begin{pmatrix} a & b \\ 0 & 1 \end{pmatrix} : a \neq 0\right\} \subset \mathrm{GL}_2\mathbb{C}.$$

Topologically this group is homeomorphic to $\mathbb{C} \times \mathbb{C}^*$. To take its universal cover, we write a general member of G_0 as

$$\begin{pmatrix} e^t & s \\ 0 & 1 \end{pmatrix}.$$

The product of two such matrices is given by

$$\begin{pmatrix} e^t & s \\ 0 & 1 \end{pmatrix} \cdot \begin{pmatrix} e^{t'} & s' \\ 0 & 1 \end{pmatrix} = \begin{pmatrix} e^{t+t'} & s + e^t s' \\ 0 & 1 \end{pmatrix},$$

so we may realize the universal cover G of G_0 as the group of pairs $(t, s) \in \mathbb{C} \times \mathbb{C}$ with group law

$$(t, s) \cdot (t', s') = (t + t', s + e^t s').$$

The center of G is just the subgroup

$$Z(G) = \{(2\pi i n, 0)\} \cong \mathbb{Z},$$

so that the connected groups with Lie algebra $\mathfrak{g}$ form a partially ordered tower

$$\begin{array}{l} G \\ \downarrow \\ \vdots \\ \downarrow \\ G_n = G/n\mathbb{Z} = \{(a, b) \in \mathbb{C}^* \times \mathbb{C};\, (a, b) \cdot (a', b') = (aa', b + a^n b')\}. \\ \downarrow \\ \vdots \\ \downarrow \\ G_0 \end{array}$$

Exercise 10.2*. Show that for $n \neq m$ the two groups G_n and G_m are not isomorphic.

Finally, in the real case things are simpler: when we exponentiate the adjoint representation as above, the Lie group we arrive at is already simply connected, and so is the unique connected real Lie group with this Lie algebra.

§10.2. Dimension Three, Rank 1

As in the case of dimension two, we look at the Lie bracket as a linear map from $\wedge^2\mathfrak{g}$ to $\mathfrak{g}$ and begin our classification by considering the rank of this map (that is, the dimension of $\mathscr{D}\mathfrak{g}$), which may be either 0, 1, 2, or 3. For the case of rank 0, we refer back to the discussion of abelian Lie groups above. We begin with the case of rank 1.

Here the kernel of the map $[\ ,\ ]: \wedge^2 \mathfrak{g} \to \mathfrak{g}$ is two dimensional, which means that for some $X \in \mathfrak{g}$ it consists of all vectors of the form $X \wedge Y$ with Y ranging over all of $\mathfrak{g}$ (X here will just be the vector corresponding to the hyperplane $\ker([\ ,\ ]) \subset \wedge^2 \mathfrak{g}$ under the natural (up to scalars) duality between a three-dimensional vector space and its exterior square). Completing X to a basis $\{X, Y, Z\}$ of $\mathfrak{g}$, we can write $\mathfrak{g}$ in the form

$$[X, Y] = [X, Z] = 0,$$
$$[Y, Z] = \alpha X + \beta Y + \gamma Z$$

for some $\alpha, \beta, \gamma \in \mathbb{C}$. If either β or γ is nonzero, we may now rechoose our basis, replacing Y by a multiple of the linear combination $\alpha X + \beta Y + \gamma Z$ and either leaving Z alone (if $\beta \neq 0$) or replacing Z by Y (if $\gamma \neq 0$). We will then have

$$[X, Y] = [X, Z] = 0,$$
$$[Y, Z] = Y$$

from which we see that $\mathfrak{g}$ is just the product of the one-dimensional abelian Lie algebra $\mathbb{C}X$ with the non-abelian two-dimensional Lie algebra $\mathbb{C}Y \oplus \mathbb{C}Z$ described in the preceding discussion. We may thus ignore this case and assume that in fact we have $\beta = \gamma = 0$; replacing X by αX we then have the Lie algebra

$$[X, Y] = [X, Z] = 0,$$
$$[Y, Z] = X.$$

How do we find the Lie groups with this Lie algebra? As before, we need to start with a faithful representation of $\mathfrak{g}$, but here the adjoint representation is useless, since X is in its kernel. We can, however, arrive at a representation of $\mathfrak{g}$ by considering the equations defining $\mathfrak{g}$: we want to find a pair of endomorphisms Y and Z on some vector space that do not commute, but that do commute with their commutator $X = [Y, Z]$; thus,

$$Y(YZ - ZY) - (YZ - ZY)Y = Y^2Z - 2YZY + ZY^2 = 0$$

and similarly for $[Z, [Y, Z]]$. One simple way to find such a pair of endomorphisms is make all three terms Y^2Z, YZY, and Z^2Y in the above equation zero, e.g., by making Y and Z both have square zero, and to have $YZ = 0$ while $ZY \neq 0$. For example, on a three-dimensional vector space with basis e_1, e_2, and e_3 we could take Y to be the map carrying e_3 to e_2 and killing e_1 and e_2, and Z the map carrying e_2 to e_1 and killing e_1 and e_3; we then have $YZ = 0$ while ZY sends e_3 to e_1. We see then that $\mathfrak{g}$ *is just the Lie algebra* $\mathfrak{n}_3$ *of strictly upper-triangular* 3×3 *matrices*. When we exponentiate we arrive at the group

$$G = \left\{ \begin{pmatrix} 1 & a & b \\ 0 & 1 & c \\ 0 & 0 & 1 \end{pmatrix}, a, b, c \in \mathbb{C} \right\}$$

which is simply connected. Now the center of G is the subgroup

$$Z(G) = \left\{ \begin{pmatrix} 1 & 0 & b \\ 0 & 1 & 0 \\ 0 & 0 & 1 \end{pmatrix}, b \in \mathbb{C} \right\} \cong \mathbb{C},$$

so the discrete subgroups of $Z(G)$ are just lattices Λ of rank 1 or 2; thus any connected group with Lie algebra $\mathfrak{g}$ is either G, $G/\mathbb{Z}$, or $G/(\mathbb{Z} \times \mathbb{Z})$—that is, an extension of $\mathbb{C} \times \mathbb{C}$ by either $\mathbb{C}$, $\mathbb{C}^*$, or a torus E.

Exercise 10.3. Show that G/Λ is determined up to isomorphism by the one-dimensional $Z(G)/\Lambda$.

A similar analysis holds in the real case: just as before, $\mathfrak{n}_3$ is the unique real Lie algebra of dimension three with commutator subalgebra of dimension one; its simply connected form is the group G of unipotent 3×3 matrices and (the center of this group being $\mathbb{R}$) the only other group with this Lie algebra is the quotient $H = G/\mathbb{Z}$.

Incidentally, the group H represents an interesting example of a group that cannot be realized as a matrix group, i.e., that admits no faithful finite-dimensional representations. One way to see this is to argue that in any irreducible finite-dimensional representation V the center S^1 of H, being compact and abelian, must be diagonalizable; and so under the corresponding representation of the Lie algebra $\mathfrak{g}$ the element X must be carried to a diagonalizable endomorphism of V. But now if $v \in V$ is any eigenvector for X with eigenvalue λ, we also have, arguing as in §9.2,

$$X(Y(v)) = Y(X(v)) = Y(\lambda v) = \lambda Y(v)$$

and similarly $X(Z(v)) = \lambda Z(v)$, i.e., both $Y(v)$ and $Z(v)$ are also eigenvectors for X with eigenvalue λ. Since Y and Z generate $\mathfrak{g}$ and the representation V is irreducible, it follows that X must act as a scalar multiple $\lambda \cdot I$ of the identity; but since $X = [Y, Z]$ is a commutator and so has trace 0, it follows that $\lambda = 0$.

Exercise 10.4*. Show that if G is a simply connected Lie group, and its Lie algebra is solvable, then G cannot contain any nontrivial compact subgroup (in particular, it contains no elements of finite order).

The group H does, however, have an important infinite-dimensional representation. This arises from the representation of the Lie algebra $\mathfrak{g}$ on the space V of $\mathscr{C}^\infty$ functions on the real line $\mathbb{R}$ with coordinate x, in which Y, Z, and X are the operators

$$Y\colon f \mapsto \pi i x \cdot f,$$

$$Z\colon f \mapsto \frac{df}{dx}$$

and $X = [Y, Z]$ is $-\pi i$ times the identity. Exponentiating, we see that e^{tY} acts on a function f by multiplying it by the function $(\cos tx + i \cdot \sin tx)$; e^{tZ} sends f to the function F_t where $F_t(x) = f(t + x)$, and e^{tX} sends f to the scalar multiple $e^{-\pi i t} \cdot f$.

§10.3. Dimension Three, Rank 2

In this case, write the commutator subalgebra $\mathscr{D}\mathfrak{g} \subset \mathfrak{g}$ as the span of two elements Y and Z. The commutator of Y and Z can then be written

$$[Y, Z] = \alpha Y + \beta Z.$$

But now the endomorphism $\mathrm{ad}(Y)$ of $\mathfrak{g}$ carries $\mathfrak{g}$ into $\mathscr{D}\mathfrak{g}$, kills Y, and sends Z to $\alpha Y + \beta Z$, and so has trace β; on the other hand, since $\mathrm{ad}(Y)$ is a commutator in $\mathrm{End}(\mathfrak{g})$, it must have trace 0. Thus, β, and similarly α, must be zero; i.e., the subalgebra $\mathscr{D}\mathfrak{g}$ must be abelian. It follows from this that for any element $X \in \mathfrak{g}$ not in $\mathscr{D}\mathfrak{g}$, the map

$$\mathrm{ad}(X)\colon \mathscr{D}\mathfrak{g} \to \mathscr{D}\mathfrak{g}$$

must be an isomorphism. We may now distinguish two possibilities: either $\mathrm{ad}(X)$ is diagonalizible or it is not.

(Note that for the first time we see a case where the classification of the real Lie algebra will be more complicated than that of the complex: in the real case we will have to deal with the third possibility that $\mathrm{ad}(X)$ is diagonalizible over $\mathbb{C}$ but not over $\mathbb{R}$, i.e., that it has two complex conjugate eigenvalues. Though we have not seen it much in these low-dimensional examples, in fact it is generally the case that the real picture is substantially more complicated than the complex one, for essentially just this reason.)

Possibility A: $\mathrm{ad}(X)$ *is diagonalizable.* In this case it is natural to use as a basis for $\mathscr{D}\mathfrak{g}$ a pair of eigenvectors Y, Z for $\mathrm{ad}(X)$; and by multiplying X by a suitable scalar we can assume that one of the eigenvalues (both of which are nonzero) is 1. We thus have the equations for $\mathfrak{g}$

$$[X, Y] = Y, \qquad [X, Z] = \alpha Z, \qquad [Y, Z] = 0 \tag{10.5}$$

for some $\alpha \in \mathbb{C}^*$.

Exercise 10.6. Show that two Lie algebras $\mathfrak{g}_\alpha, \mathfrak{g}_{\alpha'}$ corresponding to two different scalars in the structure equations (10.5) are isomorphic if and only if $\alpha = \alpha'$ or

$\alpha = 1/\alpha'$. Observe that we have for the first time a continuously varying family of nonisomorphic complex Lie algebras.

To find the groups with these Lie algebras we go to the adjoint representation, which here is faithful. Explicitly, $\mathrm{ad}(Y)$ carries X to $-Y$ and kills Y and Z; $\mathrm{ad}(Z)$ carries X to $-\alpha Z$ and also kills Y and Z; and $\mathrm{ad}(X)$ carries Y to itself, Z to αZ, and kills X. A general member $aX - bY - cZ$ of the Lie algebra is thus represented (with respect to the basis $\{Y, Z, X\}$ for $\mathfrak{g}$) by the matrix

$$\begin{pmatrix} a & 0 & b \\ 0 & \alpha a & \alpha c \\ 0 & 0 & 0 \end{pmatrix}.$$

Exponentiating, we find that a group with Lie algebra $\mathfrak{g}$ is

$$G = \left\{ \begin{pmatrix} e^t & 0 & u \\ 0 & e^{\alpha t} & v \\ 0 & 0 & 1 \end{pmatrix}, t, u, v \in \mathbb{C} \right\} \subset \mathrm{GL}_3\mathbb{C}.$$

Here we run across a very interesting circumstance. If the complex number α is not rational, then the exponential map from $\mathfrak{g}$ to G is one-to-one, and hence a homeomorphism; thus, in particular, G is simply connected. If, on the other hand, α is rational, G will have nontrivial fundamental group. To see this, observe that we always have an exact sequence of groups

$$1 \to B \to G \to A \to 1,$$

where

$$A = \left\{ \begin{pmatrix} e^t & 0 & 0 \\ 0 & e^{\alpha t} & 0 \\ 0 & 0 & 1 \end{pmatrix}, t \in \mathbb{C} \right\}$$

and

$$B = \left\{ \begin{pmatrix} 1 & 0 & u \\ 0 & 1 & v \\ 0 & 0 & 1 \end{pmatrix}, u, v \in \mathbb{C} \right\} \cong \mathbb{C} \times \mathbb{C}.$$

Now when $\alpha \notin \mathbb{Q}$, the group $A \cong \mathbb{C}$ is simply connected; but when $\alpha \in \mathbb{Q}$—whatever its denominator—we have $A \cong \mathbb{C}^*$ and correspondingly $\pi_1(G) = \mathbb{Z}$.

Exercise 10.7. Show that G has no center, and hence when $\alpha \neq \mathbb{Q}$, it is the unique connected group with Lie algebra $\mathfrak{g}$. For $\alpha \in \mathbb{Q}$, describe the universal covering of G and classify all groups with Lie algebra $\mathfrak{g}$.

Observe that in this case, even though we have a continuously varying family of Lie algebras $\mathfrak{g}_\alpha$, we have no corresponding continuously varying family of the adjoint (linear) Lie groups; the simply-connected forms do form a family, however.

Possibility B: ad(X) *is not diagonalizable.* In this case the natural thing to do is to choose a basis $\{Y, Z\}$ of $\mathscr{D}\mathfrak{g}$ with respect to which ad(X) is in Jordan normal form; replacing X by a multiple, we may assume both its eigenvalues are 1 so that we will have the Lie algebra

$$[X, Y] = Y, \qquad [X, Z] = Y + Z, \qquad [Y, Z] = 0. \tag{10.8}$$

With respect to the basis $\{Y, Z, X\}$ for $\mathfrak{g}$, then, the adjoint action of the general element $aX - bY - cZ$ of the Lie algebra is represented by the matrix

$$\begin{pmatrix} a & a & b+c \\ 0 & a & c \\ 0 & 0 & 0 \end{pmatrix}$$

and exponentiating we find that the corresponding group is

$$G = \left\{ \begin{pmatrix} e^t & te^t & u \\ 0 & e^t & v \\ 0 & 0 & 1 \end{pmatrix}, t, u, v \in \mathbb{C} \right\}.$$

Exercise 10.9. Show that this group has no center, and hence is the unique connected complex Lie group with its Lie algebra.

Note that the real Lie groups obtained by exponentiating the adjoint action of the Lie algebras given by (10.5) and (10.8) are all homeomorphic to $\mathbb{R}^3$ and have no center, and so are the only connected real Lie groups with these Lie algebras.

Exercise 10.10. Complete the analysis of real Lie groups in Case 2 by considering the third possibility mentioned above: that ad(X) acts on $\mathscr{D}\mathfrak{g}$ with distinct complex conjugate eigenvalues. Observe that in this way we arrive at our first example of two nonisomorphic real Lie algebras whose tensor products with $\mathbb{C}$ are isomorphic.

§10.4. Dimension Three, Rank 3

Our analysis of this final case begins, as in the preceding one, by looking for eigenvectors of the adjoint action of a suitable element $X \in \mathfrak{g}$. Specifically, we claim that we can find an element $H \in \mathfrak{g}$ such that ad(H): $\mathfrak{g} \to \mathfrak{g}$ has an eigenvector with nonzero eigenvalue. To see this, observe first that for any nonzero $X \in \mathfrak{g}$, the rank of ad(X) must be 2; in particular, we must have Ker(ad(X)) $= \mathbb{C}X$. Now start with any $X \in \mathfrak{g}$. Either ad(X) has an eigenvector with nonzero eigenvalue or it is nilpotent; if it is nilpotent, then there exists a vector $Y \in \mathfrak{g}$, not in the kernel of $\mathrm{ad}(X)$ but in the kernel of $\mathrm{ad}(X)^2$—that is, such that $\mathrm{ad}(X)(Y) = \alpha X$ for some nonzero $\alpha \in \mathbb{C}$. But then of course $\mathrm{ad}(Y)(X) = -\alpha X$, so that X is an eigenvector for $\mathrm{ad}(Y)$ with nonzero eigenvalue.

So: choose H and $X \in \mathfrak{g}$ so that X is an eigenvector with nonzero eigenvalue for $\mathrm{ad}(H)$, and write $[H, X] = \alpha X$. Since $H \in \mathscr{D}\mathfrak{g}$, $\mathrm{ad}(H)$ is a commutator in $\mathrm{End}(\mathfrak{g})$, and so has trace 0; it follows that $\mathrm{ad}(H)$ must have a third eigenvector Y with eigenvalue $-\alpha$. To describe the structure of $\mathfrak{g}$ completely it now remains to find the commutator of X and Y; but this follows from the Jacobi identity. We have

$$\begin{aligned}[H, [X, Y]] &= -[X, [Y, H]] - [Y, [H, X]] \\ &= -[X, \alpha Y] - [Y, \alpha X] \\ &= 0,\end{aligned}$$

from which we deduce that $[X, Y]$ must be a multiple of H; since it must be a nonzero multiple, we can multiply X or Y by a scalar to make it 1. Similarly multiplying H by a scalar we can assume α is 1 or any other nonzero scalar. Thus, there is only one possible complex Lie algebra $\mathfrak{g}$ of this type. One could look for endomorphisms H, X, and Y whose commutators satisfy these relations, as we did before. Or we may simply realize that the three-dimensional Lie algebra $\mathfrak{sl}_2\mathbb{C}$ has not yet been seen, so it must be this last possibility. In fact, a natural basis for $\mathfrak{sl}_2\mathbb{C}$ is

$$H = \begin{pmatrix} 1 & 0 \\ 0 & -1 \end{pmatrix}, \qquad X = \begin{pmatrix} 0 & 1 \\ 0 & 0 \end{pmatrix}, \qquad Y = \begin{pmatrix} 0 & 0 \\ 1 & 0 \end{pmatrix}$$

whose Lie algebra is given by

$$[H, X] = 2X, \qquad [H, Y] = -2Y, \qquad [X, Y] = H. \tag{10.11}$$

What groups other than $SL_2\mathbb{C}$ have Lie algebra $\mathfrak{sl}_2\mathbb{C}$? To begin with, the group $SL_2\mathbb{C}$ is simply connected: for example, the map $SL_2\mathbb{C} \to \mathbb{C}^2 - \{(0, 0)\}$ sending a matrix to its first row expresses the topological space $SL_2\mathbb{C}$ as a bundle with fiber $\mathbb{C}$ over $\mathbb{C}^2 - \{(0, 0)\}$. Also, it is not hard to see that the center of $SL_2\mathbb{C}$ is just the subgroup $\{\pm I\}$ of scalar matrices, so that the only other connected group with Lie algebra $\mathfrak{sl}_2\mathbb{C}$ is the quotient $PSL_2\mathbb{C} = Sl_2\mathbb{C}/\{\pm I\}$.

As in the preceding case, the analysis of real three-dimensional Lie algebras $\mathfrak{g}$ with $\mathscr{D}\mathfrak{g} = \mathfrak{g}$ involves one additional possibility. At the outset of the argument above, we started with an arbitrary $H \in \mathfrak{g}$ and said that if $\mathrm{ad}(H)$ had no eigenvector other than H itself, then it would have to be nilpotent. Of course, in the real case it is also possible that $\mathrm{ad}(H)$ has two distinct complex conjugate eigenvalues λ and $\bar{\lambda}$. Since $\mathrm{ad}(H)$ is a commutator in $\mathrm{End}(\mathfrak{g})$ and so has trace 0, λ will have to be purely imaginary in this case; and so multiplying H by a real scalar we can assume that its eigenvalues are i and $-i$. It follows then that we can find $X, Y \in \mathfrak{g}$ with

$$[H, X] = Y \quad \text{and} \quad [H, Y] = -X.$$

Using the Jacobi identity as before we may conclude that the commutator of X and Y is a multiple of H; after multiplying each of X and Y by a real scalar we can assume that it is either H or $-H$. Finally, if $[X, Y] = -H$, then we observe that we are in the case we considered before: $\mathrm{ad}(Y)$ will have $X + H$ as an eigenvector with nonzero eigenvalue, and following our previous analysis we may conclude that $\mathfrak{g} \cong \mathfrak{sl}_2\mathbb{R}$. Thus, we are left with the sole additional possibility that $\mathfrak{g}$ has structure equations

$$[H, X] = Y, \qquad [H, Y] = -X, \qquad [X, Y] = H. \tag{10.12}$$

This, finally, we may recognize as the Lie algebra $\mathfrak{su}_2$ of the real Lie group SU(2) (as you may recall, the isomorphism $\mathfrak{su}_2 \otimes \mathbb{C} \cong \mathfrak{sl}_2\mathbb{C}$ was used in the last lecture).

What are the real Lie groups with Lie algebras $\mathfrak{sl}_2\mathbb{R}$ and $\mathfrak{su}_2$? To start, the center of the group $SL_2\mathbb{R}$ is again just the scalar matrices $\{\pm I\}$, so the only group dominated by $SL_2\mathbb{R}$ is the quotient $PSL_2\mathbb{R}$. On the other hand, unlike the complex case $SL_2\mathbb{R}$ is not simply connected: now the map associating to a 2×2 matrix its first row expresses $SL_2\mathbb{R}$ as a bundle with fiber $\mathbb{R}$ over $\mathbb{R}^2 - \{(0, 0)\}$, so that $\pi_1(SL_2\mathbb{R}) = \mathbb{Z}$. More precisely $PSL_2\mathbb{R}$ maps to the real projective line $\mathbb{P}^1\mathbb{R}$, which is homeomorphic to the circle, with fiber homeomorphic to $\mathbb{R}^2$, so $\pi_1(PSL_2\mathbb{R}) = \mathbb{Z}$. We thus have a tower of covering spaces of $PSL_2\mathbb{R}$, consisting of the simply-connected group $\tilde{S}$ with center $\mathbb{Z}$ and its quotients $\tilde{S}_n = \tilde{S}/n\mathbb{Z}$ (not all of these are covers of $SL_2\mathbb{R}$, despite the diagram below).

A note: In §10.2 we encountered a real Lie group with no faithful finite-dimensional representations; only its universal cover could be represented as a matrix group. Here we find in some sense the opposite phenomenon: the groups $\tilde{S}$ and $\tilde{S}_n$ have no faithful finite-dimensional representations, all finite-dimensional representations factoring through $SL_2\mathbb{R}$ or $PSL_2\mathbb{R}$. This fact will be proved as a consequence of our discussion of the representations of the Lie algebra $\mathfrak{sl}_2\mathbb{C}$ in the next lecture.

What about groups with Lie algebra $\mathfrak{su}_2$? To begin with, there is SU(2), which (again via the map sending a matrix to its first row vector) is homeomorphic to S^3 and thus simply connected. The center of this group is again $\{\pm I\}$, so that the quotient PSU(2) is the only other group with Lie algebra $\mathfrak{su}_2$. (Alternatively, we may realize SU(2) as the group of unit quaternions, cf. Exercise 7.15.)

Finally, we remark that there are other representations of the real and complex Lie groups discussed above. As we will see, the Lie algebra $\mathfrak{so}_3\mathbb{C}$ is isomorphic to $\mathfrak{sl}_2\mathbb{C}$, which induces an isomorphism between the corresponding adjoint forms $PSL_2\mathbb{C}$ and $SO_3\mathbb{C}$ (and between the simply-connected forms $SL_2\mathbb{C}$ and the spin group $\mathrm{Spin}_3\mathbb{C}$). This in turn suggests two more real forms of this group: $SO_3\mathbb{R}$ and $SO^+(2, 1)$. In fact, it is not hard to see that $SO_3\mathbb{R} \cong$ PSU(2), while $SO^+(2, 1) \cong PSL_2\mathbb{R}$. Lastly the isomorphism $\mathfrak{su}_{1,1} \otimes \mathbb{C} \cong$

$\mathfrak{su}_2 \otimes \mathbb{C} \cong \mathfrak{sl}_2\mathbb{C}$ implies that the real Lie algebra $\mathfrak{su}_{1,1}$ is isomorphic to either $\mathfrak{su}_2$ or $\mathfrak{sl}_2\mathbb{R}$; in fact, the latter is the case and this induces an isomorphism of groups $SU_{1,1} \cong SL_2\mathbb{R}$. We summarize the isomorphisms mentioned in the diagram below:

$$\begin{array}{ccccc}
\tilde{S} & & & & \\
\downarrow & & & & \\
\tilde{S}_n & & \mathrm{Spin}_3\mathbb{C} & & \mathrm{Spin}_3\mathbb{R} \\
\downarrow & & \| & & \| \\
SU(1,1) = SL_2\mathbb{R} & \hookrightarrow & SL_2\mathbb{C} & \hookleftarrow & SU(2) = \{\text{unit quaternions}\} \\
\downarrow & & \downarrow & & \downarrow \\
PSL_2\mathbb{R} & \hookrightarrow & PSL_2\mathbb{C} & \hookleftarrow & PSU(2) = \{\text{unit quaternions}\}/\pm 1 \\
\| & & \| & & \| \\
SO^+(2,1) & \hookrightarrow & SO_3\mathbb{C} & \hookleftarrow & SO_3\mathbb{R}
\end{array} \tag{10.13}$$

Note also the coincidences:

$$Sp_2(\mathbb{C}) = SL_2(\mathbb{C}), \qquad Sp_2(\mathbb{R}) = SL_2(\mathbb{R}), \tag{10.14}$$

which follow from the fact that Sp refers to preserving a skew-symmetric bilinear form, and for 2×2 matrices the determinant is such a form.

Exercise 10.15. Identify the Lie algebras $\mathfrak{so}_3$, $\mathfrak{su}_2$, $\mathfrak{su}_{1,1}$, $\mathfrak{so}_{2,1}$, and verify the assertions made about the corresponding Lie groups in the diagram.

Exercise 10.16. For each of the Lie algebras encountered in this lecture, compute the lower central series and the derived series, and say whether the algebra is nilpotent, solvable, simple, or semisimple.

Exercise 10.17. The following are Lie groups of dimension two or three, so must appear on our list. Find them: (i) the group of affine transformations of the line ($x \mapsto ax + b$, under composition); (ii) the group of upper-triangular 2×2 matrices; (iii) the group of orientation preserving Euclidean transformations of the plane (compositions of translations and rotations).

Exercise 10.18. Locate $\mathbb{R}^3$ with the usual cross-product on our list of Lie algebras. More generally, consider the family of Lie algebras parametrized by real quadruples (a, b, c, d), each with basis X, Y, Z with bracket given by

$$[X, Y] = aZ + dY, \qquad [Y, Z] = bX, \qquad [Z, X] = cY - dZ.$$

Classify this Lie algebra as (a, b, c, d) varies in $\mathbb{R}^4$, showing in particular that every three-dimensional Lie algebra can be written in this way.

Exercise 10.19. Realize the isomorphism of $SU(1, 1)$ with $SL_2\mathbb{R}$ by identifying them with the groups of complex automorphisms of the unit disk and the upper half-plane, respectively.

Exercise 10.20. Classify all Lie algebras of dimension four and rank 1; in particular, show that they are all direct sums of Lie algebras described above.

Exercise 10.21. Show more generally that there exists a Lie algebra of dimension m and rank 1 that is not a direct sum of smaller Lie algebras if and only if m is odd; in case m is odd show that this Lie algebra is unique and realize it as a Lie subalgebra of $\mathfrak{sl}_n\mathbb{C}$.

LECTURE 11

Representations of $\mathfrak{sl}_2\mathbb{C}$

This is the first of four lectures—§11–14—that comprise in some sense the heart of the book. In particular, the naive analysis of §11.1, together with the analogous parts of §12 and §13, form the paradigm for the study of finite-dimensional representations of all semisimple Lie algebras and groups. §11.2 is less central; in it we show how the analysis carried out in §11.1 can be used to explicitly describe the tensor products of irreducible representations. §11.3 is least important; it indicates how we can interpret geometrically some of the results of the preceding section. The discussions in §11.1 and §11.2 are completely elementary (we do use the notion of symmetric powers of a vector space, but in a non-threatening way). §11.3 involves a fair amount of classical projective geometry, and can be skimmed or skipped by those not already familiar with the relevant basic notions from algebraic geometry.

§11.1. The Irreducible Representations

We start our discussion of representations of semisimple Lie algebras with the simplest case, that of $\mathfrak{sl}_2\mathbb{C}$. As we will see, while this case does not exhibit any of the complexity of the more general case, the basic idea that informs the whole approach is clearly illustrated here.

This approach is one already mentioned above, in connection with the representations of the symmetric group on three letters. The idea in that case was that given a representation of our group on a vector space V we first restrict the representation to the abelian subgroup generated by a 3-cycle τ. We obtain a decomposition

$$V = \bigoplus V_\alpha$$

of V into eigenspaces for the action of τ; the commutation relations satisfied by the remaining elements σ of the group with respect to τ implied that such σ simply permuted these subspaces V_α, so that the representation was in effect determined by the collection of eigenvalues of τ.

Of course, circumstances in the case of Lie algebra representations are quite different: to name two, it is no longer the case that the action of an abelian object on any vector space admits such a decomposition; and even if such a decomposition exists we certainly cannot expect that the remaining elements of our Lie algebra will simply permute its summands. Nevertheless, the idea remains essentially a good one, as we shall now see.

To begin with, we choose the basis for the Lie algebra $\mathfrak{sl}_2\mathbb{C}$ from the last lecture:

$$H = \begin{pmatrix} 1 & 0 \\ 0 & -1 \end{pmatrix}, \qquad X = \begin{pmatrix} 0 & 1 \\ 0 & 0 \end{pmatrix}, \qquad Y = \begin{pmatrix} 0 & 0 \\ 1 & 0 \end{pmatrix}$$

satisfying

$$[H, X] = 2X, \qquad [H, Y] = -2Y, \qquad [X, Y] = H. \tag{11.1}$$

These seem like a perfectly natural basis to choose, but in fact the choice is dictated by more than aesthetics; there is, as we will see, a nearly canonical way of choosing a basis of a semisimple Lie algebra (up to conjugation), which will yield this basis in the present circumstance and which will share many of the properties we describe below.

In any event, let V be an irreducible finite-dimensional representation of $\mathfrak{sl}_2\mathbb{C}$. We start by trotting out one of the facts that we quoted in Lecture 9, the preservation of Jordan decomposition; in the present circumstances it implies that

$$\textit{The action of } H \textit{ on } V \textit{ is diagonalizable.} \tag{11.2}$$

We thus have, as indicated, a decomposition

$$V = \bigoplus V_\alpha, \tag{11.3}$$

where the α run over a collection of complex numbers, such that for any vector $v \in V_\alpha$ we have

$$H(v) = \alpha \cdot v.$$

The next question is obviously how X and Y act on the various spaces V_α. We claim that X and Y must each carry the subspaces V_α into other subspaces $V_{\alpha'}$. In fact, we can be more specific: if we want to know where the image of a given vector $v \in V_\alpha$ under the action of X sits in relation to the decomposition (11.3), we have to know how H acts on $X(v)$; this is given by the

Fundamental Calculation (first time):

$$\begin{aligned} H(X(v)) &= X(H(v)) + [H, X](v) \\ &= X(\alpha \cdot v) + 2X(v) \\ &= (\alpha + 2) \cdot X(v); \end{aligned}$$

i.e., *if v is an eigenvector for H with eigenvalue α, then $X(v)$ is also an eigenvector for H, with eigenvalue $\alpha + 2$.* In other words, we have

$$X: V_\alpha \to V_{\alpha+2}.$$

The action of Y on each V_α is similarly calculated; we have $Y(V_\alpha) \subset V_{\alpha-2}$.

Observe that as an immediate consequence of this and the irreducibility of V, *all the complex numbers α that appear in the decomposition* (11.3) *must be congruent to one another* mod 2: for any α_0 that actually occurs, the subspace

$$\bigoplus_{n \in \mathbb{Z}} V_{\alpha_0 + 2n}$$

would be invariant under $\mathfrak{sl}_2\mathbb{C}$ and hence equal to all of V. Moreover, by the same token, the V_α that appear must form an unbroken string of numbers of the form $\beta, \beta + 2, \ldots, \beta + 2k$. We denote by n the last element in this sequence; at this point we just know n is a complex number, but we will soon see that it must be an integer.

To proceed with our analysis, we have the following picture of the action of $\mathfrak{sl}_2\mathbb{C}$ on the vector space V:

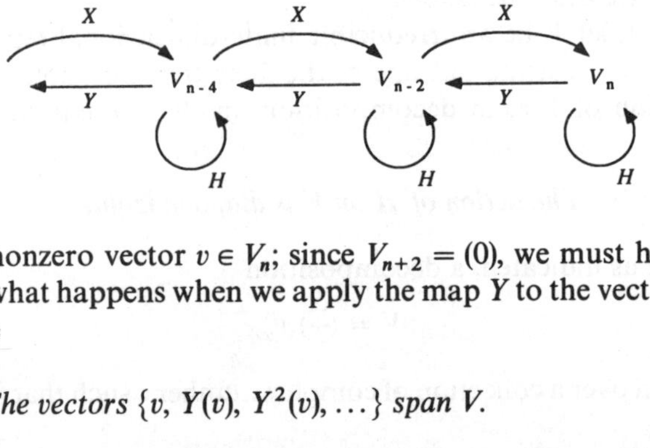

Choose any nonzero vector $v \in V_n$; since $V_{n+2} = (0)$, we must have $X(v) = 0$. We ask now what happens when we apply the map Y to the vector v. To begin with, we have

Claim 11.4. *The vectors $\{v, Y(v), Y^2(v), \ldots\}$ span V.*

PROOF. From the irreducibility of V it is enough to show that the subspace $W \subset V$ spanned by these vectors is carried into itself under the action of $\mathfrak{sl}_2\mathbb{C}$. Clearly, Y preserves W, since it simply carries the vector $Y^m(v)$ into $Y^{m+1}(v)$. Likewise, since the vector $Y^m(v)$ is in V_{n-2m}, we have $H(Y^m(v)) = (n - 2m) \cdot Y^m(v)$, so H preserves the subspace W. Thus, it suffices to check that $X(W) \subset W$, i.e., that for each m, X carries $Y^m(v)$ into a linear combination of the $Y^i(v)$. We check this in turn for $m = 0, 1, 2$, etc.

To begin with, we have $X(v) = 0 \in W$. To see what X does to $Y(v)$, we use the commutation relations for $\mathfrak{sl}_2\mathbb{C}$: we have

$$\begin{aligned} X(Y(v)) &= [X, Y](v) + Y(X(v)) \\ &= H(v) + Y(0) \\ &= n \cdot v. \end{aligned}$$

Next, we see that

$$\begin{aligned} X(Y^2(v)) &= [X, Y](Y(v)) + Y(X(Y(v))) \\ &= H(Y(v)) + Y(n \cdot v) \\ &= (n-2) \cdot Y(v) + n \cdot Y(v). \end{aligned}$$

The pattern now is clear: X carries each vector in the sequence $v, Y(v), Y^2(v), \ldots$ into a multiple of the previous vector. Explicitly, we have

$$X(Y^m(v)) = (n + (n-2) + (n-4) + \cdots + (n - 2m + 2)) \cdot Y^{m-1}(v),$$

or

$$X(Y^m(v)) = m(n - m + 1) \cdot Y^{m-1}(v), \tag{11.5}$$

as can readily be verified by induction. □

There are a number of coroliaries of the calculation in the above Claim. To begin with, we make the observation that

all the eigenspaces V_α of H are one dimensional. (11.6)

Second, since we have in the course of the proof written down a basis for V and said exactly where each of H, X, and Y takes each basis vector, the representation V is completely determined by the one complex number n that we started with; in particular, of course, we have that

V is determined by the collection of α occurring in the decomposition $V = \bigoplus V_\alpha$. (11.7)

To complete our analysis, we have to use one more time the finite dimensionality of V. This tells us that there is a lower bound on the α for which $V_\alpha \neq (0)$ as well as an upper one, so that we must have $Y^k(v) = 0$ for sufficiently large k. But now if m is the smallest power of Y annihilating v, then from the relation (11.5),

$$0 = X(Y^m(v)) = m(n - m + 1) \cdot Y^{m-1}(v),$$

and the fact that $Y^{m-1}(v) \neq 0$, we conclude that $n - m + 1 = 0$; in particular, it follows that *n is a non-negative integer*. The picture is thus that the eigenvalues α of H on V form a string of integers differing by 2 and symmetric about the origin in $\mathbb{Z}$. In sum, then, we see that there is a unique representation $V^{(n)}$ for each non-negative integer n; the representation *$V^{(n)}$ is $(n+1)$-dimensional, with H having eigenvalues $n, n-2, \ldots, -n+2, -n$.*

Note that the existence part of this statement may be deduced by checking that the actions of H, X, and Y as given above in terms of the basis v, Yv, $Y^2(v), \ldots, Y^n(v)$ for V do indeed satisfy all the commutation relations for $\mathfrak{sl}_2\mathbb{C}$. Alternatively, we will exhibit them in a moment. Note also that by the symmetry of the eigenvalues we may deduce the useful fact that *any representation V of $\mathfrak{sl}_2\mathbb{C}$ such that the eigenvalues of H all have the same parity and occur with multiplicity one is necessarily irreducible*; more generally, *the number of irreducible factors in an arbitrary representation V of $\mathfrak{sl}_2\mathbb{C}$ is exactly the sum of the multiplicities of 0 and 1 as eigenvalues of H.*

We can identify in these terms some of the standard representations of $\mathfrak{sl}_2\mathbb{C}$. To begin with, the trivial one-dimensional representation $\mathbb{C}$ is clearly just $V^{(0)}$. As for the standard representation of $\mathfrak{sl}_2\mathbb{C}$ on $V = \mathbb{C}^2$, if x and y are the standard basis for $\mathbb{C}^2$, then we have $H(x) = x$ and $H(y) = -y$, so that $V = \mathbb{C}\cdot x \oplus \mathbb{C}\cdot y = V_{-1} \oplus V_1$ is just the representation $V^{(1)}$ above. Similarly, a basis for the symmetric square $W = \mathrm{Sym}^2 V = \mathrm{Sym}^2\mathbb{C}^2$ is given by $\{x^2, xy, y^2\}$, and we have

$$H(x\cdot x) = x\cdot H(x) + H(x)\cdot x = 2x\cdot x,$$
$$H(x\cdot y) = x\cdot H(y) + H(x)\cdot y = 0,$$
$$H(y\cdot y) = y\cdot H(y) + H(y)\cdot y = -2y\cdot y,$$

so the representation $W = \mathbb{C}\cdot x^2 \oplus \mathbb{C}\cdot xy \oplus \mathbb{C}\cdot y^2 = W_{-2} \oplus W_0 \oplus W_2$ is the representation $V^{(2)}$ above. More generally, the nth symmetric power $\mathrm{Sym}^n V$ of V has basis $\{x^n, x^{n-1}y, \ldots, y^n\}$, and we have

$$\begin{aligned} H(x^{n-k}y^k) &= (n-k)\cdot H(x)\cdot x^{n-k-1}y^k + k\cdot H(y)\cdot x^{n-k}y^{k-1} \\ &= (n-2k)\cdot x^{n-k}y^k \end{aligned}$$

so that the eigenvalues of H on $\mathrm{Sym}^n V$ are exactly $n, n-2, \ldots, -n$. By the observation above that a representation for which all eigenvalues of H occur with multiplicity 1 must be irreducible, it follows that $\mathrm{Sym}^n V$ is irreducible, and hence that

$$V^{(n)} = \mathrm{Sym}^n V.$$

In sum then, we can say simply that

Any irreducible representation of $\mathfrak{sl}_2\mathbb{C}$ is a symmetric power of the standard representation $V \cong \mathbb{C}^2$. (11.8)

Observe that when we exponentiate the image of $\mathfrak{sl}_2\mathbb{C}$ under the embedding $\mathfrak{sl}_2\mathbb{C} \to \mathfrak{sl}_{n+1}\mathbb{C}$ corresponding to the representation $\mathrm{Sym}^n V$, we arrive at the group $\mathrm{SL}_2\mathbb{C}$ when n is odd and $\mathrm{PGL}_2\mathbb{C}$ when n is even. Thus, *the representations of the group $\mathrm{PGL}_2\mathbb{C}$ are exactly the even powers $\mathrm{Sym}^{2n} V$.*

Exercise 11.9. Use the analysis of the representations of $\mathfrak{sl}_2\mathbb{C}$ to prove the statement made in the previous lecture that the universal cover $\widetilde{S}$ of $\mathrm{SL}_2\mathbb{R}$ has no finite-dimensional representations.

§11.2. A Little Plethysm

Clearly, knowing the eigenspace decomposition of given representations tells us the eigenspace decomposition of all their tensor, symmetric, and alternating products and powers: for example, if $V = \bigoplus V_\alpha$ and $W = \bigoplus W_\beta$ then $V \otimes W = \bigoplus (V_\alpha \otimes W_\beta)$ and $V_\alpha \otimes W_\beta$ is an eigenspace for H with eigenvalue $\alpha + \beta$. We can use this to describe the decomposition of these products and powers into irreducible representations of the algebra $\mathfrak{sl}_2\mathbb{C}$.

For example, let $V \cong \mathbb{C}^2$ be the standard representation of $\mathfrak{sl}_2\mathbb{C}$; and suppose we want to study the representation $\mathrm{Sym}^2 V \otimes \mathrm{Sym}^3 V$; we ask in particular whether if it irreducible and, if not, how it decomposes. We have seen that the eigenvalues of $\mathrm{Sym}^2 V$ are 2, 0, and -2, and those of $\mathrm{Sym}^3 V$ are 3, 1, -1, and -3. The 12 eigenvalues of the tensor product $\mathrm{Sym}^2 V \otimes \mathrm{Sym}^3 V$ are thus 5 and -5, 3 and -3 (taken twice), and 1 and -1 (taken three times); we may represent them by the diagram

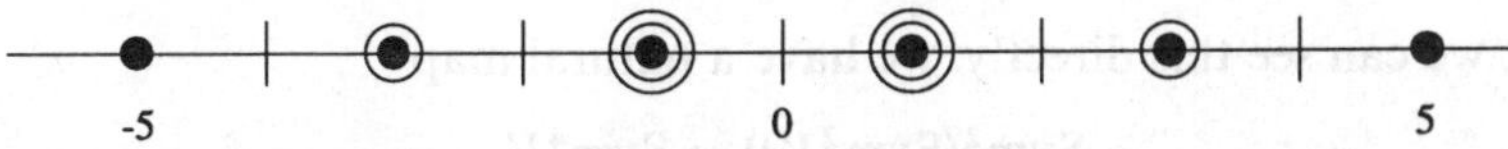

The eigenvector with eigenvalue 5 will generate a subrepresentation of the tensor product isomorphic to $\mathrm{Sym}^5 V$, which will account for one occurrence of each of the eigenvalues 5, 3, 1, -1, -3, and -5. Similarly, the complement of $\mathrm{Sym}^5 V$ in the tensor product will have eigenvalues 3 and -3, and 1 and -1 (taken twice), and so will contain a copy of the representation $\mathrm{Sym}^3 V$, which will account for one occurrence of the eigenvalues 3, 1, -1 and -3; and the complement of these two subrepresentations will be simply a copy of V. We have, thus,

$$\mathrm{Sym}^2 V \otimes \mathrm{Sym}^3 V \cong \mathrm{Sym}^5 V \oplus \mathrm{Sym}^3 V \oplus V.$$

Note that the projection map

$$\mathrm{Sym}^2 V \otimes \mathrm{Sym}^3 V \to \mathrm{Sym}^5 V$$

on the first factor is just multiplication of polynomials; the other two projections do not admit such obvious interpretations.

Exercise 11.10. Find, in a similar way, the decomposition of the tensor product $\mathrm{Sym}^2 V \otimes \mathrm{Sym}^5 V$.

Exercise 11.11*. Show, in general, that for $a \geq b$ we have

$$\mathrm{Sym}^a V \otimes \mathrm{Sym}^b V = \mathrm{Sym}^{a+b} V \oplus \mathrm{Sym}^{a+b-2} V \oplus \cdots \oplus \mathrm{Sym}^{a-b} V.$$

As indicated, we can also look at symmetric and exterior powers of given representations; in many ways this is more interesting. For example, let

$V \cong \mathbb{C}^2$ be as above the standard representation of $\mathfrak{sl}_2\mathbb{C}$, and let $W = \mathrm{Sym}^2 V$ be its symmetric square; i.e., in the notation introduced above, take $W = V^{(2)}$. We ask now whether the symmetric square of W is irreducible, and if not what its decomposition is. To answer this, observe that W has eigenvalues -2, 0, and 2, each occurring once, so that the symmetric square of W will have eigenvalues the pairwise sums of these numbers—that is, -4, -2, 0 (occurring twice), 2, and 4. We may represent $\mathrm{Sym}^2 V$ by the diagram:

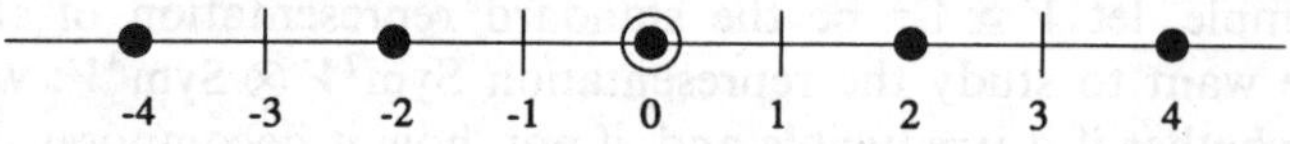

From this, it is clear that the representation $\mathrm{Sym}^2 W$ must decompose into one copy of the representation $V^{(4)} = \mathrm{Sym}^4 V$, plus one copy of the trivial (one-dimensional) representation:

$$\mathrm{Sym}^2(\mathrm{Sym}^2 V)) = \mathrm{Sym}^4 V \oplus \mathrm{Sym}^0 V. \tag{11.12}$$

Indeed, we can see this directly: we have a natural map

$$\mathrm{Sym}^2(\mathrm{Sym}^2 V)) \to \mathrm{Sym}^4 V$$

obtained simply by evaluation; this will have a one-dimensional kernel (if x and y are as above the standard basis for V we can write a generator of this kernel as $(x^2)\cdot(y^2) - (x \cdot y)^2$).

Exercise 11.13. Show that the exterior square $\wedge^2 W$ is isomorphic to W itself. Observe that this, together with the above description of $\mathrm{Sym}^2 W$, agrees with the decomposition of $W \otimes W$ given in Exercise 11.11 above.

We can, in a similar way, describe the decomposition of all the symmetric powers of the representation $W = \mathrm{Sym}^2 V$. For example, the third symmetric power $\mathrm{Sym}^3 W$ has eigenvalues given by the triple sums of the set $\{-2, 0, 2\}$; these are -6, -4, -2 (twice), 0 (twice), 2 (twice), 4, and 6; diagrammatically,

Again, there is no ambiguity about the decomposition; this collection of eigenspaces can only come from the direct sum of $\mathrm{Sym}^6 V$ with $\mathrm{Sym}^2 V$, so we must have

$$\mathrm{Sym}^3(\mathrm{Sym}^2 V) = \mathrm{Sym}^6 V \oplus \mathrm{Sym}^2 V.$$

As before, we can see at least part of this directly: we have a natural evaluation map

$$\mathrm{Sym}^3(\mathrm{Sym}^2 V) \to \mathrm{Sym}^6 V,$$

and the eigenspace decomposition tells us that the kernel is the irreducible representation $\mathrm{Sym}^2 V$.

Exercise 11.14. Use the eigenspace decomposition to establish the formula

$$\mathrm{Sym}^n(\mathrm{Sym}^2 V) = \bigoplus_{\alpha=0}^{[n/2]} \mathrm{Sym}^{2n-4\alpha} V$$

for all n.

§11.3. A Little Geometric Plethysm

We want to give some geometric interpretations of these and similar decompositions of higher tensor powers of representations of $\mathfrak{sl}_2\mathbb{C}$. One big difference is that instead of looking at the action of either the Lie algebra $\mathfrak{sl}_2\mathbb{C}$ or the groups $SL_2\mathbb{C}$ or $PGL_2\mathbb{C}$ on a representation W, we look at the action of the group $PGL_2\mathbb{C}$ on the associated projective space[1] $\mathbb{P}W$. In this context, it is natural to look at various geometric objects associated to the action: for example, we look at closures of orbits of the action, which all turn out to be algebraic varieties, i.e., definable by polynomial equations. In particular, our goal in the following will be to describe the symmetric and exterior powers of W in terms of the action of $PGL_2\mathbb{C}$ on the projective spaces $\mathbb{P}W$ and various loci in $\mathbb{P}W$.

The main point is that while the action of $PGL_2\mathbb{C}$ on the projective space $\mathbb{P}V \cong \mathbb{P}^1$ associated to the standard representation V is transitive, its action on the spaces $\mathbb{P}(\mathrm{Sym}^n V) \cong \mathbb{P}^n$ for $n > 1$ is not. Rather, the action will preserve various orbits whose closures are algebraic subvarieties of $\mathbb{P}^n$—for example, the locus of points

$$C = \{[v \cdot v \cdot \ldots \cdot v] : v \in V\} \subset \mathbb{P}(\mathrm{Sym}^n V)$$

corresponding to nth powers in $\mathrm{Sym}^n V$ will be an algebraic curve in $\mathbb{P}(\mathrm{Sym}^n V) \cong \mathbb{P}^n$, called the *rational normal curve*; and this curve will be carried into itself by any element of $PGL_2\mathbb{C}$ acting on $\mathbb{P}^n$ (more about this in a moment). Thus, a knowledge of the geometry of these subvarieties of $\mathbb{P}W$ may illuminate the representation W, and vice versa. This approach is particularly useful in describing the symmetric powers of W, since these powers can be viewed as the vector spaces of homogeneous polynomials on the projective space $\mathbb{P}(W^*)$ (or, mod scalars, as hypersurfaces in that projective space). Decomposing these symmetric powers should therefore correspond to some interesting projective geometry.

[1] $\mathbb{P}W$ here denotes the projective space of lines through the origin in W, or the quotient space of $W \setminus \{0\}$ by multiplication by nonzero scalars; we write $[w]$ for the point in $\mathbb{P}W$ determined by the nonzero vector w. For $W = \mathbb{C}^{m+1}$, $[z_0, \ldots, z_m]$ is the point in $\mathbb{P}^m = \mathbb{P}W$ determined by a point $(z_0, \ldots, z_m)$ in $\mathbb{C}^{m+1}$.

Digression on Projective Geometry

First, as we have indicated, we want to describe representations of Lie groups in terms of the corresponding actions on projective spaces. The following fact from algebraic geometry is therefore of some moral (if not logical) importance:

Fact 11.15. The group of automorphisms of projective space $\mathbb{P}^n$—either as algebraic variety or as complex manifold—is just the group $\mathrm{PGL}_{n+1}\mathbb{C}$.

For a proof, see [Ha]. (For the Riemann sphere $\mathbb{P}^1$ at least, this should be a familiar fact from complex analysis.)

For any vector space W of dimension $n+1$, $\mathrm{Sym}^k W^*$ is the space of homogeneous polynomials of degree k on the projective space $\mathbb{P}^n = \mathbb{P}W$ of lines in W; dually, $\mathrm{Sym}^k W$ will be the space of homogeneous polynomials of degree k on the projective space $\mathbb{P}^n = \mathbb{P}(W^*)$ of lines in W^*, or of hyperplanes in W. Thus, the projective space $\mathbb{P}(\mathrm{Sym}^k W)$ is the space of hypersurfaces of degree k in $\mathbb{P}^n = \mathbb{P}(W^*)$. (Because of this duality, we usually work with objects in the projective space $\mathbb{P}(W^*)$ rather than the dual space $\mathbb{P}W$ in order to derive results about symmetric powers $\mathrm{Sym}^k W$; this may seem initially more confusing, but we believe it is ultimately less so.)

For any vector space V and any positive integer n, we have a natural map, called the *Veronese embedding*

$$\mathbb{P}V^* \hookrightarrow \mathbb{P}(\mathrm{Sym}^n V^*)$$

that maps the line spanned by $v \in V^*$ to the line spanned by $v^n \in \mathrm{Sym}^n V^*$. We will encounter the Veronese embedding of higher-dimensional vector spaces in later lectures; here we are concerned just with the case where V is two dimensional, so $\mathbb{P}V^* = \mathbb{P}^1$. In this case we have a map

$$\iota_n\colon \mathbb{P}^1 \hookrightarrow \mathbb{P}^n = \mathbb{P}(\mathrm{Sym}^n V^*)$$

whose image is called the *rational normal curve $C = C_n$ of degree n*. Choosing bases $\{\alpha, \beta\}$ for V^* and $\{\ldots[n!/k!(n-k)!]\alpha^k\beta^{n-k}\ldots\}$ for $\mathrm{Sym}^n V^*$ and expanding out $(x\alpha + y\beta)^n$ we see that in coordinates this map may be given as

$$[x, y] \mapsto [x^n, x^{n-1}y, x^{n-2}y^2, \ldots, xy^{n-1}, y^n].$$

From the definition, the action of $\mathrm{PGL}_2\mathbb{C}$ on $\mathbb{P}^n$ preserves C_n; conversely, since any automorphism of $\mathbb{P}^n$ fixing C_n pointwise is the identity, from Fact 11.15 it follows that *the group G of automorphisms of $\mathbb{P}^n$ that preserve C_n is precisely* $\mathrm{PGL}_2\mathbb{C}$. (Note that conversely if W is any $(n+1)$-dimensional representation of $\mathrm{SL}_2\mathbb{C}$ and $\mathbb{P}W \cong \mathbb{P}^n$ contains a rational normal curve of degree n preserved by the action of $\mathrm{PGL}_2\mathbb{C}$, then we must have $W \cong \mathrm{Sym}^n V$; we leave this as an exercise.[2])

When $n = 2$, C is the *plane conic* defined by the equation

[2] Note that any confusion between $\mathbb{P}W$ and $\mathbb{P}W^*$ is relatively harmless for us here, since the representations $\mathrm{Sym}^n V$ are isomorphic to their duals.

$$F(Z_0, Z_1, Z_2) = Z_0Z_2 - Z_1^2 = 0.$$

For $n = 3$, C is the *twisted cubic curve* in $\mathbb{P}^3$, and is defined by three quadratic polynomials

$$Z_0Z_2 - Z_1^2, \qquad Z_0Z_3 - Z_1Z_2, \qquad \text{and} \qquad Z_1Z_3 - Z_2^2.$$

More generally, the rational normal curve is the common zero locus of the 2×2 minors of the matrix

$$M = \begin{pmatrix} Z_0 Z_1 \ldots Z_{n-1} \\ Z_1 Z_2 \ldots \; Z_n \end{pmatrix},$$

that is, the locus where the rank of M is 1.

Back to Plethysm

We start with Example (11.12). We can interpret the decomposition given there (or rather the decomposition of the representation of the corresponding Lie group $\mathrm{SL}_2\mathbb{C}$) geometrically via the Veronese embedding $\iota_2\colon \mathbb{P}^1 \hookrightarrow \mathbb{P}^2$. As noted, $\mathrm{SL}_2\mathbb{C}$ acts on $\mathbb{P}^2 = \mathbb{P}(\mathrm{Sym}^2 V^*)$ as the group of motions of $\mathbb{P}^2$ carrying the conic curve C_2 into itself. Its action on the space $\mathrm{Sym}^2(\mathrm{Sym}^2 V))$ of quadratic polynomials on $\mathbb{P}^2$ thus must preserve the one-dimensional subspace $\mathbb{C}\cdot F$ spanned by the polynomial F above that defines the conic C_2. At the same time, we see that pullback via ι_2 defines a map from the space of quadratic polynomials on $\mathbb{P}^2$ to the space of quartic polynomials on $\mathbb{P}^1$, which has kernel $\mathbb{C}\cdot F$; thus, we have an exact sequence

$$0 \to \mathbb{C} = \mathrm{Sym}^0 V \to \mathrm{Sym}^2(\mathrm{Sym}^2 V)) \to \mathrm{Sym}^4 V \to 0,$$

which implies the decomposition of $\mathrm{Sym}^2(\mathrm{Sym}^2 V))$ described above.

Note that what comes to us at first glance is not actually the direct sum decomposition (11.12) of $\mathrm{Sym}^2(\mathrm{Sym}^2 V))$, but just the exact sequence above. The splitting of this sequence of $\mathrm{SL}_2\mathbb{C}$-modules, guaranteed by the general theory, is less obvious. For example, we are saying that given a conic curve C in the plane $\mathbb{P}^2$, there is a subspace U_C of the space of all conics in $\mathbb{P}^2$, complementary to the one-dimensional subspace spanned by C itself and invariant under the action of the group of motions of the plane $\mathbb{P}^2$ carrying C into itself. Is there a geometric description of this space? Yes: the following proposition gives one.

Proposition 11.16. *The subrepresentation* $\mathrm{Sym}^4 V \subset \mathrm{Sym}^2(\mathrm{Sym}^2 V)$ *is the space of conics spanned by the family of double lines tangent to the conic* $C = C_2$.

PROOF. One way to prove this is to simply write out this subspace in coordinates: in terms of homogeneous coordinates Z_i on $\mathbb{P}^2$ as above, the tangent line to the conic C at the point $[1, \alpha, \alpha^2]$ is the line

$$L_\alpha = \{Z: \alpha^2 Z_0 - 2\alpha Z_1 + Z_2 = 0\}.$$

The double line $2L_\alpha$ is, thus, the conic with equation

$$\alpha^4 Z_0^2 - 4\alpha^3 Z_0 Z_1 + 2\alpha^2 Z_0 Z_2 + 4\alpha^2 Z_1^2 - 4\alpha Z_1 Z_2 + Z_2^2 = 0.$$

The subspace these conics generate is thus spanned by Z_0^2, $Z_0 Z_1$, $Z_1 Z_2$, Z_2^2, and $Z_0 Z_2 + 2Z_1^2$. By construction, this is invariant under the action of $\mathrm{SL}_2\mathbb{C}$, and it is visibly complementary to the trivial subrepresentation $\mathbb{C}\cdot F = \mathbb{C}\cdot(Z_0 Z_2 - Z_1^2)$.

For those familiar with some algebraic geometry, it may not be necessary to write all this down in coordinates: we could just observe that the map from the conic curve C to the projective space $\mathbb{P}(\mathrm{Sym}^2(\mathrm{Sym}^2 V))$ of conics in $\mathbb{P}^2$ sending each point $p \in C$ to the square of the tangent line to C at p is the restriction to C of the quadratic Veronese map $\mathbb{P}^2 \to \mathbb{P}^5$, and so has image a quartic rational normal curve. This spans a four-dimensional projective subspace of $\mathbb{P}(\mathrm{Sym}^2(\mathrm{Sym}^2 V))$, which must correspond to a subrepresentation isomorphic to $\mathrm{Sym}^4 V$. □

We will return to this notion in Exercise 11.26 below.

We can, in a similar way, describe the decomposition of all the symmetric powers of the representation $W = \mathrm{Sym}^2 V$; in the general setting, the geometric interpretation becomes quite handy. For example, we have seen that the third symmetric power decomposes

$$\mathrm{Sym}^3(\mathrm{Sym}^2 V) = \mathrm{Sym}^6 V \oplus \mathrm{Sym}^2 V.$$

This is immediate from the geometric description: the space of cubics in the plane $\mathbb{P}^2$ naturally decomposes into the space of cubics vanishing on the conic $C = C_2$, plus a complementary space isomorphic (via the pullback map ι_2^*) to the space of sextic polynomials on $\mathbb{P}^1$; moreover, since a cubic vanishing on C_2 factors into the quadratic polynomial F and a linear factor, the space of cubics vanishing on the conic curve $C \subset \mathbb{P}^2$ may be identified with the space of lines in $\mathbb{P}^2$.

One more special case: from the general formula (11.14), we have

$$\mathrm{Sym}^4(\mathrm{Sym}^2 V) \cong \mathrm{Sym}^8 V \oplus \mathrm{Sym}^4 V \oplus \mathrm{Sym}^0 V.$$

Again, this is easy to see from the geometric picture: the space of quartic polynomials on $\mathbb{P}^2$ consists of the one-dimensional space of quartics spanned by the square of the defining equation F of C itself, plus the space of quartics vanishing on C modulo multiples of F^2, plus the space of quartics modulo those vanishing on C. (We use the word "plus," suggesting a direct sum, but as before only an exact sequence is apparent).

Exercise 11.17. Show that, in general, the order of vanishing on C defines a filtration on the space of polynomials of degree n in $\mathbb{P}^2$, whose successive quotients are the direct sum factors on the right hand side of the decomposition of Exercise 11.14.

We can similarly analyze symmetric powers of the representation $U = \mathrm{Sym}^3 V$. For example, since U has eigenvalues $-3, -1, 1$, and 3, the symmetric square of U has eigenvalues $-6, -4, -2$ (twice), 0 (twice), 2 (twice), 4, and 6; diagrammatically, we have

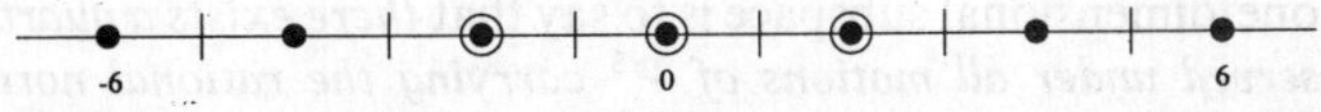

This implies that

$$\mathrm{Sym}^2(\mathrm{Sym}^3 V) \cong \mathrm{Sym}^6 V \oplus \mathrm{Sym}^2 V. \tag{11.18}$$

We can interpret this in terms of the twisted cubic $C = C_3 \subset \mathbb{P}^3$ as follows: the space of quadratic polynomials on $\mathbb{P}^3$ contains, as a subrepresentation, the three-dimensional vector space of quadrics containing the curve C itself; and the quotient is isomorphic, via the pullback map ι_3^*, to the space of sextic polynomials on $\mathbb{P}^1$.

Exercise 11.19*. By the above, the action of $\mathrm{SL}_2\mathbb{C}$ on the space of quadric surfaces containing the twisted cubic curve C is the same as its action on $\mathbb{P}(\mathrm{Sym}^2 V^*) \cong \mathbb{P}^2$. Make this explicit by associating to every quadric containing C a polynomial of degree 2 on $\mathbb{P}^1$, up to scalars.

Exercise 11.20*. The direct sum decomposition (11.18) says that there is a linear space of quadric surfaces in $\mathbb{P}^3$ preserved under the action of $\mathrm{SL}_2\mathbb{C}$ and complementary to the space of quadrics containing C. Describe this space.

Exercise 11.21. The projection map from $\mathrm{Sym}^2(\mathrm{Sym}^3 V)$ to $\mathrm{Sym}^2 V$ given by the decomposition (11.18) above may be viewed as a *quadratic* map from the vector space $\mathrm{Sym}^3 V$ to the vector space $\mathrm{Sym}^2 V$. Show that it may be given in these terms as the *Hessian*, that is, by associating to a homogeneous cubic polynomial in two variables the determinant of the 2×2 matrix of its second partials.

Exercise 11.22. The map in the preceding exercise may be viewed as associating to an unordered triple of points $\{p, q, r\}$ in $\mathbb{P}^1$ an unordered pair of points $\{s, t\} \subset \mathbb{P}^1$. Show that this pair of points is the pair of fixed points of the automorphism of $\mathbb{P}^1$ permuting the three points p, q, and r cyclically.

Exercise 11.23*. Show that

$$\mathrm{Sym}^3(\mathrm{Sym}^3 V) = \mathrm{Sym}^9 V \oplus \mathrm{Sym}^5 V \oplus \mathrm{Sym}^3 V,$$

and interpret this in terms of the geometry of the twisted cubic curve. In particular, show that the space of cubic surfaces containing the curve is the direct sum of the last two factors, and identify the subspace of cubics corresponding to the last factor.

Exercise 11.24. Analyze the representation $\mathrm{Sym}^4(\mathrm{Sym}^3 V)$ similarly. In particular, show that it contains a trivial one-dimensional subrepresentation.

The trivial subrepresentation of $\mathrm{Sym}^4(\mathrm{Sym}^3 V)$ found in the last exercise has an interesting interpretation. To say that $\mathrm{Sym}^4(\mathrm{Sym}^3 V)$ has such an invariant one-dimensional subspace is to say that *there exists a quartic surface in* $\mathbb{P}^3$ *preserved under all motions of* $\mathbb{P}^3$ *carrying the rational normal curve* $C = C_3$ *into itself*. What is this surface? The answer is simple: it is the *tangent developable* to the twisted cubic, that is, the surface given as the union of the tangent lines to C.

Exercise 11.25*. Show that the representation $\mathrm{Sym}^3(\mathrm{Sym}^4 V)$ contains a trivial subrepresentation, and interpret this geometrically.

Problem 11.26. Another way of interpreting the direct sum decomposition of $\mathrm{Sym}^2(\mathrm{Sym}^2 V)$ geometrically is to say that given a conic curve $C \subset \mathbb{P}^2$ and given four points on C, we can find a conic $C' = C'(C; p_1, \dots, p_4) \subset \mathbb{P}^2$ intersecting C in exactly these points, in a way that is preserved by the action of the group $\mathrm{PGL}_3\mathbb{C}$ of all motions of $\mathbb{P}^2$ (i.e., for any motion $A: \mathbb{P}^2 \to \mathbb{P}^2$ of the plane, we have $A(C'(C; p_1, \dots, p_4)) = C'(AC; Ap_1, \dots, Ap_4)$). What is a description of this process? In particular, show that the cross-ratio of the four points p_i on the curve C' must be a function of the cross-ratio of the p_i on C, and find this function. Observe also that this process gives an endomorphism of the pencil

$$\{C \subset \mathbb{P}^2 : p_1, \dots, p_4 \in C\} \cong \mathbb{P}^1$$

of conics passing through any four points $p_i \in \mathbb{P}^2$. What is the degree of this endomorphism?

The above questions have all dealt with the symmetric powers of $\mathrm{Sym}^n V$. There are also interesting questions about the exterior powers of $\mathrm{Sym}^n V$. To start with, consider the exterior square $\Lambda^2(\mathrm{Sym}^3 V)$. The eigenvalues of this representation are just the pairwise sums of distinct elements of $\{3, 1, -1, -3\}$, that is, 4, 2, 0 (twice), -2, and -4; we deduce that

$$\Lambda^2(\mathrm{Sym}^3 V) \cong \mathrm{Sym}^4 V \oplus \mathrm{Sym}^0 V. \tag{11.27}$$

Observe in particular that according to this there is a skew-symmetric bilinear form on the space $U = \mathrm{Sym}^3 V$ preserved (up to scalars) by the action of $\mathrm{SL}_2\mathbb{C}$. What is this form? One way of describing it would be in terms of the twisted cubic: the map from C to the dual projective space $(\mathbb{P}^3)^*$ sending each point $p \in C$ to the osculating plane to C at p extends to a skew-symmetric linear isomorphism of $\mathbb{P}^3$ with $(\mathbb{P}^3)^*$.

Exercise 11.28. Show that a line in $\mathbb{P}^3$ is isotropic for this form if and only if, viewed as an element of $\mathbb{P}(\Lambda^2 U)$, it lies in the linear span of the locus of tangent lines to the twisted cubic.

Exercise 11.29. Show that the projection on the first factor in the decomposition (11.27) is given explicitly by the map

$$F \wedge G \mapsto F \cdot dG - G \cdot dF$$

and say precisely what this means.

Exercise 11.30. Show that, in general, the representation $\wedge^2(\mathrm{Sym}^n V)$ has as a direct sum factor the representation $\mathrm{Sym}^{2n-2} V$, and that the projection on this factor is given as in the preceding exercise. Find the remaining factors of $\wedge^2(\mathrm{Sym}^n V)$, and interpret them.

More on Rational Normal Curves

Exercise 11.31. Analyze in general the representations $\mathrm{Sym}^2(\mathrm{Sym}^n V)$; show, using eigenvalues, that we have

$$\mathrm{Sym}^2(\mathrm{Sym}^n V) = \bigoplus_{\alpha \geq 0} \mathrm{Sym}^{2n-4\alpha} V.$$

Exercise 11.32*. Interpret the space $\mathrm{Sym}^2(\mathrm{Sym}^n V)$ of the preceding exercise as the space of quadrics in the projective space $\mathbb{P}^n$, and use the geometry of the rational normal curve $C = C_n \subset \mathbb{P}^n$ to interpret the decomposition of this representation into irreducible factors. In particular, show that direct sum

$$\bigoplus_{\alpha \geq 1} \mathrm{Sym}^{2n-4\alpha} V$$

is the space of quadratic polynomials vanishing on the rational normal curve; and that the direct sum

$$\bigoplus_{\alpha \geq 2} \mathrm{Sym}^{2n-4\alpha} V$$

is the space of quadrics containing the *tangential developable* of the rational normal curve, that is, the union of the tangent lines to C. Can you interpret the sums for $\alpha \geq k$ for $k > 2$?

Exercise 11.33. Note that by Exercise 11.11, the tensor power

$$\mathrm{Sym}^n V \otimes \mathrm{Sym}^n V$$

always contains a copy of the trivial representation; and that by Exercises 11.30 and 11.31, this trivial subrepresentation will lie in $\mathrm{Sym}^2(\mathrm{Sym}^n V)$ if n is even and in $\wedge^2(\mathrm{Sym}^n V)$ if n is odd. Show that in either case, the bilinear form on $\mathrm{Sym}^n V$ preserved by $\mathrm{SL}_2\mathbb{C}$ may be described as the isomorphism of $\mathbb{P}^n$ with $(\mathbb{P}^n)^*$ carrying each point p of the rational normal curve $C \subset \mathbb{P}^n$ into the osculating hyperplane to C at p.

Comparing Exercises 11.14 and 11.31, we see that $\mathrm{Sym}^2(\mathrm{Sym}^n V) \cong \mathrm{Sym}^n(\mathrm{Sym}^2 V)$; apparently coincidentally. This is in fact a special case of a more general theorem (cf. Exercise 6.18):

Exercise 11.34. (Hermite Reciprocity). Use the eigenvalues of H to prove the isomorphism

$$\mathrm{Sym}^k(\mathrm{Sym}^n V) \cong \mathrm{Sym}^n(\mathrm{Sym}^k V).$$

Can you exhibit explicitly a map between these two?

Note that in the examples of Hermite reciprocity we have seen, it seems completely coincidental: for example, the fact that the representations $\mathrm{Sym}^3(\mathrm{Sym}^4 V)$ and $\mathrm{Sym}^4(\mathrm{Sym}^3 V)$ both contain a trivial representation corresponds to the facts that the tangential developable of the twisted cubic in $\mathbb{P}^3$ has degree 4, while the chordal variety of the rational normal quartic in $\mathbb{P}^4$ has degree 3.

Exercise 11.35*. Show that $\wedge^m(\mathrm{Sym}^n V) \cong \mathrm{Sym}^m(\mathrm{Sym}^{n+1-m} V)$.

We will see in Lecture 23 that there is a unique closed orbit in $\mathbb{P}(W)$ for any irreducible representation W. For now, we can do the following special case.

Exercise 11.36. Show that the unique closed orbit of the action of $SL_2\mathbb{C}$ on the projectivization of any irreducible representation is isomorphic to $\mathbb{P}^1$ (these are the *rational normal curves* introduced above).

LECTURE 12
Representations of $\mathfrak{sl}_3\mathbb{C}$, Part I

This lecture develops results for $\mathfrak{sl}_3\mathbb{C}$ analogous to those of §11.1 (though not in exactly the same order). This involves generalizing some of the basic terms of §11 (e.g., the notions of eigenvalue and eigenvector have to be redefined), but the basic ideas are in some sense already in §11. Certainly no techniques are involved beyond those of §11.1.

We come now to a second important stage in the development of the theory: in the following, we will take our analysis of the representations of $\mathfrak{sl}_2\mathbb{C}$ and see how it goes over in the next case, the algebra $\mathfrak{sl}_3\mathbb{C}$. As we will see, a number of the basic constructions need to be modified, or at least rethought. There are, however, two pieces of good news that should be borne in mind. First, we will arrive, by the end of the following lecture, at a classification of the representations of $\mathfrak{sl}_3\mathbb{C}$ that is every bit as detailed and explicit as the classification we arrived at previously for $\mathfrak{sl}_2\mathbb{C}$. Second, once we have redone our analysis in this context, *we will need to introduce no further concepts to carry out the classification of the finite-dimensional representations of all remaining semisimple Lie algebras.*

We will proceed by analogy with the previous lecture. To begin with, we started out our analysis of $\mathfrak{sl}_2\mathbb{C}$ with the basis $\{H, X, Y\}$ for the Lie algebra; we then proceeded to decompose an arbitrary representation V of $\mathfrak{sl}_2\mathbb{C}$ into a direct sum of eigenspaces for the action of H. What element of $\mathfrak{sl}_3\mathbb{C}$ in particular will play the role of H? The answer—and this is the first and perhaps most wrenching change from the previous case—is that no one element really allows us to see what is going on.[1] Instead, we have to replace

[1] This is not literally true: as we will see from the following analysis, if H is any diagonal matrix whose entries are independent over $\mathbb{Q}$, then the action of H on any representation V of $\mathfrak{sl}_3\mathbb{C}$ determines the representation (i.e., if we know the eigenvalues of H we know V). But (as we will also see) trying to carry this out in practice would be sheer perversity.

the single element $H \in \mathfrak{sl}_2\mathbb{C}$ with a *subspace* $\mathfrak{h} \subset \mathfrak{sl}_3\mathbb{C}$, namely, the two-dimensional subspace of all diagonal matrices. The idea is a basic one; it comes down to the observation that *commuting diagonalizable matrices are simultaneously diagonalizable.* This translates in the present circumstances to the statement that any finite-dimensional representation V of $\mathfrak{sl}_3\mathbb{C}$ admits a decomposition $V = \bigoplus V_\alpha$, where every vector $v \in V_\alpha$ is an eigenvector for every element $H \in \mathfrak{h}$.

At this point some terminology is clearly in order, since we will be dealing with the action not of a single matrix H but rather a vector space $\mathfrak{h}$ of them. To begin with, by an *eigenvector* for $\mathfrak{h}$ we will mean, reasonably enough, a vector $v \in V$ that is an eigenvector for every $H \in \mathfrak{h}$. For such a vector v we can write

$$H(v) = \alpha(H)\cdot v, \tag{12.1}$$

where $\alpha(H)$ is a scalar depending linearly on H, i.e., $\alpha \in \mathfrak{h}^*$. This leads to our second notion: by an *eigenvalue* for the action of $\mathfrak{h}$ we will mean an element $\alpha \in \mathfrak{h}^*$ such that there exists a nonzero element $v \in V$ satisfying (12.1); and by the *eigenspace* associated to the eigenvalue α we will mean the subspace of all vectors $v \in V$ satisfying (12.1). Thus we may phrase the statement above as

(12.2) *Any finite-dimensional representation V of $\mathfrak{sl}_3\mathbb{C}$ has a decomposition*

$$V = \bigoplus V_\alpha,$$

where V_α is an eigenspace for $\mathfrak{h}$ and α ranges over a finite subset of $\mathfrak{h}^*$.

This is, in fact, a special case of a more general statement: for any semisimple Lie algebra $\mathfrak{g}$, we will be able to find an abelian subalgebra $\mathfrak{h} \subset \mathfrak{g}$, such that the action of $\mathfrak{h}$ on any $\mathfrak{g}$-module V will be diagonalizable, i.e., we will have a direct sum decomposition of V into eigenspaces V_α for $\mathfrak{h}$.

Having decided what the analogue for $\mathfrak{sl}_3\mathbb{C}$ of $H \in \mathfrak{sl}_2\mathbb{C}$ is, let us now consider what will play the role of X and Y. The key here is to look at the commutation relations

$$[H, X] = 2X \quad \text{and} \quad [H, Y] = -2Y$$

in $\mathfrak{sl}_2\mathbb{C}$. The correct way to interpret these is as saying that *X and Y are eigenvectors for the adjoint action of H on $\mathfrak{sl}_2\mathbb{C}$.* In our present circumstances, then, we want to look for eigenvectors (in the new sense) for the adjoint action of $\mathfrak{h}$ on $\mathfrak{sl}_3\mathbb{C}$. In other words, we apply (12.2) to the adjoint representation of $\mathfrak{sl}_3\mathbb{C}$ to obtain a decomposition

$$\mathfrak{sl}_3\mathbb{C} = \mathfrak{h} \oplus \left(\bigoplus \mathfrak{g}_\alpha\right), \tag{12.3}$$

where α ranges over a finite subset of $\mathfrak{h}^*$ and $\mathfrak{h}$ acts on each space $\mathfrak{g}_\alpha$ by scalar multiplication, i.e., for any $H \in \mathfrak{h}$ and $Y \in \mathfrak{g}_\alpha$,

$$[H, Y] = \operatorname{ad}(H)(Y) = \alpha(H)\cdot Y.$$

This is probably easier to carry out in practice than it is to say; we are being longwinded here because once this process is understood it will be straightforward to apply it to the other Lie algebras. In any case, to do it in the present circumstances, we just observe that multiplication of a matrix M on the left by a diagonal matrix D with entries a_i multiplies the ith row of M by a_i, while multiplication on the right multiplies the ith column by a_i; if the entries of M are $m_{i,j}$, the entries of the commutator $[D, M]$ are thus $(a_i - a_j)m_{i,j}$. We see then that the commutator $[D, M]$ will be a multiple of M for all D if and only if all but one entry of M are zero. Thus, if we let $E_{i,j}$ be the 3×3 matrix whose (i, j)th entry is 1 and all of whose other entries are 0, we see that the $E_{i,j}$ exactly generate the eigenspaces for the adjoint action of $\mathfrak{h}$ on $\mathfrak{g}$.

Explicitly, we have

$$\mathfrak{h} = \left\{ \begin{pmatrix} a_1 & 0 & 0 \\ 0 & a_2 & 0 \\ 0 & 0 & a_3 \end{pmatrix} : a_1 + a_2 + a_3 = 0 \right\}$$

and so we can write

$$\mathfrak{h}^* = \mathbb{C}\{L_1, L_2, L_3\}/(L_1 + L_2 + L_3 = 0)\},$$

where

$$L_i \begin{pmatrix} a_1 & 0 & 0 \\ 0 & a_2 & 0 \\ 0 & 0 & a_3 \end{pmatrix} = a_i.$$

The linear functionals $\alpha \in \mathfrak{h}^*$ appearing in the direct sum decomposition (12.3) are thus the six functionals $L_i - L_j$; the space $\mathfrak{g}_{L_i - L_j}$ will be generated by the element $E_{i,j}$. To draw a picture

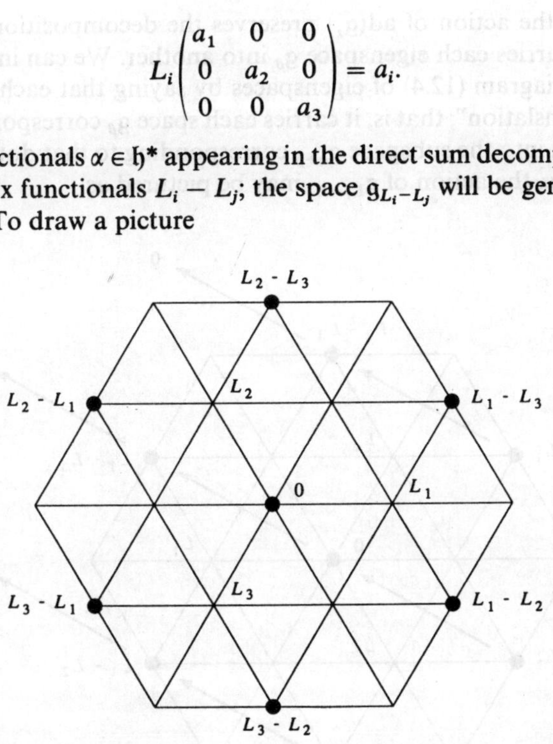

(12.4)

The virtue of this decomposition and the corresponding picture is that we can read off from it pretty much the entire structure of the Lie algebra. Of course, the action of $\mathfrak{h}$ on $\mathfrak{g}$ is clear from the picture: $\mathfrak{h}$ carries each of the subspaces $\mathfrak{g}_\alpha$ into itself, acting on each $\mathfrak{g}_\alpha$ by scalar multiplication by the linear functional represented by the corresponding dot. Beyond that, though, we can also see, much as in the case of representations of $\mathfrak{sl}_2\mathbb{C}$, how the rest of the Lie algebra acts. Basically, we let X be any element of $\mathfrak{g}_\alpha$ and ask where $\mathrm{ad}(X)$ sends a given vector $Y \in \mathfrak{g}_\beta$; the answer as before comes from knowing how $\mathfrak{h}$ acts on $\mathrm{ad}(X)(Y)$. Explicitly, we let H be an arbitrary element of $\mathfrak{h}$ and as on page 148 we make the

Fundamental Calculation (second time):

$$\begin{aligned}[H, [X, Y]] &= [X, [H, Y]] + [[H, X], Y] \\ &= [X, \beta(H)\cdot Y] + [\alpha(H)\cdot X, Y] \\ &= (\alpha(H) + \beta(H))\cdot [X, Y].\end{aligned}$$

In other words, $[X, Y] = \mathrm{ad}(X)(Y)$ *is again an eigenvector for* $\mathfrak{h}$, *with eigenvalue* $\alpha + \beta$. Thus,

$$\mathrm{ad}(\mathfrak{g}_\alpha)\colon \mathfrak{g}_\beta \to \mathfrak{g}_{\alpha+\beta};$$

in particular, the action of $\mathrm{ad}(\mathfrak{g}_\alpha)$ preserves the decomposition (12.3) in the sense that it carries each eigenspace $\mathfrak{g}_\beta$ into another. We can interpret this in terms of the diagram (12.4) of eigenspaces by saying that each $\mathfrak{g}_\alpha$ acts, so to speak, by "translation"; that is, it carries each space $\mathfrak{g}_\beta$ corresponding to a dot in the diagram into the subspace $\mathfrak{g}_{\alpha+\beta}$ corresponding to that dot translated by α. For example, the action of $\mathfrak{g}_{L_1-L_3}$ may be pictured as

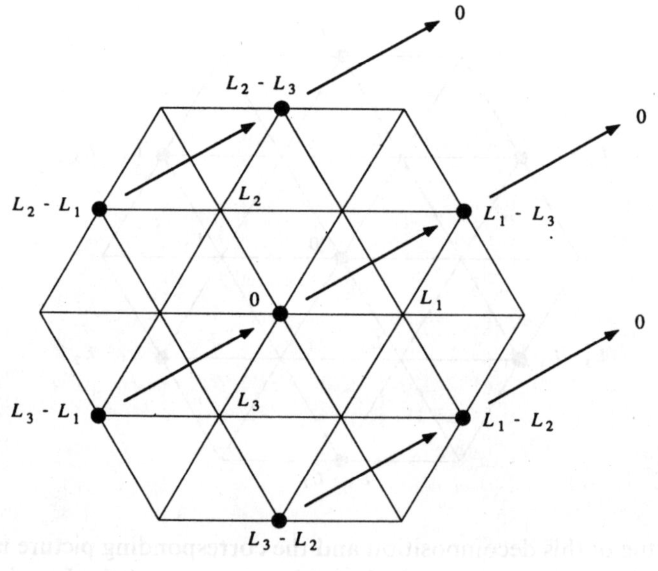

(12.5)

i.e., it carries $\mathfrak{g}_{L_2-L_1}$ into $\mathfrak{g}_{L_2-L_3}$; $\mathfrak{g}_{L_3-L_1}$ into $\mathfrak{h}$; $\mathfrak{h}$ into $\mathfrak{g}_{L_1-L_3}$, $\mathfrak{g}_{L_3-L_2}$ into $\mathfrak{g}_{L_1-L_2}$, and kills $\mathfrak{g}_{L_2-L_3}$, $\mathfrak{g}_{L_1-L_3}$, and $\mathfrak{g}_{L_1-L_2}$. Of course, not all the data can be read off of the diagram, at least on the basis on what we have said so far. For example, we do not at present see from the diagram the kernel of $\mathrm{ad}(\mathfrak{g}_{L_1-L_3})$ on $\mathfrak{h}$, though we will see later how to read this off as well. We do, however, have at least a pretty good idea of who is doing what to whom.

Pretty much the same picture applies to any representation V of $\mathfrak{sl}_3\mathbb{C}$: we start from the eigenspace decomposition $V = \bigoplus V_\alpha$ for the action of $\mathfrak{h}$ that we saw in (12.2). Next, the commutation relations for $\mathfrak{sl}_3\mathbb{C}$ tell us exactly how the remaining summands of the decomposition (12.3) of $\mathfrak{sl}_3\mathbb{C}$ act on the space V, and again we will see that each of the spaces $\mathfrak{g}_\alpha$ acts by carrying one eigenspace V_β into another. As usual, for any $X \in \mathfrak{g}_\alpha$ and $v \in V_\beta$ we can tell where X will send v if we know how an arbitrary element $H \in \mathfrak{h}$ will act on $X(v)$. This we can determine by making the

Fundamental Calculation (*third time*):

$$\begin{aligned} H(X(v)) &= X(H(v)) + [H, X](v) \\ &= X(\beta(H)\cdot v) + (\alpha(H)\cdot X)(v) \\ &= (\alpha(H) + \beta(H))\cdot X(v). \end{aligned}$$

We see from this that $X(v)$ *is again an eigenvector for the action of* $\mathfrak{h}$, *with eigenvalue* $\alpha + \beta$; in other words, the action of $\mathfrak{g}_\alpha$ carries V_β to $V_{\alpha+\beta}$. We can thus represent the eigenspaces V_α of V by dots in a plane diagram so that each $\mathfrak{g}_\alpha$ acts again "by translation," as we did for representations of $\mathfrak{sl}_2\mathbb{C}$ in the preceding lecture and the adjoint representation of $\mathfrak{sl}_3\mathbb{C}$ above. Just as in the case of $\mathfrak{sl}_2\mathbb{C}$ (page 148), we have

Observation 12.6. *The eigenvalues α occurring in an irreducible representation of $\mathfrak{sl}_3\mathbb{C}$ differ from one other by integral linear combinations of the vectors $L_i - L_j \in \mathfrak{h}^*$.*

Note that these vectors $L_i - L_j$ generate a lattice in $\mathfrak{h}^*$, which we will denote by Λ_R, and that all the α lie in some translate of this lattice.

At this point, we should begin to introduce some of the terminology that appears in this subject. The basic object here, the eigenvalue $\alpha \in \mathfrak{h}^*$ of the action of $\mathfrak{h}$ on a representation V of $\mathfrak{g}$, is called a *weight* of the representation; the corresponding eigenvectors in V_α are called, naturally enough, *weight vectors* and the spaces V_α themselves *weight spaces*. Clearly, the weights that occur in the adjoint representation are special; these are called the *roots* of the Lie algebra and the corresponding subspaces $\mathfrak{g}_\alpha \subset \mathfrak{g}$ *root spaces*; by convention, zero is not a root. The lattice $\Lambda_R \subset \mathfrak{h}^*$ generated by the roots α is called the *root lattice*.

To see what the next step should be, we go back to the analysis of representations of $\mathfrak{sl}_2\mathbb{C}$. There, at this stage we continued our analysis by going to an extremal eigenspace V_α and taking a vector $v \in V_\alpha$. The point was that since V_α was extremal, the operator X, which would carry V_α to $V_{\alpha+2}$, would have to kill v; so that v would be then both an eigenvector for H and in the kernel of X. We then saw that these two facts allowed us to completely describe the representation V in terms of images of v.

What would be the appropriately analogous setup in the case of $\mathfrak{sl}_3\mathbb{C}$? To start at the beginning, there is the question of what we mean by *extremal*: in the case of $\mathfrak{sl}_2\mathbb{C}$, since we knew that all the eigenvalues were scalars differing by integral multiples of 2, there was not much ambiguity about what we meant by this. In the present circumstance this does involve a priori a choice (though as we shall see the choice does not affect the outcome): we have to choose a direction, and look for the farthest α in that direction appearing in the decomposition (12.3). What this means is that we should choose a linear functional

$$l\colon \Lambda_R \to \mathbb{R},$$

extend it by linearity to a linear functional $l\colon \mathfrak{h}^* \to \mathbb{C}$, and then for any representation V we should go to the eigenspace V_α for which the real part of $l(\alpha)$ is maximal.[2] Of course, to avoid ambiguity we should choose l to be irrational with respect to the lattice Λ_R, that is, to have no kernel.

What is the point of this? The answer is that, just as in the case of a representation V of $\mathfrak{sl}_2\mathbb{C}$ we found in this way a vector $v \in V$ that was simultaneously in the kernel of the operator X and an eigenvector for H, in the present case what we will find is a vector $v \in V_\alpha$ that is an eigenvector for $\mathfrak{h}$, *and at the same time in the kernel of the action of* $\mathfrak{g}_\beta$ *for every* β *such that* $l(\beta) > 0$—that is, *that is killed by half the root spaces* $\mathfrak{g}_\beta$ (specifically, the root spaces corresponding to dots in the diagram (12.4) lying in a half plane). This will likewise give us a nearly complete description of the representation V.

To carry this out explicitly, choose our functional l to be given by

$$l(a_1L_1 + a_2L_2 + a_3L_3) = aa_1 + ba_2 + ca_3,$$

where $a + b + c = 0$ and $a > b > c$, so that the spaces $\mathfrak{g}_\alpha \subset \mathfrak{g}$ for which we have $l(\alpha) > 0$ are then exactly $\mathfrak{g}_{L_1-L_3}$, $\mathfrak{g}_{L_2-L_3}$, and $\mathfrak{g}_{L_1-L_2}$; they correspond to matrices with one nonzero entry above the diagonal.

[2] The real-versus-complex business is a red herring since (it will turn out very shortly) all the eigenvalues α actually occurring in any representation will in fact be in the real (in fact, the rational) linear span of Λ_R.

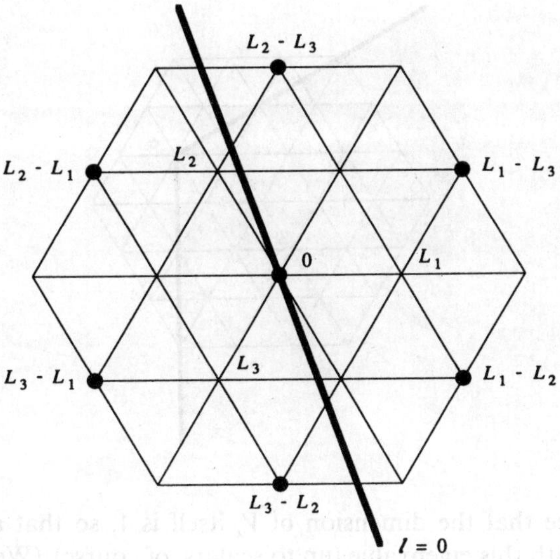

(12.7)

Thus, for $i < j$, the matrices $E_{i,j}$ *generate the positive root spaces*, and the $E_{j,i}$ *generate the negative root spaces.* We set

$$H_{i,j} = [E_{i,j}, E_{j,i}] = E_{i,i} - E_{j,j}. \tag{12.8}$$

Now let V be any irreducible, finite-dimensional representation of $\mathfrak{sl}_3\mathbb{C}$. The upshot of all the above is the

Lemma 12.9. *There is a vector $v \in V$ with the properties that*

(i) *v is an eigenvector for $\mathfrak{h}$, i.e. $v \in V_\alpha$ for some α; and*
(ii) *v is killed by $E_{1,2}$, $E_{1,3}$, and $E_{2,3}$.*

For any representation V of $\mathfrak{sl}_3\mathbb{C}$, a vector $v \in V$ with these properties is called a *highest weight vector.*

In the case of $\mathfrak{sl}_2\mathbb{C}$, having found an eigenvector v for H killed by X, we argued that the images of v under successive applications of Y generated the representation. The situation here is the same: analogous to Claim 11.4 we have

Claim 12.10. *Let V be an irreducible representation of $\mathfrak{sl}_3\mathbb{C}$, and $v \in V$ a highest weight vector. Then V is generated by the images of v under successive applications of the three operators $E_{2,1}$, $E_{3,1}$, and $E_{3,2}$.*

Before we check the claim, we note three immediate consequences. First, it says that all the eigenvalues $\beta \in \mathfrak{h}^*$ occurring in V lie in a sort of $\frac{1}{3}$-plane with corner at α:

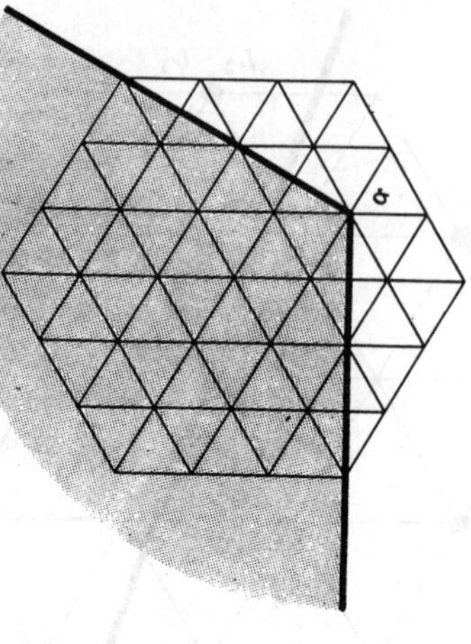

Second, we see that the dimension of V_α itself is 1, so that v is the unique eigenvector with this eigenvalue (up to scalars, of course). (We will see below that in fact v is the unique highest weight vector of V up to scalars; see Proposition 12.11.) Lastly, it says that the spaces $V_{\alpha+n(L_2-L_1)}$ and $V_{\alpha+n(L_3-L_2)}$ are all at most one dimensional, since they must be spanned by $(E_{2,1})^n(v)$ and $(E_{3,2})^n(v)$, respectively.

Proof of Claim 12.10. This is formally the same as the proof of the corresponding statement for $\mathfrak{sl}_2\mathbb{C}$: we argue that the subspace W of V spanned by images of v under the subalgebra of $\mathfrak{sl}_3\mathbb{C}$ generated by $E_{2,1}$, $E_{3,1}$, and $E_{3,2}$ is, in fact, preserved by all of $\mathfrak{sl}_3\mathbb{C}$ and hence must be all of V. To do this we just have to check that $E_{1,2}$, $E_{2,3}$, and $E_{1,3}$ carry W into itself (in fact it is enough to do this for the first two, the third being their commutator), and this is straightforward. To begin with, v itself is in the kernel of $E_{1,2}$, $E_{2,3}$, and $E_{1,3}$, so there is no problem there. Next we check that $E_{2,1}(v)$ is kept in W: we have

$$\begin{aligned} E_{1,2}(E_{2,1}(v)) &= (E_{2,1}(E_{1,2}(v)) + [E_{1,2}, E_{2,1}](v) \\ &= \alpha([E_{1,2}, E_{2,1}])\cdot v \end{aligned}$$

since $E_{1,2}(v) = 0$ and $[E_{1,2}, E_{2,1}] \in \mathfrak{h}$; and

$$\begin{aligned} E_{2,3}(E_{2,1}(v)) &= (E_{2,1}(E_{2,3}(v)) + [E_{2,3}, E_{2,1}](v) \\ &= 0 \end{aligned}$$

since $E_{2,3}(v) = 0$ and $[E_{2,3}, E_{2,1}] = 0$. A similar computation shows that $E_{3,2}(v)$ is also carried into V by $E_{1,2}$ and $E_{2,3}$.

More generally, we may argue the claim by a sort of induction: we let w_n denote any word of length n or less in the letters $E_{2,1}$ and $E_{3,2}$ and take W_n to be the vector space spanned by the vectors $w_n(v)$ for all such words; note that W is the union of the spaces W_n, since $E_{3,1}$ is the commutator of $E_{3,2}$ and $E_{2,1}$. We claim that $E_{1,2}$ and $E_{2,3}$ carry W_n into W_{n-1}. To see this, we can write w_n as either $E_{2,1} \circ w_{n-1}$ or $E_{3,2} \circ w_{n-1}$; in either case $w_{n-1}(v)$ will be an eigenvector for $\mathfrak{h}$ with eigenvalue β for some β. In the former case we have

$$\begin{aligned} E_{1,2}(w_n(v)) &= E_{1,2}(E_{2,1}(w_{n-1}(v))) \\ &= E_{2,1}(E_{1,2}(w_{n-1}(v))) + [E_{1,2}, E_{2,1}](w_{n-1}(v)) \\ &\in E_{2,1}(W_{n-2}) + \beta([E_{1,2}, E_{2,1}]) \cdot w_{n-1}(v) \\ &\subset W_{n-1} \end{aligned}$$

since $[E_{1,2}, E_{2,1}] \in \mathfrak{h}$; and

$$\begin{aligned} E_{2,3}(w_n(v)) &= E_{2,3}(E_{2,1}(w_{n-1}(v))) \\ &= E_{2,1}(E_{2,3}(w_{n-1}(v))) + [E_{2,3}, E_{2,1}](w_{n-1}(v)) \\ &\in E_{2,1}(W_{n-2}) \\ &\subset W_{n-1} \end{aligned}$$

since $[E_{2,3}, E_{2,1}] = 0$. Essentially the same calculation covers the latter case $w_n = E_{3,2} \circ w_{n-1}$, establishing the claim. □

This argument shows a little more; in fact, it proves

Proposition 12.11. *If V is any representation of $\mathfrak{sl}_3\mathbb{C}$ and $v \in V$ is a highest weight vector, then the subrepresentation W of V generated by the images of v by successive applications of the three operators $E_{2,1}$, $E_{3,1}$, and $E_{3,2}$ is irreducible.*

PROOF. Let α be the weight of v. The above shows that W is a subrepresentation, and it is clear that W_α is one dimensional. If W were not irreducible, we would have $W = W' \oplus W''$ for some representations W' and W''. But since projection to W' and W'' commute with the action of $\mathfrak{h}$, we have $W_\alpha = W'_\alpha \oplus W''_\alpha$. This shows that one of these spaces is zero, which implies that v belongs to W' or W'', and hence that W is W' or W''. □

As a corollary of this proposition we see that any irreducible representation of $\mathfrak{sl}_3\mathbb{C}$ has a unique highest weight vector, up to scalars; more generally, the set of highest weight vectors in V forms a union of linear subspaces Ψ_W corresponding to the irreducible subrepresentations W of V, with the dimension of Ψ_W equal to the number of times W appears in the direct sum decomposition of V into irreducibles.

What do we do next? Well, let us continue to look at the border vectors $(E_{2,1})^k(v)$. We call these border vectors because they live in (and, as we saw, span) a collection of eigenspaces $\mathfrak{g}_\alpha$, $\mathfrak{g}_{\alpha+L_2-L_1}$, $\mathfrak{g}_{\alpha+2(L_2-L_1)}$, ... that correspond to points on the boundary of the diagram above of possible eigenvalues of V. We also know that they span an uninterrupted string of nonzero eigenspaces $\mathfrak{g}_{\alpha+k(L_2-L_1)} \cong \mathbb{C}$, $k = 0, 1, \ldots$, until we get to the first m such that

$(E_{2,1})^m(v) = 0$; after that we have $\mathfrak{g}_{\alpha+k(L_2-L_1)} = (0)$ for all $k \geq m$. The picture is thus:

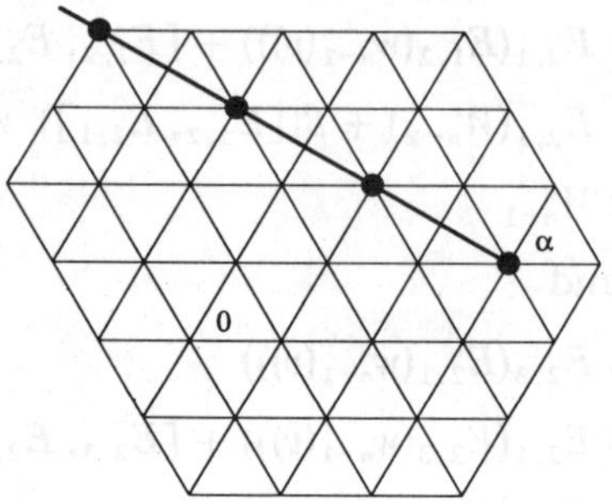

(12.12)

where we have no dots above/to the right of the bold line, and no dots on that line other than the ones marked.

The obvious question now is how long the string of dots along this line is. One way to answer this would be to make a calculation analogous to the one in the preceding lecture: use the computation made above to say explicitly for any k what multiple of $(E_{2,1})^{k-1}(v)$ the image of $(E_{2,1})^k(v)$ under the map $E_{1,2}$ is, and use the fact that $(E_{2,1})^m(v) = 0$ to determine m. It will be simpler—and more useful in general—if instead we just use what we have already learned about representations of $\mathfrak{sl}_2\mathbb{C}$. The point is, *the elements $E_{1,2}$ and $E_{2,1}$, together with their commutator $[E_{1,2}, E_{2,1}] = H_{1,2}$, span a subalgebra of $\mathfrak{sl}_3\mathbb{C}$ isomorphic to $\mathfrak{sl}_2\mathbb{C}$ via an isomorphism carrying $E_{1,2}$, $E_{2,1}$ and $H_{1,2}$ to the elements X, Y and H.* We will denote this subalgebra by $\mathfrak{s}_{L_1-L_2}$ (the notation may appear awkward, but this is a special case of a general construction). By the description we have already given of the action of $\mathfrak{sl}_3\mathbb{C}$ on the representation V in terms of the decomposition $V = \bigoplus V_\alpha$, we see that the subalgebra $\mathfrak{s}_{L_1-L_2}$ will shift eigenspaces V_α only in the direction of $L_2 - L_1$; in particular, the direct sum of the eigenspaces in question, namely the subspace

$$W = \bigoplus_k \mathfrak{g}_{\alpha+k(L_2-L_1)} \tag{12.13}$$

of V will be preserved by the action of $\mathfrak{s}_{L_1-L_2}$. In other words, W is a representation of $\mathfrak{s}_{L_1-L_2} \cong \mathfrak{sl}_2\mathbb{C}$ and we may deduce from this that *the eigenvalues of $H_{1,2}$ on W are integral, and symmetric with respect to zero.* Leaving aside the integrality for the moment, this says that the string of dots in diagram (12.12) must be symmetric with respect to the line $\langle H_{1,2}, L\rangle = 0$ in the plane $\mathfrak{h}^*$. Happily (though by no means coincidentally, as we shall see), this line is perpendicular to the line spanned by $L_1 - L_2$ in the picture we have drawn; so we can say simply that *the string of dots occurring in diagram* (12.12) *is preserved under reflection in the line $\langle H_{1,2}, L\rangle = 0$.*

In general, for any $i \neq j$ the elements $E_{i,j}$ and $E_{j,i}$, together with their commutator $[E_{i,j}, E_{j,i}] = H_{i,j}$, span a subalgebra $\mathfrak{s}_{L_i-L_j}$ of $\mathfrak{sl}_3\mathbb{C}$ isomorphic to $\mathfrak{sl}_2\mathbb{C}$ via an isomorphism carrying $E_{i,j}$, $E_{j,i}$, and $H_{i,j}$ to the elements X, Y, and H. (Note that $H_{i,j} = -H_{j,i}$.) Analyzing the action of the subalgebra $\mathfrak{s}_{L_2-L_3}$ in particular then shows that the string of dots corresponding to the eigenspaces $\mathfrak{g}_{\alpha+k}(L_3 - L_2)$ is likewise preserved under reflection in the line $\langle H_{2,3}, L\rangle = 0$ in $\mathfrak{h}^*$. The picture is thus

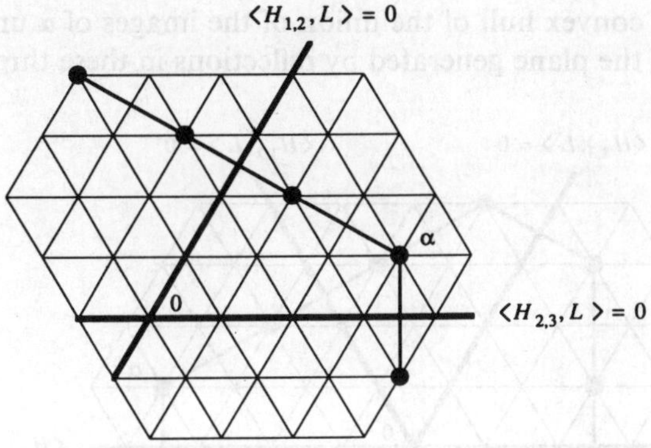

Let us now take a look at the last eigenspace in the first string, that is, V_β where m is as before the smallest integer such that $(E_{2,1})^m(v) = 0$ and $\beta = \alpha + (m-1)(L_2 - L_1)$. If $v' \in V_\beta$ is any vector, then, by definition, we have $E_{2,1}(v') = 0$; and since there are no eigenspaces V_γ corresponding to γ above the bold line in diagram (12.12), we have as well that $E_{2,3}(v') = E_{1,3}(v') = 0$. Thus, v', like v itself, satisfies the statement of Lemma 12.9, except for the exchange of the indices 2 and 1; or in other words, if we had chosen the linear functional l above differently—precisely, with coefficients $b > a > c$—then the vector whose existence is implied by Lemma 12.9 would have turned out to be v' rather than v. If, indeed, we had carried out the above analysis with respect to the vector v' instead of v, we would find that all eigenvalues of V occur below or to the right of the lines through β in the directions of $L_1 - L_2$ and $L_3 - L_1$, and that the strings of eigenvalues occurring on these two lines were symmetric about the lines $\langle H_{1,2}, L\rangle = 0$ and $\langle H_{1,3}, L\rangle = 0$, respectively. The picture now is

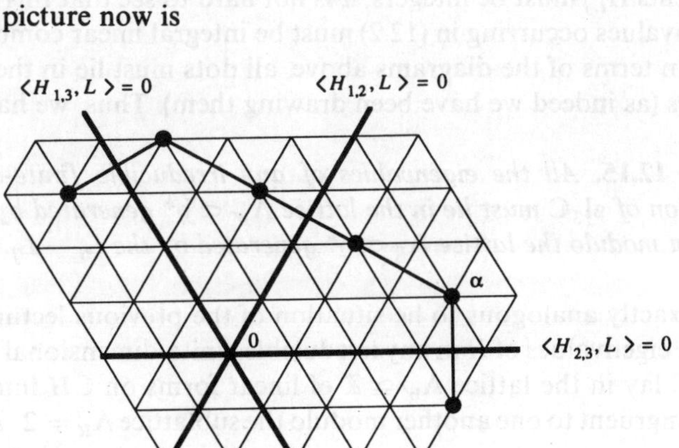

Needless to say, we can continue to play the same game all the way around: at the end of the string of eigenvalues $\{\beta + k(L_3 - L_1)\}$ we will arrive at a vector v'' that is an eigenvector for $\mathfrak{h}$ and killed by $E_{3,1}$ and $E_{2,1}$, and to which therefore the same analysis applies. In sum, then, we see that the set of eigenvalues in V will be bounded by a hexagon symmetric with respect to the lines $\langle H_{i,j}, L\rangle = 0$ and with one vertex at α; indeed, this characterizes the hexagon as the convex hull of the union of the images of α under the group of isometries of the plane generated by reflections in these three lines.

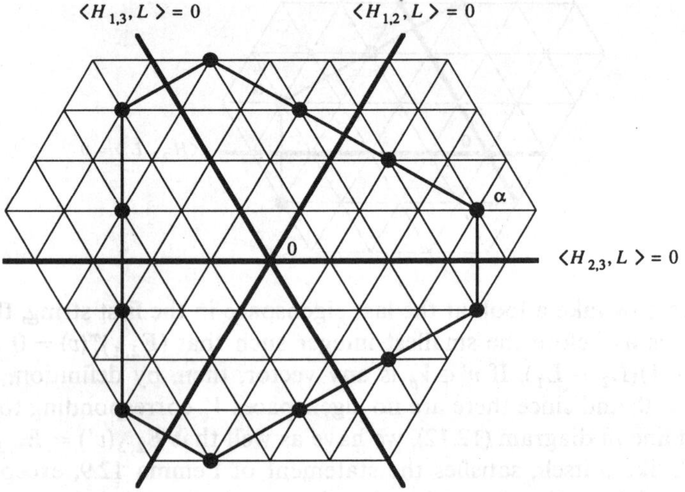

(12.14)

We will see in a moment that the set of eigenvalues will include all the points congruent to α modulo the lattice Λ_R generated by the $L_i - L_j$ lying on the boundary of this hexagon, and that each of these eigenvalues will occur with multiplicity one.

The use of the subalgebras $\mathfrak{s}_{L_i - L_j}$ does not stop here. For one thing, observe that as an immediate consequence of our analysis of $\mathfrak{sl}_2\mathbb{C}$, all the eigenvalues of the elements $H_{i,j}$ must be integers; it is not hard to see that this means that all the eigenvalues occurring in (12.2) must be integral linear combinations of the L_i, i.e., in terms of the diagrams above, all dots must lie in the lattice Λ_W of interstices (as indeed we have been drawing them). Thus, we have

Proposition 12.15. *All the eigenvalues of any irreducible finite-dimensional representation of $\mathfrak{sl}_3\mathbb{C}$ must lie in the lattice $\Lambda_W \subset \mathfrak{h}^*$ generated by the L_i and be congruent modulo the lattice $\Lambda_R \subset \mathfrak{h}^*$ generated by the $L_i - L_j$.*

This is exactly analogous to he situation of the previous lecture: there we saw that the eigenvalues of H in any irreducible, finite-dimensional representation of $\mathfrak{sl}_2\mathbb{C}$ lay in the lattice $\Lambda_W \cong \mathbb{Z}$ of linear forms on $\mathbb{C}H$ integral on H, and were congruent to one another modulo the sublattice $\Lambda_R = 2 \cdot \mathbb{Z}$ generated by the eigenvalues of H under the adjoint representation. Note that in the case of $\mathfrak{sl}_2\mathbb{C}$ we have $\Lambda_W/\Lambda_R \cong \mathbb{Z}/2$, while in the present case we have $\Lambda_W/\Lambda_R \cong \mathbb{Z}/3$; we will see later how this reflects a general pattern. The lattice Λ_W is called the *weight lattice*.

Exercise 12.16. Show that the two conditions that the eigenvalues of V are congruent to one another modulo Λ_R and are preserved under reflection in the three lines $\langle H_{i,j}, L\rangle = 0$ imply that they all lie in Λ_W, and that, in fact, this characterizes Λ_W.

To continue, we can go into the interior of the diagram (12.14) of eigenvalues of V by observing that the direct sums (12.13) are not the only visible subspaces of V preserved under the action of the subalgebras $\mathfrak{s}_{L_i-L_j}$; more generally, for any $\beta \in \mathfrak{h}^*$ appearing in the decomposition (12.2) and any i, j the direct sum

$$W = \bigoplus_k \mathfrak{g}_{\beta+k(L_i-L_j)}$$

will be a representation of $\mathfrak{s}_{L_i-L_j}$ (not necessarily irreducible, of course); in particular it follows that the values of k for which $V_{\beta+k(L_i-L_j)} \neq (0)$ form an unbroken string of integers. Observing that if β is any of the "extremal" eigenvalues pictured in diagram (12.14), then this string will include another; so that all eigenvalues congruent to the dots pictured in diagram (12.14) and lying in their convex hull must also occur. Thus, the complete diagram of eigenvalues will look like

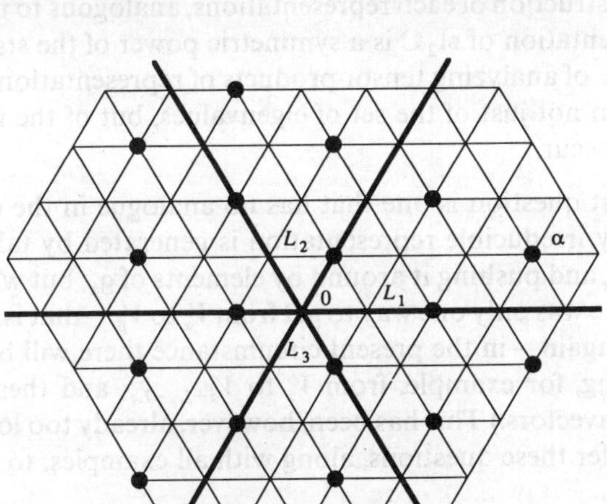

(12.17)

We can summarize this description in

Proposition 12.18. *Let V be any irreducible, finite-dimensional representation of $\mathfrak{sl}_3\mathbb{C}$. Then for some $\alpha \in \Lambda_W \subset \mathfrak{h}^*$, the set of eigenvalues occurring in V is*

exactly the set of linear functionals congruent to α *modulo the lattice* Λ_R *and lying in the hexagon with vertices the images of* α *under the group generated by reflections in the lines* $\langle H_{i,j}, L\rangle = 0$.

Remark. We did, in the analysis thus far, make one apparently arbitrary choice when we defined the notion of "extremal" eigenvalue by choosing a linear functional l on $\mathfrak{h}^*$. We remark here that, in fact, the choice was not as broad as might at first have appeared. Indeed, given the fact that the configuration of eigenvalues occurring in any irreducible finite-dimensional representation of $\mathfrak{sl}_3\mathbb{C}$ is always either a triangle or a hexagon, the "extremal" eigenvalue picked out by l will always turn out to be one of the three or six vertices of this figure; in other words, if we define the linear functional l to take $a_1L_1 + a_2L_2 + a_3L_3$ to $aa_1 + ba_2 + ca_3$, then only the ordering of the three real numbers a, b, and c matters. Indeed, in hindsight this choice was completely analogous to the choice we made (implicitly) in the case of $\mathfrak{sl}_2\mathbb{C}$ in choosing one of the two directions along the real line.

We said at the outset of this lecture that our goal was to arrive at a description of representations of $\mathfrak{sl}_3\mathbb{C}$ as complete as that for $\mathfrak{sl}_2\mathbb{C}$. We have now, certainly, as complete a description of the possible configurations of eigenvalues; but clearly much more is needed. Specifically, we should have

an existence and uniqueness theorem;

an explicit construction of each representations, analogous to the statement that every representation of $\mathfrak{sl}_2\mathbb{C}$ is a symmetric power of the standard; and

for the purpose of analyzing tensor products of representations of $\mathfrak{sl}_3\mathbb{C}$, we need a description not just of the set of eigenvalues, but of the multiplicities with which they occur.

(Note that the last question is one that has no analogue in the case of $\mathfrak{sl}_2\mathbb{C}$: in both cases, any irreducible representation is generated by taking a single eigenvector $v \in V_\alpha$ and pushing it around by elements of $\mathfrak{g}_\alpha$; but whereas in the previous case there was only one way to get from V_α to V_β—that is, by applying Y over and over again—in the present circumstance there will be more than one way of getting, for example, from V_α to $V_{\alpha+L_3-L_1}$; and these may yield independent eigenvectors.) This has been, however, already too long a lecture, and so we will defer these questions, along with all examples, to the next.

LECTURE 13

Representations of $\mathfrak{sl}_3\mathbb{C}$, Part II: Mainly Lots of Examples

In this lecture we complete the analysis of the irreducible representations of $\mathfrak{sl}_3\mathbb{C}$, culminating in §13.2 with the answers to all three of the questions raised at the end of the last lecture: we explicitly construct the unique irreducible representation with given highest weight, and in particular determine its multiplicities. The latter two sections correspond to §11.2 and §11.3 in the lecture on $\mathfrak{sl}_2\mathbb{C}$. In particular, §13.4, like §11.3, involves some projective algebraic geometry and may be skipped by those to whom this is unfamiliar.

§13.1. Examples

This lecture will be largely concerned with studying examples, giving constructions and analyzing tensor products of representations of $\mathfrak{sl}_3\mathbb{C}$. We start, however, by at least stating the basic existence and uniqueness theorem that provides the context for this analysis.

To state this, recall from the previous lecture than any irreducible, finite-dimensional representation of $\mathfrak{sl}_3\mathbb{C}$ has a vector, unique up to scalars, that is simultaneously an eigenvector for the subalgebra $\mathfrak{h}$ and killed by the three subspaces $\mathfrak{g}_{L_1-L_2}$, $\mathfrak{g}_{L_1-L_3}$, and $\mathfrak{g}_{L_2-L_3}$. We called such a vector a *highest weight vector* of the representation V; its associated eigenvalue will, of course, be called the *highest weight* of V. More generally, in any finite-dimensional representation W of $\mathfrak{sl}_3\mathbb{C}$, any vector $v \in W$ with these properties will be called a highest weight vector; we saw that it will generate an irreducible subrepresentation V of W. Finally, from the description given in the last lecture of the possible configurations of eigenvalues for a representation of $\mathfrak{sl}_3\mathbb{C}$, we see that any highest weight vector must lie in the $(\frac{1}{6})$-plane described by the inequalities $\langle H_{1,2}, L\rangle \geq 0$ and $\langle H_{2,3}, L\rangle \geq 0$, i.e., it must be of the form $(a+b)L_1 + bL_2 = aL_1 - bL_3$ for some pair of non-negative integers a and b. We can now state

Theorem 13.1. *For any pair of natural numbers a, b there exists a unique irreducible, finite-dimensional representation* $\Gamma_{a,b}$ *of* $\mathfrak{sl}_3\mathbb{C}$ *with highest weight* $aL_1 - bL_3$.

We will defer the proof of this theorem until the second section of this lecture, not so much because it is in any way difficult but simply because it is time to get to some examples. We will remark, however, that whereas in the case of $\mathfrak{sl}_2\mathbb{C}$ the analysis that led to the concept of highest weight vector immediately gave the uniqueness part of the analogous theorem, here to establish uniqueness we will be forced to resort to a more indirect trick. The proof of existence, by contrast, will be very much like that of the corresponding statement for $\mathfrak{sl}_2\mathbb{C}$: we will construct the representations $\Gamma_{a,b}$ out of the standard representation by multilinear algebra.

For the time being, though, we would like to apply the analysis of the previous lecture to some of the obvious representations of $\mathfrak{sl}_3\mathbb{C}$, partly to gain some familiarity with what goes on and partly in the hopes of seeing a general multilinear-algebraic construction.

We begin with the standard representation of $\mathfrak{sl}_3\mathbb{C}$ on $V \cong \mathbb{C}^3$. Of course, the eigenvectors for the action of $\mathfrak{h}$ are just the standard basis vectors e_1, e_2, and e_3; they have eigenvalues L_1, L_2, and L_3, respectively. The weight diagram for V is thus

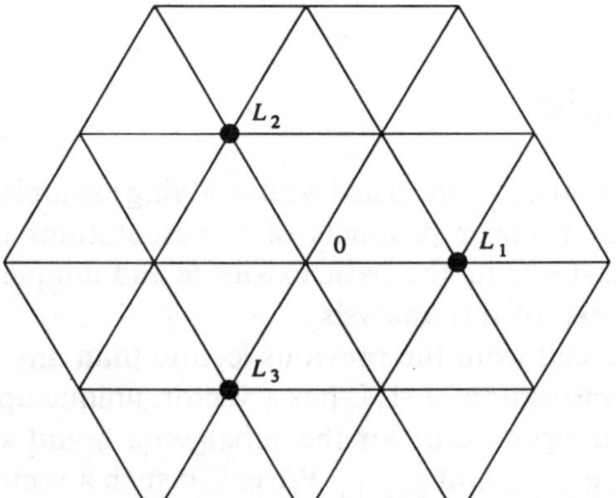

Next, consider the dual representation V^*. The eigenvalues of the dual of a representation of a Lie algebra are just the negatives of the eigenvalues of the original, so the diagram of V^* is

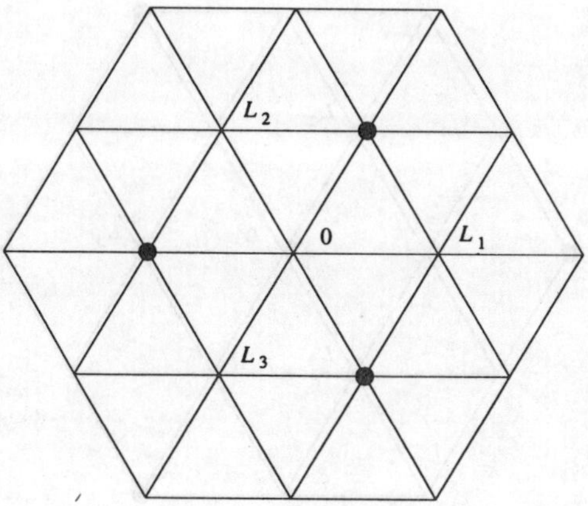

Alternatively, of course, we can just observe that the dual basis vectors e_i^* are eigenvectors with eigenvalues $-L_i$.

Note that while in the case of $\mathfrak{sl}_2\mathbb{C}$ the weights of any representation were symmetric about the origin, and correspondingly each representation was isomorphic to its dual, the same is not true here (that the diagrams for V and V^* look the same is a reflection of the fact that the two representations are carried into one another by an automorphism of $\mathfrak{sl}_3\mathbb{C}$, namely, the automorphism $X \mapsto -{}^tX$). Observe also that V^* is also isomorphic to the representation $\wedge^2 V$, whose weights are the pairwise sums of the distinct weights of V; and that likewise V is isomorphic as representation to $\wedge^2 V^*$.

Next, consider the degree 2 tensor products of V and V^*. Since the weights of the symmetric square of a representation are the pairwise sums of the weights of the original, the weight diagram of $\mathrm{Sym}^2 V$ will look like

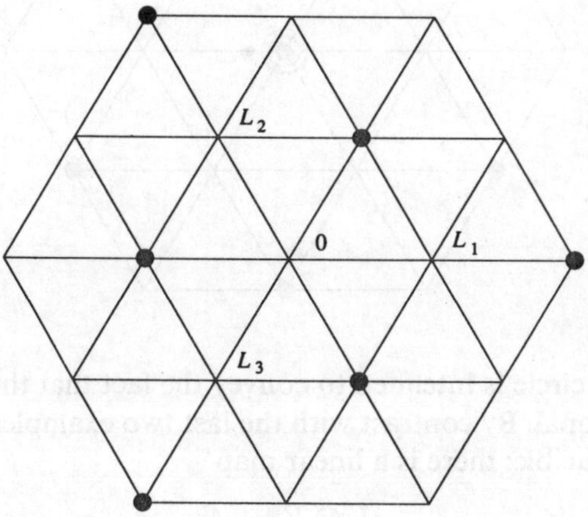

and likewise the symmetric square $\mathrm{Sym}^2 V^*$ has weights $\{-2L_i, -L_i - L_j\} = \{-2L_i - 2L_j, L_k\}$:

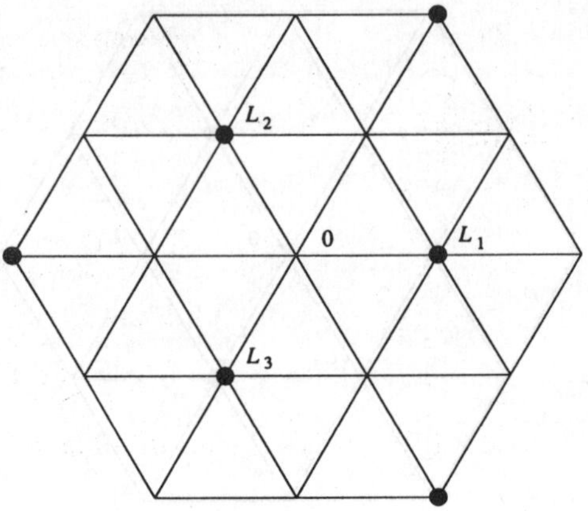

We see immediately from these diagrams that $\mathrm{Sym}^2 V$ and $\mathrm{Sym}^2 V^*$ are irreducible, since neither collection of weights is the union of two collections arising from representations of $\mathfrak{sl}_3\mathbb{C}$.

As for the tensor product $V \otimes V^*$, its weights are just the sums of the weights $\{L_i\}$ of V with those $\{-L_i\}$ of V^*, that is, the linear functionals $L_i - L_j$ (each occurring once, with weight vector $e_i \otimes e_j^*$) and 0 (occurring with multiplicity three, with weight vectors $e_i \otimes e_i^*$). We can represent these weights by the diagram

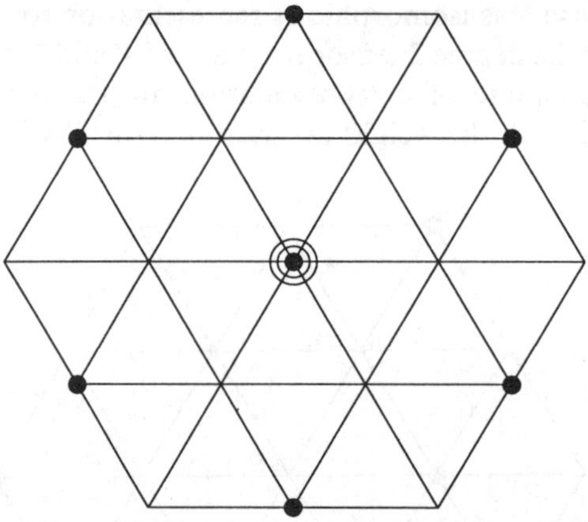

where the triple circle is intended to convey the fact that the weight space V_0 is three dimensional. By contrast with the last two examples, this representation is not irreducible: there is a linear map

$$V \otimes V^* \to \mathbb{C}$$

given simply by the contraction

$$v \otimes u^* \mapsto \langle v, u^* \rangle = u^*(v)$$

(or, in terms of the identification $V \otimes V^* \cong \mathrm{Hom}(V, V)$, by the *trace*) that is a map of $\mathfrak{sl}_3\mathbb{C}$-modules (with $\mathbb{C}$ the trivial representation, of course). The kernel of this map is then the subspace of $V \otimes V^*$ of traceless matrices, which is just the adjoint representation of the Lie algebra $\mathfrak{sl}_3\mathbb{C}$ and is irreducible (we can see this either from our explicit description of the adjoint representation—for example, $E_{1,3}$ is the unique weight vector for $\mathfrak{h}$ killed by $\mathfrak{g}_{L_1-L_2}$, $\mathfrak{g}_{L_1-L_3}$, and $\mathfrak{g}_{L_2-L_3}$—or, if we take as known the fact that $SL_3\mathbb{C}$ is simple, from the fact that a subrepresentation of the adjoint representation is an *ideal* in a Lie algebra, and exponentiates to a normal subgroup, cf. Exercise 8.43.)

(Physicists call this adjoint representation of $\mathfrak{sl}_3\mathbb{C}$ (or SU(3)) the "eightfold way," and relate its decomposition to mesons and baryons. The standard representation V is related to "quarks" and V^* to "antiquarks." See [S-W], [Mack].)

(We note that, in general, if V is any faithful representation of a Lie algebra, the adjoint representation will appear as a subrepresentation of the tensor $V \otimes V^*$.)

Let us continue now with some of the triple tensor products of V and V^*, which will be the last specific cases we look at. To begin with, we have the symmetric cubes $\mathrm{Sym}^3 V$ and $\mathrm{Sym}^3 V^*$, with weight diagrams

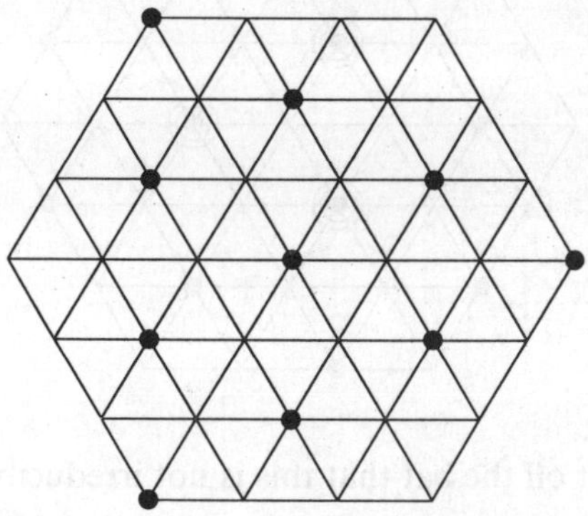

and

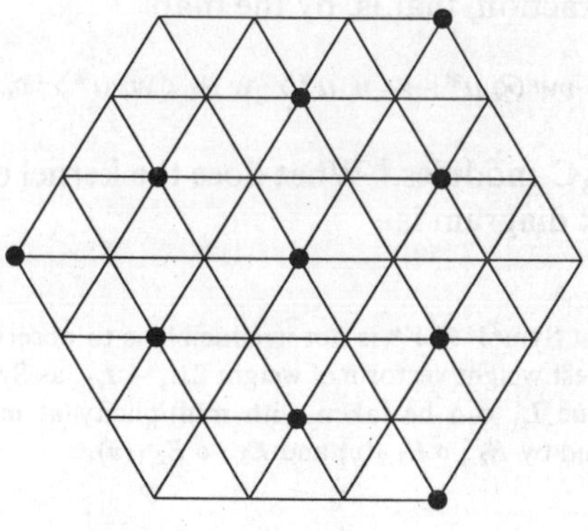

respectively. In general it is clear that, in terms of the description given in the preceding lecture of the possible weight diagrams of irreducible representations of $\mathfrak{sl}_3\mathbb{C}$, the symmetric powers of V and V^* will be exactly the representations with triangular, as opposed to hexagonal, diagrams.

It also follows from the above description and the fact that the weights of the symmetric powers $\mathrm{Sym}^n V$ occur with multiplicity 1 that $\mathrm{Sym}^n V$ and $\mathrm{Sym}^n V^*$ are all irreducible, i.e., we have, in the notation of Theorem 13.1,

$$\mathrm{Sym}^n V = \Gamma_{n,0} \quad \text{and} \quad \mathrm{Sym}^n V^* = \Gamma_{0,n}.$$

By way of notation, we will often write $\mathrm{Sym}^n V$ in place of $\Gamma_{n,0}$.

Consider now the mixed tensor $\mathrm{Sym}^2 V \otimes V^*$. Its weights are the sums of the weights of $\mathrm{Sym}^2 V$—that is, the pairwise sums of the L_i—with the weights of V^*; explicitly, these are $L_i + L_j - L_k$ and $2L_i - L_j$ (each occurring once) and the L_i themselves (each occurring three times, as $L_i + L_j - L_j$). Diagrammatically, the representation looks like

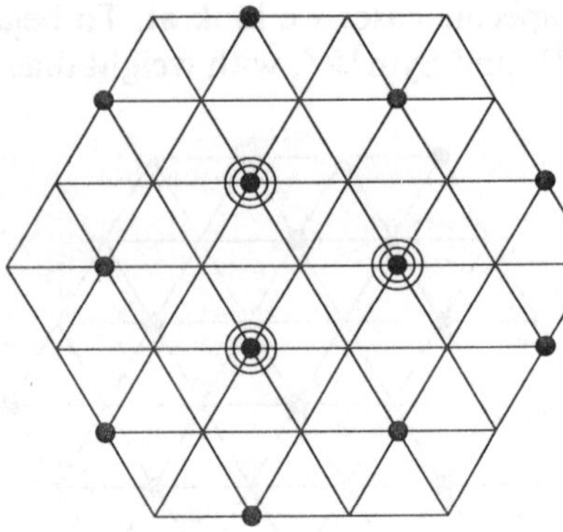

Now, we know right off the bat that this is not irreducible: we have a natural map

$$\iota\colon \mathrm{Sym}^2 V \otimes V^* \to V$$

given again by contraction, that is, by the map

$$vw \otimes u^* \mapsto \langle v, u^* \rangle \cdot w + \langle w, u^* \rangle \cdot v,$$

which is a map of $\mathfrak{sl}_3\mathbb{C}$-modules.[1] What does the kernel of this map look like? Of course, its weight diagram is

[1] Another way to see that $\mathrm{Sym}^2 V \otimes V^*$ is not irreducible is to observe that if a representation W is generated by a highest weight vector v of weight $2L_1 - L_3$, as $\mathrm{Sym}^2 V \otimes V^*$ must be if it is irreducible, the eigenvalue L_1 can be taken with multiplicity at most 2, the corresponding eigenspace being generated by $E_{2,1} \circ E_{3,2}(v)$ and $E_{3,2} \circ E_{2,1}(v)$.

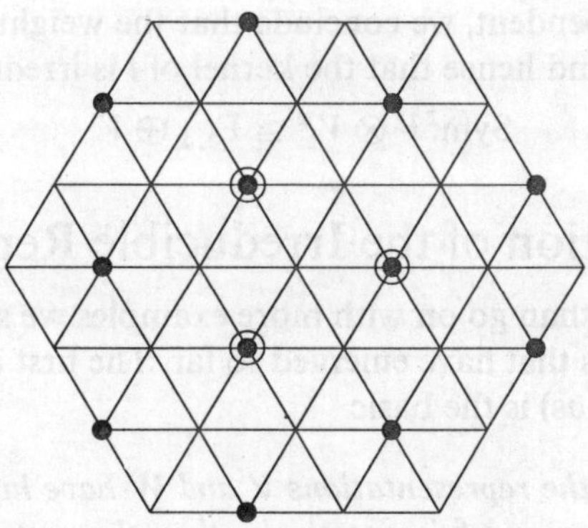

and we know one other thing: certainly any vector in the weight space of $2L_1 - L_3$—that is to say, of course, any multiple of the vector $e_1^2 \otimes e_3^*$—is killed by $\mathfrak{g}_{L_1-L_2}$, $\mathfrak{g}_{L_1-L_3}$, and $\mathfrak{g}_{L_2-L_3}$, so that the kernel of ι will contain an irreducible representation $\Gamma = \Gamma_{2,1}$ with $2L_1 - L_3$ as its highest weight. Since Γ must then assume every weight of $\mathrm{Ker}(\iota)$, there are exactly two possibilities: either $\mathrm{Ker}(\iota) = \Gamma$, which assumes the weights L_i with multiplicity 2; or all the weights of Γ occur with multiplicity one and $\mathrm{Ker}(\iota) \cong \Gamma \oplus V$.

How do we settle this issue? There are at least three ways. To begin with, we can try to analyze directly the structure of the kernel of ι. An alternative approach would be to determine a priori with what multiplicities the weights of $\Gamma_{a,b}$ are taken. Certainly it is clear that a formula giving us the latter information will be tremendously valuable—it would for one thing clear up the present confusion instantly—and indeed there exist several such, one of which, the *Weyl character formula*, we will prove later in the book. (We will also prove the *Kostant multiplicity formula*, which can be applied to deduce directly the independence statement we arrive at below.) As a third possibility, we can identify the representations $\Gamma_{a,b}$ as Weyl modules and appeal to Lecture 6. Rather than invoke such general formulas at present, however, we will take the first approach here. This is straightforward: in terms of the notation we have been using, the highest weight vector for the representation $\Gamma \subset \mathrm{Sym}^2 V \otimes V^*$ is the vector $e_1^2 \otimes e_3^*$, and so the eigenspace $\Gamma_{L_1} \subset \Gamma$ with eigenvalue L_1 will be spanned by the images of this vector under the two compositions $E_{2,1} \circ E_{3,2}$ and $E_{3,2} \circ E_{2,1}$. These are, respectively,

$$\begin{aligned} E_{2,1} \circ E_{3,2}(e_1^2 \otimes e_3^*) &= E_{2,1}(E_{3,2}(e_1^2) \otimes e_3^* + e_1^2 \otimes E_{3,2}(e_3^*)) \\ &= E_{2,1}(-e_1^2 \otimes e_2^*) \\ &= -2(e_1 \cdot e_2) \otimes e_2^* + e_1^2 \otimes e_1^* \end{aligned}$$

and

$$\begin{aligned} E_{3,2} \circ E_{2,1}(e_1^2 \otimes e_3^*) &= E_{3,2}(E_{2,1}(e_1^2) \otimes e_3^* + e_1^2 \otimes E_{2,1}(e_3^*)) \\ &= E_{3,2}((2e_1 \cdot e_2) \otimes e_3^*) \\ &= 2(e_1 \cdot e_3) \otimes e_3^* - 2(e_1 \cdot e_2) \otimes e_2^*. \end{aligned}$$

Since these are independent, we conclude that the weight L_1 does occur in Γ with multiplicity 2, and hence that the kernel of $\imath$ is irreducible, i.e.,

$$\mathrm{Sym}^2 V \otimes V^* \cong \Gamma_{2,1} \oplus V.$$

§13.2. Description of the Irreducible Representations

At this point, rather than go on with more examples we should state some of the general principles that have emerged so far. The first and most important (though pretty obvious) is the basic

Observation 13.2. *If the representations V and W have highest weight vectors v and w with weights α and β, respectively, then the vector $v \otimes w \in V \otimes W$ is a highest weight vector of weight $\alpha + \beta$.*

Of course, there are numerous generalizations of this: the vector $v^n \in \mathrm{Sym}^n V$ is a highest weight vector of weight $n\alpha$, etc.[2] Just the basic statement above, however, enables us to give the

PROOF OF THEOREM 13.1. First, the existence statement follows immediately from the observation: the representation $\mathrm{Sym}^a V \otimes \mathrm{Sym}^b V^*$ will contain an irreducible subrepresentation $\Gamma_{a,b}$ with highest weight $aL_1 - bL_3$.

The uniqueness part is only slightly harder (if less explicit): Given irreducible representations V and W with highest weight α, let $v \in V$ and $w \in W$ be highest weight vectors with weight α. Then (v, w) is again a highest weight vector in the representation $V \oplus W$ with highest weight α; let $U \subset V \oplus W$ be the irreducible subrepresentation generated by (v, w). The projection maps $\pi_1: U \to V$ and $\pi_2: U \to W$, being nonzero maps between irreducible representations of $\mathfrak{sl}_3\mathbb{C}$, must be isomorphisms, and we deduce that $V \cong W$. □

Exercise 13.3*. Let $\mathbb{S}_\lambda$ be the Schur functor introduced in Lecture 6. What can you say about the highest weight vectors in the representation $\mathbb{S}_\lambda(V)$ obtained by applying it to a given representation V?

To continue our discussion of tensor products like $\mathrm{Sym}^a V \otimes \mathrm{Sym}^b V^*$ in general, as we indicated we would like to make more explicit the construction of the representation $\Gamma_{a,b}$, which we know to be lying in $\mathrm{Sym}^a V \otimes \mathrm{Sym}^b V^*$. To begin with, we have in general a contraction map

$$\imath_{a,b}: \mathrm{Sym}^a V \otimes \mathrm{Sym}^b V^* \to \mathrm{Sym}^{a-1} V \otimes \mathrm{Sym}^{b-1} V^*$$

analogous to the map $\imath$ introduced above; we can describe this map either (in fancy language) as the dual of the map from $\mathrm{Sym}^{a-1} V \otimes \mathrm{Sym}^{b-1} V^*$ to $\mathrm{Sym}^a V \otimes \mathrm{Sym}^b V^*$ given by multiplication by the identity element in

[2] One slightly less obvious statement is this: if the weights of V are $\alpha_1, \alpha_2, \alpha_3 \ldots$ with $l(\alpha_1) > l(\alpha_2) > \ldots$, then $\bigwedge^n V$ possesses a highest weight vector weight $\alpha_1 + \cdots + \alpha_n$. Note that since the ordering of the α_i may in fact depend on the choice of l (even with the restriction $a > b > c$ on the coefficients of l as above), this may in some cases imply the existence of several subrepresentations of $\bigwedge^n V$.

$V \otimes V^* = \mathrm{Hom}(V, V)$; or, concretely, by sending

$$(v_1 \cdot \ldots \cdot v_a) \otimes (v_1^* \cdot \ldots \cdot v_b^*)$$
$$\mapsto \sum \langle v_i, v_j^* \rangle (v_1 \cdot \ldots \cdot \hat{v}_i \cdot \ldots \cdot v_a) \otimes (v_1^* \cdot \ldots \cdot \hat{v}_j^* \cdot \ldots \cdot v_b^*).$$

Clearly this map is surjective, and, since the target does not have eigenvalue $aL_1 - bL_3$, the subrepresentation $\Gamma_{a,b} \subset \mathrm{Sym}^a V \otimes \mathrm{Sym}^b V^*$ must lie in the kernel. In fact, we have, just as in the case of $\mathrm{Sym}^2 V \otimes V^*$ above,

Claim 13.4. *The kernel of the map $\iota_{a,b}$ is the irreducible representation $\Gamma_{a,b}$.*

We will defer the proof of this for a moment and consider some of its consequences. To begin with, we can deduce from this assertion the complete decomposition of $\mathrm{Sym}^a V \otimes \mathrm{Sym}^b V^*$: we must have (if, say, $b \leq a$)

$$\mathrm{Sym}^a V \otimes \mathrm{Sym}^b V^* = \bigoplus_{i=0}^{b} \Gamma_{a-i,b-i}. \tag{13.5}$$

Since we know, a priori, all the multiplicities of the eigenvalues of the tensor product $\mathrm{Sym}^a V \otimes \mathrm{Sym}^b V^*$, this will, in turn, determine (inductively at least) all the multiplicities of the representations $\Gamma_{a,b}$. In fact, the answer turns out to be very nice. To express it, observe first that if $a \geq b$, the weight diagram of either $\Gamma_{a,b}$ or $\mathrm{Sym}^a V \otimes \mathrm{Sym}^b V^*$ looks like a sequence of b shrinking concentric (not in general regular) hexagons H_i with vertices at the points $(a-i)L_1 - (b-i)L_3$ for $i = 0, 1, \ldots, b-1$, followed (after the shorter three sides of the hexagon have shrunk to points) by a sequence of $[(a-b)/3] + 1$ triangles T_j with vertices at the points $(a-b-3j)L_1$ for $j = 0, 1, \ldots, [(a-b)/3]$ (it will be convenient notationally to refer to T_0 as H_b occasionally). Diagram (13.6) shows the picture of the weights of $\mathrm{Sym}^6 V \otimes \mathrm{Sym}^2 V^*$:

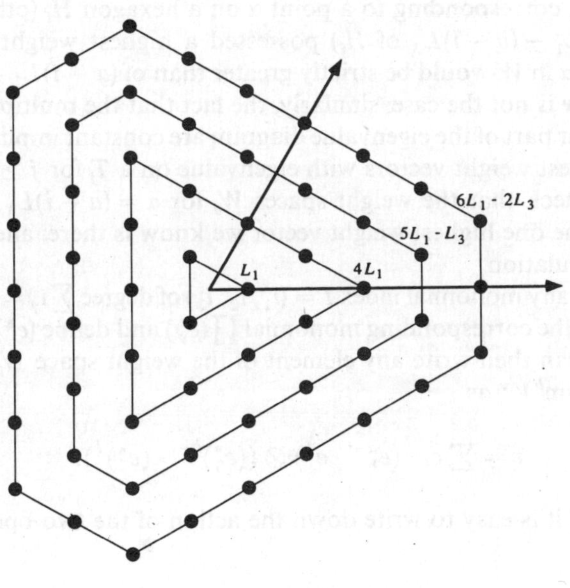

(13.6)

(Note that by the decomposition (13.5), the weights of the highest weight vectors in $\mathrm{Sym}^a V \otimes \mathrm{Sym}^b V^*$ will be $aL_1 - bL_3$, $(a-1)L_1 - (b-1)L_3, \ldots, (a-b)L_1$, as shown in the diagram.)

An examination of the representation $\mathrm{Sym}^a V \otimes \mathrm{Sym}^b V^*$ shows that it has multiplicity $(i+1)(i+2)/2$ on the hexagon H_i, and then a constant multiplicity $(b+1)(b+2)/2$ on all the triangles T_j; and it follows from the decomposition (13.5), in general, that the representation $\Gamma_{a,b}$ has multiplicity $(i+1)$ on H_i and $b+1$ on T_j. In English, *the multiplicities of $\Gamma_{a,b}$ increase by one on each of the concentric hexagons of the eigenvalue diagram and are constant on the triangles*. Note in particular that the description of $\Gamma_{2,1}$ in the preceding section is a special case of this.

PROOF OF CLAIM 13.4. We remark first that the claim will be implied by the Weyl character formula or by the description via Weyl's construction in Lecture 15; so the reader who wishes to can skip the following without dire consequences to the logical structure of the book. Otherwise, observe first that the claim is equivalent to asserting the decomposition (13.5); this, in turn, is equivalent to the statement that the representation $W = \mathrm{Sym}^a V \otimes \mathrm{Sym}^b V^*$ has exactly $b+1$ irreducible components (still assuming $a \geq b$). The irreducible factors in a representation correspond to the highest weight vectors in the representation up to scalars; so in sum the claim is equivalent to the assertion that *the eigenspace W_α of* $\mathrm{Sym}^a V \otimes \mathrm{Sym}^b V^*$ *contains a unique highest weight vector (up to scalars) if α is of the form $(a-i)L_1 - (b-i)L_3$ for $i \leq b$, and none otherwise*; this is what we shall prove.

To begin with, the "none otherwise" part of the statement follows (given the other) just from looking at the diagram: if, for example, any of the eigenspaces W_α corresponding to a point α on a hexagon H_i (other than the vertex $(a-i)L_1 - (b-i)L_3$ of H_i) possessed a highest weight vector, the multiplicity of α in W would be strictly greater than of $(a-i)L_1 - (b-i)L_3$, which we know is not the case; similarly, the fact that the multiplicities of W in the triangular part of the eigenvalue diagram are constant implies that there can be no highest weight vectors with eigenvalue on a T_j for $j \geq 1$. Thus, we just have to check that the weight spaces W_α for $\alpha = (a-i)L_1 - (b-i)L_3$ contain only the one highest weight vector we know is there; and we do this by explicit calculation.

To start, for any monomial index $I = (i_1, i_2, i_3)$ of degree $\sum i_\gamma = i$, we denote by $e^I \in \mathrm{Sym}^i V$ the corresponding monomial $\prod(e_\gamma^{i_\gamma})$ and define $(e^*)^I \in \mathrm{Sym}^i V^*$ similarly. We can then write any element of the weight space $W_{(a-i)L_1-(b-i)L_3}$ of $\mathrm{Sym}^a V \otimes \mathrm{Sym}^b V^*$ as

$$v = \sum c_I \cdot (e_1^{a-i} \cdot e^I) \otimes ((e_3^*)^{b-i} \cdot (e^*)^I).$$

In these terms, it is easy to write down the action of the two operators $E_{1,2}$

and $E_{2,3}$. First, $E_{1,2}$ kills both $e_1 \in V$ and $e_3^* \in V^*$, so that we have

$$\begin{aligned} E_{1,2}((e_1^{a-i} \cdot e^I) \otimes ((e_3^*)^{b-i} \cdot (e^*)^I)) \\ = i_2(e_1^{a-i} \cdot e^{I'}) \otimes ((e_3^*)^{b-i} \cdot (e^*)^I) \\ - i_1(e_1^{a-i} \cdot e^I) \otimes ((e_3^*)^{b-i} \cdot (e^*)^{I''}), \end{aligned}$$

where $I' = (i_1 + 1, i_2 - 1, i_3)$ and $I'' = (i_1 - 1, i_2 + 1, i_3)$ (and we adopt the convention that $e^I = 0$ if $i_\gamma < 0$ for any γ). It follows that *the vector v above is in the kernel of $E_{1,2}$ if and only if the coefficients c_I satisfy $i_2 c_I = (i_1 + 1)c_{I'}$*; and by the analogous calculation that v is in the kernel of $E_{2,3}$ if and only if $i_3 c_I = (i_2 + 1)c_J$ whenever the indices I and J are related by $j_1 = i_1$, $j_2 = i_2 + 1$, and $j_3 = i_3 - 1$. These conditions are equivalent to saying that the numbers $i_1!i_2!i_3!c_I$ are independent of I. We see, in other words, that v is a highest weight vector if and only if all the coefficients c_I are equal to $c/i_1!i_2!i_3!$ for some constant c. □

§13.3. A Little More Plethysm

We would like to consider here, as we did in the case of $\mathfrak{sl}_2\mathbb{C}$ in Lecture 11, how the tensor products and powers of the representations we have described decompose. We start with one general remark: given our knowledge of the eigenvalue diagrams of the irreducible representations of $\mathfrak{sl}_3\mathbb{C}$ (*with* multiplicities), there can be no possible ambiguity about the decomposition of any representation U given as the tensor product of representations whose eigenvalue diagrams are known. Indeed, we have an algorithm for determining the components of that decomposition, as follows:

1. Write down the eigenvalue decomposition of U.
2. Find the eigenvalue $\alpha = aL_1 - bL_3$ appearing in this diagram for which the value of $l(\alpha)$ is maximal.
3. We now know that U will contain a copy of the irreducible representation $\Gamma_\alpha = \Gamma_{a,b}$, i.e., $U \cong \Gamma_\alpha \oplus U'$ for some U'. Since we also know the eigenvalue diagram of Γ_α, we can thus write down the eigenvalue diagram of U' as well.
4. Repeat this process for U'.

To see how this goes in practice, consider some examples of tensor products of the basic irreducible representations described so far. We have already seen how the tensor products of the symmetric powers of the standard representation V of $\mathfrak{sl}_3\mathbb{C}$ and symmetric powers of its dual decompose; let us look now at an example of a more general tensor product of irreducible representations: say V itself and the representation $\Gamma_{2,1}$. We start by writing down the weights of the tensor product: since $\Gamma_{2,1}$ has weights $2L_i - L_j$, $L_i + L_j - L_k$, and L_i

(taken twice) and V has weights L_i, the tensor product will have weights $3L_i - L_j$, $2L_i + L_j - L_k$ (taken twice), $2L_i$ (taken four times), and $L_i + L_j$ (taken five times). The diagram is thus

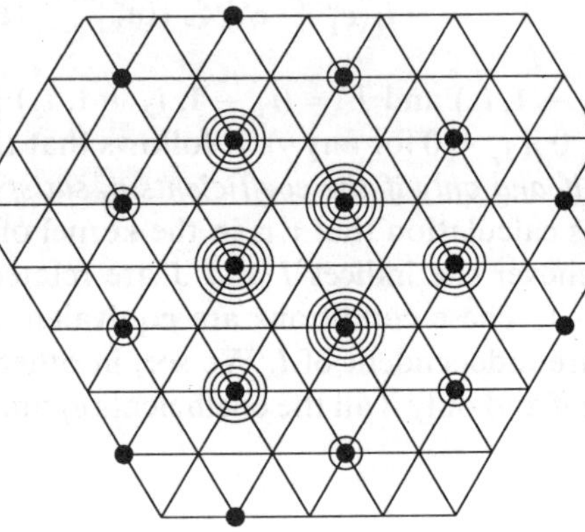

(One thing we may deduce from this diagram is that we are soon going to need a better system for presenting the data of the weights of a representation. In the future, we may simply draw one sector of the plane, and label weights with numbers to indicate multiplicities.)

We know right off the bat that the tensor product $V \otimes \Gamma_{2,1}$ contains a copy of the irreducible representation $\Gamma_{3,1}$ with highest weight $3L_1 - L_3$. By what we have said, the weight diagram of $\Gamma_{3,1}$ is

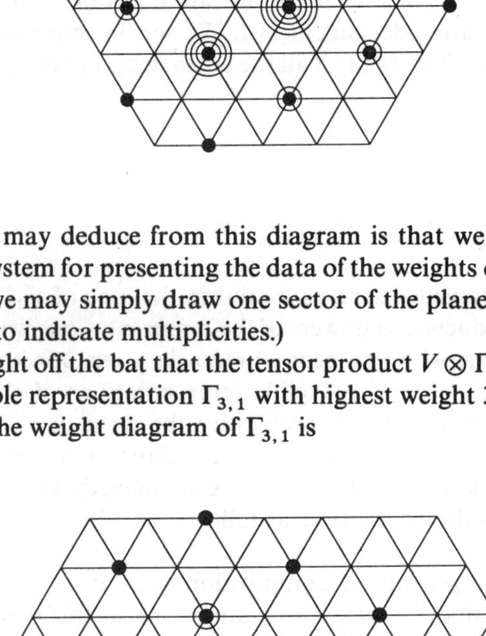

so the complement of $\Gamma_{3,1}$ in the tensor product $V \otimes \Gamma_{2,1}$ will look like

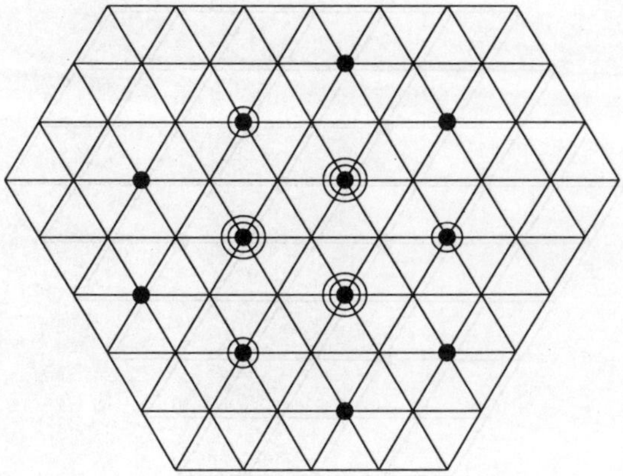

One obvious highest weight in this representation is $2L_1 + L_2 - L_3 = L_1 - 2L_3$, so that the tensor product will contain a copy of the irreducible representation $\Gamma_{1,2}$ as well; since this has weight diagram

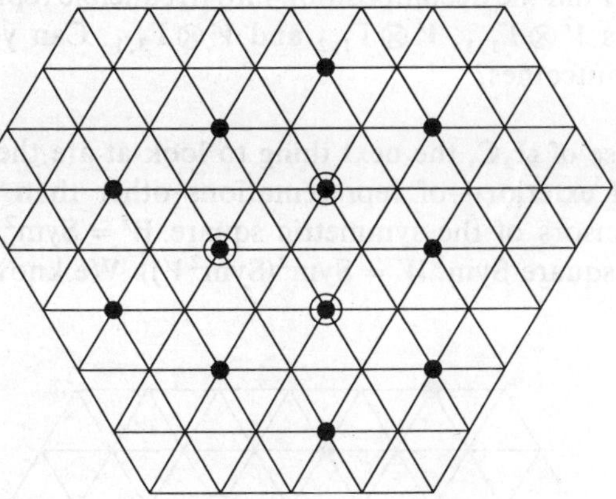

the remaining part of the tensor product will have weight diagram

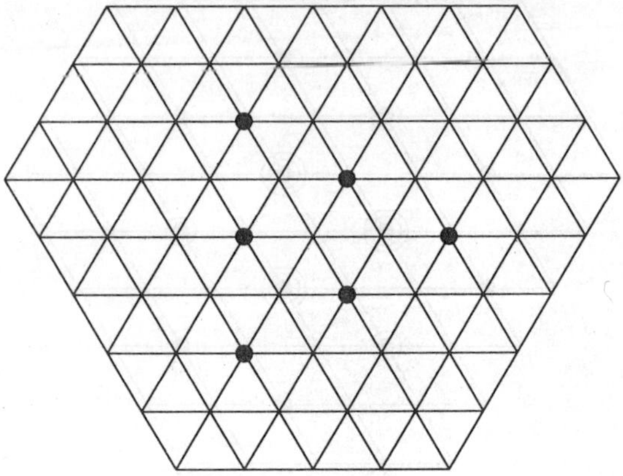

which we recognize as the weight diagram of the symmetric square $\mathrm{Sym}^2 V = \Gamma_{2,0}$ of the standard representation. We have, thus,

$$V \otimes \Gamma_{2,1} = \Gamma_{3,1} \oplus \Gamma_{1,2} \oplus \Gamma_{2,0}. \tag{13.7}$$

Exercise 13.8*. Find the decomposition into irreducible representations of the tensor products $V \otimes \Gamma_{1,1}$, $V \otimes \Gamma_{1,2}$ and $V \otimes \Gamma_{3,1}$. Can you find a general pattern to the outcomes?

As in the case of $\mathfrak{sl}_2\mathbb{C}$, the next thing to look at are the tensor powers—symmetric and exterior—of representations other than the standard; we look first at tensors of the symmetric square $W = \mathrm{Sym}^2 V$. First, consider the symmetric square $\mathrm{Sym}^2 W = \mathrm{Sym}^2(\mathrm{Sym}^2 V))$. We know the diagram for $\mathrm{Sym}^2 W$; it is

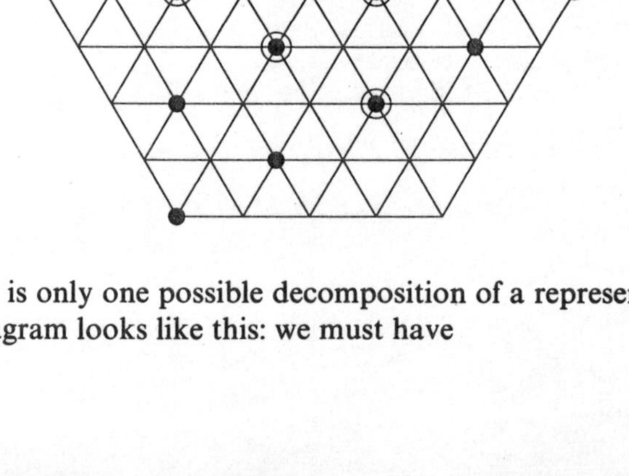

Now, there is only one possible decomposition of a representation whose eigenvalue diagram looks like this: we must have

$$\mathrm{Sym}^2(\mathrm{Sym}^2 V)) \cong \mathrm{Sym}^4 V \oplus \mathrm{Sym}^2 V^*.$$

Indeed, the presence of the $\mathrm{Sym}^4 V$ factor is clear: there is an obvious map

$$\varphi\colon \mathrm{Sym}^2(\mathrm{Sym}^2 V)) \to \mathrm{Sym}^4 V$$

obtained simply by multiplying out. The identification of the kernel of this map with the representation $\mathrm{Sym}^2 V^*$ is certainly less obvious, but can still be made explicit. We can identify V^* with $\bigwedge^2 V$ as we saw, and then define a map

$$\tau\colon \mathrm{Sym}^2(\bigwedge^2 V) \to \mathrm{Sym}^2(\mathrm{Sym}^2 V))$$

by sending the generator $(u \wedge v)\cdot(w \wedge z) \in \mathrm{Sym}^2(\bigwedge^2 V)$ to the element $(u\cdot w)\cdot(v\cdot z) - (u\cdot z)\cdot(v\cdot w) \in \mathrm{Sym}^2(\mathrm{Sym}^2 V)$, which is clearly in the kernel of φ.

Exercise 13.9. Verify that this map is well defined and that it extends linearly to an isomorphism of $\mathrm{Sym}^2(\bigwedge^2 V)$ with $\mathrm{Ker}(\varphi)$.

Exercise 13.10. Apply the techniques above to show that the representation $\bigwedge^2(\mathrm{Sym}^2 V)$ is isomorphic to $\Gamma_{2,1}$.

Exercise 13.11. Apply the same techniques to determine the irreducible factors of the representation $\bigwedge^3(\mathrm{Sym}^2 V)$. Note: we will return to this example in Exercise 13.22.

Exercise 13.12. Find the decomposition into irreducibles of the representations $\mathrm{Sym}^2(\mathrm{Sym}^3 V)$ and $\mathrm{Sym}^3(\mathrm{Sym}^2 V)$ (observe in particular that Hermite reciprocity has bitten the dust). Describe the projection maps to the various factors. Note: we will describe these examples further in the following section.

§13.4. A Little More Geometric Plethysm

Just as in the case of $\mathfrak{sl}_2\mathbb{C}$, some of these identifications can also be seen in geometric terms. To do this, recall from §11.3 the definition of the *Veronese embedding*: if $\mathbb{P}^2 = \mathbb{P}V^*$ is the projective space of one-dimensional subspaces of V^*, there is then a natural embedding of $\mathbb{P}^2$ in the projective space $\mathbb{P}^5 = \mathbb{P}(\mathrm{Sym}^2 V^*)$, obtained simply by sending the point $[v^*] \in \mathbb{P}^2$ corresponding to the vector $v^* \in V^*$ to the point $[v^{*2}] \in \mathbb{P}(\mathrm{Sym}^2 V^*)$ associated to the vector $v^{*2} = v^*\cdot v^* \in \mathrm{Sym}^2 V^*$. The image $S \subset \mathbb{P}^5$ is called the *Veronese surface*. As in the case of the rational normal curves discussed in Lecture 11, it is not hard to see that the group of automorphisms of $\mathbb{P}^5$ carrying S into itself is exactly the group $\mathrm{PGL}_3\mathbb{C}$ of automorphisms of $S = \mathbb{P}^2$.

Now, a quadratic polynomial in the homogeneous coordinates of the space $\mathbb{P}(\mathrm{Sym}^2 V^*) \cong \mathbb{P}^5$ will restrict to a quartic polynomial on the Veronese surface $S = \mathbb{P}V^*$, which corresponds to the natural evaluation map φ of the preceding section; the kernel of this map is thus the vector space of quadratic polynomials in $\mathbb{P}^5$ vanishing on the Veronese surface S, on which the group of automorphisms of $\mathbb{P}^5$ carrying S to itself obviously acts. Now, for any pair of points $P = [u^*]$, $Q = [v^*] \in S$, it is not hard to see that the cone over the Veronese surface with vertex the line $\overline{PQ} \subset \mathbb{P}^5$ (that is, the union of the 2-planes $\overline{PQR}$ as R varies over the surface S) will be a quadric hypersurface in $\mathbb{P}^5$ containing the Veronese surface; sending the generator $u^* \cdot v^* \in \mathrm{Sym}^2 V^*$ to this quadric hypersurface will then define an isomorphism of the space of such quadrics with the projective space associated to $\mathrm{Sym}^2 V^*$.

Exercise 13.13. Verify the statements made in the last paragraph: that the union of the $\overline{PQR}$ is a quadric hypersurface and that this extends to a linear isomorphism $\mathbb{P}(\mathrm{Sym}^2 V^*) \cong \mathbb{P}(\mathrm{Ker}(\varphi))$. Verify also that this isomorphism coincides with the one given in Exercise 13.9.

There is another way of representing the Veronese surface that will shed some light on this kernel. If, in terms of some coordinates e_i on V^*, we think of $\mathrm{Sym}^2 V^*$ as the vector space of symmetric 3×3 matrices, then the Veronese surface is just the locus, in the associated projective space, of rank 1 matrices up to scalars, i.e., in terms of homogeneous coordinates $Z_{i,j} = e_i \cdot e_j$ on $\mathbb{P}^5$,

$$S = \left\{ [Z]: \mathrm{rank}\begin{pmatrix} Z_{1,1} & Z_{1,2} & Z_{1,3} \\ Z_{1,2} & Z_{2,2} & Z_{2,3} \\ Z_{1,3} & Z_{2,3} & Z_{3,3} \end{pmatrix} = 1 \right\}.$$

The vector space of quadratic polynomials vanishing on S is then generated by the 2×2 minors of the matrix $(Z_{i,j})$; in particular, for any pair of linear combinations of the rows and pair of linear combinations of the columns we get a 2×2 matrix whose determinant vanishes on S.

Exercise 13.14. Show that this is exactly the isomorphism $\mathrm{Sym}^2(\wedge^2 V) \cong \mathrm{Ker}(\varphi)$ described above.

We note in passing that if indeed the space of quadrics containing the Veronese surface, with the action of the group $\mathrm{PGL}_3\mathbb{C}$ of motions of $\mathbb{P}^5$ preserving S, is the projectivization of the representation $\mathrm{Sym}^2 V^*$, then it must contain its own Veronese surface, i.e., there must be a surface $T = \mathbb{P}(V^*) \subset \mathbb{P}(\mathrm{Ker}(\varphi))$ invariant under this action. This turns out to be just the set of *quadrics of rank* 3 containing the Veronese, that is, the quadrics whose singular locus is a 2-plane. In fact, the 2-plane will be the tangent plane to S at a point, giving the identification $T = S$.

Let us consider one more example of this type, namely, the symmetric cube $\mathrm{Sym}^3(\mathrm{Sym}^2 V))$. (We promise we will stop after this one.) As before, it is easy to write down the eigenvalues of this representation; they are just the triple sums of the eigenvalues $\{2L_i, L_i + L_j\}$ of $\mathrm{Sym}^2 V$. The diagram (we will draw here only one-sixth of the plane and indicate multiplicities with numbers rather than rings) thus looks like

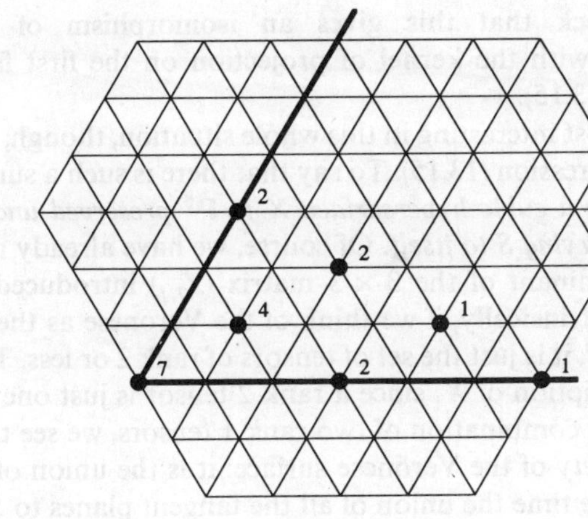

from which we see what the decomposition must be: as representations we have

$$\mathrm{Sym}^3(\mathrm{Sym}^2 V)) \cong \mathrm{Sym}^6 V \oplus \Gamma_{2,2} \oplus \mathbb{C}. \tag{13.15}$$

As before, the map to the first factor is just the obvious one; it is the identification of the kernel that is intriguing, and especially the identification of the last factor.

To see what is going on here, we should look again at the geometry of the Veronese surface $S \subset \mathbb{P}^5 = \mathbb{P}(\mathrm{Sym}^2 V^*)$. The space $\mathrm{Sym}^3(\mathrm{Sym}^2 V))$ is just the space of homogeneous cubic polynomials on the ambient space $\mathbb{P}^5$, and as before the map to the first factor of the right-hand side of (13.15) is just the restriction, so that the last two factors of (13.15) represent the vector space $I(S)_3$ of cubic polynomials vanishing on S. Note that we could in fact prove (13.15) without recourse to eigenvalue diagrams from this: since the ideal of the Veronese surface is generated by the vector space $I(S)_2$ of quadratic polynomials vanishing on it, we have a surjective map

$$I(S)_2 \otimes W = \mathrm{Sym}^2 V^* \otimes \mathrm{Sym}^2 V \to I(S)_3.$$

But we already know how the left hand side decomposes: we have

$$\mathrm{Sym}^2 V^* \otimes \mathrm{Sym}^2 V = \Gamma_{2,2} \oplus \Gamma_{1,1} \oplus \mathbb{C}, \tag{13.16}$$

so that $I(S)_3$ must be a partial direct sum of these three irreducible representations; by dimension considerations it can only be $\Gamma_{2,2} \oplus \mathbb{C}$.

This, in turn, tells us how to make the isomorphism (13.15) explicit (assuming we want to): we can define a map

$$\mathrm{Sym}^2(\wedge^2 V) \otimes \mathrm{Sym}^2 V \to \mathrm{Sym}^3(\mathrm{Sym}^2 V)$$

by sending

$$(u \wedge v)\cdot(w \wedge z) \otimes (s \cdot t) \mapsto ((u \cdot w)\cdot(v \cdot z) - (u \cdot z)\cdot(v \cdot w))\cdot(s \cdot t)$$

and then just check that this gives an isomorphism of $\Gamma_{2,2} \oplus \mathbb{C} \subset \mathrm{Sym}^2 V^* \otimes \mathrm{Sym}^2 V$ with the kernel of projection on the first factor of the right-hand side of (13.15).

What is really most interesting in this whole situation, though, is the trivial summand in the expression (13.15). To say that there is such a summand is to say that *there exists a cubic hypersurface X in $\mathbb{P}^5$ preserved under all automorphisms of $\mathbb{P}^5$ carrying S to itself.* Of course, we have already run into this one: it is the determinant of the 3×3 matrix $(Z_{i,j})$ introduced above. To express this more intrinsically, if we think of the Veronese as the set of rank 1 tensors in $\mathrm{Sym}^2 V^*$, it is just the set of tensors of rank 2 or less. This, in turn, yields another description of X: since a rank 2 tensor is just one that can be expressed as a linear combination of two rank 1 tensors, we see that X is the famous *chordal variety* of the Veronese surface: it is the union of the chords to S, and at the same time the union of all the tangent planes to S.

Exercise 13.17. Show that the only symmetric powers of $\mathrm{Sym}^2 V$ that possess trivial summands are the powers $\mathrm{Sym}^{3k}(\mathrm{Sym}^2 V))$ divisible by 3, and that the unique trivial summand in this is just the kth power of the trivial summand of $\mathrm{Sym}^3(\mathrm{Sym}^2 V))$.

Exercise 13.18. Given the isomorphism of the projectivization of the vector space $I(S)_2$—that is, the projective space of quadric hypersurfaces containing the Veronese surface—with $\mathbb{P}(\mathrm{Sym}^2 V^*)$, find the unique cubic hypersurface in $I(S)_2$ invariant under the action of $\mathrm{PGL}_3\mathbb{C}$.

Exercise 13.19. Analyze the representation $\mathrm{Sym}^2(\mathrm{Sym}^3 V))$ of $\mathfrak{sl}_3\mathbb{C}$. Interpret the direct sum factors in terms of the geometry of the Veronese embedding of $\mathbb{P}V^* = \mathbb{P}^2$ in $\mathbb{P}(\mathrm{Sym}^3 V^*) = \mathbb{P}^9$.

Exercise 13.20*. Show that the representations $\mathrm{Sym}^4(\mathrm{Sym}^3 V))$ and $\mathrm{Sym}^6(\mathrm{Sym}^3 V))$ contain trivial summands, and that the representation $\mathrm{Sym}^{12}(\mathrm{Sym}^3 V))$ contains two. Interpret these.

Exercise 13.21. Apply the techniques above to show that the representation $\Lambda^2(\mathrm{Sym}^2 V)$ is isomorphic to $\Gamma_{2,1}$.

Exercise 13.22*. Apply the techniques above to analyze the representation $\Lambda^3(\mathrm{Sym}^2 V)$, and in particular to interpret its decomposition into irreducible representations.

Exercise 13.23. If $\mathbb{P}^5 = \mathbb{P}(\mathrm{Sym}^2 V^*)$ is the ambient space of the Veronese surface, the Grassmannian $\mathbb{G}(2, 5)$ of 2-planes in $\mathbb{P}^5$ naturally embeds in the projective space $\mathbb{P}(\Lambda^3(\mathrm{Sym}^2 V))$. Describe, in terms of the decomposition in the preceding exercise, the span of the locus of tangent 2-planes to the

Veronese, and the span of the locus of 2-planes in $\mathbb{P}^5$ spanned by the images in S of lines in $\mathbb{P}V^*$.

Exercise 13.24*. Show that the unique closed orbit of the action of $SL_3\mathbb{C}$ on the representation $\Gamma_{a,b}$ is either isomorphic to $\mathbb{P}^2$ (embedded as the Veronese surface) if either a or b is zero, or to the incidence correspondence

$$\Sigma = \{(p, l): p \in l\} \subset \mathbb{P}^2 \times \mathbb{P}^{2*}$$

if neither a or b is zero.

PART III

THE CLASSICAL LIE ALGEBRAS AND THEIR REPRESENTATIONS

As we indicated at the outset, the analysis we have just carried out of the structure of $\mathfrak{sl}_2\mathbb{C}$ and $\mathfrak{sl}_3\mathbb{C}$ and their representations carries over to other semisimple complex Lie algebras. In Lecture 14 we codify this structure, using the pattern of the examples we have worked out so far to give a model for the analysis of arbitrary semisimple Lie algebras and stating some of the most important facts that are true in general. As usual, we postpone proofs of many of these facts until Part IV and the Appendices, the main point here being to introduce a unifying approach and language. The facts themselves will all be seen explicitly on a case-by-case basis for the classical Lie algebras $\mathfrak{sl}_n\mathbb{C}$, $\mathfrak{sp}_{2n}\mathbb{C}$, and $\mathfrak{so}_n\mathbb{C}$, which are studied in some detail in Lectures 15–20.

Most of the development follows the outline we developed in Lectures 11–13, the main goal being to describe the irreducible representations as explicitly as we can, and to see the decomposition of naturally occurring representations, both algebraically and geometrically. While most of the representations are found inside tensor powers of the standard representations, for the orthogonal Lie algebras this only gives half of them, and one needs new methods to construct the other "spin" representations. This is carried out using Clifford algebras in Lecture 20.

We also make the tie with Weyl's construction of representations of $GL_n\mathbb{C}$ from Lecture 6, which arose from the representation theory of the symmetric groups. We show in Lecture 15 that these are the irreducible representations of $\mathfrak{sl}_n\mathbb{C}$; in Lecture 17 we show how to use them to construct the irreducible representations of the symplectic Lie algebras, and in Lecture 19 to give the nonspin representation of the orthogonal Lie algebras. These give useful descriptions of the irreducible representations, and powerful methods for decomposing other representations, but they are not necessary for the logical progression of the book, and many of these decompositions can also be deduced from the Weyl character formula which we will discuss in Part IV.

LECTURE 14

The General Setup: Analyzing the Structure and Representations of an Arbitrary Semisimple Lie Algebra

This is the last of the four central lectures; in the body of it, §14.1, we extract from the examples of §11–13 the basic algorithm for analyzing a general semisimple Lie algebra and its representations. It is this algorithm that we will spend the remainder of Part III carrying out for the classical algebras, and the reader who finds the general setup confusing may wish to read this lecture in parallel with, for example, Lectures 15 and 16. In particular, §14.2 is less clearly motivated by what we have worked out so far; the reader may wish to skim it for now and defer a more thorough reading until after going through some more of the examples of Lectures 15–20.

§14.1. Analyzing Simple Lie Algebras in General

We said at the outset of Lecture 12 that once the analysis of the representations of $\mathfrak{sl}_3\mathbb{C}$ was understood, the analysis of the representations of any semisimple Lie algebra would be clear, at least in broad outline. Here we would like to indicate how that analysis will go in general, by providing an essentially algorithmic procedure for describing the representations of an arbitrary complex semisimple Lie algebra $\mathfrak{g}$. The process we give here is directly analogous, step for step, to that carried out in Lecture 12 for $\mathfrak{sl}_3\mathbb{C}$; the only difference is one change in the order of steps: having seen in the case of $\mathfrak{sl}_3\mathbb{C}$ the importance of the "distinguished" subalgebras $\mathfrak{s}_\alpha \cong \mathfrak{sl}_2\mathbb{C} \subset \mathfrak{g}$ and the corresponding distinguished elements $H_\alpha \in \mathfrak{s}_\alpha \subset \mathfrak{h}$, we will introduce them earlier here.

Step 0. *Verify that your Lie algebra is semisimple*; if not, none of the following will work (but see Remark 14.3). If your Lie algebra is not semisimple, pass as indicated in Lecture 9 to its semisimple part; a knowledge of the representations of this quotient algebra may not tell you everything about the representations of the original, but it will at least tell you about the irreducible representations.

Step 1. *Find an abelian subalgebra* $\mathfrak{h} \subset \mathfrak{g}$ *acting diagonally*. This is of course the analogue of looking at the specific element H in $\mathfrak{sl}_2\mathbb{C}$ and the subalgebra $\mathfrak{h}$ of diagonal matrices in the case of $\mathfrak{sl}_3\mathbb{C}$; in general, to serve an analogous function it should be an abelian subalgebra that acts diagonally on one faithful (and hence, by Theorem 9.20, on any) representation of $\mathfrak{g}$. Moreover, in order that the restriction of a representation V of $\mathfrak{g}$ to $\mathfrak{h}$ carry the greatest possible information about V, $\mathfrak{h}$ should clearly be maximal among abelian, diagonalizable subalgebras; such a subalgebra is called a *Cartan subalgebra*.

It may seem that this step is somewhat less than algorithmic; in particular, while it is certainly possible to tell when a subalgebra of a given Lie algebra is abelian, and when it is diagonalizable, it is not clear how to tell whether it is maximal with respect to these properties. This defect will, however, be largely cleared up in the next step (see Remark 14.3).

Step 2. *Let* $\mathfrak{h}$ *act on* $\mathfrak{g}$ *by the adjoint representation, and decompose* $\mathfrak{g}$ *accordingly*. By the choice of $\mathfrak{h}$, its action on any representation of $\mathfrak{g}$ will be diagonalizable; applying this to the adjoint representation we arrive at a direct sum decomposition, called a *Cartan decomposition*,

$$\mathfrak{g} = \mathfrak{h} \oplus \left(\bigoplus \mathfrak{g}_\alpha\right), \tag{14.1}$$

where the action of $\mathfrak{h}$ preserves each $\mathfrak{g}_\alpha$ and acts on it by scalar multiplication by the linear functional $\alpha \in \mathfrak{h}^*$; that is, for any $H \in \mathfrak{h}$ and $X \in \mathfrak{g}_\alpha$ we will have

$$\mathrm{ad}(H)(X) = \alpha(H) \cdot X.$$

The second direct sum in the expression (14.1) is over a finite set of eigenvalues $\alpha \in \mathfrak{h}^*$; these eigenvalues—in the language of Lecture 12, the *weights of the adjoint representation*—are called the *roots* of the Lie algebra and the corresponding subspaces $\mathfrak{g}_\alpha$ are called the *root spaces*. Of course, $\mathfrak{h}$ itself is just the eigenspace for the action of $\mathfrak{h}$ corresponding to the eigenvalue 0 (see Remark 14.3 below); so that in some contexts—such as the following paragraph, for example—it will be convenient to adopt the convention that $\mathfrak{g}_0 = \mathfrak{h}$; but we do not usually count $0 \in \mathfrak{h}^*$ as a root. The set of all roots is usually denoted $R \subset \mathfrak{h}^*$.

As in the previous cases, we can picture the structure of the Lie algebra in terms of the diagram of its roots: by the fundamental calculation of §11.1 and Lecture 12 (which we will not reproduce here for the fourth time) we see that the adjoint action of $\mathfrak{g}_\alpha$ carries the eigenspace $\mathfrak{g}_\beta$ into another eigenspace $\mathfrak{g}_{\alpha+\beta}$.

There are a couple of things we can anticipate about how the configuration of roots (and the corresponding root spaces) will look. We will simply state them here as

Facts 14.2

(i) *each root space* $\mathfrak{g}_\alpha$ *will be one dimensional.*
(ii) R *will generate a lattice* $\Lambda_R \subset \mathfrak{h}^*$ *of rank equal to the dimension of* $\mathfrak{h}$.

(iii) *R is symmetric about the origin, i.e., if $\alpha \in R$ is a root, then $-\alpha \in R$ is a root as well.*

These facts will all be proved in general in due course; for the time being, they are just things we will observe as we do the analysis of each simple Lie algebra in turn. We mention them here simply because some of what follows will make sense only given these facts. Note in particular that by (ii), the roots all lie in (and span) a real subspace of $\mathfrak{h}^*$; all our pictures clearly will be of this real subspace.

Remark 14.3. If indeed 0 does appear as an eigenvalue of the action of $\mathfrak{h}$ on $\mathfrak{g}/\mathfrak{h}$, then we may conclude from this that $\mathfrak{h}$ was not maximal to begin with: by the above, anything in the 0-eigenspace of the action of $\mathfrak{h}$ commutes with $\mathfrak{h}$ and (given the fact that the $\mathfrak{g}_\alpha$ are one dimensional) acts diagonally on $\mathfrak{g}$, so that if it not already in $\mathfrak{h}$, then $\mathfrak{h}$ could be enlarged while still retaining the properties of being abelian and diagonalizable. Similarly, the assertion in (ii) that the roots span $\mathfrak{h}^*$ follows from the fact that an element of $\mathfrak{h}$ in the annihilator of all of them would be in the center of $\mathfrak{g}$.

From what we have done so far, we get our first picture of the structure of an arbitrary irreducible finite-dimensional representation V of $\mathfrak{g}$. Specifically, V will admit a direct sum decomposition

$$V = \bigoplus V_\alpha, \tag{14.4}$$

where the direct sum runs over a finite set of $\alpha \in \mathfrak{h}^*$ and $\mathfrak{h}$ acts diagonally on each V_α by multiplication by the eigenvalue α, i.e., for any $H \in \mathfrak{h}$ and $v \in V_\alpha$ we will have

$$H(v) = \alpha(H) \cdot v.$$

The eigenvalues $\alpha \in \mathfrak{h}^*$ that appear in this direct sum decomposition are called the *weights* of V; the V_α themselves are called *weight spaces*; and the dimension of a weight space V_α will be called the *multiplicity* of the weight α in V. We will often represent V by drawing a picture of the set of its weights and thinking of each dot as representing a subspace; this picture (often with some annotation to denote the multiplicity of each weight) is called the *weight diagram* of V.

The action of the rest of the Lie algebra on V can be described in these terms: for any root β, we have

$$\mathfrak{g}_\beta \colon V_\alpha \to V_{\alpha+\beta},$$

so we can think of the action of $\mathfrak{g}_\beta$ on V as a translation in the weight diagram, shifting each of the dots over by β and mapping the weight spaces correspondingly.

Observe next that all the weights of an irreducible representation are congruent to one another modulo the root lattice Λ_R: otherwise, for any weight α of V the subspace

$$V' = \bigoplus_{\beta \in \Lambda_R} V_{\alpha+\beta}$$

would be a proper subrepresentation of V. In particular, in view of Fact 14.2(ii), this means that the weights all lie in a translate of the real subspace spanned by the roots, so that it is not so unreasonable to draw a picture of them.

Step 3. *Find the distinguished subalgebras* $\mathfrak{s}_\alpha \cong \mathfrak{sl}_2\mathbb{C} \subset \mathfrak{g}$. As we saw in the example of $\mathfrak{sl}_3\mathbb{C}$, a crucial ingredient in the analysis of an arbitrary irreducible finite-dimensional representation is the restriction of the representation to certain special copies of the algebra $\mathfrak{sl}_2\mathbb{C}$ contained in $\mathfrak{g}$, and the application of what we know from Lecture 11 about such representations. To generalize this to our arbitrary Lie algebra $\mathfrak{g}$, let $\mathfrak{g}_\alpha \subset \mathfrak{g}$ be a root space, one dimensional by (i) of Fact 14.2. Then by (iii) of Fact 14.2, there is another root space $\mathfrak{g}_{-\alpha} \subset \mathfrak{g}$; and their commutator $[\mathfrak{g}_\alpha, \mathfrak{g}_{-\alpha}]$ must be a subspace of $\mathfrak{g}_0 = \mathfrak{h}$, of dimension at most one. The adjoint action of the commutator $[\mathfrak{g}_\alpha, \mathfrak{g}_{-\alpha}]$ thus carries each of $\mathfrak{g}_\alpha$ and $\mathfrak{g}_{-\alpha}$ into itself; so that *the direct sum*

$$\mathfrak{s}_\alpha = \mathfrak{g}_\alpha \oplus \mathfrak{g}_{-\alpha} \oplus [\mathfrak{g}_\alpha, \mathfrak{g}_{-\alpha}] \tag{14.5}$$

is a subalgebra of $\mathfrak{g}$. The structure of $\mathfrak{s}_\alpha$ is not hard to describe, given two further facts that we will state here, verify in cases, and prove in general in Appendix D.

Facts 14.6.

(i) $[\mathfrak{g}_\alpha, \mathfrak{g}_{-\alpha}] \neq 0$; *and*
(ii) $[[\mathfrak{g}_\alpha, \mathfrak{g}_{-\alpha}], \mathfrak{g}_\alpha] \neq 0$.

Given these, it follows that *the subalgebra* $\mathfrak{s}_\alpha$ *is isomorphic* to $\mathfrak{sl}_2\mathbb{C}$. In particular, we can pick a basis $X_\alpha \in \mathfrak{g}_\alpha$, $Y_\alpha \in \mathfrak{g}_{-\alpha}$, and $H_\alpha \in [\mathfrak{g}_\alpha, \mathfrak{g}_{-\alpha}]$ satisfying the standard commutation relations (9.1) for $\mathfrak{sl}_2\mathbb{C}$; X_α and Y_α are not determined by this, but H_α is, being the unique element of $[\mathfrak{g}_\alpha, \mathfrak{g}_{-\alpha}]$ having eigenvalues 2 and -2 on $\mathfrak{g}_\alpha$ and $\mathfrak{g}_{-\alpha}$, respectively [i.e., H_α is uniquely characterized by the requirements that $H_\alpha \in [\mathfrak{g}_\alpha, \mathfrak{g}_{-\alpha}]$ and $\alpha(H_\alpha) = 2$.]

Step 4. *Use the integrality of the eigenvalues of the* H_α. The distinguished elements $H_\alpha \in \mathfrak{h}$ found above are important first of all because, by the analysis of the representations of $\mathfrak{sl}_2\mathbb{C}$ carried out in Lecture 9, in any representation of $\mathfrak{s}_\alpha$—and hence in any representation of $\mathfrak{g}$—*all eigenvalues of the action of* H_α *must be integers*. Thus, every eigenvalue $\beta \in \mathfrak{h}^*$ of every representation of $\mathfrak{g}$ must assume integer values on all the H_α. We correspondingly let Λ_W be the set of linear functionals $\beta \in \mathfrak{h}^*$ that are integer valued on all the H_α; Λ_W will be a lattice, called the *weight lattice* of $\mathfrak{g}$, with the property that

all weights of all representations of $\mathfrak{g}$ *will lie in* Λ_W.

Note, in particular, that $R \subset \Lambda_W$ and hence $\Lambda_R \subset \Lambda_W$; in fact, the root lattice will in general be a sublattice of finite index in the weight lattice.

Step 5. *Use the symmetry of the eigenvalues of the* H_α. The integrality of the eigenvalues of the H_α under any representation is only half the story; it is also true that they are symmetric about the origin in $\mathbb{Z}$. To express this, for any α we introduce the involution W_α on the vector space $\mathfrak{h}^*$ with $+1$-eigenspace the hyperplane

$$\Omega_\alpha = \{\beta \in \mathfrak{h}^* : \langle H_\alpha, \beta \rangle = 0\} \tag{14.7}$$

and minus 1 eigenspace the line spanned by α itself.[1] In English, W_α is the reflection in the plane Ω_α with axis the line spanned by α:

$$W_\alpha(\beta) = \beta - \frac{2\beta(H_\alpha)}{\alpha(H_\alpha)}\alpha = \beta - \beta(H_\alpha)\alpha. \tag{14.8}$$

Let $\mathfrak{W}$ be the group generated by these involutions; $\mathfrak{W}$ is called the *Weyl group* of the Lie algebra $\mathfrak{g}$.

Now suppose that V is any representation of $\mathfrak{g}$, with eigenspace decomposition $V = \bigoplus V_\beta$. The weights β appearing in this decomposition can then be broken up into equivalence classes mod α, and the direct sum

$$V_{[\beta]} = \bigoplus_{n \in \mathbb{Z}} V_{\beta + n\alpha} \tag{14.9}$$

of the eigenspaces in a given equivalence class will be a subrepresentation of V for $\mathfrak{s}_\alpha$. It follows then that the set of weights of V congruent to any given β mod α will be invariant under the involution W_α; in particular,

The set of weights of any representation of $\mathfrak{g}$ *is invariant under the Weyl group.*

To make this more explicit, the string of weights that correspond to nonzero summands in (14.9) are, possibly after replacing β by a translate by a multiple of α:

$$\beta, \beta + \alpha, \beta + 2\alpha, \ldots, \beta + m\alpha, \quad \text{with } m = -\beta(H_\alpha). \tag{14.10}$$

(Note that by our analysis of $\mathfrak{sl}_2\mathbb{C}$ this must be an uninterrupted string.) Indeed if we choose β and $m \geq 0$ so that (14.10) is the string corresponding to nonzero summands in (14.9), then the string of integers

$$\beta(H_\alpha), (\beta + \alpha)(H_\alpha) = \beta(H_\alpha) + 2, \ldots, (\beta + m\alpha)(H_\alpha) = \beta(H_\alpha) + 2m$$

must be symmetric about zero, so $\beta(H_\alpha) = -m$. In particular,

$$W_\alpha(\beta + k\alpha) = \beta + (-\beta(H_\alpha) - k)\alpha = \beta + (m - k)\alpha.$$

Note also that by the same analysis the multiplicities of the weights are invariant under the Weyl group.

We should mention one other fact about the Weyl group, whose proof we also postpone:

[1] Note that by the nondegeneracy assertion (ii) of Fact 14.6, the line $\mathbb{C} \cdot \alpha$ does not lie in the hyperplane Ω_α. Recall that $\langle\ ,\ \rangle$ is the pairing between $\mathfrak{h}$ and $\mathfrak{h}^*$, so $\langle H_\alpha, \beta \rangle = \beta(H_\alpha)$.

Fact 14.11. *Every element of the Weyl group is induced by an automorphism of the Lie algebra $\mathfrak{g}$ carrying $\mathfrak{h}$ to itself.*

We can even say what automorphism of $\mathfrak{g}$ does the trick: to get the involution W_α, take the adjoint action of the exponential $\exp(\pi i U_\alpha) \in G$, where G is any group with Lie algebra $\mathfrak{g}$ and U_α is a suitable element of the direct sum of the root spaces $\mathfrak{g}_\alpha$ and $\mathfrak{g}_{-\alpha}$. To prove that $\mathrm{Ad}(\exp(\pi i U_\alpha))$ actually does this requires more knowledge of $\mathfrak{g}$ than we currently possess; but it would be an excellent exercise to verify this assertion directly in each of the cases studied below. (For the general case see (23.20) and (26.15).)

Step 6. *Draw the picture* (optional). While there is no logical need to do so at this point, it will be much easier to think about what is going on in $\mathfrak{h}^*$ if we introduce the appropriate inner product, called the *Killing form*, on $\mathfrak{g}$ (hence by restriction on $\mathfrak{h}$, and hence on $\mathfrak{h}^*$). Since the introduction of the Killing form is, logically, a digression, we will defer until later in this lecture a discussion of its various definitions and properties. It will suffice for now to mention the characteristic property of the induced inner product on $\mathfrak{h}^*$: up to scalars it is the unique inner product on $\mathfrak{h}^*$ preserved by the Weyl group, i.e., in terms of which the Weyl group acts as a group of orthogonal transformations. Equivalently, it is the unique inner product (up to scalars) such that the line spanned by each root $\alpha \in \mathfrak{h}^*$ is actually perpendicular to the plane Ω_α (so that the involution W_α is just a reflection in that hyperplane). Indeed, in practice this is most often how we will compute it. In terms of the Killing form, then, we can say that *the Weyl group is just the group generated by the reflections in the hyperplanes perpendicular to the roots of the Lie algebra.*

Step 7. *Choose a direction in $\mathfrak{h}^*$*. By this we mean a real linear functional l on the lattice Λ_R irrational with respect to this lattice. This gives us a decomposition of the set

$$R = R^+ \cup R^-, \tag{14.12}$$

where $R^+ = \{\alpha: l(\alpha) > 0\}$ (the $\alpha \in R^+$ are called the *positive* roots, those in R^- *negative*); this decomposition is called an *ordering of the roots*. For most purposes, the only aspect of l that matters is the associated ordering of the roots.

The point of choosing a direction—and thereby an ordering of the roots $R = R^+ \cup R^-$—is, of course, to mimic the notion of highest weight vector that was so crucial in the cases of $\mathfrak{sl}_2\mathbb{C}$ and $\mathfrak{sl}_3\mathbb{C}$. Specifically, we make the

Definition. Let V be any representation of $\mathfrak{g}$. A nonzero vector $v \in V$ that is both an eigenvector for the action of $\mathfrak{h}$ and in the kernel of $\mathfrak{g}_\alpha$ for all $\alpha \in R^+$ is called a *highest weight vector* of V.

Just as in the previous cases, we then have

Proposition 14.13. *For any semisimple complex Lie algebra* $\mathfrak{g}$,

(i) *every finite-dimensional representation V of $\mathfrak{g}$ possesses a highest weight vector;*

(ii) *the subspace W of V generated by the images of a highest weight vector v under successive applications of root spaces $\mathfrak{g}_\beta$ for $\beta \in R^-$ is an irreducible subrepresentation;*

(iii) *an irreducible representation possesses a unique highest weight vector up to scalars.*

PROOF. Part (i) is immediate: we just take α to be the weight appearing in V for which the value $l(\alpha)$ is maximal and choose v any nonzero vector in the weight space V_α. Since $V_{\alpha+\beta} = (0)$ for all $\beta \in R^+$, such a vector v will necessarily be in the kernel of all root spaces $\mathfrak{g}_\beta$ corresponding to positive roots β.

Part (ii) may be proved by the same argument as in the two cases we have already discussed: we let W_n be the subspace spanned by all $w_n \cdot v$ where w_n is a word of length at most n in elements of $\mathfrak{g}_\beta$ for negative β. We then claim that for any X in any positive root space, $X \cdot W_n \subset W_n$. To see this, write a generator of W_n in the form $Y \cdot w$, $w \in W_{n-1}$, and use the commutation relation $X \cdot Y \cdot w = Y \cdot X \cdot w + [X, Y] \cdot w$; the claim follows by induction, since $[X, Y]$ is always in $\mathfrak{h}$. The subspace $W \subset V$ which is a union of all the W_n's is thus a subrepresentation; to see that it is irreducible; note that if we write $W = W' \oplus W''$, then either W' or W'' will have to contain the one-dimensional weight space W_α, and so will have to equal W.

The uniqueness of the highest weight vector of an irreducible representation follows immediately: if $v \in V_\alpha$ and $w \in V_\beta$ were two such, not scalar multiples of each other, we would have $l(\alpha) > l(\beta)$ and vice versa. □

Exercise 14.14. Show that in (ii) one need only apply those $\mathfrak{g}_\beta$ for which $\mathfrak{g}_\beta \cdot v \neq 0$. (Note: with W_n defined using only these $\mathfrak{g}_\beta$, and X in any root space, the same inductive argument shows that $X \cdot W_n \subset W_{n+1}$. On the other hand, if one uses all $\mathfrak{g}_\beta$ with β negative and primitive, as in Observation 14.16, then $X \cdot W_n \subset W_{n-1}$. One cannot combine these, however: V may *not* be generated by successively applying those $\mathfrak{g}_\beta$ with β negative, primitive, and $\mathfrak{g}_\beta \cdot v \neq 0$, e.g., the standard representation of $\mathfrak{sl}_3\mathbb{C}$.)

The weight α of the highest weight vector of an irreducible representation will be called, not unreasonably, the *highest weight* of that representation; the term *dominant weight* is also common.

We can refine part (ii) of this proposition slightly in another direction; this is not crucial but will be useful later on in estimating multiplicities of various representations. This refinement is based on

Exercise 14.15*. (a) Let $\alpha_1, \ldots, \alpha_k$ be roots of a semisimple Lie algebra $\mathfrak{g}$ and $\mathfrak{g}_{\alpha_i} \subset \mathfrak{g}$ the corresponding root spaces. Show that the subalgebra of $\mathfrak{g}$ generated by the Cartan subalgebra $\mathfrak{h}$ together with the $\mathfrak{g}_{\alpha_i}$ is exactly the direct sum $\mathfrak{h} \oplus (\bigoplus \mathfrak{g}_\alpha)$, where the direct sum is over the intersection of the set R of roots of $\mathfrak{g}$ with the semigroup $\mathbb{N}\{\alpha_1, \ldots, \alpha_k\} \subset \mathfrak{h}$ generated by the α_i.

(b) Similarly, let $\alpha_1, \ldots, \alpha_k$ be negative roots of a semisimple Lie algebra $\mathfrak{g}$ and $\mathfrak{g}_{\alpha_i} \subset \mathfrak{g}$ the corresponding root spaces. Show that the subalgebra of $\mathfrak{g}$ generated by the $\mathfrak{g}_{\alpha_i}$ is exactly the direct sum $\bigoplus \mathfrak{g}_\alpha$, where the direct sum is over the intersection of the set R of roots of $\mathfrak{g}$ with the semigroup $\mathbb{N}\{\alpha_1, \ldots, \alpha_k\} \subset \mathfrak{h}$ generated by the α_i.

(Note that by the description of the adjoint action of a Lie algebra on itself we have an obvious inclusion; the problem here is to show—given the facts above—that if $\alpha + \beta \in R$, then $[\mathfrak{g}_\alpha, \mathfrak{g}_\beta] \neq 0$.)

From this exercise, it is clear that generating a subrepresentation W of a given representation V by successive applications of root spaces $\mathfrak{g}_\beta$ for $\beta \in R^-$ to a highest weight vector v is inefficient; we need only apply the root spaces $\mathfrak{g}_\beta$ corresponding to a set of roots β generating R^- as a semigroup. We accordingly introduce another piece of terminology: we say that a positive (resp., negative) root $\alpha \in R$ is *primitive* or *simple* if it cannot be expressed as a sum of two positive (resp. negative) roots. (Note that, since there are only finitely many roots, every positive root can be written as a sum of primitive positive roots.) We then have

Observation 14.16. *Any irreducible representation V is generated by the images of its highest weight vector v under successive applications of root spaces $\mathfrak{g}_\beta$ where β ranges over the primitive negative roots.*

We have already seen one example of this in the case of $\mathfrak{sl}_3\mathbb{C}$, where we observed (in the proof of Claim 12.10 and in the analysis of $\mathrm{Sym}^2 V \otimes V^*$ in Lecture 13) that any irreducible representation was generated by applying the two elements $E_{2,1} \in \mathfrak{g}_{L_2-L_1}$ and $E_{3,2} \in \mathfrak{g}_{L_3-L_2}$ to a highest weight vector.

To return to our description of the weights of an irreducible representation V, we observe next that in fact *every vertex of the convex hull of the weights of V must be conjugate to α under the Weyl group.* To see this, note that by the above the set of weights is contained in the cone $\alpha + C_\alpha^-$, where C_α^- is the positive real cone spanned by the roots $\beta \in R^-$ such that $\mathfrak{g}_\beta(v) \neq 0$—that is, such that $\alpha(H_\beta) \neq 0$. Conversely, the weights of V will contain the string of weights

$$\alpha, \alpha + \beta, \alpha + 2\beta, \ldots, \alpha + (-\alpha(H_\beta))\beta \tag{14.17}$$

for any $\beta \in R^-$. Thus, any vertex of the convex hull of the set of weights of V adjacent to α must be of the form

$$\alpha - \alpha(H_\beta)\beta = W_\beta(\alpha)$$

for some β; applying the same analysis to each successive vertex gives the statement.

From the above, we deduce that the set of weights of V will lie in the convex hull of the images of α under the Weyl group. Since, moreover, we know that the intersection of this set with any set of weights of the form $\{\beta + n\gamma\}$ will be a connected string, it follows that *the set of weights of V will be exactly the weights that are congruent to α modulo the root lattice Λ_R and that lie in the convex hull of the images of α under the Weyl group.*

One more bit of terminology, and then we are done. By what we have seen (cf. (14.17)), the highest weight of any representation of V will be a weight α satisfying $\alpha(H_\gamma) \geq 0$ for every $\gamma \in R^+$. The locus $\mathscr{W}$, in the real span of the roots, of points satisfying these inequalities—in terms of the Killing form, making an acute or right angle with each of the positive roots—is called the (closed) *Weyl chamber* associated to the ordering of the roots. A Weyl chamber could also be described as the closure of a connected component of the complement of the union of the hyperplanes Ω_α. The Weyl group acts simply transitively on the set of Weyl chambers and likewise on the set of orderings of the roots. As usual, these statements will be easy to see in the cases we study, while the abstract proofs are postponed (to Appendix D).

Step 8. *Classify the irreducible, finite-dimensional representations of* $\mathfrak{g}$. Where all the above is leading should be pretty clear; it is expressed in the fundamental existence and uniqueness theorem:

Theorem 14.18. *For any α in the intersection of the Weyl chamber $\mathscr{W}$ associated to the ordering of the roots with the weight lattice Λ_W, there exists a unique irreducible, finite-dimensional representation Γ_α of $\mathfrak{g}$ with highest weight α; this gives a bijection between $\mathscr{W} \cap \Lambda_W$ and the set of irreducible representations of $\mathfrak{g}$. The weights of Γ_α will consist of those elements of the weight lattice congruent to α modulo the root lattice Λ_R and lying in the convex hull of the set of points in $\mathfrak{h}^*$ conjugate to α under the Weyl group.*

HALF-PROOF. We will give here just the proof of uniqueness, which is easy. The existence part we will demonstrate explicitly in each example in turn; and later on we will sketch some of the constructions that can be made in general.

The uniqueness part is exactly the same as for $\mathfrak{sl}_3\mathbb{C}$. If V and W are two irreducible, finite-dimensional representations of $\mathfrak{g}$ with highest weight vectors v and w, respectively, both having weight α, then the vector $(v, w) \in V \oplus W$ will again be a highest weight vector of weight α in that representation. Let $U \subset V \oplus W$ be the subrepresentation generated by (v, w); since U will again be irreducible the projection maps $\pi_1: U \to V$ and $\pi_2: U \to W$, being nonzero, will have to be isomorphisms. □

Another fact which we will see as we go along—and eventually prove in general—is that there are always *fundamental weights* $\omega_1, \ldots, \omega_n$ with the property that any dominant weight can be expressed uniquely as a non-negative integral linear combination of them. They can be characterized geometrically as the first weights met along the edges of the Weyl chamber, or algebraically as those elements ω_i in $\mathfrak{h}^*$ such that $\omega_i(H_{\alpha_j}) = \delta_{i,j}$, where $\alpha_1, \ldots, \alpha_n$ are the simple roots (in some order). When we have found them, we often write $\Gamma_{a_1,\ldots,a_n}$ for the irreducible representation with highest weight $a_1\omega_1 + \cdots + a_n\omega_n$; i.e.,

$$\Gamma_{a_1,\ldots,a_n} = \Gamma_{a_1\omega_1+\cdots+a_n\omega_n}.$$

As with most of the material in this section, general proofs will be found in Lecture 21 and Appendix D.

One basic point we want to repeat here (and that we hope to demonstrate in succeeding lectures) is this: that actually carrying out this process in practice is completely elementary and straightforward. Any mathematician, stranded on a desert island with only these ideas and the definition of a particular Lie algebra $\mathfrak{g}$ such as $\mathfrak{sl}_n\mathbb{C}$, $\mathfrak{so}_n\mathbb{C}$, or $\mathfrak{sp}_{2n}\mathbb{C}$, would in short order have a complete description of all the objects defined above in the case of $\mathfrak{g}$. We should say as well, however, that at the conclusion of this procedure we are left without one vital piece of information about the representations of $\mathfrak{g}$, without which we will be unable to analyze completely, for example, tensor products of known representations; this is, of course, a description of the multiplicities of the basic representations Γ_α. As we said, we will, in fact, describe and prove such a formula (the Weyl character formula); but it is of a much less straightforward character (our hypothetical shipwrecked mathematician would have to have what could only be described as a pretty good day to come up with the idea) and will be left until later. For now, we will conclude this lecture with the promised introduction to the Killing form.

§14.2. About the Killing Form

As we said, the Killing form is an inner product (symmetric bilinear form) on the Lie algebra $\mathfrak{g}$; abusing our notation, we will denote by B both the Killing form and the induced inner products on $\mathfrak{h}$ and $\mathfrak{h}^*$. B can be defined in several ways; the most common is by associating to a pair of elements $X, Y \in \mathfrak{g}$ the trace of the composition of their adjoint actions on $\mathfrak{g}$, i.e.,

$$B(X, Y) = \operatorname{Tr}(\operatorname{ad}(X) \circ \operatorname{ad}(Y): \mathfrak{g} \to \mathfrak{g}). \tag{14.19}$$

As we will see, the Killing form may be computed in practice either from this definition, or (up to scalars) by using its invariance under the group of automorphisms of $\mathfrak{g}$. We remark that this definition is not as opaque as it may seem at first. For one thing, the description of the adjoint action of the root space $\mathfrak{g}_\alpha$ as a "translation" of the root diagram—that is, carrying each root space $\mathfrak{g}_\beta$ into $\mathfrak{g}_{\alpha+\beta}$—tells us immediately that $\mathfrak{g}_\alpha$ is perpendicular to $\mathfrak{g}_\beta$ for all β other than $-\alpha$; in other words, the decomposition

$$\mathfrak{g} = \mathfrak{h} \oplus \left(\bigoplus_{\alpha \in R^+} (\mathfrak{g}_\alpha \oplus \mathfrak{g}_{-\alpha}) \right) \tag{14.20}$$

is orthogonal. As for the restriction of B to $\mathfrak{h}$, this is more subtle, but it is not hard to write down: if X, Y are in $\mathfrak{h}$, and Z_α generates $\mathfrak{g}_\alpha$, then $\operatorname{ad}(X) \circ \operatorname{ad}(Y)(Z_\alpha) = \alpha(X)\alpha(Y)Z_\alpha$, so $B(X, Y) = \sum \alpha(X)\alpha(Y)$, the sum over the roots; viewing $B|_\mathfrak{h}$ as an element of the symmetric square $\operatorname{Sym}^2(\mathfrak{h}^*)$, we have

$$B|_\mathfrak{h} = \frac{1}{2} \sum_{\alpha \in R} \alpha^2. \tag{14.21}$$

A key fact following from this—one that, if nothing else, makes picturing $\mathfrak{h}^*$ with the inner product B involve less eyestrain—is

(14.22) *B is positive definite on the real subspace of $\mathfrak{h}$ spanned by the vectors $\{H_\alpha: \alpha \in R\}$.*

Indeed, all roots take on real values on this space (since all $\alpha(H_\beta) \in \mathbb{Z} \subset \mathbb{R}$), so for H in this real subspace of $\mathfrak{h}$, $B(H, H)$ is non-negative, and is zero only when all $\alpha(H) = 0$, which implies $H = 0$, since the roots span $\mathfrak{h}^*$.

To see that the Killing form is nondegenerate on all of $\mathfrak{g}$, we need the useful identity:

$$B([X, Y], Z) = B(X, [Y, Z]) \tag{14.23}$$

for all X, Y, Z in $\mathfrak{g}$. This follows from the identity

$$\operatorname{Trace}((\bar{X}\bar{Y} - \bar{Y}\bar{X})\bar{Z}) = \operatorname{Trace}(\bar{X}(\bar{Y}\bar{Z} - \bar{Z}\bar{Y}))$$

for any endomorphisms $\bar{X}$, $\bar{Y}$, $\bar{Z}$ of a vector space. And this, in turn, follows from

$$\operatorname{Trace}(\bar{Y}\bar{X}\bar{Z} - \bar{X}\bar{Z}\bar{Y}) = \operatorname{Trace}([\bar{Y}, \bar{X}\bar{Z}]) = 0.$$

An immediate consequence of (14.23) is that if $\mathfrak{a}$ is any ideal in a Lie algebra $\mathfrak{g}$, then its orthogonal complement $\mathfrak{a}^\perp$ with respect to B is also an ideal. In particular, if $\mathfrak{g}$ is simple, the kernel of B is zero (note that the kernel cannot be $\mathfrak{g}$ since it does not contain $\mathfrak{h}$). Since the Killing form of a direct sum is the sum of the Killing forms of the factors, it follows that *the Killing form is nondegenerate on a semisimple Lie algebra* $\mathfrak{g}$.

One of the reasons the Killing form helps to picture $\mathfrak{h}^*$ is the fact mentioned above:

Proposition 14.24. *With respect to B, the line spanned by each root α is perpendicular to the hyperplane Ω_α.*

As we observed, this is equivalent to saying that the involutions W_α above are simply reflections in hyperplanes, and in turn to saying that the whole Weyl group is orthogonal. Note also that Proposition 14.24 thereby follows immediately from the Fact 14.11: from the definition of B above, it is clearly invariant under any automorphism of $\mathfrak{g}$. Nevertheless, we would prefer not to rely on this fact; and anyway giving a direct proof of the proposition is not hard, in terms of the picture we have of the adjoint action of $\mathfrak{g}$ on itself. To prove the assertion $\alpha \perp \Omega_\alpha$, it suffices to prove the dual assertion that $H \perp H_\alpha$ for all H in the annihilator of α. But now by construction H_α is the commutator $[X_\alpha, Y_\alpha]$ of an element $X_\alpha \in \mathfrak{g}_\alpha$ and an element $Y_\alpha \in \mathfrak{g}_{-\alpha}$. Using (14.23) we have for any H in $\mathfrak{h}$,

$$\begin{aligned} B(H_\alpha, H) &= B([X_\alpha, Y_\alpha], H) = B(X_\alpha, [Y_\alpha, H]) \\ &= B(X_\alpha, \alpha(H) Y_\alpha) = \alpha(H) B(X_\alpha, Y_\alpha), \end{aligned} \tag{14.25}$$

which vanishes since $\alpha(H) = 0$.

Note that as a consequence of this, we can characterize the Weyl chamber associated to an ordering of the roots as exactly those vectors in the real span of the roots forming an acute angle with all the positive roots (or, equivalently, with all the primitive ones); the Weyl chamber is thus the cone whose faces lie in the hyperplanes perpendicular to the primitive positive roots.

Equation (14.25) leads to a formula for the isomorphism of $\mathfrak{h}$ with $\mathfrak{h}^*$ determined by the Killing form. First note that for $H = H_\alpha$ it gives

$$B(H_\alpha, H_\alpha) = 2B(X_\alpha, Y_\alpha) \neq 0,$$

for if $B(X_\alpha, Y_\alpha)$ were zero we would have $B(H_\alpha, H) = 0$ for all H, contradicting the nondegeneracy of B on $\mathfrak{h}$. The element T_α of $\mathfrak{h}$ which corresponds to $\alpha \in \mathfrak{h}^*$ by the Killing form is by definition the element of $\mathfrak{h}$ that satisfies the condition

$$B(T_\alpha, H) = \alpha(H) \quad \text{for all } H \in \mathfrak{h}. \tag{14.26}$$

Looking at (14.25), we see that $T_\alpha = H_\alpha/B(X_\alpha, Y_\alpha) = 2H_\alpha/B(H_\alpha, H_\alpha)$. This proves

Corollary 14.27. *The isomorphism of $\mathfrak{h}^*$ and $\mathfrak{h}$ determined by the Killing form B carries α to $T_\alpha = (2/B(H_\alpha, H_\alpha)) \cdot H_\alpha$.*

The Killing form on $\mathfrak{h}^*$ is defined by $B(\alpha, \beta) = B(T_\alpha, T_\beta)$.

Exercise 14.28. Show that the inverse isomorphism from $\mathfrak{h}$ to $\mathfrak{h}^*$ takes H_α to $(2/B(\alpha, \alpha)) \cdot \alpha$.

The orthogonality of W_α can be expressed by the formula

$$W_\alpha(\beta) = \beta - \frac{2B(\beta, \alpha)}{B(\alpha, \alpha)} \alpha.$$

Comparing with (14.8) this says:

Corollary 14.29. *If α and β are roots, then*

$$2B(\beta, \alpha)/B(\alpha, \alpha) = \beta(H_\alpha)$$

is an integer.

By the above identification of $\mathfrak{h}$ with $\mathfrak{h}^*$, (14.22) translates to

Corollary 14.30. *The Killing form B is positive definite on the real vector space spanned by the root lattice Λ_R.*

Note that it follows immediately from (14.22) that the Weyl group $\mathfrak{W}$ is finite, being simultaneously discrete ($\mathfrak{W}$ preserves the set R of roots of $\mathfrak{g}$ and hence the lattice Λ_R; it follows that $\mathfrak{W}$ can be realized as a subgroup of $\mathrm{GL}_n\mathbb{Z}$)

and compact ($\mathfrak{W}$ preserves the Killing form, and hence is a subgroup of the orthogonal group $O_n\mathbb{R}$.) Alternatively, $\mathfrak{W}$ is a subgroup of the permutation group of the set of roots.

As we observed, the Killing form on $\mathfrak{h}^*$ is preserved by the Weyl group. In fact, in case $\mathfrak{g}$ is simple, the Killing form is, up to scalars, the unique inner product preserved by the Weyl group. This will follow from

Proposition 14.31. *The space* $\mathfrak{h}^*$ *is an irreducible representation of the Weyl group* $\mathfrak{W}$.

PROOF. Suppose that $\mathfrak{z} \subset \mathfrak{h}^*$ were preserved by the action of $\mathfrak{W}$. This means that every root $\alpha \in \mathfrak{h}^*$ of $\mathfrak{g}$ will either lie in the subspace $\mathfrak{z}$ or be perpendicular to it, i.e., for every $\alpha \in \mathfrak{z}$ and $\beta \notin \mathfrak{z}$ we will have $\beta(H_\alpha) = 0$. We claim then that *the subspace* $\mathfrak{g}'$ *of* $\mathfrak{g}$ *spanned by the subalgebras* $\{\mathfrak{s}_\alpha\}_{\alpha \in \mathfrak{z}}$ *will be an ideal in* $\mathfrak{g}$. Clearly it will be a subalgebra; the space spanned by the distinguished subalgebras $\mathfrak{s}_\alpha$ corresponding to the set of roots lying in any subspace of $\mathfrak{h}^*$ will be. To see that it is in fact an ideal, let $Y \in \mathfrak{g}_\beta$ be an element of a root space. Then for any $\alpha \in \mathfrak{z}$, we have

$$[Y, Z] \in \mathfrak{g}_{\alpha+\beta} = 0$$

since $\alpha + \beta$ is neither in $\mathfrak{z}$ nor perpendicular to it, and so cannot be a root; and

$$[Y, H_\alpha] = -[H_\alpha, Y] = \beta(H_\alpha) \cdot Y = 0.$$

Thus, $\text{ad}(Y)$ kills $\mathfrak{g}'$; since, of course, all of H itself will preserve $\mathfrak{g}'$, it follows that $\mathfrak{g}'$ is an ideal. Thus, either all the roots lie in $\mathfrak{z}$ and so $\mathfrak{z} = \mathfrak{h}^*$, or all roots are perpendicular to $\mathfrak{z}$ and correspondingly $\mathfrak{z} = (0)$. □

Note that given Fact 14.11, we can also express the last statement by saying that (in case $\mathfrak{g}$ is simple) the Killing form on $\mathfrak{h}$ is the unique form preserved by every automorphism of the Lie algebra $\mathfrak{g}$ carrying $\mathfrak{h}$ to itself. As we will see, in practice this is most often how we will first describe the Killing form.

Exercise 14.32. Find the Killing form on the Lie algebras $\mathfrak{sl}_2\mathbb{C}$ and $\mathfrak{sl}_3\mathbb{C}$ by explicit computation, and verify the statements made above in these cases.

Exercise 14.33*. If a semisimple Lie algebra is a direct sum of simple subalgebras, then its Killing form is the orthogonal sum of the Killing forms of the factors. Show that, conversely, if the roots of a semisimple Lie algebra lie in a collection of mutually perpendicular subspaces, then the Lie algebra decomposes accordingly.

Exercise 14.34*. Suppose $\mathfrak{g}$ is a Lie algebra that has an abelian subalgebra $\mathfrak{h}$ such that $\mathfrak{g}$ has a decomposition (14.1), satisfying the conditions of Facts 14.2 and 14.6. Show that $\mathfrak{g}$ is semisimple, and $\mathfrak{h}$ is a Cartan subalgebra.

The preceding exercise can be used instead of Weyl's unitary trick or any abstract theory to verify that the algebras we meet in the next few lectures are all semisimple. It is tempting to call such a Lie algebra "visibly semisimple."

The discussion of the geometry of the roots of a semisimple Lie algebra will be continued in Lecture 21 and completed in Appendix D. The Killing form becomes particularly useful in the general theory; for example, solvability and semisimplicity can both be characterized by properties of the Killing form (see Appendix C).

Exercise 14.35*. Show that $\mathfrak{b} = \mathfrak{h} \oplus \bigoplus_{\alpha > 0} \mathfrak{g}_\alpha$ is a maximal solvable subalgebra of $\mathfrak{g}$; $\mathfrak{b}$ is called a *Borel subalgebra.* Show that $\bigoplus_{\alpha > 0} \mathfrak{g}_\alpha$ is a maximal nilpotent subalgebra of $\mathfrak{g}$. These will be discussed in Lecture 25.

Exercise 14.36*. Show that the Killing form on the Lie algebra $\mathfrak{gl}_m$ is given by the formula

$$B(X, Y) = 2m \operatorname{Tr}(X \circ Y) - 2 \operatorname{Tr}(X) \operatorname{Tr}(Y).$$

Find similar formulas for $\mathfrak{sl}_m$, $\mathfrak{so}_m$, and $\mathfrak{sp}_m$, showing in each case that $B(X, Y)$ is a constant multiple of $\operatorname{Tr}(X \circ Y)$.

Exercise 14.37. If G is a real Lie group, the Killing form on its Lie algebra $\mathfrak{g} = T_e G$ may not be positive definite. When it is, it determines, by left translation, a Riemannian metric on G. Show that the Killing form is positive definite for $G = \mathrm{SO}_n\mathbb{R}$, but not for $\mathrm{SL}_n\mathbb{R}$.

LECTURE 15
$\mathfrak{sl}_4\mathbb{C}$ and $\mathfrak{sl}_n\mathbb{C}$

In this lecture, we will illustrate the general paradigm of the previous lecture by applying it to the Lie algebras $\mathfrak{sl}_n\mathbb{C}$; this is typical of the analyses of specific Lie algebras carried out in this Part. We start in §15.1 by describing the Cartan subalgebra, roots, root spaces, etc., for $\mathfrak{sl}_n\mathbb{C}$ in general. We then give in §15.2 a detailed account of the representations of $\mathfrak{sl}_4\mathbb{C}$, which generalizes directly to $\mathfrak{sl}_n\mathbb{C}$; in particular, we deduce the existence part of Theorem 14.18 for $\mathfrak{sl}_n\mathbb{C}$.

In §15.3 we give an explicit construction of the irreducible representations of $\mathfrak{sl}_n\mathbb{C}$ using the Weyl construction introduced in Lecture 6; analogous constructions of the irreducible representations of the remaining classical Lie algebras will be given in §17.3 and §19.5. This section presupposes familiarity with Lecture 6 and Appendix A, but can be skipped by those willing to forego §17.3 and 19.5 as well. Section 15.4 requires essentially the same degree of knowledge of classical algebraic geometry as §§11.3 and 13.4 (it does not presuppose §15.3), but can also be skipped. Finally, §15.5 describes representations of $GL_n\mathbb{C}$; this appears to involve the Weyl construction but in fact the main statement, Proposition 15.47 (and even its proof) can be understood without the preceding two sections.

§15.1. Analyzing $\mathfrak{sl}_n\mathbb{C}$

To begin with, we have to locate a Cartan subalgebra, and this is not hard; as in the case of $\mathfrak{sl}_2\mathbb{C}$ and $\mathfrak{sl}_3\mathbb{C}$ the subalgebra of diagonal matrices will work fine. Writing H_i for the diagonal matrix $E_{i,i}$ that takes e_i to itself and kills e_j

for $j \neq i$, we have

$$\mathfrak{h} = \{a_1H_1 + a_2H_2 + \cdots + a_nH_n: a_1 + a_2 + \cdots + a_n = 0\};$$

note that H_i is not in $\mathfrak{h}$. We can correspondingly write

$$\mathfrak{h}^* = \mathbb{C}\{L_1, L_2, \ldots, L_n\}/(L_1 + L_2 + \cdots + L_n = 0),$$

where $L_i(H_j) = \delta_{i,j}$. We often write L_i for the image of L_i in $\mathfrak{h}^*$.

We have already seen how the diagonal matrices act on the space of all traceless matrices: if $E_{i,j}$ is the endomorphism of $\mathbb{C}^n$ carrying e_j to e_i and killing e_k for all $k \neq j$, then we have

$$\mathrm{ad}(a_1H_1 + a_2H_2 + \cdots + a_nH_n)(E_{i,j}) = (a_i - a_j)\cdot E_{i,j}; \tag{15.1}$$

or, in other words, *$E_{i,j}$ is an eigenvector for the action of $\mathfrak{h}$ with eigenvalue $L_i - L_j$*; in particular, *the roots of $\mathfrak{sl}_n\mathbb{C}$ are just the pairwise differences of the L_i*.

Before we try to visualize anything taking place in $\mathfrak{h}$ or $\mathfrak{h}^*$, let us take a moment out and describe the Killing form. To this end, note that the automorphism φ of $\mathbb{C}^n$ sending e_i to e_j, e_j to $-e_i$ and fixing e_k for all $k \neq i, j$ induces an automorphism $\mathrm{Ad}(\varphi)$ of the Lie algebra $\mathfrak{sl}_n\mathbb{C}$ (or even $\mathfrak{gl}_n(\mathbb{C})$) that carries $\mathfrak{h}$ to itself, exchanges H_i and H_j, and fixes all the other H_k. Since the Killing form on $\mathfrak{h}$ must be invariant under all these automorphisms, it must satisfy $B(L_i, L_i) = B(L_j, L_j)$ for all i and j and $B(L_i, L_k) = B(L_j, L_k)$ for all i, j and $k \neq i, j$; it follows that on $\mathfrak{h}$ it must be a linear combination of the forms

$$B'(\textstyle\sum a_iH_i, \sum b_iH_i) = \sum a_ib_i$$

and

$$B''(\textstyle\sum a_iH_i, \sum b_iH_i) = \sum_{i\neq j} a_ib_j.$$

On the space $\{\sum a_iH_i: \sum a_i = 0\}$, however, we have $0 = (\sum a_i)(\sum b_j) = \sum a_ib_i + \sum a_ib_j$, so in fact these two forms are dependent; and hence we can write the Killing form simply as a multiple of B'. Similarly, the Killing form on $\mathfrak{h}^*$ must be a linear combination of the forms $B'(\sum a_iL_i, \sum b_iL_i) = \sum a_ib_i$ and $B''(\sum a_iL_i, \sum b_iL_i) = \sum_{j\neq i} a_ib_j$; the condition that $B(\sum a_iL_i, \sum b_iL_i) = 0$ whenever $a_1 = a_2 = \cdots = a_n$ or $b_1 = b_2 = \cdots = b_n$ implies that it must be a multiple of

$$\begin{aligned} B(\textstyle\sum a_iL_i, \sum b_iL_i) &= \left(\frac{n-1}{n}\right)\sum_i a_ib_i - \frac{1}{n}\sum_{i\neq j} a_ib_j \\ &= \sum_i a_ib_i - \frac{1}{n}\sum_{i,j} a_ib_j. \end{aligned} \tag{15.2}$$

We may, of course, also calculate the Killing form directly from the definition. By (14.21), since the roots of $\mathfrak{sl}_n\mathbb{C}$ are $\{L_i - L_j\}_{i\neq j}$, we have

$$\begin{aligned} B(\textstyle\sum a_iH_i, \sum b_iH_i) &= \textstyle\sum_{i\neq j} (a_i - a_j)(b_i - b_j) \\ &= \textstyle\sum_i \sum_{j\neq i} (a_ib_i + a_jb_j - a_ib_j - a_jb_i). \end{aligned}$$

Noting that $\sum_{j\neq i} a_j = -a_i$ and, similarly, $\sum_{j\neq i} b_j = -b_i$, this simplifies to

$$B(\sum a_i H_i, \sum b_i H_i) = 2n \sum a_i b_i. \tag{15.3}$$

It follows with a little calculation that the dual form on $\mathfrak{h}^*$ is

$$B(\sum a_i L_i, \sum b_i L_i) = (1/2n)(\sum_i a_i b_i - (1/n) \sum_{i,j} a_i b_j). \tag{15.4}$$

It is probably simpler just to think of this as the form, unique up to scalars, invariant under the symmetric group $\mathfrak{S}_n$ of permutations of $\{1, 2, \ldots, n\}$. The L_i, therefore, all have the same length, and the angles between all pairs are the same. To picture the roots in $\mathfrak{h}^*$, then, we should think of the points L_i as situated at the vertices of a regular $(n-1)$-simplex Δ, with the origin located at the barycenter of that simplex. This picture is easiest to visualize in the special case $n = 4$, where the L_i will be located at every other vertex of a unit cube centered at the origin:

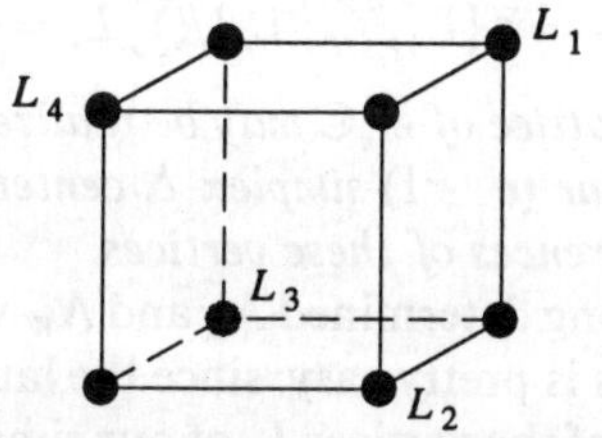

(15.5)

Now, as we said, the roots of $\mathfrak{sl}_n\mathbb{C}$ are now just the pairwise differences of the L_i. The root lattice Λ_R they generate can thus be described as

$$\Lambda_R = \{\sum a_i L_i \colon a_i \in \mathbb{Z}, \sum a_i = 0\}/(\sum L_i = 0).$$

Both the roots and the root lattice can be drawn in the case of $\mathfrak{sl}_4\mathbb{C}$: if we think of the vectors $L_i \in \mathfrak{h}^*$ as four of the vertices of a cube centered at the origin, the roots will comprise all the midpoints of the edges of a second cube whose linear dimensions are twice the dimensions of the first:

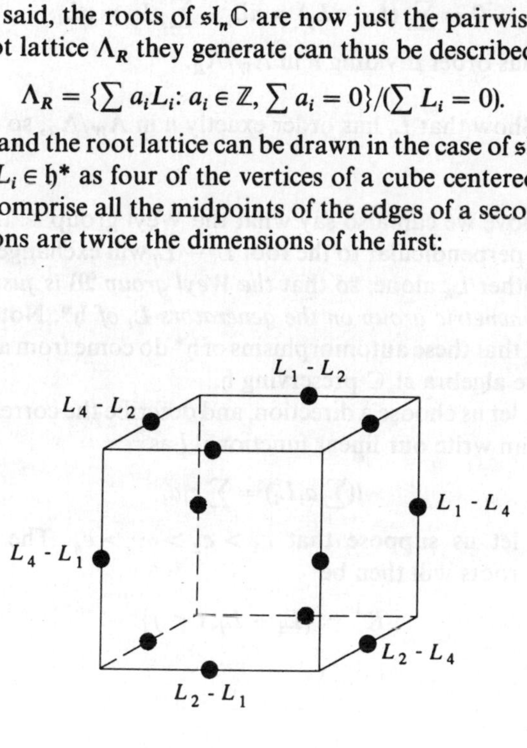

(15.6)

The next step, finding the distinguished subalgebras $\mathfrak{s}_\alpha$, is also very easy. The root space $\mathfrak{g}_{L_i-L_j}$ corresponding to the root $L_i - L_j$ is generated by $E_{i,j}$, so the subalgebra $\mathfrak{s}_{L_i-L_j}$ is generated by

$$E_{i,j}, \quad E_{j,i}, \quad \text{and} \quad [E_{i,j}, E_{j,i}] = H_i - H_j.$$

The eigenvalue of $H_i - H_j$ acting on $E_{i,j}$ is $(L_i - L_j)(H_i - H_j) = 2$, so that the corresponding distinguished element $H_{L_i-L_j}$ in $\mathfrak{h}$ must be just $H_i - H_j$. The annihilator, of course, is the hyperplane $\Omega_{L_i-L_j} = \{\sum a_i L_j : a_i = a_j\}$; note that this is indeed perpendicular to the root $L_i - L_j$ with respect to the Killing form B as described above.

Knowing the H_α we know the weight lattice: in order for a linear functional $\sum a_i L_i \in \mathfrak{h}^*$ to have integral values on all the distinguished elements, it is clearly necessary and sufficient that all the a_i be congruent to one another modulo $\mathbb{Z}$. Since $\sum L_i = 0$ in $\mathfrak{h}^*$, this means that the weight lattice is given as

$$\Lambda_W = \mathbb{Z}\{L_1, \ldots, L_n\}/(\textstyle\sum L_i = 0).$$

In sum, then, *the weight lattice of* $\mathfrak{sl}_n\mathbb{C}$ *may be realized as the lattice generated by the vertices of a regular* $(n-1)$*-simplex* Δ *centered at the origin; and the roots as the pairwise differences of these vertices.*

While we are at it, having determined Λ_R and Λ_W we might as well compute the quotient Λ_W/Λ_R. This is pretty easy: since the lattice Λ_W can be generated by Λ_R together with any of the vertices L_i of our simplex, the quotient Λ_W/Λ_R will be cyclic, generated by any L_i; since, modulo Λ_R,

$$0 = \textstyle\sum_j (L_i - L_j) = nL_i - \sum_j L_j = nL_i.$$

we see that L_i has order dividing n in Λ_W/Λ_R.

Exercise 15.7. Show that L_i has order exactly n in Λ_W/Λ_R, so that $\Lambda_W/\Lambda_R \cong \mathbb{Z}/n\mathbb{Z}$.

From the above we can also say what the Weyl group is: the reflection in the hyperplane perpendicular to the root $L_i - L_j$ will exchange L_i and $L_j \in \mathfrak{h}^*$ and leave the other L_k alone, so that *the Weyl group* $\mathfrak{W}$ *is just the group* $\mathfrak{S}_n$, *acting as the symmetric group on the generators* L_i *of* $\mathfrak{h}^*$. Note that we have already verified that these automorphisms of $\mathfrak{h}^*$ do come from automorphisms of the whole Lie algebra $\mathfrak{sl}_n\mathbb{C}$ preserving $\mathfrak{h}$.

To continue, let us choose a direction, and describe the corresponding Weyl chamber. We can write our linear functional l as

$$l(\textstyle\sum a_i L_i) = \sum c_i a_i$$

with $\sum c_i = 0$; let us suppose that $c_1 > c_2 > \cdots > c_n$. The corresponding ordering of the roots will then be

$$R^+ = \{L_i - L_j : i < j\}$$

and

$$R^- = \{L_i - L_j : j < i\}.$$

The primitive negative roots for this ordering are simply the roots $L_{i+1} - L_i$. (Note that the ordering of the roots depends only on the relative sizes of the c_i, so that the Weyl group acts simply transitively on the set of orderings.) The (closed) Weyl chamber associated to this ordering will then be the set

$$\mathscr{W} = \{\sum a_i L_i : a_1 \geq a_2 \geq \cdots \geq a_n\}.$$

One way to describe this geometrically is to say that if we take the barycentric subdivision of the faces of the simplex Δ, the Weyl chamber will be the cone over one $(n-2)$-simplex of the barycentric subdivision: e.g., in the case $n = 4$

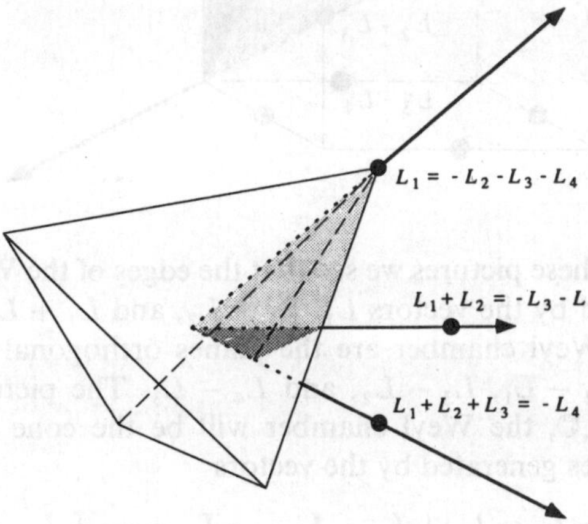

It may be easier to visualize the case $n = 4$ if we introduce the associated cubes: in terms of the cube with vertices at the points $\pm L_i$, we can draw the Weyl chamber as

(15.8)

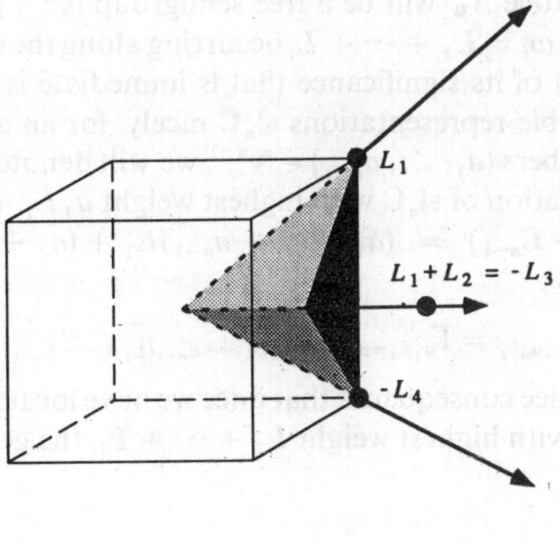

Alternatively, in terms of the slightly larger cube with vertices at the points $\pm 2L_i$, we can draw $\mathscr{W}$ as

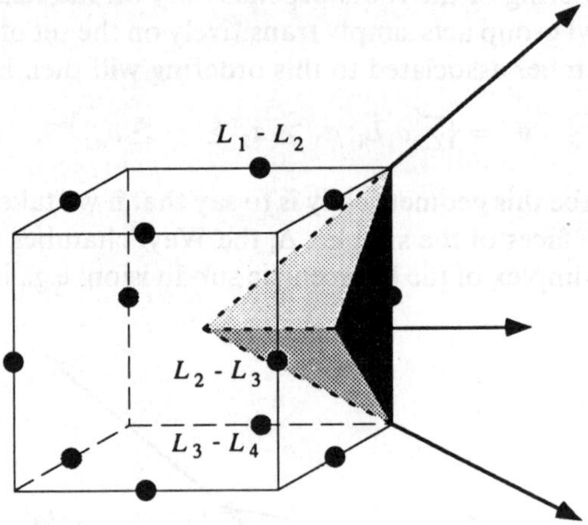

(15.9)

From the first of these pictures we see that the edges of the Weyl chamber are the rays generated by the vectors L_1, $L_1 + L_2$, and $L_1 + L_2 + L_3$; and that the faces of the Weyl chamber are the planes orthogonal to the primitive negative roots $L_2 - L_1$, $L_3 - L_2$, and $L_4 - L_3$. The picture in general is analogous: for $\mathfrak{sl}_n\mathbb{C}$, the Weyl chamber will be the cone over an $(n-2)$-simplex, with edges generated by the vectors

$$L_1, \quad L_1 + L_2, \quad L_1 + L_2 + L_3, \ldots, L_1 + \cdots + L_{n-1} = -L_n.$$

The faces of $\mathscr{W}$ will thus be the hyperplanes

$$\Omega_{L_i - L_{i+1}} = \{\sum a_j L_j \colon a_i = a_{i+1}\}$$

perpendicular to the primitive negative roots $L_{i+1} - L_i$.

Note the important phenomenon: the intersection of the closed Weyl chamber with the lattice Λ_W will be a free semigroup $\mathbb{N}^{n-1}$ generated by the fundamental weights $\omega_i = L_1 + \cdots + L_i$ occurring along the edges of the Weyl chamber. One aspect of its significance that is immediate is that it allows us to index the irreducible representations $\mathfrak{sl}_n\mathbb{C}$ nicely: for an arbitrary $(n-1)$-tuple of natural numbers $(a_1, \ldots, a_{n-1}) \in \mathbb{N}^{n-1}$ we will denote by $\Gamma_{a_1, \ldots, a_{n-1}}$ the irreducible representation of $\mathfrak{sl}_n\mathbb{C}$ with highest weight $a_1 L_1 + a_2(L_1 + L_2) + \cdots + a_{n-1}(L_1 + \cdots + L_{n-1}) = (a_1 + \cdots + a_{n-1})L_1 + (a_2 + \cdots + a_{n-1})L_2 + \cdots + a_{n-1}L_{n-1}$:

$$\Gamma_{a_1, \ldots, a_{n-1}} = \Gamma_{a_1 L_1 + a_2(L_1 + L_2) + \cdots + a_{n-1}(L_1 + \cdots + L_{n-1})}.$$

This also has the nice consequence that once we have located the irreducible representations $V^{(i)}$ with highest weight $L_1 + \cdots + L_i$, the general irreducible representation $\Gamma_{a_1,\ldots,a_{n-1}}$ with highest weight $\sum a_i(L_1 + \cdots + L_i)$ will occur inside the tensor product of symmetric powers

$$\mathrm{Sym}^{a_1}V^{(1)} \otimes \mathrm{Sym}^{a_2}V^{(2)} \otimes \cdots \otimes \mathrm{Sym}^{a_{n-1}}V^{(n-1)}$$

of these representations. Thus, the existence part of the basic Theorem 14.18 is reduced to finding the basic representations $V^{(i)}$; we will do this in due course, though at this point it is probably not too hard an exercise to guess what they are.

§15.2. Representations of $\mathfrak{sl}_4\mathbb{C}$ and $\mathfrak{sl}_n\mathbb{C}$

We begin as usual with the standard representation of $\mathfrak{sl}_4\mathbb{C}$ on $V = \mathbb{C}^4$. The standard basis vectors e_i of $\mathbb{C}^4$ are eigenvectors for the action of $\mathfrak{h}$, with eigenvalues L_i, so that the weight diagram looks like

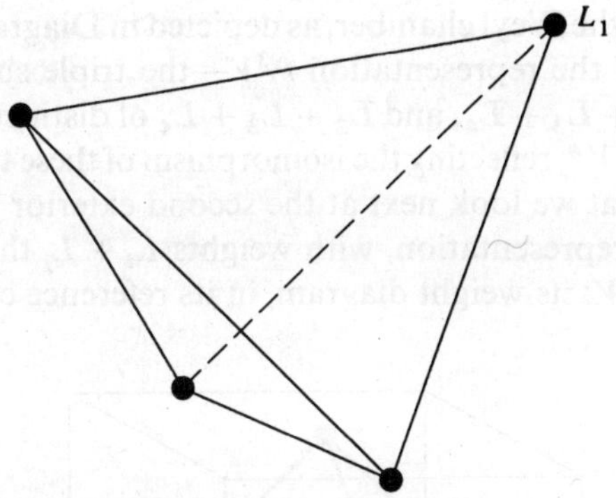

or, with the reference cube drawn as well,

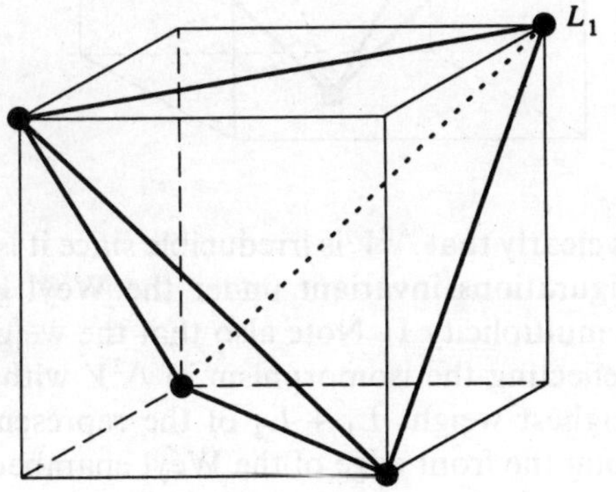

The dual representation V^* of course has weights $-L_i$ corresponding to the vectors of the dual basis e_i^* for V^*, so that the weight diagram, with its reference cube, looks like

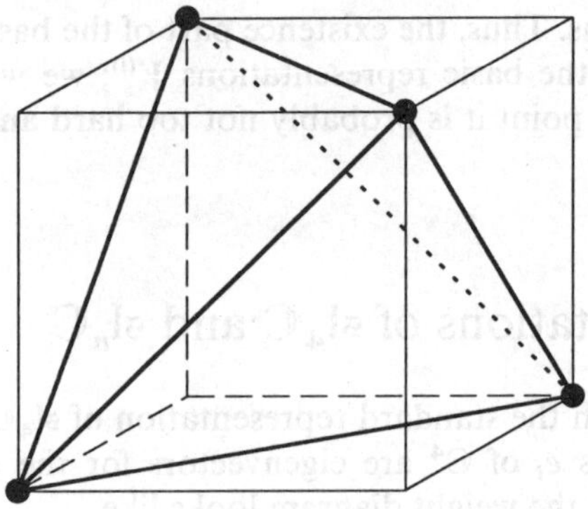

Note that the highest weight for this representation is $-L_4$, which lies along the bottom edge of the Weyl chamber, as depicted in Diagram (15.8). Note also that the weights of the representation $\wedge^3 V$—the triple sums $L_1 + L_2 + L_3$, $L_1 + L_2 + L_4$, $L_1 + L_3 + L_4$, and $L_2 + L_3 + L_4$ of distinct weights of V—are the same as those of V^*, reflecting the isomorphism of these two representations.

This suggests that we look next at the second exterior power $\wedge^2 V$. This is a six-dimensional representation, with weights $L_i + L_j$ the pairwise sums of distinct weights of V; its weight diagram, in its reference cube, looks like

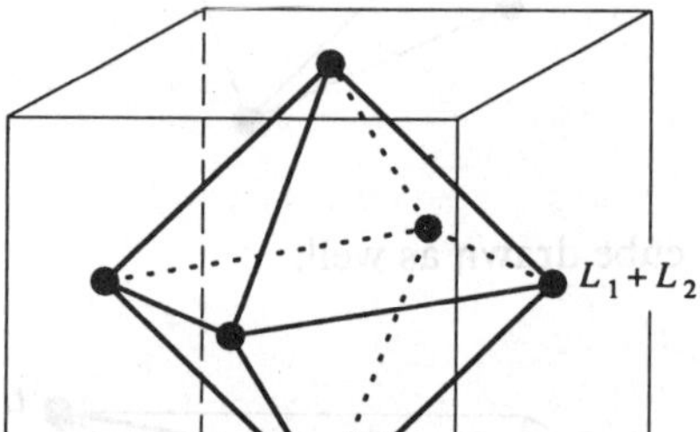

The diagram shows clearly that $\wedge^2 V$ is irreducible since it is not the nontrivial union of two configurations invariant under the Weyl group $\mathfrak{S}_4$ (and all weights occur with multiplicity 1). Note also that the weights are symmetric about the origin, reflecting the isomorphism of $\wedge^2 V$ with $(\wedge^2 V)^* = \wedge^2(V^*)$.

Note that the highest weight $L_1 + L_2$ of the representation $\wedge^2 V$ is the primitive vector along the front edge of the Weyl chamber $\mathscr{W}$ as pictured in Diagram (15.8). Now, we have already seen that the intersection of the closed

Weyl chamber with the weight lattice is a free semigroup generated by the primitive vectors along the three edges of $\mathscr{W}$—that is, every vector in $\mathscr{W} \cap \Lambda_W$ is a non-negative integral linear combination of the three vectors $L_1, L_1 + L_2$, and $L_1 + L_2 + L_3$. As we remarked at the end of the first section of this lecture, it follows that *we have proved the existence half of the general existence and uniqueness theorem* (14.18) *in the case of the Lie algebra* $\mathfrak{sl}_4\mathbb{C}$. Explicitly, since V, $\wedge^2 V$, and $\wedge^3 V = V^*$ have highest weight vectors with weights L_1, $L_1 + L_2$, and $L_1 + L_2 + L_3$, respectively, it follows that *the representation*

$$\mathrm{Sym}^a V \otimes \mathrm{Sym}^b(\wedge^2 V) \otimes \mathrm{Sym}^c(\wedge^3 V)$$

contains a highest weight vector with weight $aL_1 + b(L_1 + L_2) + c(L_1 + L_2 + L_3)$*, and hence a copy of the irreducible representation* $\Gamma_{a,b,c}$ *with this highest weight.*

Let us continue our examination of representations of $\mathfrak{sl}_4\mathbb{C}$ with a pair of tensor products of the three basic representations: $V \otimes \wedge^2 V$ and $V \otimes \wedge^3 V$. As for the first of these, its weights are easy to find: they consist of the sums $2L_i + L_j$ (which occur once, as the sum of L_i and $L_i + L_j$) and $L_i + L_j + L_k$ (which occur three times). The diagram of these weights looks like

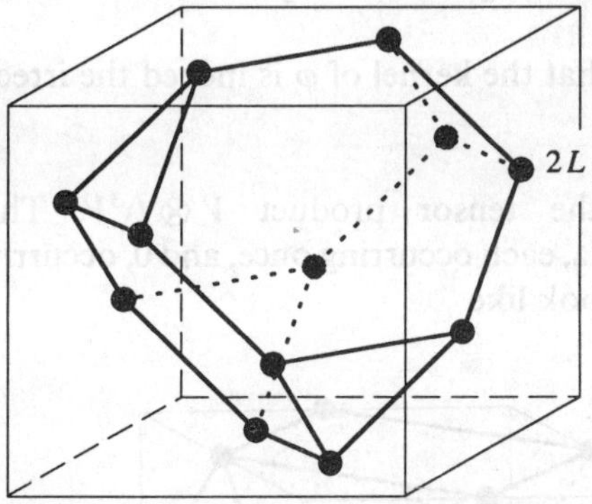

(We have drawn only the vertices of the convex hull of this diagram, thus omitting the weights $L_i + L_j + L_k$; they are located at the centers of the hexagonal faces of this polyhedron.)

Now, the representation $V \otimes \wedge^2 V$ cannot be irreducible, for at least a couple of reasons. First off, just by looking at weights, we see that the irreducible representation $W = \Gamma_{1,1,0}$ with highest weight $2L_1 + L_2$ can have multiplicity at most 2 on the weight $L_1 + L_2 + L_3$: by Observation 14.16, the weight space $W_{L_1+L_2+L_3}$ is generated by the images of the highest weight vector $v \in W_{2L_1+L_2}$ by successive applications of the primitive negative root spaces $\mathfrak{g}_{L_2-L_1}, \mathfrak{g}_{L_3-L_2}$, and $\mathfrak{g}_{L_4-L_3}$. But $L_1 + L_2 + L_3$ is uniquely expressible as a sum of $2L_1 + L_2$ and the primitive negative roots:

$$L_1 + L_2 + L_3 = 2L_1 + L_2 + (L_2 - L_1) + (L_3 - L_2);$$

so that $V_{L_1+L_2+L_3}$ is generated by the subspaces $\mathfrak{g}_{L_2-L_1}(\mathfrak{g}_{L_3-L_2}(v))$ and $\mathfrak{g}_{L_3-L_2}(\mathfrak{g}_{L_2-L_1}(v))$. We can in fact check that the representation $\Gamma_{1,1,0}$ takes on the weight $L_1 + L_2 + L_3$ with multiplicity 2 by writing out these generators explicitly and checking that they are independent: for example, we have

$$\begin{aligned}\mathfrak{g}_{L_2-L_1}(\mathfrak{g}_{L_3-L_2}(v)) &= \mathbb{C}\cdot E_{2,1}(E_{3,2}(e_1 \otimes (e_1 \wedge e_2))) \\ &= \mathbb{C}\cdot E_{2,1}(e_1 \otimes (e_1 \wedge e_3)) \\ &= \mathbb{C}\cdot(e_2 \otimes (e_1 \wedge e_3) + e_1 \otimes (e_2 \wedge e_3)).\end{aligned}$$

This is in fact what is called for in Exercise 15.10.

Alternatively, forgetting weights entirely, we can see from standard multilinear algebra that the representation $V \otimes \wedge^2 V$ cannot be irreducible: we have a natural map of representations

$$\varphi: V \otimes \wedge^2 V \to \wedge^3 V$$

which is obviously surjective. The kernel of this map is a representation with the same set of weights as $V \otimes \wedge^2 V$ (but taking on the weights $L_i + L_j + L_k$ with multiplicity 2 rather than 3), and so must contain the irreducible representation $\Gamma_{1,1,0}$ with highest weight $2L_1 + L_2$.

Exercise 15.10. Prove that the kernel of φ is indeed the irreducible representation $\Gamma_{1,1,0}$.

Finally, consider the tensor product $V \otimes \wedge^3 V$. This has weights $2L_i + L_k + L_l = L_i - L_j$, each occurring once, and 0, occurring four times. Its weight diagrams thus look like

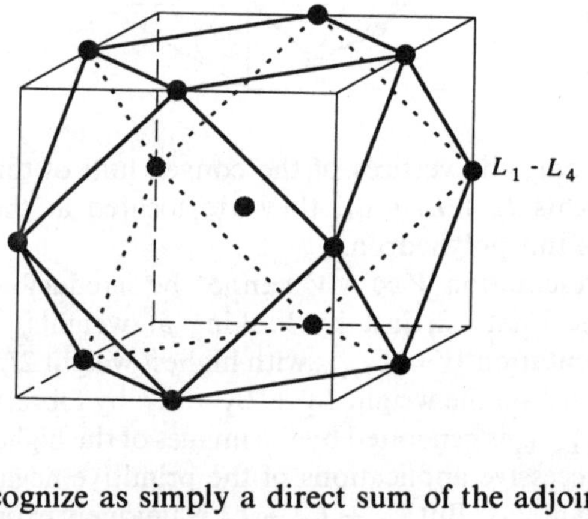

This we may recognize as simply a direct sum of the adjoint representation with a copy of the trivial; this corresponds to the kernel and image of the obvious contraction (or trace) map

$$V \otimes \Lambda^3 V = V \otimes V^* = \mathrm{Hom}(V, V) \to \mathbb{C}.$$

(Note that the adjoint representation is the irreducible representation with highest weight $2L_1 + L_2 + L_3$, or in other words the representation $\Gamma_{1,0,1}$.)

Exercise 15.11. Describe the weights of the representations $\mathrm{Sym}^n V$, and deduce that they are all irreducible.

Exercise 15.12. Describe the weights of the representations $\mathrm{Sym}^n(\Lambda^2 V)$, and deduce that they are not irreducible. Describe maps

$$\varphi_n\colon \mathrm{Sym}^n(\Lambda^2 V) \to \mathrm{Sym}^{n-2}(\Lambda^2 V)$$

and show that the kernel of φ_n is the irreducible representation with highest weight $n(L_1 + L_2)$.

Exercise 15.13. The irreducible representation $\Gamma_{1,1,1}$ with highest weight $3L_1 + 2L_2 + L_3$ occurs as a subrepresentation of the tensor product $V \otimes \Lambda^2 V \otimes \Lambda^3 V$ lying in the kernel of each of the three maps

$$V \otimes \Lambda^2 V \otimes \Lambda^3 V \to \Lambda^3 V \otimes \Lambda^3 V$$

$$V \otimes \Lambda^2 V \otimes \Lambda^3 V \to \Lambda^2 V \otimes \Lambda^4 V \cong \Lambda^2 V$$

$$V \otimes \Lambda^2 V \otimes \Lambda^3 V \cong V \otimes \Lambda^2 V^* \otimes V^* \to V \otimes \Lambda^3 V^* \cong V \otimes V$$

obtained by wedging two of the three factors. Is it equal to the intersection of these kernels? To test your graphic abilities, draw a diagram of the weights (ignoring multiplicities) of this representation.

Representations of $\mathfrak{sl}_n\mathbb{C}$

Once the case of $\mathfrak{sl}_4\mathbb{C}$ is digested, the case of the special linear group in general offers no surprises; the main difference in the general case is just the absence of pictures. Of course, the standard representation V of $\mathfrak{sl}_n\mathbb{C}$ has highest weight L_1, and similarly the exterior power $\Lambda^k V$ is irreducible with highest weight $L_1 + \cdots + L_k$. It follows that the irreducible representation $\Gamma_{a_1,\ldots,a_{n-1}}$ with highest weight $(a_1 + \cdots + a_{n-1})L_1 + \cdots + a_{n-1}L_{n-1}$ will appear inside the tensor product

$$\mathrm{Sym}^{a_1} V \otimes \mathrm{Sym}^{a_2}(\Lambda^2 V) \otimes \cdots \otimes \mathrm{Sym}^{a_{n-1}}(\Lambda^{n-1} V),$$

demonstrating the existence theorem (14.18) for representations of $\mathfrak{sl}_n\mathbb{C}$.

Exercise 15.14. Verify that the exterior powers of the standard representations of $\mathfrak{sl}_n\mathbb{C}$ are indeed irreducible (though this is not necessary for the truth of the last sentence).

§15.3. Weyl's Construction and Tensor Products

At the end of the preceding section, we saw that the irreducible representation $\Gamma_{a_1,\ldots,a_{n-1}}$ of $\mathfrak{sl}_n\mathbb{C}$ with highest weight $(a_1 + \cdots + a_{n-1})L_1 + \cdots + a_{n-1}L_{n-1}$ will appear as a subspace of the tensor product

$$\mathrm{Sym}^{a_1}V \otimes \mathrm{Sym}^{a_2}(\wedge^2 V) \otimes \cdots \otimes \mathrm{Sym}^{a_{n-1}}(\wedge^{n-1}V),$$

or equivalently as a subspace of the dth tensor power $V^{\otimes d}$ of the standard representation V. The natural question is, how can we describe this subspace? We have seen the answer in one case already (two cases, if you count the trivial answer $\Gamma_a = \mathrm{Sym}^a V$ in the case $n = 2$): the representation $\Gamma_{a,b}$ of $\mathfrak{sl}_3\mathbb{C}$ can be realized as the kernel of the contraction map

$$\mathrm{Sym}^a V \otimes \mathrm{Sym}^b(\wedge^2 V) \to \mathrm{Sym}^{a-1}V \otimes \mathrm{Sym}^{b-1}(\wedge^2 V).$$

This raises the question of whether the representation $\Gamma_{\mathbf{a}}$ can in general be described as a subspace of the tensor power $\bigotimes(\mathrm{Sym}^{a_i}(\wedge^i V))$ by intersecting kernels of such contraction/wedge product maps. Specifically, for i and j with $i + j \leq n$ we can define maps

$$\begin{aligned}\mathrm{Sym}^{a_1}V \otimes \mathrm{Sym}^{a_2}(\wedge^2 V) &\otimes \cdots \otimes \mathrm{Sym}^{a_{n-1}}(\wedge^{n-1}V)\\ &\to \wedge^i V \otimes \wedge^j V \otimes \mathrm{Sym}^{a_1}V \otimes \cdots \otimes \mathrm{Sym}^{a_i-1}(\wedge^i V) \otimes \cdots\\ &\qquad \otimes \mathrm{Sym}^{a_j-1}(\wedge^j V) \otimes \cdots \otimes \mathrm{Sym}^{a_{n-1}}(\wedge^{n-1}V)\end{aligned}$$

and we have similar maps for $i < j$ with $i + j \geq n$ and i even with $2i \geq n$; there are likewise analogously defined maps in which we split off three or more factors. The representation $\Gamma_{a_1,\ldots,a_{n-1}}$ is in the kernel of all such maps; and we may ask whether the intersection of all such kernels is equal to $\Gamma_{\mathbf{a}}$.

The answer, it turns out, is no. (It is a worthwhile exercise to find an example of a representation $\Gamma_{\mathbf{a}}$ that cannot be realized in this way.) There is, however, another way of describing $\Gamma_{\mathbf{a}}$ as a subspace of $V^{\otimes d}$: in fact, we have already met these representations in Lecture 6, under the guise of *Schur functors* or *Weyl modules*. In fact, at the end of this lecture we will see how to describe them explicitly as subspaces of the above spaces $\bigotimes(\mathrm{Sym}^{a_i}(\wedge^i V))$. Recall that for $V = \mathbb{C}^n$ an n-dimensional vector space, and any partition

$$\lambda\colon \lambda_1 \geq \lambda_2 \geq \cdots \geq \lambda_n \geq 0,$$

we can apply the Schur functor $\mathbb{S}_\lambda$ to V to obtain a representation $\mathbb{S}_\lambda V = \mathbb{S}_\lambda(\mathbb{C}^n)$ of $\mathrm{GL}(V) = \mathrm{GL}_n(\mathbb{C})$. If $d = \sum \lambda_i$, this was realized as

$$\mathbb{S}_\lambda V = V^{\otimes d} \cdot c_\lambda = V^{\otimes d} \otimes_{\mathbb{C}\mathfrak{S}_d} V_\lambda,$$

where c_λ is the Young symmetrizer corresponding to λ, and V_λ is the irreducible representation of $\mathfrak{S}_d$ corresponding to λ.

We saw in Lecture 6 that $\mathbb{S}_\lambda V$ is an irreducible representation of $\mathrm{GL}_n\mathbb{C}$. It follows immediately that $\mathbb{S}_\lambda V$ remains irreducible as a representation of $\mathrm{SL}_n\mathbb{C}$, since any element of $GL_n\mathbb{C}$ is a scalar multiple of an element of $SL_n\mathbb{C}$. In particular, it determines an irreducible representation of the Lie algebra $\mathfrak{sl}_n\mathbb{C}$.

Proposition 15.15. *The representation $\mathbb{S}_\lambda(\mathbb{C}^n)$ is the irreducible representation of $\mathfrak{sl}_n\mathbb{C}$ with highest weight $\lambda_1 L_1 + \lambda_2 L_2 + \cdots + \lambda_n L_n$.*

In particular, $\mathbb{S}_\lambda(\mathbb{C}^n)$ and $\mathbb{S}_\mu(\mathbb{C}^n)$ are isomorphic representations of $\mathfrak{sl}_n\mathbb{C}$ if and only if $\lambda_i - \mu_i$ is constant, independent of i. To relate this to our earlier notation, we may say that the irreducible representation $\Gamma_{a_1,\ldots,a_{n-1}}$ of $\mathfrak{sl}_n\mathbb{C}$ with highest weight $a_1 L_1 + a_2(L_1 + L_2) + \cdots + a_{n-1}(L_1 + \cdots + L_{n-1})$ is obtained by applying the Schur functor $\mathbb{S}_\lambda$ to the standard representation V, where

$$\lambda = (a_1 + \cdots + a_{n-1}, a_2 + \cdots + a_{n-1}, \ldots, a_{n-1}, 0).$$

(If we want a unique Schur functor for each representation, we can restrict to those λ with $\lambda_n = 0$.) In terms of the Young diagram for λ, the coefficients $a_i = \lambda_i - \lambda_{i+1}$ are the differences of lengths of rows. For example, if $n = 6$,

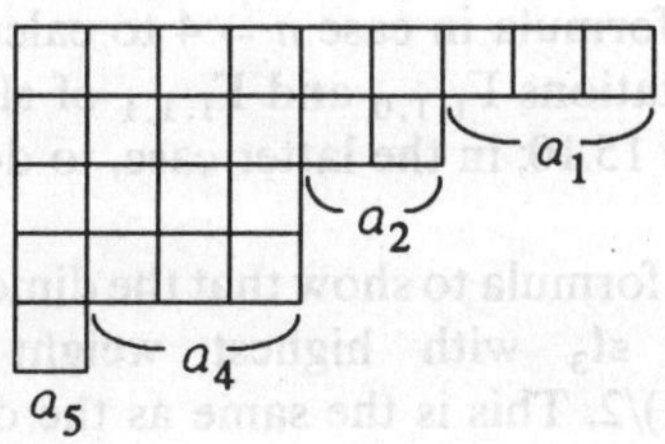

is the Young diagram corresponding to $\Gamma_{3,2,0,3,1}$.

Proof of the Proposition. In Theorem 6.3 we calculated that the trace of a diagonal matrix with entries $x_1, \ldots, x_n$ on $\mathbb{S}_\lambda(\mathbb{C}^n)$ is the Schur polynomial $S_\lambda(x_1, \ldots, x_n)$. By Equation (A.19), when the Schur polynomial is written out it takes the form

$$S_\lambda(x_1, \ldots, x_n) = M_\lambda + \sum_{\mu < \lambda} K_{\lambda\mu} M_\mu, \tag{15.16}$$

where M_μ is the sum of the monomial $X^\mu = x_1^{\mu_1} x_2^{\mu_2} \cdot \ldots \cdot x_n^{\mu_n}$ and all distinct monomials obtained from it by permuting the variables, and the $K_{\lambda\mu}$ are certain non-negative integers called Kostka numbers. When $\mathbb{S}_\lambda(\mathbb{C}^n)$ is diagonalized with respect to the group of diagonal matrices in $GL_n(\mathbb{C})$, it is also diagonalized with respect to $\mathfrak{h} \subset \mathfrak{sl}_n(\mathbb{C})$. There is one monomial in the displayed equation for each one-dimensional eigenspace. The weights of $\mathbb{S}_\lambda(\mathbb{C}^n)$ as a representation of $\mathfrak{sl}_n(\mathbb{C})$ therefore consist of all

$$\mu_1 L_1 + \mu_2 L_2 + \cdots + \mu_n L_n,$$

each occurring as often as it does in the monomial X^μ in the polynomial

$S_\lambda(x_1, \ldots, x_n)$. Since the sum is over those partitions μ for which the first nonzero $\lambda_i - \mu_i$ is positive, the highest weight that appears is $\lambda_1 L_1 + \lambda_2 L_2 + \cdots + \lambda_n L_n$, which concludes the proof. [In fact one can describe an explicit basis of eigenvectors for $\mathbb{S}_\lambda(\mathbb{C}^n)$ which correspond to the monomials that appear in (15.16), cf. Problem 6.15 or Proposition 15.55.] □

In particular, we have (by Theorem 6.3) formulas for the dimension of the representation with given highest weight. Explicitly, one formula says that

$$\dim(\Gamma_{a_1,\ldots,a_{n-1}}) = \prod_{1\le i<j\le n} \frac{(a_i + \cdots + a_{j-1}) + j - i}{j - i}. \tag{15.17}$$

As we saw in the proof, this proposition also gives the multiplicities of all weight spaces as the integers $K_{\lambda\mu}$ that appear in (15.16), which have a simple combinatorial description (p. 456): *the dimension of the weight space with weight μ in the representation $\mathbb{S}_\lambda(\mathbb{C}^n)$ is the number of ways one can fill the Young diagram of λ with μ_1 1's, μ_2 2's, ..., μ_n n's, in such a way that the entries in each row are nondecreasing and those in each column are strictly increasing.*

Exercise 15.18. Use the formula in case $n = 4$ to calculate the dimensions of the irreducible representations $\Gamma_{1,1,0}$ and $\Gamma_{1,1,1}$ of $\mathfrak{sl}_4\mathbb{C}$. In the former case, use this to redo Exercise 15.10; in the latter case, to do Exercise 15.13.

Exercise 15.19*. Use this formula to show that the dimension of the irreducible representation $\Gamma_{a,b}$ of $\mathfrak{sl}_3$ with highest weight $aL_1 + b(L_1 + L_2)$ is $(a + b + 1)(a + 1)(b + 1)/2$. This is the same as the dimension of the kernel of the contraction map

$$\iota_{a,b}\colon \mathrm{Sym}^a V \otimes \mathrm{Sym}^b V^* \to \mathrm{Sym}^{a-1} V \otimes \mathrm{Sym}^{b-1} V^*.$$

Use this to give another proof of the assertion made in Claim 13.4 that $\Gamma_{a,b}$ is this kernel.

Exercise 15.20*. As an application of the above formula, show that if V is the standard representation of $\mathfrak{sl}_n\mathbb{C}$, then the kernel of the wedge product map

$$V \otimes \wedge^k V \to \wedge^{k+1} V$$

is the irreducible representation $\Gamma_{1,0,\ldots,0,1,0,\ldots}$ with highest weight $2L_1 + L_2 + \cdots + L_k$; and that the irreducible representation $\Gamma_{k-1,1,0,\ldots}$ with highest weight $k \cdot L_1 + L_2$ is the kernel of the product map

$$V \otimes \mathrm{Sym}^k V \to \mathrm{Sym}^{k+1} V.$$

Exercise 15.21*. Show that the only nontrivial irreducible representations of $\mathfrak{sl}_n\mathbb{C}$ of dimension less than or equal to n are V and V^*.

One important consequence of the fact that the irreducible representations of $\mathfrak{sl}_n\mathbb{C}$ are obtained by applying Schur functors to the standard representation is that *identities among the Schur–Weyl functors give rise to identities among representations of* GL_n (and hence SL_n and $\mathfrak{sl}_n$), as we saw in Lecture 6. For example, the representation

$$\mathrm{Sym}^{\lambda_1}(V) \otimes \mathrm{Sym}^{\lambda_2}(V) \otimes \cdots \otimes \mathrm{Sym}^{\lambda_n}(V) \tag{15.22}$$

is a direct sum of representations $\mathbb{S}_\lambda(V) \oplus \bigoplus_\mu K_{\mu\lambda} \mathbb{S}_\mu(V)$, where $K_{\mu\lambda}$ is the coefficient described above. The particular application of this principle that we will use most frequently in the sequel, however, is the consequence that *one knows the decomposition of a tensor product of any two irreducible representations of* $\mathfrak{sl}_n\mathbb{C}$: specifically, the tensor power $\mathbb{S}_\lambda(V) \otimes \mathbb{S}_\mu(V)$ decomposes into a direct sum of irreducible representations

$$\mathbb{S}_\lambda(V) \otimes \mathbb{S}_\mu(V) = \bigoplus_\nu N_{\lambda\mu\nu} \mathbb{S}_\nu(V), \tag{15.23}$$

where the coefficients $N_{\lambda\mu\nu}$ are given by the *Littlewood–Richardson rule*, which is a formula in terms of the number of ways to fill the Young diagram between λ and ν with μ_1 1's, μ_2 2's, ..., μ_n n's, satisfying a certain combinatorial condition described in (A.8).

Exercise 15.24. Use the Littlewood–Richardson rule to show that the representation $\Gamma_{a_1+b_1,\ldots,a_{n-1}+b_{n-1}}$ occurs exactly once in the tensor product $\Gamma_{a_1,\ldots,a_{n-1}} \otimes \Gamma_{b_1,\ldots,b_{n-1}}$.

A special case of this is the analogue of Pieri's formula, which allows us to decompose the tensor product of an arbitrary irreducible representation with either $\mathrm{Sym}^k V = \Gamma_{k,0,\ldots,0}$ or the fundamental representation $\bigwedge^k V = \Gamma_{0,\ldots,1,0,\ldots,0}$, (where the 1 occurs in the kth place):

Proposition 15.25. (i) *The tensor product of* $\Gamma_{a_1,\ldots,a_{n-1}}$ *with* $\mathrm{Sym}^k V = \Gamma_{k,0,\ldots,0}$ *decomposes into a direct sum*:

$$\Gamma_{a_1,\ldots,a_{n-1}} \otimes \Gamma_{k,\ldots,0} = \bigoplus \Gamma_{b_1,\ldots,b_{n-1}},$$

the sum over all $(b_1,\ldots,b_{n-1})$ *for which there are non-negative integers* $c_1,\ldots,c_n$ *whose sum is* k, *with* $c_{i+1} \le a_i$ *for* $1 \le i \le n-1$, *and with* $b_i = a_i + c_i - c_{i+1}$ *for* $1 \le i \le n-1$.

(ii) *The tensor product of* $\Gamma_{a_1,\ldots,a_{n-1}}$ *with* $\bigwedge^k V = \Gamma_{0,\ldots,0,1,0,\ldots,0}$ *decomposes into a direct sum*:

$$\Gamma_{a_1,\ldots,a_{n-1}} \otimes \Gamma_{0,\ldots,0,1,0,\ldots,0} = \bigoplus \Gamma_{b_1,\ldots,b_{n-1}},$$

the sum over all $(b_1,\ldots,b_{n-1})$ *for which there is a subset* S *of* $\{1,\ldots,n\}$ *of cardinality* k, *such that if* $i \notin S$ *and* $i+1 \in S$, *then* $a_i > 0$, *with*

$$b_i = \begin{cases} a_i - 1 & \text{if } i \notin S \text{ and } i+1 \in S \\ a_i + 1 & \text{if } i \in S \text{ and } i+1 \notin S \\ a_i & \text{otherwise.} \end{cases}$$

PROOF. This is simply a matter of translating the prescriptions of (6.8) and (6.9), which describe the decompositions in terms of adding boxes to the Young diagrams. In (i), the c_i are the number of boxes added to the ith row, and in (ii), S is the set of rows to which a box is added. □

Exercise 15.26. Verify the descriptions in Section 2 of this lecture of $V \otimes \wedge^2 V$ and $V \otimes \wedge^3 V$, where V is the standard representation of $\mathfrak{sl}_4\mathbb{C}$.

Exercise 15.27. Use Pieri's formula (with $n = 4$) twice to find the decomposition into irreducibles of $V \otimes \wedge^2 V \otimes \wedge^3 V$, where V is the standard representation of $\mathfrak{sl}_4\mathbb{C}$. Use this to redo Exercise 15.13.

Exercise 15.28. Use Pieri's formula to prove (13.5). You may also want to look around in Lecture 13 to see which other of the decompositions found there by hand may be deduced from these formulas.

Exercise 15.29. Verify that the statement of Exercise 15.20 follows directly from Pieri's formula.

In the following exercises, $V = \mathbb{C}^n$ is the standard representation of $\mathfrak{sl}_n\mathbb{C}$.

Exercise 15.30. Consider now tensor products of the form $\wedge^k V \otimes \wedge^l V$, with, say, $k \geq l$. Show that there is a natural map

$$\wedge^k V \otimes \wedge^l V \to \wedge^{k+1} V \otimes \wedge^{l-1} V$$

given by contraction with the element "trace" (or "identity") in $V \otimes V^* = \operatorname{End}(V)$. Explicitly, this map may be given by

$$(v_1 \wedge \cdots \wedge v_k) \otimes (w_1 \wedge \cdots \wedge w_l)$$
$$\mapsto \sum_{i=1}^{l} (-1)^i (v_1 \wedge \cdots \wedge v_k \wedge w_i) \otimes (w_1 \wedge \cdots \wedge \widehat{w_i} \wedge \cdots \wedge w_l).$$

What is the image of this map? Show that the kernel is the irreducible representation $\Gamma_{0,\ldots,0,1,0,\ldots,0,1,0,\ldots}$ with highest weight $2L_1 + \cdots + 2L_l + L_{l+1} + \cdots + L_k$.

Exercise 15.31*. Carry out an analysis similar to that of the preceding exercise for the maps

$$\operatorname{Sym}^k V \otimes \operatorname{Sym}^l V \to \operatorname{Sym}^{k+1} V \otimes \operatorname{Sym}^{l-1} V$$

defined analogously.

Exercise 15.32*. As a special case of Pieri's formula, we see that if V is the standard representation of $\mathfrak{sl}_n\mathbb{C}$, the tensor product

$$\textstyle\bigwedge^k V \otimes \bigwedge^k V = \bigoplus \mathbb{S}_{(2,\ldots,2,1,\ldots,1,0,\ldots)}(V)$$
$$= \bigoplus \Gamma_{0,\ldots,0,1,0,\ldots,0,1,0,\ldots},$$

where in the ith factor the 1's occur in the $(k-i)$th and $(k+i)$th places. At the same time, of course, we know that

$$\textstyle\bigwedge^k V \otimes \bigwedge^k V = \mathrm{Sym}^2(\bigwedge^k V) \oplus \bigwedge^2(\bigwedge^k V).$$

If we denote the ith term on the right-hand side of the first displayed equation for $\bigwedge^k V \otimes \bigwedge^k V$ by Θ_i, show that

$$\textstyle\mathrm{Sym}^2(\bigwedge^k V) = \bigoplus \Theta_{2i} \quad \text{and} \quad \bigwedge^2(\bigwedge^k V) = \bigoplus \Theta_{2i+1}.$$

Exercise 15.33*. As another special case of Pieri's formula, we see that the tensor product

$$\mathrm{Sym}^k V \otimes \mathrm{Sym}^k V = \bigoplus \mathbb{S}_{(k+i,k-i)}(V)$$
$$= \bigoplus \Gamma_{2i,k-i,0\ldots0}.$$

At the same time, of course, we know that

$$\mathrm{Sym}^k V \otimes \mathrm{Sym}^k V = \mathrm{Sym}^2(\mathrm{Sym}^k V) \oplus \textstyle\bigwedge^2(\mathrm{Sym}^k V).$$

Which of the factors appearing in the first decomposition lie in $\mathrm{Sym}^2(\mathrm{Sym}^k V)$, and which in $\bigwedge^2(\mathrm{Sym}^k V)$?

It follows from the Littlewood–Richardson rule that if λ, μ, and ν all have at most two rows, then the coefficient $N_{\lambda\mu\nu}$ is zero or one (and it is easy to say which occurs). In particular, for the Lie algebras $\mathfrak{sl}_2\mathbb{C}$ and $\mathfrak{sl}_3\mathbb{C}$, the decomposition of the tensor product of two irreducible representations is always multiplicity free. Groups whose representations have this property, such as SU(2), SU(3), and SO(3) which are so important in physics, are called "simply reducible," cf. [Mack].

§15.4. Some More Geometry

Let V be an n-dimensional vector space, and $G(k, n) = G(k, V) = \mathrm{Grass}_k V$ the Grassmannian of k-planes in V. $\mathrm{Grass}_k V$ is embedded as a subvariety of the projective space $\mathbb{P}(\bigwedge^k V)$ by the *Plücker embedding*:

$$\rho\colon \mathrm{Grass}_k V \hookrightarrow \mathbb{P}(\textstyle\bigwedge^k V)$$

sending the plane W spanned by vectors $v_1, \ldots, v_k$ to the alternating tensor $v_1 \wedge \cdots \wedge v_k$. Equivalently, noting that if $W \subset V$ is a k-dimensional subspace, then $\bigwedge^k W$ is a line in $\bigwedge^k V$, we may write this simply as

$$\rho\colon W \mapsto \textstyle\bigwedge^k W.$$

This embedding is compatible with the action of the general linear group:

$$\mathrm{PSL}_n\mathbb{C} = \mathrm{Aut}(\mathbb{P}(V)) = \{\sigma \in \mathrm{Aut}(\mathbb{P}(\wedge^k V)) : \sigma(G(k, V)) = G(k, V)\}^\circ.$$

This follows from a fact in algebraic geometry ([Ha]): all automorphisms of the Grassmannian are induced by automorphisms of V, unless $n = 2k$, in which case we can choose an arbitrary isomorphism of V with V^* and compose these with the automorphism that takes W to $(\mathbb{C}^n/W)^*$. Here the superscript $\circ$ denotes the connected component of the identity. As in previous lectures, if we want symmetric powers to correspond to homogeneous polynomials on projective space, we should consider the dual situation: $G = \mathrm{Grass}^k V$ is the Grassmannian of k-dimensional *quotient* spaces of V, and the Plücker embedding embeds G in the projective space $\mathbb{P}(\wedge^k V^*)$ of one-dimensional quotients of $\wedge^k V$.

The space of all homogeneous polynomials of degree m on $\mathbb{P}(\wedge^k V^*)$ is naturally the symmetric power $\mathrm{Sym}^m(\wedge^k V)$. Let $I(G)_m$ denote the subspace of those polynomials of degree m on $\mathbb{P}(\wedge^k V^*)$ that vanish on G. Each $I(G)_m$ is a representation of $\mathfrak{sl}_n\mathbb{C}$:

$$0 \to I(G)_m \to \mathrm{Sym}^m(\wedge^k V) \to W_m \to 0,$$

where W_m denotes the restrictions to G of the polynomials of degree m on the ambient space $\mathbb{P}(\wedge^k V^*)$. We shall see later that W_m is the irreducible representation $\Gamma_{0,\ldots,0,m,0,\ldots}$ with highest weight $m(L_1 + \cdots + L_k)$ (the case $m = 2$ will be dealt with below). In the following discussion, we consider the problem of describing the quadratic part $I(G)_2$ of the ideal as a representation of $\mathfrak{sl}_n\mathbb{C}$.

Exercise 15.34. Consider the first case of a Grassmannian that is not a projective space, that is, $k = 2$. The ideal of the Grassmannian $G(2, V)$ of 2-planes in a vector space is easy to describe: a tensor $\varphi \in \wedge^2 V$ is decomposable if and only if $\varphi \wedge \varphi = 0$ (equivalently, if we think of φ as given by a skew-symmetric $n \times n$ matrix, if and only if the Pfaffians of symmetric 4×4 minors all vanish); and indeed the quadratic relations we get in this way generate the ideal of the Grassmannian. We, thus, have an isomorphism

$$I(G)_2 \cong \wedge^4 V$$

and correspondingly a decomposition into irreducibles

$$\mathrm{Sym}^2(\wedge^2 V) \cong \wedge^4 V \oplus \Gamma_{0,2,0,\ldots,0},$$

where $\Gamma_{0,2,0,\ldots,0}$ is, as above, the irreducible representation with highest weight $2(L_1 + L_2)$, cf. Exercise 15.32.

Exercise 15.35. When $k = 2$ and $n = 4$, G is a quadric hypersurface in $\mathbb{P}^5$, so polynomials vanishing on G are simply those divisible the quadratic polynomial that defines G. Deduce an isomorphism.

$$I(G)_m = \mathrm{Sym}^{m-2}(\wedge^2 V).$$

The first case of a Grassmannian that is not a projective space or of the form $G(2, V)$ is, of course, $G(3, 6)$, and this yields an interesting example.

Exercise 15.36. Let V be six dimensional. By examining weights, show that the space $I(G)_2$ of quadratic polynomials vanishing on the Grassmannian $G(3, V) \subset \mathbb{P}(\wedge^3 V)$ is isomorphic to the adjoint representation of $\mathfrak{sl}_6\mathbb{C}$, i.e., that we have a map

$$\varphi\colon \mathrm{Sym}^2(\wedge^3 V) \to V \otimes V^*$$

with image the space of traceless matrices.

Exercise 15.37. Find explicitly the map φ of the preceding exercise.

Exercise 15.38. Again, let V be six dimensional. Show that the representation $\mathrm{Sym}^4(\wedge^3 V)$ has a trivial direct summand, corresponding to the hypersurface in $\mathbb{P}(\wedge^3 V^*)$ dual to the Grassmannian $G = G(3, V) \subset \mathbb{P}(\wedge^3 V)$.

In general, the ideal $I(G) = \bigoplus I(G)_m$ is generated by the famous *Plücker equations*. These are homogeneous polynomials of degree two, and may be written down explicitly, cf. (15.53), [H-P], or [Ha]. In the following exercises, we will give a more intrinsic description of these relations, which will allow us to identify the space $I(G)_2$ they span as a representation on $\mathfrak{sl}_n\mathbb{C}$ (and to see the general pattern of which the above are special cases).

Exercise 15.39. For a given tensor $\Lambda \in \wedge^k V$, we introduce two associated subspaces:

$$W = \{v \in V\colon v \wedge \Lambda = 0\} \subset V$$

and

$$W^* = \{v^* \in V^*\colon v^* \wedge \Lambda^* = 0\} \subset V^*,$$

where, abusing notation slightly, Λ^* is the tensor Λ viewed as an element of $\wedge^k V = \wedge^{n-k} V^*$. Show that the dimensions of W and W^* are at most k and $n - k$, respectively, and that Λ is decomposable if and only if W has dimension k or W^* has dimension $n - k$; and deduce that Λ is decomposable if and only if the annihilator W' of W^* is equal to W.

Exercise 15.40. Now let $\Xi \in \wedge^{k+1} V^* = \wedge^{n-k-1} V$. Wedge product gives a map

$$\iota_\Xi\colon \wedge^k V \to \wedge^{n-1} V = V^*.$$

Using the preceding exercise, show that Λ is decomposable if and only if

$$\iota_\Xi(\Lambda) \wedge \Lambda = 0 \in \wedge^{k-1} V$$

for all $\Xi \in \wedge^{k+1} V^*$.

Exercise 15.41. Observe that in the preceding exercise we construct a map

$$\wedge^{k+1}V^* \otimes \mathrm{Sym}^2(\wedge^k V) \to \wedge^{k-1}V,$$

or, by duality, a map

$$\wedge^{k+1}V^* \otimes \wedge^{k-1}V^* \to \mathrm{Sym}^2(\wedge^k V^*) \tag{15.42}$$

whose image is a vector space of quadrics on $\mathbb{P}(\wedge^k V)$ whose common zeros are exactly the locus of decomposable vectors, that is, the Grassmannian $G(k, V)$. Show that this image is exactly the span of the Plücker relations above.

Exercise 15.43. Show that the map (15.42) of the preceding exercise is just the dual of the map constructed in Exercise 15.30, with $k = l$ and restricted to the symmetric product. Combining this with the result of Exercise 15.32 (and assuming the statement that the Plücker relations do indeed span $I(G)_2$), deduce that in terms of the description

$$\mathrm{Sym}^2(\wedge^k V) = \bigoplus \Theta_{2i}$$

of the symmetric square of $\wedge^k V$, we have

$$W_2 = \Theta_0 = \Gamma_{0,\ldots,0,2,0,\ldots}$$

(the irreducible representation with highest weight $2(L_1 + \cdots + L_k)$), and

$$I(G)_2 = \bigoplus_{i \geq 1} \Theta_{2i}.$$

Hard Exercise 15.44. Show that in the last equation the sub-direct sum

$$I(l) = \bigoplus_{i \geq l} \Theta_{2i}$$

is just the quadratic part of the ideal of the *restricted chordal variety* of the Grassmannian: that is, the union of the chords $\overline{LM}$ joining pairs of points in G corresponding to pairs of planes L and M meeting in a subspace of dimension at least $k - 2l + 1$. (Question: What is the actual zero locus of these quadrics?)

Exercise 15.45. Carry out an analysis similar to the above to relate the ideal of a Veronese variety $\mathbb{P}V^* \subset \mathbb{P}(\mathrm{Sym}^k V^*)$ to the decomposition given in Exercise 15.33 of $\mathrm{Sym}^2(\mathrm{Sym}^k V)$. For which k do the quadratic polynomials vanishing the Veronese give an irreducible representation?

Exercise 15.46. (For algebraic geometers and/or commutative algebraists.) Just as the group $\mathrm{PGL}_n\mathbb{C}$ acts on the ring S of polynomials on projective space $\mathbb{P}^N$, preserving the ideal of the Veronese variety, so it acts on that space of relations on the ideal (that is, inasmuch as the ideal is generated by quadrics, the kernel of the multiplication map $I_X(2) \otimes S \to S$), and likewise on the entire minimal resolution of the ideal of X. Show that this resolution has the form

$$\cdots \to R_2 \otimes S \to R_1 \otimes S \to I_X(2) \otimes S,$$

where all the R_i are finite-dimensional representations of $\mathrm{PGL}_n\mathbb{C}$, and identify the representations R_i in the specific cases of

(i) the rational normal curve in $\mathbb{P}^3$,
(ii) the rational normal curve in $\mathbb{P}^4$, and
(iii) the Veronese surface in $\mathbb{P}^5$.

§15.5. Representations of $\mathrm{GL}_n\mathbb{C}$

We have said that there is little difference between representations of $\mathrm{GL}_n\mathbb{C}$ and those of the subgroup $\mathrm{SL}_n\mathbb{C}$ of matrices of determinant 1. Our object here is to record the difference, which, naturally enough, comes from the determinant: if $V = \mathbb{C}^n$ is the standard representation, $\wedge^n V$ is trivial for $\mathrm{SL}_n\mathbb{C}$ but not for $\mathrm{GL}_n\mathbb{C}$. Similarly, V and $\wedge^{n-1}V^*$ are isomorphic for $\mathrm{SL}_n\mathbb{C}$ but not for $\mathrm{GL}_n\mathbb{C}$.

To relate representations of $\mathrm{SL}_n\mathbb{C}$ and $\mathrm{GL}_n\mathbb{C}$, we first need to define some representations of $\mathrm{GL}_n\mathbb{C}$. To begin with, let D_k denote the one-dimensional representation of $\mathrm{GL}_n\mathbb{C}$ given by the kth power of the determinant. When k is non-negative, $D_k = (\wedge^n V)^{\otimes k}$; D_{-k} is the dual $(D_k)^*$ of D_k. Next, note that the irreducible representations of $\mathrm{SL}_n\mathbb{C}$ may be lifted to representations of $\mathrm{GL}_n\mathbb{C}$ in two ways. First, for any index $\mathbf{a} = (a_1, \ldots, a_n)$ of length n we may take $\Phi_{\mathbf{a}}$ to be the subrepresentation of the tensor product

$$\mathrm{Sym}^{a_1}V \otimes \cdots \otimes \mathrm{Sym}^{a_{n-1}}(\wedge^{n-1}V) \otimes \mathrm{Sym}^{a_n}(\wedge^n V)$$

spanned by the highest weight vector with weight $a_1 L_1 + a_2(L_1 + L_2) + \cdots + a_{n-1}(L_1 + \cdots + L_{n-1})$—that is, the vector

$$v = (e_1)^{a_1} \cdot (e_1 \wedge e_2)^{a_2} \cdot \ldots \cdot (e_1 \wedge \cdots \wedge e_n)^{a_n}.$$

This restricts to $\mathrm{SL}_n\mathbb{C}$ to give the representation $\Gamma_{a'}$, where $a' = (a_1, \ldots, a_{n-1})$; taking different values of a_n amounts to tensoring the representation with different factors $\mathrm{Sym}^{a_n}(\wedge^n V) = (\wedge^n V)^{\otimes a_n} = D_{a_n}$. In particular, we have

$$\Phi_{a_1, \ldots, a_n + k} = \Phi_{a_1, \ldots, a_n} \otimes D_k,$$

which allows us to extend the definition of $\Phi_{\mathbf{a}}$ to indices a with $a_n < 0$: we simply set

$$\Phi_{a_1, \ldots, a_n} = \Phi_{a_1, \ldots, a_n + k} \otimes D_{-k}$$

for large k.

Alternatively, we may consider the Schur functor $\mathbb{S}_\lambda$ applied to the standard representation V of $\mathrm{GL}_n\mathbb{C}$, where

$$\lambda = (a_1 + \cdots + a_n, a_2 + \cdots + a_n, \ldots, a_{n-1} + a_n, a_n).$$

We will denote this representation $\mathbb{S}_\lambda V$ of $\mathrm{GL}_n\mathbb{C}$ by Ψ_λ; note that

$$\Psi_{\lambda_1 + k, \ldots, \lambda_n + k} = \Psi_{\lambda_1, \ldots, \lambda_n} \otimes D_k$$

which likewise allows us to define Ψ_λ for any index λ with $\lambda_1 \geq \lambda_2 \geq \cdots \geq \lambda_n$, even if some of the λ_i are negative: we simply take

$$\Psi_{\lambda_1,\ldots,\lambda_n} = \Psi_{\lambda_1+k,\ldots,\lambda_n+k} \otimes D_{-k}$$

for any sufficiently large k.

As is not hard to see, *the two representations $\Phi_{\mathbf{a}}$ and Ψ_λ are isomorphic as representations of* $\mathrm{GL}_n\mathbb{C}$: by §15.3 their restrictions to $\mathrm{SL}_n\mathbb{C}$ agree, so it suffices to check their restrictions to the center $\mathbb{C}^* \subset \mathrm{GL}_n\mathbb{C}$, where each acts by multiplication by $z^{\sum\lambda_i} = z^{\sum i a_i}$). It is even clearer that there are no coincidences among the $\Phi_{\mathbf{a}}$ (i.e., $\Phi_{\mathbf{a}}$ will be isomorphic to $\Phi_{\mathbf{a}'}$ if and only if $\mathbf{a} = \mathbf{a}'$): if $\Phi_{\mathbf{a}} \cong \Phi_{\mathbf{a}'}$, we must have $a_i = a_i'$ for $i = 1, \ldots, n-1$, so the statement follows from the nontriviality of D_k for $k \neq 0$. Thus, to complete our description of the irreducible finite-dimensional representations of $\mathrm{GL}_n\mathbb{C}$, we just have to check that we have found them all. We may then express the completed result as

Proposition 15.47. *Every irreducible complex representation of* $\mathrm{GL}_n\mathbb{C}$ *is isomorphic to Ψ_λ for a unique index $\lambda = \lambda_1, \ldots, \lambda_n$ with $\lambda_1 \geq \lambda_2 \geq \cdots \geq \lambda_n$ (equivalently, to $\Phi_{\mathbf{a}}$ for a unique index $\mathbf{a} = a_1, \ldots, a_n$ with $a_1, \ldots, a_{n-1} \geq 0$).*

PROOF. We start by going back to the corresponding Lie algebras. The scalar matrices form a one-dimensional ideal $\mathbb{C}$ in $\mathfrak{gl}_n\mathbb{C}$, and in fact $\mathfrak{gl}_n\mathbb{C}$ is a product of Lie algebras:

$$\mathfrak{gl}_n\mathbb{C} = \mathfrak{sl}_n\mathbb{C} \times \mathbb{C}. \tag{15.48}$$

In particular, $\mathbb{C}$ is the radical of $\mathfrak{gl}_n\mathbb{C}$, and $\mathfrak{sl}_n\mathbb{C}$ is the semisimple part. It follows from Proposition 9.17 that every irreducible representation of $\mathfrak{gl}_n\mathbb{C}$ is a tensor product of an irreducible representation of $\mathfrak{sl}_n\mathbb{C}$ and a one-dimensional representation. More precisely, let $W_\lambda = \mathbb{S}_\lambda(\mathbb{C}^n)$ be the representation of $\mathfrak{sl}_n\mathbb{C}$ determined by the partition λ (extended to $\mathfrak{sl}_n\mathbb{C} \times \mathbb{C}$ by making the second factor act trivially). For $w \in \mathbb{C}$, let $L(w)$ be the one-dimensional representation of $\mathfrak{sl}_n\mathbb{C} \times \mathbb{C}$ which is zero on the first factor and multiplication by w on the second; the proof of Proposition 9.17 shows that any irreducible representation of $\mathfrak{sl}_n\mathbb{C} \times \mathbb{C}$ is isomorphic to a tensor product $W_\lambda \otimes L(w)$. The same is therefore true for the simply connected[1] group $\mathrm{SL}_n\mathbb{C} \times \mathbb{C}$ with this Lie algebra.

We write $\mathrm{GL}_n\mathbb{C}$ as a quotient modulo a discrete subgroup of the center of $\mathrm{SL}_n\mathbb{C} \times \mathbb{C}$:

$$1 \to \mathrm{Ker}(\rho) \to \mathrm{SL}_n\mathbb{C} \times \mathbb{C} \xrightarrow{\rho} \mathrm{GL}_n\mathbb{C} \to 1, \tag{15.49}$$

where $\rho(g \times z) = e^z \cdot g$, so the kernel of ρ is generated by $e^s \cdot I \times (-s)$, where $s = 2\pi i/n$.

Our task is simply to see which of the representations $W_\lambda \otimes L(w)$ of $\mathrm{SL}_n\mathbb{C} \times \mathbb{C}$ are trivial on the kernel of ρ. Now $e^s \cdot I$ acts on $\mathbb{S}_\lambda\mathbb{C}^n$ by multi-

[1] For a proof that $\mathrm{SL}_n\mathbb{C}$ is simply connected, see §23.1.

plication by e^{sd}, where $d = \sum \lambda_i$; indeed, this is true on the entire representation $(\mathbb{C}^n)^{\otimes d}$ which contains $\mathbb{S}_\lambda \mathbb{C}^n$. And $-s$ acts on $L(w)$ by multiplication by e^{-sw}, so $e^s \cdot I \times (-s)$ acts on the tensor product by multiplication by e^{sd-sw}. The tensor product is, therefore, trivial on the kernel of ρ precisely when $sd - sw \in 2\pi i\mathbb{Z}$, i.e., when

$$w = \sum \lambda_i + kn$$

for some integer k.

We claim finally that any representation $W_\lambda \otimes L(w)$ satisfying this condition is the pullback via ρ of a representation Ψ on $GL_n\mathbb{C}$. In fact, it is not hard to see that it is the pullback of the representation $\Psi_{\lambda_1+k,\ldots,\lambda_n+k}$: the two clearly restrict to the same representation on $SL_n\mathbb{C}$, and their restrictions to $\mathbb{C}$ are just multiplication by $e^{wz} = e^{(\sum \lambda_i + nk)z}$. □

Exercise 15.50. Show that the dual of the representation Ψ_λ which is isomorphic to $\mathbb{S}_\lambda(V^*)$ is the representation $\Psi_{(-\lambda_n,\ldots,-\lambda_1)}$.

Exercise 15.51*. Show that if $\rho: GL_n\mathbb{C} \to GL(W)$ is a representation (assumed to be holomorphic), then W decomposes into a direct sum of irreducible representations.

Exercise 15.52*. Show that the Hermite reciprocity isomorphism of Exercise 11.34 is an isomorphism over $GL_2\mathbb{C}$, not just over $SL_2\mathbb{C}$.

More Remarks on Weyl's Construction

We close out this lecture by looking once more at the Weyl construction of these representations of $GL(V)$. This will include a realization "by generators and relations," as well as giving a natural basis for each representation. First, it may be illuminating—and it will be useful later—to look more closely at how $\mathbb{S}_\lambda V$ sits in $V^{\otimes d}$. We want to realize $\mathbb{S}_\lambda V$ as a subspace of the subspace

$$\mathrm{Sym}^{a_k}(\wedge^k V) \otimes \mathrm{Sym}^{a_{k-1}}(\wedge^{k-1} V) \otimes \cdots \otimes \mathrm{Sym}^{a_1}(V) \subset V^{\otimes d},$$

where a_i is the number of columns of the Young diagram of λ of length i (and k is the number of rows). This space is embedded in $V^{\otimes d}$ in the natural way: from left to right, a factor $\mathrm{Sym}^a(\wedge^b V)$ is embedded in the corresponding $V^{\otimes ab}$ by mapping a symmetric product of exterior products

$$(v_{1,1} \wedge v_{2,1} \wedge \cdots \wedge v_{b,1}) \cdot (v_{1,2} \wedge v_{2,2} \wedge \cdots \wedge v_{b,2}) \cdot \ldots \cdot (v_{1,a} \wedge v_{2,a} \wedge \cdots \wedge v_{b,a})$$

to

$$\sum \mathrm{sgn}(q)(v_{q_1(1),p(1)} \otimes \cdots \otimes v_{q_1(b),p(1)}) \otimes \cdots \otimes (v_{q_a(1),p(a)} \otimes \cdots \otimes v_{q_a(b),p(a)}),$$

the sum over $p \in \mathfrak{S}_a$ and $q = (q_1, \ldots, q_a) \in \mathfrak{S}_b \times \cdots \times \mathfrak{S}_b$. In other words, one first symmetrizes by permuting columns of the same length, and then performs an alternating symmetrizer on each column.

Letting $\mathbf{a} = (a_1, \ldots, a_k)$, let $A^{\mathbf{a}}(V)$ denote this tensor product of symmetric powers of exterior powers, i.e., set

$$A^{\mathbf{a}}V = \mathrm{Sym}^{a_k}(\wedge^k V) \otimes \mathrm{Sym}^{a_{k-1}}(\wedge^{k-1} V) \otimes \cdots \otimes \mathrm{Sym}^{a_1}(V).$$

We want to realize $\mathbb{S}_\lambda V$ as a subspace of $A^{\mathbf{a}}V$. To do this we use the construction of $\mathbb{S}_\lambda V$ as $V^{\otimes d} \cdot c_\lambda$, where c_λ is a Young symmetrizer; to get compatibility with the embedding of $A^{\mathbf{a}}V$ we have just made, we use the tableau which numbers the columns from top to bottom, then left to right.

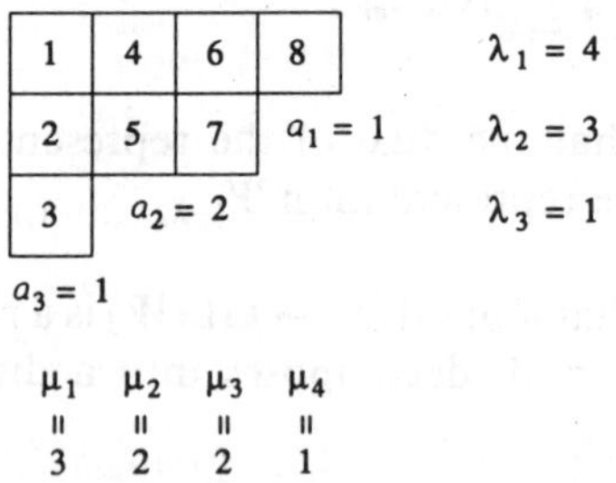

We take $\mu = \lambda' = (\mu_1 \geq \cdots \geq \mu_l > 0)$ to be the conjugate of λ. The symmetrizer c_λ is a product $a_\lambda \cdot b_\lambda$, where $a_\lambda = \sum e_p$, the sum over all p in the subgroup $P = \mathfrak{S}_{\lambda_1} \times \cdots \times \mathfrak{S}_{\lambda_k}$ of $\mathfrak{S}_d$ preserving the rows, $b_\lambda = \sum \mathrm{sgn}(q)q$, the sum over the subgroup $Q = \mathfrak{S}_{\mu_1} \times \cdots \times \mathfrak{S}_{\mu_l}$ preserving the columns, as described in Lecture 4. The symmetrizing by rows can be done in two steps as follows. There is a subgroup

$$R = \mathfrak{S}_{a_k} \times \cdots \times \mathfrak{S}_{a_1}$$

of P, which consists of permutations that move all entries of each column to the same position in some column of the same length; in other words, permutations in R are determined by permuting columns which have the same length. (In the illustration, $R = \{1, (46)(57)\}$.) Set

$$a'_\lambda = \sum_{r \in R} e_r \quad \text{in } \mathbb{C}\mathfrak{S}_d.$$

Now if we define a''_λ to be $\sum e_p$, where the sum is over any set of representatives in P for the left cosets P/R, then the row symmetrizer a_λ is the product of a''_λ and a'_λ. So

$$\mathbb{S}_\lambda(V) = (V^{\otimes d} \cdot a''_\lambda) \cdot a'_\lambda \cdot b_\lambda.$$

The point is that, by what we have just seen,

$$V^{\otimes d} \cdot a'_\lambda \cdot b_\lambda = A^{\mathbf{a}}V.$$

Since $V^{\otimes d} \cdot a''_\lambda$ is a subspace of $V^{\otimes d}$, its image $\mathbb{S}_\lambda(V)$ by $a'_\lambda \cdot b_\lambda$ is a subspace of $A^{\mathbf{a}}(V)$, as we claimed.

There is a simple way to construct all the representations $\mathbb{S}_\lambda V$ of $GL(V)$ at once. In fact, the direct sum of all the representations $\mathbb{S}_\lambda V$, over all (nonnegative) partitions λ, can be made into a commutative, graded ring, which we denote by $\mathbb{S}^\bullet$ or $\mathbb{S}^\bullet(V)$, with simple generators and relations. This is similar to the fact that the symmetric algebra $\mathrm{Sym}^\bullet V = \bigoplus \mathrm{Sym}^k V$ and the exterior algebra $\bigwedge^\bullet V = \bigoplus \bigwedge^k V$ are easier to describe than the individual graded pieces, and it has some of the similar advantages for studying all the representations at once. This algebra has appeared and reappeared frequently, cf. [H-P]; the construction we give is essentially that of Towber [Tow1].

To construct $\mathbb{S}^\bullet(V)$, start with the symmetric algebra on the sum of all the positive exterior products of V: set

$$\begin{aligned} A^\bullet(V) &= \mathrm{Sym}^\bullet(V \oplus \textstyle\bigwedge^2 V \oplus \bigwedge^3 V \oplus \cdots \oplus \bigwedge^n V) \\ &= \bigoplus_{a_1, \ldots, a_n} \mathrm{Sym}^{a_n}(\textstyle\bigwedge^n V) \otimes \cdots \otimes \mathrm{Sym}^{a_2}(\bigwedge^2 V) \otimes \mathrm{Sym}^{a_1}(V), \end{aligned}$$

the sum over all n-tuples $a_1, \ldots, a_n$ of non-negative integers. So $A^\bullet(V)$ is the direct sum of the $A^{\mathbf{a}}(V)$ just considered. The ring $\mathbb{S}^\bullet = \mathbb{S}^\bullet(V)$ is defined to be the quotient of this ring $A^\bullet(V)$ modulo the graded, two-sided ideal $I^\bullet$ generated by all elements ("Plücker relations") of the form

$$\begin{aligned} &(v_1 \wedge \cdots \wedge v_p) \cdot (w_1 \wedge \cdots \wedge w_q) \\ &\quad - \sum_{i=1}^{p} (v_1 \wedge \cdots \wedge v_{i-1} \wedge w_1 \wedge v_{i+1} \wedge \cdots \wedge v_p) \cdot (v_i \wedge w_2 \wedge \cdots \wedge w_q) \end{aligned} \tag{15.53}$$

for all $p \geq q \geq 1$ and all $v_1, \ldots, v_p, w_1, \ldots, w_q \in V$. (If $p = q$, this is an element of $\mathrm{Sym}^2(\bigwedge^p V)$; if $p > q$, it is in $\bigwedge^p V \otimes \bigwedge^q V = \mathrm{Sym}^1(\bigwedge^p V) \otimes \mathrm{Sym}^1(\bigwedge^q V)$. Note that the multiplication in $\mathbb{S}^\bullet(V)$ comes entirely from its being a symmetric algebra and does not involve the wedge products in $\bigwedge^\bullet V$.)

Exercise 15.54*. Show that $I^\bullet$ contains all elements of the form

$$\begin{aligned} &(v_1 \wedge \cdots \wedge v_p) \cdot (w_1 \wedge \cdots \wedge w_q) \\ &\quad - \sum (v_1 \wedge \cdots \wedge w_1 \wedge \cdots \wedge w_r \wedge \cdots \wedge v_p) \\ &\qquad \cdot (v_{i_1} \wedge v_{i_2} \wedge \cdots \wedge v_{i_r} \wedge w_{r+1} \wedge \cdots \wedge w_q) \end{aligned}$$

for all $p \geq q \geq r \geq 1$ and all $v_1, \ldots, v_p, w_1, \ldots, w_q \in V$, where the sum is over all $1 \leq i_1 < i_2 < \cdots < i_r \leq p$, and the elements $w_1, \ldots, w_r$ are inserted at the corresponding places in $v_1 \wedge \cdots \wedge v_p$.

Remark. You can avoid this exercise by simply taking the elements in the exercise as defining generators for the ideal $I^\bullet$. When $p = q = r$, the calculation of Exercise 15.54 shows that the relation $(v_1 \wedge \cdots \wedge v_p)\cdot(w_1 \wedge \cdots \wedge w_p) = (w_1 \wedge \cdots \wedge w_p)\cdot(v_1 \wedge \cdots \wedge v_p)$ follows from the generating equations for $I^{\bullet}$. In particular, this commutativity shows that one could define $\mathbb{S}^{\bullet}(V)$ to be the full tensor algebra on $V \oplus \wedge^2 V \oplus \cdots \oplus \wedge^n V$ modulo the ideal generated by the same generators.

The algebra $\mathbb{S}^{\bullet}(V)$ is the direct sum of the images $\mathbb{S}^{\mathbf{a}}(V)$ of the summands $A^{\mathbf{a}}(V)$. Let $e_1, \ldots, e_n$ be a basis for V. We will construct a basis for $\mathbb{S}^{\mathbf{a}}(V)$, with a basis element e_T for every semistandard tableau T on the partition λ which corresponds to $\mathbf{a}$. Recall that a semistandard tableau is a numbering of the boxes of the Young diagram with the integers $1, \ldots, n$, in such a way that the entries in each row are nondecreasing, and the entries in each column are strictly increasing. Let $T(i, j)$ be the entry of T in the ith row and the jth column. Define e_T to be the image in $\mathbb{S}^{\mathbf{a}}(V)$ of the element

$$\prod_{j=1}^{l} e_{T(1,j)} \wedge e_{T(2,j)} \wedge \cdots \wedge e_{T(\mu_j,j)} \in \operatorname{Sym}^{a_n}(\wedge^n V) \otimes \cdots \otimes \operatorname{Sym}^{a_1}(V),$$

i.e., wedge together the basis elements corresponding to the entries in the columns, and multiply the results in $\mathbb{S}^{\bullet}(V)$.

Proposition 15.55. (1) *The projection from $A^{\mathbf{a}}(V)$ to $\mathbb{S}^{\mathbf{a}}(V)$ maps the subspace $\mathbb{S}_\lambda(V)$ isomorphically onto $\mathbb{S}^{\mathbf{a}}(V)$.*

(2) *The e_T for T a semistandard tableau on λ form a basis for $\mathbb{S}^{\mathbf{a}}(V)$.*

PROOF. We show first that the elements e_T span $\mathbb{S}^{\mathbf{a}}(V)$. It is clear that the e_T span if we allow all tableaux T that number the boxes of λ with integers between 1 and n with strictly increasing columns, for such elements span before dividing by the ideal $I^{\bullet}$. We order such tableaux by listing their entries column by column, from left to right and top to bottom, and using the reverse lexicographic order: $T' > T$ if the last entry where they differ has a larger entry for T' than for T. If T is not semistandard, there will be two successive columns of T, say the jth and $(j + 1)$st, in which we have $T(r, j) > T(r, j + 1)$ for some r. It suffices to show how to use relations in $I^{\bullet}$ to write e_T as a linear combination of elements $e_{T'}$ with $T' > T$. For this we use the relation in Exercise 15.54, with $v_i = e_{T(i,j)}$ for $1 \le i \le p = \mu_j$, and $w_i = e_{T(i,j+1)}$ for $1 \le i \le q = \mu_{j+1}$, to interchange the first r of the $\{w_i\}$ with subsets of r of the $\{v_i\}$. The terms on the right-hand side of the relation will all correspond to tableaux T' in which the r first entries in the $(j + 1)$st column of T are replaced by r of the enties in the jth column, and are not otherwise changed beyond the jth column. All of these are larger than T in the ordering, which proves the assertion.

It is possible to give a direct proof that the e_T corresponding to semistandard tableaux T are linearly independent (see [Tow1]), but we can get by with less. Among the semistandard tableaux on λ there is a smallest one T_0 whose ith row is filled with the integer i. We need to know that e_{T_0} is not zero in $\mathbb{S}^{\bullet}$. This is easy to see directly. In fact, the relations among the e_T in $I^{\bullet} \cap A^{\mathbf{a}}(V)$ are spanned by those obtained by substituting r elements from some column of some T to an earlier column, as in the preceding paragraph. Such will never involve the generator e_{T_0} unless the T that is used is T_0, and in this case, the resulting element of $I^{\bullet}$ is zero. Since e_{T_0} occurs in no nontrivial relation, its image in $\mathbb{S}^{\bullet}$ cannot vanish.

Since e_{T_0} comes from $\mathbb{S}_\lambda(V)$, it follows that the projection from $\mathbb{S}_\lambda(V)$ to $\mathbb{S}^{\mathbf{a}}(V)$ is not zero. Since this projection is a mapping of representations of $SL(V)$, it follows that $\mathbb{S}^{\mathbf{a}}(V)$ must contain a copy of the irreducible representation $\mathbb{S}_\lambda(V)$. We know from Theorem 6.3 and Exercise A.31 that the dimension of $\mathbb{S}_\lambda(V)$ is the number of semistandard tableaux on λ. Since we have proved that the dimension of $\mathbb{S}^{\mathbf{a}}(V)$ is at most this number, the projection from $\mathbb{S}_\lambda(V)$ to $\mathbb{S}^{\mathbf{a}}(V)$ must be surjective, and since $\mathbb{S}_\lambda(V)$ is irreducible, it must be injective as well, and the e_T for T a semistandard tableau on λ must form a basis, as asserted. □

Note that this proposition gives another description of the representations $\mathbb{S}_\lambda(V)$, as the quotient of the space $A^{\mathbf{a}}(V)$ by the subspace generated by the "Plücker" relations (15.53).

Exercise 15.56. Show that, if the factor $\bigwedge^n V$ is omitted from the construction, the resulting algebra is the direct sum of all irreducible representations of $SL(V) = SL_n\mathbb{C}$.

It is remarkable that all the representations $\mathbb{S}_\lambda(\mathbb{C}^n)$ of $GL_n\mathbb{C}$ were written down by Deruyts (following Clebsch) a century ago, before representation theory was born, as in the following exercise.

Exercise 15.57*. Let $X = (x_{i,j})$ be an $n \times n$ matrix of indeterminates. The group $G = GL_n\mathbb{C}$ acts on the polynomial ring $\mathbb{C}[x_{i,j}]$ by $g \cdot x_{i,j} = \sum_{k=1}^{n} a_{k,i} x_{k,j}$ for $g = (a_{i,j}) \in GL_n\mathbb{C}$. For any tableau T on the Young diagram of λ consisting of the integers from 1 to n, strictly increasing in the columns, let e_T be the product of minors constructed from X, one for each column, as follows: if the column of T has length μ_j, form the minor using the first μ_j columns, and use the rows that are numbered by the entries of the column of T. Let D_λ be the subspace of $\mathbb{C}[x_{i,j}]$ spanned by these e_T, where d is the number partitioned by λ. Show that: (i) D_λ is preserved by $GL_n\mathbb{C}$; (ii) the e_T, where T is semistandard, form a basis for D_λ; (iii) D_λ is isomorphic to $\mathbb{S}_\lambda(\mathbb{C}^n)$.

LECTURE 16

Symplectic Lie Algebras

In this lecture we do for the symplectic Lie algebras exactly what we did for the special linear ones in §15.1 and most of §15.2: we will first describe in general the structure of a symplectic Lie algebra (that is, give a Cartan subalgebra, find the roots, describe the Killing form, and so on). We will then work out in some detail the representations of the specific algebra $\mathfrak{sp}_4\mathbb{C}$. As in the case of the corresponding analysis of the special linear Lie algebras, this is completely elementary.

§16.1. The Structure of $\mathrm{Sp}_{2n}\mathbb{C}$ and $\mathfrak{sp}_{2n}\mathbb{C}$

Let V be a $2n$-dimensional complex vector space, and

$$Q: V \times V \to \mathbb{C},$$

a nondegenerate, skew-symmetric bilinear form on V. The symplectic Lie group $\mathrm{Sp}_{2n}\mathbb{C}$ is then defined to be the group of automorphisms A of V preserving Q—that is, such that $Q(Av, Aw) = Q(v, w)$ for all $v, w \in V$—and the symplectic Lie algebra $\mathfrak{sp}_{2n}\mathbb{C}$ correspondingly consists of endomorphisms $A: V \to V$ satisfying

$$Q(Av, w) + Q(v, Aw) = 0$$

for all v and $w \in V$. Clearly, the isomorphism classes of the abstract group and Lie algebra do not depend on the particular choice of Q; but in order to be able to write down elements of both explicitly we will, for the remainder of our discussion, take Q to be the bilinear form given, in terms of a basis $e_1, \ldots,$

e_{2n} for V, by

$$Q(e_i, e_{i+n}) = 1,$$
$$Q(e_{i+n}, e_i) = -1,$$

and

$$Q(e_i, e_j) = 0 \quad \text{if } j \neq i \pm n.$$

The bilinear form Q may be expressed as

$$Q(x, y) = {}^t x \cdot M \cdot y,$$

where M is the $2n \times 2n$ matrix given in block form as

$$M = \begin{pmatrix} 0 & I_n \\ -I_n & 0 \end{pmatrix};$$

the group $Sp_{2n}\mathbb{C}$ is thus the group of $2n \times 2n$ matrices A satisfying

$$M = {}^t A \cdot M \cdot A$$

and the Lie algebra $\mathfrak{sp}_{2n}\mathbb{C}$ correspondingly the space of matrices X satisfying the relation

$$ {}^t X \cdot M + M \cdot X = 0. \tag{16.1}$$

Writing a $2n \times 2n$ matrix X in block form as

$$X = \begin{pmatrix} A & B \\ C & D \end{pmatrix}$$

we have

$$ {}^t X \cdot M = \begin{pmatrix} -{}^t C & {}^t A \\ -{}^t D & {}^t B \end{pmatrix}$$

and

$$M \cdot X = \begin{pmatrix} C & D \\ -A & -B \end{pmatrix}$$

so that this relation is equivalent to saying that *the off-diagonal blocks B and C of X are symmetric, and the diagonal blocks A and D of X are negative transposes of each other.*

With this said, there is certainly an obvious candidate for Cartan subalgebra $\mathfrak{h}$ in $\mathfrak{sp}_{2n}\mathbb{C}$, namely the subalgebra of matrices diagonal in this representation; in fact, this works, as we shall see shortly. The subalgebra $\mathfrak{h}$ is thus spanned by the n $2n \times 2n$ matrices $H_i = E_{i,i} - E_{n+i,n+i}$ whose action on V is to fix e_i, send e_{n+i} to its negative, and kill all the remaining basis vectors; we will correspondingly take as basis for the dual vector space $\mathfrak{h}^*$ the dual basis L_j, where $\langle L_j, H_i \rangle = \delta_{i,j}$.

We have already seen how the diagonal matrices act on the algebra of all matrices, so that it is easy to describe the action of $\mathfrak{h}$ on $\mathfrak{g}$. For example, for

$1 \leq i, j \leq n$ the matrix $E_{i,j} \in \mathfrak{gl}_{2n}\mathbb{C}$ is carried into itself under the adjoint action of H_i, into minus itself by the action of H_j, and to 0 by all the other H_k; and the same is true of the matrix $E_{n+j,n+i}$. The element

$$X_{i,j} = E_{i,j} - E_{n+j,n+i} \in \mathfrak{sp}_{2n}\mathbb{C}$$

is thus an eigenvector for the action of $\mathfrak{h}$, with eigenvalue $L_i - L_j$. Similarly, for $i \neq j$ we see that the matrices $E_{i,n+j}$ and $E_{j,n+i}$ are carried into themselves by H_i and H_j and killed by all the other H_k; and likewise $E_{n+i,j}$ and $E_{n+j,i}$ are each carried into their negatives by H_i and H_j and killed by the others. Thus, the elements

$$Y_{i,j} = E_{i,n+j} + E_{j,n+i}$$

and

$$Z_{i,j} = E_{n+i,j} + E_{n+j,i}$$

are eigenvectors for the action of $\mathfrak{h}$, with eigenvalues $L_i + L_j$ and $-L_i - L_j$, respectively. Finally, when $i = j$ the same calculation shows that $E_{i,n+i}$ is doubled by H_i and killed by all other H_j; and likewise $E_{n+i,i}$ is sent to minus twice itself by H_i and to 0 by the others. Thus, the elements

$$U_i = E_{i,n+i}$$

and

$$V_i = E_{n+i,i}$$

are eigenvectors with eigenvalues $2L_i$ and $-2L_i$, respectively. In sum, then, *the roots of the Lie algebra* $\mathfrak{sp}_{2n}\mathbb{C}$ *are the vectors* $\pm L_i \pm L_j \in \mathfrak{h}^*$.

In the first case $n = 1$, of course we just get the root diagram of $\mathfrak{sl}_2\mathbb{C}$, which is the same algebra as $\mathfrak{sp}_2\mathbb{C}$. In case $n = 2$, we have the diagram

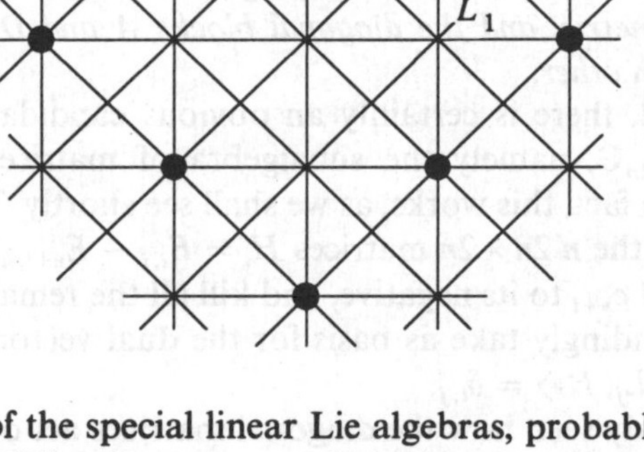

(16.2)

As in the case of the special linear Lie algebras, probably the easiest way to determine the Killing form on $\mathfrak{sp}_{2n}\mathbb{C}$ (at least up to scalars) is to use its invariance under the automorphisms of $\mathfrak{sp}_{2n}\mathbb{C}$ preserving $\mathfrak{h}$. For example, we have the automorphisms of $\mathfrak{sp}_{2n}\mathbb{C}$ induced by permutations of the basis vectors e_i of V: for any permutation σ of $\{1, 2, \ldots, n\}$ we can define an automorphism of V preserving Q by sending e_i to $e_{\sigma(i)}$ and e_{n+i} to $e_{n+\sigma(i)}$, and this induces an automorphism of $\mathfrak{sp}_{2n}\mathbb{C}$ preserving $\mathfrak{h}$ and carrying H_i to $H_{\sigma(i)}$. Also, for any i we can define an involution of V—and thereby of $\mathfrak{sp}_{2n}\mathbb{C}$—by sending e_i to e_{n+i}, e_{n+i} to $-e_i$, and all the other basis vectors to themselves; this will have the effect of sending H_i to $-H_i$ and preserving all the other H_j. Now, the Killing form on $\mathfrak{h}$ must be invariant under these automorphisms; from the first batch it follows that for some pair of constants α and β we must have

$$B(H_i, H_i) = \alpha$$

and

$$B(H_i, H_j) = \beta \quad \text{for } i \neq j;$$

from the second batch it follows that, in fact, $\beta = 0$. Thus, B is just a multiple of the standard quadratic form $B(H_i, H_j) = \delta_{i,j}$, and the dual form correspondingly a multiple of $B(L_i, L_j) = \delta_{i,j}$; so that the angles in the diagram above are correct.

Also as in the case of $\mathfrak{sl}_n\mathbb{C}$, one can also compute the Killing form directly from the definition: $B(H, H') = \sum \alpha(H)\alpha(H')$, the sum over all roots α. For $H = \sum a_i H_i$ and $H' = \sum b_i H_i$, this gives $B(H, H')$ as a sum

$$\sum_{i \neq j} (a_i + a_j)(b_i + b_j) + 2 \sum_i (2a_i)(2bi) + \sum_{i \neq j} (a_i - a_j)(b_i - b_j)$$

which simplifies to

$$B(H, H') = (4n + 4)(\textstyle\sum a_i b_i). \tag{16.3}$$

Our next job is to locate the distinguished copies $\mathfrak{s}_\alpha$ of $\mathfrak{sl}_2\mathbb{C}$, and the corresponding elements $H_\alpha \in \mathfrak{h}$. This is completely straightforward. We start with the eigenvalues $L_i - L_j$ and $L_j - L_i$ corresponding to the elements $X_{i,j}$ and $X_{j,i}$; we have

$$\begin{aligned}
[X_{i,j}, X_{j,i}] &= [E_{i,j} - E_{n+j,n+i}, E_{j,i} - E_{n+i,n+j}] \\
&= [E_{i,j}, E_{j,i}] + [E_{n+j,n+i}, E_{n+i,n+j}] \\
&= E_{i,i} - E_{j,j} + E_{n+j,n+j} - E_{n+i,n+i} \\
&= H_i - H_j.
\end{aligned}$$

Thus, the distinguished element $H_{L_i - L_j}$ is a multiple of $H_i - H_j$. To see what multiple, recall that $H_{L_i - L_j}$ should act on $X_{i,j}$ by multiplication by 2 and on $X_{j,i}$ by multiplication by -2; since we have

$$\begin{aligned}
\mathrm{ad}(H_i - H_j)(X_{i,j}) &= ((L_i - L_j)(H_i - H_j)) \cdot X_{i,j} \\
&= 2X_{i,j},
\end{aligned}$$

we conclude that

$$H_{L_i-L_j} = H_i - H_j.$$

Next consider the pair of opposite eigenvalues $L_i + L_j$ and $-L_i - L_j$, corresponding to the eigenvectors $Y_{i,j}$ and $Z_{i,j}$. We have

$$\begin{aligned}[Y_{i,j}, Z_{i,j}] &= [E_{i,n+j} + E_{j,n+i}, E_{n+i,j} + E_{n+j,i}] \\ &= [E_{i,n+j}, E_{n+j,i}] + [E_{j,n+i}, E_{n+i,j}] \\ &= E_{i,i} - E_{n+j,n+j} + E_{j,j} - E_{n+i,n+i} \\ &= H_i + H_j.\end{aligned}$$

We calculate then

$$\begin{aligned}\operatorname{ad}(H_i + H_j)(Y_{i,j}) &= ((L_i + L_j)(H_i + H_j)) \cdot Y_{i,j} \\ &= 2 \cdot Y_{i,j},\end{aligned}$$

so we have

$$H_{L_i+L_j} = H_i + H_j$$

and similarly

$$H_{-L_i-L_j} = -H_i - H_j.$$

Finally, we look at the pair of eigenvalues $\pm 2L_i$ coming from the eigenvectors U_i and V_i. To complete the span of U_i and V_i to a copy of $\mathfrak{sl}_2\mathbb{C}$ we add

$$\begin{aligned}[U_i, V_i] &= [E_{i,n+i}, E_{n+i,i}] \\ &= E_{i,i} - E_{n+i,n+i} \\ &= H_i.\end{aligned}$$

Since

$$\begin{aligned}\operatorname{ad}(H_i)(U_i) &= (2L_i(H_i)) \cdot U_i \\ &= 2 \cdot U_i,\end{aligned}$$

we conclude that the distinguished element H_{2L_i} is H_i, and likewise $H_{-2L_i} = -H_i$. Thus, the distinguished elements $\{H_\alpha\} \subset \mathfrak{h}$ are $\{\pm H_i \pm H_j, \pm H_i\}$; in particular, the weight lattice Λ_W of linear forms on $\mathfrak{h}$ integral on all the H_α is exactly the lattice of integral linear combinations of the L_i. In Diagram (16.2), for example, this is just the lattice of intersections of the horizontal and vertical lines drawn; observe that for all n the index $[\Lambda_W : \Lambda_R]$ of the root lattice in the weight lattice is just 2.

Next we consider the group of symmetries of the weights of an arbitrary representation of $\mathfrak{sp}_{2n}\mathbb{C}$. For each root α we let W_α be the involution in $\mathfrak{h}^*$ fixing the hyperplane Ω_α given by $\langle H_\alpha, L\rangle = 0$ and acting as $-I$ on the line spanned by α; we observe in this case that, as we claimed will be true in general, the line generated by α is perpendicular to the hyperplane Ω_α, so that the involution is just a reflection in this plane. In the case $n = 2$, for example, we get the dihedral group generated by reflections around the four lines drawn through the origin:

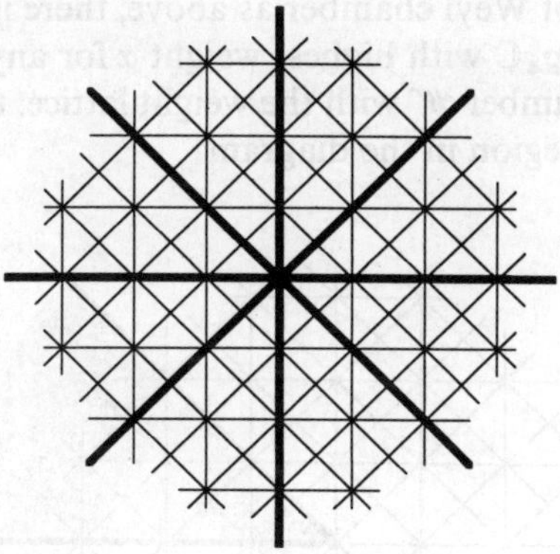

so that the weight diagram of a representation of $\mathfrak{sp}_4\mathbb{C}$ will look like an octagon in general, or (in some cases) a square.

In general, reflection in the plane Ω_{2L_i} given by $\langle H_i, L\rangle = 0$ will simply reverse the sign of L_i while leaving the other L_j fixed; reflection in the plane $\langle H_i - H_j, L\rangle = 0$ will exchange L_i and L_j and leave the remaining L_k alone. The Weyl group $\mathfrak{W}$ acts as the full automorphism group of the lines spanned by the L_i and fits into a sequence

$$1 \to (\mathbb{Z}/2\mathbb{Z})^n \to \mathfrak{W} \to \mathfrak{S}_n \to 1.$$

Note that the sequence splits: $\mathfrak{W}$ is a semidirect product of $\mathfrak{S}_n$ and $(\mathbb{Z}/2\mathbb{Z})^n$. (This is a special case of a *wreath product*.) In particular the order of $\mathfrak{W}$ is $2^n n!$.

We can choose a positive direction as before:

$$l(\sum a_i L_i) = c_1 a_1 + \cdots + c_n a_n, \qquad c_1 > c_2 > \cdots > c_n > 0.$$

The positive roots are then

$$R^+ = \{L_i + L_j\}_{i \le j} \cup \{L_i - L_j\}_{i<j}, \tag{16.4}$$

with primitive positive roots $\{L_i - L_{i+1}\}_{i=1,\ldots,n-1}$ and $2L_n$. The corresponding (closed) Weyl chamber is

$$\mathscr{W} = \{a_1 L_1 + a_2 L_2 + \cdots + a_n L_n : a_1 \ge a_2 \ge \cdots \ge a_n \ge 0\}; \tag{16.5}$$

note that the walls of this chamber—the cones

$$\{\sum a_i L_i : a_1 > \cdots > a_i = a_{i+1} > \cdots > a_n > 0\}$$

and

$$\{\sum a_i L_i : a_1 > a_2 > \cdots > a_n = 0\}$$

lie in the hyperplanes $\Omega_{L_i - L_{i+1}}$ and Ω_{2L_n} perpendicular to the primitive positive or negative roots, as expected.

§16.2. Representations of $\mathfrak{sp}_4\mathbb{C}$

Let us consider now the representations of the algebra $\mathfrak{sp}_4\mathbb{C}$ specifically. Recall that, with the choice of Weyl chamber as above, there is a unique irreducible representation Γ_α of $\mathfrak{sp}_4\mathbb{C}$ with highest weight α for any α in the intersection of the closed Weyl chamber $\mathscr{W}$ with the weight lattice: that is, for each lattice vector in the shaded region in the diagram

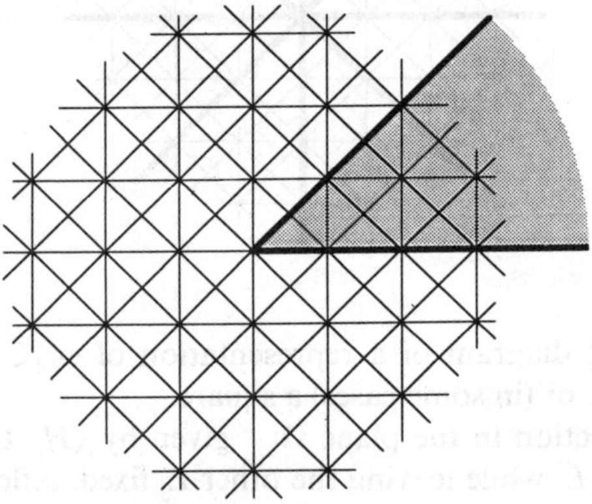

Any such highest weight vector can be written as a non-negative integral linear combination of L_1 and $L_1 + L_2$; for simplicity we will just write $\Gamma_{a,b}$ for the irreducible representation $\Gamma_{aL_1+b(L_1+L_2)}$ with highest weight $aL_1 + b(L_1 + L_2) = (a + b)L_1 + bL_2$.

To begin with, we have the standard representation as the algebra of endomorphisms of the four-dimensional vector space V; the four standard basis vectors e_1, e_2, e_3, and e_4 are eigenvectors with eigenvalues $L_1, L_2, -L_1$, and $-L_2$, respectively, so that the weight diagram of V is

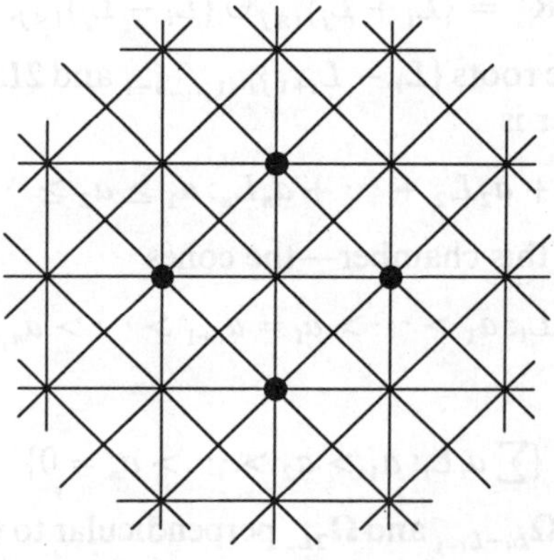

V is just the representation $\Gamma_{1,0}$ in the notation above. Note that the dual of this representation is isomorphic to it, which we can see either from the symmetry of the weight diagram, or directly from the fact that the corresponding group representation preserves a bilinear form $V \times V \to \mathbb{C}$ giving an identification of V with V^*.

The next representation to consider is the exterior square $\bigwedge^2 V$. The weights of $\bigwedge^2 V$, the pairwise sums of distinct weights of V, are just the linear forms $\pm L_i \pm L_j$ (each appearing once) and 0 (appearing twice, as $L_1 - L_1$ and $L_2 - L_2$), so that its weight diagram looks like

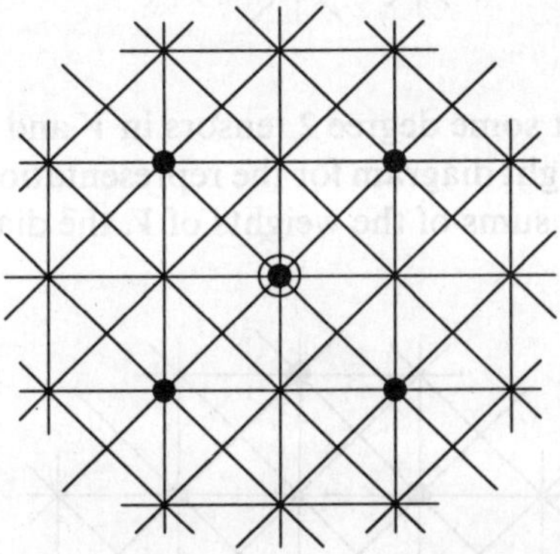

Clearly this representation is not irreducible. We can see this from the weight diagram, using Observation 14.16: there is only one way of getting to the weight space 0 from the highest weight $L_1 + L_2$ by successive applications of the primitive negative root spaces $\mathfrak{g}_{-L_1+L_2}$ (spanned by $X_{2,1} = E_{2,1} - E_{3,4}$) and $\mathfrak{g}_{-2L_2}$ (spanned by $V_2 = E_{4,2}$)—that is, by applying first V_2, which takes you to the weight space of $L_1 - L_2$, and then $X_{2,1}$—and so the dimension of the zero weight space in the irreducible representation $\Gamma_{0,1}$ with highest weight $L_1 + L_2$ must be one. Of course, we know in any event that $\bigwedge^2 V$ cannot be irreducible: the corresponding group action of $\mathrm{Sp}_4\mathbb{C}$ on V by definition preserves the skew form $Q \in \bigwedge^2 V^* \cong \bigwedge^2 V$. Either way, we conclude that we have a direct sum decomposition

$$\textstyle\bigwedge^2 V = W \oplus \mathbb{C},$$

where W is the irreducible, five-dimensional representation of $\mathfrak{sp}_4\mathbb{C}$ with highest weight $L_1 + L_2$—in our notation, $\Gamma_{0,1}$—and weight diagram

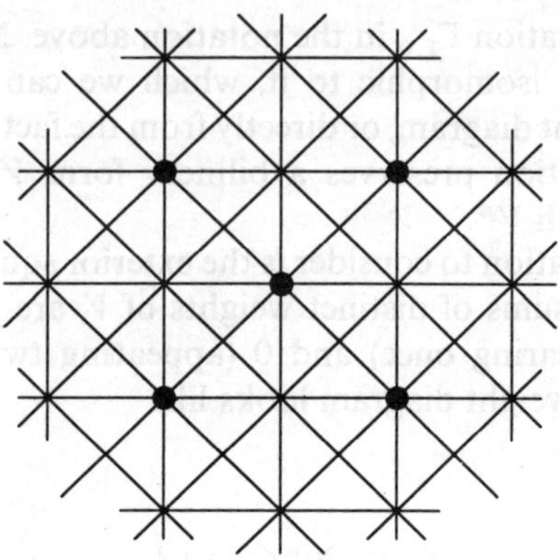

Let us consider next some degree 2 tensors in V and W. To begin with, we can write down the weight diagram for the representation $\mathrm{Sym}^2 V$; the weights being just the pairwise sums of the weights of V, the diagram is

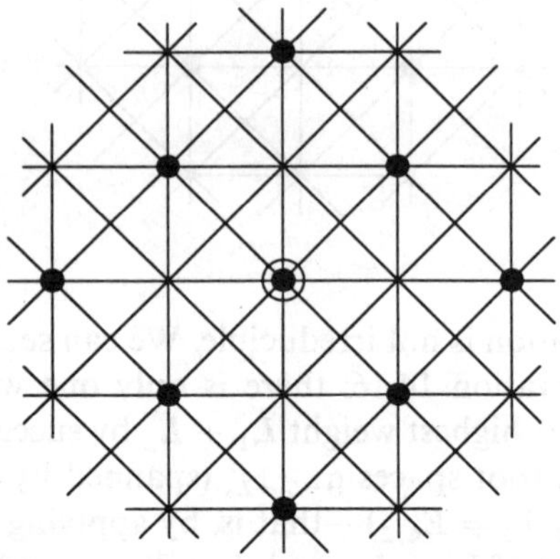

This looks like the weight diagram of the adjoint representation, and indeed that is what it is: in terms of the identification of V and V^* given by the skew form Q, the relation (16.1) defining the symplectic Lie algebra says that the subspace

$$\mathfrak{sp}_4\mathbb{C} \subset \mathrm{Hom}(V, V) = V \otimes V^* = V \otimes V$$

is just the subspace $\mathrm{Sym}^2 V \subset V \otimes V$. In particular, $\mathrm{Sym}^2 V$ is the irreducible representation $\Gamma_{2,0}$ with highest weight $2L_1$.

Next, consider the symmetric square $\mathrm{Sym}^2 W$, which has weight diagram

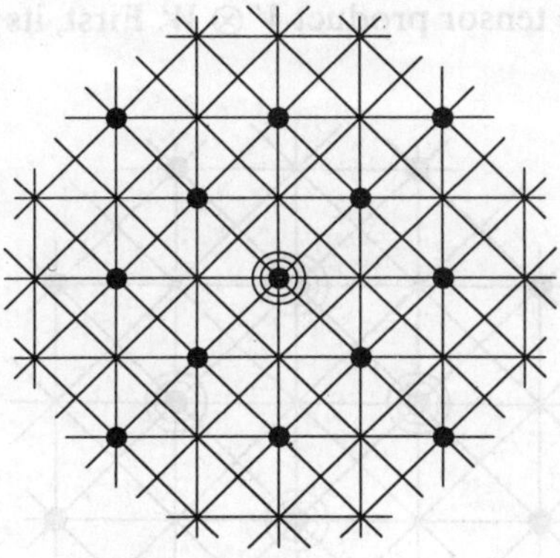

To see if this is irreducible we first look at the weight diagram: this time there are three ways of getting from the weight space with highest weight $2L_1 + 2L_2$ to the space of weight 0 by successively applying $X_{2,1} = E_{2,1} - E_{3,4}$ and $V_2 = E_{4,2}$, so if we want to proceed by this method we are forced to do a little calculation, which we leave as Exercise 16.7.

Alternatively, we can see directly that $\mathrm{Sym}^2 W$ decomposes: the natural map given by wedge product

$$\wedge^2 V \otimes \wedge^2 V \to \wedge^4 V = \mathbb{C}$$

is symmetric, and so factors to give a map

$$\mathrm{Sym}^2(\wedge^2 V)) \to \mathbb{C}.$$

Moreover, since this map is well defined up to scalars—in particular, it does not depend on the choice of skew form Q—it cannot contain the subspace $\mathrm{Sym}^2 W \subset \mathrm{Sym}^2(\wedge^2 V))$ in its kernel, so that it restricts to give a surjection

$$\varphi\colon \mathrm{Sym}^2 W \to \mathbb{C}.$$

This approach would appear to leave two possibilities open: either the kernel of this map is irreducible, or it is the direct sum of an irreducible representation and a further trivial summand. In fact, however, from the principle that an irreducible representation cannot have two independent invariant bilinear forms, we see that $\mathrm{Sym}^2 W$ can contain at most one trivial summand, and so the former alternative must hold, i.e., we have

$$\mathrm{Sym}^2 W = \Gamma_{0,2} \oplus \mathbb{C}. \qquad (16.6)$$

Exercise 16.7*. Prove (16.6) directly, by showing that if v is a highest weight vector, then the three vectors $X_{2,1} V_2 X_{2,1} V_2 v$, $X_{2,1} X_{2,1} V_2 V_2 v$, and $V_2 X_{2,1} X_{2,1} V_2 v$ span a two-dimensional subspace of the kernel of φ.

Exercise 16.8. Verify that $\wedge^2 W \cong \mathrm{Sym}^2 V$. The significance of this isomorphism will be developed further in Lecture 18.

Lastly, consider the tensor product $V \otimes W$. First, its weight diagram:

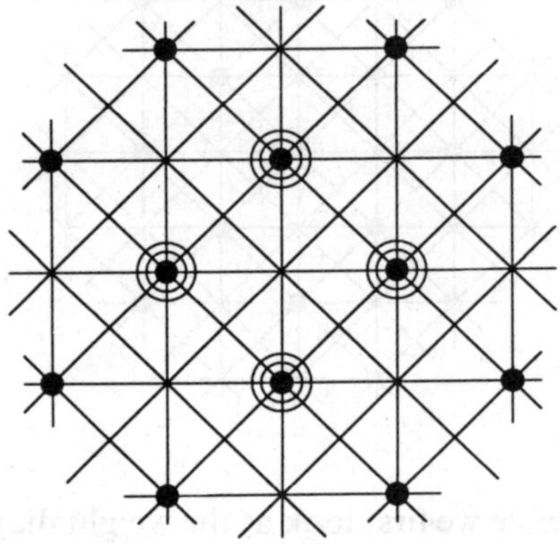

This obviously must contain the irreducible representation $\Gamma_{1,1}$ with highest weight $2L_1 + L_2$; but it cannot be irreducible, for either of two reasons. First, looking at the weight diagram, we see that $\Gamma_{1,1}$ can take on the eigenvalues $\pm L_i$ with multiplicity at most 2, so that $V \otimes W$ must contain at least one copy of the representation V. Alternatively, we have a natural map given by wedge product

$$\wedge: V \otimes \Lambda^2 V \to \Lambda^3 V = V^* = V;$$

and since this map does not depend on the choice of skew form Q, it must restrict to give a nonzero (and hence surjective) map

$$\varphi: V \otimes W \to V.$$

Exercise 16.9. Show that the kernel of this map is irreducible, and hence that we have

$$V \otimes W = \Gamma_{1,1} \oplus V.$$

What about more general tensors? To begin with, note that we have established the existence half of the standard existence and uniqueness theorem (14.18) in the case of $\mathfrak{sp}_4\mathbb{C}$: the irreducible representation $\Gamma_{a,b}$ may be found somewhere in the tensor product $\mathrm{Sym}^a V \otimes \mathrm{Sym}^b W$. The question that remains is, where? In other words, we would like to be able to say how these tensor products decompose. This will be, as it was in the case of $\mathfrak{sl}_3\mathbb{C}$, nearly tantamount (modulo the combinatorics needed to count the multiplicity with which the tensor product $\mathrm{Sym}^a V \otimes \mathrm{Sym}^b W$ assumes each of its eigenvalues) to specifying the multiplicities of the irreducible representations $\Gamma_{a,b}$.

Let us start with the simplest case, namely, the representations $\mathrm{Sym}^a V$. These have weight diagram a sequence of nested diamonds D_i with vertices at aL_1, $(a-2)L_1$, etc.:

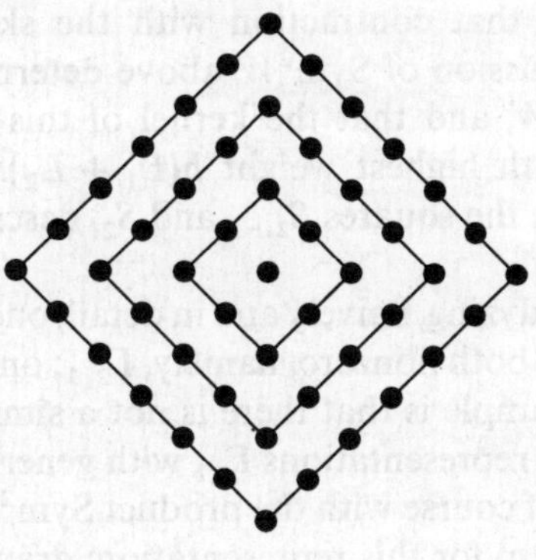

Moreover, it is not hard to calculate the multiplicities of $\mathrm{Sym}^a V$: the multiplicity on the outer diamond D_1 is one, of course; and then the multiplicities will increase by one on successive rings, so that the multiplicity along the diamond D_i will be i.

Exercise 16.10. Using the techniques of Lecture 13, show that the representations $\mathrm{Sym}^a V$ are irreducible.

The next simplest representations, naturally enough, are the symmetric powers $\mathrm{Sym}^b W$ of W. These have eigenvalue diagrams in the shape of a sequence of squares S_i with vertices at $b(L_1 + L_2)$, $(b - 1)(L_1 + L_2)$, and so on:

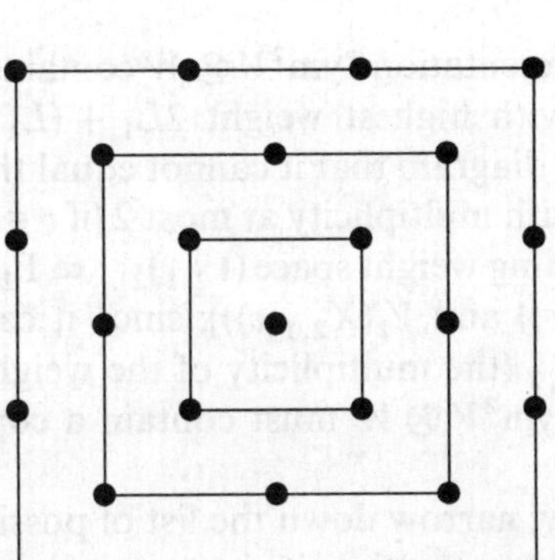

Here, however, the multiplicities increase in a rather strange way: they grow quadratically, but only on every other ring. Explicitly, the multiplicity will be one on the outer two rings, then 3 on the next two rings, 6 on the next two; in general, it will be $i(i + 1)/2$ on the $(2i - 1)$st and $(2i)$th squares S_{2i-1} and S_{2i}.

Exercise 16.11. Show that contraction with the skew form $\varphi \in \mathrm{Sym}^2 W^*$ introduced in the discussion of $\mathrm{Sym}^2 W$ above determines a surjection from $\mathrm{Sym}^b W$ onto $\mathrm{Sym}^{b-2} W$, and that the kernel of this map is the irreducible representation $\Gamma_{0,b}$ with highest weight $b(L_1 + L_2)$. Show that the multiplicities of $\Gamma_{0,b}$ are i on the squares S_{2i-1} and S_{2i} described above.

We will finish by analyzing, naively and in detail, one example of a representation $\Gamma_{a,b}$ with a and b both nonzero, namely, $\Gamma_{2,1}$; one thing we may observe on the basis of this example is that there is not a similarly simple pattern to the multiplicities of the representations $\Gamma_{a,b}$ with general a and b. To carry out our analysis, we start of course with the product $\mathrm{Sym}^2 V \otimes W$. We can readily draw the weight diagram for this representation; drawing only one-eighth of the plane and indicating multiplicities by numbers, it is

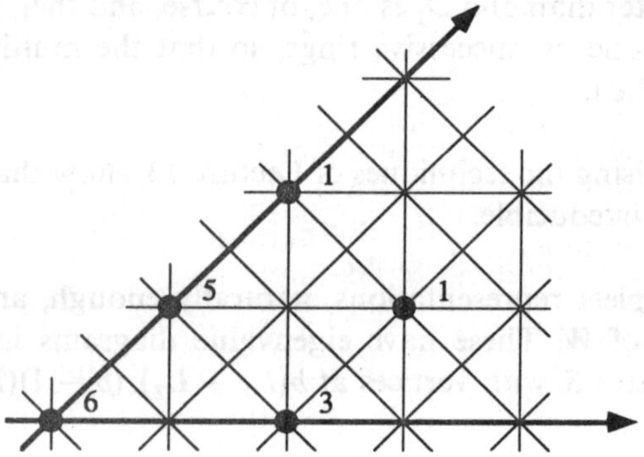

We know that the representation $\mathrm{Sym}^2 V \otimes W$ contains a copy of the irreducible representation $\Gamma_{2,1}$ with highest weight $2L_1 + (L_1 + L_2)$; and we can see immediately from the diagram that it cannot equal this: for example, $\Gamma_{2,1}$ can take the weight $2L_1$ with multiplicity at most 2 (if $v \in \Gamma_{2,1}$ is its highest weight vector, the corresponding weight space $(\Gamma_{2,1})_{2L_1} \subset \Gamma_{2,1}$ will be spanned by the two vectors $X_{2,1}(V_2(v))$ and $V_2(X_{2,1}(v))$); since it cannot contain a copy of the representation $\Gamma_{0,2}$ (the multiplicity of the weight $2(L_1 + L_2)$ being just one) it follows that $\mathrm{Sym}^2 V \otimes W$ must contain a copy of the representation $\Gamma_{2,0} = \mathrm{Sym}^2 V$.

We can, in this way, narrow down the list of possibilities a good deal. For example, $\Gamma_{2,1}$ cannot have multiplicity just one at each of the weights $2L_1$ and $L_1 + L_2$: if it did, $\mathrm{Sym}^2 V \otimes W$ would have to contain two copies of $\mathrm{Sym}^2 V$ and a further two copies of W to make up the multiplicity at $L_1 + L_2$; but since 0 must appear as a weight of $\Gamma_{2,1}$, this would give a total multiplicity of at least 7 for the weight 0 in $\mathrm{Sym}^2 V \otimes W$. Similarly, $\Gamma_{2,1}$ cannot have multiplicity 1 at $2L_1$ and 2 at $L_1 + L_2$: we would then have two copies of $\mathrm{Sym}^2 V$ and one of W in $\mathrm{Sym}^2 V \otimes W$; and since the multiplicity of 0 in $\Gamma_{2,1}$ will in this case be at least 2 (being greater than or equal to the multiplicity of $L_1 + L_2$), this would again imply a multiplicity of at least 7 for the weight 0 in $\mathrm{Sym}^2 V \otimes W$. It follows that $\mathrm{Sym}^2 V \otimes W$ must contain exactly one copy of $\mathrm{Sym}^2 V$; and since the multiplicity of $L_1 + L_2$ in $\Gamma_{2,1}$ is at most 3, it follows that $\mathrm{Sym}^2 V \otimes W$ will contain at least one copy of $\Gamma_{0,1} = W$ as well.

Exercise 16.12. Prove, independently of the above analysis, that $\mathrm{Sym}^2 V \otimes W$ must contain a copy of $\mathrm{Sym}^2 V$ and a copy of W by looking at the map

$$\varphi\colon \mathrm{Sym}^2 V \otimes W \to V \otimes V$$

obtained by sending

$$u \cdot v \otimes (w \wedge z) \mapsto u \otimes \tilde{Q}(v \wedge w \wedge z) + v \otimes \tilde{Q}(u \wedge w \wedge z),$$

where we are identifying $\bigwedge^3 V$ with the dual space V^* and denoting by $\tilde{Q}\colon V^* \to V$ the isomorphism induced by the skew form Q on V. Specifically, show that the image of this map is complementary to the line spanned by the element $Q \in \bigwedge^2 V^* = \bigwedge^2 V \subset V \otimes V$.

The above leaves us with exactly two possibilities for the weights of $\Gamma_{2,1}$: we know that the multiplicity of $2L_1$ in $\Gamma_{2,1}$ is exactly 2; so either the multiplicities of $L_1 + L_2$ and 0 in $\Gamma_{2,1}$ are both 3 and we have

$$\mathrm{Sym}^2 V \otimes W = \Gamma_{2,1} \oplus \mathrm{Sym}^2 V \oplus W;$$

or the multiplicities of $L_1 + L_2$ and 0 in $\Gamma_{2,1}$ are both 2 and we have

$$\mathrm{Sym}^2 V \otimes W = \Gamma_{2,1} \oplus \mathrm{Sym}^2 V \oplus W^{\oplus 2}.$$

Exercise 16.13. Show that the former of these two possibilities actually occurs, by

(a) Showing that if v is the highest weight vector in $\Gamma_{2,1} \subset \mathrm{Sym}^2 V \otimes W$, then the images $(X_{2,1})^2 V_2(v)$, $X_{2,1} V_2 X_{2,1}(v)$, and $V_2(X_{2,1})^2 v$ are independent; and (redundantly)

(b) Showing that the representation $\mathrm{Sym}^2 V \otimes W$ contains only one highest weight vector of weight $L_1 + L_2$.

The weight diagram of $\Gamma_{2,1}$ is therefore

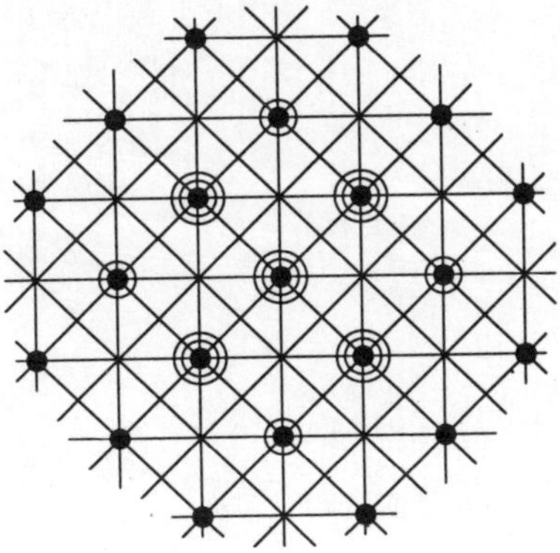

We see from all this that, in particular, the weights of the irreducible representations of $\mathfrak{sp}_4\mathbb{C}$ are not constant on the rings of their weight diagrams.

Exercise 16.14. Analyze the representation $V \otimes \mathrm{Sym}^2 W$ of $\mathfrak{sp}_4\mathbb{C}$. Find in particular the multiplicities of the representation $\Gamma_{1,2}$.

Exercise 16.15. Analyze the representation $\mathrm{Sym}^2 V \otimes \mathrm{Sym}^2 W$ of $\mathfrak{sp}_4\mathbb{C}$. Find in particular the multiplicities of the representation $\Gamma_{2,2}$.

LECTURE 17

$\mathfrak{sp}_6\mathbb{C}$ and $\mathfrak{sp}_{2n}\mathbb{C}$

In the first two sections of this lecture we complete our classification of the representations of the symplectic Lie algebras: we describe in detail the example of $\mathfrak{sp}_6\mathbb{C}$, then sketch the representation theory of symplectic Lie algebras in general, in particular proving the existence part of Theorem 14.18 for $\mathfrak{sp}_{2n}\mathbb{C}$. In the final section we describe an analog for the symplectic algebras of the construction given in §15.3 of the irreducible representations of the special linear algebras via Weyl's construction, though we postpone giving analogous formulas for the decomposition of tensor products of irreducible representations. Sections 17.1 and 17.2 are completely elementary, given the by now standard multilinear algebra of Appendix B. Section 17.3, like §15.3, requires familiarity with the contents of Lecture 6 and Appendix A; but, like that section, it can be skipped without affecting most of the rest of the book.

§17.1. Representations of $\mathfrak{sp}_6\mathbb{C}$

As we have seen, the Cartan algebra $\mathfrak{h}$ of $\mathfrak{sp}_6\mathbb{C}$ is three-dimensional, with the linear functionals L_1, L_2, and L_3 forming an orthonormal basis in terms of the Killing form; and the roots of $\mathfrak{sp}_6\mathbb{C}$ are then the 18 vectors $\pm L_i \pm L_j$. We can draw this in terms of a "reference cube" in $\mathfrak{h}^*$ with faces centered at the points $\pm L_i$; the vectors $\pm L_i \pm L_j$ with $i \neq j$ are then the midpoints of edges of this reference cube and the vectors $\pm 2L_i$ the midpoints of the faces of a cube twice as large. Alternatively, we can draw a reference octahedron with vertices at the vectors $\pm 2L_i$; the roots $\pm L_i \pm L_j$ with $i \neq j$ will then be the midpoints of the edges of this octahedron:

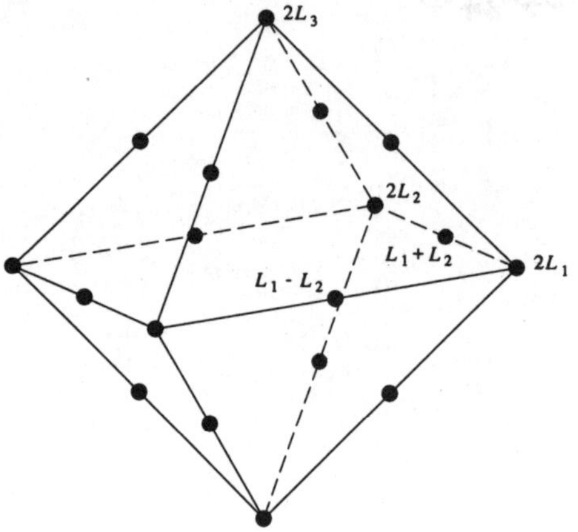

or, if we include the reference cube as well, as

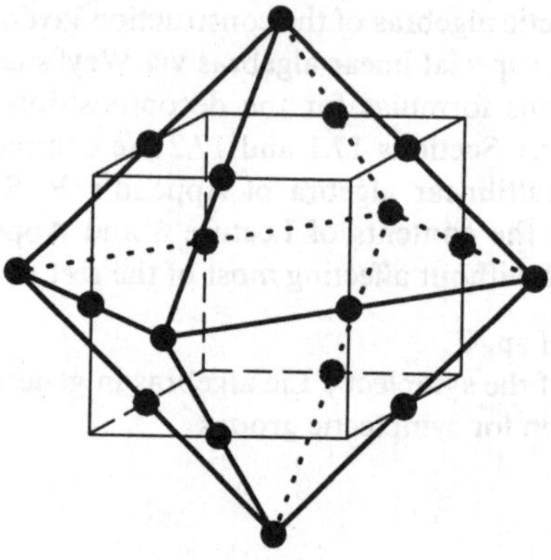

(17.1)

This last diagram suggests a comparison with the root diagram of $\mathfrak{sl}_4\mathbb{C}$; in fact the 12 roots of $\mathfrak{sp}_6\mathbb{C}$ of the form $\pm L_i \pm L_j$ for $i \neq j$ are congruent to the 12 roots of $\mathfrak{sl}_4\mathbb{C}$. In particular, the Weyl group of $\mathfrak{sp}_6\mathbb{C}$ will be generated by the Weyl group of $\mathfrak{sl}_4\mathbb{C}$, plus any of the additional three reflections in the planes perpendicular to the L_i (i.e., the planes parallel to the faces of the reference cube in the root diagram of either Lie algebra). We can indicate the planes perpendicular to the roots of $\mathfrak{sp}_6\mathbb{C}$ by drawing where they cross the visible part of the reference cube:

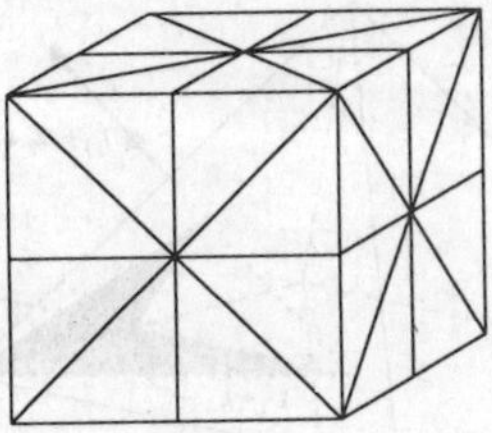

We see from this that the effect of the additional reflections in the Weyl group of $\mathfrak{sp}_6\mathbb{C}$ on the Weyl chamber of $\mathfrak{sl}_4\mathbb{C}$ is simply to cut it in half; whereas the Weyl chamber of $\mathfrak{sl}_4\mathbb{C}$ looked like

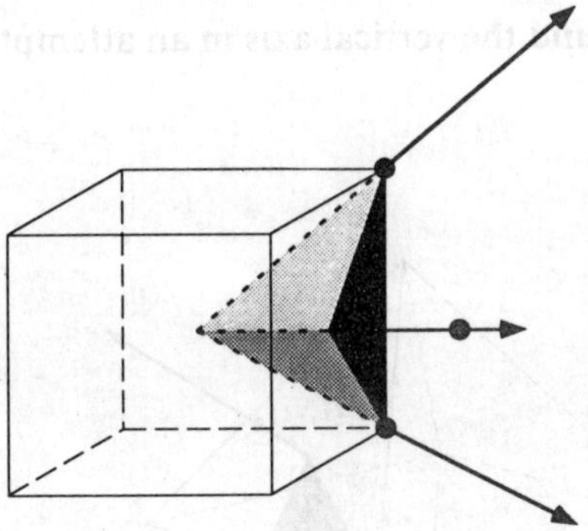

the Weyl chamber of $\mathfrak{sp}_6\mathbb{C}$ will look like just the upper half of this region:

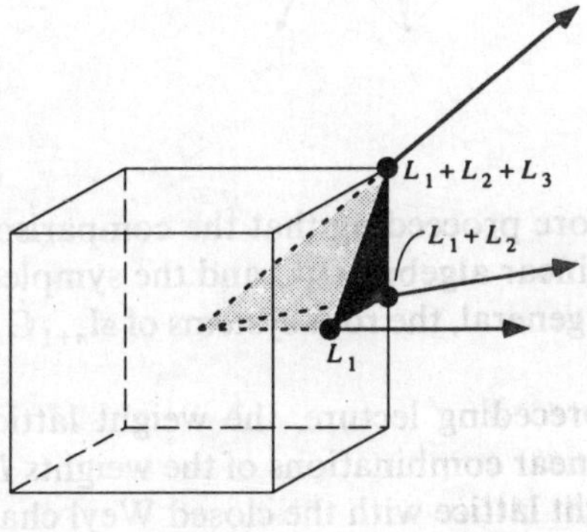

In terms of the reference octahedron, this is the cone over one part of the barycentric subdivision of a face:

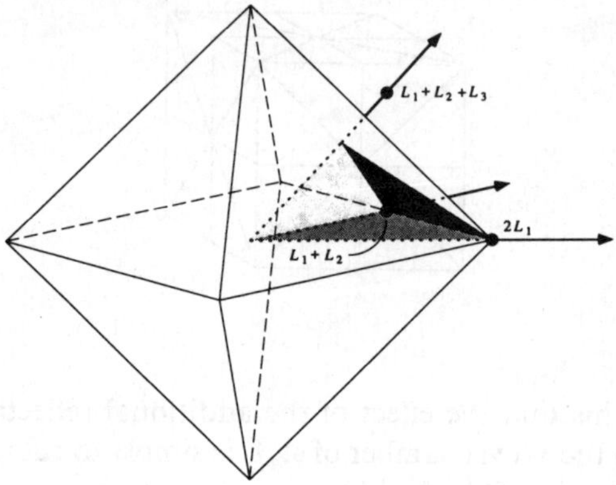

or, if we rotate 90° around the vertical axis in an attempt to make the picture clearer,

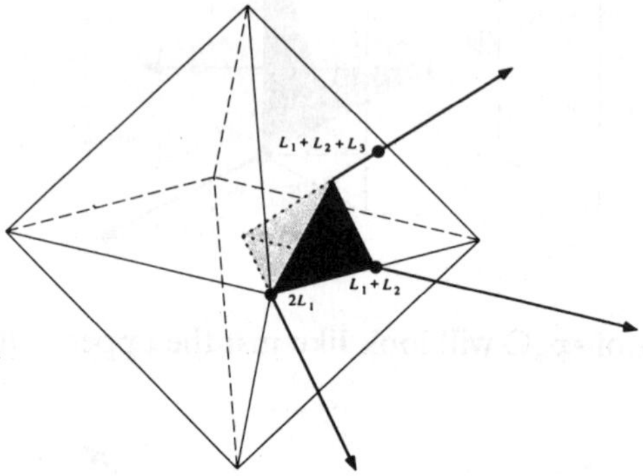

(17.2)

We should remark before proceeding that the comparison between the root systems of the special linear algebra $\mathfrak{sl}_4\mathbb{C}$ and the symplectic algebra $\mathfrak{sp}_6\mathbb{C}$ is peculiar to this case; in general, the root systems of $\mathfrak{sl}_{n+1}\mathbb{C}$ and $\mathfrak{sp}_{2n}\mathbb{C}$ will bear no such similarity.

As we saw in the preceding lecture, the weight lattice of $\mathfrak{sp}_6\mathbb{C}$ consists simply of the integral linear combinations of the weights L_i. In particular, the intersection of the weight lattice with the closed Weyl chamber chosen above will consist exactly of integral linear combinations $a_1L_1 + a_2L_2 + a_3L_3$ with $a_1 \geq a_2 \geq a_3 \geq 0$. By our general existence and uniqueness theorem, then, for every triple (a, b, c) of non-negative integers there will exist a unique irreducible representation of $\mathfrak{sp}_6\mathbb{C}$ with highest weight $aL_1 + b(L_1 + L_2) +$

$c(L_1 + L_2 + L_3) = (a + b + c)L_1 + (b + c)L_2 + cL_3$; we will denote this representation by $\Gamma_{a,b,c}$ and will demonstrate its existence in the following.

We start by considering the standard representation of $\mathfrak{sp}_6\mathbb{C}$ on $V = \mathbb{C}^6$. The eigenvectors of the action of $\mathfrak{h}$ on V are just the standard basis vectors e_i, and these have eigenvalues $\pm L_i$, so that the weight diagram of V looks like the midpoints of the faces of the reference cube (or the vertices of an octahedron one-half the size of the reference octahedron):

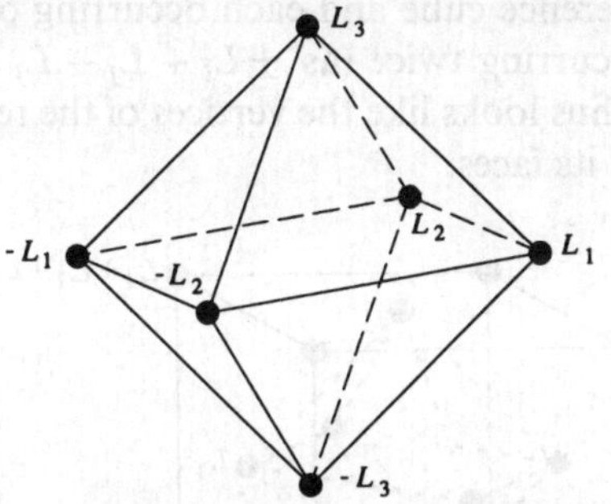

In particular, V is the representation $\Gamma_{1,0,0}$.

Since we are going to want to find a representation with highest weight $L_1 + L_2$, the natural thing to look at next is the second exterior power $\wedge^2 V$ of the standard representation. This will have weights the pairwise sum of distinct weights of V, or in other words the 12 weights $\pm L_i \pm L_j$ with $i \neq j$, and the weight 0 taken three times. This is not irreducible: by definition the action of $\mathfrak{sp}_6\mathbb{C}$ on the standard representation preserves a skew form, so that the representation on $\wedge^2 V$ will have a trivial summand. On the other hand, the skew form on V preserved by $\mathfrak{sp}_6\mathbb{C}$, and hence that trivial summand of $\wedge^2 V$, is unique; and since all the nonzero weights of $\wedge^2 V$ occur with multiplicity 1 and are conjugate under the Weyl group, it follows that the complement W of the trivial representation in $\wedge^2 V$ is irreducible. So $W = \Gamma_{0,1,0}$.

As in previous examples, we can also see that $\wedge^2 V$ is not irreducible by using the fact (Observation 14.16) that the irreducible representation $\Gamma_{0,1,0}$ with highest weight $L_1 + L_2$ will be generated by applying to a single highest weight vector v the root spaces $\mathfrak{g}_{L_2 - L_1}$, $\mathfrak{g}_{L_3 - L_2}$, and $\mathfrak{g}_{-2L_3}$ corresponding to primitive negative roots. We can then verify that in the irreducible representation W with highest weight $L_1 + L_2$, there are only three ways of going from the highest weight space to the zero weight space by successive application of these roots spaces: we can go

$$\begin{array}{ccccccc} L_1 + L_2 \to L_1 + L_3 & \to & L_1 - L_3 & \to & L_1 - L_2 & & \\ & \searrow & & \searrow & & \searrow & \\ & & L_2 + L_3 & \to & L_2 - L_3 & \to & 0 \end{array}$$

Exercise 17.3. Verify this, and also verify that the lower two routes to the zero-weight space in $\wedge^2 V$ yield the same nonzero vector, and that the upper route yields an independent element of $\wedge^2 V$, so that 0 does indeed occur with multiplicity 2 as a weight of $\Gamma_{0,1,0}$.

To continue, we look next at the third exterior power $\wedge^3 V$ of the standard representation; we know that this will contain a copy of the irreducible representation $\Gamma_{0,0,1}$ with highest weight $L_1 + L_2 + L_3$. The weights of $\wedge^3 V$ are of two kinds: we have the eight sums $\pm L_1 \pm L_2 \pm L_3$, corresponding to the vertices of the reference cube and each occurring once; and we have the weights $\pm L_i$ each occurring twice (as $\pm L_i + L_j - L_j$ and $\pm L_i + L_k - L_k$). The weight diagram thus looks like the vertices of the reference cube together with the midpoints of its faces:

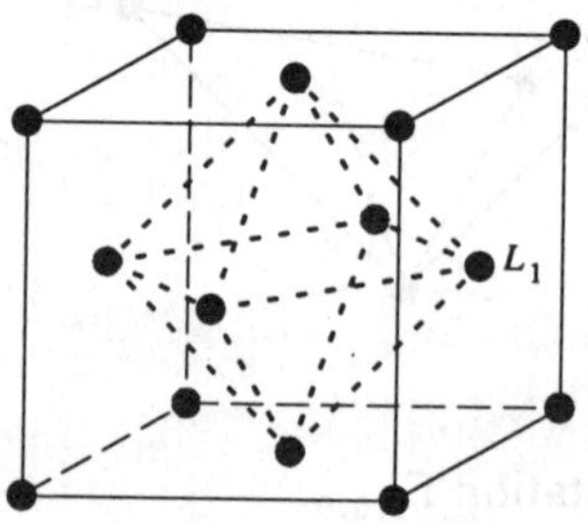

Now, the weights $\pm L_i$ must occur in the representation $\Gamma_{0,0,1}$ with highest weight $L_1 + L_2 + L_3$, since they are congruent to $L_1 + L_2 + L_3$ modulo the root lattice and lie in the convex hull of the translates of $L_1 + L_2 + L_3$ under the Weyl group (that is, they lie in the closed reference cube). But they cannot occur with multiplicity greater than 1: for example, the only way to get from the point $L_1 + L_2 + L_3$ to the point L_1 by translations by the basic vectors $L_2 - L_1$, $L_3 - L_2$, and $-2L_3$ pictured in Diagram (17.1) above (while staying inside the reference cube) is by translation by $-2L_3$ first, and then by $L_3 - L_2$. it follows that the multiplicities of the weights $\pm L_i$ in $\Gamma_{0,0,1}$ are 1. On the other hand, we have a natural map

$$\wedge^3 V \to V$$

obtained by contracting with the element of $\wedge^2 V^*$ preserved by the action of $\mathfrak{sp}_6\mathbb{C}$, and the kernel of this map, which must contain the representation $\Gamma_{0,0,1}$, will have exactly these weights. The kernel of φ is thus the irreducible representation with highest weight $L_1 + L_2 + L_3$; we will call this representation U for now.

At this point, we have established the existence theorem for representations of $\mathfrak{sp}_6\mathbb{C}$: the irreducible representation $\Gamma_{a,b,c}$ with highest weight $(a + b + c)L_1 + (a + b)L_2 + cL_3$ will occur inside the representation

$$\mathrm{Sym}^a V \otimes \mathrm{Sym}^b W \otimes \mathrm{Sym}^c U.$$

For example, suppose we want to find the irreducible representation $\Gamma_{1,1,0}$ with highest weight $2L_1 + L_2$. The weights of this representation will be the 24 weights $\pm 2L_i \pm L_j$, each taken with multiplicity 1; the 8 weights $\pm L_1 \pm L_2 \pm L_3$, taken with a multiplicity we do not a priori know (but that the reader can verify must be either 1 or 2), and the weights $\pm L_i$ taken with some other multiplicity. At the same time, the representation $V \otimes W$, which contains $\Gamma_{1,1,0}$, will take on these weights, with multiplicities 1, 3, and 6, respectively. In particular, it follows that $V \otimes W$ will contain a copy of the irreducible representation U with highest weight $L_1 + L_2 + L_3$ as well; alternatively, we can see this directly by observing that the wedge product map

$$V \otimes \Lambda^2 V \to \Lambda^3 V$$

factors to give a map

$$V \otimes W \to U$$

and that $\Gamma_{1,1,0}$ must lie in the kernel of this map. To say more about the location of $\Gamma_{1,1,0}$ inside $V \otimes W$, and its exact weights, would require either explicit calculation or something like the Weyl character formula. We will see in Lecture 24 how the latter can be used to solve the problem; for the time being we leave this as

Exercise 17.4. Verify by direct calculation that the multiplicities of the weights of $\Gamma_{1,1,0}$ are 1, 2, and 5, and hence that the kernel of the map φ above is exactly the representation $\Gamma_{1,1,0}$.

§17.2. Representations of $\mathfrak{sp}_{2n}\mathbb{C}$ in General

The general picture for representations of the symplectic Lie algebras offers no further surprises. As we have seen, the weight lattice consists simply of integral linear combinations of the L_i. And our typical Weyl chamber is a cone over a simplex in n-space, with edges the rays defined by

$$a_1 = a_2 = \cdots = a_i > a_{i+1} = \cdots = a_n = 0.$$

The primitive lattice element on the ith ray is the weight $\omega_i = L_1 + \cdots + L_i$, and we may observe that, similarly to the case of the special linear Lie algebras, these n fundamental weights generate as a semigroup the intersection of the closed Weyl chamber with the lattice. Thus, our basic existence and uniqueness theorem asserts that for an arbitrary n-tuple of natural numbers $(a_1, \ldots, a_n) \in \mathbb{N}^n$ there will be a unique irreducible representation with highest weight

$$\begin{aligned} &a_1\omega_1 + a_2\omega_2 + \cdots + a_n\omega_n \\ &\quad = (a_1 + \cdots + a_n)L_1 + (a_2 + \cdots + a_n)L_2 + \cdots + a_nL_n. \end{aligned}$$

As before, we denote this by $\Gamma_{a_1,\ldots,a_n}$:

$$\Gamma_{a_1,\ldots,a_n} = \Gamma_{a_1L_1+a_2(L_1+L_2)+\cdots+a_n(L_1+\cdots+L_n)}.$$

These exhaust all irreducible representations of $\mathfrak{sp}_{2n}\mathbb{C}$.

We can find the irreducible representation $V^{(k)} = \Gamma_{0,\ldots,1,\ldots,0}$ with highest weight $L_1 + \cdots + L_k$ easily enough. Clearly, it will be contained in the kth exterior power $\wedge^k V$ of the standard representation. Moreover, we have a natural contraction map

$$\varphi_k: \wedge^k V \to \wedge^{k-2} V$$

defined by

$$\varphi_k(v_1 \wedge \cdots \wedge v_k) = \sum_{i<j} Q(v_i, v_j)(-1)^{i+j-1} v_1 \wedge \cdots \wedge \hat{v}_i \wedge \cdots \wedge \hat{v}_j \wedge \cdots \wedge v_k$$

(see §B.3 of Appendix B for an intrinsic definition and explanation). Since the representation $\wedge^{k-2}V$ does not have the weight $L_1 + \cdots + L_k$, the irreducible representation with this highest weight will have to be contained in the kernel of this map. We claim now that conversely

Theorem 17.5. *For $1 \leq k \leq n$, the kernel of the map φ_k is exactly the irreducible representation $V^{(k)} = \Gamma_{0,\ldots,0,1,0,\ldots,0}$ with highest weight $L_1 + \cdots + L_k$.*

PROOF. Clearly, it is enough to show that the kernel of φ_k is an irreducible representation of $\mathfrak{sp}_{2n}\mathbb{C}$. We will do this by restricting to a subalgebra of $\mathfrak{sp}_{2n}\mathbb{C}$ isomorphic to $\mathfrak{sl}_n\mathbb{C}$, and using what we have learned about representations of $\mathfrak{sl}_n\mathbb{C}$.

To describe this copy of $\mathfrak{sl}_n\mathbb{C}$ inside $\mathfrak{sp}_{2n}\mathbb{C}$, consider the subgroup $G \subset \mathrm{Sp}_{2n}\mathbb{C}$ of transformations of the space $V = \mathbb{C}^{2n}$ preserving the skew form Q introduced in Lecture 16 and preserving as well the decomposition $V = \mathbb{C}\{e_1, \ldots, e_n\} \oplus \mathbb{C}\{e_{n+1}, \ldots, e_{2n}\}$. These can act arbitrarily on the first factor, as long as they do the opposite on the second; in coordinates, they are the matrices

$$G = \left\{\begin{pmatrix} X & 0 \\ 0 & {}^tX^{-1} \end{pmatrix}, X \in \mathrm{GL}_n\mathbb{C}\right\}.$$

We have, correspondingly, a subalgebra

$$\mathfrak{s} = \left\{\begin{pmatrix} A & 0 \\ 0 & -{}^tA \end{pmatrix}, A \in \mathfrak{sl}_n\mathbb{C}\right\} \subset \mathfrak{sp}_{2n}\mathbb{C}$$

isomorphic to $\mathfrak{sl}_n\mathbb{C}$.

Now, denote by W the standard representation of $\mathfrak{sl}_n\mathbb{C}$. The restriction of the representation V of $\mathfrak{sp}_{2n}\mathbb{C}$ to the subalgebra $\mathfrak{s}$ then splits

$$V = W \oplus W^*$$

into a direct sum of W and its dual; and we have, correspondingly,

$$\wedge^k V = \bigoplus_{a+b=k} (\wedge^a W \otimes \wedge^b W^*).$$

How does the tensor product $\wedge^a W \otimes \wedge^b W^*$ decompose as a representation of $\mathfrak{sl}_n\mathbb{C}$? We know the answer to this from the discussion in Lecture 15 (see Exercise 15.30): we have contraction maps

$$\Psi_{a,b}: \wedge^a W \otimes \wedge^b W^* \to \wedge^{a-1} W \otimes \wedge^{b-1} W^*;$$

and the kernel of $\Psi_{a,b}$ is the irreducible representation $W^{(a,b)} = \Gamma_{0,\ldots,0,1,0,\ldots,0,1,0,\ldots}$ with (if, say, $a \leq n - b$) highest weight $2L_1 + \cdots + 2L_a + L_{a+1} + \cdots + L_{n-b}$. The restriction of $\wedge^k V$ to $\mathfrak{s}$ is thus given by

$$\wedge^k V = \bigoplus_{\substack{a+b \leq k \\ a+b \equiv k(2)}} W^{(a,b)}$$

and by the same token,

$$\operatorname{Ker}(\varphi_k) = \bigoplus_{a+b=k} W^{(a,b)}.$$

Note that the actual highest weight factor in the summand $W^{(a,b)} \subset \operatorname{Ker}(\varphi_k) \subset \wedge^k V$ is the vector

$$\begin{aligned} w^{(a,b)} &= e_1 \wedge \cdots \wedge e_a \wedge e_{2n-b+1} \wedge \cdots \wedge e_{2n} \\ &= e_1 \wedge \cdots \wedge e_a \wedge e_{2n-k+a+1} \wedge \cdots \wedge e_{2n}. \end{aligned}$$

Exercise 17.6. Show that more generally the highest weight vector in any summand $W^{(a,b)} \subset \wedge^k V$ is the vector

$$\begin{aligned} w^{(a,b)} &= e_1 \wedge \cdots \wedge e_a \wedge e_{2n-k+a+1} \wedge \cdots \wedge e_{2n} \wedge Q^{(k-a-b)/2} \\ &= e_1 \wedge \cdots \wedge e_a \wedge e_{2n-k+a+1} \wedge \cdots \wedge e_{2n} \wedge \left(\sum (e_i \wedge e_{n+i})\right)^{(k-a-b)/2}. \end{aligned}$$

By the above, any subspace of $\operatorname{Ker}(\varphi_k)$ invariant under $\mathfrak{sp}_{2n}\mathbb{C}$ must be a direct sum, over a subset of pairs (a, b) with $a + b = k$, of subspaces $W^{(a,b)}$. But now (supposing for the moment that $k < n$) we observe that the element

$$Z_{a,n-b} = E_{2n-b,a} + E_{n+a,n-b} \in \mathfrak{sp}_{2n}\mathbb{C}$$

carries the vector $w^{(a,b)}$ into $w^{(a-1,b+1)}$ and, likewise,

$$Y_{a+1,n-b+1} = E_{a+1,2n-b+1} + E_{n-b+1,n+a+1} \in \mathfrak{sp}_{2n}\mathbb{C}$$

carries $w^{(a,b)}$ to $w^{(a+1,b-1)}$. In case $a + b = k = n$, we see similarly that

$$V_a = E_{n+a,a} \in \mathfrak{sp}_{2n}\mathbb{C}$$

carries the vector $w^{(a,b)}$ into $w^{(a-1,b+1)}$, and

$$U_{a+1} = E_{a+1,n+a+1} \in \mathfrak{sp}_{2n}\mathbb{C}$$

carries $w^{(a,b)}$ to $w^{(a+1,b-1)}$. Thus, any representation of $\mathfrak{sp}_{2n}\mathbb{C}$ contained in $\operatorname{Ker}(\varphi_k)$ and containing any one of the factors $W^{(a,b)}$ will contain them all, and we are done. □

Exercise 17.7. Another way to conclude this proof would be to remark that, inasmuch as all the $w^{(a,b)}$ above are eigenvectors of different weights, any highest weight vector for the action of $\mathfrak{sp}_{2n}\mathbb{C}$ on $\ker(\varphi_k) \subset \bigwedge^k V$ would have to be (up to scalars) one of the $w^{(a,b)}$. It would thus be sufficient to find, for each (a, b) with $a + b = k$ other than $(a, b) = (k, 0)$, a positive root α such that $\mathfrak{g}_\alpha(w^{(a,b)}) \neq 0$. Do this.

Note that, having found the irreducible representations $V^{(k)} = \Gamma_{0,\dots,1,\dots,0}$ with highest weight $L_1 + \cdots + L_k$, any other representation of $\mathfrak{sp}_{2n}\mathbb{C}$ will occur in a tensor product of these; specifically, the irreducible representation $\Gamma_{a_1,\dots,a_n}$ with highest weight $a_1 L_1 + \cdots + a_n(L_1 + \cdots + L_n)$ will occur in the product $\mathrm{Sym}^{a_1} V \otimes \mathrm{Sym}^{a_2} V^{(2)} \otimes \cdots \otimes \mathrm{Sym}^{a_n} V^{(n)}$.

One further remark is that there exist geometric interpretations of the action of $\mathfrak{sl}_{2n}\mathbb{C}$ on the fundamental representations $V^{(k)}$. We have said before that the group $\mathrm{PSp}_{2n}\mathbb{C}$ may be characterized as the subgroup of $\mathrm{PGL}_{2n}\mathbb{C}$ carrying isotropic subspaces of V into isotropic subspaces. At the same time, $\mathrm{PGL}_{2n}\mathbb{C}$ acts on the projective space $\mathbb{P}(\bigwedge^k V)$ as the connected component of the identity in the group of motions of this space carrying the Grassmannian $G = G(k, V) \subset \mathbb{P}(\bigwedge^k V)$ into itself. Now, the subset $G_L \subset G$ of k-dimensional isotropic subspaces of V is exactly the intersection of the Grassmannian G with the subspace $\mathbb{P}(V^{(k)})$ associated to the kernel of the map φ above; so that $\mathrm{PSp}_{2n}\mathbb{C}$ will act on $\mathbb{P}(V^{(k)})$ carrying G_L into itself and indeed when $1 < k \leq n$ may be characterized as the connected component of the identity in the group of motions of $\mathbb{P}(V^{(k)})$ preserving the variety G_L.

Exercise 17.8. Show that if $k > n$ the contraction φ_k is injective.

§17.3. Weyl's Construction for Symplectic Groups

We have just seen how the basic representations for $\mathfrak{sp}_{2n}\mathbb{C}$ can be obtained by taking certain basic representations of the larger Lie algebra $\mathfrak{sl}_{2n}\mathbb{C}$—in this case, $\bigwedge^k V$ for $k \leq n$—and intersecting with the kernel of a contraction constructed from the symplectic form. In fact, *all* the representations of the symplectic Lie algebras can be given a similar conrete realization, by intersecting certain of the irreducible representations of $\mathfrak{sl}_{2n}\mathbb{C}$ with the intersections of the kernels of all such contractions.

Recall from Lectures 6 and 15 that the irreducible representations of $\mathfrak{sl}_{2n}\mathbb{C}$ are given by Schur functors $\mathbb{S}_\lambda V$, where $\lambda = (\lambda_1 \geq \cdots \geq \lambda_{2n} \geq 0)$ is a partition of some integer $d = \sum \lambda_i$, and $V = \mathbb{C}^{2n}$. This representation is realized as the image of a corresponding Young symmetrizer c_λ acting on the d-fold tensor product space $V^{\otimes d}$. For each pair $I = \{p < q\}$ of integers between 1 and d, the symplectic form Q determines a contraction

$$\Phi_I \colon V^{\otimes d} \to V^{\otimes(d-2)},$$
$$v_1 \otimes \cdots \otimes v_d \mapsto Q(v_p, v_q) v_1 \otimes \cdots \otimes \hat{v}_p \otimes \cdots \otimes \hat{v}_q \otimes \cdots \otimes v_d. \tag{17.9}$$

Let $V^{\langle d\rangle} \subset V^{\otimes d}$ denote the intersection of the kernels of all these contractions. These subspaces is mapped to itself by permutations, so $V^{\langle d\rangle}$ is a subrepresentation of $V^{\otimes d}$ as a representation of the symmetric group $\mathfrak{S}_d$. Now let[1]

$$\mathbb{S}_{\langle\lambda\rangle}V = V^{\langle d\rangle} \cap \mathbb{S}_\lambda V. \tag{17.10}$$

This space is a representation of the symplectic group $\mathrm{Sp}_{2n}\mathbb{C}$ of Q, since $V^{\langle d\rangle}$ and $\mathbb{S}_\lambda(V)$ are subrepresentations of $V^{\otimes d}$ over $\mathrm{Sp}_{2n}\mathbb{C}$.

Theorem 17.11. *The space $\mathbb{S}_{\langle\lambda\rangle}(V)$ is nonzero if and only if the Young diagram of λ has at most n rows, i.e., $\lambda_{n+1} = 0$. In this case, $\mathbb{S}_{\langle\lambda\rangle}(V)$ is the irreducible representation of $\mathfrak{sp}_{2n}\mathbb{C}$ with highest weight $\lambda_1 L_1 + \cdots + \lambda_n L_n$.*

In other words, for an n-tuple $(a_1, \ldots, a_n)$ of non-negative integers

$$\Gamma_{a_1,\ldots,a_n} = \mathbb{S}_{\langle\lambda\rangle}V,$$

where λ is the partition $(a_1 + a_2 + \cdots + a_n, a_2 + \cdots + a_n, \ldots, a_n)$.

The proof follows the pattern for the general linear group given in §6.2, but we will have to call on a basic result from invariant theory in place of the simple Lemma 6.23. We first show how to find a complement to $V^{\langle d\rangle}$ in $V^{\otimes d}$ For example, if $d = 2$, then

$$V^{\otimes 2} = V^{\langle 2\rangle} \oplus \mathbb{C}\cdot\psi,$$

where ψ is the element of $V \otimes V$ corresponding to the quadratic form Q. In terms of our canonical basis, $\psi = \sum (e_i \otimes e_{n+i} - e_{n+i} \otimes e_i)$. In general, for any $I = \{p < q\}$ define

$$\Psi_I \colon V^{\otimes(d-2)} \to V^{\otimes d}$$

by inserting ψ in the p, q factors. Note that $\Phi_I \circ \Psi_I$ is multiplication by $2n = \dim V$ on $V^{\otimes(d-2)}$. We claim that

$$V^{\otimes d} = V^{\langle d\rangle} \oplus \textstyle\sum_I \Psi_I(V^{\otimes(d-2)}). \tag{17.12}$$

To prove this, put the standard Hermitian metric (,) on $V = \mathbb{C}^{2n}$, using the given e_i as a basis, so that $(ae_i, be_j) = \delta_{ij}\bar{a}b$. This extends to give a Hermitian metric on each $V^{\otimes d}$. We claim that the displayed equation is a perpendicular direct sum. This follows from the following exercise.

Exercise 17.13. (i) Verify that for $v, w \in V$, $(\psi, v \otimes w) = Q(v, w)$.

(ii) Use (i) to show that $\mathrm{Ker}(\Phi_I) = \mathrm{Im}(\Psi_I)^\perp$ for each I.

Now define $F_r^d \subset V^{\otimes d}$ to be the intersection of the kernels of all r-fold contractions $\Phi_{I_1} \circ \cdots \circ \Phi_{I_r}$, and set

$$V_{d-2r}^{\langle d\rangle} = \sum \Psi_{I_1} \circ \cdots \circ \Psi_{I_r}(V^{\otimes\langle d-2r\rangle}). \tag{17.14}$$

[1] This follows a classical notation of using ⟨ ⟩ for the symplectic group and [ ] for the orthogonal group (although we have omitted the corresponding notation { } for the general linear group).

Lemma 17.15. *The tensor power $V^{\otimes d}$ decomposes into a direct sum*

$$V^{\otimes d} = V^{\langle d \rangle} \oplus V_{d-2}^{\langle d \rangle} \oplus V_{d-4}^{\langle d \rangle} \oplus \cdots \oplus V_{d-2p}^{\langle d \rangle},$$

with $p = [d/2]$, and, for all $r \geq 1$,

$$F_r^d = V^{\langle d \rangle} \oplus V_{d-2}^{\langle d \rangle} \oplus \cdots \oplus V_{d-2r+2}^{\langle d \rangle}.$$

Exercise 17.16. (i) Show as in the preceding exercise that there is a perpendicular decomposition

$$V^{\otimes d} = F_r^d \oplus \sum \Psi_{I_1} \circ \cdots \circ \Psi_{I_r}(V^{\otimes(d-2r)}).$$

(ii) Verify that $\Psi_I(F_p^{d-2}) \subset F_{p+1}^d$.

(iii) Show by induction that $V^{\otimes d}$ is the sum of the spaces $V_{d-2r}^{\langle d \rangle}$.

(iv) Finish the proof of the lemma, using (i) and (ii) to deduce that both sums are orthogonal splittings. □

All the subspaces in these splittings are invariant by the action of the symplectic group $\mathrm{Sp}_{2n}\mathbb{C}$, as well as the action of the symmetric group $\mathfrak{S}_d$. In particular, we see that

$$\mathbb{S}_{\langle \lambda \rangle} V = V^{\langle d \rangle} \cdot c_\lambda = \mathrm{Im}(c_\lambda \colon V^{\langle d \rangle} \to V^{\langle d \rangle}). \tag{17.17}$$

Exercise 17.18*. (i) Show that if $s > n$, then $\wedge^s V \otimes V^{\otimes(d-s)}$ is contained in $\sum_I \Psi_I(V^{\otimes(d-2)})$, and deduce that $\mathbb{S}_{\langle \lambda \rangle}(V) = 0$ if λ_{n+1} is not 0.

(ii) Show that $\mathbb{S}_{\langle \lambda \rangle}(V)$ is not zero if $\lambda_{n+1} = 0$.

For any pair of integers I from $\{1, \ldots, d\}$, define

$$\vartheta_I = \Psi_I \circ \Phi_I \colon V^{\otimes d} \to V^{\otimes d}.$$

From what we have seen, $V^{\langle d \rangle}$ is the intersection of the kernels of all these endomorphisms. Note that the endomorphism of $V^{\otimes d}$ determined by any symplectic automorphism of V not only commutes with all permutations of the factors $\mathfrak{S}_d$ but also commutes with the operators ϑ_I. We need a fact which is proved in Appendix F.2:

Invariant Theory Fact 17.19. *Any endomorphism of $V^{\otimes d}$ that commutes with all permutations in $\mathfrak{S}_d$ and all the operators ϑ_I is a finite $\mathbb{C}$-linear combination of operators of the form $A \otimes \cdots \otimes A$, for $A \in \mathrm{Sp}_{2n}\mathbb{C}$.*

Now let B be the algebra of all endomorphisms of the space $V^{\langle d \rangle}$ that are $\mathbb{C}$-linear combinations of operators of the form $A \otimes \cdots \otimes A$, for $A \in \mathrm{Sp}_{2n}\mathbb{C}$.

Proposition 17.20. *The algebra B is precisely the algebra of all endomorphisms of $V^{\langle d \rangle}$ commuting with all permutations in $\mathfrak{S}_d$.*

PROOF. If F is an endomorphism of $V^{\langle d \rangle}$ commuting with all permutations of factors, then the endomorphism $\tilde{F}$ of $V^{\otimes d}$ that is F on the factor $V^{\langle d \rangle}$ and zero on the complementary summand $\sum_I \Psi_I(V^{\otimes(d-2)})$ is an endomorphism that commutes with all permutations and all operators ϑ_I. The fact that $\tilde{F}$ is a linear combination of operators from the symplectic group (which we know from Fact 17.19) implies the same for F. □

Corollary 17.21. *The representations* $\mathbb{S}_{\langle\lambda\rangle}(V)$ *are irreducible representations of* $\mathrm{Sp}_{2n}\mathbb{C}$.

PROOF. Since B is the commutator algebra to $A = \mathbb{C}[\mathfrak{S}_d]$ acting on the space $V^{\langle d\rangle}$, Lemma 6.22 implies that $(V^{\langle d\rangle})\cdot c_\lambda$ is an irreducible B-module. But we have seen that $(V^{\langle d\rangle})\cdot c_\lambda = \mathbb{S}_{\langle\lambda\rangle}V$, and the proposition shows that being irreducible over B is the same as being irreducible over $\mathrm{Sp}_{2n}\mathbb{C}$. □

Exercise 17.22*. Show that the multiplicity with which $\mathbb{S}_{\langle\lambda\rangle}(V)$ occurs in $V^{\langle d\rangle}$ is the dimension m_λ of the corresponding representation V_λ of $\mathfrak{S}_d$.

As was the case for the Weyl construction over $\mathrm{GL}_n\mathbb{C}$, there are general formulas for decomposing tensor products of these representations, as well as restrictions to subgroups $\mathrm{Sp}_{2n-2}\mathbb{C}$, and for their dimensions and multiplicities of weight spaces. We postpone these questions to Lecture 25, when we will have the Weyl character formula at our disposal.

As we saw in Lecture 15 for $\mathrm{GL}_n\mathbb{C}$, it is possible to make a commutative algebra which we denote by $\mathbb{S}^{\langle\cdot\rangle} = \mathbb{S}^{\langle\cdot\rangle}(V)$ out of the sum of all the irreducible representations of $\mathrm{Sp}_{2n}\mathbb{C}$, where $V = \mathbb{C}^{2n}$ is the standard representation. Probably the simplest way to do this, given what we have proved so far, is to start with the ring

$$\begin{aligned} A^\cdot(V, n) &= \mathrm{Sym}^\cdot(V \oplus \Lambda^2 V \oplus \Lambda^3 V \oplus \cdots \oplus \Lambda^n V) \\ &= \bigoplus_{a_1,\ldots,a_n} \mathrm{Sym}^{a_n}(\Lambda^n V) \otimes \cdots \otimes \mathrm{Sym}^{a_2}(\Lambda^2 V) \otimes \mathrm{Sym}^{a_1}(V), \end{aligned}$$

the sum over all n-tuples $\mathbf{a} = (a_1, \ldots, a_n)$ of non-negative integers. Define a ring $\mathbb{S}^\cdot(V, n)$ to be the quotient of $A^\cdot(V, n)$ by the ideal generated by the same relations as in (15.53). By the argument in §15.5, the ring $\mathbb{S}^\cdot(V, n)$ is the direct sum of all the representations $\mathbb{S}_\lambda(V)$ of $\mathrm{GL}(V)$, as λ varies over all partitions with at most n parts.

The decomposition $V^{\otimes d} = V^{\langle d\rangle} \oplus W^{\langle d\rangle}$ of (17.12) determines a decomposition $V^{\otimes d}\cdot c_\lambda = V^{\langle d\rangle}\cdot c_\lambda \oplus W^{\langle d\rangle}\cdot c_\lambda$, which is a decomposition

$$\mathbb{S}_\lambda(V) = \mathbb{S}_{\langle\lambda\rangle}(V) \oplus J_{\langle\lambda\rangle}(V)$$

of representations of $\mathrm{Sp}_{2n}\mathbb{C}$. We claim that the sum $J^{\langle\cdot\rangle} = \bigoplus_\lambda J_{\langle\lambda\rangle}(V)$ is an ideal in $\mathbb{S}^\cdot(V, n) = \bigoplus_\lambda \mathbb{S}_\lambda(V)$. This is easy to see using weights, since $J_{\langle\lambda\rangle}(V)$ is the sum of all the representations in $\mathbb{S}_\lambda(V)$ whose highest weight is strictly smaller than λ. This implies that the image of $J_{\langle\lambda\rangle}(V) \otimes \mathbb{S}_\mu(V)$ in $\mathbb{S}_{\lambda+\mu}(V)$ is a sum of representations whose highest weights are less than $\lambda + \mu$, so they must be in $J_{\langle\lambda+\mu\rangle}(V)$.

The quotient ring is, therefore, the ring $\mathbb{S}^{\langle\cdot\rangle}(V)$ we were looking for:

$$\mathbb{S}^{\langle\bullet\rangle} = \mathbb{S}^{\bullet}(V, n)/J^{\langle\bullet\rangle} = \bigoplus_{\lambda} \mathbb{S}_{\langle\lambda\rangle}(V).$$

In fact, the ideal $J^{\langle\bullet\rangle}$ is generated by elements of the form $x \wedge \psi$, where $x \in \bigwedge^i V$, $i \leq n-2$, and ψ is the element in $\bigwedge^2 V$ corresponding to the skew form Q. An outline of the proof is sketched at the end of Lecture 25. The calculations, as well as other constructions of the ring, can be found in [L-T], where one can also find a discussion of functorial properties of the construction. For bases, see [DC-P], [L-M-S], and [M-S].

LECTURE 18
Orthogonal Lie Algebras

In this and the following two lectures we carry out for the orthogonal Lie algebras what we have already done in the special linear and symplectic cases. As in those cases, we start by working out in general the structure of the orthogonal Lie algebras, describing the roots, root spaces, Weyl group, etc., and then go to work on low-dimensional examples. There is one new phenomenon here: as it turns out, all three of the Lie algebras we deal with in §18.2 are isomorphic to symplectic or special linear Lie algebras we have already analyzed (this will be true of $\mathfrak{so}_6\mathbb{C}$ as well, but of no other orthogonal Lie algebra). As in the previous cases, the analysis of the Lie algebras and their representation theory will be completely elementary. Algebraic geometry does intrude into the discussion, however: we have described the isomorphisms between the orthogonal Lie algebras discussed and special linear and symplectic ones in terms of projective geometry, since that is what seems to us most natural. This should not be a problem; there are many other ways of describing these isomorphisms, and readers who disagree with our choice can substitute their own.

§18.1. $\mathrm{SO}_m\mathbb{C}$ and $\mathfrak{so}_m\mathbb{C}$

We will take up now the analysis of the Lie algebras of orthogonal groups. Here there is, as we will see very shortly, a very big difference in behavior between the so-called "even" orthogonal Lie algebras $\mathfrak{so}_{2n}\mathbb{C}$ and the "odd" orthogonal Lie algebras $\mathfrak{so}_{2n+1}\mathbb{C}$. Interestingly enough, the latter seem at first glance to be more complicated, especially in terms of notation; but when we analyze their representations we see that in fact they behave more regularly than the even ones. In any event, we will try to carry out the analysis in parallel fashion for as long as is feasible; when it becomes necessary to split up into cases, we will usually look at the even orthogonal Lie algebras first and then consider the odd.

Let V be a m-dimensional complex vector space, and

$$Q: V \times V \to \mathbb{C}$$

a nondegenerate, symmetric bilinear form on V. The orthogonal group $\mathrm{SO}_m\mathbb{C}$ is then defined to be the group of automorphisms A of V of determinant 1 preserving Q—that is, such that $Q(Av, Aw) = Q(v, w)$ for all $v, w \in V$—and the orthogonal Lie algebra $\mathfrak{so}_m\mathbb{C}$ correspondingly consists of endomorphisms $A: V \to V$ satisfying

$$Q(Av, w) + Q(v, Aw) = 0 \tag{18.1}$$

for all v and $w \in V$. As in the case of the symplectic Lie algebras, to carry out our analysis we want to write Q explicitly in terms of a basis for V, and here is where the cases of even and odd m first separate. In case $m = 2n$ is even, we will choose a basis for V in terms of which the quadratic form Q is given by

$$Q(e_i, e_{i+n}) = Q(e_{i+n}, e_i) = 1$$

and

$$Q(e_i, e_j) = 0 \quad \text{if } j \neq i \pm n.$$

The bilinear form Q may be expressed as

$$Q(x, y) = {}^t x \cdot M \cdot y,$$

where M is the $2n \times 2n$ matrix given in block form as

$$M = \begin{pmatrix} 0 & I_n \\ I_n & 0 \end{pmatrix};$$

the group $\mathrm{SO}_{2n}\mathbb{C}$ is thus the group of $2n \times 2n$ matrices A with $\det(A) = 1$ and

$$M = {}^t A \cdot M \cdot A,$$

and the Lie algebra $\mathfrak{so}_{2n}\mathbb{C}$ correspondingly the space of matrices X satisfying the relation

$${}^t X \cdot M + M \cdot X = 0.$$

Writing a $2n \times 2n$ matrix X in block form as

$$X = \begin{pmatrix} A & B \\ C & D \end{pmatrix}$$

we have

$${}^t X \cdot M = \begin{pmatrix} {}^t C & {}^t A \\ {}^t D & {}^t B \end{pmatrix}$$

and

$$M \cdot X = \begin{pmatrix} C & D \\ A & B \end{pmatrix}$$

so that this relation is equivalent to saying that *the off-diagonal blocks B and C of X are skew-symmetric, and the diagonal blocks A and D of X are negative transposes of each other.*

Exercise 18.2. Show that with this choice of basis,

$$SO_2(\mathbb{C}) = \left\{ \begin{pmatrix} a & 0 \\ 0 & a^{-1} \end{pmatrix} \right\} \cong \mathbb{C}^*,$$

and $\mathfrak{so}_2\mathbb{C} = \mathbb{C}$.

The situation in case the dimension m of V is odd is similar, if a little messier. To begin with, we will take Q to be expressible, in terms of a basis $e_1, \ldots, e_{2n+1}$ for V, by

$$Q(e_i, e_{i+n}) = Q(e_{i+n}, e_i) = 1 \quad \text{for } 1 \le i \le n;$$

$$Q(e_{2n+1}, e_{2n+1}) = 1;$$

and

$$Q(e_i, e_j) = 0 \quad \text{for all other pairs } i, j.$$

The bilinear form Q may be expressed as

$$Q(x, y) = {}^t x \cdot M \cdot y,$$

where M is the $(2n+1) \times (2n+1)$ matrix

$$M = \left(\begin{array}{c|c|c} 0 & I_n & 0 \\ \hline I_n & 0 & 0 \\ \hline 0 & 0 & 1 \end{array}\right)$$

(the diagonal blocks here having widths n, n, and 1). The Lie algebra $\mathfrak{so}_{2n+1}\mathbb{C}$ is correspondingly the space of matrices X satisfying the relation ${}^tX \cdot M + M \cdot X = 0$; if we write X in block form as

$$X = \left(\begin{array}{c|c|c} A & B & E \\ \hline C & D & F \\ \hline G & H & J \end{array}\right),$$

then this is equivalent to saying that, as in the previous case, *B and C are skew-symmetric and A and D negative transposes of each other*; *and in addition $E = -{}^tH$, $F = -{}^tG$, and $J = 0$.*

With these choices, we may take as Cartan subalgebra—in both the even and odd cases—the subalgebra of matrices diagonal in this representation.[1]

[1] Note that if we had taken the simpler choice of Q, with M the identity matrix, the Lie algebra would have consisted of skew-symmetric matrices, and there would have been no nonzero diagonal matrices in the Lie algebra.

The subalgebra $\mathfrak{h}$ is thus generated by the n $2n \times 2n$ matrices $H_i = E_{i,i} - E_{n+i,n+i}$ whose action on V is to fix e_i, send e_{n+i} to its negative, and kill all the remaining basis vectors; note that this is the same whether $m = 2n$ or $2n + 1$. We will correspondingly take as basis for the dual vector space $\mathfrak{h}^*$ the dual basis L_j, where $\langle L_j, H_i \rangle = \delta_{i,j}$.

Given that the Cartan subalgebra of $\mathfrak{so}_{2n}\mathbb{C}$ coincides, as a subspace of $\mathfrak{sl}_{2n}\mathbb{C}$, with the Cartan subalgebra of $\mathfrak{sp}_{2n}\mathbb{C}$, we can use much of the description of the roots of $\mathfrak{sp}_{2n}\mathbb{C}$ to help locate the roots and root spaces of $\mathfrak{so}_{2n}\mathbb{C}$. For example, we saw in Lecture 16 that the endomorphism

$$X_{i,j} = E_{i,j} - E_{n+j,n+i} \in \mathfrak{sp}_{2n}\mathbb{C}$$

is an eigenvector for the action of $\mathfrak{h}$ with eigenvalue $L_i - L_j$. Since $X_{i,j}$ is also an element of $\mathfrak{so}_{2n}\mathbb{C}$, we see that $L_i - L_j$ is likewise a root of $\mathfrak{so}_{2n}\mathbb{C}$, with root space generated by $X_{i,j}$. Less directly but using the same analysis, we find that the endomorphisms

$$Y_{i,j} = E_{i,n+j} - E_{j,n+i}$$

and

$$Z_{i,j} = E_{n+i,j} - E_{n+j,i}$$

are eigenvectors for the action of $\mathfrak{h}$, with eigenvalues $L_i + L_j$ and $-L_i - L_j$, respectively (note that $Y_{i,j}$ and $Z_{i,j}$ do not coincide with their definitions in Lecture 16). In sum, then, *the roots of the Lie algebra* $\mathfrak{so}_{2n}\mathbb{C}$ *are the vectors* $\{\pm L_i \pm L_j\}_{i \neq j} \subset \mathfrak{h}^*$.

The case of the algebra $\mathfrak{so}_{2n+1}\mathbb{C}$ is similar; indeed, all the eigenvectors for the action of $\mathfrak{h}$ found above in $\mathfrak{so}_{2n}\mathbb{C}$, viewed as endomorphisms of $\mathbb{C}^{2n+1}$, are likewise eigenvectors for the action of $\mathfrak{h}$ on $\mathfrak{so}_{2n+1}\mathbb{C}$. In addition, we have the endomorphisms

$$U_i = E_{i,2n+1} - E_{2n+1,n+i}$$

and

$$V_i = E_{n+i,2n+1} - E_{2n+1,i}$$

which are eigenvectors with eigenvalues $+L_i$ and $-L_i$, respectively. The roots of $\mathfrak{so}_{2n+1}\mathbb{C}$ are thus the roots $\pm L_i \pm L_j$ of $\mathfrak{so}_{2n}\mathbb{C}$, together with additional roots $\pm L_i$.

We note that we could have arrived at these statements without decomposing the Lie algebras $\mathfrak{so}_m\mathbb{C}$: the description (18.1) of the orthogonal Lie algebra may be interpreted as saying that, in terms of the identification of V with V^* given by the form Q, $\mathfrak{so}_m\mathbb{C}$ is just the Lie algebra of skew-symmetric endomorphisms of V (an endomorphism being skew-symmetric if it is equal to minus its transpose). That is, the adjoint representation of $\mathfrak{so}_m\mathbb{C}$ is isomorphic to the wedge product $\wedge^2 V$. In the even case $m = 2n$, since the weights of V are $\pm L_i$ (inasmuch as the subalgebras $\mathfrak{h} \subset \mathrm{End}(V)$ coincide, the weights of V must likewise be the same for $\mathfrak{so}_{2n}\mathbb{C}$ as for $\mathfrak{sp}_{2n}\mathbb{C}$), it follows that the roots of $\mathfrak{so}_{2n}\mathbb{C}$

are just the pairwise distinct sums $\pm L_i \pm L_j$. In the odd case $m = 2n + 1$, we see that $e_{2n+1} \in V$ is an eigenvector for the action of $\mathfrak{h}$ with eigenvalue 0, so that the weights of the standard representation V are $\{\pm L_i\} \cup \{0\}$ and the weights of the adjoint representation correspondingly $\{\pm L_i \pm L_j\} \cup \{\pm L_i\}$.

Exercise 18.3. Use a similar analysis to find the roots of $\mathfrak{sp}_{2n}\mathbb{C}$ without explicit calculation.

To make a comparison with the Lie algebra $\mathfrak{sp}_{2n}\mathbb{C}$, we can say that the root diagram of $\mathfrak{so}_{2n}\mathbb{C}$ looks like that of $\mathfrak{sp}_{2n}\mathbb{C}$ with the roots $\pm 2L_i$ removed, whereas the root diagram of $\mathfrak{so}_{2n+1}\mathbb{C}$ looks like that of $\mathfrak{sp}_{2n}\mathbb{C}$ with the roots $\pm 2L_i$ replaced by $\pm L_i$. Note that this immediately tells us what the Weyl groups are: first, in the case of $\mathfrak{so}_{2n+1}\mathbb{C}$, the Weyl group is the same as that of $\mathfrak{sp}_{2n}\mathbb{C}$:

$$1 \to (\mathbb{Z}/2)^n \to \mathfrak{W}_{\mathfrak{so}_{2n+1}\mathbb{C}} \to \mathfrak{S}_n \to 1.$$

In the case of $\mathfrak{so}_{2n}\mathbb{C}$, the Weyl group is the subgroup of the Weyl group of $\mathfrak{sp}_{2n}\mathbb{C}$ generated by reflection in the hyperplanes perpendicular to the roots $\pm L_i \pm L_j$, without the additional generator given by reflection in the roots $\pm L_i$. This subgroup still acts as the full symmetric group on the set of coordinate axes in $\mathfrak{h}^*$; but the kernel of this action, instead of acting as $\pm I$ on each of the coordinate axes independently, will consist of transformations of determinant 1; i.e., will act as -1 on an even number of axes. (That every such transformation is indeed in the Weyl group is easy to see: for example, reflection in the plane perpendicular to $L_i + L_j$ followed by reflection in the plane perpendicular to $L_i - L_j$ will send L_i to $-L_i$, L_j to $-L_j$, and L_k to L_k for $k \neq i, j$.) Another way to say this is that the Weyl group is the subgroup of the Weyl group of $\mathfrak{sp}_{2n}\mathbb{C}$ consisting of transformations whose determinant agrees with the sign of the induced permutation of the coordinate axes; so that while the Weyl group of $\mathfrak{sp}_{2n}\mathbb{C}$ fits into the exact sequence

$$1 \to (\mathbb{Z}/2)^n \to \mathfrak{W}_{\mathfrak{sp}_{2n}\mathbb{C}} \to \mathfrak{S}_n \to 1,$$

the Weyl group of $\mathfrak{so}_{2n}\mathbb{C}$ has instead the sequence

$$1 \to (\mathbb{Z}/2)^{n-1} \to \mathfrak{W}_{\mathfrak{so}_{2n}\mathbb{C}} \to \mathfrak{S}_n \to 1.$$

We can likewise describe the Weyl chambers of $\mathfrak{so}_{2n}\mathbb{C}$ and $\mathfrak{so}_{2n+1}\mathbb{C}$ by direct comparison with $\mathfrak{sp}_{2n}\mathbb{C}$. To start, to choose an ordering of the roots we take as linear functional on $\mathfrak{h}^*$ a form $l = c_1 H_1 + \cdots + c_n H_n$, where $c_1 > c_2 > \cdots > c_n > 0$. The positive roots in the case of $\mathfrak{so}_{2n+1}\mathbb{C}$ are then

$$R^+ = \{L_i + L_j\}_{i<j} \cup \{L_i - L_j\}_{i<j} \cup \{L_i\}_i,$$

whereas in the case of $\mathfrak{so}_{2n}\mathbb{C}$ we have

$$R^+ = \{L_i + L_j\}_{i<j} \cup \{L_i - L_j\}_{i<j}.$$

The primitive positive roots are

$$L_1 - L_2, L_2 - L_3, \ldots, L_{n-1} - L_n, L_n \qquad \text{for } \mathfrak{so}_{2n+1}\mathbb{C};$$
$$L_1 - L_2, L_2 - L_3, \ldots, L_{n-1} - L_n, L_{n-1} + L_n \qquad \text{for } \mathfrak{so}_{2n}\mathbb{C}.$$

In the first case, the Weyl chamber is exactly the same as for $\mathfrak{sp}_{2n}\mathbb{C}$, namely, for $m = 2n + 1$,

$$\mathscr{W} = \{\sum a_i L_i \colon a_1 \geq a_2 \geq \cdots \geq a_n \geq 0\}$$

since the roots are the same except for the factor of 2 on some. In the case of $\mathfrak{so}_{2n}\mathbb{C}$, since there is no root along the line spanned by the L_i, the equality $a_n = 0$ does not describe a face of the Weyl chamber; however, since $L_{n-1} + L_n$ is still a root (and a positive one) we still have the inequality $a_{n-1} + a_n \geq 0$ in $\mathscr{W}$, so that we can write, for $m = 2n$,

$$\mathscr{W} = \{\sum a_i L_i \colon a_1 \geq a_2 \geq \cdots \geq a_{n-1} \geq |a_n|\}.$$

(Note that in the case of $\mathfrak{so}_{2n}\mathbb{C}$ we could have chosen our linear functional $l = c_1 H_1 + \cdots + c_n H_n$ with $c_1 > c_2 > \cdots > -c_n > 0$; the ordering of the roots, and consequently the Weyl chamber, would still be the same.)

As for the Killing form, the same considerations as for the symplectic case show that it must be, up to scalars, the standard quadratic form: $B(H_i, H_j) = \delta_{i,j}$. (This was implicit in the above description of the Weyl group.) The explicit calculation is no more difficult, and we leave it as an exercise:

$$B(\sum a_i H_i, \sum b_i H_i) = \begin{cases} (4n-2)\sum a_i b_i & \text{if } m = 2n+1 \\ (4n-4)\sum a_i b_i & \text{if } m = 2n. \end{cases}$$

Next, to describe the representations of the orthogonal Lie algebras we have to determine the weight lattice in $\mathfrak{h}^*$; and to do this we must, as before, locate the copies $\mathfrak{s}_\alpha$ of $\mathfrak{sl}_2\mathbb{C}$ corresponding to the root pairs $\pm\alpha$, and the corresponding distinguished elements H_α of $\mathfrak{h}$. This is so similar to the case of $\mathfrak{sp}_{2n}\mathbb{C}$ that we will leave the actual calculations as an exercise; we will simply state here the results that in $\mathfrak{so}_m\mathbb{C}$ for any m,

(i) the distinguished copy $\mathfrak{s}_{L_i - L_j}$ of $\mathfrak{sl}_2\mathbb{C}$ associated to the root $L_i - L_j$ is the span of the root spaces $\mathfrak{g}_{L_i - L_j} = \mathbb{C}\cdot X_{i,j}$, $\mathfrak{g}_{-L_i - L_j} = \mathbb{C}\cdot X_{j,i}$ and their commutator $[X_{i,j}, X_{j,i}] = E_{i,i} - E_{j,j} + E_{n+j,n+j} - E_{n+i,n+i}$, with distinguished element $H_{L_i - L_j} = H_i - H_j$ (this is exactly as in the case of $\mathfrak{sp}_{2n}\mathbb{C}$);

(ii) the distinguished copy $\mathfrak{s}_{L_i + L_j}$ of $\mathfrak{sl}_2\mathbb{C}$ associated to the root $L_i + L_j$ is the span of the root spaces $\mathfrak{g}_{L_i + L_j} = \mathbb{C}\cdot Y_{i,j}$, $\mathfrak{g}_{-L_i - L_j} = \mathbb{C}\cdot Z_{i,j}$ and their commutator $[Y_{i,j}, Z_{i,j}] = -E_{i,i} + E_{j,j} - E_{n+j,n+j} + E_{n+i,n+i} = -H_i - H_j$, with distinguished element $H_{L_i + L_j} = H_i + H_j$ (so that we have also $H_{-L_i - L_j} = -H_i - H_j$); and in the case of $\mathfrak{so}_{2n+1}\mathbb{C}$,

(iii) the distinguished copy $\mathfrak{s}_{L_i}$ of $\mathfrak{sl}_2\mathbb{C}$ associated to the root L_i is the span of the root spaces $\mathfrak{g}_{L_i} = \mathbb{C}\cdot U_i$, $\mathfrak{g}_{-L_i} = \mathbb{C}\cdot V_i$ and their commutator $[U_i, V_i] = [E_{i,2n+1} - E_{2n+1,n+i}, E_{n+i,2n+1} - E_{2n+1,i}] = -H_i$, with distinguished element $H_{L_i} = 2H_i$ (so that $H_{-L_i} = -2H_i$ as well).

Exercise 18.4. Verify the computations made here.

Again, the configuration of distinguished elements resembles that of $\mathfrak{sp}_{2n}\mathbb{C}$ closely; that of $\mathfrak{so}_{2n+1}\mathbb{C}$ differs from it by the substitution of $\pm 2H_i$ for $\pm H_i$, whereas that of $\mathfrak{so}_{2n}\mathbb{C}$ differs by the removal of the $\pm H_i$. The effect on the weight lattice is the same in either case: *for both even and odd orthogonal Lie algebras, the weight lattice Λ_W is the lattice generated by the L_i together with the element $(L_1 + \cdots + L_n)/2$.*

Exercise 18.5. Show that

$$\Lambda_W/\Lambda_R = \begin{cases} \mathbb{Z}/2 & \text{if } m = 2n+1 \\ \mathbb{Z}/4 & \text{if } m = 2n \text{ and } n \text{ is odd} \\ \mathbb{Z}/2 \oplus \mathbb{Z}/2 & \text{if } m = 2n \text{ and } n \text{ is even.} \end{cases}$$

§18.2. Representations of $\mathfrak{so}_3\mathbb{C}$, $\mathfrak{so}_4\mathbb{C}$, and $\mathfrak{so}_5\mathbb{C}$

To give some examples, start with the case $n = 1$. Of course, $\mathfrak{so}_2\mathbb{C} \cong \mathbb{C}$ is not semisimple. The root system of $\mathfrak{so}_3\mathbb{C}$, on the other hand, looks like that of $\mathfrak{sl}_2\mathbb{C}$:

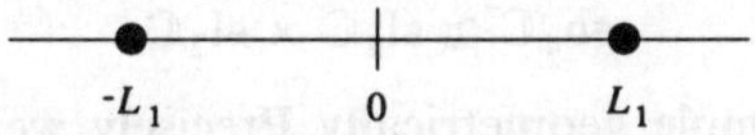

This is because, in fact, the two Lie algebras are isomorphic. Indeed, like the symplectic group, the quotient $PSO_m\mathbb{C}$ of the orthogonal group by its center can be realized as the motions of the projective space $\mathbb{P}V$ preserving isotropic subspaces for the quadratic form Q; in particular, this means we can realize $PSO_m\mathbb{C}$ as the group of motions of $\mathbb{P}V = \mathbb{P}^{m-1}$ carrying the quadric hypersurface

$$\bar{Q} = \{[v]: Q(v, v) = 0\}$$

into itself. In the first case of this, we see that the group $PSO_3\mathbb{C}$ is the group of motions of the projective plane $\mathbb{P}^2$ carrying a conic curve $C \subset \mathbb{P}^2$ into itself. But we have seen before that this group is also $PGL_2\mathbb{C}$ (the conic curve is itself isomorphic to $\mathbb{P}^1$, and the group acts as its full group of automorphisms), giving us the isomorphism $\mathfrak{so}_3\mathbb{C} \cong \mathfrak{sl}_2\mathbb{C}$. One thing to note here is that the "standard" representation of $\mathfrak{so}_3\mathbb{C}$ is not the standard representation of $\mathfrak{sl}_2\mathbb{C}$, but rather its symmetric square. In fact, the irreducible representation with highest weight $\frac{1}{2}L_1$ is not contained in tensor powers of the standard representation of $\mathfrak{so}_3\mathbb{C}$. This will turn out to be significant: the standard representation of $\mathfrak{sl}_2\mathbb{C}$, viewed as a representation of $\mathfrak{so}_3\mathbb{C}$, is the first example of a *spin* representation of an orthogonal Lie algebra.

The next examples involve two-dimensional Cartan algebras. First we have $\mathfrak{so}_4\mathbb{C}$, whose root diagram looks like

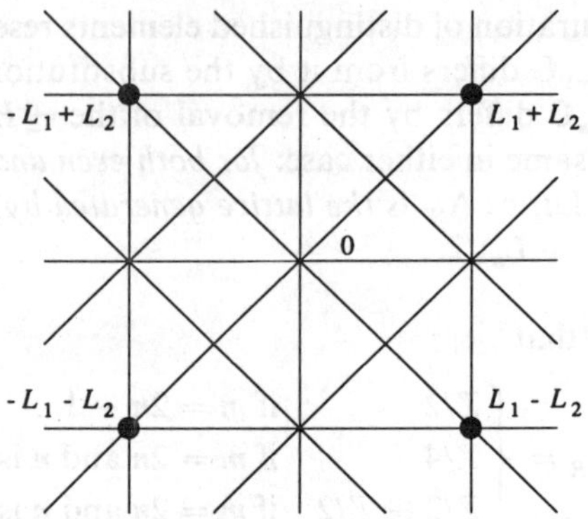

Note one thing about this diagram: the roots are located on the union of two complementary lines. This says, by Exercise 14.33, that the Lie algebra $\mathfrak{so}_4\mathbb{C}$ is decomposable, and in fact should be the sum of two algebras each of whose root diagrams looks like that of $\mathfrak{sl}_2\mathbb{C}$; explicitly, $\mathfrak{so}_4\mathbb{C}$ is the direct sum of the two algebras $\mathfrak{s}_\alpha$, for $\alpha = L_1 + L_2$ and $\alpha = L_1 - L_2$. In fact, we can see this isomorphism

$$\mathfrak{so}_4\mathbb{C} \cong \mathfrak{sl}_2\mathbb{C} \times \mathfrak{sl}_2\mathbb{C}, \tag{18.6}$$

as in the previous example, geometrically. Precisely, we may realize the group $PSO_4\mathbb{C} = SO_4\mathbb{C}/\{\pm I\}$ as the connected component of the identity in the group of motions of projective three-space $\mathbb{P}^3$ carrying a quadric hypersurface Q into itself. But a quadric hypersurface in $\mathbb{P}^3$ has two rulings by lines, and these two rulings give an isomorphism of Q with a product $\mathbb{P}^1 \times \mathbb{P}^1$

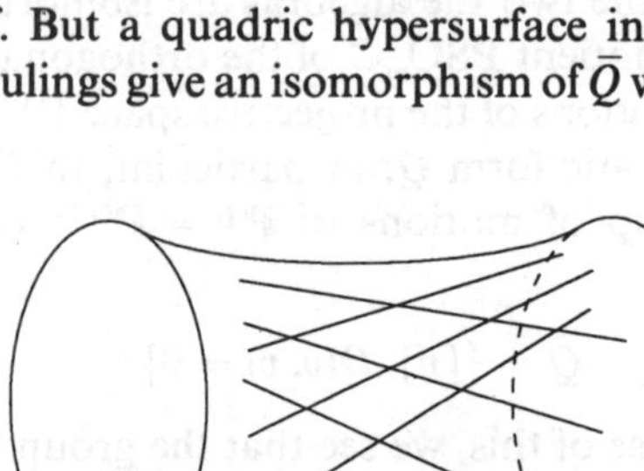

$PSO_4\mathbb{C}$ thus acts on the product $\mathbb{P}^1 \times \mathbb{P}^1$; and since the connected component of the identity in the automorphism group of this variety is just the product $PGL_2\mathbb{C} \times PGL_2\mathbb{C}$, we get an inclusion

$$PSO_4\mathbb{C} \to PGL_2\mathbb{C} \times PGL_2\mathbb{C}.$$

Another way of saying this is to remark that $PSO_4\mathbb{C}$ acts on the variety of isotropic 2-planes for the quadratic form Q on V; and this variety is just the disjoint union of two copies of $\mathbb{P}^1$. To see in this case that the map is an isomorphism, consider the tensor product $V = U \otimes W$ of the pullbacks to $\mathfrak{sl}_2\mathbb{C} \times \mathfrak{sl}_2\mathbb{C}$ of the standard representations of the two factors. Clearly the action on $\mathbb{P}(U \otimes W)$ will preserve the points corresponding to decomposable tensors (that is, points of the form $[u \otimes w]$); but the locus of such points is just a quadric hypersurface, giving us the inverse inclusion of $PGL_2\mathbb{C} \times PGL_2\mathbb{C}$ in $PSO_4\mathbb{C}$.

In fact, all of this will fall out of the analysis of the representations of $\mathfrak{so}_4\mathbb{C}$, if we just pursue it as usual. To begin with, the Weyl chamber we have selected looks like

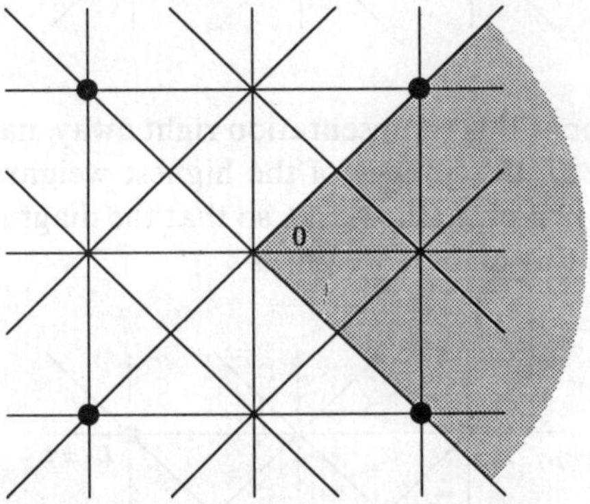

Now, the standard representation has, as noted above, weight diagram

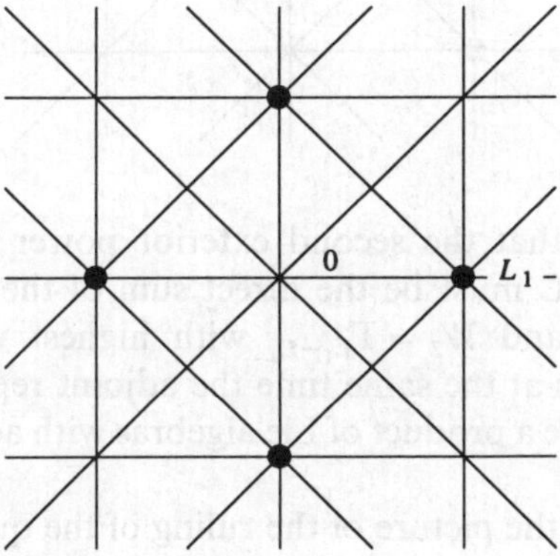

with highest weight L_1 (note that the highest weight of the standard representation lies in this case in the interior of the Weyl chamber, something of an anomaly). Its second exterior power will have weights $\pm L_1 \pm L_2$ and 0 (occurring with multiplicity 2), i.e., diagram

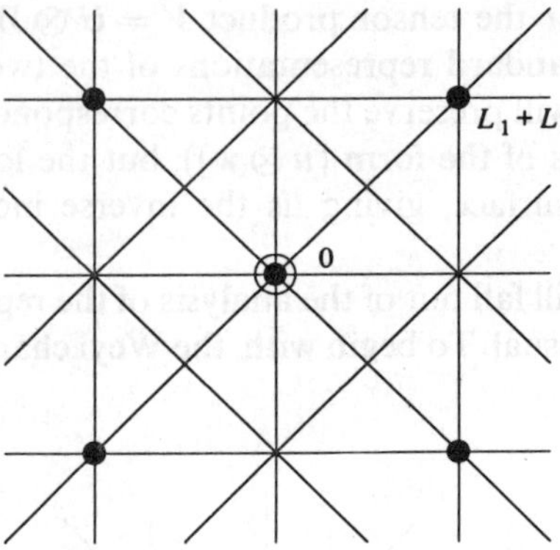

We see one thing about this representation right away, namely, that it cannot be irreducible. Indeed, the images of the highest weight $L_1 + L_2$ under the Weyl group consist just of $\pm(L_1 + L_2)$, so that the diagram of the irreducible representation with this highest weight is

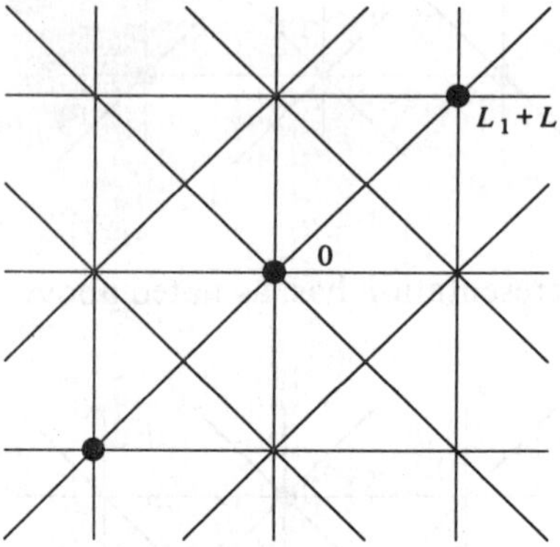

We see from this that the second exterior power $\wedge^2 V$ of the standard representation of $\mathfrak{so}_4\mathbb{C}$ must be the direct sum of the irreducible representations $W_1 = \Gamma_{L_1+L_2}$ and $W_2 = \Gamma_{L_1-L_2}$ with highest weights $L_1 + L_2$ and $L_1 - L_2$. Since $\wedge^2 V$ is at the same time the adjoint representation, this says that $\mathfrak{so}_4\mathbb{C}$ itself must be a product of Lie algebras with adjoint representations $\Gamma_{L_1+L_2}$ and $\Gamma_{L_1-L_2}$.

One way to derive the picture of the ruling of the quadric in $\mathbb{P}^3$ from this decomposition is to view $\mathfrak{so}_4\mathbb{C}$ as a subalgebra of $\mathfrak{sl}_4\mathbb{C}$, and the action of $\mathrm{PSO}_4\mathbb{C}$ on $\mathbb{P}(\wedge^2 V)$ as a subgroup of the group of motions of $\mathbb{P}^2(\wedge^2 V) = \mathbb{P}^5$ preserving the Grassmannian $G = G(2, V)$ of lines in $\mathbb{P}^3$. In fact, we see from the above that the action of PSO_4 on $\mathbb{P}^5$ will preserve a pair of complementary 2-planes $\mathbb{P}W_1$ and $\mathbb{P}W_2$; it follows that this action must carry into themselves the intersections of these 2-planes with the Grassmannian. These intersections are conic curves, corresponding to one-parameter families of lines sweeping out a quadric surface (necessarily the same quadric, since the action of $SO_4\mathbb{C}$ on V preserves a unique quadratic form); thus, the two rulings of the quadric.

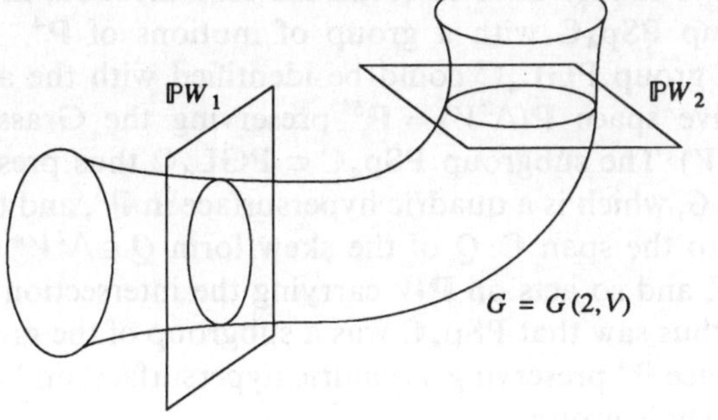

Note one more aspect of this example: as in the case of $\mathfrak{so}_3\mathbb{C} \cong \mathfrak{sl}_2\mathbb{C}$, the weights of the standard representation of $\mathfrak{so}_4\mathbb{C}$ do not generate the weight lattice, but rather a sublattice $\mathbb{Z}\{L_1, L_2\}$ of index 2 in Λ_W. Thus, there is no way of constructing all the representations of $\mathfrak{so}_4\mathbb{C}$ by applying linear- or multilinear-algebraic constructions to the standard representation; it is only after we are aware of the isomorphism $\mathfrak{so}_4\mathbb{C} \cong \mathfrak{sl}_2\mathbb{C} \times \mathfrak{sl}_2\mathbb{C}$ that we can construct, for example, the representation $\Gamma_{(L_1+L_2)/2}$ with highest weight $(L_1 + L_2)/2$ (of course, this is just the pullback from the first factor of $\mathfrak{sl}_2\mathbb{C} \times \mathfrak{sl}_2\mathbb{C}$ of the standard representation of $\mathfrak{sl}_2\mathbb{C}$).

We come now to the case of $\mathfrak{so}_5\mathbb{C}$, which is more interesting. The root diagram in this case looks like

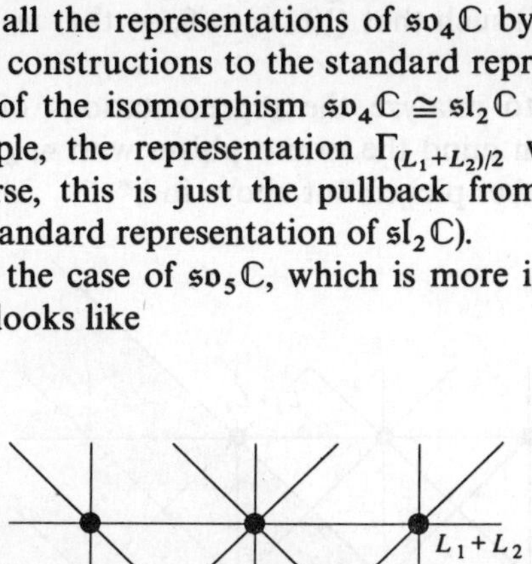

(as in the preceding example, the weight lattice is the lattice of intersections of all the lines drawn). The first thing we should notice about this diagram is that it is isomorphic to the weight diagram of the Lie algebra $\mathfrak{sp}_4\mathbb{C}$; the diagram just appears here rotated through an angle of $\pi/4$. Indeed, this is not accidental; the two Lie algebras $\mathfrak{sp}_4\mathbb{C}$ and $\mathfrak{so}_5\mathbb{C}$ are isomorphic, and it is not hard to construct this isomorphism explicitly. To see the isomorphism geometrically, we simply have to recall the identification, made in Lecture 14, of the group $\mathrm{PSp}_4\mathbb{C}$ with a group of motions of $\mathbb{P}^4$. There, we saw that the larger group $\mathrm{PGL}_4\mathbb{C}$ could be identified with the automorphisms of the projective space $\mathbb{P}(\wedge^2 V) = \mathbb{P}^5$ preserving the Grassmannian $G = G(2, 4) \subset \mathbb{P}(\wedge^2 V)$. The subgroup $\mathrm{PSp}_4\mathbb{C} \subset \mathrm{PGL}_4\mathbb{C}$ thus preserves both the Grassmannian G, which is a quadric hypersurface in $\mathbb{P}^5$, and the decomposition of $\wedge^2 V$ into the span $\mathbb{C}\cdot Q$ of the skew form $Q \in \wedge^2 V^* \cong \wedge^2 V$ and its complement W, and so acts on $\mathbb{P}W$ carrying the intersection $G_L = G \cap \mathbb{P}W$ into itself. We thus saw that $\mathrm{PSp}_4\mathbb{C}$ was a subgroup of the group of motions of projective space $\mathbb{P}^4$ preserving a quadric hypersurface, and asserted that in fact it was the whole group.

(To see the reverse inclusion directly, we can invoke a little algebraic geometry, which tells us that the locus of isotropic lines for a quadric in $\mathbb{P}^4$ is isomorphic to $\mathbb{P}^3$, so that $\mathrm{PSO}_5\mathbb{C}$ acts on $\mathbb{P}^3$. Moreover, this action preserves the subset of pairs of points in $\mathbb{P}^3$ whose corresponding lines in $\mathbb{P}^4$ intersect, which, for a suitably defined skew-symmetric bilinear form $\tilde{Q}$, is exactly the set of pairs $([v], [w])$ such that $\tilde{Q}(v, w) = 0$, so that we have an inclusion of $\mathrm{PSO}_5\mathbb{C}$ in $\mathrm{PSp}_4\mathbb{C}$.)

Let us proceed to analyze the representations of $\mathfrak{so}_5\mathbb{C}$ as we would ordinarily, bearing in mind the isomorphism with $\mathfrak{sp}_4\mathbb{C}$. To begin with, we draw the Weyl chamber picked out above in $\mathfrak{h}^*$:

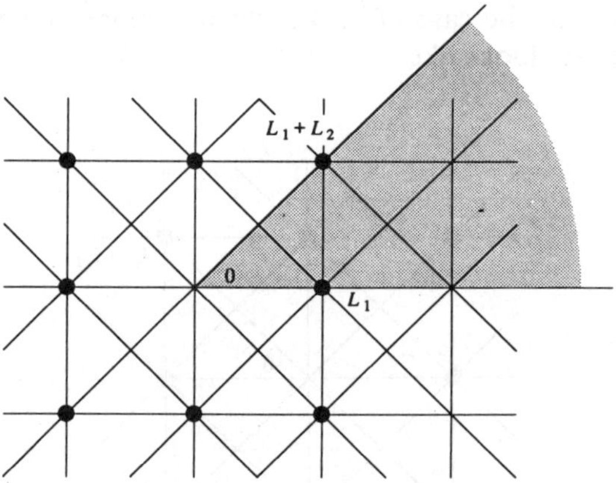

As for the representations of $\mathfrak{so}_5\mathbb{C}$, we have to begin with the standard, which has weight diagram

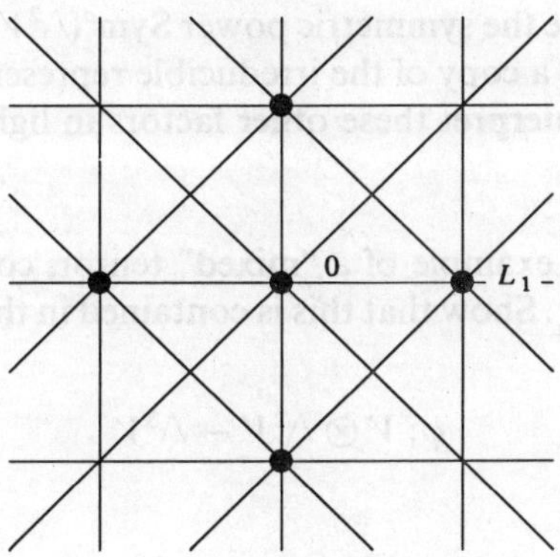

This we see corresponds to the representation $W = \Lambda^2 V/\mathbb{C}\cdot Q$ of $\mathfrak{sp}_4\mathbb{C}$. Next, the second exterior power of the standard representation of $\mathfrak{so}_5\mathbb{C}$ has weights

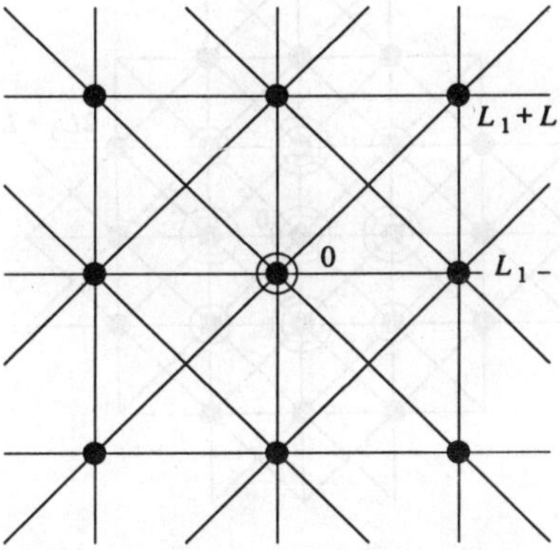

This is of course the adjoint representation of $\mathfrak{so}_5\mathbb{C}$; it is the irreducible representation with highest weight $L_1 + L_2$. Note that it corresponds to the symmetric square $\mathrm{Sym}^2 V$ of the standard representation of $\mathfrak{sp}_4\mathbb{C}$ (see Exercise 16.8).

Exercise 18.7. Show that contraction with the quadratic form $Q \in \mathrm{Sym}^2 V^*$ preserved by the action of $\mathfrak{so}_5\mathbb{C}$ induces maps

$$\varphi\colon \mathrm{Sym}^a V \to \mathrm{Sym}^{a-2} V.$$

Show that the kernel of this contraction is exactly the irreducible representation with highest weight $a \cdot L_1$. Compare this with the analysis in Exercise 16.11.

Exercise 18.8. Examine the symmetric power $\mathrm{Sym}^a(\wedge^2 V)$ of the representation $\wedge^2 V$. This will contain a copy of the irreducible representation $\Gamma_{a(L_1+L_2)}$; what else will it contain? Interpret these other factors in light of the isomorphism $\mathfrak{so}_5\mathbb{C} \cong \mathfrak{sp}_4\mathbb{C}$.

Exercise 18.9. For an example of a "mixed" tensor, consider the irreducible representation $\Gamma_{2L_1+L_2}$. Show that this is contained in the kernels of the wedge product map

$$\varphi: V \otimes \wedge^2 V \to \wedge^3 V$$

and the composition

$$\varphi': V \otimes \wedge^2 V \to V^* \otimes \wedge^2 V \to V,$$

where the first map is induced by the isomorphism $\tilde{Q}: V \to V^*$ and the second is the contraction $V^* \otimes \wedge^2 V \to V$. Is it equal to the intersection of these kernels? Show that the weight diagram of this representation is

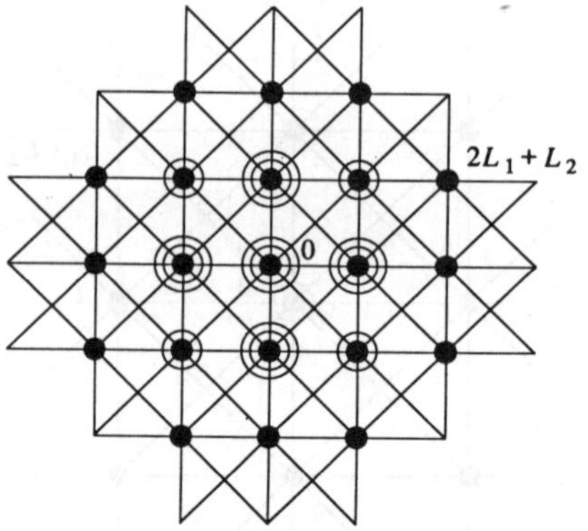

After you are done with this analysis, compare with the analysis given of the corresponding representation in Lecture 16.

Note that, as in the case of the other orthogonal Lie algebras studied so far (and as is the case for all $\mathfrak{so}_m\mathbb{C}$), *the weights of the standard representation do not generate the weight lattice, but only the sublattice of index two generated by the L_i*. Thus, the tensor algebra of the standard representation will contain only one-half of all the irreducible representations of $\mathfrak{so}_5\mathbb{C}$. Now, we do know that there are others, and even something about them—for example, we see in the following exercise that the irreducible representation of $\mathfrak{so}_5\mathbb{C}$ with highest weight $(L_1 + L_2)/2$ is a sort of "symmetric square root" of the adjoint representation:

Exercise 18.10. Show, using only root and weight diagrams for $\mathfrak{so}_5\mathbb{C}$, that the exterior square $\wedge^2 V$ of the standard representation of $\mathfrak{so}_5\mathbb{C}$ is actually the symmetric square of an irreducible representation.

We can also describe this irreducible representation via the isomorphism of $\mathfrak{so}_5\mathbb{C}$ with $\mathfrak{sp}_4\mathbb{C}$: it is just the standard representation of $\mathfrak{sp}_4\mathbb{C}$ on $\mathbb{C}^4$. We do not at this point have, however, a way of constructing this representation without invoking the isomorphism. This representation, the representation of $\mathfrak{so}_3\mathbb{C}$ with highest weight $L_1/2$, and the representation of $\mathfrak{so}_4\mathbb{C}$ with highest weight $(L_1 + L_2)/2$ discussed above are called *spin* representations of the corresponding Lie algebras and will be the subject matter of Lecture 20.

LECTURE 19
$\mathfrak{so}_6\mathbb{C}$, $\mathfrak{so}_7\mathbb{C}$, and $\mathfrak{so}_m\mathbb{C}$

This lecture is analogous in content (and prerequisites) to Lecture 17: we do some more low-dimensional examples and then describe the general picture of the representations of the orthogonal Lie algebras. One difference is that only half the irreducible representations of $\mathfrak{so}_m\mathbb{C}$ lie in the tensor algebra of the standard; to complete the picture of the representation theory we have to construct the spin representations, which is the subject matter of the following lecture. The first four sections are completely elementary (except possibly for the discussion of the isomorphism $\mathfrak{so}_6\mathbb{C} \cong \mathfrak{sl}_4\mathbb{C}$ in §19.1); the last section assumes a knowledge of Lecture 6 and §15.3, but can be skipped by those who did not read those sections.

§19.1. Representations of $\mathfrak{so}_6\mathbb{C}$

We continue our discussion of orthogonal Lie algebras with the example of $\mathfrak{so}_6\mathbb{C}$. First, its root diagram:

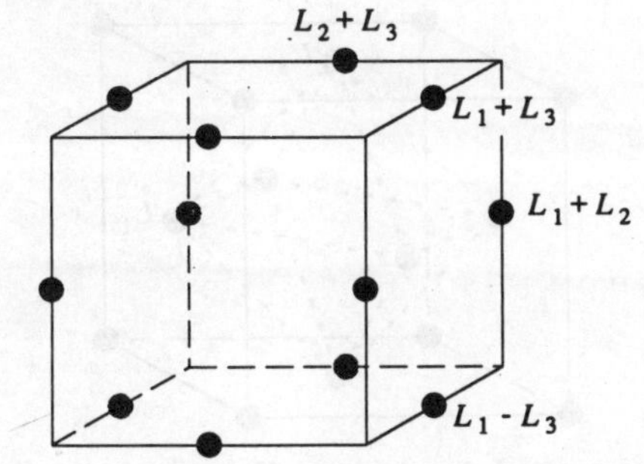

Once more (and for the last time), we notice a coincidence between this and the root diagram of a Lie algebra already studied, namely, $\mathfrak{sl}_4\mathbb{C}$. In fact, the two Lie algebras are isomorphic. The isomorphism is one we have already observed, in a sense: in the preceding lecture we noted that if V is a four-dimensional vector space, then the group $PGL_4\mathbb{C}$ may be realized as the connected component of the identity in the group of motions of $\mathbb{P}(\wedge^2 V) = \mathbb{P}^5$ carrying the Grassmannian $G = G(2, 4) \subset \mathbb{P}(\wedge^2 V)$ into itself, and $PSp_4\mathbb{C} \subset PGL_4\mathbb{C}$ the subgroup fixing a hyperplane $\mathbb{P}W = \mathbb{P}^4 \subset \mathbb{P}^5$. We used this to identify the subgroup $PSp_4\mathbb{C}$ with the orthogonal group $PSO_5\mathbb{C}$; at the same time it gives an identification of the larger group $PGL_4\mathbb{C}$ with the orthogonal group $PSO_6\mathbb{C}$.

Even though $\mathfrak{so}_6\mathbb{C}$ is isomorphic to a Lie algebra we have already examined, it is worth going through the analysis of its representations for what amounts to a second time, partly so as to understand the isomorphism better, but mainly because we will see clearly in the case of $\mathfrak{so}_6\mathbb{C}$ a number of phenomena that will hold true of the even orthogonal groups in general. To start, we draw the Weyl chamber in $\mathfrak{h}^*$:

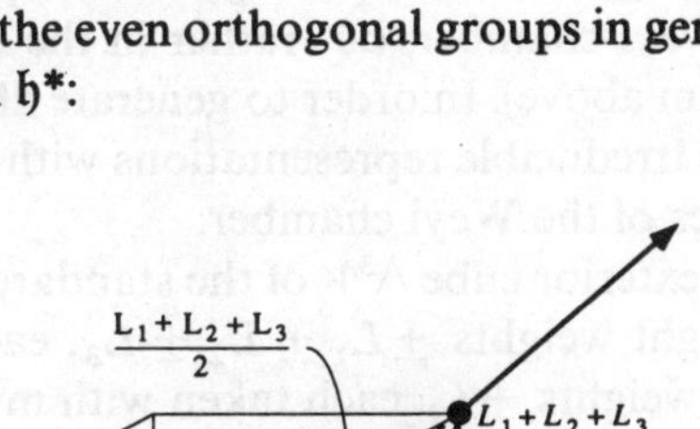

As usual, we begin with the standard representation, which has weights $\pm L_i$, corresponding to the centers of the faces of the cube:

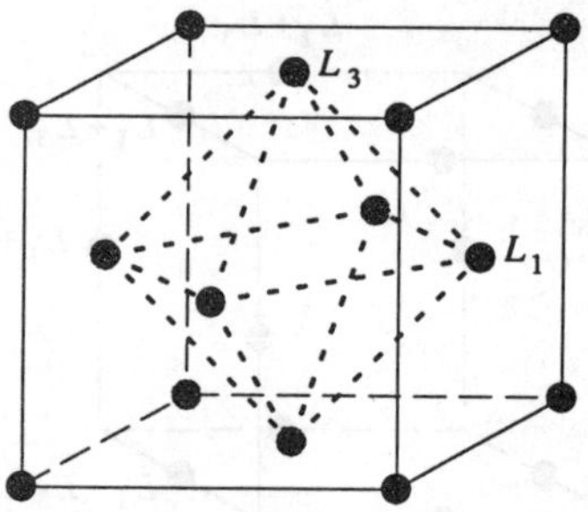

Note that the highest weight L_1 once more lies on an edge of the Weyl chamber (the front edge, in the diagram on the preceding page). Observe that the standard representation of $\mathfrak{so}_6\mathbb{C}$ corresponds, as we have already pointed out, to the exterior square of the standard representation of $\mathfrak{sl}_4\mathbb{C}$.

Next, we look at the exterior square $\wedge^2 V$ of the standard representation of $\mathfrak{so}_6\mathbb{C}$. This will have weights $\pm L_i \pm L_j$ (of course, it is the adjoint representation) and so will have weight diagram

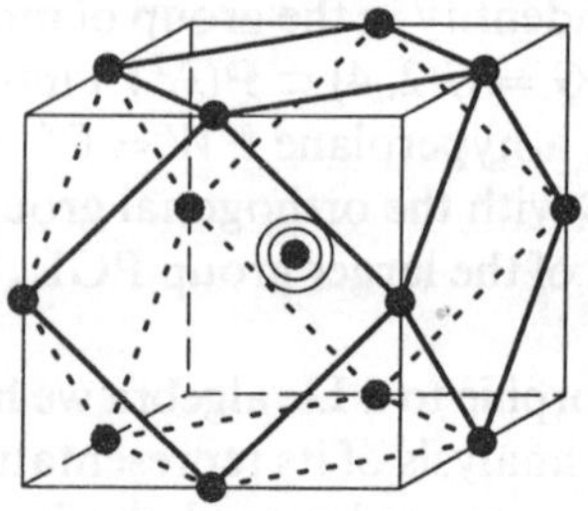

Note that the highest weight vector $L_1 + L_2$ of this representation does not lie on an edge of the Weyl chamber, but rather in the interior of a face (the back face, in the diagram above). In order to generate all the representations, we still need to find the irreducible representations with highest weight along the remaining two edges of the Weyl chamber.

We look next at the exterior cube $\wedge^3 V$ of the standard representation. The weights here are the eight weights $\pm L_1 \pm L_2 \pm L_3$, each taken with multiplicity one, and the six weights $\pm L_i$, each taken with multiplicity 2, as in the diagram

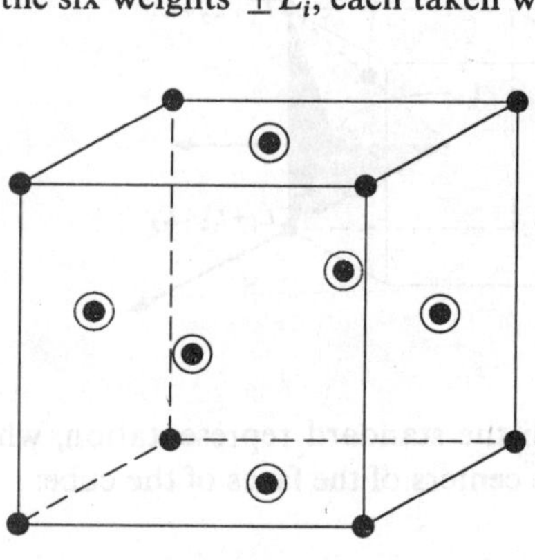

Now, we notice something very interesting: this cannot be an irreducible representation. We can see this in a number of ways: the images of the weight $L_1 + L_2 + L_3$ under the Weyl group, for example, consist of every other vertex of the reference cube; in particular, their convex hull does not contain the remaining four vertices including $L_1 + L_2 - L_3$. Equivalently, there is no way to go from $L_1 + L_2 + L_3$ to $L_1 + L_2 - L_3$ by translation by negative root vectors. The representation $\wedge^3 V$ will thus contain copies of the irreducible representations $\Gamma_{L_1+L_2+L_3}$ and $\Gamma_{L_1+L_2-L_3}$ with highest weights $L_1 + L_2 + L_3$ and $L_1 + L_2 - L_3$, with weight diagrams

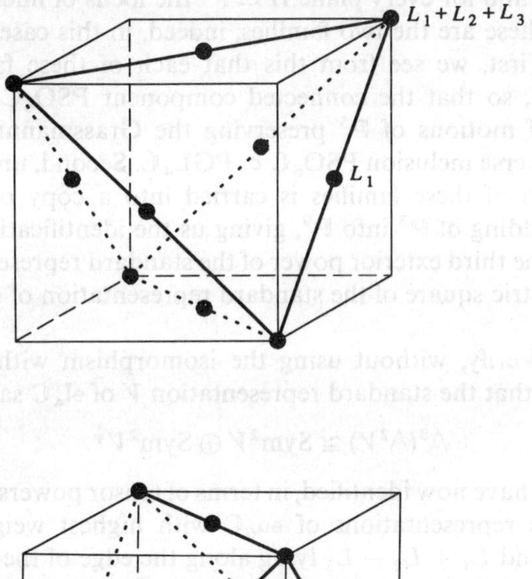

and

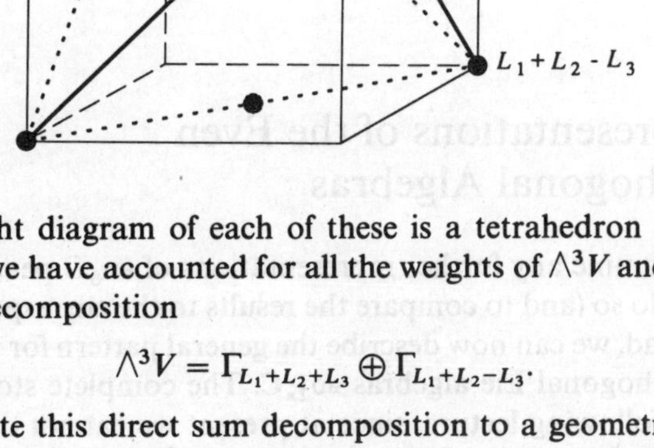

Since the weight diagram of each of these is a tetrahedron containing the weights $\pm L_i$, we have accounted for all the weights of $\wedge^3 V$ and so must have a direct sum decomposition

$$\wedge^3 V = \Gamma_{L_1+L_2+L_3} \oplus \Gamma_{L_1+L_2-L_3}.$$

We can relate this direct sum decomposition to a geometric feature of a quadric hypersurface in $\mathbb{P}^5$, analogous to the presence of two rulings on a quadric in $\mathbb{P}^3$. We saw before that the locus of lines lying on a quadric surface in $\mathbb{P}^3$ turns out to be disconnected, consisting of two components each isomorphic to $\mathbb{P}^1$ (and embedded, via the Plücker embedding of the Grassmannian $G = G(2, 4)$ of lines in $\mathbb{P}^3$ in $\mathbb{P}(\wedge^2\mathbb{C}) = \mathbb{P}^5$, as a pair of conic curves lying in complementary 2-planes in $\mathbb{P}^5$). In a similar fashion, the variety of 2-planes lying on a quadric hypersurface in $\mathbb{P}^5$ turns out to be disconnected, consisting of two components that, under the Plücker embedding of $G(3, 6)$ in $\mathbb{P}(\wedge^3\mathbb{C}^6) = \mathbb{P}^{19}$, span two complementary 9-planes $\mathbb{P}W_1$ and $\mathbb{P}W_2$; these two planes give the direct sum decomposition of $\wedge^3 V$ as an $\mathfrak{so}_6\mathbb{C}$-module.

In fact, if we think of a quadric hypersurface in $\mathbb{P}^5$ as the Grassmannian $G = G(2, 4)$ of lines in $\mathbb{P}^3$, we can see explicitly what these two families of 2-planes are: for every point $p \in \mathbb{P}^3$ the locus of lines passing through p forms a 2-plane on G, and for every plane $H \subset \mathbb{P}^3$ the locus of lines lying in H is a 2-plane in G. These are the two families; indeed, in this case we can go two steps further. First, we see from this that each of these families is parametrized by $\mathbb{P}^3$, so that the connected component $\mathrm{PSO}_6\mathbb{C}$ of the identity in the group of motions of $\mathbb{P}^5$ preserving the Grassmannian acts on $\mathbb{P}^3$, giving us the inverse inclusion $\mathrm{PSO}_6\mathbb{C} \subset \mathrm{PGL}_4\mathbb{C}$. Second, under the Plücker embedding each of these families is carried into a copy of the quadratic Veronese embedding of $\mathbb{P}^3$ into $\mathbb{P}^9$, giving us the identification of the direct sum factors of the third exterior power of the standard representation of $\mathfrak{so}_6\mathbb{C}$ with the symmetric square of the standard representation of $\mathfrak{sl}_4\mathbb{C}$.

Exercise 19.1. Verify, without using the isomorphism with $\mathfrak{so}_6\mathbb{C}$ and the analysis above, that the standard representation V of $\mathfrak{sl}_4\mathbb{C}$ satisfies

$$\wedge^3(\wedge^2 V) \cong \mathrm{Sym}^2 V \oplus \mathrm{Sym}^2 V^*.$$

Note that we have now identified, in terms of tensor powers of the standard one, irreducible representations of $\mathfrak{so}_6\mathbb{C}$ with highest weight vectors L_1, $L_1 + L_2 + L_3$ and $L_1 + L_2 - L_3$ lying along the edge of the Weyl chamber, as well as one with highest weight $L_1 + L_2$ lying in a face. We can thus find irreducible representations with highest weight γ, if not for every γ in $\Lambda_W \cap \mathscr{W}$, at least for every weight γ in the intersection of $\mathscr{W}$ with a sublattice of index 2 in Λ_W.

§19.2. Representations of the Even Orthogonal Algebras

We will not examine any further representations of $\mathfrak{so}_6\mathbb{C}$ per se, leaving it as an exercise to do so (and to compare the results to the corresponding analysis for $\mathfrak{sl}_4\mathbb{C}$). Instead, we can now describe the general pattern for representations of the even orthogonal Lie algebras $\mathfrak{so}_{2n}\mathbb{C}$. The complete story will have to wait until the following lecture, since at present we cannot construct all the representations of $\mathfrak{so}_{2n}\mathbb{C}$ (as we have pointed out, we have been able to do so in the cases $n = 2$ and 3 studied so far only by virtue of isomorphisms with other Lie algebras; and there are no more such isomorphisms from this point on). We will nonetheless give as much of the picture as we can.

To begin with, recall that the weight lattice of $\mathfrak{so}_{2n}\mathbb{C}$ is generated by $L_1, \ldots, L_n$ together with the further vector $(L_1 + \cdots + L_n)/2$. The Weyl chamber, on the other hand, is the cone

$$\mathscr{W} = \{\sum a_i L_i : a_1 \geq a_2 \geq \cdots \geq \pm a_n\}.$$

Note that the Weyl chamber is a simplicial cone, with faces corresponding to the n planes $a_1 = a_2, \ldots, a_{n-1} = a_n$ and $a_{n-1} = -a_n$; the edges of the Weyl chamber are thus the rays generated by the vectors L_1, $L_1 + L_2, \ldots,$ $L_1 + \cdots + L_{n-2}$, $L_1 + \cdots + L_n$ and $L_1 + \cdots + L_{n-1} - L_n$ (note that $L_1 + \cdots + L_{n-1}$ is not on an edge of the Weyl chamber). We see from this that, as in every previous case, the intersection of the weight lattice with the closed Weyl cone is a free semigroup generated by fundamental weights, in this case the vectors L_1, $L_1 + L_2, \ldots, L_1 + \cdots + L_{n-2}$ and the vectors[1]

$$\alpha = (L_1 + \cdots + L_n)/2 \quad \text{and} \quad \beta = (L_1 + \cdots + L_{n-1} - L_n)/2.$$

As before, the obvious place to start to look for irreducible representations is among the exterior powers of the standard representation. This almost works: we have

Theorem 19.2. (i) *The exterior powers $\wedge^k V$ of the standard representation V of $\mathfrak{so}_{2n}\mathbb{C}$ are irreducible for $k = 1, 2, \ldots, n-1$; and* (ii) *The exterior power $\wedge^n V$ has exactly two irreducible factors.*

PROOF. The proof will follow the same lines as that of the analogous theorem for the symplectic Lie algebras in Lecture 17; in particular, we will start by considering the restriction to the same subalgebra as in the case of $\mathfrak{sp}_{2n}\mathbb{C}$.

Recall that the group $\mathrm{Sp}_{2n}\mathbb{C} \subset \mathrm{SL}_{2n}\mathbb{C}$ of automorphisms preserving the skew form Q introduced in Lecture 16 contains the subgroup G of automorphisms of the space $V = \mathbb{C}^{2n}$ preserving the decomposition $V = \mathbb{C}\{e_1, \ldots, e_n\} \oplus \mathbb{C}\{e_{n+1}, \ldots, e_{2n}\}$, acting as an arbitrary automorphism on the first factor and as the inverse transpose of that automorphism on the second factor; in matrices

$$G = \left\{ \begin{pmatrix} X & 0 \\ 0 & {}^t X^{-1} \end{pmatrix}, X \in \mathrm{GL}_n\mathbb{C} \right\}.$$

In fact, the subgroup $\mathrm{SO}_{2n}\mathbb{C} \subset \mathrm{SL}_{2n}\mathbb{C}$ also contains the same subgroup; we have, correspondingly a subalgebra

[1] To conform to standard conventions, with simple roots $\alpha_i = L_i - L_{i+1}$ for $1 \leq i \leq n-1$, and $\alpha_n = L_{n-1} + L_n$, to have $\omega_i(H_{\alpha_j}) = \delta_{i,j}$, the fundamental weights ω_i should be put in the order: $\omega_i = L_1 + \cdots + L_i$ for $1 \leq i \leq n-2$, and

$$\omega_{n-1} = \beta = (L_1 + \cdots + L_{n-1} - L_n)/2, \qquad \omega_n = \alpha = (L_1 + \cdots + L_n)/2.$$

$$\mathfrak{s} = \left\{ \begin{pmatrix} A & 0 \\ 0 & -{}^tA \end{pmatrix}, A \in \mathfrak{sl}_n\mathbb{C} \right\} \subset \mathfrak{so}_{2n}\mathbb{C}$$

isomorphic to $\mathfrak{sl}_n\mathbb{C}$.

Denote by W the standard representation of $\mathfrak{sl}_n\mathbb{C}$. As in the previous case, the restriction of the standard representation V of $\mathfrak{so}_{2n}\mathbb{C}$ to the subalgebra $\mathfrak{s}$ then splits

$$V = W \oplus W^*$$

into a direct sum of W and its dual; and we have, correspondingly,

$$\wedge^k V = \bigoplus_{a+b=k} (\wedge^a W \otimes \wedge^b W^*).$$

We also can say how each factor on the right-hand side of this expression decomposes as a representation of $\mathfrak{sl}_n\mathbb{C}$: we have contraction maps

$$\Psi_{a,b}\colon \wedge^a W \otimes \wedge^b W^* \to \wedge^{a-1} W \otimes \wedge^{b-1} W^*;$$

and the kernel of $\Psi_{a,b}$ is the irreducible representation $W^{(a,b)}$ with highest weight $2L_1 + \cdots + 2L_a + L_{a+1} + \cdots + L_{n-b}$. The restriction of $\wedge^k V$ to $\mathfrak{s}$ is thus given by

$$\wedge^k V = \bigoplus_{\substack{a+b \le k \\ a+b \equiv k(2)}} W^{(a,b)},$$

where the actual highest weight factor in the summand $W^{(a,b)} \subset \wedge^k V$ is the vector

$$\begin{aligned} w^{(a,b)} &= e_1 \wedge \cdots \wedge e_a \wedge e_{2n-b+1} \wedge \cdots \wedge e_{2n} \wedge Q^{(k-a-b)/2} \\ &= e_1 \wedge \cdots \wedge e_a \wedge e_{2n-b+1} \wedge \cdots \wedge e_{2n} \wedge \left(\sum (e_i \wedge e_{n+i})\right)^{(k-a-b)/2}. \end{aligned}$$

Now, all the vectors $w^{(a,b)}$ have distinct weights; and it follows, as in Exercise 17.7, that *any highest weight vector for the action of* $\mathfrak{so}_{2n}\mathbb{C}$ *on* $\wedge^k V$ *will be a scalar multiple of one of the* $w^{(a,b)}$. It will thus suffice, in order to show that $\wedge^k V$ is irreducible as representation of $\mathfrak{so}_{2n}\mathbb{C}$ for $k < n$, to exhibit for each (a, b) with $a + b \le k < n$ other than $(k, 0)$ a positive root α such that the image $\mathfrak{g}_\alpha(w^{(a,b)}) \ne 0$. This is simplest in the case $a + b = k < n$ (so there is no factor of Q in $w^{(a,b)}$): just as in the case of $\mathfrak{sp}_{2n}\mathbb{C}$ we have

$$\begin{aligned} &Y_{a+1,n-b+1}(w^{(a,b)}) \\ &= (E_{a+1,2n-b+1} - E_{n-b+1,n+a+1})(e_1 \wedge \cdots \wedge e_a \wedge e_{2n-b+1} \wedge \cdots \wedge e_{2n}) \\ &= w^{(a+1,b-1)} \\ &\ne 0 \end{aligned}$$

and $Y_{i,j}$ is the generator of the positive root space $\mathfrak{g}_{L_i+L_j}$.

In case $a + b < k < n$, we observe first that for any i and j

$$\begin{aligned} Y_{i,j}(Q) &= (E_{i,n+j} - E_{j,n+i})(\textstyle\sum (e_p \wedge e_{n+p})) \\ &= 2 \cdot e_j \wedge e_i \\ &\neq 0 \end{aligned}$$

so that whenever $a < i, j \leq n - b$,

$$\begin{aligned} &Y_{i,j}(w^{(a,b)}) \\ &= Y_{i,j}(e_1 \wedge \cdots \wedge e_a \wedge e_{2n-b+1} \wedge \cdots \wedge e_{2n} \wedge Q^{(k-a-b)/2}) \\ &= e_1 \wedge \cdots \wedge e_a \wedge e_{2n-b+1} \wedge \cdots \wedge e_{2n} \wedge Y_{i,j}((\textstyle\sum (e_p \wedge e_{n+p}))^{(k-a-b)/2}) \\ &= (k-a-b) \cdot (e_1 \wedge \cdots \wedge e_a \wedge e_j \wedge e_i \wedge e_{2n-b+1} \wedge \cdots \wedge e_{2n} \wedge Q^{(k-a-b-2)/2}) \\ &\neq 0. \end{aligned}$$

It is always possible to find a pair (i, j) satisfying the conditions $a < i, j \leq n - b$ since we are assuming $a + b < k < n$; this concludes the proof of part (i).

The proof of part (ii) requires only one further step: we have to check the vectors $w^{(a,b)}$ with $a + b = k = n$ to see if any of them might be highest weight vectors for $\mathfrak{so}_{2n}\mathbb{C}$. In fact (as the statement of the theorem implies), two of them are: It is not hard to check that, in fact, $w^{(n,0)}$ and $w^{(n-1,1)}$ are killed by every positive root space $\mathfrak{g}_{L_i+L_j}$. To see that no other vector $w^{(a,n-a)}$ is, look at the action of $Y_{a+1,a+2} \in \mathfrak{g}_{L_{a+1}+L_{a+2}}$: we have

$$\begin{aligned} &Y_{a+1,a+2}(w^{(a,n-a)}) \\ &\quad = (E_{a+1,n+a+2} - E_{a+2,n+a+1})(e_1 \wedge \cdots \wedge e_a \wedge e_{n+a+1} \wedge \cdots \wedge e_{2n}) \\ &\quad = e_1 \wedge \cdots \wedge e_a \wedge e_{a+1} \wedge e_{n+a+1} \wedge e_{n+a+3} \wedge \cdots \wedge e_{2n} \\ &\qquad - e_1 \wedge \cdots \wedge e_a \wedge e_{a+2} \wedge e_{n+a+2} \wedge \cdots \wedge e_{2n} \\ &\quad \neq 0. \end{aligned}$$

□

Remarks. (i) This theorem will be a consequence of the Weyl character formula, which will tell us a priori that the dimension of the irreducible representation of $\mathfrak{so}_{2n}\mathbb{C}$ with highest weight $L_1 + \cdots + L_k$ has dimension $\binom{2n}{k}$ if $k < n$, and half that if $k = n$.

(ii) Note also that by the above, $\wedge^n V$ is the direct sum of the two irreducible representations $\Gamma_{2\alpha}$ and $\Gamma_{2\beta}$ with highest weights $2\alpha = L_1 + \cdots + L_n$ and $2\beta = L_1 + \cdots + L_{n-1} - L_n$. Indeed, the inclusion $\Gamma_{2\alpha} \oplus \Gamma_{2\beta} \subset \wedge^n V$ can be seen just from the weight diagram: $\wedge^n V$ possesses a highest weight vector with highest weight $L_1 + \cdots + L_n$, and so contains a copy of $\Gamma_{2\alpha}$; but this representation does not possess the weight 2β, and so $\wedge^n V$ must contain $\Gamma_{2\beta}$ as well. (Alternatively, we observed in the preceding lecture that in choosing an ordering of the roots we could have chosen our linear functional $l = c_1 H_1 + \cdots + c_n H_n$ with $c_1 > c_2 > \cdots > -c_n > 0$ without altering the positive roots or the Weyl chamber; in this case the weight λ of $\wedge^n V$ with $l(\lambda)$ maximal would be 2β, showing that $\Gamma_{2\beta} \subset \wedge^n V$.)

(iii) If we want to avoid weight diagrams altogether, we can still see that $\wedge^n V$ must be reducible, because the action of $\mathfrak{so}_{2n}\mathbb{C}$ preserves two bilinear forms: first, we have the bilinear form induced on $\wedge^n V$ by the form Q on V; and second we have the wedge product

$$\varphi: \wedge^n V \times \wedge^n V \to \wedge^{2n} V = \mathbb{C},$$

the last map taking $e_1 \wedge \cdots \wedge e_{2n}$ to 1. It follows that $\wedge^n V$ is reducible; indeed, if we want to see the direct sum decomposition asserted in the statement of the theorem we can look at the composition

$$\tau: \wedge^n V \to \wedge^n V^* \to \wedge^n V,$$

where the first map is the isomorphism given by Q and the second is the isomorphism given by φ. The square of this map is the identity, and decomposing $\wedge^n V$ into $+1$ and -1 eigenspaces for this map gives two subrepresentations.

Exercise 19.3*. Part (i) of Theorem 19.2 can also be proved by showing that for any nonzero vector $w \in \wedge^k V$, the linear span of the vectors $X(w)$, for $X \in \mathfrak{so}_m\mathbb{C}$, is all of $\wedge^k V$. For these purposes take, instead of the basis we have been using, an orthonormal basis $v_1, \ldots, v_m$ for $V = \mathbb{C}^m$, $m = 2n$, so $Q(v_i, v_j) = \delta_{i,j}$. The vectors $v_I = v_{i_1} \wedge \cdots \wedge v_{i_k}$, $I = \{i_1 < \cdots < i_k\}$, form a basis for $\wedge^k V$, and $\mathfrak{so}_m\mathbb{C}$ has a basis consisting of endomorphisms $V_{p,q}$, $p < q$, which takes v_q to v_p, v_p to $-v_q$, and takes the other v_i to zero. Compute the images $V_{p,q}(v_I)$, and prove the claim, first, when $w = v_I$ for some I, and then by induction on the number of nonzero coefficients in the expression $w = \sum a_I v_I$. For (ii) a similar argument shows that $\wedge^n V$ is an irreducible representation of the group $\mathrm{O}_n\mathbb{C}$, and the ideas of §5.1 (cf. §19.5) can be used to see how it decomposes over the subgroup $\mathrm{SO}_n\mathbb{C}$ of index two.

We return now to our analysis of the representations of $\mathfrak{so}_{2n}\mathbb{C}$. By the theorem, the exterior powers $V, \wedge^2 V, \ldots, \wedge^{n-2} V$ provide us with the irreducible representations with highest weight the fundamental weight along the first $n-2$ edges of the Weyl chamber (of course, the exterior power $\wedge^{n-1} V$ is irreducible as well, but as we have observed, $L_1 + \cdots + L_{n-1}$ is not on an edge of the Weyl chamber, and so $\wedge^{n-1} V$ is not as useful for our purposes). For the remaining two edges, we have found irreducible representations with highest weights located there, namely the two direct sum factors of $\wedge^n V$; but the highest weights of these two representations are not primitive ones; they are divisible by 2. Thus, given the theorem above, we see that we have constructed exactly one-half the irreducible representations of $\mathfrak{so}_{2n}\mathbb{C}$, namely, those whose highest weight lies in the sublattice $\mathbb{Z}\{L_1, \ldots, L_n\} \subset \Lambda_W$. Explicitly, any weight γ in the closed Weyl chamber can be expressed (uniquely) in the form

$$\gamma = a_1L_1 + \cdots + a_{n-2}(L_1 + \cdots + L_{n-2}) \\ + a_{n-1}(L_1 + \cdots + L_{n-1} - L_n)/2 + a_n(L_1 + \cdots + L_n)/2$$

with $a_i \in \mathbb{N}$. If $a_{n-1} + a_n$ is even, with $a_{n-1} \geq a_n$ we see that the representation

$$\mathrm{Sym}^{a_1}V \otimes \cdots \otimes \mathrm{Sym}^{a_{n-2}}(\wedge^{n-2}V) \otimes \mathrm{Sym}^{a_n}(\wedge^{n-1}V) \otimes \mathrm{Sym}^{(a_{n-1}-a_n)/2}(\Gamma_{2\beta})$$

will contain an irreducible representation Γ_γ with highest weight γ; whereas if $a_n \geq a_{n-1}$, we will find Γ_γ inside

$$\mathrm{Sym}^{a_1}V \otimes \cdots \otimes \mathrm{Sym}^{a_{n-2}}(\wedge^{n-2}V) \otimes \mathrm{Sym}^{a_{n-1}}(\wedge^{n-1}V) \otimes \mathrm{Sym}^{(a_n-a_{n-1})/2}(\Gamma_{2\alpha}).$$

There remains the problem of constructing irreducible representations Γ_γ whose highest weight γ involves an odd number of α's and β's. To do this, we clearly have to exhibit irreducible representations Γ_α and Γ_β with highest weights α and β. These exist, and are called the *spin representations* of $\mathfrak{so}_{2n}\mathbb{C}$; we will study them in detail in the following lecture. We see from the above that once we exhibit the two representations Γ_α and Γ_β, we will have constructed all the representations of $\mathfrak{so}_{2n}\mathbb{C}$. The representation Γ_γ with highest weight γ written above will be found in the tensor product

$$\mathrm{Sym}^{a_1}V \otimes \cdots \otimes \mathrm{Sym}^{a_{n-2}}(\wedge^{n-2}V) \otimes \mathrm{Sym}^{a_{n-1}}(\Gamma_\beta) \otimes \mathrm{Sym}^{a_n}(\Gamma_\alpha).$$

For the time being, we will assume the existence of the spin representations of $\mathfrak{so}_{2n}\mathbb{C}$; there is a good deal we can say about these representations just on the basis of their weight diagrams.

Exercise 19.4*. Find the weights (with multiplicities) of the representations $\wedge^k V$, and also of $\Gamma_{2\alpha}$, $\Gamma_{2\beta}$, Γ_α, and Γ_β.

Exercise 19.5. Using the above, show that Γ_α and Γ_β are dual to one another when n is odd, and that they are self-dual when n is even.

Exercise 19.6. Give the complete decomposition into irreducible representations of $\mathrm{Sym}^2\Gamma_\alpha$ and $\wedge^2\Gamma_\alpha$. Show that

$$\Gamma_\alpha \otimes \Gamma_\alpha = \Gamma_{2\alpha} \oplus \wedge^{n-2}V \oplus \wedge^{n-4}V \oplus \wedge^{n-6}V \otimes \cdots.$$

Exercise 19.7. Show that

$$\Gamma_\alpha \otimes \Gamma_\beta = \wedge^{n-1}V \oplus \wedge^{n-3}V \oplus \wedge^{n-5}V \oplus \cdots.$$

Exercise 19.8. Verify directly the above statements in the case of $\mathfrak{so}_6\mathbb{C}$, using the isomorphism with $\mathfrak{sl}_4\mathbb{C}$.

Exercise 19.9. Show that the automorphism of $\mathbb{C}^{2n}$ that interchanges e_n and e_{2n}, leaving the other e_i fixed, determines an automorphism of $\mathfrak{so}_{2n}\mathbb{C}$ that preserves the $n-2$ roots $L_1 - L_2, \ldots, L_{n-2} - L_{n-1}$ and interchanges $L_{n-1} - L_n$ and $L_{n-1} + L_n$. This automorphism takes the representation V to itself, but interchanges Γ_α and Γ_β.

§19.3. Representations of $\mathfrak{so}_7\mathbb{C}$

While we might reasonably be apprehensive about the prospect of a family of Lie algebras even more strangely behaved than the even orthogonal algebras, there is some good news: even though the roots systems of the odd Lie algebras appear more complicated than those of the even, the representation theory of the odd algebras is somewhat tamer. We will describe these representations, starting with the example of $\mathfrak{so}_7\mathbb{C}$; we begin, as always, with a picture of the root diagram:

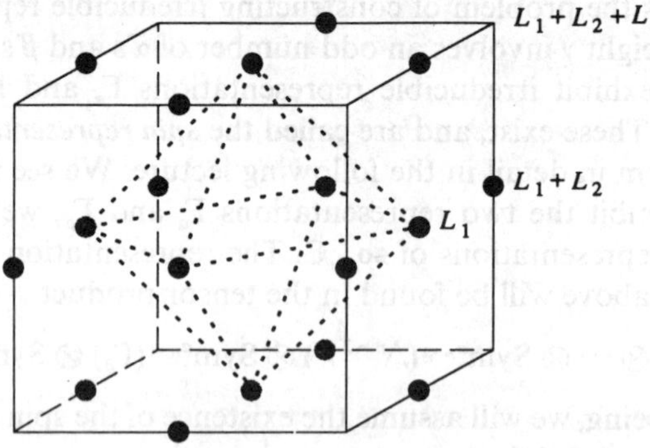

As we said, this looks like the root diagram for $\mathfrak{sp}_6\mathbb{C}$, except that the roots $\pm 2L_i$ have been shortened to $\pm L_i$. Unlike the case of $\mathfrak{so}_5\mathbb{C}$, however, where the long and short roots could be confused and the root diagram was correspondingly congruent to that of $\mathfrak{sp}_4\mathbb{C}$, in the present circumstance the root diagram is not similar to any other; the Lie algebra $\mathfrak{so}_7\mathbb{C}$, in fact, is *not* isomorphic to any of the others we have studied. Next, the Weyl chamber:

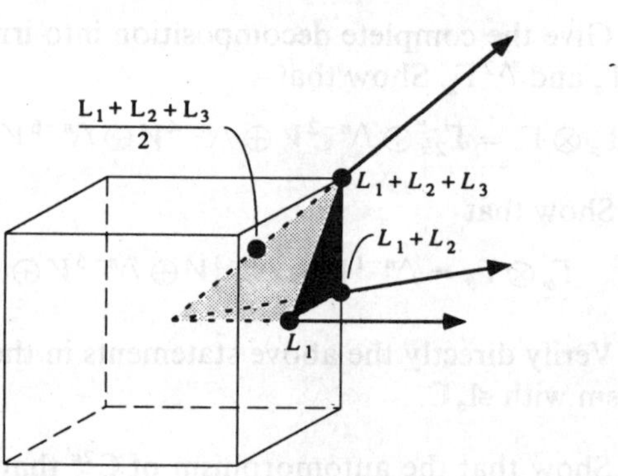

Again, the Weyl chamber itself looks just like that of $\mathfrak{sp}_6\mathbb{C}$; the difference in this picture is in the weight lattice, which contains the additional vector $(L_1 + L_2 + L_3)/2$.

As usual, we start our study of the representations of $\mathfrak{so}_7\mathbb{C}$ with the standard representation, whose weights are $\pm L_i$ and 0:

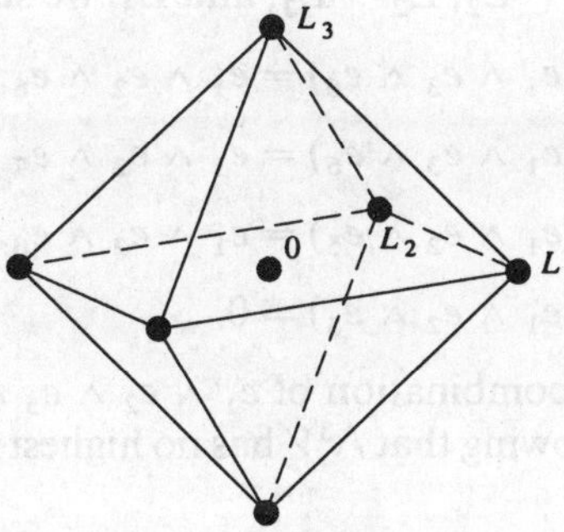

Note that the highest weight L_1 of this representation lies along the front edge of the Weyl chamber. Next, the weights of the exterior square $\wedge^2 V$ are $\pm L_i \pm L_j$, $\pm L_i$, and 0 (taken three times); this, of course, is just the adjoint representation. Note that the highest weight $L_1 + L_2$ of this representation is the same as that of the exterior square of the standard representation for $\mathfrak{so}_6\mathbb{C}$, but because of the smaller Weyl chamber this weight does indeed lie on an edge of the chamber.

Next, consider the third exterior power $\wedge^3 V$ of the standard. This has weights $\pm L_1 \pm L_2 \pm L_3$, $\pm L_i \pm L_j$, $\pm L_i$ (with multiplicity 2) and 0 (with multiplicity 3), i.e., at the midpoints of all the vertices, edges, and faces of the cube:

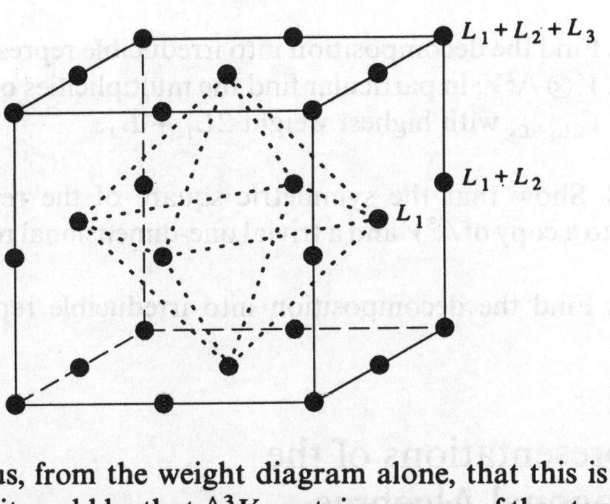

It is not obvious, from the weight diagram alone, that this is an irreducible representation; it could be that $\wedge^3 V$ contains a copy of the standard representation V and that the irreducible representation $\Gamma_{L_1+L_2+L_3}$ thus has multiplicity 1 on the weights $\pm L_i$ and multiplicity 2 (or 1) at 0. We can rule out this possibility by direct calculation: for example, if this were the case, then $\wedge^3 V$ would contain a highest weight vector with weight L_1. The weight space with eigenvalue L_1 in $\wedge^3 V$ is spanned by the tensors $e_1 \wedge e_2 \wedge e_5$ and $e_1 \wedge e_3 \wedge e_6$, however, and if we apply to these the generators $X_{1,2} = E_{1,2} - E_{5,4}$, $X_{2,3} = E_{2,3} - E_{6,5}$, and $U_3 = E_{3,7} - E_{7,6}$ of the root spaces corresponding to the positive roots $L_1 - L_2$, $L_2 - L_3$, and L_3, we see that

$$X_{2,3}(e_1 \wedge e_3 \wedge e_6) = e_1 \wedge e_2 \wedge e_6,$$

$$U_3(e_1 \wedge e_3 \wedge e_6) = e_1 \wedge e_3 \wedge e_7 \neq 0;$$

$$X_{2,3}(e_1 \wedge e_2 \wedge e_5) = e_1 \wedge e_2 \wedge e_6,$$

$$U_3(e_1 \wedge e_2 \wedge e_5) = 0.$$

There is thus no linear combination of $e_1 \wedge e_2 \wedge e_5$ and $e_1 \wedge e_3 \wedge e_6$ killed by both U_3 and $X_{2,3}$, showing that $\wedge^3 V$ has no highest weight vector of weight L_1.

Exercise 19.10. Verify that $\wedge^3 V$ does not contain the trivial representation.

We have thus found irreducible representations of $\mathfrak{so}_7\mathbb{C}$ with highest weight vectors along the three edges of the Weyl chamber, and as in the case of $\mathfrak{so}_6\mathbb{C}$ we have thereby established the existence of the irreducible representations of $\mathfrak{so}_7\mathbb{C}$ with highest weight in the sublattice $\mathbb{Z}\{L_1, L_2, L_3\}$. To complete the description, we need to know that the representation Γ_α with highest weight $\alpha = (L_1 + L_2 + L_3)/2$ exists, and what it looks like, and this time there is no isomorphism to provide this; we will have to wait until the following lecture. In the meantime, we can still have fun playing around both with the representations we do know exist, and also with those whose existence is simply asserted.

Exercise 19.11. Find the decomposition into irreducible representations of the tensor product $V \otimes \wedge^2 V$; in particular find the multiplicities of the irreducible representation $\Gamma_{2L_1+L_2}$ with highest weight $2L_1 + L_2$.

Exercise 19.12. Show that the symmetric square of the representation Γ_α decomposes into a copy of $\wedge^3 V$ and a trivial one-dimensional representation.

Exercise 1913. Find the decomposition into irreducible representations of $\wedge^2 \Gamma_\alpha$.

§19.4. Representations of the Odd Orthogonal Algebras

We will now describe as much as we can of the general pattern for representations of the odd orthogonal Lie algebras $\mathfrak{so}_{2n+1}\mathbb{C}$. As in the case of the even orthogonal Lie algebras, the proof of the existence part of the basic theorem (14.18) (that is, the construction of the irreducible representation with given highest weight) will not be complete until the following lecture, but we can work around this pretty well.

To begin with, recall that the weight lattice of $\mathfrak{so}_{2n+1}\mathbb{C}$ is, like that of $\mathfrak{so}_{2n}\mathbb{C}$, generated by $L_1, \ldots, L_n$ together with the further vector $(L_1 + \cdots + L_n)/2$. The Weyl chamber, on the other hand, is the cone

$$\mathscr{W} = \{\sum a_i L_i : a_1 \geq a_2 \geq \cdots \geq a_n \geq 0\}.$$

The Weyl chamber is as we have pointed out the same as for $\mathfrak{sp}_{2n}\mathbb{C}$, that is, it is a simplicial cone with faces corresponding to the n planes $a_1 = a_2, \ldots, a_{n-1} = a_n$ and $a_n = 0$. The edges of the Weyl chamber are thus the rays generated by the vectors $L_1, L_1 + L_2, \ldots, L_1 + \cdots + L_{n-1}$ and $L_1 + \cdots + L_n$ (note that $L_1 + \cdots + L_{n-1}$ *is* on an edge of the Weyl chamber). Again, the intersection of the weight lattice with the closed Weyl cone is a free semigroup, in this case generated by the fundamental weights $\omega_1 = L_1$, $\omega_2 = L_1 + L_2$, $\ldots$, $\omega_{n-1} = L_1 + \cdots + L_{n-1}$ and the weight $\omega_n = \alpha = (L_1 + \cdots + L_n)/2$. Moreover, as we saw in the cases of $\mathfrak{so}_5\mathbb{C}$ and $\mathfrak{so}_7\mathbb{C}$, the exterior powers of the standard representation do serve to generate all the irreducible representations whose highest weights are in the sublattice $\mathbb{Z}\{L_1, \ldots, L_n\}$: in general we have the following theorem.

Theorem 19.14. *For $k = 1, \ldots, n$, the exterior power $\bigwedge^k V$ of the standard representation V of $\mathfrak{so}_{2n+1}\mathbb{C}$ is the irreducible representation with highest weight $L_1 + \cdots + L_k$.*

Proof. We will leave this as an exercise; the proof is essentially the same as in the case of $\mathfrak{so}_{2n}\mathbb{C}$, with enough of a difference to make it interesting. □

We have thus constructed one-half of the irreducible representations of $\mathfrak{so}_{2n+1}\mathbb{C}$: any weight γ in the closed Weyl chamber can be written

$$\gamma = a_1 L_1 + a_2(L_1 + L_2) + \cdots + a_{n-1}(L_1 + \cdots + L_{n-1}) + a_n(L_1 + \cdots + L_n)/2$$

with $a_i \in \mathbb{N}$; and if a_n is even, the representation

$$\mathrm{Sym}^{a_1} V \otimes \cdots \otimes \mathrm{Sym}^{a_{n-1}}(\textstyle\bigwedge^{n-1} V) \otimes \mathrm{Sym}^{a_n/2}(\bigwedge^n V)$$

will contain an irreducible representation Γ_γ with highest weight γ. We are still missing, however, any representation whose weights involve odd multiples of α; to construct these, we clearly have to exhibit an irreducible representation Γ_α with highest weight α. This exists and is called (as in the case of the even orthogonal Lie algebras) the *spin representation* of $\mathfrak{so}_{2n+1}\mathbb{C}$. We see from the above that once we exhibit the spin representation Γ_α, we will have constructed all the representations of $\mathfrak{so}_{2n+1}\mathbb{C}$; for any γ as above the tensor

$$\mathrm{Sym}^{a_1} V \otimes \cdots \otimes \mathrm{Sym}^{a_{n-1}}(\textstyle\bigwedge^{n-1} V) \otimes \mathrm{Sym}^{a_n}(\Gamma_\alpha)$$

will contain a copy of Γ_γ.

As in the case of the spin representation Γ_α of the even orthogonal Lie algebras, we can say some things about Γ_α even in advance of its explicit construction; for example, we can do the following exercises.

Exercise 19.15. Find the weights (with multiplicities) of the representations $\bigwedge^k V$, and also of Γ_α.

Exercise 19.16. Give the complete decomposition into irreducible representations of $\mathrm{Sym}^2\Gamma_\alpha$ and $\bigwedge^2\Gamma_\alpha$. Show that

$$\Gamma_\alpha \otimes \Gamma_\alpha = \bigwedge^n V \oplus \bigwedge^{n-1} V \oplus \bigwedge^{n-2} V \oplus \cdots \oplus \bigwedge^1 V \oplus \bigwedge^0 V.$$

Exercise 19.17. Verify directly the above statements in the case of $\mathfrak{so}_5\mathbb{C}$, using the isomorphism with $\mathfrak{sp}_4\mathbb{C}$.

§19.5. Weyl's Construction for Orthogonal Groups

The same procedure we saw in the symplectic case can be used to construct representations of the orthogonal groups, this time generalizing what we saw directly for $\bigwedge^k V$ in §§19.2 and 19.4. For the symmetric form Q on $V = \mathbb{C}^m$, the same formula (17.9) determines contractions from $V^{\otimes d}$ to $V^{\otimes(d-2)}$. Denote the intersection of the kernels of all these contractions by $V^{[d]}$. For any partition $\lambda = (\lambda_1 \geq \cdots \geq \lambda_m \geq 0)$ of d, let

$$\mathbb{S}_{[\lambda]}V = V^{[d]} \cap \mathbb{S}_\lambda V. \tag{19.18}$$

As before, this is a representation of the orthogonal group $\mathrm{O}_m\mathbb{C}$ of Q.

Theorem 19.19. *The space $\mathbb{S}_{[\lambda]}V$ is an irreducible representation of $\mathrm{O}_m\mathbb{C}$; $\mathbb{S}_{[\lambda]}V$ nonzero if and only if the sum of the lengths of the first two columns of the Young diagram of λ is at most m.*

The tensor power $V^{\otimes d}$ decomposes exactly as in Lemma 17.15, with everything the same but replacing the symbol $\langle d\rangle$ by $[d]$. In particular,

$$\mathbb{S}_{[\lambda]}V = V^{[d]} \cdot c_\lambda = \mathrm{Im}(c_\lambda\colon V^{[d]} \to V^{[d]}).$$

Exercise 19.20. Verify that $\mathbb{S}_{[\lambda]}V$ is zero when the sum of the lengths of the first two columns is greater than m by showing that $\bigwedge^a V \otimes \bigwedge^b V \otimes V^{(d-a-b)}$ is contained in $\sum_I \Psi_I(V^{\otimes(d-2)})$ when $a + b > m$. Show that $\mathbb{S}_{[\lambda]}V$ is not zero when the sum of the lengths of the first two columns is at most m.

Exercise 19.21*. (i) Show that the kernel of the contraction from $\mathrm{Sym}^d V$ to $\mathrm{Sym}^{d-2}V$ is the irreducible representation $\mathbb{S}_{[d]}V$ of $\mathfrak{so}_m\mathbb{C}$ with highest weight dL_1.

(ii) Show that

$$\mathrm{Sym}^d V = \mathbb{S}_{[d]}V \oplus \mathbb{S}_{[d-2]}V \oplus \cdots \oplus \mathbb{S}_{[d-2p]}V,$$

where p is the largest integer $\leq d/2$.

The proof of the theorem proceeds exactly as in §17.3. The fundamental fact from invariant theory is the same statement as (17.19), with, of course, the operators $\vartheta_I = \Psi_I \circ \Phi_I$ defined using the given symmetric form, and the group $\mathrm{Sp}_{2n}\mathbb{C}$ replaced by $\mathrm{O}_m\mathbb{C}$ (and the same reference to Appendix F.2 for the proof). The theorem then follows from Lemma 6.22 in exactly the same way as for the symplectic group.

To find the irreducible representations over $\mathrm{SO}_m\mathbb{C}$ one can proceed as in §5.1. Weyl calls two partitions (each with the sum of the first two column lengths at most m) *associated* if the sum of the lengths of their first columns is m and the other columns of their Young diagrams have the same lengths. Representations of associated partitions restrict to isomorphic representations of $\mathrm{SO}_m\mathbb{C}$. Note that at least one of each pair of associated partitions will have a Young diagram with at most $\frac{1}{2}m$ rows. If $m = 2n + 1$ is odd, no λ is associated to itself, but if $m = 2n$ is even, any λ with a Young diagram with n nonzero rows will be associated to itself, and its restriction will be the sum of two conjugate representations of $\mathrm{SO}_m\mathbb{C}$ of the same dimension. The final result is:

Theorem 19.22. (i) *If $m = 2n + 1$, and $\lambda = (\lambda_1 \geq \cdots \geq \lambda_n \geq 0)$, then $\mathbb{S}_{[\lambda]}V$ is the irreducible representation of $\mathfrak{so}_m\mathbb{C}$ with highest weight $\lambda_1 L_1 + \cdots + \lambda_n L_n$.*

(ii) *If $m = 2n$, and $\lambda = (\lambda_1 \geq \cdots \geq \lambda_{n-1} \geq 0)$, then $\mathbb{S}_{[\lambda]}V$ is the irreducible representation of $\mathfrak{so}_m\mathbb{C}$ with highest weight $\lambda_1 L_1 + \cdots + \lambda_n L_n$.*

(iii) *If $m = 2n$, and $\lambda = (\lambda_1 \geq \cdots \geq \lambda_{n-1} \geq \lambda_n > 0)$, then $\mathbb{S}_{[\lambda]}V$ is the sum of two irreducible representations of $\mathfrak{so}_m\mathbb{C}$ with highest weights $\lambda_1 L_1 + \cdots + \lambda_n L_n$ and $\lambda_1 L_1 + \cdots + \lambda_{n-1}L_{n-1} - \lambda_n L_n$.*

Exercise 19.23. When m is odd, show that $\mathrm{O}_m\mathbb{C} = \mathrm{SO}_m\mathbb{C} \times \{\pm I\}$. Show that if λ and μ are associated, then $\mu = \lambda \otimes \varepsilon$, where ε is the sign of the determinant.

We postpone to Lecture 25 all discussion of multiplicities of weight spaces, or decomposing tensor products or restrictions to subgroups.

As we saw in Lecture 15 for $\mathrm{GL}_n\mathbb{C}$ and in Lecture 17 for $\mathrm{Sp}_{2n}\mathbb{C}$, it is possible to make a commutative algebra $\mathbb{S}^{[\bullet]} = \mathbb{S}^{[\bullet]}(V)$ out of the sum of all the irreducible representations of $\mathrm{SO}_m\mathbb{C}$, where $V = \mathbb{C}^m$ is the standard representation. First suppose $m = 2n + 1$ is odd. Define the ring $\mathbb{S}^\bullet(V, n)$ as in §15.5, which is a sum of all the representations $\mathbb{S}_\lambda(V)$ of $\mathrm{GL}(V)$ where λ runs over all partitions with at most n parts. As in the symplectic case, there is a canonical decomposition

$$\mathbb{S}_\lambda(V) = \mathbb{S}_{[\lambda]}(V) \oplus J_{[\lambda]}(V),$$

and the direct sum $J^{[\bullet]} = \bigoplus_\lambda J_{[\lambda]}(V)$ is an ideal in $\mathbb{S}^\bullet(V, n)$. The quotient ring

$$\mathbb{S}^{[\bullet]}(V) = A^\bullet(V, n)/J^{[\bullet]} = \bigoplus_\lambda \mathbb{S}_{[\lambda]}(V)$$

is a commutative graded ring which contains each irreducible representation of $\mathrm{SO}_{2n+1}\mathbb{C}$ once.

If $m = 2n$ is even, the above quotient will contain each representation $\mathbb{S}_{[\lambda]}(V)$ twice if λ has n rows. To cut it down so there is only one of each, one can add to $J^{[\cdot]}$ relations of the form $x - \tau(x)$, for $x \in \bigwedge^n V$, where $\tau: \bigwedge^n V \to \bigwedge^n V$ is the isomorphism described in the remark (iii) after the proof of Theorem 19.2. For a detailed discussion, with explicit generators for the ideals, see [L-T].

LECTURE 20

Spin Representations of $\mathfrak{so}_m\mathbb{C}$

In this lecture we complete the picture of the representations of the orthogonal Lie algebras by constructing the spin representations $S^{\pm}$ of $\mathfrak{so}_m\mathbb{C}$; this also yields a description of the spin groups $\mathrm{Spin}_m\mathbb{C}$. Since the representation-theoretic analysis of the spaces $S^{\pm}$ was carried out in the preceding lecture, we are concerned here primarily with the algebra involved in their construction. Thus, §20.1 and §20.2, while elementary, involve some fairly serious algebra. Section 20.3, where we briefly sketch the notion of triality, may seem mysterious to the reader (this is at least in part because it is so to the authors); if so, it may be skipped. Finally, we should say that the subject of the spin representations of $\mathfrak{so}_m\mathbb{C}$ is a very rich one, and one that accommodates many different points of view; the reader who is interested is encouraged to try some of the other approaches that may be found in the literature.

§20.1: Clifford algebras and spin representations of $\mathfrak{so}_m\mathbb{C}$
§20.2: The spin groups $\mathrm{Spin}_m\mathbb{C}$ and $\mathrm{Spin}_m\mathbb{R}$
§20.3: $\mathrm{Spin}_8\mathbb{C}$ and triality

§20.1. Clifford Algebras and Spin Representations of $\mathfrak{so}_m\mathbb{C}$

We begin this section by trying to motivate the definition of Clifford algebras. We may begin by asking, why were we able to find all the representations of $\mathrm{SL}_n\mathbb{C}$ or $\mathrm{Sp}_{2n}\mathbb{C}$ inside tensor powers of the standard representation, but only half the representations of $\mathrm{SO}_m\mathbb{C}$ arise this way? One difference that points in this direction lies in the topology of these groups: $\mathrm{SL}_n\mathbb{C}$ and $\mathrm{Sp}_{2n}\mathbb{C}$ are simply connected, while $\mathrm{SO}_m\mathbb{C}$ has fundamental group $\mathbb{Z}/2$ for $m > 2$ (for proofs see §23.1). Therefore $\mathrm{SO}_m\mathbb{C}$ has a double covering, the *spin group* $\mathrm{Spin}_m\mathbb{C}$. (For $m \leq 6$, these coverings could also be extracted from our identifications of the adjoint group $PSO_m\mathbb{C}$ with the adjoint group of other simply connected groups; e.g. the double cover of $SO_3\mathbb{C}$ is $SL_2\mathbb{C}$.) We will see that the missing representations are those representations of $Spin_m\mathbb{C}$ that do not come from representations of $SO_m\mathbb{C}$.

This double covering may be most readily visible, and probably familiar, for the case of the real subgroup $SO_3\mathbb{R}$ of rotations: a rotation is specified by an axis to rotate about, given by a unit vector u, and an angle of rotation about u; the two choices $\pm u$ of unit vector give a two-sheeted covering. In other words, if D^3 is the unit ball in $\mathbb{R}^3$, there is a double covering

$$S^3 = D^3/\partial D^3 \to SO_3\mathbb{R},$$

which sends a vector v in D^3 to rotation by the angle $2\pi\|v\|$ about the unit vector $v/\|v\|$ (the origin and the unit sphere ∂D^3 are sent to the identity transformation).

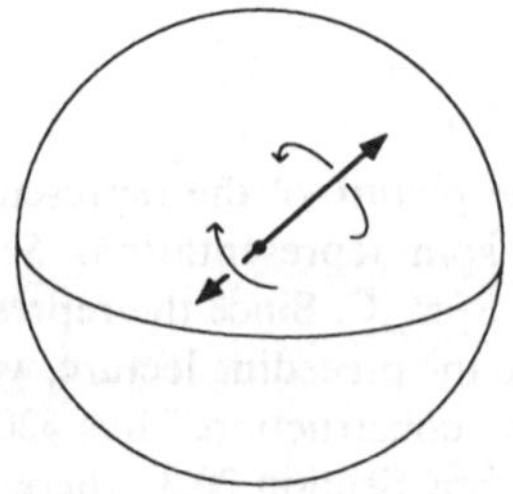

This covering is even easier to see for the entire orthogonal group $O_3\mathbb{R}$, which is generated by reflections R_v in unit vectors v (with $\pm v$ determining the same reflection): we can describe the double cover of $O_3\mathbb{R}$ as the group generated by unit vectors v, with relations

$$v_1 \cdot \ldots \cdot v_n = w_1 \cdot \ldots \cdot w_m$$

whenever the compositions of the corresponding reflections are equal, i.e., whenever

$$R_{v_1} \circ \cdots \circ R_{v_n} = R_{w_1} \circ \cdots \circ R_{w_m};$$

and also relations

$$(-v)\cdot(-w) = v \cdot w$$

for all pairs of unit vectors v and w. (Note that if we restricted ourselves to products of even numbers of the generators $v \in \partial D^3$ we would get back the double cover of the special orthogonal group $SO_3\mathbb{C}$.)

How should we generalize this? The answer is not obvious. For one thing, for various reasons we will not try to construct directly a group that covers the orthogonal group in general. Instead, given a vector space V (real or complex) and a quadratic form Q on V, we will first construct an algebra $\mathrm{Cliff}(V, Q)$, called the *Clifford algebra*. The algebra $\mathrm{Cliff}(V, Q)$ will then turn out to contain in its multiplicative group a subgroup which is a double cover of the orthogonal group $O(V, Q)$ of automorphisms of V preserving Q.

By analogy with the construction of the double cover of $SO_3\mathbb{R}$, the Clifford algebra $\mathrm{Cliff}(V, Q)$ associated to the pair (V, Q) is an associative algebra containing and generated by V. (When we want to describe the spin group inside $\mathrm{Cliff}(V, Q)$ we will restrict ourselves to products of even numbers of elements of V having a fixed norm $Q(v, v)$; if odd products are allowed as well, we get a group called "Pin" which is a double covering of the whole orthogonal group.) To motivate the definition, we would like $\mathrm{Cliff}(V, Q)$ to be the algebra generated by V subject to relations analogous to those above for the double cover of the orthogonal group. In particular, for any vector v with $Q(v, v) = 1$, since the reflection R_v in the hyperplane perpendicular to v is an involution, we want

$$v \cdot v = 1$$

in $\mathrm{Cliff}(V, Q)$. By polarization, this is the same as imposing the relation

$$v \cdot w + w \cdot v = 2Q(v, w)$$

for all v and w in V. In particular, $w \cdot v = -v \cdot w$ if v and w are perpendicular. In fact, the Clifford algebra[1] will be defined below to be the associative algebra generated by V and subject to the equation $v \cdot v = Q(v, v)$.

Looking ahead, we will see later in this section that each complex Clifford algebra contains an orthogonal Lie algebra as a subalgebra. The key theorem is then that $\mathrm{Cliff}(V, Q)$ *is isomorphic either to a matrix algebra or to a sum of two matrix algebras*. This in turn determines either one or two representations of the orthogonal Lie algebras, which turn out to be the representations which were needed to complete the story in the last lecture. Just as in the special linear and symplectic cases, the corresponding Lie groups are not really needed to construct the representations; they can be written down directly from the Lie algebra. In this section we do this, using the Clifford algebras to construct these representations of $\mathfrak{so}_m\mathbb{C}$ directly, and verify that they give the missing spin representations. In the second section of this lecture we will show how the spin groups sit as subgroups in their multiplicative groups.

Clifford Algebras

Given a symmetric bilinear form Q on a vector space V, the *Clifford algebra* $C = C(Q) = \mathrm{Cliff}(V, Q)$ is an associative algebra with unit 1, which contains and is generated by V, with $v \cdot v = Q(v, v) \cdot 1$ for all $v \in V$. Equivalently, we have the equation

$$v \cdot w + w \cdot v = 2Q(v, w), \tag{20.1}$$

[1] The mathematical world seems to be about evenly divided about the choice of signs here, and one must translate from Q to $-Q$ to go from one side to the other.

for all v and w in V. The Clifford algebra can be defined to be the universal algebra with this property: if E is any associative algebra with unit, and a linear mapping $j: V \to E$ is given such that $j(v)^2 = Q(v, v)\cdot 1$ for all $v \in V$, or equivalently

$$j(v)\cdot j(w) + j(w)\cdot j(v) = 2Q(v, w)\cdot 1 \tag{20.2}$$

for all $v, w \in V$, then there should be a unique homomorphism of algebras from $C(Q)$ to E extending j. The Clifford algebra can be constructed quickly by taking the tensor algebra

$$T^{\cdot}(V) = \bigoplus_{n\geq 0} V^{\otimes n} = \mathbb{C} \oplus V \oplus (V \otimes V) \oplus (V \otimes V \otimes V) \oplus \cdots,$$

and setting $C(Q) = T^{\cdot}(V)/I(Q)$, where $I(Q)$ is the two-sided ideal generated by all elements of the form $v \otimes v - Q(v, v)\cdot 1$. It is automatic that this $C(Q)$ satisfies the required universal property.

The facts that the dimension of C is 2^m, where $m = \dim(V)$, and that the canonical mapping from V to C is an embedding, are part of the following lemma:

Lemma 20.3. *If $e_1, \ldots, e_m$ form a basis for V, then the products $e_I = e_{i_1}\cdot e_{i_2}\cdot \ldots \cdot e_{i_k}$, for $I = \{i_1 < i_2 < \cdots < i_k\}$, and with $e_\phi = 1$, form a basis for $C(Q) = \mathrm{Cliff}(V, Q)$.*

Proof. From the equations $e_i\cdot e_j + e_j\cdot e_i = 2Q(e_i, e_j)$ it follows immediately that the elements e_I generate $C(Q)$. Their independence is not hard to verify directly; it also follows by seeing that the images in the matrix algebras under the mappings constructed below are independent. For another proof, note that when $Q \equiv 0$, the Clifford algebra is just the exterior algebra $\bigwedge^{\cdot} V$. In general, the Clifford algebra can be filtered by subspaces F_k, consisting of those elements which can be written as sums of at most k products of elements in V; one checks that the associated graded space F_k/F_{k+1} is $\bigwedge^k V$. For a third proof, one can verify that the Clifford algebra of the direct sum of two orthogonal spaces is the skew commutative tensor product of the Clifford algebras of the two spaces (cf. Exercise B.9), which reduces one to the trivial case where $\dim V = 1$. □

Since the ideal $I(Q) \subset T(V)$ is generated by elements of even degree, the Clifford algebra inherits a $\mathbb{Z}/2\mathbb{Z}$ grading:

$$C = C^{\text{even}} \oplus C^{\text{odd}} = C^+ \oplus C^-,$$

with $C^+\cdot C^+ \subset C^+, C^+\cdot C^- \subset C^-, C^-\cdot C^+ \subset C^-, C^-\cdot C^- \subset C^+$; C^+ is spanned by products of an even number of elements in V and C^- is spanned by products of an odd number. In particular, C^{even} is a subalgebra of dimension 2^{m-1}.

Since $C(Q)$ is an associative algebra, it determines a Lie algebra, with bracket $[a, b] = a\cdot b - b\cdot a$. From now on we assume Q is nondegenerate. The new representations of $\mathfrak{so}_m\mathbb{C}$ will be found in two steps:

(i) embedding the Lie algebra $\mathfrak{so}(Q) = \mathfrak{so}_m\mathbb{C}$ inside the Lie algebra of the even part of the Clifford algebra $C(Q)$;
(ii) identifying the Clifford algebras with one or two copies of matrix algebras.

To carry out the first step we make explicit the isomorphism of $\bigwedge^2 V$ with $\mathfrak{so}(Q)$ that we have discussed before. Recall that

$$\mathfrak{so}(Q) = \{X \in \operatorname{End}(V)\colon Q(Xv, w) + Q(v, Xw) = 0 \text{ for all } v, w \text{ in } V\}.$$

The isomorphism is given by

$$\textstyle\bigwedge^2 V \overset{\cong}{\rightrightarrows} \mathfrak{so}(Q) \subset \operatorname{End}(V), \qquad a \wedge b \mapsto \varphi_{a\wedge b},$$

for a and b in V, where $\varphi_{a\wedge b}$ is defined by

$$\varphi_{a\wedge b}(v) = 2(Q(b, v)a - Q(a, v)b). \tag{20.4}$$

It is a simple verification that $\varphi_{a\wedge b}$ is in $\mathfrak{so}(Q)$. One sees that the natural bases correspond up to scalars, e.g., $e_i \wedge e_{n+j}$ maps to $2(E_{i,j} - E_{n+j,n+i})$, so the map is an isomorphism. (The choice of scalar factor is unimportant here; it was chosen to simplify later formulas.) One calculates what the bracket on $\bigwedge^2 V$ must be to make this an isomorphism of Lie algebras:

$$\begin{aligned}
[\varphi_{a\wedge b}, \varphi_{c\wedge d}](v) &= \varphi_{a\wedge b} \circ \varphi_{c\wedge d}(v) - \varphi_{c\wedge d} \circ \varphi_{a\wedge b}(v)\\
&= 2\varphi_{a\wedge b}(Q(d, v)c - Q(c, v)d) - 2\varphi_{c\wedge d}(Q(b, v)a - Q(a, v)b)\\
&= 4Q(d, v)(Q(b, c)a - Q(a, c)b)\\
&\quad - 4Q(c, v)(Q(b, d)a - Q(a, d)b)\\
&\quad - 4Q(b, v)(Q(d, a)c - Q(c, a)d)\\
&\quad + 4Q(a, v)(Q(d, b)c - Q(c, b)d)\\
&= 2Q(b, c)\varphi_{a\wedge d}(v) - 2Q(b, d)\varphi_{a\wedge c}(v)\\
&\quad - 2Q(a, d)\varphi_{c\wedge b}(v) + 2Q(a, c)\varphi_{d\wedge b}(v).
\end{aligned}$$

This gives an explicit formula for the bracket on $\bigwedge^2 V$:

$$\begin{aligned}
[a \wedge b, c \wedge d] = 2Q(b, c)a \wedge d - 2Q(b, d)a \wedge c\\
- 2Q(a, d)c \wedge b + 2Q(a, c)d \wedge b.
\end{aligned} \tag{20.5}$$

On the other hand, the bracket in the Clifford algebra satisfies

$$\begin{aligned}
[a\cdot b, c\cdot d] &= a\cdot b\cdot c\cdot d - c\cdot d\cdot a\cdot b\\
&= (2Q(b, c)a\cdot d - a\cdot c\cdot b\cdot d) - (2Q(a, d)c\cdot b - c\cdot a\cdot d\cdot b)\\
&= 2Q(b, c)a\cdot d - (2Q(b, d)a\cdot c - a\cdot c\cdot d\cdot b)\\
&\quad - 2Q(a, d)c\cdot b + (2Q(a, c)\cdot d\cdot b - a\cdot c\cdot d\cdot b)\\
&= 2Q(b, c)a\cdot d - 2Q(b, d)a\cdot c - 2Q(a, d)c\cdot b + 2Q(a, c)\cdot d\cdot b.
\end{aligned}$$

It follows that the map $\psi: \wedge^2 V \to \mathrm{Cliff}(V, Q)$ defined by

$$\psi(a \wedge b) = \tfrac{1}{2}(a \cdot b - b \cdot a) = a \cdot b - Q(a, b) \tag{20.6}$$

is a map[2] of Lie algebras, and by looking at basis elements again one sees that it is an embedding. This proves:

Lemma 20.7. *The mapping $\psi \circ \varphi^{-1}: \mathfrak{so}(Q) \to C(Q)^{\text{even}}$ embeds $\mathfrak{so}(Q)$ as a Lie subalgebra of $C(Q)^{\text{even}}$.*

Exercise 20.8. Show that the image of ψ is

$$F_2 \cap C(Q)^{\text{even}} \cap \mathrm{Ker}(\mathrm{trace}),$$

where F_2 is the subspace of $C(Q)$ spanned by products of at most two elements of V, and the trace of an element of $C(Q)$ is the trace of left multiplication by that element on $C(Q)$.

We consider first the *even* case: write $V = W \oplus W'$, where W and W' are n-dimensional isotropic spaces for Q. (Recall that a space is isotropic when Q restricts to the zero form on it.) With our choice of standard Q on $V = \mathbb{C}^{2n}$, W can be taken to be the space spanned by the first n basis vectors, W' by the last n.

Lemma 20.9. *The decomposition $V = W \oplus W'$ determines an isomorphism of algebras*

$$C(Q) \cong \mathrm{End}(\wedge^{\bullet} W),$$

where $\wedge^{\bullet} W = \wedge^0 W \oplus \cdots \oplus \wedge^n W$.

Proof. Mapping $C(Q)$ to the algebra $E = \mathrm{End}(\wedge^{\bullet} W)$ is the same as defining a linear mapping from V to E, satisfying (20.2). We must construct maps $l: W \to E$ and $l': W' \to E$ such that

$$l(w)^2 = 0, \qquad l'(w')^2 = 0, \tag{20.10}$$

and

$$l(w) \circ l'(w') + l'(w') \circ l(w) = 2Q(w, w')I$$

for any $w \in W$, $w' \in W'$. For each $w \in W$, let $L_w \in E$ be left multiplication by w on the exterior algebra $\wedge^{\bullet} W$:

$$L_w(\xi) = w \wedge \xi, \quad \xi \in \wedge^{\bullet} W.$$

For $\vartheta \in W^*$, let $D_\vartheta \in E$ be the derivation of $\wedge^{\bullet} W$ such that $D_\vartheta(1) = 0$, $D_\vartheta(w) = \vartheta(w) \in \wedge^0 W = \mathbb{C}$ for $w \in W = \wedge^1 W$, and

[2] Note that the bilinear form ψ given by (20.6) is alternating since $\psi(a \wedge a) = 0$, so it defines a linear map on $\wedge^2 V$.

$$D_\vartheta(\zeta \wedge \xi) = D_\vartheta(\zeta) \wedge \xi + (-1)^{\deg(\zeta)}\zeta \wedge D_\vartheta(\zeta).$$

Explicitly, $D_\vartheta(w_1 \wedge \cdots \wedge w_r) = \sum(-1)^{i-1}\vartheta(w_i)(w_1 \wedge \cdots \wedge \hat{w}_i \wedge \cdots \wedge w_r)$. Now set

$$l(w) = L_w, \qquad l'(w') = D_\vartheta, \tag{20.11}$$

where $\vartheta \in W^*$ is defined by the identity $\vartheta(w) = 2Q(w, w')$ for all $w \in W$. The required equations (20.10) are straightforward verifications: one checks directly on elements in $W = \wedge^1 W$, and then that, if they hold on ζ and ξ, they hold on $\zeta \wedge \xi$. Finally, one may see that the resulting map is an isomorphism by looking at what happens to a basis. □

Exercise 20.12. The left $C(Q)$-module $\wedge^\bullet W$ is isomorphic to a left ideal in $C(Q)$. Show that if f is a generator for $\wedge^n W'$, then $C(Q)\cdot f = \wedge^\bullet W \cdot f$, and the map $\zeta \mapsto \zeta \cdot f$ gives an isomorphism

$$\wedge^\bullet W \to \wedge^\bullet W \cdot f = C(Q)\cdot f$$

of left $C(Q)$-modules.

Now we have a decomposition $\wedge^\bullet W = \wedge^{\text{even}} W \oplus \wedge^{\text{odd}} W$ into the sum of even and odd exterior powers, and $C(W)^{\text{even}}$ respects this splitting. We deduce from Lemma 20.9 an isomorphism

$$C(Q)^{\text{even}} \cong \operatorname{End}(\wedge^{\text{even}} W) \oplus \operatorname{End}(\wedge^{\text{odd}} W). \tag{20.13}$$

Combining with Lemma 20.7, we now have an embedding of Lie algebras:

$$\mathfrak{so}(Q) \subset C(Q)^{\text{even}} \cong \mathfrak{gl}(\wedge^{\text{even}} W) \oplus \mathfrak{gl}(\wedge^{\text{odd}} W), \tag{20.14}$$

and hence we have two representations of $\mathfrak{so}(Q) = \mathfrak{so}_{2n}\mathbb{C}$, which we denote by

$$S^+ = \wedge^{\text{even}} W \quad \text{and} \quad S^- = \wedge^{\text{odd}} W.$$

Proposition 20.15. *The representations $S^\pm$ are the irreducible representations of $\mathfrak{so}_{2n}\mathbb{C}$ with highest weights $\alpha = \frac{1}{2}(L_1 + \cdots + L_n)$ and $\beta = \frac{1}{2}(L_1 + \cdots + L_{n-1} - L_n)$. More precisely,*

$$S^+ = \Gamma_\alpha \quad \textit{and} \quad S^- = \Gamma_\beta \quad \textit{if } n \textit{ is even;}$$

$$S^+ = \Gamma_\beta \quad \textit{and} \quad S^- = \Gamma_\alpha \quad \textit{if } n \textit{ is odd.}$$

PROOF. We show that the natural basis vectors $e_I = e_{i_1} \wedge \cdots \wedge e_{i_k}$ for $\wedge^\bullet W$ are weight vectors. Tracing through the isomorphisms established above, we see that $H_i = E_{i,i} - E_{n+i,n+i}$ in $\mathfrak{h} \subset \mathfrak{so}_{2n}\mathbb{C}$ corresponds to $\frac{1}{2}(e_i \wedge e_{n+i})$ in $\wedge^2 V$, which corresponds to $\frac{1}{2}(e_i \cdot e_{n+i} - 1)$ in $C(Q)$, which maps to

$$\tfrac{1}{2}(L_{e_i} \circ D_{2e_i^*} - I) = L_{e_i} \circ D_{e_i^*} - \tfrac{1}{2}I \in \operatorname{End}(\wedge^\bullet W).$$

A simple calculation shows that

$$L_{e_i} \circ D_{e_i^*}(e_I) = \begin{cases} e_I & \text{if } i \in I \\ 0 & \text{if } i \notin I. \end{cases}$$

Therefore, e_I spans a weight space with weight $\frac{1}{2}(\sum_{i \in I} L_i - \sum_{j \notin I} L_j)$. All such weights with given $|I|$ mod 2 are congruent by the Weyl group, so each of $S^+ = \bigwedge^{\text{even}} W^+$ and $S^- = \bigwedge^{\text{odd}} W$ must be an irreducible representation. The highest weights are easy to read off. For example, the highest weight for $\bigwedge^{\text{even}} W$ is $\frac{1}{2}\sum L_i = \alpha$ if n is even, while if n is odd, its highest weight is β. □

These two representations S^+ and S^- are usually called the *half-spin representations* of $\mathfrak{so}_{2n}\mathbb{C}$, while their sum $S = S^+ \oplus S^- = \bigwedge^{\bullet} W$ is called the *spin representation*. Frequently, especially when we speak of the even and odd cases together, we call them all simply "spin representations." Elements of S are called *spinors*. For other proofs of the proposition see Exercises 20.34 and 20.35.

For the *odd* case, write $V = W \oplus W' \oplus U$, where W and W' are n-dimensional isotropic subspaces, and U is a one-dimensional space perpendicular to them. For our standard Q on $\mathbb{C}^{2n+1}$, these are spanned by the first n, the second n, and the last basis vector.

Lemma 20.16. *The decomposition $V = W \oplus W' \oplus U$ determines an isomorphism of algebras*

$$C(Q) \cong \operatorname{End}(\textstyle\bigwedge^{\bullet} W) \oplus \operatorname{End}(\bigwedge^{\bullet} W').$$

Proof. Proceeding as in the even case, to map V to $E = \operatorname{End}(\bigwedge^{\bullet} W)$, map $w \in W$ to L_w, $w' \in W'$ to D_ϑ, where $\vartheta(w) = 2Q(w, w')$ as before. Let u_0 be the element in U such that $Q(u_0, u_0) = 1$, and send u_0 to the endomorphism that is the identity on $\bigwedge^{\text{even}} W$, and minus the identity on $\bigwedge^{\text{odd}} W$. Since this involution skew commutes with all L_w and D_ϑ, the resulting map from $V = W \oplus W' \oplus U$ to E determines an algebra homomorphism from $C(Q)$ to E. The map to $\operatorname{End}(\bigwedge^{\bullet} W')$ is defined similarly, reversing the roles of W and W'. Again one checks that the map is an isomorphism by looking at bases. □

Exercise 20.17*. Find a generator for a left ideal of $C(Q)$ that is isomorphic to $\bigwedge^{\bullet} W$.

The subalgebra $C(Q)^{\text{even}}$ of $C(Q)$ is mapped isomorphically onto either of the factors by the isomorphism of the lemma, so we have an isomorphism in the odd case:

$$C(Q)^{\text{even}} \cong \operatorname{End}(\textstyle\bigwedge^{\bullet} W). \tag{20.18}$$

As before, this gives a representation $S = \bigwedge^{\bullet} W$ of Lie algebras:

$$\mathfrak{so}_{2n+1}\mathbb{C} = \mathfrak{so}(Q) \subset C(Q)^{\text{even}} \cong \mathfrak{gl}(\textstyle\bigwedge^{\bullet} W) = \mathfrak{gl}(S). \tag{20.19}$$

Proposition 20.20. *The representation $S = \bigwedge^{\bullet} W$ is the irreducible representation of $\mathfrak{so}_{2n+1}\mathbb{C}$ with highest weight*

$$\alpha = \tfrac{1}{2}(L_1 + \cdots + L_n).$$

PROOF. Exactly as in the even case, each e_I is an eigenvector with with weight $\frac{1}{2}(\sum_{i \in I} L_i - \sum_{j \notin I} L_j)$. This time all such weights are congruent by the Weyl group, so this must be an irreducible representation, and the highest weight is clearly $\frac{1}{2}(L_1 + \cdots + L_n)$. □

As we saw in Lecture 19, the construction of this *spin representation S* finishes the proof of the existence theorem for representations of $\mathfrak{so}_m\mathbb{C}$, and hence for all of the classical complex semisimple Lie algebras.

Exercise 20.21*. Use the above identification of the Clifford algebras with matrix algebras (or direct calculation) to compute their centers. In particular, show that the intersection of the center of C with the even subalgebra C^{even} is always the one-dimensional space of scalars. Show similarly that if x is in C^{odd} and $x \cdot v = -v \cdot x$ for all v in V, then $x = 0$.

Exercise 20.22*. For $X \in \mathfrak{so}(Q)$ and $v \in V$, we have $X \cdot v \in V$ by the standard action of $\mathfrak{so}(Q)$ on V. On the other hand, we have identified $\mathfrak{so}(Q)$ and V as subspaces of the Clifford algebra C, so we can compute the commutator $[X, v]$. Show that these agree:

$$X \cdot v = [X, v] \in V \subset C.$$

Problem 20.23*. Let $C(p, q)$ be the real Clifford algebra corresponding to the quadratic form with p positive and q negative eigenvalues. Lemmas 20.9 and 20.16 actually construct isomorphisms of $C(n, n)$ with a real matrix algebra, and of $C(n + 1, n)$ with a product of two real matrix algebras. Compute $C(p, q)$ for other p and q. All are products of one or two matrix algebras over $\mathbb{R}$, $\mathbb{C}$, or $\mathbb{H}$.

§20.2. The Spin Groups $\text{Spin}_m\mathbb{C}$ and $\text{Spin}_m\mathbb{R}$

The Clifford algebra $C = C(Q)$ is generated by the subspace $V = \mathbb{C}^m$, and C has an anti-involution $x \mapsto x^*$, determined by

$$(v_1 \cdot \ldots \cdot v_r)^* = (-1)^r v_r \cdot \ldots \cdot v_1$$

for any $v_1, \ldots, v_r$ in V. This operation $*$, sometimes called the *conjugation*, is the composite of:

the *main antiautomorphism* or reversing map $\tau\colon C \to C$ determined by

$$\tau(v_1 \cdot \ldots \cdot v_r) = v_r \cdot \ldots \cdot v_1 \tag{20.24}$$

for $v_1, \ldots, v_r$ in V, and

the *main involution* α which is the identity on C^{even} and minus the identity on C^{odd}, i.e.,

$$\alpha(v_1 \cdot \ldots \cdot v_r) = (-1)^r v_1 \cdot \ldots \cdot v_r. \tag{20.25}$$

Note that $(x \cdot y)^* = y^* \cdot x^*$, which comes from the identities $\tau(x \cdot y) = \tau(y) \cdot \tau(x)$ and $\alpha(x \cdot y) = \alpha(x) \cdot \alpha(y)$.

Exercise 20.26. Use the universal property for C to verify that these are well defined: show that α is a homomorphism from C to C and τ is a well-defined homomorphism from C to the opposite algebra of C (the algebra with the same vector space structure, but with reversed multiplication: $x \tilde{\cdot} y = y \cdot x$).

Instead of defining the spin group as the set of products of certain elements of V, it will be convenient to start with a more abstract definition. Set

$$\mathrm{Spin}(Q) = \{x \in C(Q)^{\text{even}}: x \cdot x^* = 1 \text{ and } x \cdot V \cdot x^* \subset V\}. \tag{20.27}$$

We see from this definition that $\mathrm{Spin}(Q)$ forms a closed subgroup of the group of units in the (even) Clifford algebra. Any x in $\mathrm{Spin}(Q)$ determines an endomorphism $\rho(x)$ of V by

$$\rho(x)(v) = x \cdot v \cdot x^*, \quad v \in V.$$

Proposition 20.28. *For* $x \in \mathrm{Spin}(Q)$, $\rho(x)$ *is in* $\mathrm{SO}(Q)$. *The mapping*

$$\rho: \mathrm{Spin}(Q) \to \mathrm{SO}(Q)$$

is a homomorphism, making $\mathrm{Spin}(Q)$ *a connected two-sheeted covering of* $\mathrm{SO}(Q)$. *The kernel of* ρ *is* $\{1, -1\}$.

PROOF. We will prove something more. Define a larger subgroup, this time of the multiplicative group of $C(Q)$, by

$$\mathrm{Pin}(Q) = \{x \in C(Q): x \cdot x^* = 1 \text{ and } x \cdot V \cdot x^* \subset V\}, \tag{20.29}$$

and define a homomorphism

$$\rho: \mathrm{Pin}(Q) \to \mathrm{O}(Q), \qquad \rho(x)(v) = \alpha(x) \cdot v \cdot x^*, \tag{20.30}$$

where $\alpha: C(Q) \to C(Q)$ is the main involution.

To see that $\rho(x)$ preserves the quadratic form Q, we use the fact that for w in V, $Q(w, w) = w \cdot w = -w \cdot w^*$, and calculate:

$$\begin{aligned} Q(\rho(x)(v), \rho(x)(v)) &= -\alpha(x) \cdot v \cdot x^* \cdot (\alpha(x) \cdot v \cdot x^*)^* \\ &= -\alpha(x) \cdot v \cdot x^* \cdot x \cdot v^* \cdot \alpha(x)^* \end{aligned}$$

$$\begin{aligned} &= -\alpha(x)\cdot v\cdot v^*\cdot\alpha(x^*) \\ &= Q(v, v)\alpha(x)\cdot\alpha(x^*) \\ &= Q(v, v)\alpha(x\cdot x^*) = Q(v, v). \end{aligned}$$

We claim next that ρ is surjective. This follows from the standard fact (see Exercise 20.32) that the orthogonal group $O(Q)$ is generated by reflections. Indeed, if R_w is the reflection in the hyperplane perpendicular to a vector w, normalized so that $Q(w, w) = -1$, it is easy to see that w is in $\text{Pin}(Q)$ and $\rho(w) = R_w$; in fact,

$$w\cdot w^* = w\cdot(-w) = -Q(w, w) = 1,$$

and so

$$\rho(w)(w) = \alpha(w)\cdot w\cdot w^* = -w\cdot 1 = -w;$$

and if $Q(w, v) = 0$,

$$\rho(w)(v) = \alpha(w)\cdot v\cdot w^* = -w\cdot v\cdot w^* = v\cdot w\cdot w^* = v.$$

The next claim is that the kernel of ρ on the larger group $\text{Pin}(Q)$ is ± 1. Suppose x is in the kernel, and write $x = x_0 + x_1$ with $x_0 \in C^{\text{even}}$ and $x_1 \in C^{\text{odd}}$. Then $x_0\cdot v = v\cdot x_0$ for all $v \in V$, so x_0 is in the center of C. And $x_1\cdot v = -v\cdot x_1$ for all $v \in V$. By Exercise 20.21, x_0 is in $\mathbb{C}\cdot 1$, and $x_1 = 0$. So $x = x_0$ is in $\mathbb{C}$ and $x^2 = 1$; so $x = \pm 1$.

It follows that if $R \in O(Q)$ is written as a product of reflections $R_{w_1}\circ\ldots\circ R_{w_r}$, then the two elements in $\rho^{-1}(R)$ are $\pm w_1\cdot\ldots\cdot w_r$. In particular, we get another description of the spin groups:

$$\begin{aligned} \text{Spin}(Q) &= \text{Pin}(Q)\cap C(Q)^{\text{even}} = \rho^{-1}(\text{SO}(Q)) \\ &= \{\pm w_1\cdot\ldots\cdot w_{2k}\colon w_i \in V, Q(w_i, w_i) = -1\}. \qquad (20.31) \end{aligned}$$

Since $-1 = v\cdot v$ for any v with $Q(v, v) = -1$, we see that the spin group consists of even products of such elements.

To complete the proof, we must check that $\text{Spin}(Q)$ is connected or, equivalently, that the two elements in the kernel of ρ can be connected by a path. We leave this now as an exercise, since much more will be seen shortly. □

Exercise 20.32*. Let Q be a nondegenerate symmetric bilinear form on a real or complex vector space V.

(a) Show that if v and w are vectors in V with $Q(v, v) = Q(w, w) \neq 0$, then there is either a reflection or a product of two reflections that takes v into w.

(b) Deduce that every element of the orthogonal group of Q can be written as the product of at most $2\cdot\dim(V)$ reflections.

Exercise 20.33*. Since $\text{Spin}(Q)$ is a subgroup of the multiplicative group of $C(Q)$, its Lie algebra is a subalgebra of $C(Q)$ with its usual bracket. Verify that this subalgebra is the subalgebra $\mathfrak{so}(Q)$ that was constructed in §20.1.

Exercise 20.34. The fact that $\wedge^{\bullet}W$ (and $\wedge^{\bullet}W'$ in the odd case) is an irreducible module over $C(Q)$ is equivalent to the fact that it is an irreducible representation of the group $\mathrm{Pin}(Q)$ since the linear span of $\mathrm{Pin}(Q)$ is dense in $C(Q)$.

(a) Apply the analysis of §5.1 to the subgroup

$$\mathrm{Spin}(Q) \subset \mathrm{Pin}(Q)$$

of index two. In the odd case, $\wedge^{\bullet}W$ and $\wedge^{\bullet}W'$ are conjugate representations, so their restrictions to $\mathrm{Spin}(Q)$ are isomorphic and irreducible: this is the spin representation. In the even case, $\wedge^{\bullet}W$ is self-conjugate, and its restriction to $\mathrm{Spin}(Q)$ is a sum of two conjugate irreducible representations, which are the two half-spin representations.

(b) Of the representations of $\mathrm{Spin}(Q)$ (i.e., the representations of $\mathfrak{so}_m\mathbb{C}$), which induce irreducible representations of $\mathrm{Pin}(Q)$ and which are restrictions of irreducible representations of $\mathrm{Pin}(Q)$?

Exercise 20.35. Deduce the irreducibility of the spin and half-spin representations from the fact that their restrictions to the 2-groups of Exercise 3.9 are irreducible representations of these finite groups.

Exercise 20.36*. Show that the center of $\mathrm{Spin}_m(\mathbb{C})$ is $\rho^{-1}(1) = \{\pm 1\}$ if m is odd. If m is even show that the center is

$$\rho^{-1}(\pm 1) = \{\pm 1, \pm\omega\},$$

where, in terms of our standard basis,

$$\omega = \frac{ie_1 \cdot e_{n+1} - ie_{n+1} \cdot e_1}{2} \cdot \ldots \cdot \frac{ie_n \cdot e_{2n} - ie_{2n} \cdot e_n}{2}.$$

Exercise 20.37*. Show that the spin representation $\mathrm{Spin}(Q) \to \mathrm{GL}(S)$ maps into the special linear group $\mathrm{SL}(S)$. Show that for $m = 2n$ and n even, the half-spin representations also map into the special linear groups $\mathrm{SL}(S^+)$ and $\mathrm{SL}(S^-)$.

Exercise 20.38*. Construct a nondegenerate bilinear pairing β on the spinor space $S = \wedge^{\bullet}W$ by choosing an isomorphism of $\wedge^n W$ with $\mathbb{C}$ and letting $\beta(s, t)$ be the image of $\tau(s) \wedge t \in \wedge^{\bullet}W$ by the projection to $\wedge^n W = \mathbb{C}$, where τ is the main antiautomorphism).

(a) When $m = 2n$, show that β can also be defined by the identity $\beta(s, t)f = \tau(s \cdot f) \cdot t \cdot f$ for an appropriate generator f of $\wedge^n W'$. Deduce that the action of $\mathrm{Spin}(Q)$ on S respects the bilinear form β.

(b) Show that β is symmetric if n is congruent to 0 or 3 modulo 4, and skew-symmetric otherwise. So the spin representation is a homomorphism

$$\mathrm{Spin}_{2n+1}\mathbb{C} \to \mathrm{SO}_{2n}\mathbb{C} \quad \text{if } n \equiv 0, 3\ (4),$$

$$\mathrm{Spin}_{2n+1}\mathbb{C} \to \mathrm{Sp}_{2n}\mathbb{C} \quad \text{if } n \equiv 1, 2\ (4).$$

(c) If $m = 2n$, the restrictions of β to S^+ and S^- are zero if n is odd. For n even, deduce that the half-spin representations are homomorphisms

$$\mathrm{Spin}_{2n}\mathbb{C} \to \mathrm{SO}_{2^{n-1}}\mathbb{C} \quad \text{if } n \equiv 0\ (4),$$
$$\mathrm{Spin}_{2n}\mathbb{C} \to \mathrm{Sp}_{2^{n-1}}\mathbb{C} \quad \text{if } n \equiv 2\ (4).$$

Note in particular that $\mathrm{Spin}_8\mathbb{C}$ has two maps to $\mathrm{SO}_8\mathbb{C}$ in addition to the original covering. "Triality," which we discuss in the next section, describes the relation among these three homomorphisms.

Exercise 20.39. Show that the spin and half-spin representations give the isomorphisms we have seen before:

$$\mathrm{Spin}_2\mathbb{C} \cong \mathrm{GL}(S^+) = \mathrm{GL}_1\mathbb{C} = \mathbb{C}^*,$$
$$\mathrm{Spin}_3\mathbb{C} \cong \mathrm{SL}(S) = \mathrm{SL}_2\mathbb{C},$$
$$\mathrm{Spin}_4\mathbb{C} \cong \mathrm{SL}(S^+) \times \mathrm{SL}(S^-) = \mathrm{SL}_2\mathbb{C} \times \mathrm{SL}_2\mathbb{C},$$
$$\mathrm{Spin}_5\mathbb{C} \cong \mathrm{Sp}(S) = \mathrm{Sp}_4\mathbb{C},$$
$$\mathrm{Spin}_6\mathbb{C} \cong \mathrm{SL}(S^+) = \mathrm{SL}_4\mathbb{C}.$$

Exercise 20.40. Let C_m denote the Clifford algebra of the vector space $\mathbb{C}^m$ with our standard quadratic form Q_m.

(a) The embedding of $\mathbb{C}^{2n} = W \oplus W'$ in $\mathbb{C}^{2n+1} = W \oplus W' \oplus U$ as indicated induces an embedding of C_{2n} in C_{2n+1}, and corresponding embedding of $\mathrm{Spin}_{2n}\mathbb{C}$ in $\mathrm{Spin}_{2n+1}\mathbb{C}$ and of $\mathrm{SO}_{2n}\mathbb{C}$ in $\mathrm{SO}_{2n+1}\mathbb{C}$. Show that the spin representation S of $\mathrm{Spin}_{2n+1}\mathbb{C}$ restricts to the spin representation $S^+ \oplus S^-$ of $\mathrm{Spin}_{2n}\mathbb{C}$.

(b) Similarly there is an embedding of $\mathrm{Spin}_{2n+1}\mathbb{C}$ in $\mathrm{Spin}_{2n+2}\mathbb{C}$ coming from an embedding of $\mathbb{C}^{2n+1} = W \oplus W' \oplus U$ in $\mathbb{C}^{2n+2} = W \oplus W' \oplus U_1 \oplus U_2$; here $U_1 \oplus U_2 = \mathbb{C} \oplus \mathbb{C}$ with the quadratic form $\begin{pmatrix} 0 & 1 \\ 1 & 0 \end{pmatrix}$, and $U = \mathbb{C}$ is embedded in $U_1 \oplus U_2$ by sending 1 to $\left(\frac{1}{\sqrt{2}}, \frac{1}{\sqrt{2}}\right)$. Show that each of the half-spin representations of $\mathrm{Spin}_{2n+2}\mathbb{C}$ restricts to the spin representation of $\mathrm{Spin}_{2n+1}\mathbb{C}$.

Very little of the above discussion needs to be changed to construct the real spin groups $\mathrm{Spin}_m(\mathbb{R})$, which are double coverings of the real orthogonal groups $\mathrm{SO}_m(\mathbb{R})$. One uses the real Clifford algebra $\mathrm{Cliff}(\mathbb{R}^m, Q)$ associated to the real quadratic form $Q = -Q_m$, where Q_m is the standard positive definite quadratic form on $\mathbb{R}^m$. If v_i are an orthonormal basis, the products in this Clifford algebra are given by

$$v_i \cdot v_j = -v_j \cdot v_i \quad \text{if } i \neq j, \quad \text{and} \quad v_i \cdot v_i = -1.$$

The same definitions can be given as in the complex case, giving rise to coverings $\text{Pin}_m(\mathbb{R})$ of $O_m(\mathbb{R})$ and $\text{Spin}_m(\mathbb{R})$ of $SO_m(\mathbb{R})$.

Exercise 20.41. Show that $\text{Spin}_m\mathbb{R}$ is connected by showing that if v and w are any two perpendicular elements in V with $Q(v, v) = Q(w, w) = -1$, the path

$$t \mapsto (\cos(t)v + \sin(t)w)\cdot(\cos(t)v - \sin(t)w), \quad 0 \le t \le \pi/2$$

connects -1 to 1.

Exercise 20.42. Show that $i \mapsto v_2 \cdot v_3$, $j \mapsto v_3 \cdot v_1$, $k \mapsto v_1 \cdot v_2$ determines an isomorphism of the quaternions $\mathbb{H}$ onto the even part of $\text{Cliff}(\mathbb{R}^3, -Q_3)$, such that conjugation $^-$ in $\mathbb{H}$ corresponds to the conjugation $*$ in the Clifford algebra. Show that this maps $\text{Sp}(2) = \{q \in \mathbb{H} \mid q\bar{q} = 1\}$ isomorphically onto $\text{Spin}_3\mathbb{R}$, and that this isomorphism is compatible with the map to $SO_3\mathbb{R}$ defined in Exercise 7.15.

More generally, if Q is a quadratic form on $\mathbb{R}^m$ with p positive and q negative eigenvalues, we get a group $\text{Spin}^+(p, q)$ in the Clifford algebra $C(p, q) = \text{Cliff}(\mathbb{R}^m, Q)$, with double coverings

$$\text{Spin}^+(p, q) \to SO^+(p, q).$$

Exercise 20.43*. Show that $\text{Spin}^+(p, q)$ is connected if p and q are positive, except for the case $p = q = 1$, when it has two components. Show that if in the definition of spin groups one relaxes the condition $x \cdot x^* = 1$ to the condition $x \cdot x^* = \pm 1$, one gets coverings $\text{Spin}(p, q)$ of $SO(p, q)$.

§20.3. $\text{Spin}_8\mathbb{C}$ and Triality

When m is even, there is always an outer automorphism of $\text{Spin}_m(\mathbb{C})$ that interchanges the two spin representations S^+ and S^-, while preserving the basic representation $V = \mathbb{C}^m$ (cf. Exercise 19.9). In case $m = 8$, all three of these representations V, S^+, and S^- are eight dimensional. One basic expression of *triality* is the fact that there are automorphisms of $\text{Spin}_8\mathbb{C}$ or $\mathfrak{so}_8\mathbb{C}$ that permute these three representations arbitrarily. (In fact, the group of outer automorphisms modulo inner automorphisms is the symmetric group on three elements.) We give a brief discussion of this phenomenon in this section, in the form of an extended exercise.

To see where these automorphisms might come from, consider the four simple roots:

$$\alpha_1 = L_1 - L_2, \quad \alpha_2 = L_2 - L_3, \quad \alpha_3 = L_3 - L_4, \quad \alpha_4 = L_3 + L_4.$$

Note that α_1, α_3, and α_4 are mutually perpendicular, and that each makes an angle of 120° with α_2:

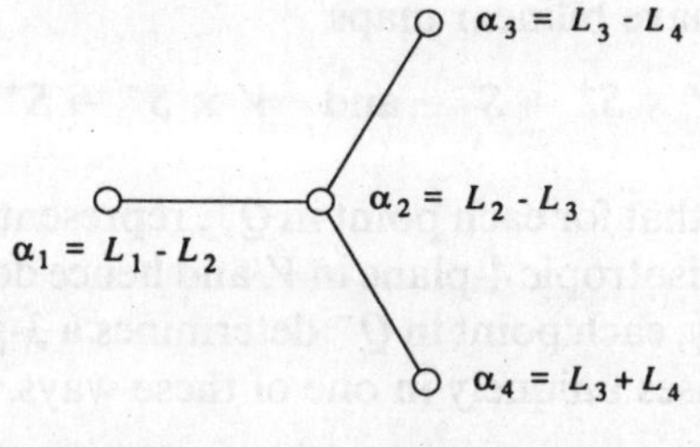

Exercise 20.44*. For each of the six permutations of $\{\alpha_1, \alpha_3, \alpha_4\}$ find the orthogonal automorphism of the root space which fixes α_2 and realizes the permutation of α_1, α_3, and α_4.

Each automorphism of this exercise corresponds to an automorphism of the Cartan subalgebra $\mathfrak{h}$. In the next lecture we will see that such automorphisms can be extended (nonuniquely) to automorphisms of the Lie algebra $\mathfrak{so}_8(\mathbb{C})$. (For explicit formulas see [Ca2].)

There is also a purely geometric notion of triality. Recall that an even-dimensional quadric Q can contain linear spaces Λ of at most half the dimension of Q, and that there are two families of linear spaces of this maximal dimension (cf. [G-H], [Ha]). In case Q is six-dimensional, each of these families can themselves be realized as six-dimensional quadrics, which we may denote by Q^+ and Q^- (see below). Moreover, there are correspondences that assign to a point of any one of these quadrics a 3-plane in each of the others:

$$\begin{array}{ccccc} & \text{Point in } Q & \longrightarrow & \text{3-plane in } Q^+ & \\ \swarrow & & & & \nwarrow \\ \text{3-plane in } Q^- & & & & \text{Point in } Q^- \\ \nwarrow & & & & \swarrow \\ & \text{Point in } Q^+ & \longrightarrow & \text{3-plane in } Q & \end{array} \tag{20.45}$$

Given $P \in Q$, $\{\Lambda \in Q^+ : \Lambda \text{ contains } P\}$ is a 3-plane in Q^+, and $\{\Lambda \in Q^- : \Lambda \text{ contains } P\}$ is a 3-plane in Q^-.

Given $\Lambda \in Q^+$, Λ itself is a 3-plane in Q, and $\{\Gamma \in Q^- : \Gamma \cap \Lambda \text{ is a 2-plane}\}$ is a 3-plane in Q^-.

Given $\Lambda \in Q^-$, Λ itself is a 3-plane in Q, and $\{\Gamma \in Q^+ : \Gamma \cap \Lambda \text{ is a 2-plane}\}$ is a 3-plane in Q^+.

To relate these two notions of triality, take Q to be our standard quadric in $\mathbb{P}^7 = \mathbb{P}(V)$, with $V = W \oplus W'$ with our usual quadratic space, and let $S^+ = \wedge^{\text{even}} W$ and $S^- = \wedge^{\text{odd}} W$ be the two spin representations. In Exercise 20.38 we constructed quadratic forms on S^+ and S^-, by choosing an isomorphism of $\wedge^4 W$ with $\mathbb{C}$. This gives us two quadrics Q^+ and Q^- in $\mathbb{P}(S^+)$ and $\mathbb{P}(S^-)$.

To identify Q^+ and Q^- with the families of 3-planes in Q, recall the action of V on $S = \wedge^\bullet W = S^+ \oplus S^-$ which gave rise to the isomorphism of the Clifford algebra with $\mathrm{End}(S)$ (cf. Lemma 20.9). This in fact maps S^+ to S^-

and S^- to S^+; so we have bilinear maps

$$V \times S^+ \dashrightarrow S^- \quad \text{and} \quad V \times S^- \dashrightarrow S^+. \tag{20.46}$$

Exercise 20.47. Show that for each point in Q^+, represented by a vector $s \in S^+$, $\{v \in V: v \cdot s = 0\}$ is an isotropic 4-plane in V, and hence determines a projective 3-plane in Q. Similarly, each point in Q^- determines a 3-plane in Q. Show that every 3-plane in Q arises uniquely in one of these ways.

Let $\langle\ ,\ \rangle_V$ denote the symmetric form corresponding to the quadratic form in V, and similarly for S^+ and S^-. Define a product

$$S^+ \times S^- \dashrightarrow V, \qquad s \times t \mapsto s \cdot t, \tag{20.48}$$

by requiring that $\langle v, s \cdot t \rangle_V = \langle v \cdot s, t \rangle_{S^-}$ for all $v \in V$.

Exercise 20.49. Use this product, together with those in (20.46), to show that the other four arrows in the hexagon (20.45) for geometric triality can be described as in the preceding exercise.

This leads to an algebraic version of triality, which we sketch following [Ch2]. The above products determine a commutative but nonassociative product on the direct sum $A = V \oplus S^+ \oplus S^-$. The operation

$$(v, s, t) \mapsto \langle v \cdot s, t \rangle_{S^-}$$

determines a cubic form on A, which by polarization determines a symmetric trilinear form Φ on A.

Exercise 20.50*. One can construct an automorphism J of A of order three that sends V to S^+, S^+ to S^-, and S^- to V, preserving their quadratic forms, and compatible with the cubic form. The definition of J depends on the choice of an element $v_1 \in V$ and $s_1 \in S^+$ with $\langle v_1, v_1 \rangle_V = \langle s_1, s_1 \rangle_{S^+} = 1$; set $t_1 = v_1 \cdot s_1$, so that $\langle t_1, t_1 \rangle_{S^-} = 1$ as well. The map J is defined to be the composite $\mu \circ \nu$ of two involutions μ and ν, which are determined by the following:

(i) μ interchanges S^+ and S^-, and maps V to itself, with $\mu(s) = v_1 \cdot s$ for $s \in S^+$; $\mu(v) = 2\langle v, v_1 \rangle_V v_1 - v$ for $v \in V$.
(ii) ν interchanges V and S^-, maps S^+ to itself, with $\nu(v) = v \cdot s_1$ for $v \in V$; $\nu(s) = 2\langle s, s_1 \rangle_{S^+} s_1 - s$ for $s \in S^+$.

Show that this J satisfies the asserted properties.

Exercise 20.51*. In this algebraic form, triality can be expressed by the assertion that there is an automorphism j of $\mathrm{Spin}_8\mathbb{C}$ of order 3 compatible with J, i.e., such that for all $x \in \mathrm{Spin}_8\mathbb{C}$, the following diagrams commute:

$$\begin{array}{ccccccc}
V & \xrightarrow{J} & S^+ & \xrightarrow{J} & S^- & \xrightarrow{J} & V \\
\Big\downarrow{\scriptstyle \rho(x)} & & \Big\downarrow{\scriptstyle \rho^+(j(x))} & & \Big\downarrow{\scriptstyle \rho^-(j^2(x))} & & \Big\downarrow{\scriptstyle \rho(x)} \\
V & \xrightarrow[J]{} & S^+ & \xrightarrow[J]{} & S^- & \xrightarrow[J]{} & V
\end{array}$$

If $j'\colon \mathfrak{so}_8\mathbb{C} \to \mathfrak{so}_8\mathbb{C}$ is the map induced by j, the fact that j is compatible with the trilinear form Φ (cf. Exercise 20.49) translates to the "local triality" equation

$$\Phi(Xv, s, t) + \Phi(v, Ys, t) + \Phi(v, s, Zt) = 0$$

for $X \in \mathfrak{so}_8\mathbb{C}$, $Y = j'(X)$, $Z = j'(Y)$.

PART IV

LIE THEORY

The purpose of this final part of the book is threefold.

First of all, we want to complete the program stated in the introduction to Part II. We have completed the first two steps of this program, showing in Part II how the analysis of representations of Lie groups could be reduced to the study of representations of complex Lie algebras, of which the most important are the semisimple; and carrying out in Part III such an analysis for the classical Lie algebras $\mathfrak{sl}_n\mathbb{C}$, $\mathfrak{sp}_{2n}\mathbb{C}$, and $\mathfrak{so}_m\mathbb{C}$. To finish the story, we want now to translate our answers back into the terms of the original problem. In particular, we want to deal with representations of Lie groups as well as Lie algebras, and real groups and algebras as well as complex. The passage back to groups is described in Lecture 21, and the analysis of the real case in Lecture 26.

Another goal of this Part is to establish a framework for some of the results of the preceding lectures—to describe the general theory of semisimple Lie algebras and Lie groups. The key point here is the introduction of the Dynkin diagram and its use in classifying all semisimple Lie algebras over $\mathbb{C}$. From one point of view, the impact of the classification theorem is not great: it just tells us that we have in fact already analyzed all but five of the simple Lie algebras in existence. Beyond that, however, it provides a picture and a language for the description of the general Lie algebra. This both yields a description of the five remaining simple Lie algebras and allows us to give uniform descriptions of associated objects: for example, the compact homogeneous spaces associated to simple Lie groups, or the characters of their representations. The classification theory of semisimple Lie algebras is given in Lecture 21; the description in these terms of their representations and characters is given in Lecture 23. The five exceptional simple Lie algebras, whose existence is revealed from the Dynkin diagrams, are studied in Lecture

22; we give a fairly detailed account of one of them ($\mathfrak{g}_2$), with only brief descriptions of the others.

Third, all this general theory makes it possible to answer the main outstanding problem left over from Part III: a description of the multiplicities of the weights in the irreducible representations of the simple Lie algebras. We give in Lectures 24 and 25 a number of formulas for these multiplicities.

This, it should be said, represents in some ways a shift in style. In the previous lectures we would typically analyze special cases first and deduce general patterns from these cases; here, for example, the Weyl character formula is stated and proved in general, then specialized to the various individual cases (this is the approach more often taken in the literature on the subject). In some ways, this is a fourth goal of Part IV: to provide a bridge between the naive exploration of Lie theory undertaken in Parts II and III, and the more general theory readers will find elsewhere when they pursue the subject further.

Finally, we should repeat here the disclaimer made in the Preface. This part of the book, to the extent that it is successful, will introduce the reader to the rich and varied world of Lie theory; but it certainly undertakes no serious exploration of that world. We do not, for example, touch on such basic constructions as the universal enveloping algebra, Verma modules, Tits buildings; and we do not even hint at the fascinating subject of (infinite-dimensional) unitary representations. The reader is encouraged to sample these and other topics, as well as those included here, according to background and interest.

LECTURE 21

The Classification of Complex Simple Lie Algebras

In the first section of this lecture we introduce the Dynkin diagram associated to a semisimple Lie algebra $\mathfrak{g}$. This is an amazingly efficient way of conveying the structure of $\mathfrak{g}$: it is a simple diagram that not only determines $\mathfrak{g}$ up to isomorphism in theory, but in practice exhibits many of the properties of $\mathfrak{g}$. The main use of Dynkin diagrams in this lecture, however, will be to provide a framework for the basic classification theorem, which says that with exactly five exceptions the Lie algebras discussed so far in these lectures are all the simple Lie algebras. To do this, in §21.2 we show how to list all diagrams that arise from semisimple Lie algebras. In §21.3 we show how to recover such a Lie algebra from the data of its diagram, completing the proof of the classification theorem. All three sections are completely elementary, though §21.3 gets a little complicated; it may be useful to read it in conjunction with §22.1, where the process described is carried out in detail for the exceptional algebra $\mathfrak{g}_2$. (Note that neither §21.3 or §22.1 is a prerequisite for §22.3, where another description of $\mathfrak{g}_2$ will be given.)

§21.1. Dynkin Diagrams Associated to Semisimple Lie Algebras

For the following, we will let $\mathfrak{g}$ be a semisimple Lie algebra; as usual, a Cartan subalgebra $\mathfrak{h}$ of $\mathfrak{g}$ will be fixed throughout. As we have seen, the roots R of $\mathfrak{g}$ span a real subspace of $\mathfrak{h}^*$ on which the Killing form is positive definite. We denote this Euclidean space here by $\mathbb{E}$, and the Killing form on $\mathbb{E}$ simply by

$(\ , \ )$ instead of $B(\ , \ )$. The geometry of how R sits in $\mathbb{E}$ is very rigid, as indicated by the pictures we have seen for the classical Lie algebras. In this section we will classify the possible configurations, up to rotation and multiplication by a positive scalar in $\mathbb{E}$. In the next section we will see that this geometry completely determines the Lie algebra.

The following four properties of the root system are all that are needed:

(1) *R is a finite set spanning $\mathbb{E}$.*
(2) *$\alpha \in R \Rightarrow -\alpha \in R$, but $k \cdot \alpha$ is not in R if k is any real number other than ± 1.*
(3) *For $\alpha \in R$, the reflection W_α in the hyperplane $\alpha^\perp$ maps R to itself.*
(4) *For $\alpha, \beta \in R$, the real number*

$$n_{\beta\alpha} = 2\frac{(\beta, \alpha)}{(\alpha, \alpha)}$$

is an integer.

Except perhaps for the second part of (2), these properties have been seen in Lecture 14. For example, (4) is Corollary 14.29. Note that $n_{\beta\alpha} = \beta(H_\alpha)$, and

$$W_\alpha(\beta) = \beta - n_{\beta\alpha}\alpha. \tag{21.1}$$

For (2), consider the representation $\mathfrak{i} = \bigoplus_k \mathfrak{g}_{k\alpha}$ of the Lie algebra $\mathfrak{s}_\alpha \cong \mathfrak{sl}_2\mathbb{C}$. Note that all the nonzero factors but $\mathfrak{h} = \mathfrak{g}_0$ are one dimensional. We may assume α is the smallest nonzero root that appears in the string. Now, decompose $\mathfrak{i}$ as an $\mathfrak{s}_\alpha$-module:

$$\mathfrak{i} = \mathfrak{s}_\alpha \oplus \mathfrak{i}'.$$

By the hypothesis that α is the smallest nonzero root that appears in the string, $\mathfrak{i}'$ is a representation of $\mathfrak{s}_\alpha$ having no eigenspace with eigenvalue 1 or 2 for H_α. It follows that $\mathfrak{i}'$ must be trivial, i.e., $\mathfrak{g}_{k\alpha} = (0)$ for $k \neq 0$ or ± 1.

Any set R of elements in a Euclidean space $\mathbb{E}$ satisfying conditions (1) to (4) may be called an (*abstract*) *root system.*

Property (4) puts very strong restrictions on the geometry of the roots. If ϑ is the angle between α and β, we have

$$n_{\beta\alpha} = 2\cos(\vartheta)\frac{\|\beta\|}{\|\alpha\|}. \tag{21.2}$$

In particular,

$$n_{\alpha\beta}n_{\beta\alpha} = 4\cos^2(\vartheta) \tag{21.3}$$

is an integer between 0 and 4. The case when this integer is 4 occurs when $\cos(\vartheta) = \pm 1$, i.e. $\beta = \pm\alpha$. Omitting this trivial case, the only possibilities are therefore those given in the following table. Here we have ordered the two roots so that $\|\beta\| \geq \|\alpha\|$, or $|n_{\beta\alpha}| \geq |n_{\alpha\beta}|$.

Table 21.4

$\cos(\vartheta)$	$\sqrt{3}/2$	$\sqrt{2}/2$	$1/2$	0	$-1/2$	$-\sqrt{2}/2$	$-\sqrt{3}/2$
ϑ	$\pi/6$	$\pi/4$	$\pi/3$	$\pi/2$	$2\pi/3$	$3\pi/4$	$5\pi/6$
$n_{\beta\alpha}$	3	2	1	0	-1	-2	-3
$n_{\alpha\beta}$	1	1	1	0	-1	-1	-1
$\frac{\\|\beta\\|}{\\|\alpha\\|}$	$\sqrt{3}$	$\sqrt{2}$	1	*	1	$\sqrt{2}$	$\sqrt{3}$

In other words, the relation of any two roots α and β is one of

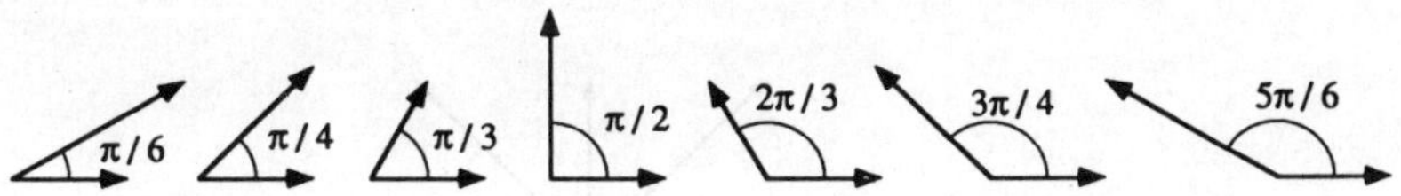

The dimension $n = \dim_{\mathbb{R}} \mathbb{E} = \dim_{\mathbb{C}} \mathfrak{h}$ is called the *rank* (of the Lie algebra, or the root system). It is easy to find all those of smallest ranks. As we write them down, we will label them by the labels (A_n), (B_n), ... that have become standard.

Rank 1. The only possibility is

(A_1)

which is the root system of $\mathfrak{sl}_2\mathbb{C}$.

Rank 2. Note first that by Property (3), the angle between two roots must be the same for any pair of adjacent roots in a two-dimensional root system. As we will see, any of the four angles $\pi/2$, $\pi/3$, $\pi/4$, and $\pi/6$ can occur; once this angle is specified the relative lengths of the roots are determined by Property (4), except in the case of right angles. Thus, up to scalars there are exactly four root systems of dimension two. First we have the case $\vartheta = \pi/2$,

$(A_1 \times A_1)$

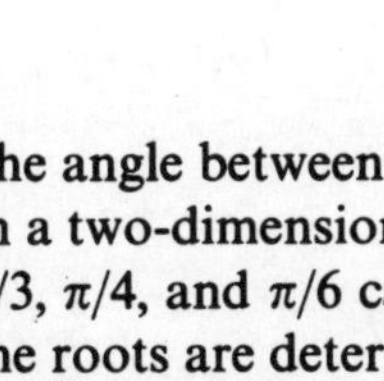

which is the root system of $\mathfrak{sl}_2\mathbb{C} \times \mathfrak{sl}_2\mathbb{C} \cong \mathfrak{so}_4\mathbb{C}$.

(In general, the orthogonal direct sum of two root systems is a root system;

a root system that is not such a sum is called *irreducible*. Our task will be to classify all irreducible root systems.)

The other root systems of rank 2 are

(A_2)

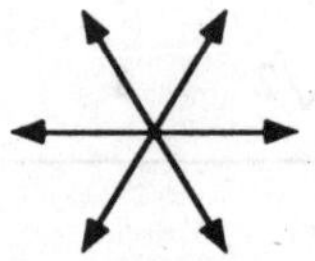

the root system of $\mathfrak{sl}_3\mathbb{C}$;

(B_2)

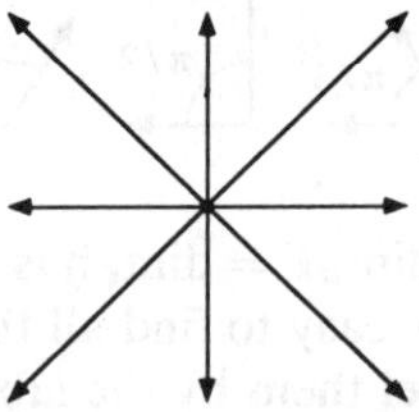

the root system of $\mathfrak{so}_5\mathbb{C} \cong \mathfrak{sp}_4\mathbb{C}$; and

(G_2)

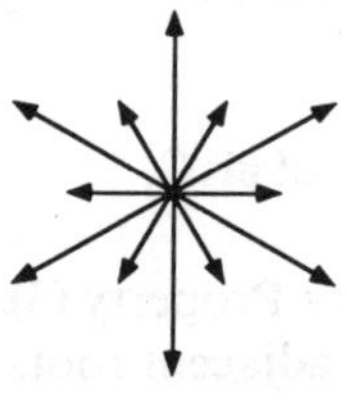

Although we have not yet seen a Lie algebra with this root system, we will see that there is one.

Exercise 21.5. Show that these are all the root systems of rank 2.

Exercise 21.6. Show that a semisimple Lie algebra is simple if and only if its root system is irreducible.

Rank 3. Besides the direct sums of (A_1) with one of those of rank 2, we have the irreducible root systems we have seen; we draw only dots at the ends of the vectors, the origins being in the centers of the reference cubes:

(A_3)

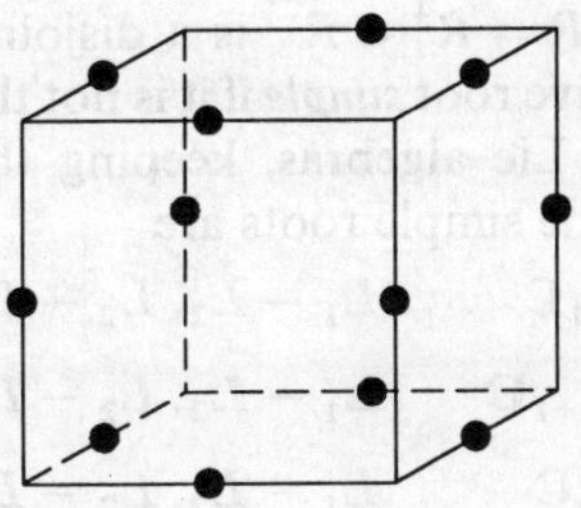

which is the root system of $\mathfrak{sl}_4\mathbb{C} \cong \mathfrak{so}_6\mathbb{C}$;

(B_3)

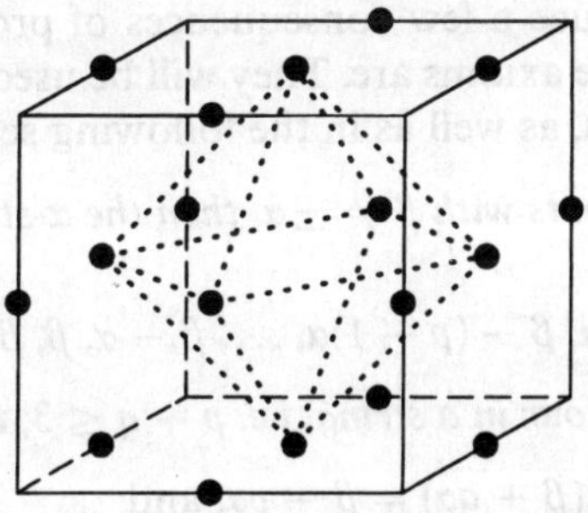

the root system of $\mathfrak{so}_7\mathbb{C}$;

(C_3)

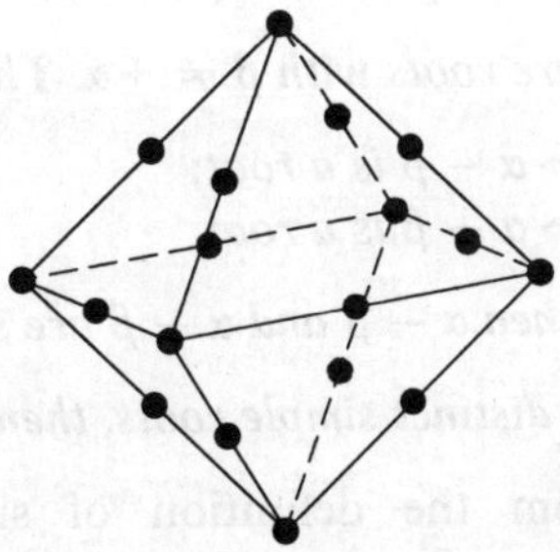

the root system of $\mathfrak{sp}_6\mathbb{C}$.

Exercise 21.7. Show that there are no other root systems of rank 3.

We can further reduce the data of a root system by introducing a subset of the roots, called the simple roots. First, choose as in Lecture 14 a direction

$l: \mathbb{E} \to \mathbb{R}$, so that $R = R^+ \cup R^-$ is a disjoint union of positive and negative roots. Call a positive root *simple* if it is not the sum of two other positive roots. For the classical Lie algebras, keeping the notations and conventions of Lectures 15–20, the simple roots are

(A_n)	$\mathfrak{sl}_{n+1}\mathbb{C}$	$L_1 - L_2, L_2 - L_3, \ldots, L_{n-1} - L_n, L_n - L_{n+1}$,
(B_n)	$\mathfrak{so}_{2n+1}\mathbb{C}$	$L_1 - L_2, L_2 - L_3, \ldots, L_{n-1} - L_n, L_n$,
(C_n)	$\mathfrak{sp}_{2n}\mathbb{C}$	$L_1 - L_2, L_2 - L_3, \ldots, L_{n-1} - L_n, 2L_n$,
(D_n)	$\mathfrak{so}_{2n}\mathbb{C}$	$L_1 - L_2, L_2 - L_3, \ldots, L_{n-1} - L_n, L_{n-1} + L_n$.

Exercise 21.8. Verify this list, and find two simple roots for (G_2).

We next deduce a few consequences of properties (1)–(4), which indicate how strong these axioms are. They will be used in the present classification of abstract systems, as well as in the following section.

(5) *If α, β are roots with $\beta \neq \pm\alpha$, then the α-string through β, i.e., the roots of the form*

$$\beta - p\alpha, \beta - (p-1)\alpha, \ldots, \beta - \alpha, \beta, \beta + \alpha, \beta + 2\alpha, \ldots, \beta + q\alpha$$

has at most four in a string, i.e. $p + q \leq 3$; in addition, $p - q = n_{\beta\alpha}$.

Indeed, since $W_\alpha(\beta + q\alpha) = \beta - p\alpha$, and

$$W_\alpha(\beta + q\alpha) = (\beta - n_{\beta\alpha}\alpha) - q\alpha,$$

we must have $p = n_{\beta\alpha} + q$, which is the second equality. For the first, we may take $p = 0$, and then $q = -n_{\beta\alpha}$, which we have seen is an integer no larger than three. As a consequence of (5) we have

(6) *Suppose α, β are roots with $\beta \neq \pm\alpha$. Then*

$$(\beta, \alpha) > 0 \Rightarrow \alpha - \beta \text{ is a root;}$$
$$(\beta, \alpha) < 0 \Rightarrow \alpha + \beta \text{ is a root.}$$

If $(\beta, \alpha) = 0$, then $\alpha - \beta$ and $\alpha + \beta$ are simultaneously roots or nonroots.

(7) *If α and β are distinct simple roots, then $\alpha - \beta$ and $\beta - \alpha$ are not roots.*

This follows from the definition of simple, since from the equation $\alpha = \beta + (\alpha - \beta)$, $\alpha - \beta$ cannot be in R^+, and similarly $-(\alpha - \beta) = \beta - \alpha$ cannot be in R^+. From (6) and (7) we deduce that $(\alpha, \beta) \leq 0$, i.e.,

(8) *The angle between two distinct simple roots cannot be acute.*

(9) *The simple roots are linearly independent.*

This follows from (8) by

Exercise 21.9*. If a set of vectors lies on one side of a hyperplane, with all mutual angles at least 90°, show that they must be linearly independent.

(10) *There are precisely n simple roots. Each positive root can be written uniquely as a non-negative integral linear combination of simple roots.*

Since R spans $\mathbb{E}$, the first statement follows from (9), as does the uniqueness of the second statement. The fact that any positive root can be written as a positive sum of simple roots follows readily from the definition, for if α were a positive root with minimal $l(\alpha)$ that could not be so written, then α is not simple, so $\alpha = \beta + \gamma$, with β and γ positive roots with $l(\beta), l(\gamma) < l(\alpha)$.

Note that as an immediate corollary of (10) it follows that *no root is a linear combination of the simple roots α_i with coefficients of mixed sign.* For example, (7) is just a special case of this.

The *Dynkin diagram* of the root system is drawn by drawing one node ○ for each simple root and joining two nodes by a number of lines depending on the angle ϑ between them:

no lines		if $\vartheta = \pi/2$
one line		if $\vartheta = 2\pi/3$
two lines	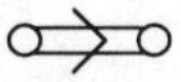	if $\vartheta = 3\pi/4$
three lines	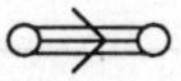	if $\vartheta = 5\pi/6$.

When there is one line, the roots have the same length; if two or three lines, an arrow is drawn pointing from the *longer* to the *shorter* root.

Exercise 21.10. Show that a root system is irreducible if and only if its Dynkin diagram is connected.

We will see later that the Dynkin diagram of a root system is independent of the choice of direction, i.e., of the decomposition of R into R^+ and R^-.

§21.2. Classifying Dynkin Diagrams

The wonderful thing about Dynkin diagrams is that from this very simple picture one can reconstruct the entire Lie algebra from which it came. We will see this in the following section; for now, we ask the complementary question of which diagrams arise from Lie algebras. Our goal is the following classification theorem, which is a result in pure Euclidean geometry. (The subscripts on the labels (A_n), ... are the number of nodes.)

Theorem 21.11. *The Dynkin diagrams of irreducible root systems are precisely:*

(A_n)  $(n \geq 1)$

(B_n)  $(n \geq 2)$

(C_n)  $(n \geq 3)$

(D_n) 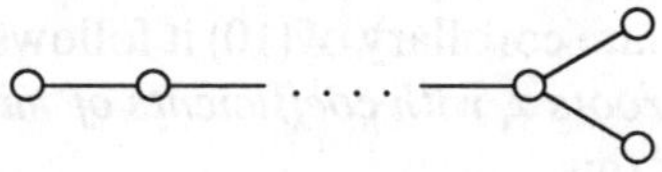 $(n \geq 4)$

(E_6) 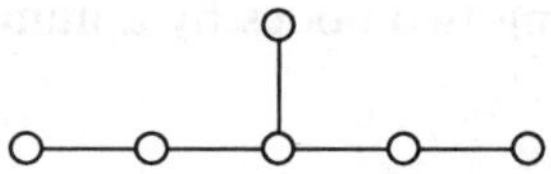

(E_7) 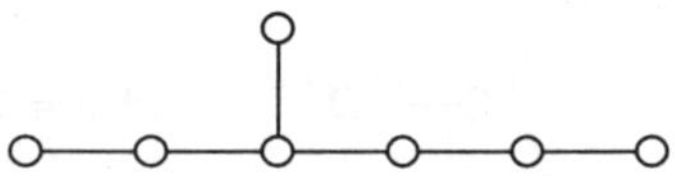

(E_8) 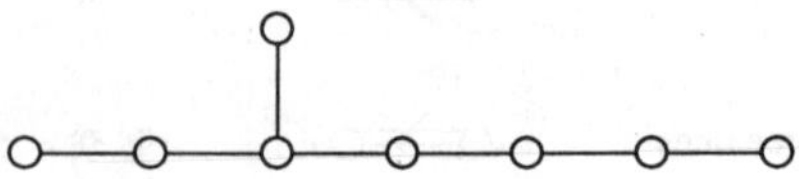

(F_4) 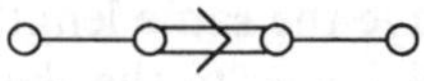

(G_2) 

The first four are those belonging to the classical series we have been studying:

$$
\begin{array}{ll}
(A_n) & \mathfrak{sl}_{n+1}\mathbb{C} \\
(B_n) & \mathfrak{so}_{2n+1}\mathbb{C} \\
(C_n) & \mathfrak{sp}_{2n}\mathbb{C} \\
(D_n) & \mathfrak{so}_{2n}\mathbb{C}
\end{array}
$$

The restrictions on n in these series are to avoid repeats, as well as degenerate cases. Indeed, the diagrams can be used to recall all the coincidences we have seen:

When $n = 1$, all four of the diagrams become one node. The case (D_1) is degenerate, since $\mathfrak{so}_2\mathbb{C}$ is not semisimple, while the coincidences $(C_1) = (B_1) = (A_1)$ correspond to the isomorphisms

$$\mathfrak{sp}_2\mathbb{C} \cong \mathfrak{so}_3\mathbb{C} \cong \mathfrak{sl}_2\mathbb{C} \qquad \circ .$$

For $n = 2$, $(D_2) = (A_1) \times (A_1)$ consists of two disjoint nodes, corresponding to the isomorphism

$$\mathfrak{so}_4\mathbb{C} \cong \mathfrak{sl}_2\mathbb{C} \times \mathfrak{sl}_2\mathbb{C}$$ 

The coincidence $(C_2) = (B_2)$ corresponds to the isomorphism

$$\mathfrak{sp}_4\mathbb{C} \cong \mathfrak{so}_5\mathbb{C}$$ 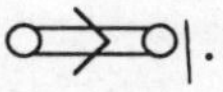

For $n = 3$, the fact that $(D_3) = (A_3)$ reflects the isomorphism

$$\mathfrak{so}_6\mathbb{C} \cong \mathfrak{sl}_4\mathbb{C}$$ 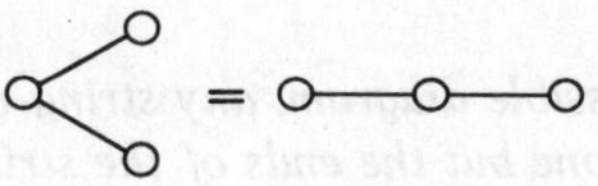

Proof of the Theorem. Our desert-island reader would find this a pleasant pastime. For example, if there are two simple roots with angle $5\pi/6$, the plane of these roots must contain the G_2 configuration of 12 roots. It is not hard to see that one cannot add another root that is not perpendicular to this plane, without some of the 12 angles and lengths being wrong. This shows that (G_2) is the only connected diagram containing a triple line. At the risk of spoiling your fun, we give the general proof of a slightly stronger result.

In fact, the angles alone determine the possible diagrams. Such diagrams, without the arrows to indicate relative lengths, are often called *Coxeter diagrams* (or Coxeter graphs). Define a diagram of n nodes, with each pair connected by 0, 1, 2, or 3 lines, to be *admissible* if there are n independent unit vectors $e_1, \ldots, e_n$ in a Euclidean space $\mathbb{E}$ with the angle between e_i and e_j being $\pi/2$, $2\pi/3$, $3\pi/4$, or $5\pi/6$, according as the number of lines between corresponding nodes is 0, 1, 2, or 3. The claim is that the diagrams of the above Dynkin diagrams, ignoring the arrows, are the only connected admissible diagrams. Note that

$$(e_i, e_j) = 0, -1/2, -\sqrt{2}/2, \text{or } -\sqrt{3}/2, \tag{21.12}$$

according as the number of lines between them is 0, 1, 2, or 3; equivalently,

$$4(e_i, e_j)^2 = \text{number of lines between } e_i \text{ and } e_j. \tag{21.13}$$

The steps of the proof are as follows:

(i) *Any subdiagram of an admissible diagram, obtained by removing some nodes and all lines to them, will also be admissible.*

(ii) *There are at most $n - 1$ pairs of nodes that are connected by lines. The diagram has no cycles (loops).*

Indeed, if e_i and e_j are connected, $2(e_i, e_j) \le -1$, and

$$0 < \left(\sum e_i, \sum e_i\right) = n + 2 \sum_{i<j} (e_i, e_j),$$

which proves the first statement of (ii). The second follows from the first and (i).

(iii) *No node has more than three lines to it.*

By (i), we may assume that e_1 is connected to each of the other nodes; by (ii), no other nodes are connected to each other. We must show that $\sum_{j=2}^{n} 4(e_1, e_j)^2 < 4$. Since $e_2, \ldots, e_n$ are perpendicular unit vectors, and e_1 is not in their span,

$$1 = (e_1, e_1)^2 > \sum_{j=2}^{n} (e_1, e_j)^2,$$

as required.

(iv) *In an admissible diagram, any string of nodes connected to each other by one line, with none but the ends of the string connected to any other nodes, can be collapsed to one node, and resulting diagram remains admissible*:

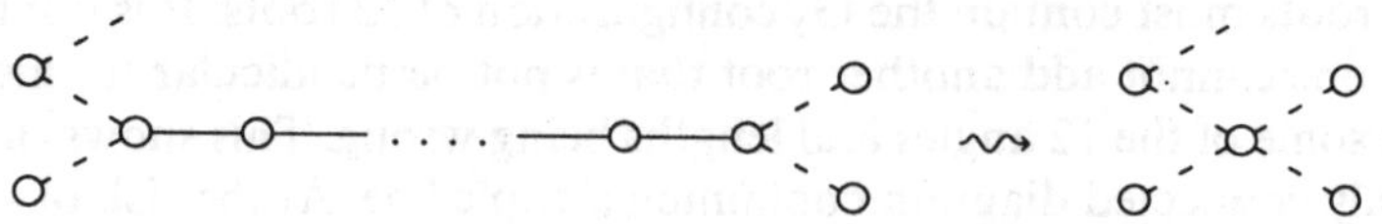

If $e_1, \ldots, e_r$ are the unit vectors corresponding to the string of nodes, then $e' = e_1 + \cdots + e_r$ is a unit vector, since

$$\begin{aligned}(e', e') &= r + 2((e_1, e_2) + (e_2, e_3) + \cdots + (e_{r-1}, e_r)) \\ &= r - (r - 1).\end{aligned}$$

Moreover, e' satisfies the same conditions with respect to the other vectors since (e', e_j) is either (e_1, e_j) or (e_r, e_j).

Now we can rule out the other admissible connected diagrams not on our list. First, from (iii) we see that the diagram (G_2) has the only triple edge. Next, there cannot be two double lines, or we could find a subdiagram of the form:

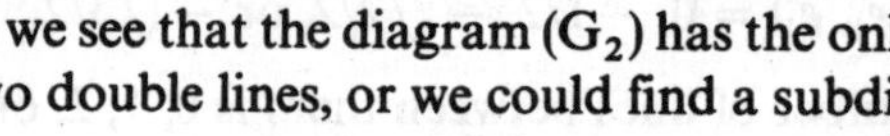

and then collapse the middle to get , contradicting (iii). Similarly there can be at most one triple node, i.e., a node with single lines to three other nodes, by

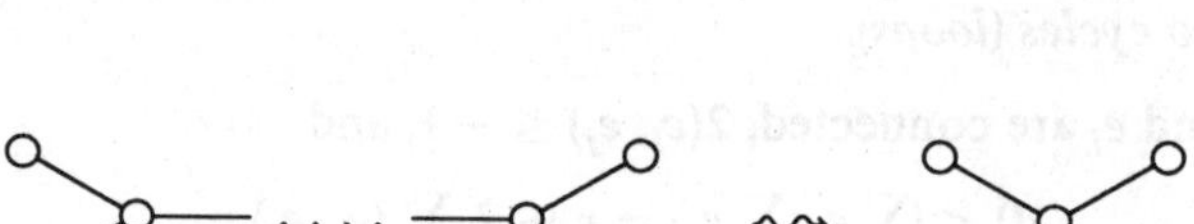

By the same reasoning, there cannot be a triple node together with a double line:

To finish the case with double lines, we must simply verify that

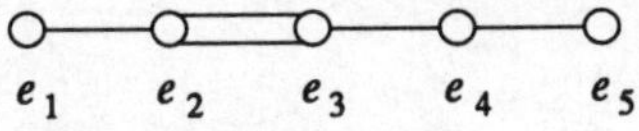

is not admissible. Consider general vectors $v = a_1e_1 + a_2e_2$, and $w = a_3e_3 + a_4e_4 + a_5e_5$. We have

$$\|v\|^2 = a_1^2 + a_2^2 - a_1a_2, \qquad \|w\|^2 = a_3^2 + a_4^2 + a_5^2 - a_3a_4 - a_4a_5,$$

and $(v, w) = -a_2a_3/\sqrt{2}$. We want to choose v and w to contradict the Cauchy–Schwarz inequality $(v, w)^2 < \|v\|^2\|w\|^2$. For this we want $|a_2|/\|v\|$ and $|a_3|/\|w\|$ to be as large as possible.

Exercise 21.14. Show that these maxima are achieved by taking $a_2 = 2a_1$ and $a_3 = 3a_5$, $a_4 = 2a_5$.

In fact, $v = e_1 + 2e_2$, $w = 3e_3 + 2e_4 + e_5$ do give the contradictory

$$(v, w)^2 = 18, \quad \|v\|^2 = 3, \quad \text{and} \quad \|w\|^2 = 6.$$

Finally, we must show that the strings coming out from a triple node cannot be longer that those specified in types (D_n), (E_6), (E_7), or (E_8). First, we rule out

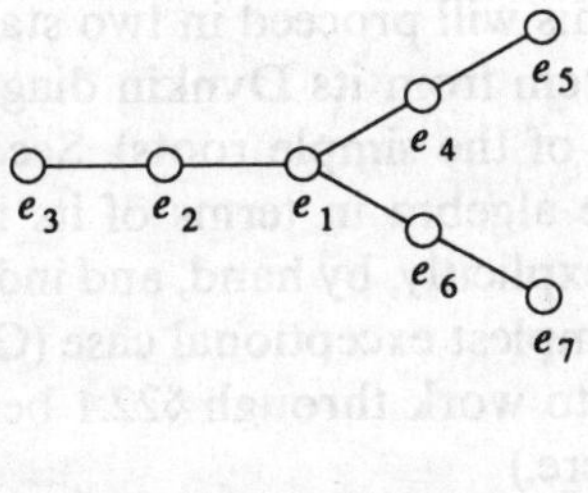

Consider the three perpendicular unit vectors:

$$u = (2e_2 + e_3)/\sqrt{3}, \qquad v = (2e_4 + e_5)/\sqrt{3}, \qquad w = (2e_6 + e_7)/\sqrt{3}.$$

Then as in (iii), since e_1 is not in the span of them,

$$1 = \|e_1\|^2 > (e_1, u)^2 + (e_1, v)^2 + (e_1, w)^2 = 1/3 + 1/3 + 1/3 = 1,$$

a contradiction.

Exercise 21.15*. Similarly, rule out

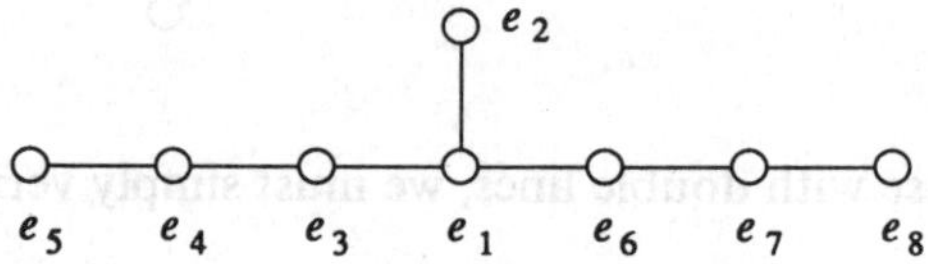

and

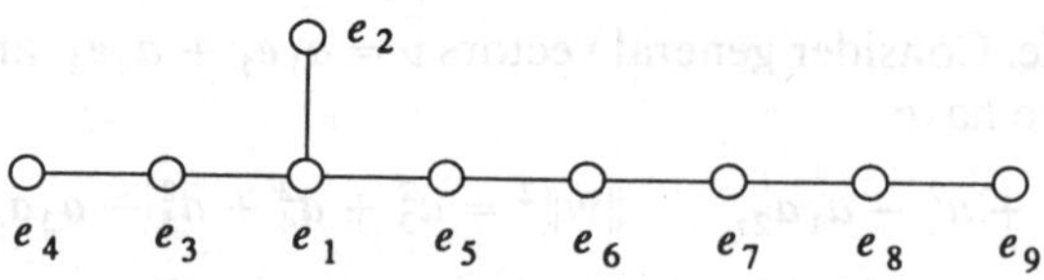

(The last few arguments can be amalgamated, by showing that if the legs from a triple node have lengths p, q, and r, then $1/p + 1/q + 1/r$ must be greater than 1.)

This finishes the proof of the theorem. □

§21.3. Recovering a Lie Algebra from Its Dynkin Diagram

In this section we will complete the classification theorem for simple Lie algebras by showing how one may recover a simple Lie algebra from the data of its Dynkin diagram. This will proceed in two stages: first, we will see how to reconstruct a root system from its Dynkin diagram (which a priori only tells us the configuration of the simple roots). Secondly, we will show how to describe the entire Lie algebra in terms of its root system. (In the next lecture we will do all this explicitly, by hand, and independently of the general discussion here, for the simplest exceptional case (G_2); as we have noted, the reader may find it useful to work through §22.1 before or while reading the general story described here.)

To begin with, to recover the root system from the Dynkin diagram, let $\alpha_1, \ldots, \alpha_n$ be the simple roots corresponding to the nodes of a connected Dynkin diagram. We must show which non-negative integral linear combinations $\sum m_i \alpha_i$ are roots. Call $\sum m_i$ the *level* of $\sum m_i \alpha_i$. Those of level one are the simple roots. For level two, we see from Property (2) that no $2\alpha_i$ is a root, and by

Property (6) that $\alpha_i + \alpha_j$ is a root precisely when $(\alpha_i, \alpha_j) < 0$, i.e., when the corresponding nodes are joined by a line.

Suppose we know all positive roots of level at most m, and let $\beta = \sum m_i \alpha_i$ be any positive root of level m. We next determine for each simple root $\alpha = \alpha_j$, whether $\beta + \alpha$ is also a root. Look at the α-string through β:

$$\beta - p\alpha, \ldots, \beta, \ldots, \beta + q\alpha.$$

We know p by induction (no root is a linear combination of the simple roots α_i with coefficients of mixed sign, so $p \leq m_j$ and $\beta - p\alpha$ is a positive root). By Property (5), $q = p - n_{\beta\alpha}$. So $\beta + \alpha$ is a root exactly when

$$p > n_{\beta\alpha} = 2\frac{(\beta, \alpha)}{(\alpha, \alpha)} = \sum_{i=1}^{n} m_i n_{\alpha_i \alpha}.$$

In effect, the additional roots we will find in this way are those obtained by reflecting a known positive root in the hyperplane perpendicular to a simple root α_i (and filling in the string if necessary).

To finish the proof, we must show that we get all the positive roots in this way. This will follow once from the fact that any positive root of level $m + 1$ can be written in at least one way as a sum of a positive root of level m and a simple root. If $\gamma = \sum r_i \alpha_i$ has level $m + 1$, from

$$0 < (\gamma, \gamma) = \sum r_i (\gamma, \alpha_i),$$

some (γ, α_i) must be positive, with $r_i > 0$. By property (6), $\gamma - \alpha_i$ is a root, as required.

By way of example, consider the rank 2 root systems. In the case of $\mathfrak{sl}_3\mathbb{C}$, we start with a pair of simple roots α_1, α_2 with $n_{\alpha_1, \alpha_2} = -1$, i.e., at an angle of $2\pi/3$; as always, we know that $\beta = \alpha_1 + \alpha_2$ is a root as well.

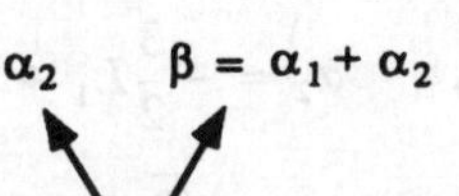

On the other hand, since $\beta - 2\alpha_1 = \alpha_2 - \alpha_1$ is not a root, $\beta + \alpha_1$ cannot be either, and likewise $\beta + \alpha_2$ is not; so we have all the positive roots.

In the case of $\mathfrak{sp}_4\mathbb{C}$, we have two simple roots α_1 and α_2 at an angle of $3\pi/4$; in terms of an orthonormal basis L_1 and L_2 these may be taken to be L_1 and $L_2 - L_1$, respectively.

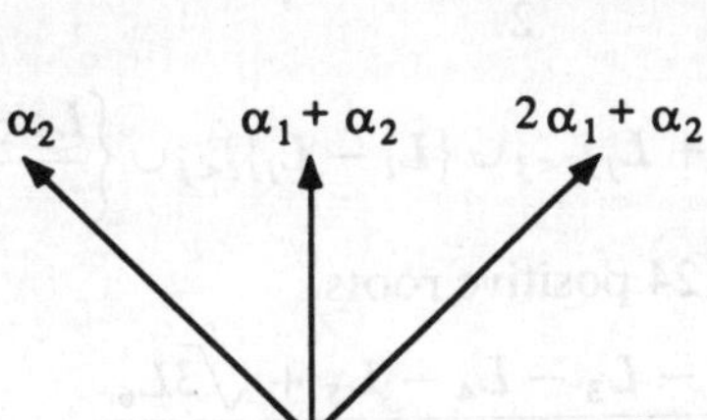

We then see that in addition to $\beta = \alpha_1 + \alpha_2$, the sum $\beta + \alpha_1 = 2\alpha_1 + \alpha_2$ is a root—it is just the reflection of α_2 in the plane perpendicular to α_1—but $\beta + \alpha_2 = \alpha_1 + 2\alpha_2$ and $3\alpha_1 + \alpha_2$ are not because $\alpha_1 - \alpha_2$ and $\alpha_2 - \alpha_1$ are not respectively (alternatively, we could note that they would form inadmissible angles with α_1 and α_2 respectively).

Finally, in the case of (G_2), we have two simple roots α_1, α_2 at an angle of $5\pi/6$, which in terms of an orthonormal basis for $\mathbb{E}$ may be taken to be L_1 and $(-3L_1 + \sqrt{3}L_2)/2$ respectively.

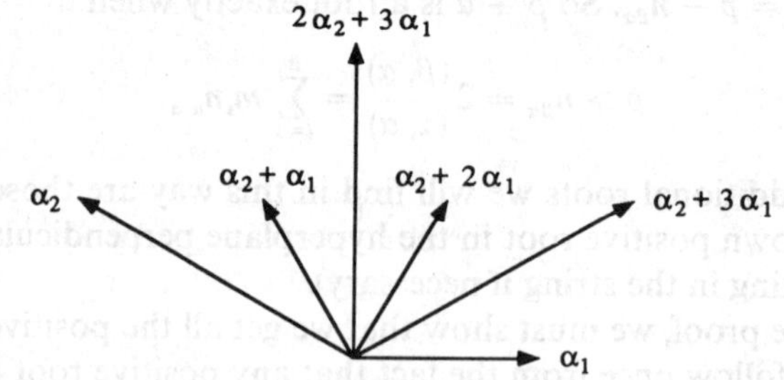

Reflecting α_2 in the plane perpendicular to α_1 yields a string of roots $\alpha_2 + \alpha_1$, $\alpha_2 + 2\alpha_1$ and $\alpha_2 + 3\alpha_2$. Moreover, reflecting the last of these in the plane perpendicular to α_2 yields one more root, $2\alpha_2 + 3\alpha_3$. Finally, these are all the positive roots, giving us the root system for the diagram (G_2).

We state here the results of applying this process to the exceptional diagrams (F_4), (E_6), (E_7), and (E_8) (in addition to (G_2)). In each case, $L_1, \ldots, L_n$ is an orthogonal basis for $\mathbb{E}$, the simple roots α_i can be taken to be as follows, and the corresponding root systems are given:

(G_2)
$$\alpha_1 = L_1, \quad \alpha_2 = -\frac{3}{2}L_1 + \frac{\sqrt{3}}{2}L_2;$$

$$R^+ = \left\{L_1, \sqrt{3}L_2, \pm L_1 + \frac{\sqrt{3}}{2}L_2, \pm\frac{3}{2}L_1 + \frac{\sqrt{3}}{2}L_2\right\}.$$

(G_2) thus has 6 positive roots.

(F_4)
$$\alpha_1 = L_2 - L_3, \quad \alpha_2 = L_3 - L_4, \quad \alpha_3 = L_4,$$
$$\alpha_4 = \frac{L_1 - L_2 - L_3 - L_4}{2};$$

$$R^+ = \{L_i\} \cup \{L_i + L_j\}_{i<j} \cup \{L_i - L_j\}_{i<j} \cup \left\{\frac{L_1 \pm L_2 \pm L_3 \pm L_4}{2}\right\}.$$

In particular, (F_4) has 24 positive roots.

(E_6)
$$\alpha_1 = \frac{L_1 - L_2 - L_3 - L_4 - L_5 + \sqrt{3}L_6}{2}, \qquad \alpha_2 = L_1 + L_2,$$

$$\alpha_3 = L_2 - L_1, \quad \alpha_4 = L_3 - L_2,$$
$$\alpha_5 = L_4 - L_3, \quad \alpha_6 = L_5 - L_4;$$
$$R^+ = \{L_i + L_j\}_{i<j\le 5} \cup \{L_i - L_j\}_{j<i\le 5}$$
$$\cup \left\{\frac{\pm L_1 \pm L_2 \pm L_3 \pm L_4 \pm L_5 + \sqrt{3}L_6}{2}\right\}_{\text{number of minus signs even}}$$

(E_6) has 36 positive roots.

(E_7) $$\alpha_1 = \frac{L_1 - L_2 - \cdots - L_6 + \sqrt{2}L_7}{2}, \quad \alpha_2 = L_1 + L_2,$$
$$\alpha_3 = L_2 - L_1, \quad \alpha_4 = L_3 - L_2, \quad \alpha_5 = L_4 - L_3,$$
$$\alpha_6 = L_5 - L_4, \quad \alpha_7 = L_6 - L_5;$$
$$R^+ = \{L_i + L_j\}_{i<j\le 6} \cup \{L_i - L_j\}_{j<i\le 6} \cup \{\sqrt{2}L_7\}$$
$$\cup \left\{\frac{\pm L_1 \pm L_2 \pm \cdots \pm L_6 + \sqrt{2}L_7}{2}\right\}_{\text{number of minus signs odd}}$$

Thus, (E_7) has 63 positive roots.

(E_8) $$\alpha_1 = \frac{L_1 - L_2 - \cdots - L_7 + L_8}{2}, \quad \alpha_2 = L_1 + L_2,$$
$$\alpha_3 = L_2 - L_1, \quad \alpha_4 = L_3 - L_2, \quad \alpha_5 = L_4 - L_3,$$
$$\alpha_6 = L_5 - L_4, \quad \alpha_7 = L_6 - L_5, \quad \alpha_8 = L_7 - L_6.$$
$$R^+ = \{L_i + L_j\}_{i<j\le 8} \cup \{L_i - L_j\}_{j<i\le 8}$$
$$\cup \left\{\frac{\pm L_1 \pm L_2 \pm \cdots \pm L_7 + L_8}{2}\right\}_{\text{number of minus signs even}}$$

(E_8) has 120 positive roots.

For (G_2) and (F_4) the simple roots are listed in order reading from left to right in their Dynkin diagrams

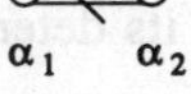

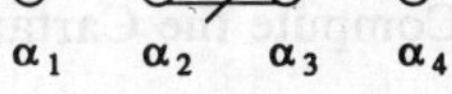

as in the classical series (A_n)–(D_n). For (E_8), the numbering is

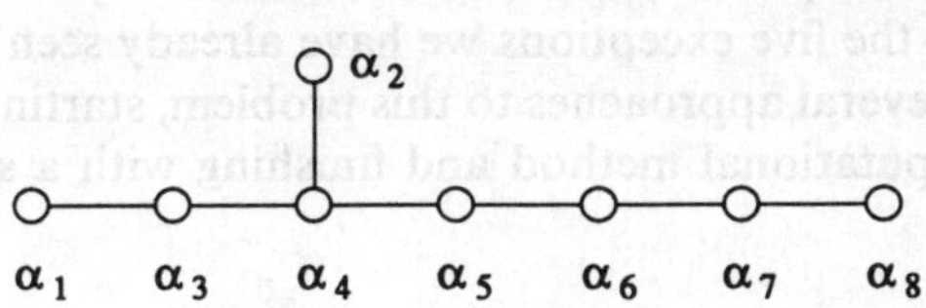

while those for (E_7) and (E_6) are obtained by removing the last one or two nodes. Note that, given the root system of (E_8), we can find the root system of (E_7) or (E_6) by taking the subspace spanned by the first seven or six simple roots.

Exercise 21.16*. (a) Verify the above lists of roots.
(b) In each case, calculate the corresponding fundamental weights.

Exercise 21.17*. Show that no two of the root systems of (A_n)–(E_8) are isomorphic, and deduce that the Dynkin diagram of a root system is independent of choice of positive roots.

A more satisfying reason for the last fact is the observation that any two choices of positive roots differ by an element of the Weyl group—the group generated by reflections W_α in the simple roots. This can be seen directly for each of the diagrams (A_n)–(E_8); for a general proof that two choices differ by an element of the Weyl group, see Proposition D.29.

We should mention here another way of conveying the data of a Dynkin diagram. This is simply the $n \times n$ matrix of integers $(n_{i,j} = n_{\alpha_i \alpha_j})$, where we take $n_{i,i} = 2$; it is called the *Cartan matrix* of the Dynkin diagram (or of the Lie algebra). Thus, for example, the Cartan matrix of (A_n) is

$$\begin{pmatrix} 2 & -1 & 0 & \cdot & \cdot & \cdot & 0 \\ -1 & 2 & -1 & 0 & \cdot & \cdot & 0 \\ 0 & -1 & 2 & -1 & \cdot & \cdot & 0 \\ \cdot & \cdot & \cdot & \cdot & \cdot & \cdot & \cdot \\ \cdot & \cdot & \cdot & \cdot & \cdot & \cdot & \cdot \\ 0 & 0 & \cdot & \cdot & -1 & 2 & -1 \\ 0 & 0 & \cdot & \cdot & \cdot & -1 & 2 \end{pmatrix}.$$

These matrices pop up remarkably often, in a variety of seemingly unrelated areas of mathematics. They will not play a major role in the present text, but the reader has probably encountered them already in one form or another, and will probably do so again.

Exercise 21.18*. Compute the Cartan matrix, and its determinant, for each Dynkin diagram.

The next task is to see how the root system determines the Lie algebra. We concentrate on the uniqueness, since there are other ways to see the existence; indeed, for all but the five exceptions we have already seen the Lie algebras. We will describe several approaches to this problem, starting with a straightforward and computational method and finishing with a slick but abstract approach.

Assume as before that $\mathfrak{g}$ is a simple Lie algebra, with a chosen Cartan subalgebra $\mathfrak{h}$ and decomposition of the roots R into positive and negative roots; let $\alpha_1, \ldots, \alpha_n$ be the simple roots. The Dynkin diagram information is the knowledge of (α_i, α_j) for all $i \neq j$. Let $H_i = H_{\alpha_i}$ be the corresponding basis of $\mathfrak{h}$, defined by the rule we have seen in Lecture 14: if $\{T_i\}$ is the basis corresponding via the Killing form to $\{\alpha_i\}$, set $H_i = 2T_i/(\alpha_i, \alpha_i)$.

Choose any nonzero element X_i in the root space $\mathfrak{g}_{\alpha_i}$, for $1 \leq i \leq n$. This determines elements Y_i in $\mathfrak{g}_{-\alpha_i}$ such that $[X_i, Y_i] = H_i$. We claim first that these $3n$ elements $\{H_i, X_i, Y_i\}$ generate $\mathfrak{g}$ as a Lie algebra. This follows from

Claim 21.19. *If α, β, and $\alpha + \beta$ are roots, then $[\mathfrak{g}_\alpha, \mathfrak{g}_\beta] = \mathfrak{g}_{\alpha+\beta}$.*

Proof. Again look at the α-string through $\mathfrak{g}_\beta$, i.e., $\bigoplus_{k \in \mathbb{Z}} \mathfrak{g}_{\beta+k\alpha}$. This is an irreducible representation of $\mathfrak{s}_\alpha \cong \mathfrak{sl}_2\mathbb{C}$, since all the terms are one dimensional (this follows from the fact that no $\beta + k\alpha$ can be zero, given that $\beta \neq \pm\alpha$). But now if $[\mathfrak{g}_\alpha, \mathfrak{g}_\beta] = 0$, $\bigoplus_{k \leq 0} \mathfrak{g}_{\beta+k\alpha}$ would be a nontrivial subrepresentation. □

For each positive root β, we have seen that can write β as a sum of simple roots $\beta = \alpha_{i_1} + \cdots + \alpha_{i_r}$ such that each of the sums $\alpha_{i_1} + \cdots + \alpha_{i_s}$ is a root, $1 \leq s \leq r$. If we choose such a presentation for each β, and set

$$X_\beta = [X_{i_r}, [X_{i_{r-1}}, \ldots, [X_{i_2}, X_{i_1}] \ldots]]$$

and

$$Y_\beta = [Y_{i_r}, [Y_{i_{r-1}}, \ldots, [Y_{i_2}, Y_{i_1}] \ldots]],$$

then the collection

$$\{H_i, 1 \leq i \leq n; X_\beta, Y_\beta, \beta \in R^+\} \tag{21.20}$$

forms a basis for $\mathfrak{g}$. Note that if β is not simple, there is no reason to expect $[X_\beta, Y_\beta]$ to be the distinguished element H_β in $\mathfrak{h}$.

We want to show that the multiplication table for these basis elements is completely determined by the Dynkin diagram. The main difficulty is that the ordering of the simple roots in the above expression for β may not be unique. For example, suppose

$$\beta = (\alpha_1 + \alpha_2) + \alpha_3 = (\alpha_2 + \alpha_3) + \alpha_1,$$

with $\alpha_1 + \alpha_2$ and $\alpha_2 + \alpha_3$ roots. We must compare $[X_3, [X_2, X_1]]$ with $[X_1, [X_3, X_2]]$. In fact, they must be negatives of each other. For, by Jacobi, we have

$$[X_1, [X_3, X_2]] = -[X_3, [X_2, X_1]] - [X_2, [X_1, X_3]] = -[X_3, [X_2, X_1]],$$

noting that $[X_1, X_3] = 0$ since $\alpha_1 + \alpha_3$ cannot be a root, e.g., by step (ii) of the preceding section.

For any sequence $I = (i_1, \ldots, i_r)$, $1 \leq i_j \leq n$, set

$$\alpha_I = \alpha_{i_1} + \cdots + \alpha_{i_r},$$
$$X_I = [X_{i_r}, [X_{i_{r-1}}, \ldots, [X_{i_2}, X_{i_1}] \ldots]],$$
$$Y_I = [Y_{i_r}, [Y_{i_{r-1}}, \ldots, [Y_{i_2}, Y_{i_1}] \ldots]].$$

Call I *admissible* if each partial sum $\alpha_{i_1} + \cdots + \alpha_{i_s}$ is a root, $1 \le s \le r$; note that I is admissible exactly when X_I is not zero.

Lemma 21.21. *If I and J are two admissible sequences for which $\alpha_I = \alpha_J$, then there is a nonzero rational number q determined by I, J, and the Dynkin diagram, such that $X_J = q \cdot X_I$.*

PROOF. Let $k = i_r$ be the last entry in I. If $j_r = k$ as well, the result follows by induction on r. We reduce the general case to this case, by maneuvering to replace j_r by k. We have first

$$X_J = q_1 \cdot [X_k, [Y_k, X_J]],$$

with q_1 a nonzero rational number depending only on J, k, and the Dynkin diagram, since $\alpha_J - \alpha_k = \alpha_I - \alpha_k$ is a root; the point is that we know how $\mathfrak{s}_{\alpha_k} \cong \mathfrak{sl}_2$ acts on the α_k-string through α_J as soon as we know the length of the string, and this is Dynkin diagram information. Next, let s be the largest integer such that $j_s = k$. Then

$$[Y_k, X_J] = [X_{j_r}, \ldots [X_{j_{s+1}}, [Y_k, [X_k, X_K]]] \ldots],$$

where $K = (j_1, \ldots, j_{s-1})$, since $[Y_k, [X_i, Z]] = [X_i, [Y_k, Z]]$ when $i \neq k$. Finally,

$$[Y_k, [X_k, X_K]] = q_2 \cdot X_K,$$

with q_2 a nonzero rational number depending only on K, k, and the Dynkin diagram, since $\alpha_K + \alpha_k$ is a root. Combining these three equations, we get

$$X_J = q_1 q_2 \cdot [X_k, [X_{j_r}, \ldots [X_{j_{s+1}}, X_K] \ldots]],$$

which suffices since the sequence for the term on the right ends in the same integer k as I. □

Proposition 21.22. *The bracket of any two basis elements in (21.20) is a rational multiple of another basis element, that multiple determined from the Dynkin diagram.*

PROOF. This is clear for brackets of an H_i with any basis element. Lemma 21.21 handles brackets of the form $[X_I, X_J]$, and those involving only Y's are similar. For brackets $[Y_I, X_J]$, it suffices inductively to compute $[Y_k, X_J]$ as a rational multiple of some X_K, with K shorter than J (or of H_k if J has one term); but this was worked out in the proof of the lemma. □

Exercise 21.23*. (i) Show that in (G_2) each positive root can be written in only one way as a sum of simple roots, up to the order of the first two roots.

(ii) Work out the multiplication table from the Dynkin diagram. (iii) Verify that the result is indeed a Lie algebra, which is (visibly) simple.

This exercise will be worked out in detail to start the next lecture. Of course, there is nothing but lack of time to keep us from verifying that the other four exceptional Dynkin diagrams do lead, by the same prescription, to honest Lie algebras, but doing it by hand gets pretty laborious, and we will describe some of the other methods available.

The fact that the multiplication table can be defined with rational coefficients becomes important when one wants to reduce them modulo prime numbers, which we will not discuss here. The fact that they can be taken to be real, on the other hand, will come up later, when we discuss real forms of complex Lie algebras and groups.

There is a more general and elegant way to proceed, given by Serre [Se3]. Write n_{ij} in place of $n_{\alpha_i\alpha_j}$. Form the free Lie algebra on generators

$$H_1, \ldots, H_n, X_1, \ldots, X_n, Y_1, \ldots, Y_n,$$

i.e., form the free (tensor) algebra with this basis, and divide modulo by the relations $[A, B] + [B, A] = 0$ and the Jacobi relation. Then take this free Lie algebra, and divide by the relations

$$[H_i, H_j] = 0 \text{ (all } i, j); \qquad [X_i, Y_i] = H_i \text{ (all } i); \qquad [X_i, Y_j] = 0 \; (i \neq j);$$

$$[H_i, X_j] = n_{ji} X_j \text{ (all } i, j); \qquad [H_i, Y_j] = -n_{ji} Y_j \text{ (all } i, j);$$

and, for all $i \neq j$,

$$[X_i, X_j] = 0, \qquad [Y_i, Y_j] = 0 \qquad \text{if } n_{ij} = 0;$$

$$[X_i, [X_i, X_j]] = 0, \qquad [Y_i, [Y_i, Y_j]] = 0 \qquad \text{if } n_{ij} = -1;$$

$$[X_i, [X_i, [X_i, X_j]]] = 0, \qquad [Y_i, [Y_i, [Y_i, Y_j]]] = 0 \qquad \text{if } n_{ij} = -2;$$

$$[X_i, [X_i, [X_i, [X_i, X_j]]]] = 0, \qquad [Y_i, [Y_i, [Y_i, [Y_i, Y_j]]]] = 0 \quad \text{if } n_{ij} = -3.$$

Exercise 21.24. Verify that if one starts with a semisimple Lie algebra with a given Dynkin diagram, the above equations must hold.

Serre shows ([Se3, Chap. VI App.], cf. [Hu1 §18]) that the resulting Lie algebra is a finite-dimensional semisimple Lie algebra, with Cartan subalgebra generated by $H_1, \ldots, H_n$ and given root system. In particular, this includes a proof of the existence of all the simple Lie algebras.

Here is a third approach to uniqueness. Suppose $\mathfrak{g}$ and $\mathfrak{g}'$, with given Cartan subalgebras $\mathfrak{h}$ and $\mathfrak{h}'$, and choice of positive roots, have isomorphic root systems. There is an isomorphism $\mathfrak{h} \to \mathfrak{h}'$, taking corresponding H_i to H_i'. Choose arbitrarily nonzero vectors X_i and X_i' in the root spaces of $\mathfrak{g}$ and $\mathfrak{g}'$ corresponding to the simple roots.

Claim 21.25. *There is a unique isomorphism from* $\mathfrak{g}$ *to* $\mathfrak{g}'$ *extending the isomorphism of* $\mathfrak{h}$ *with* $\mathfrak{h}'$, *and mapping* X_i *to* X_i' *for all i.*

PROOF. The uniqueness of the isomorphism is easy: the resulting map is determined on the Y_i by $\mathfrak{sl}_2$ considerations, and the H_i, X_i, and Y_i generate $\mathfrak{g}$. For the existence of the isomorphism consider the subalgebra $\tilde{\mathfrak{g}}$ of $\mathfrak{g} \oplus \mathfrak{g}'$ generated by $\tilde{H}_i = H_i \oplus H_i'$, $\tilde{X}_i = X_i \oplus X_i'$, and $\tilde{Y}_i = Y_i \oplus Y_i'$. It suffices to prove that the two projections from $\tilde{\mathfrak{g}}$ to $\mathfrak{g}$ and $\mathfrak{g}'$ are isomorphisms. The kernel of the second projection is $\mathfrak{k} \oplus 0$, where $\mathfrak{k}$ is an ideal in $\mathfrak{g}$. Since $\mathfrak{g}$ is simple, $\mathfrak{k}$ is either 0, as required, or $\mathfrak{k} = \mathfrak{g}$. In the latter case, we must have $\tilde{\mathfrak{g}} = \mathfrak{g} \oplus \mathfrak{g}'$.

To see that this is impossible, consider a maximal positive root β, take nonzero vectors X_β, X_β' in the corresponding root spaces, and set $\tilde{X}_\beta = X_\beta \oplus X_\beta'$, a highest weight vector in $\tilde{\mathfrak{g}}$. Let W be the subspace of $\tilde{\mathfrak{g}}$ obtained by successively applying all $\tilde{Y}_i$'s. Then W is a proper subspace of $\tilde{\mathfrak{g}}$, since its weight space W_β corresponding to β is one dimensional. By the argument we have seen several times, $\tilde{\mathfrak{g}}$ preserves W. Now if $\tilde{\mathfrak{g}} = \mathfrak{g} \oplus \mathfrak{g}'$, W would be an ideal in $\mathfrak{g} \oplus \mathfrak{g}'$, and this would force $X_\beta \oplus 0$ to belong to W, making W_β two dimensional again. □

To finish this story, we should show that the simple Lie algebras corresponding to two different Dynkin diagrams cannot be isomorphic, i.e., that the two choices made in going from a semisimple Lie algebra to Dynkin diagram do not change the answer. The general facts are:

(1) Any two Cartan subalgebras of a semisimple Lie algebra are conjugate, i.e., there is an inner automorphism by an element in the corresponding adjoint group, which takes one into the other.
(2) Any two decompositions of a root system into positive and negative roots differ by an element of the Weyl group.

These are standard facts which are proved in Appendix D. Both statements are subsumed in the fact that any two *Borel subalgebras* of a semisimple Lie algebra are conjugate, a Borel subalgebra being the subspace spanned by the Cartan subalgebra and the root spaces $\mathfrak{g}_\alpha$ for positive α. For those readers who crave logical completeness but do not want to go through so much general theory, we observe that most possible coincidences can be ruled out by such simple considerations as computing dimensions, and others can be ruled out by simple ad hoc methods, cf. Exercise 21.17.

Finally, we must also prove the "existence theorem": that there is a simple Lie algebra for each Dynkin diagram. Serre's theorem quoted above gives a unified proof of existence. But we have seen and studied the Lie algebras for the classical cases (A_n)–(D_n), and it is more in keeping with the spirit of these lectures to at least try to see the five exceptions explicitly. This is the subject of the next lecture.

LECTURE 22

$\mathfrak{g}_2$ and Other Exceptional Lie Algebras

This lecture is mainly about $\mathfrak{g}_2$, with just enough discussion of the algebraic constructions of the other exceptional Lie algebras to give the reader a sense of their complexity. $\mathfrak{g}_2$, being only 14-dimensional, is different: we can reasonably carry out in practice the process described in §21.3 to arrive at an explicit description of the algebra by specifying a basis and all pairwise products; we do this in §22.1 and verify in §22.2 that the result really is a Lie algebra. In §22.3 we analyze the representations of $\mathfrak{g}_2$, and arrive in particular at another description of $\mathfrak{g}_2$: it is the algebra of endomorphisms of a seven-dimensional vector space preserving a general trilinear form. (Note that §22.3 may be read independently of either §22.1, §21.2, or §21.3.) Finally, in the fourth section we will sketch some of the more abstract (i.e., coordinate free) approaches to the construction of the five exceptional Lie algebras. While the first two sections are completely elementary, the constructions given in §22.4 involve some fairly serious algebra.

§22.1. Construction of $\mathfrak{g}_2$ from Its Dynkin Diagram

In this section we will carry out explicitly the process described in the preceding section for the Dynkin diagram (G_2), constructing in this way a Lie algebra $\mathfrak{g}_2$ with diagram (G_2) (and in particular proving its existence).

The first step is to find the root system from the Dynkin diagram. In the case of $\mathfrak{g}_2$ this is immediate; we may draw the root system $R \subset \mathfrak{h}^*$ associated to the diagram G_2 as follows:

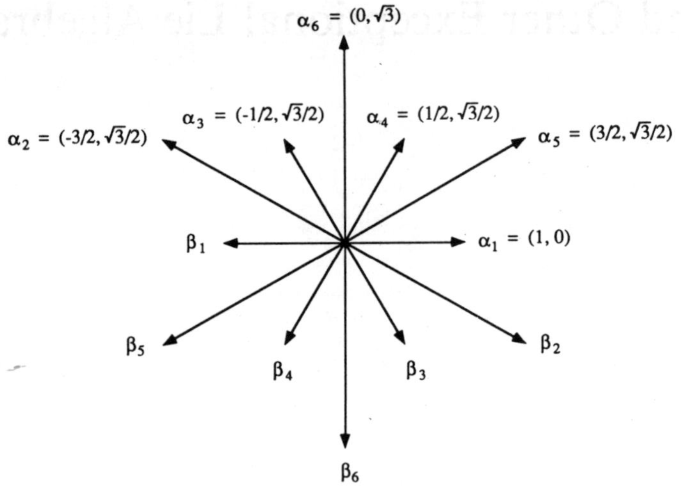

Here the positive roots are denoted α_i, with α_1 and α_2 the simple roots. The coordinate system here has no particular significance (in particular, recall that the configuration of roots α_i and β_i is determined only up to a real scalar), but is convenient for calculating inner products. Note that the Weyl group is the dihedral group generated by rotation through an angle of $\pi/3$ and reflection in the horizontal; the Weyl chamber associated to the choice of ordering of the roots given is the cone between the roots α_6 and α_4.

As indicated in the preceding section, we start by letting X_1 be any eigenvector for the action of $\mathfrak{h}$ with eigenvalue α_1, and X_2 any eigenvector for the action of $\mathfrak{h}$ with eigenvalue α_2. We similarly let Y_1 and Y_2 be eigenvectors with eigenvalues β_1 and β_2 and set

$$H_1 = [X_1, Y_1] \quad \text{and} \quad H_2 = [X_2, Y_2].$$

We can choose Y_1 and Y_2 so that the elements $H_i \in \mathfrak{h}$ satisfy $\alpha_1(H_1) = \alpha_2(H_2) = 2$, i.e.,

$$[H_1, X_1] = 2 \cdot X_1 \quad \text{and} \quad [H_2, X_2] = 2 \cdot X_2.$$

It follows that

$$[H_1, Y_1] = -2 \cdot Y_1 \quad \text{and} \quad [H_2, Y_2] = -2 \cdot Y_2,$$

i.e., H_i, X_i, and Y_i span a subalgebra $\mathfrak{s}_{\alpha_i} \cong \mathfrak{sl}_2\mathbb{C}$, with H_i, X_i, and Y_i a normalized basis for this copy of $\mathfrak{sl}_2\mathbb{C}$.

Now, it is clear from the diagram above that there is a unique way of writing each positive root α_i as a sum of simple roots $\alpha_{i_1} + \cdots + \alpha_{i_k}$ so that the partial sums $\alpha_{i_1} + \cdots + \alpha_{i_l}$ are roots for each $l \leq k$ (modulo exchanging the first two terms): we go through the root system by the path

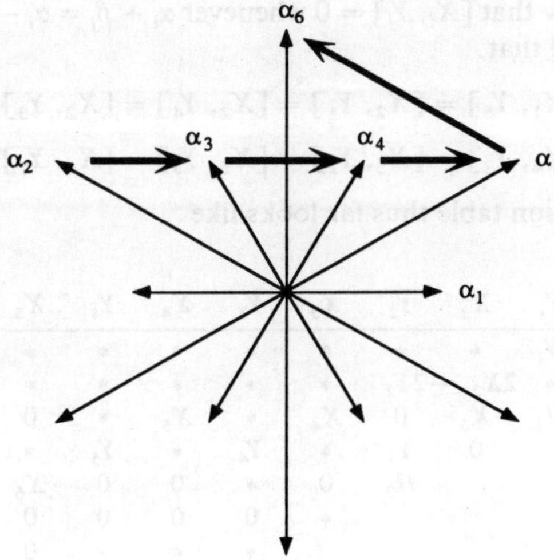

i.e., we write

$$\alpha_3 = \alpha_1 + \alpha_2,$$

$$\alpha_4 = \alpha_1 + \alpha_3 = \alpha_1 + \alpha_1 + \alpha_2,$$

$$\alpha_5 = \alpha_1 + \alpha_4 = \alpha_1 + \alpha_1 + \alpha_1 + \alpha_2,$$

$$\alpha_6 = \alpha_2 + \alpha_5 = \alpha_2 + \alpha_1 + \alpha_1 + \alpha_1 + \alpha_2.$$

According to the general recipe, this means we now set

$$X_3 = [X_1, X_2], \qquad X_4 = [X_1, X_3],$$

$$X_5 = [X_1, X_4], \qquad X_6 = [X_2, X_5],$$

and define $Y_3, \ldots, Y_6$ similarly. The elements $H_1, H_2, X_1, \ldots, X_6, Y_1, \ldots, Y_6$ then form a basis for the 14-dimensional $\mathfrak{g}_2$, with H_1 and H_2 a basis for $\mathfrak{h}$, X_i a generator of the eigenspace $\mathfrak{g}_{\alpha_i}$, and Y_i a generator of $\mathfrak{g}_{\beta_i}$ for $i = 1, \ldots, 6$.

The task at hand now is to write down the multiplication table for $\mathfrak{g}_2$ in terms of this basis. Of course, some products are already known: we know, for example, that H_i, X_i, and Y_i form a normalized basis for $\mathfrak{sl}_2\mathbb{C}$ for $i = 1, 2$, and we have the relations defining $X_3, \ldots, X_6$ and $Y_1, \ldots, Y_6$ above. In addition, since we know that the product $[X_i, X_j]$ lies in the root space $\mathfrak{g}_{\alpha_i+\alpha_j}$ for each i and j, we see immediately that $[X_i, X_j] = 0$ whenever $\alpha_i + \alpha_j$ is not a root. We deduce that

$$[X_1, X_5] = [X_1, X_6] = [X_2, X_3] = [X_2, X_4] = [X_2, X_6] = [X_3, X_5]$$
$$= [X_3, X_6] = [X_4, X_5] = [X_4, X_6] = [X_5, X_6] = 0,$$

and likewise

$$[Y_1, Y_5] = [Y_1, Y_6] = [Y_2, Y_3] = [Y_2, Y_4] = [Y_2, Y_6] = [Y_3, Y_5]$$
$$= [Y_3, Y_6] = [Y_4, Y_5] = [Y_4, Y_6] = [Y_5, Y_6] = 0.$$

Similarly, we know that $[X_i, Y_j] = 0$ whenever $\alpha_i + \beta_j = \alpha_i - \alpha_j$ is not a root; this tells us as well that

$$\begin{aligned}[X_1, Y_2] = [X_1, Y_6] = [X_2, Y_1] = [X_2, Y_4] = [X_2, Y_5] = [X_3, Y_5] \\ = [X_4, Y_2] = [X_5, Y_2] = [X_5, Y_3] = [X_6, Y_1] = 0.\end{aligned}$$

The multiplication table thus far looks like

	H_2	X_1	Y_1	X_2	Y_2	X_3	Y_3	X_4	Y_4	X_5	Y_5	X_6	Y_6
H_1	0	$2X_1$	$-2Y_1$	*	*	*	*	*	*	*	*	*	*
H_2		*	*	$2X_2$	$-2Y_2$	*	*	*	*	*	*	*	*
X_1			H_1	X_3	0	X_4	*	X_5	*	0	*	0	0
Y_1				0	Y_3	*	Y_4	*	Y_5	*	0	0	0
X_2					H_2	0	*	0	0	X_6	0	0	*
Y_2						*	0	0	0	0	Y_6	*	0
X_3							*	*	*	0	0	0	*
Y_3								*	*	0	0	*	0
X_4									*	0	*	0	*
Y_4										*	0	*	0
X_5											*	0	*
Y_5												*	0
X_6													*

The next thing to do is to describe the action of H_1 and H_2 on the various vectors X_i and Y_i. This can be done using the inner product on $\mathfrak{h}$, but it is perhaps simpler to go back to the basic idea of restriction to the subalgebras $\mathfrak{s}_{\alpha_1}$ and $\mathfrak{s}_{\alpha_2}$. For example, if we want to determine the action of H_1 on the various X_i, consider how the algebra $\mathfrak{g} = \mathfrak{h} \bigoplus (\mathfrak{g}_{\alpha_i} \oplus \mathfrak{g}_{\beta_i})$ decomposes as a representation of $\mathfrak{s}_{\alpha_1}$:

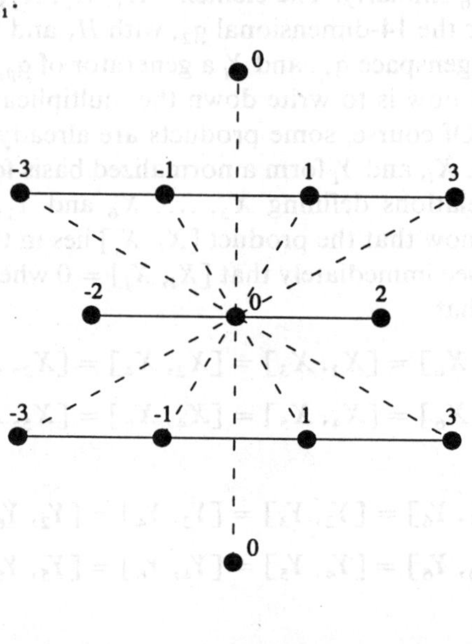

We get two trivial representations (the spans of X_6 and Y_6, as already noted); one copy of the adjoint representation $\mathrm{Sym}^2 V$ (the subalgebra $\mathfrak{s}_{\alpha_1}$ itself) spanned by X_1, Y_1, and H_1; and two copies of the irreducible four-dimensional representation $\mathrm{Sym}^3 V$ spanned by X_2, X_3, X_4, and X_5 and Y_5, Y_4, Y_3, and Y_2. In particular, it follows that X_2, X_3, X_4, and X_5 are eigenvectors for the action of H_1 with eigenvalues of $-3, -1, 1$, and 3, respectively; and likewise Y_5, Y_4, Y_3, and Y_2 are eigenvectors with eigenvalues -3, -1, 1, and 3. In similar fashion, we consider the decomposition of $\mathfrak{g}$ under the action of $\mathfrak{s}_{\alpha_2} = \mathbb{C}\{H_2, X_2, Y_2\}$: diagrammatically, this looks like

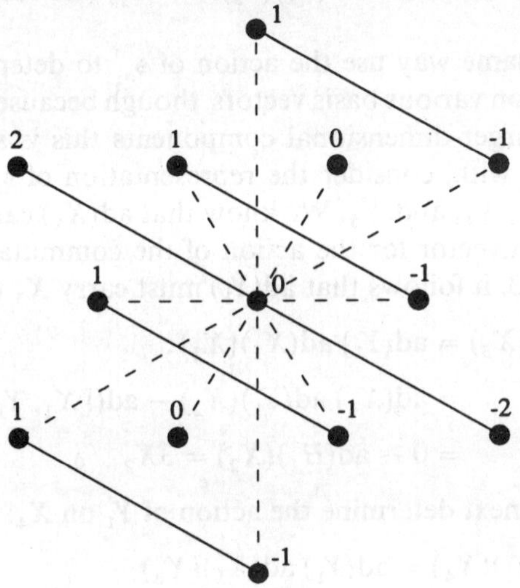

Here we have two trivial representations, spanned by X_4 and Y_4, one adjoint ($\mathfrak{s}_{\alpha_2}$ itself), and four copies of the standard two-dimensional representation V, spanned by X_6 and X_5, X_3 and X_1, Y_1 and Y_3, and Y_5 and Y_6. It follows that X_6, X_3, Y_1, and Y_5 are eigenvectors for the action of H_2 with eigenvalue 1, and likewise X_5, X_1, Y_3, and Y_6 are eigenvectors with eigenvalue -1.

Including this information, we can fill in the top two rows of the multiplication table:

	H_2	X_1	Y_1	X_2	Y_2	X_3	Y_3	X_4	Y_4	X_5	Y_5	X_6	Y_6
H_1	0	$2X_1$	$-2Y_1$	$-3X_2$	$3Y_2$	$-X_3$	Y_3	X_4	$-Y_4$	$3X_5$	$-3Y_5$	0	0
H_2		$-X_1$	Y_1	$2X_2$	$-2Y_2$	X_3	$-Y_3$	0	0	$-X_5$	Y_5	X_6	$-Y_6$

Decomposing $\mathfrak{g}_2$ according to the action of $\mathfrak{s}_{\alpha_1}$ and $\mathfrak{s}_{\alpha_2}$ gives us information about the action of X_1, X_2, Y_1, and Y_2 on the other basis vectors as well. For example, we saw a moment ago that X_5 and X_6 together span a subrepresentation of $\mathfrak{g}_2$ under the action of $\mathfrak{s}_{\alpha_2}$, with $\mathrm{ad}(X_2)$ carrying X_5 to X_6. It follows from this that $\mathrm{ad}(Y_2)$ must carry X_6 back to X_5: we have

$$\begin{aligned}\mathrm{ad}(Y_2)(X_6) &= \mathrm{ad}(Y_2)\,\mathrm{ad}(X_2)(X_5)\\ &= \mathrm{ad}(X_2)\,\mathrm{ad}(Y_2)(X_5) - \mathrm{ad}([X_2, Y_2])(X_5)\\ &= 0 - \mathrm{ad}(H_2)(X_5) = X_5.\end{aligned}$$

Similarly, since $\mathrm{ad}(X_2)$ carries X_1 into $-X_3$, which together with X_1 spans a copy of the standard two-dimensional representation of $\mathfrak{s}_{\alpha_2} \cong \mathfrak{sl}_2\mathbb{C}$, it follows that $\mathrm{ad}(Y_2)$ will carry $-X_3$ back to X_1. Likewise from the fact that $\mathrm{ad}(Y_2)$ carries Y_1 to $-Y_3$ we see that $\mathrm{ad}(Y_2)(Y_3) = -Y_1$, and since $\mathrm{ad}(Y_2)\colon Y_5 \mapsto Y_6$, $\mathrm{ad}(X_2)\colon Y_6 \mapsto Y_5$.

We can in the same way use the action of $\mathfrak{s}_{\alpha_1}$ to determine the values of $\mathrm{ad}(X_1)$ and $\mathrm{ad}(Y_2)$ on various basis vectors, though because the representation of $\mathfrak{s}_{\alpha_1}$ on $\mathfrak{g}_2$ has larger-dimensional components this is slightly more complicated. To begin with, consider the representation of $\mathfrak{s}_{\alpha_1}$ on the subspace spanned by X_2, X_3, X_4, and X_5. We know that $\mathrm{ad}(X_1)$ carries X_2 to X_3, and since X_2 is an eigenvector for the action of the commutator $[X_1, Y_1] = H_1$ with eigenvalue -3, it follows that $\mathrm{ad}(Y_1)$ must carry X_3 to $3X_2$: we have

$$\begin{aligned}\mathrm{ad}(Y_1)(X_3) &= \mathrm{ad}(Y_1)\,\mathrm{ad}(X_1)(X_2)\\ &= \mathrm{ad}(X_1)\,\mathrm{ad}(Y_1)(X_2) - \mathrm{ad}([X_1, Y_1])(X_2)\\ &= 0 - \mathrm{ad}(H_1)(X_2) = 3X_2.\end{aligned}$$

Using this, we can next determine the action of Y_1 on X_4:

$$\begin{aligned}\mathrm{ad}(Y_1)(X_4) &= \mathrm{ad}(Y_1)\,\mathrm{ad}(X_1)(X_3)\\ &= \mathrm{ad}(X_1)\,\mathrm{ad}(Y_1)(X_3) - \mathrm{ad}(H_1)(X_3)\\ &= \mathrm{ad}(X_1)(3X_2) + X_3 = 4X_3,\end{aligned}$$

and we calculate likewise that $\mathrm{ad}(Y_1)(X_5) = 3X_4$. Analogously, knowing that $\mathrm{ad}(Y_1)$ carries Y_2 to Y_3 to Y_4 to Y_5 yields the information that $\mathrm{ad}(X_1)$ must carry Y_3, Y_4, and Y_5 to $3Y_2$, $4Y_3$ and, $3Y_4$, respectively. Including all this information in the chart, the next four rows of our multiplication table are

	H_2	X_1	Y_1	X_2	Y_2	X_3	Y_3	X_4	Y_4	X_5	Y_5	X_6	Y_6
X_1			H_1	X_3	0	X_4	$3Y_2$	X_5	$4Y_3$	0	$3Y_4$	0	0
Y_1				0	Y_3	$3X_2$	Y_4	$4X_3$	Y_5	$3X_4$	0	0	0
X_2					H_2	0	$-Y_1$	0	0	X_6	0	0	Y_5
Y_2						$-X_1$	0	0	0	0	Y_6	X_5	0

We next have to find the commutators of the basis elements X_i and Y_j for $i, j \geq 3$. We cannot do this by looking at the action of the subalgebras generated by X_i and Y_i, since for $i \geq 3$ we do not know the commutator $[X_i, Y_i]$. Rather, the way to do this is outlined in the general proof in the preceding section: we just use the expression of the X_i and Y_j as brackets of the generators X_1, X_2, Y_1, and Y_2 to reduce the problem to brackets with these generators, which we now know. Thus, for example, the first unknown entry in the table at present is the bracket $[X_3, Y_3]$. We calculate this by writing X_3 as $[X_1, X_2]$, so that

$$\begin{aligned}\operatorname{ad}(X_3)(Y_3) &= \operatorname{ad}([X_1, X_2])(Y_3)\\ &= \operatorname{ad}(X_1)\operatorname{ad}(X_2)(Y_3) - \operatorname{ad}(X_2)\operatorname{ad}(X_1)(Y_3)\\ &= \operatorname{ad}(X_1)(-Y_1) - \operatorname{ad}(X_2)(3Y_2)\\ &= -H_1 - 3H_2.\end{aligned}$$

Likewise, to evaluate $[X_3, X_4]$ we have

$$\begin{aligned}\operatorname{ad}(X_3)(X_4) &= \operatorname{ad}([X_1, X_2])(X_4)\\ &= \operatorname{ad}(X_1)\operatorname{ad}(X_2)(X_4) - \operatorname{ad}(X_2)\operatorname{ad}(X_1)(X_4)\\ &= -\operatorname{ad}(X_2)(X_5) = -X_6.\end{aligned}$$

In this way, we can evaluate all brackets with X_3; knowing these, we can reduce any bracket with X_4 to one involving X_1 and X_3 by writing $X_4 = [X_1, X_3]$, and so on. Continuing in this way, we may complete our multiplication table:

	H_2	X_1	Y_1	X_2	Y_2	X_3	Y_3	X_4	Y_4	X_5	Y_5	X_6	Y_6
H_1	0	$2X_1$	$-2Y_1$	$-3X_2$	$3Y_2$	$-X_3$	Y_3	X_4	$-Y_4$	$3X_5$	$-3Y_5$	0	0
H_2		$-X_1$	Y_1	$2X_2$	$-2Y_2$	X_3	$-Y_3$	0	0	$-X_5$	Y_5	X_6	$-Y_6$
X_1			H_1	X_3	0	X_4	$3Y_2$	X_5	$4Y_3$	0	$3Y_4$	0	0
Y_1				0	Y_3	$3X_2$	Y_4	$4X_3$	Y_5	$3X_4$	0	0	0
X_2					H_2	0	$-Y_1$	0	0	X_6	0	0	Y_5
Y_2						$-X_1$	0	0	0	0	Y_6	X_5	0
X_3							$-H_1 -3H_2$	$-X_6$	$4Y_1$	0	0	0	$3Y_4$
Y_3								$4X_1$	$-Y_6$	0	0	$3X_4$	0
X_4									$8H_1 +12H_2$	0	$-12Y_1$	0	$12Y_3$
Y_4										$-12X_1$	0	$12X_3$	0
X_5											$-36H_1 -36H_2$	0	$36Y_2$
Y_5												$36X_2$	0
X_6													$36H_1 +72H_2$

Of course, in retrospect we see that the basis we have chosen is far from the most symmetric one possible: for example, if we divided X_4 and Y_4 by 2 and X_5, X_6, Y_5, and Y_6 by 6, and changed the signs of X_5 and Y_3, the form of the table would be

Table 22.1

	H_2	X_1	Y_1	X_2	Y_2	X_3	Y_3	X_4	Y_4	X_5	Y_5	X_6	Y_6
H_1	0	$2X_1$	$-2Y_1$	$-3X_2$	$3Y_2$	$-X_3$	Y_3	X_4	$-Y_4$	$3X_5$	$-3Y_5$	0	0
H_2		$-X_1$	Y_1	$2X_2$	$-2Y_2$	X_3	$-Y_3$	0	0	$-X_5$	Y_5	X_6	$-Y_6$
X_1			H_1	X_3	0	$2X_4$	$-3Y_2$	$-3X_5$	$-2Y_3$	0	Y_4	0	0
Y_1				0	$-Y_3$	$3X_2$	$-2Y_4$	$2X_3$	$3Y_5$	$-X_4$	0	0	0
X_2					H_2	0	Y_1	0	0	$-X_6$	0	0	Y_5
Y_2						$-X_1$	0	0	0	0	Y_6	$-X_5$	0
X_3							H_1+3H_2	$-3X_6$	$2Y_1$	0	0	0	Y_4
Y_3								$-2X_1$	$3Y_6$	0	0	$-X_4$	0
X_4									$2H_1+3H_2$	0	$-Y_1$	0	$-Y_3$
Y_4										X_1	0	X_3	0
X_5											H_1+H_2	0	$-Y_2$
Y_5												X_2	0
X_6													H_1+2H_2

There was another good reason for these changes: now each of the brackets $[X_i, Y_i]$ will be the distinguished element of $\mathfrak{h}$ corresponding to the root α_i. If we denote this element by H_i, then we read off from the table that

$$H_3 = H_1 + 3H_2, \qquad H_4 = 2H_1 + 3H_2, \qquad H_5 = H_1 + H_2, \qquad H_6 = H_1 + 2H_2, \tag{22.2}$$

and

$$H_i = [X_i, Y_i], \qquad [H_i, X_i] = 2X_i, \qquad [H_i, Y_i] = -2Y_i, \tag{22.3}$$

for $i = 1, 2, 3, 4, 5, 6$.

§22.2. Verifying That $\mathfrak{g}_2$ Is a Lie Algebra

The calculation of the preceding section gives a complete description of what the Lie algebra $\mathfrak{g}_2$ must look like, but there is still some work to be done: unless we know that there is a Lie algebra with diagram (G_2), we do not know that the above multiplication table defines a Lie algebra, let alone a simple one. In fact, the simplicity is not much of a problem (cf. Exercise 14.34), but to know that it is a Lie algebra requires knowing that the Jacobi identity is valid. One could simply check this from the table for all $\binom{14}{3}$ triples of elements from the basis, a rather uninviting task.

There is another way, which gives more structure to the preceding calculations, and which will give a clue for possible constructions of other Lie algebras. The root diagram for (G_2) is made up of two hexagons, one with long arrows, the other with short. This suggests that we should find a copy of the corresponding Lie algebra $\mathfrak{sl}_3\mathbb{C}$ inside $\mathfrak{g}_2$. The subspace spanned by $\mathfrak{h}$ and the root spaces corresponding to the six longer roots is clearly closed under brackets, so is the obvious candidate. The long roots are α_5, α_2, and $\alpha_6 = \alpha_5 + \alpha_2$, and their inverses. So we define $\mathfrak{g}_0$ to be the subspace spanned by the corresponding vectors:

$$\mathfrak{g}_0 = \mathbb{C}\{H_5, H_2, X_5, Y_5, X_2, Y_2, X_6, Y_6\}.$$

The multiplication table for $\mathfrak{g}_0$ is read off from Table 22.1:

	H_2	X_5	Y_5	X_2	Y_2	X_6	Y_6
H_5	0	$2X_5$	$-2Y_5$	$-X_2$	Y_2	X_6	$-Y_6$
H_2		$-X_5$	Y_5	$2X_2$	$-2Y_2$	X_6	$-Y_6$
X_5			H_5	X_6	0	0	$-Y_2$
Y_5				0	$-Y_6$	X_2	0
X_2					H_2	0	Y_5
Y_2						$-X_5$	0
X_6							$H_5 + H_2$

This is exactly the multiplication table for $\mathfrak{sl}_3\mathbb{C}$, with its standard basis (in the same order):

$$\mathfrak{sl}_3\mathbb{C} = \mathbb{C}\{E_{1,1} - E_{2,2}, E_{2,2} - E_{3,3}, E_{1,2}, E_{2,1}, E_{2,3}, E_{3,2}, E_{1,3}, E_{3,1}\}.$$

So we have determined an isomorphism

$$\mathfrak{g}_0 \cong \mathfrak{sl}_3\mathbb{C}.$$

(Note right away that this verifies the Jacobi identity for triples taken from $\mathfrak{g}_0$.)

The rest of the Lie algebra must be a representation of the subalgebra $\mathfrak{g}_0 \cong \mathfrak{sl}_3\mathbb{C}$, and we know what this must be: the smaller hexagon is the union of the two triangles which are the weight diagrams for the standard representation of $\mathfrak{sl}_3$ and its dual, which we denote here by W and W^*; W is the sum of the root spaces for α_4, β_1, and β_3, while W^* is the sum of those for β_4, α_1, and α_3.

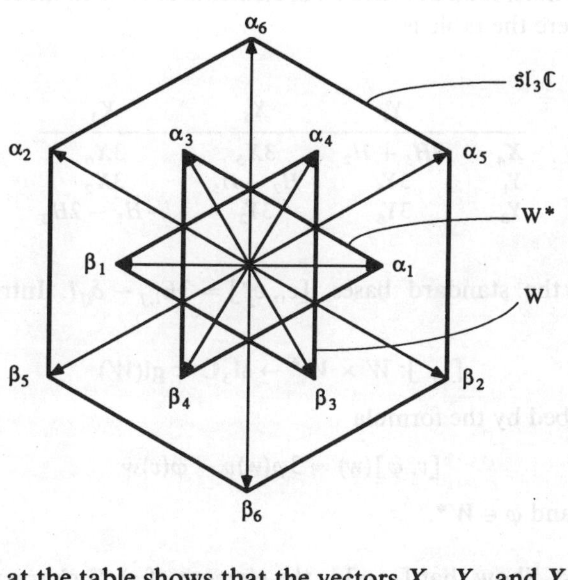

Again, a look at the table shows that the vectors X_4, Y_1, and Y_3 form a basis for $W = \mathbb{C}^3$ that corresponds to the standard basis e_1, e_2, and e_3, and similarly Y_4, X_1, and X_3 form a basis for $W^* = (\mathbb{C}^3)^*$ that corresponds to the dual basis e_1^*, e_2^*, and e_3^*: we have

$$W = \mathbb{C}\{X_4, Y_1, Y_3\}; \qquad W^* = \mathbb{C}\{Y_4, X_1, X_3\};$$

$$\mathfrak{g}_2 = \mathfrak{g}_0 \oplus W \oplus W^*.$$

With these isomorphisms, the brackets

$$\mathfrak{g}_0 \times W \to W \quad \text{and} \quad \mathfrak{g}_0 \times W^* \to W^*$$

correspond to the standard operations of $\mathfrak{sl}_3\mathbb{C}$ on $\mathbb{C}^3$ and $(\mathbb{C}^3)^*$.

Next we look at brackets of elements in W. Note that $[W, W]$ is contained in W^*, either by weights or by looking at the table. The table is

	Y_1	Y_3			e_2	e_3
X_4	$-2X_3$	$2X_1$	or	e_1	$-2e_3^*$	$2e_2^*$
Y_1	0	$-2Y_4$		e_2	0	$-2e_1^*$

Identifying $W = \mathbb{C}^3$, $W^* = (\mathbb{C}^3)^*$ as above, we see that the bracket $W \times W \to W^*$ becomes the map

$$W \times W \to W^* = \Lambda^2 W, \qquad v \times w \mapsto -2 \cdot v \wedge w.$$

Similarly for W^*, we have $[W^*, W^*] \subset W$, and the bracket is identified with the map

$$W^* \times W^* \to W = \Lambda^2 W^*, \qquad \varphi \times \psi \mapsto 2 \cdot \varphi \wedge \psi.$$

Finally we must look at brackets of elements of W with those of W^*, which land in $\mathfrak{g}_0$. Here the table is

	Y_4	X_1	X_3
X_4	$2H_5 + H_2$	$3X_5$	$3X_6$
Y_1	$3Y_5$	$H_2 - H_5$	$3X_2$
Y_3	$3Y_6$	$3Y_2$	$-H_5 - 2H_2$

In terms of the standard bases, $[e_i, e_j^*] = 3E_{i,j} - \delta_{ij}I$. Intrinsically, this mapping

$$[\ ,\ ]\colon W \times W^* \to \mathfrak{sl}_3\mathbb{C} \subset \mathfrak{gl}(W)$$

can be described by the formula

$$[v, \varphi](w) = 3\varphi(w)v - \varphi(v)w \tag{22.4}$$

for $v, w \in W$ and $\varphi \in W^*$.

Exercise 22.5*. Show that $[v, \varphi]$ is the element of $\mathfrak{sl}_3\mathbb{C}$ characterized by the formula

$$B([v, \varphi], Z) = 18\varphi(Z \cdot v) \quad \text{for all } Z \in \mathfrak{sl}_3\mathbb{C},$$

where B is the Killing form on $\mathfrak{g}_0 = \mathfrak{sl}_3\mathbb{C}$. In other words, if we write $v * \varphi$ for the element in $\mathfrak{g}_0 = \mathfrak{sl}_3$ satisfying the identity

$$B(v * \varphi, Z) = \varphi(Z \cdot v) \quad \text{for all } Z \in \mathfrak{g}_0 = \mathfrak{sl}_3\mathbb{C}, \tag{22.6}$$

then the bracket $[v, \varphi]$ can be written in the form

$$[v, \varphi] = 18 \cdot v * \varphi. \tag{22.7}$$

It is now a relatively painless task to verify the Jacobi identity, since, rather than having to check it for triples from a basis, it suffices to check it on triples of arbitrary elements of the three spaces $\mathfrak{g}_0$, W, and W^* using the above linear algebra descriptions for the brackets. We will write out this exercise, since the same reasoning will be used later. For example, for three or two elements from $\mathfrak{g}_0$, this amounts to the fact that $\mathfrak{g}_0 = \mathfrak{sl}_3\mathbb{C}$ is a Lie algebra and W and W^* are representations.

For one element Z in $\mathfrak{g}_0$, and two elements v and w in W, the Jacobi identity for these three elements is equivalent to the identity

$$Z \cdot (v \wedge w) = (Z \cdot v) \wedge w + v \wedge (Z \cdot w),$$

which we know for the action of a Lie algebra on an exterior product; and similarly for one element in $\mathfrak{g}_0$ and two in W^*.

The Jacobi identity for $Z \in \mathfrak{g}_0$, $v \in W$, and $\varphi \in W^*$ amounts to

$$[Z, v * \varphi] = (Z \cdot v) * \varphi + v * (Z \cdot \varphi).$$

Applying $B(Y, —)$ to both sides, and using the identity $B(Y, [Z, X]) = B([Y, Z], X)$, this becomes

$$\varphi([Y, Z] \cdot v) = \varphi(Y \cdot (Z \cdot v)) + (Z \cdot \varphi)(Y \cdot v).$$

Since $\varphi([Y, Z] \cdot v) = \varphi(Y \cdot (Z \cdot v)) - \varphi(Z \cdot (Y \cdot v))$, this reduces to

$$(Z \cdot \varphi)(w) = -\varphi(Z \cdot w),$$

for $w = Y \cdot v$, which comes from the fact that W and W^* are dual representations.

For triples u, v, w in W, the Jacobi identity is similarly reduced to the identity

$$(u \wedge v)(Z \cdot w) + (v \wedge w)(Z \cdot u) + (w \wedge u)(Z \cdot v) = 0$$

for all $z \in \mathfrak{g}_0$, which amounts to

$$u \wedge v \wedge (Z \cdot w) + u \wedge (Z \cdot v) \wedge w + (Z \cdot u) \wedge v \wedge w$$
$$= Z \cdot (u \wedge v \wedge w) = 0 \quad \text{in } \textstyle\bigwedge^3 W = \mathbb{C};$$

and similarly for triples from W^*.

For $v, w \in W$, and $\varphi \in W^*$, noting that

$$[[v, w], \varphi] = -2 \cdot [v \wedge w, \varphi] = -4 \cdot (v \wedge w) \wedge \varphi = -4 \cdot (\varphi(v)w - \varphi(w)v),$$

the Jacobi identity for these elements reads

$$-4\cdot(\varphi(v)w - \varphi(w)v) = -[w, \varphi](v) + [v, \varphi](w). \tag{22.8}$$

The right-hand side is

$$-[w, \varphi](v) + [v, \varphi](w) = -(3\varphi(v)w - \varphi(w)v) + (3\varphi(w)v - \varphi(v)w),$$

which proves this case. (This last line was the only place where we needed to use the definition (22.4) in place of the fancier (22.7).)

The last case is for one element v in W and two elements φ and ψ in W^*. This time identity to be proved comes down to

$$-4\cdot(\psi(v)\varphi - \varphi(v)\psi) = [v, \varphi]\cdot\psi - [v, \psi]\cdot\varphi.$$

Applying both sides to an element w in W, this becomes

$$-4\cdot(\psi(v)\varphi(w) - \varphi(v)\psi(w)) = \varphi([v, \psi]\cdot w) - \psi([v, \varphi]\cdot w).$$

If we apply ψ to the previous case (22.8) we have

$$-4\cdot(\varphi(v)\psi(w) - \varphi(w)\psi(v)) = -\psi([w, \varphi]\cdot v) + \psi([v, \varphi]\cdot w).$$

And these are the same, using the symmetry of the Killing form:

$$18\cdot\varphi([v, \psi]\cdot w) = B([v, \psi], [w, \varphi]) = B([w, \varphi], [v, \psi]) = 18\psi([w, \varphi]\cdot v).$$

This completes the proof that the algebra with multiplication table (22.1) is a Lie algebra. With the hindsight derived from working all this out, of course, we see that there is a quicker way to construct $\mathfrak{g}_2$, without any multiplication table: simply start with $\mathfrak{sl}_3\mathbb{C} \oplus W \oplus W^*$, and define products according to the above rules.

§22.3. Representations of $\mathfrak{g}_2$

We would now like to use the standard procedure, outlined in Lecture 14 (and carried out for the classical Lie algebras in Lectures 15–20) to say something about the representations of $\mathfrak{g}_2$. One nice aspect of this is that, working simply from the root system of $\mathfrak{g}_2$ and analyzing its representations, we will arrive at what is perhaps the simplest description of the algebra: we will see that $\mathfrak{g}_2$ is the algebra of endomorphisms of a seven-dimensional vector space preserving a general trilinear form.

The first step is to find the weight lattice for $\mathfrak{g}_2$. This is the lattice $\Lambda_W \subset \mathfrak{h}^*$ dual to the lattice $\Gamma_W \subset \mathfrak{h}$ generated by the six distinguished elements H_i. By (22.2), Γ_W is generated by H_1 and H_2. Since the values of the eigenvalues α_1 and α_2 on H_1 and H_2 are given by

$$\alpha_1(H_1) = 2, \qquad \alpha_1(H_2) = -1,$$

$$\alpha_2(H_1) = -3, \qquad \alpha_2(H_2) = 2,$$

it follows that the weight lattice is generated by the eigenvalues α_1 and α_2 (and in particular the weight lattice Λ_W is equal to the root lattice Λ_R). The picture is thus

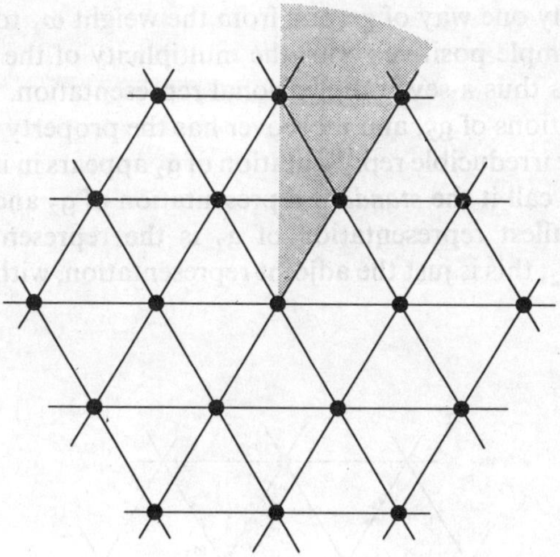

As in the case of the classical Lie algebras, the intersection of the (closed) Weyl chamber $\mathscr{W}$ with the weight lattice is a free semigroup on the two fundamental weights

$$\omega_1 = 2\alpha_1 + \alpha_2 \quad \text{and} \quad \omega_2 = 3\alpha_1 + 2\alpha_2.$$

Any irreducible representation of $\mathfrak{g}_2$ will thus have a highest weight vector λ which is a non-negative linear combination of these two. As usual, we write $\Gamma_{a,b}$ for the irreducible representation with highest weight $a\omega_1 + b\omega_2$.

Let us consider first the representation $\Gamma_{1,0}$ with highest weight ω_1. Translating ω_1 around by the action of the Weyl group, we see that the weight diagram of $\Gamma_{1,0}$ looks like

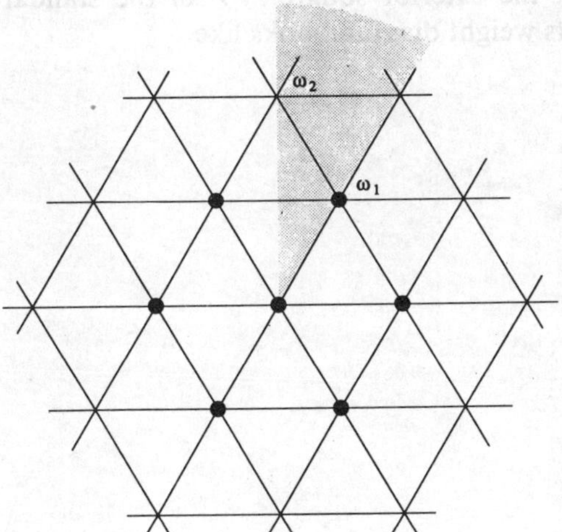

Since there is only one way of getting from the weight ω_1 to the weight 0 by subtraction of simple positive roots, the multiplicity of the weight 0 in $\Gamma_{1,0}$ must be 1. $\Gamma_{1,0}$ is thus a seven-dimensional representation. It is the smallest of the representations of $\mathfrak{g}_2$, and moreover has the property (as we will verify below) that every irreducible representation of $\mathfrak{g}_2$ appears in its tensor algebra; we will therefore call it the *standard* representation of $\mathfrak{g}_2$ and denote it V.

The next smallest representation of $\mathfrak{g}_2$ is the representation $\Gamma_{0,1}$ with highest weight ω_2; this is just the adjoint representation, with weight diagram

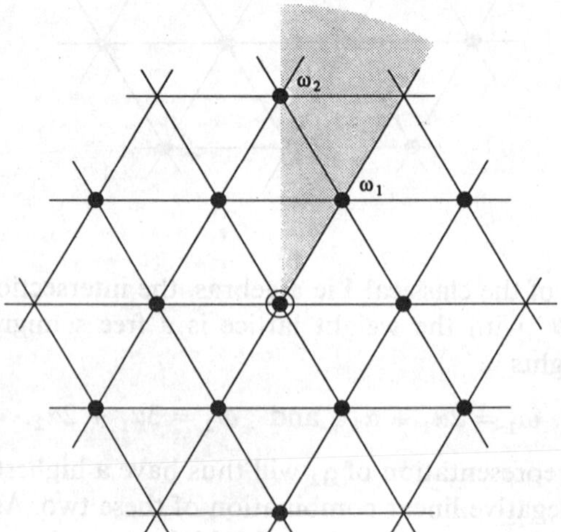

Note that the multiplicity of 0 as a weight of $\Gamma_{0,1}$ is 2, and the dimension of $\Gamma_{0,1}$ is 14.

Consider next the exterior square $\wedge^2 V$ of the standard representation $V = \Gamma_{1,0}$ of $\mathfrak{g}_2$. Its weight diagram looks like

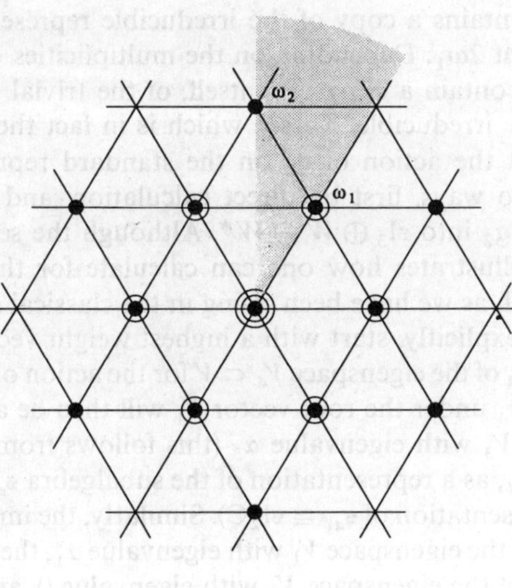

from which we may deduce that

$$\bigwedge^2 V \cong \Gamma_{0,1} \oplus V.$$

In particular, since the adjoint representation $\Gamma_{0,1}$ of $\mathfrak{g}_2$ is contained in $\bigwedge^2 V$, and the irreducible representation $\Gamma_{a,b}$ with highest weight $a\omega_1 + b\omega_2$ is contained in the tensor product $\mathrm{Sym}^a V \otimes \mathrm{Sym}^b \Gamma_{0,1}$, we see that *every irreducible representation of* $\mathfrak{g}_2$ *appears in some tensor power* $V^{\otimes m}$ *of the standard representation*, as stated above.

Next, look at the symmetric square $\mathrm{Sym}^2 V$ of the standard representation. It has weight diagram

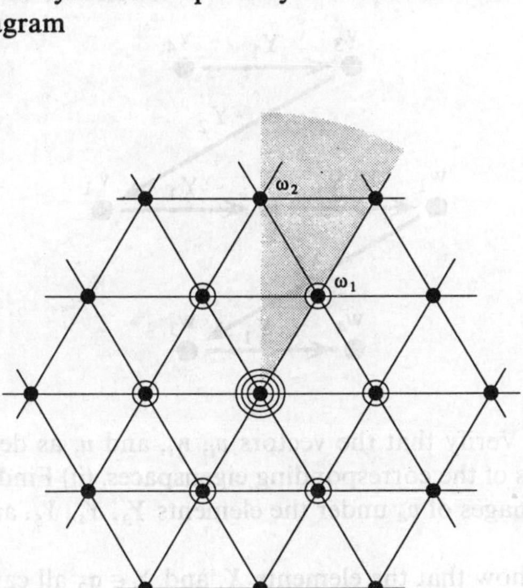

Clearly, this contains a copy of the irreducible representation $\Gamma_{2,0}$ of $\mathfrak{g}_2$ with highest weight $2\omega_1$. Depending on the multiplicities of this representation, it may also contain a copy of V itself, of the trivial representation, or both; or it may be irreducible. To see which is in fact the case, we need to know more about the action of $\mathfrak{g}_2$ on the standard representation V. We will do this in two ways, first by direct calculation, and second using the decomposition of $\mathfrak{g}_2$ into $\mathfrak{sl}_3 \oplus W \oplus W^*$. Although the second approach is shorter, the first illustrates how one can calculate for the exceptional Lie algebras very much as we have been doing in the classical cases.

To describe V explicitly, start with a highest weight vector for V, i.e., any nonzero element v_4 of the eigenspace $V_4 \subset V$ for the action of $\mathfrak{h}$ with eigenvalue α_4. The image of v_4 under the root vector Y_1 will then be a nonzero element of the eigenspace V_3 with eigenvalue α_3 (this follows from the fact that the direct sum $V_3 \oplus V_4$, as a representation of the subalgebra $\mathfrak{s}_{\alpha_1} \subset \mathfrak{g}$, is a copy of the standard representation of $\mathfrak{s}_{\alpha_1} \cong \mathfrak{sl}_2\mathbb{C}$). Similarly, the image of v_3 under Y_2 is a generator v_1 of the eigenspace V_1 with eigenvalue α_1, the image of v_1 under Y_1 is a generator of the eigenspace V_0 with eigenvalue 0, and so on. We may thus choose as a basis for V the vectors

$$v_4, \qquad v_3 = Y_1(v), \qquad v_1 = -Y_2(v_3), \qquad u = Y_1(v_1),$$

$$w_1 = \tfrac{1}{2}Y_1(u), \qquad w_3 = Y_2(w_1), \qquad \text{and} \qquad w_4 = -Y_1(w_3),$$

where v_i (resp. w_i) is an eigenvector with eigenvalue α_i (resp. β_i). (The signs and coefficient $\frac{1}{2}$ in the definition of w_1 are there for reasons of symmetry—see Exercise 22.10.) Diagrammatically, the action of $\mathfrak{g}_2$ may be represented by the arrows

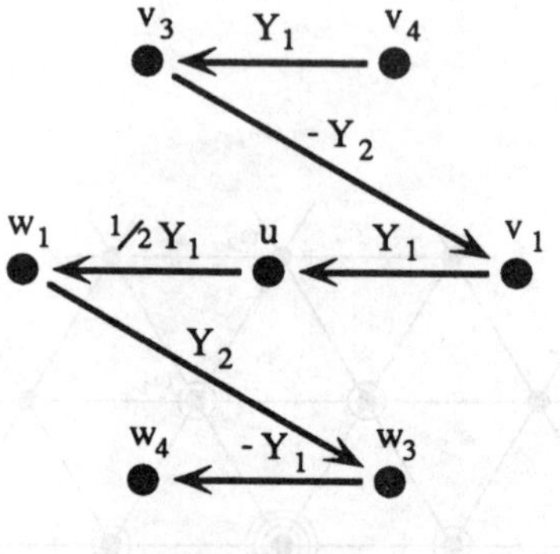

Exercise 22.9. (i) Verify that the vectors v_i, w_i, and u, as defined above, are indeed generators of the corresponding eigenspaces. (ii) Find, in terms of this basis for V, the images of v_4 under the elements Y_3, Y_4, Y_5, and Y_6.

Exercise 22.10. Show that the elements X_i and $Y_i \in \mathfrak{g}_2$ all carry basis vectors v_j and w_j into other basis vectors, up to sign (or to zero, of course), and carry u to twice basis vectors, that is, $X_i u = 2v_i$ and $Y_i u = 2w_i$ for $i = 1, 3, 4$.

Now, the representation $\mathrm{Sym}^2 V$ has, as basis, the pairwise products of the basis vectors for V; and the subrepresentation $\Gamma_{2,0}$ is just the subspace generated by the images of the highest weight vector v_4^2 under (repeated applications of) the generators Y_1, Y_2 of the negative root spaces of $\mathfrak{g}_2$. Thus, for example, the eigenspace in $\mathrm{Sym}^2 V$ with eigenvalue α_4 is the span of the products $u \cdot v_4$ and $v_3 \cdot v_1$; the part of this lying in $\Gamma_{2,0}$ will be the span of the two vectors $Y_2 Y_1 Y_1(v_4^2)$ and $Y_1 Y_2 Y_1(v_4^2)$. We calculate:

$$\begin{aligned} Y_2 Y_1 Y_1(v_4^2) &= Y_2 Y_1(2v_3 \cdot v_4) = Y_2(2v_3^2) \\ &= -4v_1 \cdot v_3 \end{aligned}$$

and

$$\begin{aligned} Y_1 Y_2 Y_1(v_4^2) &= Y_1 Y_2(2v_3 \cdot v_4) = -Y_1(2v_1 \cdot v_4) \\ &= -2v_1 \cdot v_3 - 2u \cdot v_4. \end{aligned}$$

We see, in other words, that $\Gamma_{2,0}$ assumes the weight α_4 with multiplicity 2, so that in particular $\mathrm{Sym}^2 V$ does not contain a copy of V.

Similarly, to see whether or not $\mathrm{Sym}^2 V$ contains a copy of the trivial representation, we have to calculate the multiplicity of the weight 0 in $\Gamma_{2,0}$. Since any path in the weight lattice from the eigenvalue $2\alpha_4$ to 0 obtained by subtracting α_1 and α_2 must pass through α_4, we can do this by evaluating the products of Y_1 and Y_2 on the generators $v_1 \cdot v_3$ and $u \cdot v_4$ of the eigenspace with eigenvalue α_4: we have

$$\begin{aligned} Y_1 Y_1 Y_2(v_1 v_3) &= -Y_1 Y_1(v_1^2) = -Y_1(2u \cdot v_1) \\ &= -4w_1 \cdot v_1 - 2u^2; \\ Y_1 Y_1 Y_2(u \cdot v_4) &= 0; \\ Y_1 Y_2 Y_1(v_1 v_3) &= Y_1 Y_2(u \cdot v_3) = -Y_1(u \cdot v_1) \\ &= -2w_1 \cdot v_1 - u^2; \\ Y_1 Y_2 Y_1(u \cdot v_4) &= Y_1 Y_2(u \cdot v_3 + 2w_1 \cdot v_4) \\ &= Y_1(-u \cdot v_1 + 2w_3 \cdot v_4) \\ &= -2w_1 \cdot v_1 - u^2 - 2w_4 \cdot v_4 + 2w_3 v_3; \\ Y_2 Y_1 Y_1(v_1 v_3) &= Y_2 Y_1(u \cdot v_3) = Y_2(2w_1 \cdot v_3) \\ &= -2w_1 \cdot v_1 + 2w_3 \cdot v_3; \end{aligned}$$

and

$$\begin{aligned} Y_2 Y_1 Y_1(u \cdot v_4) &= Y_2 Y_1(u \cdot v_3 + 2w_1 \cdot v_4) = Y_2(4w_1 \cdot v_3) \\ &= -4w_1 \cdot v_1 + 4w_3 \cdot v_3. \end{aligned}$$

We see from this that the 0-eigenspace of $\Gamma_{2,0}$ is three dimensional; we thus have the decomposition

$$\mathrm{Sym}^2 V \cong \Gamma_{2,0} \oplus \mathbb{C}.$$

In particular, we deduce that *the action of* $\mathfrak{g}_2$ *on the standard representation* $V = \mathbb{C}^7$ *preserves a quadratic form*; and correspondingly that the subalgebra $\mathfrak{g}_2 \subset \mathfrak{sl}(V) = \mathfrak{sl}_7\mathbb{C}$ is actually contained in the algebra $\mathfrak{so}_7\mathbb{C}$. We will see this again in the following section, where we will give alternative descriptions of the exceptional Lie algebras, and again in §23.3 where we describe compact homogeneous spaces for Lie groups.

Exercise 22.11. Analyze in general the symmetric powers $\mathrm{Sym}^k V$ of the standard representation V of $\mathfrak{g}_2$.

Finally, consider the exterior cube $\bigwedge^3 V$ of the standard representation. The weight diagram is

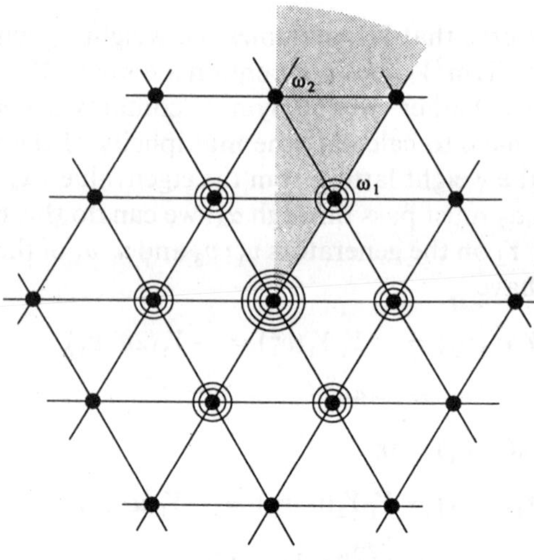

and after we remove one copy of the representation $\Gamma_{2,0}$ with highest weight $2\omega_1$ (this is the sum of the three highest weights α_4, α_3, and α_1 of V), we are left with

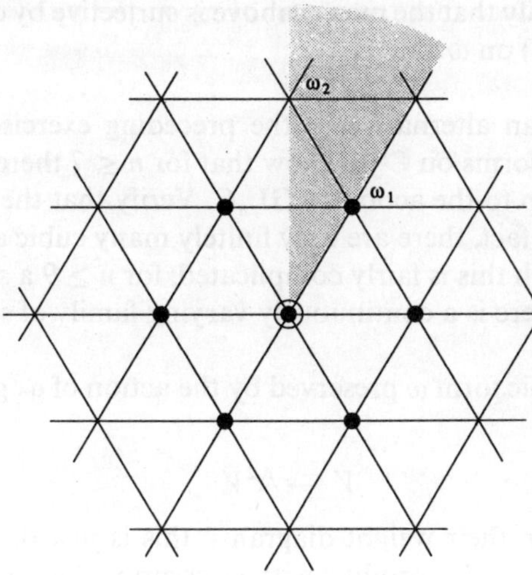

This, by what we have seen, can only be the direct sum of the standard representation V with the trivial representation $\mathbb{C}$. In sum, then, we conclude that

$$\Lambda^3 V \cong \Gamma_{2,0} \oplus V \oplus \mathbb{C}.$$

Note in particular that, as a corollary, *the action of $\mathfrak{g}_2$ on the standard representation preserves a skew-symmetric trilinear form ω on V.* It is not hard to write down this form: it is a linear combination of the five vectors $w_3 \wedge u \wedge v_3$, $v_4 \wedge u \wedge w_4$, $w_1 \wedge u \wedge v_1$, $v_1 \wedge v_3 \wedge w_4$, and $w_1 \wedge w_3 \wedge v_4$; and the fact that it is preserved by X_1 and X_2 is enough to determine the coefficients: we have

$$\omega = w_3 \wedge u \wedge v_3 + v_4 \wedge u \wedge w_4 + w_1 \wedge u \wedge v_1$$
$$+ 2v_1 \wedge v_3 \wedge w_4 + 2w_1 \wedge w_3 \wedge v_4.$$

The fact that the action of $\mathfrak{g}_2$ on V preserves the skew-symmetric cubic form ω takes on additional significance when we make a naive dimension count. The space $\Lambda^3 V$ of all such alternating forms has dimension 35, while the algebra $\mathfrak{gl}(V)$ of endomorphisms of V has dimension 49; the difference is exactly the dimension of the algebra $\mathfrak{g}_2$. In fact, we can check directly that the linear map

$$\varphi\colon \mathfrak{gl}(V) \to \Lambda^3 V$$

sending $A \in \operatorname{End}(V)$ to $A(\omega)$ is surjective. We deduce that ω is a general cubic alternating form [i.e., an open dense subset of $\Lambda^3 V$ corresponds to forms equivalent to ω under $\operatorname{Aut}(V)$], and hence that

Proposition 22.12. *The algebra $\mathfrak{g}_2$ is exactly the algebra of endomorphisms of a seven-dimensional vector space V preserving a general skew-symmetric cubic form ω on V.*

Exercise 22.13*. Verify that the map φ above is surjective by direct calculation of the action of $\mathfrak{gl}(V)$ on $\omega \in \wedge^3 V$.

Exercise 22.14. As an alternative to the preceding exercise, analyze skew-symmetric trilinear forms on $\mathbb{C}^n$ to show that for $n \leq 7$ there are only finitely many such forms, up to the action of $GL_n\mathbb{C}$. Verify that the form ω above is general in $\wedge^3\mathbb{C}^7$. (In fact, there are only finitely many cubic alternating forms on $\mathbb{C}^8$ as well, though this is fairly complicated; for $n \geq 9$ a simple dimension count shows that there is a continuously varying family of such forms.)

Note that the cubic form ω preserved by the action of $\mathfrak{g}_2$ gives us explicitly the inclusion

$$V \hookrightarrow \wedge^2 V$$

deduced earlier from their weight diagrams: this is just the map $V^* \to \wedge^2 V$ given by contraction/wedge product with ω, composed with the isomorphism of V with V^*.

Exercise 22.15*. Find the algebra of endomorphisms of a six-dimensional vector space preserving a general skew-symmetric trilinear form.

We will see the form ω again when we describe $\mathfrak{g}_2$ in the following section.

These calculations using the table amount to using all the information that can be extracted from the subalgebras $\mathfrak{s}_\alpha \cong \mathfrak{sl}_2\mathbb{C}$ of $\mathfrak{g}_2$. Using the copy of $\mathfrak{sl}_3\mathbb{C}$ that we found in the second section can make some of this more transparent. Make the identification

$$\mathfrak{g}_2 = \mathfrak{g}_0 \oplus W \oplus W^* = \mathfrak{sl}_3\mathbb{C} \oplus W \oplus W^*.$$

As a representation of $\mathfrak{sl}_3\mathbb{C}$, the seven-dimensional representation V must be the sum of W, W^*, and the trivial representation $\mathbb{C}$. If we make this identification,

$$V = W \oplus W^* \oplus \mathbb{C},$$

it is not hard to work out how the rest of $\mathfrak{g}_2$ acts. This is given in the following table:

		W	W^*	$\mathbb{C}$
		w	ψ	z
$\mathfrak{g}_0$	X	$X \cdot w$	$X \cdot \psi$	0
W	v	$-v \wedge w$	$\psi(v)$	$2z \cdot v$
W^*	φ	$\varphi(w)$	$\varphi \wedge \psi$	$2z \cdot \varphi$

With this identification, we have $u = 1$ in $\mathbb{C}$, and

$$v_4 = e_1, \qquad w_1 = e_2, \qquad w_3 = e_3 \quad \text{in } W = \mathbb{C}^3;$$

$$w_4 = e_1^*, \qquad v_1 = e_2^*, \qquad v_3 = e_3^* \quad \text{in } W^* = (\mathbb{C}^3)^*.$$

Conversely, it is not hard to verify that the above table defines a representation of $\mathfrak{g}_2$, by checking the various cases of the identity $[\xi, \eta] \cdot y = \xi \cdot (\eta \cdot y) - \eta \cdot (\xi \cdot y)$ for ξ, η in $\mathfrak{g}_2$ and y in V. Note that the cubic form ω becomes

$$\omega = \sum_{i=1}^{3} e_i \wedge u \wedge e_i^* + 2(e_1 \wedge e_2 \wedge e_3 + e_1^* \wedge e_2^* \wedge e_3^*).$$

This description of V can be used to verify the calculations made earlier, and also to study its symmetric and exterior powers. For example, $\mathrm{Sym}^2 V$ decomposes over $\mathfrak{sl}_3\mathbb{C}$ into

$$\begin{aligned}&\mathrm{Sym}^2 W \oplus \mathrm{Sym}^2 W^* \oplus \mathrm{Sym}^2 \mathbb{C} \oplus W \otimes \mathbb{C} \oplus W^* \otimes \mathbb{C} \oplus W \otimes W^* \\ &= \mathrm{Sym}^2 W \oplus \mathrm{Sym}^2 W^* \oplus \mathbb{C} \oplus W \oplus W^* \oplus \mathfrak{sl}_3\mathbb{C} \oplus \mathbb{C}.\end{aligned}$$

To get the weights around the outside ring, the irreducible representation $\Gamma_{2,0}$ must include $\mathrm{Sym}^2 W$, $\mathrm{Sym}^2 W^*$, and $\mathfrak{sl}_3\mathbb{C}$. Checking that $W \subset \mathfrak{g}_2$ maps $\mathrm{Sym}^2 W^*$ nontrivially to W^* shows that it must also include W and W^*. To finish it suffices to compute the part killed by $\mathfrak{g}_2$, which must lie in the sum of the two components which are trivial for $\mathfrak{sl}_3\mathbb{C}$; checking that this is one dimensional, one recovers the decomposition

$$\mathrm{Sym}^2 V = \Gamma_{2,0} \oplus \mathbb{C}.$$

Exercise 22.16. Use this method to decompose $\bigwedge^3 V$ and $\mathrm{Sym}^3 V$.

§22.4. Algebraic Constructions of the Exceptional Lie Algebras

In this section we will sketch a few of the abstract approaches to the construction of the five exceptional Lie algebras. The constructions are not as easy as you might wish: although the exceptional Lie groups and their Lie algebras have a remarkable way of showing up unexpectedly in many areas of mathematics and physics, they do not have such simple descriptions as the classical series. Indeed, they were not discovered until the classification theorem forced mathematicians to look for them.

To begin with, the method we used to construct $\mathfrak{g}_2$ in the second section of this lecture can be generalized to construct other Lie algebras. This is the construction of Freudenthal, which we do first. It can be used to construct the Lie algebra $\mathfrak{e}_8$ for the diagram (E_8). From $\mathfrak{e}_8$ it is possible to construct $\mathfrak{e}_7$ and $\mathfrak{e}_6$ and $\mathfrak{f}_4$. Then we will present (or at least sketch) several other approaches to their construction. Since it is a rather technical subject, probably not really suited for a first course, we will touch on several approaches rather than give a detailed discussion of one.

The construction of $\mathfrak{g}_2$ as a sum $\mathfrak{g}_0 \oplus W \oplus W^*$ that we found in the second section works more generally, with very little change. Suppose $\mathfrak{g}_0$ is a semi-simple Lie algebra, and W is a representation of $\mathfrak{g}_0$; let W^* be the dual representation, and set

$$\mathfrak{g} = \mathfrak{g}_0 \oplus W \oplus W^*.$$

We also need maps

$$\wedge : \Lambda^2 W \to W^* \quad \text{and} \quad \wedge : \Lambda^2 W^* \to W$$

of representations of $\mathfrak{g}_0$. We assume these are given by trilinear maps of $\mathfrak{g}_0$-representations $T: \Lambda^3 W \to \mathbb{C}$ and $T': \Lambda^3 W^* \to \mathbb{C}$, which means that

$$(u \wedge v)(w) = T(u, v, w) \quad \text{and} \quad \vartheta(\varphi \wedge \psi) = T'(\varphi, \psi, \vartheta).$$

We can then define a bracket on $\mathfrak{g}$ by the same rules as in the second section. To describe it, we let $X, Y, Z, \ldots$ denote arbitrary elements of $\mathfrak{g}_0$, $u, v, w, \ldots$ elements of W, and $\varphi, \psi, \vartheta, \ldots$ elements of W^*. The bracket in $\mathfrak{g}$ is determined by setting:

(i) $[X, Y] = [X, Y]$	(the given bracket in $\mathfrak{g}_0$),
(ii) $[X, v] = X \cdot v$	(the action of $\mathfrak{g}_0$ on W),
(iii) $[X, \varphi] = X \cdot \varphi$	(the canonical action of $\mathfrak{g}_0$ on W^*),
(iv) $[v, w] = a \cdot (v \wedge w)$	(for a scalar a to be determined),
(v) $[\varphi, \psi] = b \cdot (\varphi \wedge \psi)$	(for a scalar b to be determined)
(vi) $[v, \varphi] = c \cdot (v * \varphi)$	(for a scalar c to be determined).

As before, $v * \varphi$ is the element of $\mathfrak{g}_0$ such that

$$B(v * \varphi, Z) = \varphi(Z \cdot v) \quad \text{for all } Z \in \mathfrak{g}_0,$$

where B is the Killing form on $\mathfrak{g}_0$. The rules (i)–(vi) determine a bilinear product [,] on all of $\mathfrak{g}$, and the fact that it is skew follows from the facts that $[X, X] = 0$, $[v, v] = 0$, and $[\varphi, \varphi] = 0$.

The argument that we gave showing that $\mathfrak{g}_2$ satisfies the Jacobi identity works in this general case without essential change, except for the last two cases, where explicit calculation is needed. For $v, w \in W$, and $\varphi \in W^*$, the Jacobi identity is equivalent to the identity

$$ab((v \wedge w) \wedge \varphi) = c((v * \varphi) \cdot w - (w * \varphi) \cdot v). \tag{22.17}$$

For $v \in W$, $\varphi, \psi \in W^*$, the Jacobi identity amounts to

$$ab((\varphi \wedge \psi) \wedge v) = c((v * \psi) \cdot \varphi - (v * \varphi) \cdot \psi). \tag{22.18}$$

We will see in Exercise 22.20 that (22.17) and (22.18) are equivalent. Again, the simplicity of the resulting Lie algebra is easy to see, provided all the weight spaces are one dimensional, using Exercise 14.34, so we have:

Proposition 22.19 (Freudenthal). *Given a representation W of a semisimple Lie algebra $\mathfrak{g}_0$ and trilinear forms T and T' inducing maps $\wedge^2 W \to W^*$ and $\wedge^2 W^* \to W$, such that (22.17) and (22.18) are satisfied, the above products make*

$$\mathfrak{g} = \mathfrak{g}_0 \oplus W \oplus W^*$$

into a Lie algebra. If the weight spaces of W are all one dimensional, and the weights of W, W^*, and the roots of $\mathfrak{g}_0$ are all distinct, and $abc \neq 0$, then $\mathfrak{g}$ is semisimple, with the same Cartan subalgebra as $\mathfrak{g}_0$.

Exercise 22.20*. (a) Show that the trilinear map T determines a map $\wedge$: $\wedge^2 W \to W^*$ of representations if and only if it satisfies the identity

$$T(X \cdot u, v, w) + T(u, X \cdot v, w) + T(u, v, X \cdot w) = 0 \qquad \forall X \in \mathfrak{g}_0,$$

and similarly for T'.

(b) Show that each of (22.17) and (22.18) is equivalent to the identity

$$ab \cdot (v \wedge w)(\varphi \wedge \psi) = c \cdot (B(w * \psi, v * \varphi) - B(w * \varphi, v * \psi)).$$

The Lie algebra $\mathfrak{e}_8$ for (E_8) can be constructed by this method. This time $\mathfrak{g}_0$ is taken to be the Lie algebra $\mathfrak{sl}_9\mathbb{C}$; if $V = \mathbb{C}^9$ is the standard representation of $\mathfrak{sl}_9\mathbb{C}$, let $W = \wedge^3 V$, so $W^* = \wedge^3 V^*$; the trilinear map is the usual wedge product

$$\wedge^3 V \otimes \wedge^3 V \otimes \wedge^3 V \to \wedge^9 V = \mathbb{C},$$

and similarly for $\wedge^3 V^*$. We leave the verifications to the reader:

Exercise 22.21*. (i) Verify the conditions on the roots of $\mathfrak{sl}_9$ and the weights of $\wedge^3 V$ and $\wedge^3 V^*$. (ii) Use the fact that $B(X, Y) = 18 \cdot \mathrm{Tr}(XY)$ for $\mathfrak{sl}_9$ to show that (22.17) holds precisely if $c = -18ab$. (iii) Show that the Dynkin diagram of the resulting Lie algebra is (E_8).

Note that the dimension of $\mathfrak{sl}_9\mathbb{C}$ is 80, and that of W and W^* is 84, so the sum has dimension 248, as predicted by the root system of (E_8).

Once the Lie algebra $\mathfrak{e}_8$ is constructed, $\mathfrak{e}_7$ and $\mathfrak{e}_6$ can be found as subalgebras, as follows. Note that removing one or two nodes from the long arm of the Dynkin diagram of (E_8) leads to the Dynkin diagrams (E_7) and (E_6).

In general, if $\mathfrak{g}$ is a simple Lie algebra, with Dynkin diagram D, consider a subdiagram D° of D obtained by removing some subset of nodes, together with all the lines meeting these nodes.[1] Then we can construct a semisimple subalgebra $\mathfrak{g}^\circ$ of $\mathfrak{g}$ with D° as its Dynkin diagram. In fact, $\mathfrak{g}^\circ$ is the subalgebra generated by all the root spaces $\mathfrak{g}_{\pm\alpha}$, where α is a root in D°.

[1] If there are double or triple lines between two nodes, both nodes should be removed or kept together.

Exercise 22.22. (a) Prove this by verifying that the positive roots of $\mathfrak{g}^\circ$ are the positive roots β of $\mathfrak{g}$ that are sums of the roots in D°, and the Cartan subalgebra $\mathfrak{h}^\circ$ is spanned by the corresponding vectors $H_\beta \in \mathfrak{h}$.

(b) Carry this out for $\mathfrak{e}_7$ and $\mathfrak{e}_6$; in particular, show again that $\mathfrak{e}_7$ has 63 positive roots, so dimension $7 + 2(63) = 133$, and $\mathfrak{e}_6$ has 36 positive roots, so dimension $6 + 2(36) = 78$.

Exercise 22.23. For each of the simple Lie algebras, find the subalgebras obtained by removing one node from an end of its Dynkin diagram.

The last exceptional Lie algebra $\mathfrak{f}_4$ can be constructed by taking an invariant subalgebra of $\mathfrak{e}_6$ by an involution. This involution corresponds to the evident symmetry in the Dynkin diagram:

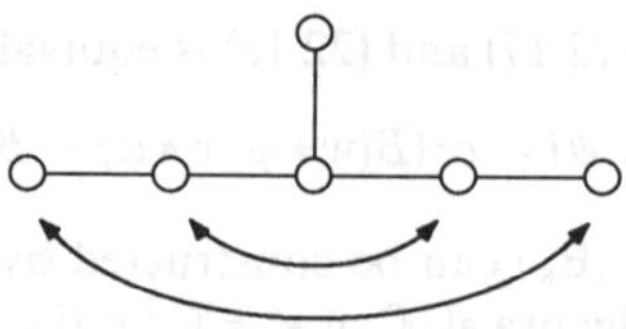

In general, an automorphism of a Dynkin diagram arises from an automorphism of the corresponding semisimple Lie algebra, as follows from the fact that the multiplication table is determined by the Dynkin diagram, cf. Proposition 21.22 and Claim 21.25.

Exercise 22.24*. (a) Show that the invariant subalgebra for the indicated involution of $\mathfrak{e}_6$ is a simple Lie algebra $\mathfrak{f}_4$ with Dynkin diagram (F_4).

(b) Find the invariant subalgebra for the involutions of (A_n) and (D_n), and for an automorphism of order three of (D_4).

Exercise 22.25*. For each automorphism of the Dynkin diagrams (A_n) and (D_n), find an explicit automorphism of $\mathfrak{sl}_{n+1}\mathbb{C}$ and $\mathfrak{so}_{2n}\mathbb{C}$ that induces it.

The exceptional Lie algebras can also be realized as the Lie algebras of derivations of certain nonassociative algebras. This also gives realizations of corresponding Lie groups as groups of automorphism of these algebras (see Exercise 8.28). Some examples of this for associative algebras should be familiar. The group of automorphisms of the algebra $\mathbb{H}$ of (real) quaternions is $O(3)$, so the Lie algebra of derivations is $\mathfrak{so}_3\mathbb{R}$. The Lie algebra of derivations of the complexification $\mathbb{H}_\mathbb{C}$ is $\mathfrak{so}_3\mathbb{C} \cong \mathfrak{sl}_2\mathbb{C}$.

The exceptional group G_2 can be realized as the group of automorphisms of the complexification of the eight-dimensional *Cayley algebra*, or algebra of *octonions*. Recall that the quaternions $\mathbb{H} = \mathbb{C} \oplus \mathbb{C}j$ can be constructed as the set of pairs (a, b) of complex numbers. In a similar way the Cayley algebra, which we denote by $\mathbb{O}$, can be constructed as the set of pairs (a, b), with a and b quaternions. The addition is componentwise, with multiplication

$$(a, b) \circ (c, d) = (ac - \bar{d}b, da + b\bar{c}),$$

where $^-$ denotes conjugation in $\mathbb{H}$. This algebra $\mathbb{O}$ also has a conjugation, which takes (a, b) to $(\bar{a}, -b)$. It has a basis $1 = (1, 0)$, together with seven elements $e_1, \ldots, e_7$:

$$(i, 0), (j, 0), (k, 0), (0, 1), (0, i), (0, j), (0, k).$$

These satisfy $e_p \circ e_p = -1$ and $e_p \circ e_q = -e_q \circ e_p$ for $p \neq q$, and the conjugate $\bar{e}_p$ of e_p is $-e_p$. The multiplication table can be encoded in the diagram:

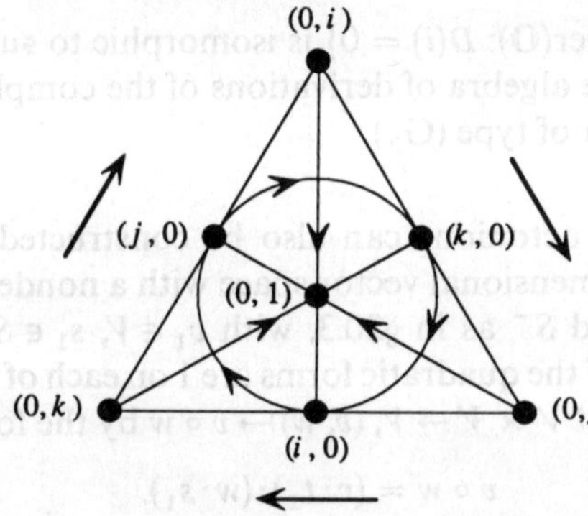

Here, if e_p, e_q, and e_r appear on a line in the order shown by the arrow, then

$$e_p \circ e_q = e_r, \qquad e_q \circ e_r = e_p, \qquad e_q \circ e_p = e_r.$$

Note in particular that any two of these basic elements generate a subalgebra of $\mathbb{O}$ isomorphic to $\mathbb{H}$.

Exercise 22.26. Show that the subalgebra of $\mathbb{O}$ generated by any two elements is isomorphic to $\mathbb{R}$, $\mathbb{C}$, or $\mathbb{H}$. Deduce that, although $\mathbb{O}$ is noncommutative and nonassociative, it is "alternative," i.e., it satisfies the identities $(x \circ x) \circ y = x \circ (x \circ y)$ and $y \circ (x \circ x) = (y \circ x) \circ x$.

A trace and norm can be defined on $\mathbb{O}$ by

$$\mathrm{Tr}(x) = \tfrac{1}{2}(x + \bar{x}), \qquad N(x) = x \circ \bar{x};$$

these satisfy the relation $x^2 - 2\,\mathrm{Tr}(x) + N(x) = 0$. Let $\beta(x, y) = \frac{1}{2}(x \circ \bar{y} + y \circ \bar{x})$ be the bilinear form associated to N; note that the above basis is an orthonormal basis for this inner product.

Let G be the group of algebra automorphisms of the real algebra $\mathbb{O}$. The next exercise sketches a proof that the complexification of G is a Lie group of type (G_2).

Exercise 22.27*. The center of $\mathbb{O}$ is $\mathbb{R}\cdot 1$, which is preserved by G. Let Y be orthogonal space to $\mathbb{R}\cdot 1$ with respect to the quadratic form N. Then G is imbedded in the group $SO(Y)$ of orthogonal transformations of Y.

(a) Define a "cross product" $\times$ on Y by the formula $v \times w = v \cdot w + \beta(v, w)\cdot 1$. Show that G can be identified with the group of orthogonal transformations of Y that preserve the cross product.
(b) Show that $G = \mathrm{Aut}(\mathbb{O})$ acts transitively on the 6-sphere
$$S^6 = \{\sum r_i e_i : \sum r_i^2 = 1\},$$
and the subgroup K that fixes $i = e_1$ is mapped onto the 5-sphere in $e_1^\perp$ by the map $g \mapsto g\cdot j$. Conclude from this that G is 14-dimensional and simply connected.
(c) Show that $\{D \in \mathrm{Der}(\mathbb{O}) : D(i) = 0\}$ is isomorphic to $\mathfrak{su}_3$.
(d) Verify that the Lie algebra of derivations of the complex octonians is the simple Lie algebra of type (G_2).

Exercise 22.28*. The octonions can also be constructed from the Clifford algebra of an eight-dimensional vector space with a nondegenerate quadratic form. With V, S^+, and S^- as in §20.3, with $v_1 \in V$, $s_1 \in S^+$, $t_1 = v_1 \cdot s_1 \in S^-$ chosen so the values of the quadratic forms are 1 on each of them as in Exercise 20.50, define a product $V \times V \to V$, $(v, w) \mapsto v \circ w$ by the formula
$$v \circ w = (v \cdot t_1)\cdot(w \cdot s_1).$$
Note that $v \cdot t_1 \in S^+$, $w \cdot s_1 \in S^-$, so their product $(v \cdot t_1)\cdot(w \cdot s_1)$ is back in V.

(a) Show that V with this product is isomorphic to the complex octonians $\mathbb{O}$, with unit v_1, with the map $v \mapsto -\rho(v_1)(v)$ corresponding to conjugation in $\mathbb{O}$.

Conversely, starting with the complex octonians $\mathbb{O}$, one can reconstruct the algebra of §20.3: define $A = \mathbb{O} \oplus \mathbb{O} \oplus \mathbb{O}$, define an automorphism J of order 3 of A by $J(x, y, z) = (z, x, y)$, and define a product $\cdot$ from each succession of two factors to the third by the formulas $x \cdot y = \bar{x} \circ \bar{y}$, $y \cdot z = \bar{y} \circ \bar{z}$, $z \cdot x = \bar{z} \circ \bar{x}$.

(b) Show that A is isomorphic to the algebra described in §20.3.

(c) Identifying $\mathfrak{so}_8\mathbb{C}$ with the space of skew linear transformations of $\mathbb{O}$, show that for each A in $\mathfrak{so}_8\mathbb{C}$ there are unique B and C in $\mathfrak{so}_8\mathbb{C}$ such that
$$A(x \circ y) = B(x) \circ y + x \circ C(y)$$
for all complex octonions x and y. Equivalently, if one defines a trilinear form $(\ ,\ ,\ )$ on the octonions by $(x, y, z) = \mathrm{Tr}((x \circ y) \circ z) = \mathrm{Tr}(x \circ (y \circ z))$,
$$(Ax, y, z) + (x, By, z) + (x, y, Cz) = 0$$
for all x, y, z. Show that this trilinear form agrees with that defined in Exercise 20.49, and the mapping $A \mapsto B$ determines the triality automorphism j' of $\mathfrak{so}_8\mathbb{C}$ of order three described in Exercise 20.51.

Exercise 22.29. Define three homomorphisms from the real Clifford algebra $C_7 = C(0, 7)$ to $\mathrm{End}_{\mathbb{R}}(\mathbb{O})$ by sending $v \in \mathbb{R}^7 = \sum \mathbb{R}e_i$ to the maps L_v, R_v, and T_v defined by $L_v(x) = v \circ x$, $R_v(x) = x \circ v$, and $T_v(x) = v \circ (x \circ v) = (v \circ x) \circ v$.

(a) Show that these do determine maps of the Clifford algebra, and that the induced maps

$$\mathrm{Spin}_8 \mathbb{R} \hookrightarrow C_8^{\mathrm{even}} = C_7 \to \mathrm{End}_{\mathbb{R}}(\mathbb{O})$$

are the two spin representations and the standard representation, respectively.

(b) Verify that $T_v(x \circ y) = L_v(x) \cdot L_v(y)$ for all v, x, y, and use this to verify the triality formula in (c) of the preceding exercise.

The algebra $\mathfrak{f}_4$ can be realized as the derivation algebra of the complexification of a 27-dimensional *Jordan algebra* $\mathbb{J}$. This can be constructed as the set of matrices of the form

$$\begin{pmatrix} a & \alpha & \beta \\ \bar{\alpha} & b & \gamma \\ \bar{\beta} & \bar{\gamma} & c \end{pmatrix},$$

with a, b, c scalars, and α, β, γ in $\mathbb{O}$. The product $\circ$ in $\mathbb{J}$ is given by

$$x \circ y = \tfrac{1}{2}(xy + yx),$$

where the products on the right-hand side are defined by usual matrix multiplication. This algebra is commutative but not associative, and satisfies the identity $((x \circ x) \circ y) \circ x = (x \circ x) \circ (y \circ x)$. In fact, (F_4) is the group of automorphisms of this 27-dimensional space that preserve the scalar product $(x, y) = \mathrm{Tr}(x \circ y)$ and the scalar triple product $(x, y, z) = \mathrm{Tr}((x \circ y) \circ z)$. The kernel of the trace map is an irreducible 26-dimensional representation of $\mathfrak{f}_4$. For details see [Ch-S], [To], [Pos].

In addition, there is a cubic form "det" on $\mathbb{J}$ such that the linear automorphisms of $\mathbb{J}$ that preserve this form is a group of type (E_6). This again shows $\mathfrak{f}_4$ as a subalgebra of $\mathfrak{e}_6$.

The other exceptional Lie algebras can also be constructed as derivations of appropriate algebras. We refer for this to [Ti2], [Dr], [Fr2], [Jac2], and the references found in these sources. Other constructions were given by Witt, cf. [Wa]. The simple Lie algebras are also constructed explicitly in [S-K, §1]. See also [Ch-S], [Fr1], and [Sc].

What little we will have to say about the representations of the four exceptional Lie algebras besides $\mathfrak{g}_2$ can wait until we have the Weyl character formula.

LECTURE 23
Complex Lie Groups; Characters

This lecture serves two functions. First and foremost, we make the transition back from Lie algebras to Lie groups: in §23.1 we classify the groups having a given semisimple Lie algebra, and say which representations of the Lie algebra, as described in the preceding lectures, lift to which groups. Secondly, we introduce in §23.2 the notion of *character* in the context of Lie theory; this gives us another way of describing the representations of the classical groups, and also provides a necessary framework for the results of the following two lectures. Then in §23.3 we sketch the beautiful interrelationships among Dynkin diagrams, compact homogeneous spaces and the irreducible representations of a Lie group. The first two sections are elementary modulo a little topology needed to calculate the fundamental groups of the classical groups in §23.1. The third section, by contrast, may appear impossible: it involves, at various points, projective algebraic geometry, holomorphic line bundles, and their cohomology. In fact, a good deal of §23.3 can be understood without these notions; the reader is encouraged to read as much of the section as seems intelligible. A final section §23.4 gives a very brief introduction to the related Bruhat decomposition, which is included because of its ubiquity in the literature.

§23.1. Representations of Complex Simple Lie Groups

In Lecture 21 we classified all simple Lie algebras over $\mathbb{C}$. This in turn yields a classification of simple complex Lie groups: as we saw in Lecture 7, for any Lie algebra $\mathfrak{g}$ there is a unique simply connected group G, and all other (connected) complex Lie groups with Lie algebra $\mathfrak{g}$ are quotients of G by discrete subgroups of the center $Z(G)$. In this section, we will first describe the groups associated to the classical Lie algebras, and then proceed to describe which of the representations of the classical algebras we have described in Part III lift to which of the groups. We start with

Proposition 23.1. *For all $n \geq 1$, the Lie groups $\mathrm{SL}_n\mathbb{C}$ and $\mathrm{Sp}_{2n}\mathbb{C}$ are connected and simply connected. For $n \geq 1$, $\mathrm{SO}_n\mathbb{C}$ is connected, with $\pi_1(\mathrm{SO}_2\mathbb{C}) = \mathbb{Z}$, and $\pi_1(\mathrm{SO}_n\mathbb{C}) = \mathbb{Z}/2$ for $n \geq 3$.*

PROOF. The main tool needed from topology is the long exact homotopy sequence of a fibration. If the Lie group G acts transitively on a manifold M, and H is the isotropy group of a point P_0 of M, then $G/H = M$, and the map $G \to M$ by $g \mapsto g \cdot P_0$ is a fibration with fiber H. The resulting long exact sequence is, assuming the spaces are connected,

$$\cdots \to \pi_2(M) \to \pi_1(H) \to \pi_1(G) \to \pi_1(M) \to \{1\}. \tag{23.2}$$

(The base points, which are omitted in this notation, can be taken to be the identity elements of H and G, and the point P_0 in M.) In practice we will know M and H are connected, from which it follows that G is also connected. From this exact sequence, if M and H are also simply connected, the same follows for G.

To apply the long exact homotopy sequence in our present circumstance we argue by induction, noting first that $\mathrm{SL}_1\mathbb{C} = \mathrm{SO}_1\mathbb{C} = \{1\}$. Now consider the action of $G = \mathrm{SL}_n\mathbb{C}$ on the manifold $M = \mathbb{C}^n \backslash \{0\}$. The subgroup H fixing the vector $P_0 = (1, 0, \ldots, 0)$ consists of matrices whose first column is $(1, 0, \ldots, 0)$ and whose lower right $(n-1)$ by $(n-1)$ matrix is in $\mathrm{SL}_{n-1}\mathbb{C}$; it follows that as topological spaces $H \cong \mathrm{SL}_{n-1}\mathbb{C} \times \mathbb{C}^{n-1}$. Since M is simply connected for $n \geq 2$ (having the sphere S^{2n-1} as a deformation retract), and H has $\mathrm{SL}_{n-1}\mathbb{C}$ as a deformation retract, the claim for $\mathrm{SL}_n\mathbb{C}$ follows from (23.2) by induction on n.

The group $\mathrm{SO}_2\mathbb{C}$ is isomorphic to the multiplicative group $\mathbb{C}^*$, which has the circle as a deformation retract, so $\pi_1(\mathrm{SO}_2\mathbb{C}) = \mathbb{Z}$. The group $G = \mathrm{SO}_n\mathbb{C}$ acts transitively on $M = \{v \in \mathbb{C}^n \colon Q(v, v) = 1\}$, where Q is the symmetric bilinear form preserved by G. (The transitivity of the action is more or less equivalent to knowing that all nondegenerate symmetric bilinear forms are equivalent.) For explicit calculations take the standard Q for which the standard basis $\{e_i\}$ of $\mathbb{C}^n$ is an orthonormal basis. This time the subgroup H fixing e_1 is $\mathrm{SO}_{n-1}\mathbb{C}$. From the following exercise, it follows that M has the sphere S^{n-1} as a deformation retract. By (23.2) the map

$$\pi_1(\mathrm{SO}_{n-1}\mathbb{C}) \to \pi_1(\mathrm{SO}_n\mathbb{C})$$

is an isomorphism for $n \geq 4$. So it suffices to look at $\mathrm{SO}_3\mathbb{C}$. This could be done by looking at the maps in the sme exact sequence, but we saw in Lecture 10 that $\mathrm{SO}_3\mathbb{C}$ has a two-sheeted covering by $\mathrm{SL}_2\mathbb{C}$, which is simply connected by the preceding paragraph, so $\pi_1(\mathrm{SO}_3\mathbb{C}) = \mathbb{Z}/2$, as required.

The group $G = \mathrm{Sp}_{2n}\mathbb{C}$ acts transitively on

$$M = \{(v, w) \in \mathbb{C}^{2n} \times \mathbb{C}^{2n}: Q(v, w) = 1\},$$

where Q is the skew form preserved by G, and the isotropy group is $\mathrm{Sp}_{2n-2}\mathbb{C}$. Since $\mathrm{Sp}_2\mathbb{C} = \mathrm{SL}_2\mathbb{C}$, the first case is known. By the following exercise, since M is defined in $\mathbb{C}^{4n}$ by a nondegenerate quadratic form, M has S^{4n-1} as a deformation retract, so we conclude again by induction. □

Exercise 23.3*. Show that $\{(z_1, \ldots, z_n) \in \mathbb{C}^n: \sum z_i^2 = 1\}$ is homeomorphic to the tangent bundle to the $(n-1)$-sphere, i.e., to

$$T_{S^{n-1}} = \{(u, v) \in S^{n-1} \times \mathbb{R}^n: u \cdot v = 0\}.$$

Using the exact sequence $\{1\} \to \mathrm{SL}_n\mathbb{C} \to \mathrm{GL}_n\mathbb{C} \to \mathbb{C}^* \to \{1\}$ we deduce from the proposition and (23.2) that

$$\pi_1(\mathrm{GL}_n\mathbb{C}) = \mathbb{Z}. \tag{23.4}$$

Exercise 23.5. Show that for all the above groups G, the second homotopy groups $\pi_2(G)$ are trivial.

We digress a moment here to mention a famous fact. Each of the above groups G has an associated compact subgroup: $\mathrm{SU}(n) \subset \mathrm{SL}_n\mathbb{C}$, $\mathrm{Sp}(n) \subset \mathrm{Sp}_{2n}\mathbb{C}$, and $\mathrm{SO}(n) \subset \mathrm{SO}_n\mathbb{C}$. In fact, each of these subgroups is connected, and these inclusions induce isomorphisms of their fundamental groups.

Exercise 23.6. Prove these assertions by finding compatible actions of the subgroups on appropriate manifolds. Alternatively, observe that in each case the compact subgroup in question is just the subgroup of G preserving a Hermitian form on $\mathbb{C}^n$ or $\mathbb{C}^{2n}$, and use Gram–Schmidt to give a retraction of G onto the subgroup.

Now, by Proposition 23.1 the simply-connected complex Lie groups corresponding to the Lie algebras $\mathfrak{g} = \mathfrak{sl}_n\mathbb{C}$, $\mathfrak{sp}_{2n}\mathbb{C}$, and $\mathfrak{so}_m\mathbb{C}$ are

$$\tilde{G} = \mathrm{SL}_n\mathbb{C}, \quad \mathrm{Sp}_{2n}\mathbb{C}, \quad \text{and } \mathrm{Spin}_m\mathbb{C}.$$

We also know the center $Z(\tilde{G})$ of each of these groups. From Lecture 7 we also know the other connected groups with these Lie algebras:

- The complex Lie groups with Lie algebra $\mathfrak{sl}_n\mathbb{C}$ are $\mathrm{SL}_n\mathbb{C}$ and quotients of $\mathrm{SL}_n\mathbb{C}$ by subgroups of the form $\{e^{2\pi li/m} \cdot I\}_l$ for m dividing n (in particular, if n is prime the only such groups are $\mathrm{SL}_n\mathbb{C}$ and $\mathrm{PSL}_n\mathbb{C}$).
- The complex Lie groups with Lie algebra $\mathfrak{sp}_{2n}\mathbb{C}$ are $\mathrm{Sp}_{2n}\mathbb{C}$ and $\mathrm{PSp}_{2n}\mathbb{C}$.
- The complex Lie groups with Lie algebra $\mathfrak{so}_{2n+1}\mathbb{C}$ are $\mathrm{Spin}_{2n+1}\mathbb{C}$ and $\mathrm{SO}_{2n+1}\mathbb{C}$.
 and
- The complex Lie groups with Lie algebra $\mathfrak{so}_{2n}\mathbb{C}$ are $\mathrm{Spin}_{2n}\mathbb{C}$, $\mathrm{SO}_{2n}\mathbb{C}$ and $\mathrm{PSO}_{2n}\mathbb{C}$; in addition, if n is even, there are two other groups covered doubly by $\mathrm{Spin}_{2n}\mathbb{C}$ and covering doubly $\mathrm{PSO}_{2n}\mathbb{C}$ [cf. Exercise 20.36].

These are called the *classical groups*. In the cases where we have observed coincidences of Lie algebras, we have the following isomorphisms of groups:

$$\mathrm{Spin}_3\mathbb{C} \cong \mathrm{SL}_2\mathbb{C} \quad \text{and} \quad \mathrm{SO}_3\mathbb{C} \cong \mathrm{PSL}_2\mathbb{C};$$

$$\mathrm{Spin}_4\mathbb{C} \cong \mathrm{SL}_2\mathbb{C} \times \mathrm{SL}_2\mathbb{C} \quad \text{and} \quad \mathrm{PSO}_4\mathbb{C} \cong \mathrm{PSL}_2\mathbb{C} \times \mathrm{PSL}_2\mathbb{C};$$

$$\mathrm{Spin}_5\mathbb{C} \cong \mathrm{Sp}_4\mathbb{C} \quad \text{and} \quad \mathrm{SO}_5\mathbb{C} \cong \mathrm{PSp}_4\mathbb{C};$$

and

$$\mathrm{Spin}_6\mathbb{C} \cong \mathrm{SL}_4\mathbb{C} \quad \text{and} \quad \mathrm{PSO}_6\mathbb{C} \cong \mathrm{PSL}_4\mathbb{C}.$$

Note that in the first case $n = 4$ where there is an intermediate subgroup between $\mathrm{SL}_n\mathbb{C}$ and $\mathrm{PSL}_n\mathbb{C}$, the subgroup in question is interesting: it turns out to be $\mathrm{SO}_6\mathbb{C}$. In general, however, these intermediate groups seldom arise.

Consider now representations of these classical groups. According to the basic result of Lecture 7, representations of a complex Lie algebra $\mathfrak{g}$ will correspond exactly to representations of the associated simply connected Lie group $\tilde{G}$: specifically, for any representation

$$\rho\colon \mathfrak{g} \to \mathfrak{gl}(V)$$

of $\mathfrak{g}$, setting

$$\tilde{\rho}(\exp(X)) = \exp(\rho(X))$$

determines a well-defined homomorphism

$$\tilde{\rho}\colon \tilde{G} \to \mathrm{GL}(V).$$

For any other group with algebra $\mathfrak{g}$, given as the quotient $\tilde{G}/C$ of $\tilde{G}$ by a subgroup $C \subset Z(\tilde{G})$, the representations of G are simply the representations of $\tilde{G}$ trivial on C. It is therefore enough to see which of the representations of the classical Lie algebras described in Part III are trivial on which subgroups $C \subset Z(\tilde{G})$.

This turns out to be very straightforward. To begin with, we observe that the center of each group G with Lie algebra $\mathfrak{g}$ lies in the image of the chosen Cartan subalgebra $\mathfrak{h} \subset \mathfrak{g}$ under the exponential map. It will therefore be enough to know when $\exp(\rho(X)) = I$ for $X \in \mathfrak{h}$; and since the representations ρ of $\mathfrak{g}$ are particularly simple on $\mathfrak{h}$ this presents no difficulty.

What we do have to do first is to describe the restriction of the exponential map to $\mathfrak{h}$, so that we can say which elements of $\mathfrak{h}$ exponentiate to elements of $Z(\tilde{G})$. For the groups that are given as matrix groups, this will all be perfectly obvious, but for the spin groups we will need to do a little calculation. We will also want to describe the *Cartan subgroup* H of each of the classical groups G, which is the connected subgroup whose Lie algebra is the Cartan subalgebra $\mathfrak{h}$ of $\mathfrak{g}$. For $G = \mathrm{SL}_n\mathbb{C}$, H is just the diagonal matrices in G, i.e.,

$$H = \{\mathrm{diag}(z_1, \ldots, z_n)\colon z_1 \cdot \ldots \cdot z_n = 1\}.$$

Similarly in $\mathrm{Sp}_{2n}\mathbb{C}$ or $\mathrm{SO}_{2n}\mathbb{C}$, $H = \{\mathrm{diag}(z_1, \ldots, z_n, z_1^{-1}, \ldots, z_n^{-1})\}$, whereas in $\mathrm{SO}_{2n+1}\mathbb{C}$, $H = \{\mathrm{diag}(z_1, \ldots, z_n, z_1^{-1}, \ldots, z_n^{-1}, 1)\}$. In each of these cases the exponential mapping from $\mathfrak{h}$ to H is just the usual exponentiation of diagonal matrices.

To calculate the exponential mapping for $\mathrm{Spin}_m\mathbb{C}$, we need to describe the elements in $\mathrm{Spin}_m\mathbb{C}$ that lie over the diagonal matrices in $\mathrm{SO}_m\mathbb{C}$. This is not a difficult task. Calculating as in §20.2, we find that for any nonzero complex number z and any $1 \leq j \leq n$, and with $m = 2n + 1$ or $m = 2n$, the elements

$$w_j(z) = \frac{1}{2}(ze_j \cdot e_{n+j} + z^{-1}e_{n+j} \cdot e_j) = z^{-1} + \left(\frac{z - z^{-1}}{2}\right)e_j \cdot e_{n+j} \tag{23.7}$$

in the Clifford algebra are in fact elements of $\mathrm{Spin}_m\mathbb{C}$. Moreover, if $\rho\colon \mathrm{Spin}_m\mathbb{C} \to \mathrm{SO}_m\mathbb{C}$ is the covering, the image $\rho(w_j(z))$ is the diagonal matrix whose jth entry is z^2, $(n + j)$th entry is z^{-2}, and other diagonal entries are 1. These elements $w_j(z)$ also commute with each other, so for any nonzero complex numbers $z_1, \ldots, z_n$ we can define

$$w(z_1, \ldots, z_n) = w_1(z_1) \cdot w_2(z_2) \cdot \ldots \cdot w_n(z_n). \tag{23.8}$$

Then $\rho(w(z_1, \ldots, z_n)) = \mathrm{diag}(z_1^2, \ldots, z_n^2, z_1^{-2}, \ldots, z_n^{-2})$ if $m = 2n$, while if $m = 2n + 1$, we get the same diagonal matrix but with a 1 at the end.

Let $H_i = E_{i,i} - E_{n+i,n+i}$, the usual basis for $\mathfrak{h} \subset \mathfrak{so}_m\mathbb{C}$.

Lemma 23.9. *For any complex numbers* $a_1, \ldots, a_n$,

$$\exp(a_1 H_1 + \cdots + a_n H_n) = w(e^{a_1/2}, \ldots, e^{a_n/2})$$

in $\mathrm{Spin}_m\mathbb{C}$.

PROOF. Since the map $\exp\colon \mathfrak{h} \to \mathrm{Spin}_m\mathbb{C}$ is determined by the facts that it is continuous, it takes 0 to 1, and its composite with ρ is the exponential for $\mathrm{SO}_m\mathbb{C}$, this follows from the preceding formulas. □

Exercise 23.10*. Show that $\exp(\sum a_j H_j) = 1$ if and only if each a_j is in $2\pi i\mathbb{Z}$ and $\sum a_j \in 4\pi i\mathbb{Z}$.

We see also that $\exp(\mathfrak{h})$ contains the center of $\mathrm{Spin}_m\mathbb{C}$. Indeed, $-1 = w(-1, 1, \ldots, 1)$, and if m is even, the other central elements are $\pm\omega$, with $\omega = w(i, \ldots, i)$, as we calculated in Exercise 20.36. (This, of course, also contains the fact that there is a path between 1 and -1, proving again that $\mathrm{Spin}_m\mathbb{C}$ is connected.)

Exercise 23.11*. Verify for all the classical groups G that: (i) $H = \exp(\mathfrak{h})$ is a closed subgroup of G that contains the center of G; (ii) the map of fundamental groups $\pi_1(H, e) \to \pi_1(G, e)$ is surjective; (iii) for any connected covering $\pi\colon G' \to G$, $\pi^{-1}(H)$ is connected and is the Cartan subgroup of G'.

Now let $G = \tilde{G}/C$ be a semisimple Lie group with Lie algebra $\mathfrak{g}$ and Cartan subalgebra $\mathfrak{h}$. Choose an ordering of the roots, and let Γ_λ be the irreducible representation of $\mathfrak{g}$ with highest weight λ. The basic fact that we need is

Lemma 23.12. *The representation Γ_λ is a representation of $G = \tilde{G}/C$ if and only if*

$$\lambda(X) \in 2\pi i\mathbb{Z} \quad \textit{whenever } \exp(X) \in C.$$

PROOF. The representation Γ_λ is a representation of G when $g \cdot v = v$ for all $g \in C$, where v is a highest weight vector in Γ_λ. Since $\exp(\mathfrak{h})$ contains C, this says $\exp(X) \cdot v = v$ for all $X \in \mathfrak{h}$ such that $\exp(X) \in C$. Now by the naturality of the exponential map, and since $X \cdot v = \lambda(X)v$ for $X \in \mathfrak{h}$, we have $\exp(X) \cdot v = e^{\lambda(X)}v$. Hence the condition is that $e^{\lambda(X)}v = v$, or that $e^{\lambda(X)} = 1$ if $\exp(X) \in C$, which is the displayed criterion. □

Let us work this out explicitly for each of the classical groups. It may help to introduce a notation for the irreducible representations which, among other virtues, allows some common terminology in the various cases. Note that for each of $\mathfrak{sl}_{n+1}$, $\mathfrak{sp}_{2n}$, $\mathfrak{so}_{2n}$, and $\mathfrak{so}_{2n+1}$ the root space $\mathfrak{h}^*$ is spanned by weights we have called $L_1, \ldots, L_n$, so a weight can be written uniquely in form $\lambda_1 L_1 + \cdots + \lambda_n L_n$. We may sometimes write λ in place of the weight $\lambda_1 L_1 + \cdots + \lambda_n L_n$. In the rest of this lecture at least, we write Γ_λ for the irreducible representation with highest weight $\lambda_1 L_1 + \cdots + \lambda_n L_n$. Note that by our choice of Weyl chambers the highest weights $\lambda = (\lambda_1, \ldots, \lambda_n)$ that arise satisfy

$$\lambda_1 \geq \lambda_2 \geq \cdots \geq \lambda_n \geq 0 \quad \text{for } \mathfrak{sl}_{n+1}, \mathfrak{sp}_{2n}, \text{ and } \mathfrak{so}_{2n+1},$$

where the λ_i are all integers in the first two cases, and for $\mathfrak{so}_{2n+1}$ they are either all integers or all half-integers; and

$$\lambda_1 \geq \lambda_2 \geq \cdots \geq \lambda_{n-1} \geq |\lambda_n| \geq 0 \quad \text{for } \mathfrak{so}_{2n},$$

with the λ_i all integers or all half-integers.

Proposition 23.13. *For each subgroup C of the center of $\tilde{G}$, the representation Γ_λ is a representation of $\tilde{G}/C$ precisely under the following conditions*:

(i) $\tilde{G} = \mathrm{SL}_{n+1}\mathbb{C}$, *$C$ has order m dividing $n+1$*: $\sum \lambda_j \equiv 0 \bmod(m)$.
(ii) $\tilde{G} = \mathrm{Sp}_{2n}\mathbb{C}$, $C = \{\pm 1\}$: *$\sum \lambda_j$ is even.*
(iii) $\tilde{G} = \mathrm{Spin}_{2n}\mathbb{C}$ *or* $\mathrm{Spin}_{2n+1}\mathbb{C}$, $C = \{\pm 1\}$: *all λ_i are integers.*
(iv) $\tilde{G} = \mathrm{Spin}_{2n}\mathbb{C}$, $C = \{\pm 1, \pm\omega\}$: *all λ_i are integers, $\sum \lambda_j$ is even.*
(v) $\tilde{G} = \mathrm{Spin}_{2n}\mathbb{C}$, *$n$ even*, $C = \{1, \omega\}$: *$\sum \lambda_j$ is an even integer; and for* $C = \{1, -\omega\}$: *$\sum \lambda_j - n/2$ is an odd integer.*

In particular, representations of $\mathrm{PSL}_{n+1}\mathbb{C}$ are given by partitions λ with $\sum \lambda_j \equiv 0 \bmod(n+1)$, and those for $\mathrm{PSp}_{2n}\mathbb{C}$ have $\sum \lambda_j$ even. Case (iii) verifies what we saw in Lecture 19 about representations of $\mathrm{SO}_m\mathbb{C}$. Representations of $\mathrm{PSO}_m\mathbb{C}$ correspond to integral partitions λ with $\sum \lambda_j$ even.

PROOF. With the preceding lemma and the explicit description of everything in sight, the calculations are routine. In case (i), for example, a generator for

C is of the form $\exp(X)$, with

$$X = (2\pi i/m)\left(\sum_{j=1}^{n} E_{j,j} - nE_{n+1,n+1}\right),$$

and so $\lambda(X) = (2\pi i/m)(\sum \lambda_j)$ will be a multiple of $2\pi i$ exactly when $\sum \lambda_j$ is divisible by m. For $\mathrm{Sp}_{2n}\mathbb{C}$, $\exp(X) = -1$ when $X = \pi i(\sum H_j)$, so $\lambda(X) = \pi i \sum \lambda_j$, and (ii) follows. The calculations are similar for $\mathrm{Spin}_m\mathbb{C}$, noting that $\exp(2\pi i(H_1) = -1$ and $\exp(\pi i(\sum H_j)) = \omega$. □

By way of an example, recall that any irreducible representation of $\mathfrak{sl}_2\mathbb{C}$ is of the form $\mathrm{Sym}^k V$, where V is the standard two-dimensional representation. Any such representation, of course, lifts to the group $\mathrm{SL}_2\mathbb{C}$; but it lifts to $\mathrm{PSL}_2\mathbb{C} \cong \mathrm{SO}_3\mathbb{C}$ if and only if k is even (in particular, the "standard" representation of $\mathrm{SO}_3\mathbb{C}$ on $\mathbb{C}^3$ is the symmetric square $\mathrm{Sym}^2 V$). For another example, we have seen that any irreducible representation of $\mathfrak{sp}_4\mathbb{C}$ may be found in a tensor product $\mathrm{Sym}^k V \otimes \mathrm{Sym}^l W$, where V is the standard four-dimensional representation of $\mathfrak{sp}_4\mathbb{C}$ and $W \subset \wedge^2 V$ the complement of the trivial one-dimensional representation. All such representations lift to $\mathrm{Sp}_4\mathbb{C}$, but they lift to $\mathrm{PSp}_4\mathbb{C} \cong \mathrm{SO}_5\mathbb{C}$ if and only if k is even—equivalently, if they are contained in a representation of the form $\mathrm{Sym}^l W \otimes \mathrm{Sym}^k(\wedge^2 W)$, where W is the "standard" representation of $\mathrm{SO}_5\mathbb{C}$.

Exercise 23.14. Show that each of these semisimple complex Lie groups G has a finite-dimensional faithful representation.

The result of the proposition can be put in a more formal setting, which brings out a feature that our alert reader has surely noticed: the center of the simply-connected form of $\mathfrak{g}$ is isomorphic to the quotient group Λ_W/Λ_R of the weight lattice modulo the root lattice. We note first that this abelian group Λ_W/Λ_R is *finite*. We have seen this for the classical Lie algebras. In general, we have

Lemma 23.15. *The group Λ_W/Λ_R is finite, of order equal to the determinant of the Cartan matrix.*

Proof. The simple roots α form a basis for the root lattice Λ_R. The corresponding elements H_α form a basis for

$$\Gamma_R = \mathbb{Z}\{H_\gamma : \gamma \in R\},$$

a lattice in $\mathfrak{h}$; this is proved in Appendix D.4. Since Λ_W is defined to be the lattice of elements of $\mathfrak{h}^*$ that take integral values on Γ_R, the determinant

$$\det(\alpha(H_\beta)) = \det(n_{\alpha\beta})$$

is the index $[\Lambda_W : \Lambda_R]$. □

In particular, for the exceptional groups, Λ_W/Λ_R is trivial for (G_2), (F_4), and (E_8), and cyclic of order two for (E_7) and order three for (E_6).

In fact, the center of the simply-connected group is naturally isomorphic to the *dual* of Λ_W/Λ_R. To express this, consider the natural dual of this last group. The lattice Γ_R defined in the preceding proof is a sublattice of the lattice

$$\Gamma_W = \{X \in \mathfrak{h}: \alpha(X) \in \mathbb{Z} \text{ for all } \alpha \in R\}.$$

Note that Λ_W was defined to be the lattice of elements of $\mathfrak{h}^*$ that take integral values on Γ_R. It follows formally from the definitions and the fact that Λ_W/Λ_R is finite that we have a perfect pairing

$$\Gamma_W/\Gamma_R \times \Lambda_W/\Lambda_R \to \mathbb{Q}/\mathbb{Z}, \qquad (X, \alpha) \mapsto \alpha(X).$$

The claim is that there is a natural isomorphism from Γ_W/Γ_R to the center of $\tilde{G}$, which is given by the exponential. More precisely, let $e_G: \mathfrak{h} \to H \subset G$ be the homomorphism defined by

$$e_G(X) = \exp(2\pi i X).$$

We claim that when $G = \tilde{G}$ is the simply-connected group, $\operatorname{Ker}(e_{\tilde{G}}) = \Gamma_R$ and $e_{\tilde{G}}(\Gamma_W)$ is the center of $\tilde{G}$, from which it follows that $e_{\tilde{G}}$ induces an isomorphism

$$\Gamma_W/\Gamma_R \cong Z(\tilde{G}).$$

More generally, for any $G = \tilde{G}/C$, define a lattice $\Gamma(G)$ between Γ_R and Γ_W by

$$\Gamma(G) = \operatorname{Ker}(e_G).$$

Then e_G determines an isomorphism

$$\Gamma_W/\Gamma(G) \cong Z(G).$$

We may thus state our result as

Theorem 23.16. *There is a one-to-one correspondence between connected Lie groups G with the Lie algebra $\mathfrak{g}$ and lattices $\Lambda \subset \mathfrak{h}^*$ such that*

$$\Lambda_R \subset \Lambda \subset \Lambda_W.$$

The correspondence is given by associating to a group G the lattice dual to the kernel of the exponential map $\exp: \mathfrak{g} \to G$; *in particular, the largest lattice Λ_W corresponds to the simply-connected group, the smallest Λ_R to the adjoint group with no center. In terms of this correspondence, the irreducible representation V_λ of $\mathfrak{g}$ with highest weight $\lambda \in \mathfrak{h}^*$ will lift to a representation of the group G corresponding to $\Lambda \subset \mathfrak{h}^*$ if and only if $\lambda \in \Lambda$.*

Note also that

$$H = \mathfrak{h}/\Gamma(G) \cong \mathbb{C}^* \times \cdots \times \mathbb{C}^*,$$

with $n = \dim_{\mathbb{C}} \mathfrak{h}$ copies of $\mathbb{C}^*$.

Exercise 23.17*. Show that these claims follow formally from what we have seen: that the image of the exponential map contains the center, and that for any weight α there is a representation V of $\mathfrak{g}$ whose weight space V_α is not zero. Show also that e_G determines an isomorphism $\Gamma(G)/\Gamma_R \cong \pi_1(G)$. In diagram form,

$$\begin{array}{lcc}
\Gamma_W & & G_0\\
\cup & \Big\}\ \text{Center}(G) & \uparrow \\
\Gamma(G) & & G\\
\cup & \Big\}\ \pi_1(G) & \uparrow\\
\Gamma_R & & \tilde{G}
\end{array}$$

Exercise 23.18. Find the kernels of each of the spin and half-spin representations $\mathrm{Spin}_m\mathbb{C} \to \mathrm{GL}(S)$ and $\mathrm{Spin}_m\mathbb{C} \to \mathrm{GL}(S^{\pm})$.

Exercise 23.19*. Classify the irreducible representations of the full orthogonal group $\mathrm{O}_m\mathbb{C}$.

Note that by our analysis of the Lie algebra $\mathfrak{g}_2$ there is a unique group G_2 with this Lie algebra, which is simultaneously the simply-connected and adjoint forms; the representations of this group are exactly those of the algebra $\mathfrak{g}_2$. The same is true for the Lie algebras of type (F_4) and (E_8), while (E_7) and (E_6) each have two associated groups, an adjoint one with fundamental group $\mathbb{Z}/2$ and $\mathbb{Z}/3$, and a simply-connected form with center $\mathbb{Z}/2$ and $\mathbb{Z}/3$ respectively.

It may be worth pointing out that each complex simple Lie group G can be realized as a closed subgroup defined by polynomial equations in some general linear group, i.e., that G is an *affine algebraic group*. Every irreducible representation $G \to \mathrm{GL}(V)$ is also defined by polynomials in appropriate coordinates. This explains why the whole subject can be developed from the point of view of algebraic groups, as in [Bor1] and [Hu2].

The Weyl group $\mathfrak{W}$, which we defined as a subgroup of $\mathrm{Aut}(\mathfrak{h}^*)$, can be interpreted in terms of any connected Lie group G with Lie algebra $\mathfrak{g}$. Let H be the Cartan subgroup corresponding to $\mathfrak{h}$, and let $N(H)$ be the normalizer:

$$N(H) = \{g \in G: gHg^{-1} = H\}.$$

We have homomorphisms:

$$N(H) \to \mathrm{Aut}(H) \to \mathrm{Aut}(\mathfrak{h}) \to \mathrm{Aut}(\mathfrak{h}^*),$$

the first defined by conjugation, the second by differentiation at the identity, and the third using the identification of $\mathfrak{h}$ and $\mathfrak{h}^*$ via the Killing form. Fact 14.11 can be sharpened to the claim that this map determines an *isomorphism*

$$N(H)/N \cong \mathfrak{W}. \tag{23.20}$$

When G is the adjoint form of the Lie algebra, this isomorphism is proved in Appendix D. The general case follows, using:

Exercise 23.21. Show that if $\pi: G' \to G$ is a connected covering, with Cartan subgroups $H' = \pi^{-1}(H)$, then the induced map $N(H')/H' \to N(H)/H$ is an isomorphism.

Exercise 23.22. For each of the classical groups, and each simple root α, find an element in $N(H)$ that maps to the reflection W_α in $\mathfrak{W}$.

§23.2. Representation Rings and Characters

Just as with finite groups, we can form the representation ring R of a semisimple Lie algebra or Lie group: take the free abelian group on the isomorphism classes $[V]$ of finite-dimensional representations V, and divide by the relations $[V] = [V'] + [V'']$ whenever $V \cong V' \oplus V''$. By the complete reducibility of representations, it follows as before that R is a free abelian group on the classes $[V]$ of irreducible representations. Again, the tensor product of representations makes R into a ring: $[V]\cdot[W] = [V \otimes W]$. Many of our questions about decomposing representations and tensor products of representations can be nicely encoded by describing R more fully. We do this first for the Lie algebras.

For a semisimple Lie algebra $\mathfrak{g}$, let $\Lambda = \Lambda_W$ be the weight lattice, and let $\mathbb{Z}[\Lambda]$ be the integral group ring on the abelian group Λ. We write $e(\lambda)$ for the basis element of $\mathbb{Z}[\Lambda]$ corresponding to the weight λ; for now at least these are just formal symbols, having nothing to do with exponentials (but see (23.40)). Elements of $\mathbb{Z}[\Lambda]$ are expressions of the form $\sum n_\lambda e(\lambda)$, i.e., they assign an integer n_λ to each weight λ, with all but a finite number being zero. So $\mathbb{Z}[\Lambda]$ is a natural carrier for the information about multiplicities of representations. Define a *character homomorphism*

$$\text{Char}: R(\mathfrak{g}) \to \mathbb{Z}[\Lambda] \tag{23.23}$$

by the formula $\text{Char}[V] = \sum \dim(V_\lambda)e(\lambda)$, where V_λ is the weight space of V for the weight λ and $\dim(V_\lambda)$ its multiplicity. This is clearly an additive homomorphism.

The first assertion about this character map is that it is *injective*. This comes down to the fact that a representation is determined by the multiplicities of its weight spaces, which is something we saw in Lecture 14.

The product in the group ring $\mathbb{Z}[\Lambda]$ is determined by $e(\alpha)\cdot e(\beta) = e(\alpha + \beta)$. We claim next that Char is a *ring homomorphism*. This comes from the familiar fact that

$$(V \otimes W)_\lambda = \bigoplus_{\mu+\nu=\lambda} V_\mu \otimes W_\nu.$$

The Weyl group $\mathfrak{W}$ acts on $\mathbb{Z}[\Lambda]$, and a third simple claim is that the image of Char is contained in the ring of invariants $\mathbb{Z}[\Lambda]^{\mathfrak{W}}$. This comes down to the fact that, for an irreducible (and hence for any) representation V, the weight spaces obtained by reflecting in walls of the Weyl chambers all have the same dimension.

Let $\omega_1, \ldots, \omega_n$ be a set of fundamental weights; as we have seen, these are the first weights along edges of a Weyl chamber, and they are free generators for the lattice Λ. Let $\Gamma_1, \ldots, \Gamma_n$ be the classes in $R(\mathfrak{g})$ of the irreducible representations with highest weights $\omega_1, \ldots, \omega_n$.

Theorem 23.24. (a) *The representation ring $R(\mathfrak{g})$ is a polynomial ring on the variables $\Gamma_1, \ldots, \Gamma_n$.*

(b) *The homomorphism $R(\mathfrak{g}) \to \mathbb{Z}[\Lambda]^{\mathfrak{W}}$ is an isomorphism.*

In particular, this says that $\mathbb{Z}[\Lambda]^{\mathfrak{W}}$ is a polynomial ring on the variables $\operatorname{Char}(\Gamma_1), \ldots, \operatorname{Char}(\Gamma_n)$. In fact, the theorem is equivalent to this assertion, since if we take variables $U_1, \ldots, U_n$ and map the polynomial ring on the U_i to $R(\mathfrak{g})$ by sending U_i to Γ_i, we have

$$\mathbb{Z}[U_1, \ldots, U_n] \to R(\mathfrak{g}) \to \mathbb{Z}[\Lambda]^{\mathfrak{W}}.$$

If the composite is an isomorphism, the second being injective, both must be isomorphisms, which is what the theorem says.

In spite of its fancy appearance, we will see that the theorem follows quite easily from what we know about the action of the Weyl group $\mathfrak{W}$ on the weights.

For any $P \in \mathbb{Z}[\Lambda]$ let us say that α is a *highest weight* for P if the coefficient of $e(\alpha)$ in P is nonzero, and, with a chosen ordering of weights as before, α is the largest such weight. We first observe that if P is invariant under $\mathfrak{W}$, then the highest weight for P is in $\mathscr{W} \cap \Lambda$, where $\mathscr{W}$ is our chosen (closed) Weyl chamber. In general, weights in $\mathscr{W} \cap \Lambda$ are often referred to as *dominant weights*.

Now suppose $\{P_\lambda\}$ is any collection of elements in $\mathbb{Z}[\Lambda]^{\mathfrak{W}}$, one for each dominant weight λ, such that P_λ has highest weight λ and the coefficient of $e(\lambda)$ is 1. We claim that the P_λ form an additive basis for $\mathbb{Z}[\Lambda]^{\mathfrak{W}}$ over $\mathbb{Z}$. This is easy to see and is the same argument used in the theory of symmetric polynomials in any algebra text: given P with highest weight λ, if the coefficient of $e(\lambda)$ is m, then $P - mP_\lambda$ is invariant whose highest weight is lower, and one continues inductively until one reaches weight zero, i.e., the constants.

Let $P_i = \operatorname{Char}(\Gamma_i)$, which has highest weight ω_i, and suppose the coefficient of $e(\omega_i)$ is 1. Since any weight $\lambda \in \mathscr{W} \cap \Lambda$ can be uniquely expressed in the form $\lambda = \sum m_i \omega_i$, for some non-negative integers m_i, and the highest weight of $\prod (P_i)^{m_i}$ is $\sum m_i \omega_i$, it follows that the monomials $\prod (P_i)^{m_i}$ in $P_1, \ldots, P_n$ form an additive basis for $\mathbb{Z}[\Lambda]^{\mathfrak{W}}$. This says precisely that $\mathbb{Z}[P_1, \ldots, P_n] = \mathbb{Z}[\Lambda]^{\mathfrak{W}}$, and completes the proof. □

Let us work this out concretely for each of our cases $\mathfrak{sl}_{n+1}\mathbb{C}$, $\mathfrak{sp}_n\mathbb{C}$, $\mathfrak{so}_{2n+1}\mathbb{C}$, and $\mathfrak{so}_{2n}\mathbb{C}$. Each lattice Λ contains weights we have called $L_1, \ldots, L_n$; in the first case we also have L_{n+1} with $L_1 + \cdots + L_{n+1} = 0$. We set

$$x_i = e(L_i), \qquad x_i^{-1} = e(-L_i) \in \mathbb{Z}[\Lambda]. \tag{23.25}$$

Note that in case $L_1, \ldots, L_n$ is a basis for Λ, then

$$\mathbb{Z}[\Lambda] = \mathbb{Z}[x_1, \ldots, x_n, x_1^{-1}, \ldots, x_n^{-1}] = \mathbb{Z}[x_1, \ldots, x_n, (x_1 \cdot \ldots \cdot x_n)^{-1}]$$

as a subring of the field $\mathbb{Q}(x_1, \ldots, x_n)$.

(A_n) For $\mathfrak{sl}_{n+1}\mathbb{C}$, fundamental weights are

$$L_1, \quad L_1 + L_2, \quad L_1 + L_2 + L_3, \ldots, L_1 + \cdots + L_n,$$

corresponding to the irreducible representations $V, \wedge^2 V, \ldots, \wedge^n V$, with $V = \mathbb{C}^{n+1}$ the standard representation. The character of $\wedge^k V$ is $\sum e(\alpha)$, the sum over all α that are sums of k different L_i for $1 \le i \le n+1$. So $\mathrm{Char}(\wedge^k V) = A_k$, where A_k is the kth elementary symmetric function of $x_1, \ldots, x_{n+1}$. The Weyl group is the symmetric group $\mathfrak{S}_{n+1}$, acting by permutation on the indices, so the theorem in this case says that

$$R(\mathfrak{sl}_{n+1}) = \mathbb{Z}[\Lambda]^{\mathfrak{S}_{n+1}} = \mathbb{Z}[A_1, \ldots, A_n]. \tag{23.26}$$

Note that $\mathbb{Z}[\Lambda] = \mathbb{Z}[x_1, \ldots, x_n, x_{n+1}]/(x_1 \cdot \ldots \cdot x_{n+1} - 1)$, so $\mathbb{Z}[\Lambda]$ has an additive basis consisting of all monomials x^α, with α an n-tuple of non-negative integers, but with not all α_i positive.

(C_n) For $\mathfrak{sp}_{2n}\mathbb{C}$, the lattice Λ and fundamental weights have the same description as in the preceding case. The corresponding irreducible representations are the kernels $V^{(k)}$ of the contraction maps $\wedge^k V \to \wedge^{k-2} V$, with now $V = \mathbb{C}^{2n}$ the standard representation, $k = 1, \ldots, n$. The character of $\wedge^k V$ is $\sum e(\alpha)$, the sum over all α that are sums of k different $\pm L_i$ for $1 \le i \le n$. The character $\mathrm{Char}(\wedge^k V)$ is thus the elementary symmetric polynomial C_k in the variables $x_1, x_1^{-1}, x_2, x_2^{-1}, \ldots, x_n, x_n^{-1}$. The theorem then says that

$$\begin{aligned} R(\mathfrak{sp}_{2n}\mathbb{C}) &= \mathbb{Z}[\Lambda]^{\mathfrak{W}} = \mathbb{Z}[C_1, C_2 - 1, C_3 - C_1, \ldots, C_n - C_{n-2}] \\ &= \mathbb{Z}[C_1, C_2, C_3, \ldots, C_n]. \end{aligned} \tag{23.27}$$

(B_n) For $\mathfrak{so}_{2n+1}\mathbb{C}$, Λ is spanned by the L_i together with $\frac{1}{2}(L_1 + \cdots + L_n)$. The fundamental representations are $V, \wedge^2 V, \ldots, \wedge^{n-1} V$, and the spin representation S. The character of $\wedge^k V$ is the kth elementary symmetric function of the $2n+1$ elements $x_1, x_1^{-1}, \ldots, x_n, x_n^{-1}$, and 1; denote this by B_k. The character of S, which we denote by B, is the sum $\sum x_1^{\pm 1/2} \cdot \ldots \cdot x_n^{\pm 1/2}$, where

$$x_i^{+1/2} = e(L_i/2), \qquad x_i^{-1/2} = e(-L_i/2). \tag{23.28}$$

So B is the nth elementary symmetric polynomial in the variables $x_i^{+1/2} + x_i^{-1/2}$. Therefore,

$$R(\mathfrak{so}_{2n+1}\mathbb{C}) = \mathbb{Z}[\Lambda]^{\mathfrak{W}} = \mathbb{Z}[B_1, \ldots, B_{n-1}, B]. \tag{23.29}$$

(D_n) For $\mathfrak{so}_{2n}\mathbb{C}$, Λ and $\mathbb{Z}[\Lambda]$ are the same as in the preceding case. The fundamental representations are $V, \wedge^2 V, \ldots, \wedge^{n-2} V$, and the half-spin representations S^+ and S^-. The character of $\wedge^k V$, denoted D_k, is the kth elementary symmetric function of the $2n$ elements $x_1, x_1^{-1}, \ldots, x_n, x_n^{-1}$. The character $D^{\pm}$ of $S^{\pm}$ is the sum $\sum x_1^{\pm 1/2} \cdot \ldots \cdot x_n^{\pm 1/2}$, where the number of plus signs is even or odd according to the sign. We have

$$R(\mathfrak{so}_{2n}\mathbb{C}) = \mathbb{Z}[\Lambda]^{\mathfrak{W}} = \mathbb{Z}[D_1, \ldots, D_{n-2}, D^+, D^-]. \tag{23.30}$$

Exercise 23.31*. (a) Prove the following relation in $R(\mathfrak{so}_{2n+1}\mathbb{C})$:

$$B^2 = B_n + \cdots + B_1 + 1,$$

corresponding to the isomorphism

$$S \otimes S \cong \wedge^n V \oplus \cdots \oplus \wedge^1 V \oplus \wedge^0 V.$$

This describes $R(\mathfrak{so}_{2n+1}\mathbb{C})$ as a quadratic extension of the ring $\mathbb{Z}[B_1, \ldots, B_n]$.

(b) Let D_n^+ (respectively, D_n^-) be the character of the representation whose highest weight is twice that of D^+ (resp., D^-), so that, for example, the sum of the representations D_n^+ and D_n^- is $\wedge^n V$. Prove the relations in $R(\mathfrak{so}_{2n}\mathbb{C})$:

$$D^+ \cdot D^+ = D_n^+ + D_{n-2} + D_{n-4} + \cdots,$$

$$D^- \cdot D^- = D_n^- + D_{n-2} + D_{n-4} + \cdots,$$

$$D^+ \cdot D^- = D_{n-1} + D_{n-3} + D_{n-5} + \cdots.$$

We can likewise describe the representation ring for $\mathfrak{g}_2$. Here, we may take as generators for the weight lattice the weights L_1 and L_2 as pictured in the diagram

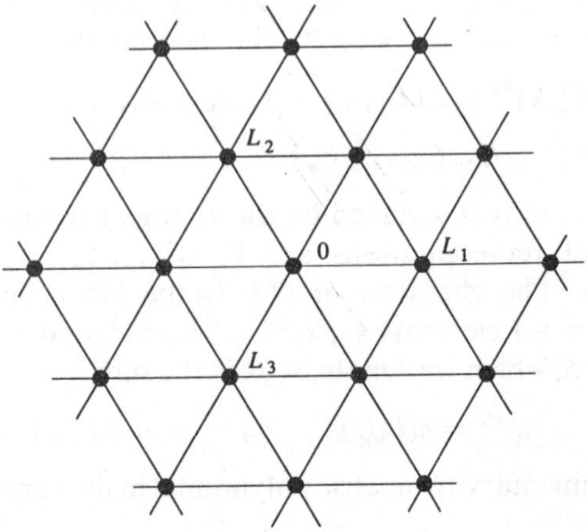

and correspondingly write $\mathbb{Z}[\Lambda]$ as $\mathbb{Z}[x_1, x_1^{-1}, x_2, x_2^{-1}]$, where $x_i = e(L_i)$. It will be a little more symmetric to introduce $L_3 = -L_1 - L_2$ as pictured and $x_3 = x_1^{-1} \cdot x_2^{-1} = e(L_3)$, and write

$$\mathbb{Z}[\Lambda] = \mathbb{Z}[x_1, x_2, x_3]/(x_1 x_2 x_3 - 1).$$

In these terms the Weyl group is the group $\mathfrak{W}$ generated by the symmetric group $\mathfrak{S}_3$ permuting the variables x_i and the involution sending each x_i to x_i^{-1}. The standard representation has weights $\pm L_i$ and 0, and so has character

$$A = A(x_1, x_2, x_3) = 1 + x_1 + x_1^{-1} + x_2 + x_2^{-1} + x_3 + x_3^{-1}.$$

Similarly, the adjoint representation has weights $\pm L_i$, $\pm(L_i - L_j)$, and 0 (taken twice); its character is

$$B = A(x_1, x_2, x_3) + A(x_1/x_2, x_2/x_3, x_3/x_1).$$

The theorem thus implies in this case the equality

$$R(\mathfrak{g}_2) = \mathbb{Z}[\Lambda]^{\mathfrak{W}} = \mathbb{Z}[A, B]. \tag{23.32}$$

Exercise 23.33. Verify directly the statement that any element of $\mathbb{Z}[x_1, x_2, x_3]/(x_1x_2x_3 - 1)$ invariant under the group $\mathfrak{W}$ as described is in fact a polynomial in A and B.

Similarly we can define the representation ring $R(G)$ of a semisimple group G. When G is the simply-connected form of its Lie algebra $\mathfrak{g}$, $R(G) = R(\mathfrak{g})$, so $R(\mathrm{SL}_n\mathbb{C})$, $R(\mathrm{Sp}_{2n}\mathbb{C})$, $R(\mathrm{Spin}_{2n+1}\mathbb{C})$, and $R(\mathrm{Spin}_{2n}\mathbb{C})$ are given by (23.26), (23.27), (23.29), and (23.30). In general, $R(G)$ is a subring of $R(\mathfrak{g})$; we can read off which subring by looking at Proposition 23.13. We have, in fact,

$$R(\mathrm{SO}_{2n+1}\mathbb{C}) = \mathbb{Z}[B_1, \ldots, B_n]; \tag{23.34}$$

$$R(\mathrm{SO}_{2n}\mathbb{C}) = \mathbb{Z}[D_1, \ldots, D_{n-1}, D_n^+, D_n^-], \tag{23.35}$$

with D_n^+ and D_n^- as in Exercise 23.31. But this time there is one relation:

$$(D_n^+ + D_{n-2} + D_{n-4} + \cdots + 1)(D_n^- + D_{n-2} + D_{n-4} + \cdots + 1)$$
$$= (D_{n-1} + D_{n-3} + \cdots +)^2.$$

Exercise 23.36*.

(a) Prove (23.34).
(b) Show that the relation in (23.35) comes from Exercise 23.31(b). Show that $R(\mathrm{SO}_{2n}\mathbb{C})$ is the polynomial ring in the $n + 1$ generators shown, modulo the ideal generated by the one polynomial indicated.
(c) Describe the representation rings for the other groups with these simple Lie algebras.
(d) Prove the isomorphism

$$R(\mathrm{GL}_n\mathbb{C}) = \mathbb{Z}[E_1, \ldots, E_n, E_n^{-1}],$$

where the E_k are the elementary symmetric functions of $x_1, \ldots, x_n$.

Exercise 23.37*. (a) Show that the image of $R(\mathrm{O}_m\mathbb{C})$ in $R(\mathrm{SO}_m\mathbb{C})$ is the polynomial ring $\mathbb{Z}[B_1, \ldots, B_n]$ if $m = 2n + 1$, and $\mathbb{Z}[D_1, \ldots, D_n]$ if $m = 2n$.

(b) Show that

$$R(O_{2n+1}\mathbb{C}) = R(SO_{2n+1}\mathbb{C}) \otimes R(\mathbb{Z}/2)$$
$$= \mathbb{Z}[B_1, \ldots, B_n, B_{2n+1}]/((B_{2n+1})^2 - 1)$$

and

$$R(O_{2n}\mathbb{C}) = \mathbb{Z}[D_1, \ldots, D_n, D_{2n}]/I,$$

where I is the ideal generated by $(D_{2n})^2 - 1$ and $D_n D_{2n} - D_n$.

Exercise 23.38*. The mapping that takes a representation V to its dual V^* induces an involution of the representation ring: $[V]^* = [V^*]$. The ring $\mathbb{Z}[\Lambda]$ has an involution determined by $(e(\lambda))^* = e(-\lambda)$. Show that the character homomorphism commutes with these involutions. Show that for $\mathfrak{sl}_{n+1}$, $(A_k)^* = A_{n+1-k}$; for $\mathfrak{so}_{2n+1}\mathbb{C}$, and $\mathfrak{sp}_{2n}\mathbb{C}$, and $\mathfrak{so}_{2n}\mathbb{C}$ for n even, the involution is the identity; while for $\mathfrak{so}_{2n}\mathbb{C}$ with n odd, $(D_k)^* = D_k$, $(D^+)^* = D^-$, $(D^-)^* = D^+$ Deduce that all representations of all symplectic and orthogonal groups are self-dual. Note that when $*$ is the identity, all representations are self-dual. In the other cases, compute the duals of irreducible representations with given highest weight.

The following exercise deals with a special property of the representation rings of semisimple Lie groups and algebras.

Exercise 23.39*. The representation rings $R = R(\mathfrak{g})$ and $R(G)$ have another important structure: they are *λ-rings*. There are operators

$$\lambda^i: R(G) \to R(G), \quad i = 0, 1, 2, \ldots,$$

determined by $\lambda^i([V]) = [\wedge^i V]$ for any representation V.

(a) Show that this determines well-defined maps, satisfying $\lambda^0 = 1, \lambda^1 = \mathrm{Id}$, and

$$\lambda^i(x + y) = \sum_{i+j=k} \lambda^i(x) \cdot \lambda^j(y)$$

for any x and y in R. In fact, R is what is called a *special* λ-ring: there are formulas for $\lambda^i(x \cdot y)$ and $\lambda^i(\lambda^j(x))$, valid as if x and y could be written as sums of one-dimensional representations (see, e.g., [A-T]).

(b) Show that λ^i extends to $\mathbb{Z}[\Lambda]$, and use this to verify that $R(G)$ is a special λ-ring.

Define *Adams operators* $\psi^k: R \to R$ by $\psi^k(x) = P_k(\lambda^1 x, \ldots, \lambda^n x)$, where P_k is the expression for the kth power sum (cf. Exercise A.32) in terms of the elementary symmetric functions, $n \geq k$. Equivalently,

$$\psi^k(x) - \psi^{k-1}(x)\lambda^1(x) + \cdots + (-1)^k k\lambda^k(x) = 0.$$

(c) Show that, regarding R as the ring of functions on the group G, $(\psi^k x)(g) = x(g^k)$. Equivalently, $\psi^k(e(\lambda)) = e(k\lambda)$.

(d) Show that each ψ^k is a ring homomorphism, and $\psi^k \circ \psi^l = \psi^{k+l}$.
(e) Show that for a representation V,

$$\mathrm{Char}(\mathrm{Sym}^2 V) = \tfrac{1}{2}\,\mathrm{Char}(V)^2 + \tfrac{1}{2}\psi^2(\mathrm{Char}(V)),$$
$$\mathrm{Char}(\wedge^2 V) = \tfrac{1}{2}\,\mathrm{Char}(V)^2 - \tfrac{1}{2}\psi^2(\mathrm{Char}(V)).$$

Show that $\mathrm{Char}(\mathrm{Sym}^d V)$ and $\mathrm{Char}(\wedge^d V)$ can be written as polynomials in $\psi^k(\mathrm{Char}(V))$, $1 \le k \le d$.

Formal Characters and Actual Characters

Let G be a Lie group with Lie algebra $\mathfrak{g}$. For any representation V of $\mathfrak{g}$, the image of $[V] \in R(\mathfrak{g})$ in $\mathbb{Z}[\Lambda]$ is called the *formal character* of V. As it turns out, this formal character can be identified with the honest character of the corresponding representation of the group G, restricted to the Cartan subgroup H:

(23.40) *If* $\mathrm{Char}(V) = \sum m_\alpha e(\alpha)$ *is the formal character, and* $\exp(X)$ *is an element of H, then the trace of* $\exp(X)$ *on V is* $\sum m_\alpha e^{\alpha(X)}$.

This is simply because $\exp(X)$ acts on the weight space V_μ by multiplication by $e^{\mu(X)}$, as we have seen. In particular, a representation is determined by the character of its restriction to a Cartan subgroup.

Another common notation for this is to set $e(X) = \exp(2\pi i X)$, and $e(z) = \exp(2\pi i z)$. Then the trace of $e(X)$ is $\sum m_\alpha e(\alpha(X))$.

Exercise 23.41. As a function on H, the character of a representation is invariant under the Weyl group $\mathfrak{W} = N(H)/H$. Describe $R(G)$ as a ring of $\mathfrak{W}$-invariant functions on H.

This is also compatible with our descriptions of elements of $\mathbb{Z}[\Lambda]^{\mathfrak{W}}$ as Laurent polynomials in variables x_i or $x_i^{1/2}$. For $\mathrm{SL}_{n+1}\mathbb{C}$, for example, if the character $\mathrm{Char}(W)$ of a representation W is $P(x_1, \ldots, x_{n+1})$, the trace of the matrix $\mathrm{diag}(z_1, \ldots, z_{n+1})$ on V is $P(z_1, \ldots, z_{n+1})$. Similarly for the other groups, using the diagonal matrices described in the first section of this lecture. For the spin groups, the element $w(z_1, \ldots, z_n)$ defined in (23.8) has trace given by substituting z_i for $x_i^{1/2}$, and z_i^{-1} for $x_i^{-1/2}$ in the corresponding Laurent polynomial.

Exercise 23.42*. If $\mathfrak{g}_1$ and $\mathfrak{g}_2$ are two semisimple Lie algebras, show that

$$R(\mathfrak{g}_1 \times \mathfrak{g}_2) = R(\mathfrak{g}_1) \otimes R(\mathfrak{g}_2).$$

Exercise 23.43*. (a) For the natural inclusion $\mathfrak{sl}_n\mathbb{C} \subset \mathfrak{sl}_{n+1}\mathbb{C}$, restriction of representations gives a homomorphism $R(\mathfrak{sl}_{n+1}\mathbb{C}) \to R(\mathfrak{sl}_n\mathbb{C})$, which can be described by saying what happens to the polynomial generators. Since $\wedge^k(\mathbb{C}^n \oplus \mathbb{C}) = \wedge^k(\mathbb{C}^n) \oplus \wedge^{k-1}(\mathbb{C}^n)$, this is

$$A_k \mapsto A_k + A_{k-1}.$$

Give the analogous descriptions for the following inclusions:

$$\mathfrak{sp}_{2n-2}\mathbb{C} \subset \mathfrak{sp}_{2n}\mathbb{C}, \qquad \mathfrak{so}_{2n}\mathbb{C} \subset \mathfrak{so}_{2n+1}\mathbb{C}, \qquad \mathfrak{so}_{2n-1}\mathbb{C} \subset \mathfrak{so}_{2n}\mathbb{C};$$

$$\mathfrak{sl}_n\mathbb{C} \subset \mathfrak{sp}_{2n}\mathbb{C}, \qquad \mathfrak{sl}_n\mathbb{C} \subset \mathfrak{so}_{2n+1}\mathbb{C}, \qquad \mathfrak{sl}_n\mathbb{C} \subset \mathfrak{so}_{2n}\mathbb{C};$$

$$\mathfrak{sp}_{2n}\mathbb{C} \subset \mathfrak{sl}_{2n}\mathbb{C}, \qquad \mathfrak{so}_{2n+1}\mathbb{C} \subset \mathfrak{sl}_{2n+1}\mathbb{C}, \qquad \mathfrak{so}_{2n}\mathbb{C} \subset \mathfrak{sl}_{2n}\mathbb{C}.$$

(b) The inclusion $\mathfrak{sl}_n\mathbb{C} \times \mathfrak{sl}_m\mathbb{C} \subset \mathfrak{sl}_{n+m}\mathbb{C}$ determines a restriction homomorphism $R(\mathfrak{sl}_{n+m}\mathbb{C}) \to R(\mathfrak{sl}_n\mathbb{C} \times \mathfrak{sl}_m\mathbb{C}) = R(\mathfrak{sl}_n\mathbb{C}) \otimes R(\mathfrak{sl}_m\mathbb{C})$, which takes polynomial generators A_k to $A_k \otimes 1 + A_{k-1} \otimes A_1 + \cdots + 1 \otimes A_k$. Compute analogously for

$$\mathfrak{sp}_{2n}\mathbb{C} \times \mathfrak{sp}_{2m}\mathbb{C} \subset \mathfrak{sp}_{2n+2m}\mathbb{C}, \qquad \mathfrak{so}_n\mathbb{C} \times \mathfrak{so}_m\mathbb{C} \subset \mathfrak{so}_{n+m}\mathbb{C}.$$

Which of these inclusions correspond to removing nodes from the Dynkin diagrams?

Exercise 23.44. Compute the isomorphisms of representation rings corresponding to the isomorphisms $\mathfrak{sl}_2\mathbb{C} \cong \mathfrak{so}_3\mathbb{C}$, $\mathfrak{so}_5\mathbb{C} \cong \mathfrak{sp}_4\mathbb{C}$, and $\mathfrak{sl}_4\mathbb{C} \cong \mathfrak{so}_6\mathbb{C}$.

§23.3. Homogeneous Spaces

In this section we will introduce and describe the compact homogeneous spaces associated to the classical groups. As we will see, these are classified neatly in terms of Dynkin diagrams, and are, in turn, closely related to the representation theory of the groups acting on them. Unfortunately, we are unable to give here more than the barest outline of this beautiful subject; but we will at least try to say what the principal objects are, and what connections among them exist. In particular, we give at the end of the section a diagram (23.58) depicting these objects and correspondences to which the reader can refer while reading this section.

We begin by introducing the notion of Borel subalgebras and Borel subgroups. Recall first that a choice of Cartan subalgebra $\mathfrak{h}$ in a semisimple Lie algebra $\mathfrak{g}$ determines, as we have seen, a decomposition $\mathfrak{g} = \mathfrak{h} \oplus \bigoplus_{\alpha \in R} \mathfrak{g}_\alpha$. To each choice of ordering of the root system $R = R^+ \cup R^-$, we can associate a subalgebra

$$\mathfrak{b} = \mathfrak{h} \oplus \bigoplus_{\alpha \in R^+} \mathfrak{g}_\alpha,$$

called a *Borel subalgebra.* Note that $\mathfrak{b}$ is solvable, since $\mathscr{D}\mathfrak{b} \subset \bigoplus \mathfrak{g}_\alpha$, $\mathscr{D}^2\mathfrak{b} \subset \bigoplus \mathfrak{g}_{\alpha+\beta}$, etc. In fact, $\mathfrak{b}$ is a maximal solvable subalgebra (Exercise 14.35).

If G is a Lie group with semisimple Lie algebra $\mathfrak{g}$, the connected subgroup B of G with Lie algebra $\mathfrak{b}$ is called a *Borel subgroup*.

Claim 23.45. *B is a closed subgroup of G, and the quotient G/B is compact.*

PROOF. Consider the adjoint representation of G on $\mathfrak{g}$. The action of the Borel subalgebra $\mathfrak{b}$ obviously preserves the subspace $\mathfrak{b} \subset \mathfrak{g}$, and, in fact, $\mathfrak{b}$ is just the inverse image of the subalgebra of $\mathfrak{gl}(\mathfrak{g})$ preserving this subspace: if $X = \sum X_\alpha$ is any element of $\mathfrak{g}$ with $X_\alpha \in \mathfrak{g}_\alpha$ and $X_\alpha \neq 0$ for some $\alpha \in R^-$, we could find an element H of $\mathfrak{h} \subset \mathfrak{b}$ with $\mathrm{ad}(X)(H) \notin \mathfrak{b}$—any H not in the annihilator of $\alpha \in \mathfrak{h}^*$ would do. B is thus (the connected component of the identity in) the inverse image in G of the subgroup of $GL(\mathfrak{g})$ carrying $\mathfrak{b}$ into itself. It follows that B is closed; and the quotient G/B is contained in a Grassmannian and hence compact. (Alternatively, we could consider the action of G on the projective space $\mathbb{P}(\wedge^m \mathfrak{g})$, where m is the number of positive roots, and observe that B is the stabilizer of the point corresponding to the exterior product of the positive root spaces.)

In fact, in the case of the classical groups, it is easy to describe the Borel subgroups and the corresponding quotients.

For $G = SL_{n+1}\mathbb{C}$, B is the group of all upper-triangular matrices in G, i.e., those automorphisms preserving the standard flag. It follows that G/B is the usual (complete) flag manifold, i.e., the variety of all flags

$$G/B = \{0 \subset V_1 \subset \cdots \subset V_n \subset \mathbb{C}^{n+1}\}$$

of subspaces with $\dim(V_r) = r$.

For $G = SO_{2n+1}\mathbb{C}$ the orthogonal group of automorphisms of $\mathbb{C}^{2n+1}$ preserving a quadratic form Q, B is the subgroup of automorphisms which preserve a fixed flag $V_1 \subset \cdots \subset V_n$ of isotropic subspaces with $\dim(V_r) = r$. All such flags being conjugate, G/B is the variety of all such flags, i.e.,

$$G/B = \{0 \subset V_1 \subset \cdots \subset V_n \subset \mathbb{C}^{2n+1} \colon Q(V_n, V_n) \equiv 0\}.$$

Note that B automatically preserves the flag of orthogonal subspaces, so that we could also characterize G/B as the space of complete flags equal to their orthogonal complements, i.e.,

$$G/B = \{V_1 \subset \cdots \subset V_{2n} \subset \mathbb{C}^{2n+1} \colon Q(V_i, V_{2n+1-i}) = 0\}.$$

The same holds for $Sp_{2n}\mathbb{C}$: the Borel subgroups $B \subset Sp_{2n}\mathbb{C}$ are just the subgroups preserving a half-flag of isotropic subspaces, or equivalently a full flag of pairwise complementary subspaces; and the quotient G/B is correspondingly the variety of all such flags.

For $G = SO_{2n}\mathbb{C}$, B fixes an isotropic flag $V_1 \subset \cdots \subset V_{n-1}$, and

$$G/B = \{0 \subset V_1 \subset \cdots \subset V_{n-1} \subset \mathbb{C}^{2n} \colon Q(V_{n-1}, V_{n-1}) = 0\}.$$

Exercise 23.46. With our choice of basis $\{e_i\}$, let V_r be the subspace spanned by the first r basic vectors. If B is defined to be the subgroup that preserves V_r for $1 \leq r \leq n$, verify that the Lie algebra of B is spanned by the Cartan subalgebra and the positive root spaces described in Lectures 17 and 19.

We now want to consider more general quotients of a semisimple complex group G. To begin with, we say that a (closed, complex analytic, and connected[1]) subgroup P of G is *parabolic* if the quotient G/P can be realized as the orbit of the action of G on $\mathbb{P}(V)$ for some representation V of G. In particular, G/P is a projective algebraic variety. It follows from the proof of Claim 23.45 that any Borel subgroup B of G is parabolic. The following two claims characterize parabolic subgroups as those containing a Borel subgroup, i.e., the Borel subgroups are exactly the *minimal* parabolic subgroups.

Claim 23.47. *If B is a Borel subgroup and P a parabolic subgroup of G, then there is an $x \in G$ with*

$$B \subset xPx^{-1}.$$

Claim 23.48. *If a subgroup P of G contains a Borel subgroup B, then P is parabolic.*

The first claim is deduced from a version of *Borel's fixed point theorem*: if B is a connected solvable group, V a representation of B and $X \subset \mathbb{P}V$ a projective variety carried into itself under the action of B on $\mathbb{P}V$, then B must have a fixed point on X. This is straightforward: we observe (by Lie's theorem (9.11)) that the action of the solvable group B on V must preserve a flag of subspaces

$$0 \subset V_1 \subset \cdots \subset V_n = V$$

with $\dim(V_i) = i$. We can thus find a subspace $V_i \subset V$ fixed by B such that X intersects $\mathbb{P}V_i$ in a finite collection of points, which must then be fixed points for the action of B on X. As for Claim 23.48 we will soon see directly how G/P is a projective variety whenever P is a subgroup containing B.

We can now completely classify the parabolic subgroups of a simple group, up to conjugacy. By the above, we may assume that P contains a Borel subgroup B. Correspondingly, its Lie algebra $\mathfrak{p}$ is a subspace of $\mathfrak{g}$ containing $\mathfrak{b}$ and invariant under the action of B on $\mathfrak{g}$; i.e., it is a direct sum

$$\mathfrak{p} = \mathfrak{h} \oplus \bigoplus_{\alpha \in T} \mathfrak{g}_\alpha$$

for some subset T of R that contains all positive roots. Now, in order for $\mathfrak{p}$ to be a subalgebra of $\mathfrak{g}$, the subset T must be closed under addition (that is, if two roots are in T, then either their sum is in T or is not a root). Since, in addition, T contains all the positive roots, we may observe that if α, β, and γ are positive roots with $\alpha = \beta + \gamma$, then we must have

$$-\alpha \in T \Rightarrow -\beta \in T \text{ and } -\gamma \in T.$$

[1] It is a general fact that P must be connected if G/P is a projective variety.

Clearly, any such subset T must be generated by R^+ together with the negatives of a subset Σ of the set of simple roots. Thus, if for each subset Σ of the set of simple roots we let $T(\Sigma)$ consist of all roots which can be written as sums of negatives of the roots in Σ, together with all positive roots, and form the subalgebra

$$\mathfrak{p}(\Sigma) = \mathfrak{h} \oplus \bigoplus_{\alpha \in T(\Sigma)} \mathfrak{g}_\alpha, \tag{23.49}$$

then $\mathfrak{p}(\Sigma)$ is a parabolic subalgebra, the corresponding Lie group $P(\Sigma)$ is a parabolic subgroup containing B, and we obtain in this way all the parabolic subgroups of G. We can express this as the observation that, up to conjugacy, *parabolic subgroups of the simple group G are in one-to-one correspondence with subsets of the nodes of the Dynkin diagram, i.e., with subsets of the set of simple roots.*

Examples. In the case of $\mathfrak{sl}_3\mathbb{C}$, there is a symmetry in the Dynkin diagram, so that there is only one parabolic subgroup other than the Borel, corresponding to the diagram

This, in turn, gives the subset of the root system

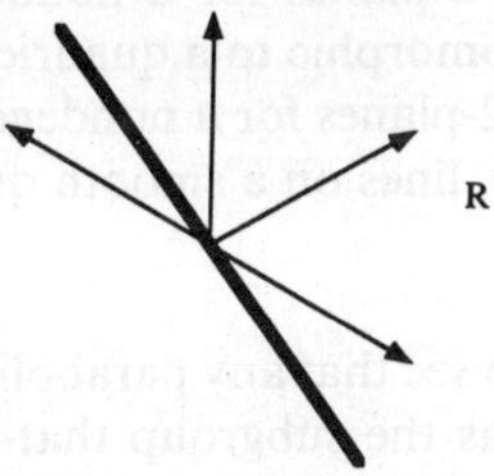

corresponding to the subgroup

$$P = \left\{ \begin{pmatrix} * & * & * \\ * & * & * \\ 0 & 0 & * \end{pmatrix} \right\}$$

and the homogeneous space

$$G/P = \mathbb{P}^2.$$

In the case of $\mathfrak{sp}_4\mathbb{C}$, there are two subdiagrams of the Dynkin diagram:

 and

these correspond to the subsets of the root system

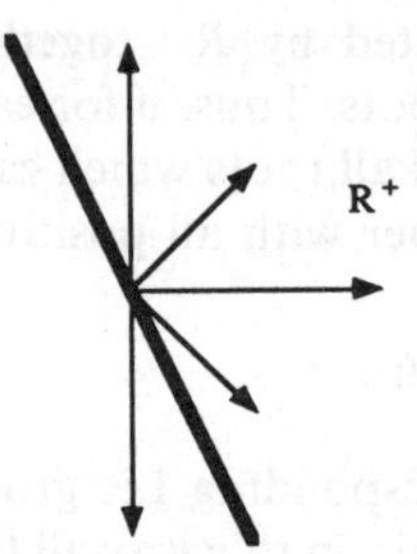

and

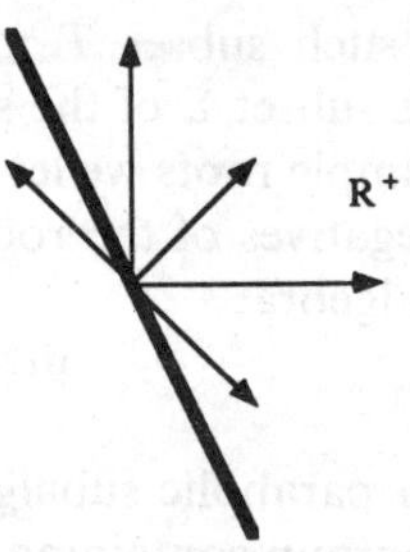

(Here we are using a black dot to indicate an omitted simple root, a white dot to indicate an included one.) The corresponding subgroups of $Sp_4\mathbb{C}$ are those preserving the vector e_1, and preserving the subspace spanned by e_1 and e_2, respectively. The quotients G/B are thus the variety of one-dimensional isotropic subspaces (i.e., the variety $\mathbb{P}^3$ of all the one-dimensional spaces) and the variety of two-dimensional isotropic subspaces.

Exercise 23.50. Interpret the diagrams above as giving rise to parabolic subgroups of the group $SO_5\mathbb{C}$ of automorphisms of $\mathbb{C}^5$ preserving a symmetric bilinear form. Show that the corresponding homogeneous spaces are the variety of isotropic planes and lines in $\mathbb{C}^5$, respectively. In particular, deduce the classical algebraic geometry facts that:

(i) The variety of isotropic 2-planes for a nondegenerate skew-symmetric bilinear form on $\mathbb{C}^4$ is isomorphic to a quadric hypersurface in $\mathbb{P}^4$.
(ii) The variety of isotropic 2-planes for a nondegenerate symmetric bilinear form on $\mathbb{C}^5$ (equivalently, lines on a smooth quadric hypersurface in $\mathbb{P}^4$) is isomorphic to $\mathbb{P}^3$.

In general, it is not hard to see that any parabolic subgroup P in a classical group G may be described as the subgroup that preserves a partial flag in the standard representation. In particular, a maximal parabolic subgroup, corresponding to omitting one node of the Dynkin diagram, may be described as the subgroup of G preserving a single subspace. Thus, for $G = SL_m\mathbb{C}$, the kth node of the Dynkin diagram

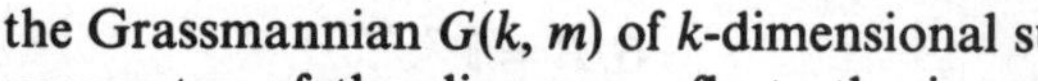

corresponds to the Grassmannian $G(k, m)$ of k-dimensional subspaces of $\mathbb{C}^m$. (Note that the symmetry of the diagram reflects the isomorphism of the Grassmannians $G(k, m)$ and $G(m - k, m)$.)

For $Sp_{2n}\mathbb{C}$, the kth node of the Dynkin diagram corresponds to the *Lagrangian Grassmannian* of isotropic k-planes, for $k = 1, 2, \ldots, n$. Similarly, for $G = SO_{2n+1}\mathbb{C}$, the kth node of the Dynkin diagram corresponds to the *orthogonal Grassmannian* of isotropic k-planes in $\mathbb{C}^{2n+1}$. Finally, for $SO_{2n}\mathbb{C}$, for $k = 1, 2, \ldots, n-2$ the kth node of the Dynkin diagram

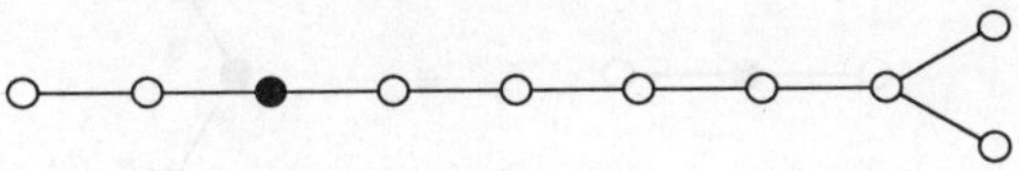

yields the orthogonal Grassmannian of isotropic k-planes in $\mathbb{C}^{2n}$, but there is one anomaly: either of the last two nodes

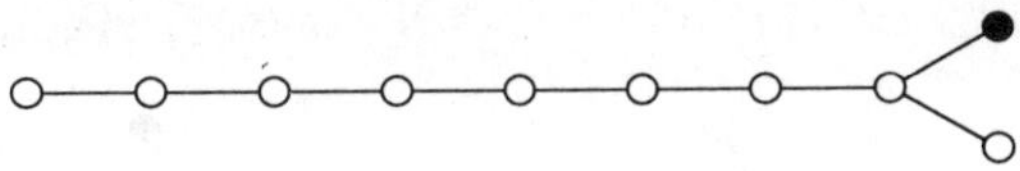

gives one of the two connected components of the Grassmannian of isotropic n-planes.

Exercise 23.51*. Compute $\mathfrak{p}(\Sigma)$ directly for each of the classical groups, and verify the above statements. Why is the orthogonal Grassmannian of isotropic $(n-1)$-planes in $\mathbb{C}^{2n}$ not included on the list?

As we saw already in Exercise 23.50, the low-dimensional coincidences between Dynkin diagrams can be used to recover some facts we have seen before. For example, the coincidence $(D_2) = (A_1) \times (A_1)$ identifies the two family of lines on a quadratic surface in $\mathbb{P}^3$ with two copies of $\mathbb{P}^1$. The coincidence $(A_3) = (D_3)$

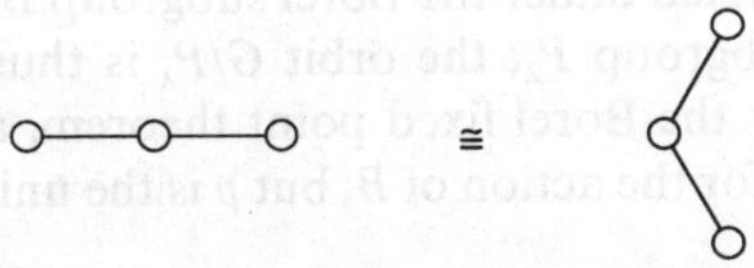

gives rise to two identifications of marked diagrams: we have

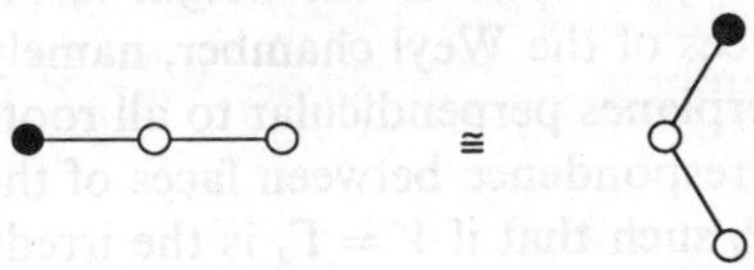

corresponding to the isomorphism between the Grassmann varieties $\mathbb{P}^3 = G(1, 4)$, $\check{\mathbb{P}}^3 = G(3, 4)$ and the two components of the family of 2-planes on a quadric hypersurface Q in $\mathbb{P}^5$; and

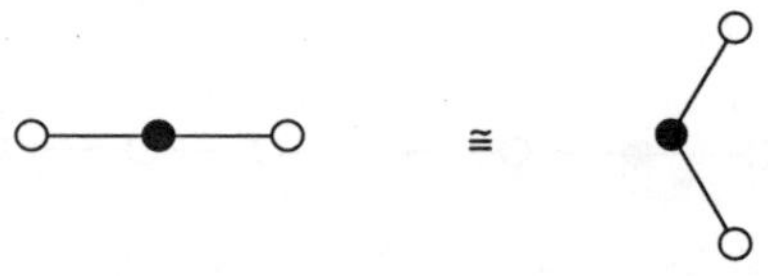

corresponding to the isomorphism of the Grassmannian $G(2, 4)$ with the quadric hypersurface Q itself. Finally, an observation that is not quite so elementary, but which we saw in §20.3: the identification of the diagrams

says that *either connected component of the variety of* 3-*planes on a smooth quadric hypersurface* Q *in* $\mathbb{P}^7$ *is isomorphic to the quadric* Q *itself.*

There is another way to realize the compact homogeneous spaces associated to a simple group G. Let $V = \Gamma_\lambda$ be an irreducible representation of G with highest weight λ, and consider the action of G on the projective space $\mathbb{P}V$. Let $p \in \mathbb{P}V$ be the point corresponding to the eigenspace with eigenvalue λ. We have then

Claim 23.52. *The orbit* $G \cdot p$ *is the unique closed orbit of the action of* G *on* $\mathbb{P}V$.

PROOF. The point p is fixed under the Borel subgroup B, so that the stabilizer of p is a parabolic subgroup P_λ; the orbit G/P_λ is thus compact and hence closed. Conversely, by the Borel fixed point theorem, any closed orbit of G contains a fixed point for the action of B; but p is the unique point in $\mathbb{P}V$ fixed by B. □

In fact, it is not hard to say which parabolic subgroup P_λ is, in terms of the classification above: *it is the parabolic subgroup corresponding to the subset of simple roots that are perpendicular to the weight* λ. Now, sets Σ of simple roots correspond to faces of the Weyl chamber, namely, the face that is the intersection of all hyperplanes perpendicular to all roots in Σ.

We thus have a correspondence between faces of the Weyl chamber and parabolic subgroups P, such that if $V = \Gamma_\lambda$ is the irreducible representation with highest weight λ, then the unique closed orbit of the action of G on $\mathbb{P}V$

is of the form G/P, where P is the parabolic subgroup corresponding to the open face of $\mathscr{W}$ containing λ. In particular, weights in the interior of the Weyl chamber correspond to $P_\lambda = B$, and so determine the full flag manifold G/B, whereas weights on the edges give rise to the quotients of G by maximal parabolics. Note that we do obtain in this way all compact homogeneous spaces for G.

For example, we have the representations of $\mathrm{SL}_3\mathbb{C}$: as we have seen, the representations $\mathrm{Sym}^k V$ and $\mathrm{Sym}^k V^*$, with highest weights on the boundaries of the Weyl chamber, have closed orbits $\{v^k\}_{v\in V}$ and $\{l^k\}_{l\in V^*}$, isomorphic to $\mathbb{P}V$ and $\mathbb{P}V^*$. By contrast, the adjoint representation—the complement of the trivial representation in $\mathrm{Hom}(V, V) = V \otimes V^*$—has as closed orbit the variety of traceless rank 1 homomorphisms, which is isomorphic to the flag manifold via the map sending a homomorphism φ to the pair ($\mathrm{Im}\ \varphi$, $\mathrm{Ker}\ \varphi$). The picture is

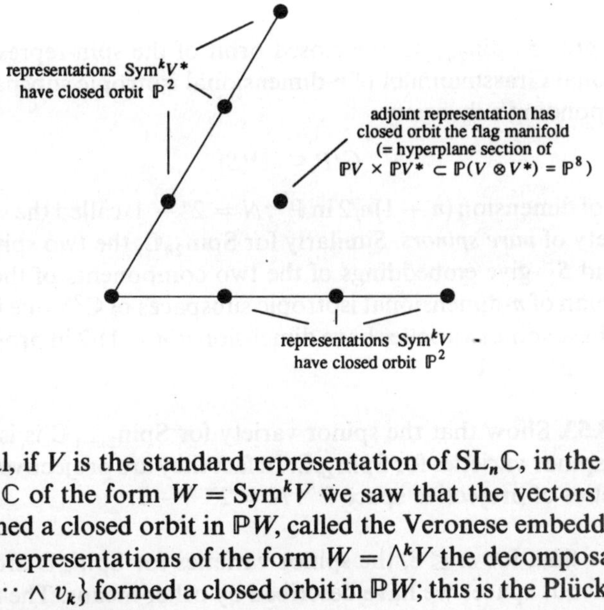

In general, if V is the standard representation of $\mathrm{SL}_n\mathbb{C}$, in the representations of $\mathrm{SL}_n\mathbb{C}$ of the form $W = \mathrm{Sym}^k V$ we saw that the vectors of the form $\{v^k\}_{v\in V}$ formed a closed orbit in $\mathbb{P}W$, called the Veronese embedding of $\mathbb{P}^{n-1}$. Likewise, in representations of the form $W = \bigwedge^k V$ the decomposable vectors $\{v_1 \wedge v_2 \wedge \cdots \wedge v_k\}$ formed a closed orbit in $\mathbb{P}W$; this is the Plücker embedding of the Grassmannian.

Similarly, we may identify the closed orbits in representations of $\mathrm{Sp}_4\mathbb{C}$. Recall here that the basic representations of $\mathrm{Sp}_4\mathbb{C}$ are the standard representation $V \cong \mathbb{C}^4$ and the complement W of the trivial representation in the exterior square $\bigwedge^2 V$; all other representations are contained in a tensor product of symmetric powers of these. Now, $\mathrm{Sp}_4\mathbb{C}$ acts transitively on $\mathbb{P}V$; the closed orbit is all of $\mathbb{P}^3$. In general, in $\mathbb{P}(\mathrm{Sym}^k V)$ the closed orbit is just the set of vectors $\{v^k\}_{v\in V} \cong \mathbb{P}^3$. By contrast, the closed orbit in $\mathbb{P}W$ is just the intersection of the hyperplane $\mathbb{P}W \subset \mathbb{P}(\bigwedge^2 V)$ with the locus of decomposable vectors $\{v \wedge w\}_{v,w\in V}$; this is the variety

$$X = \{v \wedge w: Q(v, w) = 0\}$$

of isotropic 2-planes $\Lambda \subset V$ for the skew form Q.

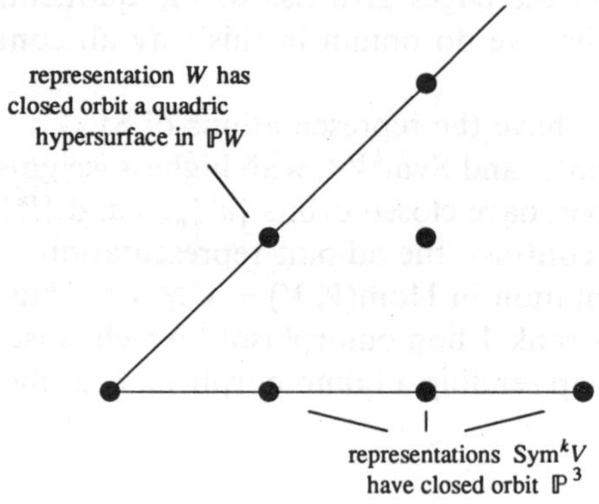

For the group $\mathrm{Spin}_{2n+1}\mathbb{C}$, the closed orbit of the spin representation S is the orthogonal Grassmannian of n-dimensional isotropic subspaces of $\mathbb{C}^{2n+1}$ The corresponding subvariety

$$G/P \hookrightarrow \mathbb{P}(S)$$

is a variety of dimension $(n+1)n/2$ in $\mathbb{P}^N$, $N = 2^n - 1$, called the *spinor variety*, or the variety of *pure spinors*. Similarly for $\mathrm{Spin}_{2n}\mathbb{C}$, the two spin representations S^+ and S^- give embeddings of the two components of the orthogonal Grassmannian of n-dimensional isotropic subspaces of $\mathbb{C}^{2n}$, one in $\mathbb{P}(S^+)$, one in $\mathbb{P}(S^-)$. These spinor varieties have dimension $n(n-1)/2$ in projective spaces of dimension $2^{n-1} - 1$.

Exercise 23.53. Show that the spinor variety for $\mathrm{Spin}_{2n-1}\mathbb{C}$ is isomorphic to each of the spinor varieties for $\mathrm{Spin}_{2n}\mathbb{C}$. In fact they are projectively equivalent as subvarieties of projective space $\mathbb{P}^N$, $N = 2^{n-1} - 1$.

It follows that, for $m \leq 8$, the spinor varieties for $\mathrm{Spin}_m\mathbb{C}$ are isomorphic to homogeneous spaces we have described by other means. The first new one is the 10-dimensional variety in $\mathbb{P}^{15}$, which comes from $\mathrm{Spin}_9\mathbb{C}$ or $\mathrm{Spin}_{10}\mathbb{C}$.

It is worth going back to interpret some of the "geometric plethysm" of earlier lectures (e.g., Exercises 11.36 and 13.24) in this light.

Finally, we can describe (at least one of) the compact homogeneous spaces for the group G_2 in this way. To begin with, G_2 has two maximal parabolic subgroups, corresponding to the diagrams

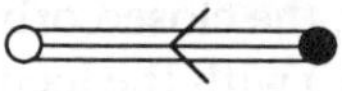 and 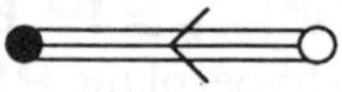

These are the groups whose Lie algebras are the parabolic subalgebras spanned by the Cartan subalgebra $\mathfrak{h} \subset \mathfrak{g}$ together with the root spaces corresponding to the roots in the diagrams

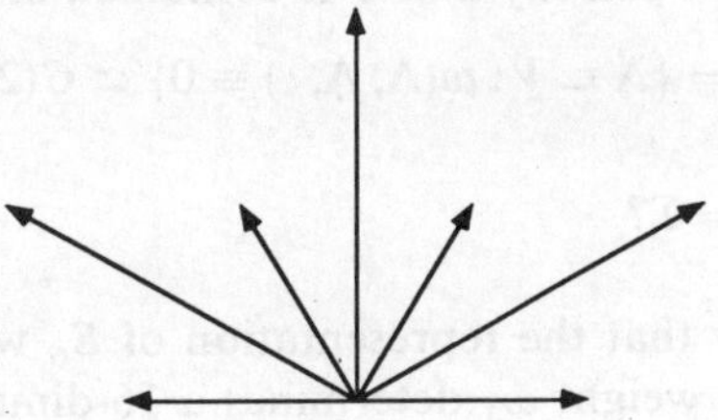

and

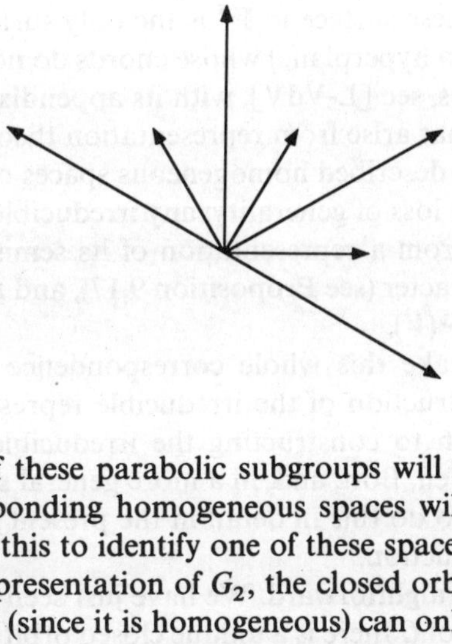

In particular, each of these parabolic subgroups will have dimension 9, so that both the corresponding homogeneous spaces will be five-dimensional varieties. We can use this to identify one of these spaces: if V is the standard seven-dimensional representation of G_2, the closed orbit in $\mathbb{P}V \cong \mathbb{P}^6$ will be a hypersurface, which (since it is homogeneous) can only be a quadric hypersurface. Thus, the homogeneous space for G_2 corresponding to the diagram

is a quadric hypersurface in $\mathbb{P}^6$. In particular, we see again that the action of G_2 on V preserves a nondegenerate bilinear form, i.e., we have an inclusion

$$G_2 \hookrightarrow \mathrm{SO}_7\mathbb{C}.$$

The other homogeneous space Y of $\mathfrak{g}_2$ is less readily described. One way to describe it is to use the fact that the adjoint representation W of $\mathfrak{g}_2$ is contained in the exterior square $\wedge^2 V$ of the standard. Since the Grassmannian $\mathbb{G}(1, 7) \subset \mathbb{P}(\wedge^2 V)$ of lines in $\mathbb{P}V$ is closed and invariant in $\mathbb{P}(\wedge^2 V)$, it follows that Y is contained in the intersection of $\mathbb{G}$ with the subspace $\mathbb{P}W \subset \mathbb{P}(\wedge^2 V)$. In other words, in terms of the skew-symmetric trilinear form ω on V preserved by the action of G_2, we can say that Y is contained in the locus

$$\Sigma = \{\Lambda \subset V: \omega(\Lambda, \Lambda, \cdot) \equiv 0\} \subset G(2, V).$$

Problem 23.54. Is $Y = \Sigma$?

Exercise 23.55. Show that the representation of E_6 whose highest weight is the first fundamental weight ω_1 determines a 16-dimensional homogeneous space in $\mathbb{P}^{26}$

These homogeneous spaces have an amazing way of showing up as extremal examples of subvarieties of projective spaces, starting with a discovery of Severi that the Veronese surface in $\mathbb{P}^5$ is the only surface in $\mathbb{P}^5$ (nonsingular and not contained in a hyperplane) whose chords do not fill up $\mathbb{P}^5$. For recent work along these lines, see [L-VdV], with its appendix by Zak on interesting projective varieties that arise from representation theory.

Although we have described homogeneous spaces only for semisimple Lie groups, this is no real loss of generality: any irreducible representation V of a Lie group G comes from a representation of its semisimple quotient, up to multiplying by a character (see Proposition 9.17), and this character does not change the orbits in $\mathbb{P}(V)$.

It is possible to take this whole correspondence one step further and use it to give a construction of the irreducible representations of G; this is the modern approach to constructing the irreducible representations, due primarily to Borel, Weil, Bott, and, in a more general setting, Schmid. We do not have the means to do this in detail in the present circumstances, but we will sketch the construction.

The idea is very straightforward. We have just seen that for every irreducible representation V of G there is a unique closed orbit $X = G/P$ of the action of G on $\mathbb{P}V$. We obtain in this way from V a projective variety X together with a line bundle L on X invariant under the action of G (the restriction of the universal bundle from $\mathbb{P}V$). In fact, we may recover V from this data simply as the vector space of holomorphic sections of the line bundle L on X. What ties this all together is the fact that this gives us a one-to-one correspondence between irreducible representations of G and ample (positive) line bundles on compact homogeneous spaces G/P. More generally, using the projection maps $G/B \to G/P$, we may pull back all these line bundles to line bundles on G/B. This then extends to give an isomorphism between the weight lattice of $\mathfrak{g}$ and the group of line bundles on G/B, with the wonderful property that for dominant weights λ, the space of holomorphic sections of the associated line bundle L_λ is the irreducible representation of G with highest weight λ.

The point of all this, apart from its intrinsic beauty, is that we can go backward: starting with just the group G, we can construct the homogeneous space G/B, and then realize all the irreducible representations of G as cohomology groups of line bundles on G/B. To carry this out, start with a weight $\lambda \in \mathfrak{h}^*$ for $\mathfrak{g}$. We have seen that λ exponentiates to a homomorphism $H \to \mathbb{C}^*$, i.e., it gives a one-dimensional representation $\mathbb{C}_\lambda$ of H. We want to induce this representation from H to G. If $H \subset B \subset G$ is a Borel subgroup, the representation extends trivially to B, since B is a semidirect product of H and the nilpotent subgroup N whose Lie algebra is the direct sum of those $\mathfrak{g}_\alpha$ for positive roots α. Then we can form

$$\begin{aligned} L_\lambda &= G \times_B \mathbb{C}_\lambda \\ &= (G \times \mathbb{C}_\lambda)/\{(g, v) \sim (gx, x^{-1}v), x \in B\}, \end{aligned}$$

which, with its natural projection to G/B, is a holomorphic line bundle on the projective variety G/B. The cohomology groups of such a line bundle are finite dimensional, and since G acts on L_λ, these cohomology groups are representations of G.

We have Bott's theorem for the vanishing of the cohomology of this line bundle:

Claim 23.56. $H^i(G/B, L_\lambda) = 0$ *for* $i \neq i(\lambda)$,

where $i(\lambda)$ is an integer depending on which Weyl chamber λ belongs to. If λ is a dominant weight (i.e., belongs to the closure of the positive Weyl chamber for the choice of positive roots used in defining B), then $i(-\lambda) = 0$. In this case the sections $H^0(G/B, L_{-\lambda})$ are a finite-dimensional vector space, on which G acts.

Claim 23.57. *For λ a dominant weight, the space of sections $H^0(G/B, L_{-\lambda})$ is the irreducible representation with highest weight λ.*

In this context the Riemann–Roch theorem can be applied to give a formula for the dimension of the irreducible representation. In fact, the dimension part of Weyl's character formula can be proved this way. More refined analysis, using the Woods Hole fixed point theorem, can be used to get the full character formula (cf. [A-B]). For a very readable introduction to this, see [Bot].

We conclude this discussion by giving a diagram showing the relationships among the various objects associated to an irreducible representation of a semisimple Lie algebra $\mathfrak{g}$. The objects and maps in diagram (23.58) are explained next.

First of all, as we have indicated, the term "Grassmannians" means the ordinary Grassmannians in the case of the groups $SL_n\mathbb{C}$, and the Lagrangian Grassmannians and the orthogonal Grassmannians of isotropic subspaces in the cases of $Sp_{2n}\mathbb{C}$ and $SO_m\mathbb{C}$, respectively. Likewise, "flag manifolds" refers to the spaces parametrizing nested sequences of such subspaces. In the cases

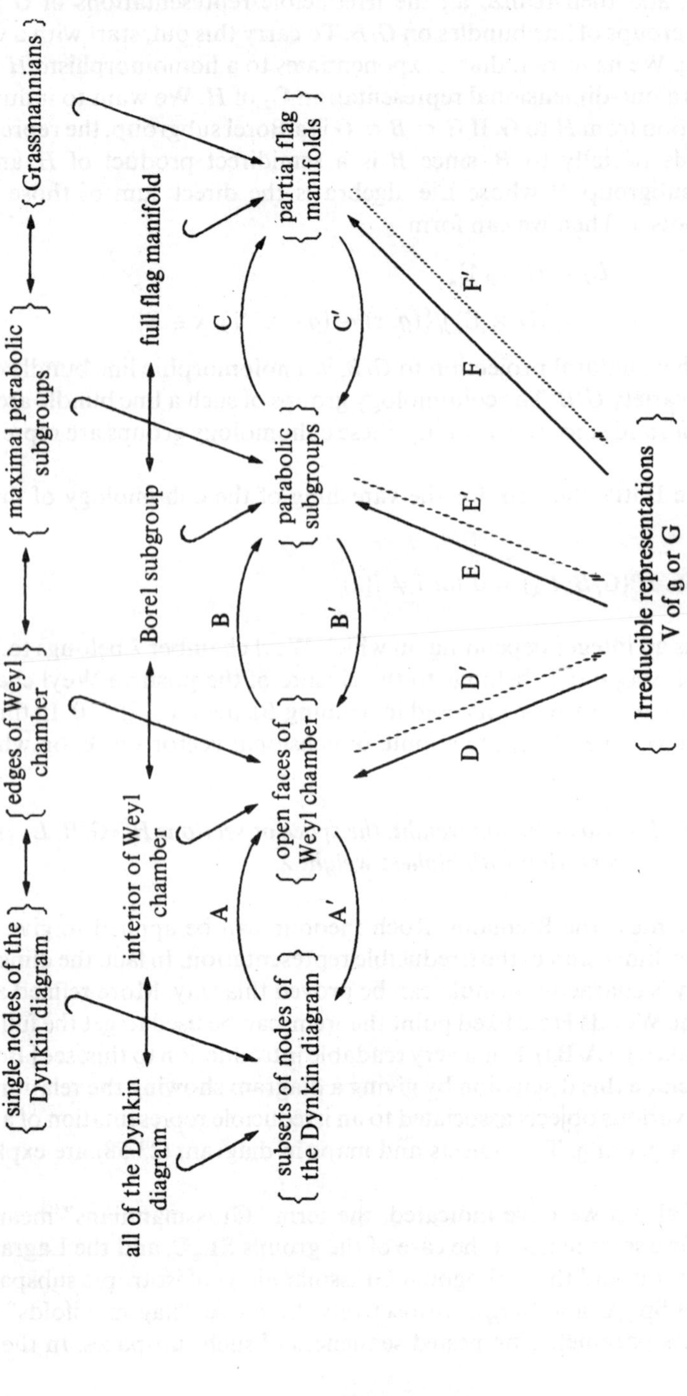

Diagram 23.58

of the exceptional Lie algebras, the term "Grassmannian" should just be ignored; except for the quotient of G_2 by one of its two maximal parabolic subgroups, the homogeneous spaces for the exceptional groups are not varieties with which we are likely to be a priori familiar.

With this said, we may describe the maps A, B, etc., as follows:

A, A': the map A associates to a subset of the nodes of the Dynkin diagram (equivalently, a subset S of the set of simple roots) the face of the Weyl chamber described by

$$\mathscr{W}_S = \left\{\lambda : \begin{matrix} (\lambda, \alpha) > 0, \forall \alpha \in S; \\ (\lambda, \alpha) = 0, \forall \alpha \notin S \end{matrix}\right\},$$

where (,) is the Killing form; the inverse is clear.

B, B': the map B associates to a face $\mathscr{W}_S$ of the Weyl chamber the subalgebra $\mathfrak{g}_S$ spanned by the Cartan subalgebra $\mathfrak{h}$, the positive root spaces $\mathfrak{g}_\alpha$, $\alpha \in R^+$, and the root spaces $\mathfrak{g}_{-\alpha}$ corresponding to those positive roots α perpendicular to $\mathscr{W}_S$. Equivalently, in terms of the corresponding subset S of the simple roots, $\mathfrak{g}_S$ will be generated by the Borel subalgebra, together with the root spaces $\mathfrak{g}_{-\alpha}$ for $\alpha \notin S$. Again, since every parabolic subalgebra is conjugate to one of this form, the inverse map is clear.

C, C': The map C simply associates to a parabolic subalgebra $\mathfrak{p} \subset \mathfrak{g}$ the quotient G/P of G by the corresponding parabolic subgroup $P \subset G$. In the other direction, given the homogeneous space $X = G/P$, with the action of G, the group P is just the stabilizer of a point in X. Note that the connected component of the identity in the automorphism group of G/P may be strictly larger: for example, $\mathbb{P}^{2n-1}$ is a compact homogeneous space for $Sp_{2n}\mathbb{C}$, and we have seen that a quadric hypersurface in $\mathbb{P}^6$ is a homogeneous space for G_2.

D, D': The map D associates to the irreducible representation V of $\mathfrak{g}$ with highest weight λ the open face of the Weyl chamber containing λ. In the other direction, given an open face $\mathscr{W}_S$ of $\mathscr{W}$, choose a lattice point $\lambda \in \mathscr{W}_S \cap \Lambda_W$ and take $V = \Gamma_\lambda$.

E: We send the representation V to the subalgebra or subgroup fixing the highest weight vector $v \in V$.

F, F': We associate to the representation V the (unique) closed orbit of the corresponding action of the group G on the projective space $\mathbb{P}V$. Going in the other direction, we have to choose an ample line bundle L on the space G/P, and then take its vector space of holomorphic sections.

§23.4. Bruhat Decompositions

We end this lecture with a brief introduction to the *Bruhat decomposition* of a semisimple complex Lie group G, and the related *Bruhat cells* in the flag manifold G/B. These ideas are not used in this course, but they appear so often elsewhere that it may be useful to describe them in the language we have developed in this lecture. We will give the general statements, but verify them only for the classical groups. General proofs can be found in [Bor1] or [Hu2].

As we have seen, a choice of positive roots determines a Borel subgroup B and Cartan subgroup H, with normalizer $N(H)$, so $N(H)/H$ is identified with the Weyl group $\mathfrak{W}$. For each $W \in \mathfrak{W}$ fix a representative n_W in $N(H)$. The double coset $B \cdot n_W \cdot B$ is clearly independent of choice of n_W, and will be denoted $B \cdot W \cdot B$.

Theorem 23.59 (Bruhat Decomposition). *The group G is a disjoint union of the $|\mathfrak{W}|$ double cosets $B \cdot W \cdot B$, as W varies over the Weyl group.*

Let us first see this explicitly for $G = \mathrm{SL}_m\mathbb{C}$. Here $N(H)$ consists of all monomial matrices in $\mathrm{SL}_m\mathbb{C}$, i.e., matrices with exactly one nonzero entry in each row and each column, and $\mathfrak{W} = \mathfrak{S}_m$; a monomial matrix with nonzero entry in the $\sigma(j)$th row of the jth column maps to the permutation σ. To see that the double cosets cover G, given $g \in G$, use elementary row operations by left multiplication by elements in B to get an element $b \cdot g^{-1}$, with $b \in B$ chosen so that the total number of zeros appearing at the left in the rows in $b \cdot g^{-1}$ is as large as possible. If two rows of $b \cdot g^{-1}$ had the same number of zeros at the left, one could increase the total by an elementary row operation. Since all the rows of $b \cdot g^{-1}$ start with different numbers of zeros, this matrix can be put in upper-triangular form by left multiplication by a monomial matrix; therefore, there is a permutation σ so that $b' = n_\sigma \cdot b \cdot g^{-1}$ is upper triangular, i.e., $g = (b')^{-1} \cdot n_\sigma \cdot b$ is in $B \cdot \sigma \cdot B$. To see that the double cosets are disjoint, suppose $n_{\sigma'} = b' \cdot n_\sigma \cdot b$ for some b and b' in B. From the equation $b = (n_\sigma)^{-1} \cdot (b')^{-1} \cdot n_{\sigma'}$ one sees that b must have nonzero entries in each place where $(n_\sigma)^{-1} \cdot n_{\sigma'}$ does, from which it follows that $\sigma' = \sigma$.

In fact, this can be strengthened as follows. Let U (resp. U^-) be the subgroup of G whose Lie algebra is the sum of all root spaces $\mathfrak{g}_\alpha$ for all positive (resp. negative) roots α. For $G = \mathrm{SL}_m\mathbb{C}$, U (resp. U^-) consists of upper- (resp. lower-) triangular matrices with 1's on the diagonal. For W in the Weyl group, define subgroups

$$U(W) = U \cap n_W \cdot U^- \cdot n_W^{-1}, \qquad U(W)' = U \cap n_W \cdot U \cdot n_W^{-1}$$

of U, which are again independent of the choice of representative n_W for W.

Corollary 23.60. *Every element in $B \cdot W \cdot B$ can be written $u \cdot n_W \cdot b$ for unique elements u in $U(W)$ and b in B.*

To see the existence of such an expression, note first that the Lie algebra of $U(W)$ is the sum of all root spaces $\mathfrak{g}_\alpha$ for which α is positive and $W^{-1}(\alpha)$ is negative; and the Lie algebra of $U(W)'$ is the sum of all root spaces $\mathfrak{g}_\alpha$ for which α and $W^{-1}(\alpha)$ are positive. One sees from this that $U(W) \cdot U(W)' \cdot H$ is the entire Borel group B. Since $H \cdot n_W = n_W \cdot H$ and $U(W)' \cdot n_W = n_W \cdot U$, and H and U are subgroups of B,

$$\begin{aligned} B\cdot n_W\cdot B &= U(W)\cdot U(W)'\cdot H\cdot n_W\cdot B \\ &= U(W)\cdot U(W)'\cdot n_W\cdot B \\ &= U(W)\cdot n_W\cdot B. \end{aligned}$$

To see the uniqueness, suppose that $n_W = u\cdot n_W\cdot b$ for some u in $U(W)$ and b in B. Then $n_W^{-1}\cdot u\cdot n_W$ is in $U^-\cap B = \{1\}$, so $u = 1$, as required.

Note in particular that the dimension of $U(W)$ is the cardinality of $R^+\cap W(R^-)$, where R^+ and R^- are the positive and negative roots; this is also the minimum number $l(W)$ of reflections in simple roots whose product is W, cf. Exercise D.30. It is a general fact, which we will see for the classical groups, that $U(W)$ is isomorphic to an affine space $\mathbb{C}^{l(W)}$.

It follows from the Bruhat decomposition that G/B is a disjoint union of the cosets $X_W = B\cdot n_W\cdot B/B$, again with W varying over the Weyl group. These X_W are called *Bruhat cells*. From the corollary we see that X_W is isomorphic to the affine space $U(W)\cong\mathbb{C}^{l(W)}$.

For $G = \mathrm{SL}_m\mathbb{C}$ and σ in $\mathfrak{W} = \mathfrak{S}_m$, the group $U(\sigma)$ consists of matrices with 1's on the diagonal, and zero entry in the i, j place whenever either $i > j$ or $\sigma^{-1}(i) < \sigma^{-1}(j)$, which is an affine space of dimension $l(\sigma) = \#\{(i, j)\colon i > j$ and $\sigma(i) < \sigma(j)\}$.

Exercise 23.61. Identifying $\mathrm{SL}_m\mathbb{C}/B$ with the space of all flags, show that X_σ consists of those flags $0\subset V_1\subset V_2\subset\cdots$ such that the dimensions of intersections with the standard flag are governed by σ, in the following sense: for each $1\le k\le m$, the set of k numbers d such that $V_k\cap\mathbb{C}^{d-1}\subsetneqq V_k\cap\mathbb{C}^d$ is precisely the set $\{\sigma(1), \sigma(2), \ldots, \sigma(k)\}$.

We will verify the Bruhat decomposition for $\mathrm{Sp}_{2n}\mathbb{C}$ by regarding it as a subgroup of $\mathrm{SL}_{2n}\mathbb{C}$ and using what we have just seen for $\mathrm{SL}_{2n}\mathbb{C}$, following [Ste2]. Our description of $\mathrm{Sp}_{2n}\mathbb{C}$ in Lecture 16 amounts to saying that it is the fixed point set of the automorphism φ of $\mathrm{SL}_{2n}\mathbb{C}$ given by $\varphi(A) = M^{-1}\cdot{}^tA^{-1}\cdot M$, with $M = \begin{pmatrix} 0 & I_n \\ -I_n & 0\end{pmatrix}$. The Borel subgroup of $\mathrm{Sp}_{2n}\mathbb{C}$ will be the intersection of the Borel subgroup B of $\mathrm{SL}_{2n}\mathbb{C}$ with $\mathrm{Sp}_{2n}\mathbb{C}$, provided we change the order of the basis of $\mathbb{C}^{2n}$ to $e_1, \ldots, e_n, e_{2n}, \ldots, e_{n+1}$, so that B consists of matrices whose upper left block is upper triangular, whose lower left block is zero, and whose lower right block is lower triangular. The automorphism φ maps this B to itself, and also preserves the diagonal subgroup H and its normalizer $N(H)$, and the groups U and U^-. The Weyl group of $\mathrm{Sp}_{2n}\mathbb{C}$ can be identified with the permutations in $\mathfrak{S}_{2n}$ such that $\sigma(n+i) = \sigma(i)\pm n$ for all $1\le i\le n$, and it is exactly for these σ for which one can choose a monomial representative n_σ in $\mathrm{Sp}_{2n}\mathbb{C}$. Now if g is any element in $\mathrm{Sp}_{2n}\mathbb{C}$, write $g = u\cdot n_\sigma\cdot b$ according to the above corollary. Then

$$g = \varphi(g) = \varphi(u)\cdot\varphi(n_\sigma)\cdot\varphi(b),$$

and by uniqueness of the decomposition we must have $\varphi(u) = u$, $\varphi(n_\sigma) = n_\sigma \cdot h$, $h \in H$, and $\varphi(b) = h^{-1} \cdot b$. It follows that σ belongs to the Weyl group of $\mathrm{Sp}_{2n}\mathbb{C}$. This gives the Bruhat decomposition, and, moreover, a unique decomposition of $g \in \mathrm{Sp}_{2n}\mathbb{C}$ into $u \cdot n_\sigma \cdot b$, with u in $U(\sigma) \cap \mathrm{Sp}_{2n}\mathbb{C}$. Since this latter is an affine space, this shows that the corresponding Bruhat cell in the symplectic flag manifold is an affine space.

Exactly the same idea works for the orthogonal groups $\mathrm{SO}_m\mathbb{C}$, by realizing them as fixed points of automorphisms of $\mathrm{SL}_m\mathbb{C}$ of the form $A \mapsto M^{-1} \cdot {}^tA^{-1} \cdot M$, with M the matrix giving the quadratic form.

Note finally that if W' is the element in the Weyl group that takes each Weyl chamber to its negative, then $B \cdot W' \cdot B$ is a dense open subset of G, a fact which is evident for the classical groups by the above discussion. The corresponding Bruhat cell $X_{W'}$ is the image of U^- in G/B, which is also a dense open set. It follows that a function or section of a line bundle on G/B is determined by its values on U^-. For treatises developing representation theory via functions on U^-, see [N-S] or [Žel].

The following exercise uses these ideas to sketch a proof of Claim 23.57 that the sections of the bundle $L_{-\lambda}$ on G/B form the irreducible representation with highest weight λ:

Exercise 23.62*. (a) Show that sections s of $L_{-\lambda}$ are all of the form $s(gB) = (g, f(g))$, where f is a holomorphic function on G satisfying

$$f(g \cdot x) = \lambda(x) f(g) \quad \text{for all } x \in B.$$

(b) Let $n' \in N(H)$ be a representative of the element W' in the Weyl group which takes each element to its negative. Show that f is determined by its value at n'.

(c) Show that any highest weight for f must be λ, and conclude that $H^0(G/B, L_{-\lambda})$ is the irreducible representation Γ_λ with highest weight λ.

The holomorphic functions f of this exercise are functions on the space G/U. In other words, all irreducible representations of G can be found in spaces of functions on G/U. This is one common approach to the study of representations, especially by the Soviet school, cf. [N-S], [Žel].

Functions on G/U form a commutative ring, which indicates how to make the sum of all the irreducible representations into a commutative ring. In fact, for the classical groups, these rings are the algebras $\mathbb{S}^{\bullet}$, $\mathbb{S}^{\langle\bullet\rangle}$, and $\mathbb{S}^{[\bullet]}$ constructed in Lectures 15, 17, and 19, cf. [L-T]. They are also coordinate rings for natural embeddings of flag manifolds in products of projective spaces.

LECTURE 24

Weyl Character Formula

This lecture is pretty straightforward: we simply state the Weyl character formula in §24.1, then show how it may be worked out in specific examples in §24.2. In particular, we derive in the case of the classical algebras formulas for the character of a given irreducible representation as a polynomial in the characters of certain basic ones (either the alternating or the symmetric powers of the standard representation for $\mathfrak{sl}_n\mathbb{C}$ and their analogues for $\mathfrak{sp}_{2n}\mathbb{C}$ and $\mathfrak{so}_m\mathbb{C}$). The proofs of the formula are deferred to the following two lectures. The techniques involved here are elementary, though the determinantal formulas are fairly complex, involving all the algebra of Appendix A.

§24.1. The Weyl Character Formula

We have already seen the Weyl character formula in the case of $\mathfrak{sl}_n\mathbb{C}$, and it is one reason why we were able to calculate so many more representations in that case. We saw in Lectures 6 and 15 that for the representation $\Gamma_\lambda = \mathbb{S}_\lambda\mathbb{C}^n$ of $\mathrm{SL}_n\mathbb{C}$ with highest weight $\lambda = \sum \lambda_i L_i$, the trace of the action of a diagonal matrix $A \in \mathrm{SL}_n\mathbb{C}$ with entries $x_1, \ldots, x_n$ is the symmetric function called the Schur polynomial $S_\lambda(x_1, \ldots, x_n)$. This included a formula for the multiplicities, which are the coefficients of the monomials in these variables.

In order to extend this formula to the other Lie algebras, let us try to rewrite this Schur polynomial in a way that may generalize. The Schur polynomial is defined to be a quotient of two alternating polynomials:

$$S_\lambda(x_1, \ldots, x_n) = \frac{|x_j^{\lambda_i+n-i}|}{|x_j^{n-i}|}.$$

These determinants can be expanded as usual as a sum over the symmetric group $\mathfrak{S}_n$, which is the Weyl group $\mathfrak{W}$. Writing $x_i = e(L_i)$ in $\mathbb{Z}[\Lambda]$ as in the preceding lecture, and writing $(-1)^W$ for $\operatorname{sgn}(W) = \det(W)$ for W in the Weyl group, the numerator may be expanded in the form

$$\begin{aligned}\sum_{W \in \mathfrak{W}} (-1)^W x_{W(1)}^{\lambda_1+n-1} \cdot \ldots \cdot x_{W(n)}^{\lambda_n} &= \sum_{W \in \mathfrak{W}} (-1)^W e(W(\Sigma(\lambda_i + n - i)L_i)) \\ &= \sum_{W \in \mathfrak{W}} (-1)^W e(W(\lambda + \rho)),\end{aligned}$$

where we write λ for $\Sigma\lambda_i L_i$ and we set $\rho = \Sigma(n - i)L_i$. Our formula therefore takes the form

$$\operatorname{Char}(\Gamma_\lambda) = \frac{\Sigma(-1)^W e(W(\lambda + \rho))}{\Sigma(-1)^W e(W(\rho))}.$$

The denominator is the discriminant

$$\Delta(x_1, \ldots, x_n) = \prod_{i<j} (x_i - x_j) = \prod_{i<j} (e(L_i) - e(L_j)).$$

This can be written in terms of the positive roots $L_i - L_j$, $i < j$, as

$$\Delta(x_1, \ldots, x_n) = \prod_{i<j} (e(\tfrac{1}{2}(L_i - L_j)) - e(-\tfrac{1}{2}(L_i - L_j))).$$

Note also that

$$\begin{aligned}\rho = \Sigma(n - i)L_i &= L_1 + (L_1 + L_2) + \cdots + (L_1 + \cdots + L_{n-1}) \\ &= \frac{1}{2}\sum_{i<j} (L_i - L_j),\end{aligned}$$

which is the *sum of the fundamental weights*, and *half the sum of the positive roots*.

These are the formulas that generalize to the other semisimple Lie algebras: For any weight μ, define $A_\mu \in \mathbb{Z}[\Lambda]$ by

$$A_\mu = \sum_{W \in \mathfrak{W}} (-1)^W e(W(\mu)). \tag{24.1}$$

Note that A_μ is not invariant by the Weyl group, but is *alternating*: $W(A_\mu) = (-1)^W A_\mu$ for $W \in \mathfrak{W}$. The ratio of two alternating polynomials will be invariant.

Theorem 24.2 (Weyl Character Formula). *Let ρ be half the sum of the positive roots. Then ρ is a weight, and $A_\rho \neq 0$. The character of the irreducible representation Γ_λ with highest weight λ is*

$$\operatorname{Char}(\Gamma_\lambda) = \frac{A_{\lambda+\rho}}{A_\rho}. \tag{WCF}$$

The assertions about ρ are part of the following lemma and exercise, which will also be useful in the applications:

Lemma 24.3. *The denominator A_ρ of Weyl's formula is*

$$\begin{aligned} A_\rho &= \prod_{\alpha \in R^+} (e(\alpha/2) - e(-\alpha/2)) \\ &= e(\rho) \prod_{\alpha \in R^+} (1 - e(-\alpha)) \\ &= e(-\rho) \prod_{\alpha \in R^+} (e(\alpha) - 1). \end{aligned}$$

PROOF. Since $e(\rho) = e(\sum \alpha/2) = \prod e(\alpha/2)$, the equality of the three displayed expressions is evident; denote these expressions temporarily by A. The key point is to see that A is alternating. For this, it suffices to see that A changes sign when a reflection in a hyperplane perpendicular to one of the simple roots is applied to it, since these reflections generate the Weyl group. This follows immediately from the first expression for A and (a) in Exercise 24.4 below.

Now, by the second displayed expression, the highest weight term that appears in A is $e(\rho)$, which is the same as that appearing in A_ρ. Calculating $1/A$ formally as in (24.5) below, we see that A_ρ/A is a formal sum $\sum m_\mu e(\mu)$ that is invariant by the Weyl group, and, using part (c) of the following exercise, it has weight 0. As in Theorem 23.24 it follows that A_ρ/A is constant; and, since A and A_ρ have the same leading term $e(\rho)$, we must have $A_\rho = A$. □

Exercise 24.4*. (a) If $W = W_{\alpha_i}$ is the reflection in the hyperplane perpendicular to a simple root α_i, show that $W(\alpha_i) = -\alpha_i$, and W permutes the other positive roots.

(b) With W as in (a), show that $W(\rho) = \rho - \alpha_i$. Deduce that ρ is the element in $\mathfrak{h}^*$ such that $\rho(H_{\alpha_i}) = 2(\rho, \alpha_i)/(\alpha_i, \alpha_i) = 1$ for each simple root α_i. Equivalently, ρ is the sum of the fundamental weights. In particular, ρ is a weight.

(c) For any $W \neq 1$ in the Weyl group, show that $\rho - W(\rho)$ is a sum of distinct positive roots. Deduce that $W(\rho)$ is not in the closure of the positive Weyl chamber.

Proofs of the character formula will be given in §25.2 and again in §26.2. For now we should at least verify that it is plausible, i.e., that $A_{\lambda+\rho}/A_\rho$ is in $\mathbb{Z}[\Lambda]^{\mathfrak{W}}$ and that the highest weight that occurs is λ. Note that since the numerator and denominator are alternating, the ratio is invariant. The fact that A_ρ is not zero follows from the second expression in the preceding lemma. To see that the ratio is actually in $\mathbb{Z}[\Lambda]$, however, we must verify that it has only a finite number of nonzero coefficients. Write

$$\frac{1}{A_\rho} = e(-\rho) \prod_{\alpha \in R^+} (1 - e(-\alpha))^{-1} = e(-\rho) \prod_{\alpha} \sum_{n=0}^{\infty} e(-n\alpha). \qquad (24.5)$$

When this is multiplied by $A_{\lambda+\rho} = \sum (-1)^W e(W(\lambda + \rho))$, we get a formal sum where the highest weight that occurs is the weight λ. This means in particular that there are only a finite number of nonzero terms corresponding to weights in the fundamental (positive) Weyl chamber $\mathscr{W}$. But since the ratio is invariant by the Weyl group, the same is true for all Weyl chambers, so $A_{\lambda+\rho}/A_\rho$ is in $\mathbb{Z}[\Lambda]^{\mathfrak{W}}$, and has highest weight λ. It follows in particular that the $A_{\lambda+\rho}/A_\rho$, as λ varies over $\mathscr{W} \cap \Lambda$, form an additive basis for $\mathbb{Z}[\Lambda]^{\mathfrak{W}}$.

Before considering the proof or any other special cases, we apply (WCF) to give a formula for the dimension of Γ_λ:

Corollary 24.6. *The dimension of the irreducible representation Γ_λ is*

$$\dim \Gamma_\lambda = \prod_{\alpha \in R^+} \frac{\langle \lambda + \rho, \alpha \rangle}{\langle \rho, \alpha \rangle} = \prod_{\alpha \in R^+} \frac{(\lambda + \rho, \alpha)}{(\rho, \alpha)},$$

where $\langle \alpha, \beta \rangle = \alpha(H_\beta) = 2(\alpha, \beta)/(\beta, \beta)$ and $(\ ,\ )$ is the Killing form.

Proof. The dimension of Γ_λ is obtained by adding the coefficients of all $e(\alpha)$ in $\operatorname{Char}(\Gamma_\lambda)$, i.e., computing the image of $\operatorname{Char}(\Gamma_\lambda)$ by the homomorphism from $\mathbb{Z}[\Lambda]$ to $\mathbb{C}$ which sends each $e(\alpha)$ to 1. However, as in the case of the Schur polynomial, the denominator vanishes if we try to do this directly. To get around this, we factor this homomorphism through the ring of power series:

$$\mathbb{Z}[\Lambda] \xrightarrow{\Psi} \mathbb{C}[[t]] \to \mathbb{C},$$

where the second homomorphism sets the variable t equal to zero, i.e., picks off the constant term of the power series, and the first homomorphism Ψ takes $e(\alpha)$ to $e^{(\rho, \alpha)t}$. More generally, for any weight μ define a homomorphism

$$\Psi_\mu \colon \mathbb{Z}[\Lambda] \to \mathbb{C}[[t]], \qquad e(\alpha) \mapsto e^{(\mu, \alpha)t}.$$

We claim that $\Psi_\mu(A_\lambda) = \Psi_\lambda(A_\mu)$ for all λ and μ. This is a simple consequence of the invariance of the metric $(\ ,\ )$ under the Weyl group:

$$\begin{aligned} \Psi_\mu(A_\lambda) &= \sum (-1)^W e^{(\mu, W(\lambda))t} \\ &= \sum (-1)^W e^{(W^{-1}(\mu), \lambda)t} \\ &= \sum (-1)^W e^{(W(\mu), \lambda)t} \\ &= \Psi_\lambda(A_\mu). \end{aligned}$$

Therefore,

$$\begin{aligned} \Psi(A_\lambda) &= \Psi_\rho(A_\lambda) = \Psi_\lambda(A_\rho) \\ &= \prod_{\alpha \in R^+} (e^{(\lambda, \alpha)t/2} - e^{-(\lambda, \alpha)t/2}) \end{aligned}$$

$$= \left(\prod_{\alpha \in R^+} (\lambda, \alpha) \right) t^{\#(R^+)} + \text{terms of higher degree in } t.$$

Hence,

$$\begin{aligned} \Psi(A_{\lambda+\rho}/A_\rho) &= \Psi(A_{\lambda+\rho})/\Psi(A_\rho) \\ &= \frac{\prod (\lambda + \rho, \alpha)}{\prod (\rho, \alpha)} + \text{terms of positive degree in } t, \end{aligned}$$

which finishes the proof. □

Exercise 24.7. In the case of $\mathfrak{sl}_n\mathbb{C}$, verify that the above corollary gives the dimension we found in Lecture 6.

Exercise 24.8. Verify directly that the right-hand side of the formula for the dimension is positive.

Since $\chi_\lambda = A_{\lambda+\rho}/A_\rho$ is the character of a virtual representation which takes on a positive value at the identity, as in the case of finite groups, to prove that it is the character of an irreducible representation, it suffices to show that $\int_G \chi_\lambda \bar{\chi}_\lambda = 1$ for an appropriate compact group G. This was the original approach of Weyl, which we will describe in the last lecture. Since the highest weight appearing is λ, we will know then that this irreducible representation must be Γ_λ.

Exercise 24.9. Use Corollary 24.6 to show that if λ is a dominant weight (i.e., in the closure of the positive Weyl chamber), and ω is a fundamental weight, then the dimension of $\Gamma_{\lambda+\omega}$ is greater than the dimension of Γ_λ. Conclude that the nontrivial representations of smallest dimension must be among the n representations Γ_ω with ω a fundamental weight.

§24.2. Applications to Classical Lie Algebras and Groups

In the case of the general linear group $GL_n\mathbb{C}$, the character[1] of the representation Γ_λ is the Schur polynomial

$$S_\lambda(x_1, \ldots, x_n) = \frac{|x_j^{\lambda_i+n-i}|}{|x_j^{n-i}|},$$

[1] We use the representation of $GL_n\mathbb{C}$ instead of its restriction to $SL_n\mathbb{C}$, since the latter would require the product of the variables x_i to be 1.

which has several expressions in terms of simpler symmetric functions. Note that the character of the dth symmetric power of the standard representation is the dth complete symmetric polynomial H_d in n variables (Appendix A.1):

$$H_d = \mathrm{Char}(\mathrm{Sym}^d(\mathbb{C}^n)).$$

The first "Giambelli" or determinantal formula (A.5) of Appendix A gives the character of the representation with highest weight $\lambda = (\lambda_1 \geq \cdots \geq \lambda_r > 0)$ as an $r \times r$ determinant:

$$\mathrm{Char}(\Gamma_\lambda) = |H_{\lambda_i+j-i}| = \begin{vmatrix} H_{\lambda_1} & H_{\lambda_1+1} \cdots H_{\lambda_1+k-1} \\ H_{\lambda_2-1} H_{\lambda_2} \cdots & \\ \vdots & \\ H_{\lambda_k-k+1} \cdots & H_{\lambda_k} \end{vmatrix}. \tag{24.10}$$

Equivalently, this expresses a general element $\Gamma_\lambda \in R(G)$ of the representation ring as a polynomial in the representations $\mathrm{Sym}^d(\mathbb{C}^n)$. A second determinantal formula, from (A.6), expresses Γ_λ in terms of the basic representations $\wedge^d(\mathbb{C}^n)$, whose characters are the elementary symmetric polynomials

$$E_d = \mathrm{Char}(\wedge^d(\mathbb{C}^n)).$$

This formula is, with μ the conjugate partition to λ,

$$\mathrm{Char}(\Gamma_\lambda) = |E_{\mu_i+j-i}| = \begin{vmatrix} E_{\mu_1} & E_{\mu_1+1} \cdots E_{\mu_1+l-1} \\ E_{\mu_2-1} E_{\mu_2} \cdots & \\ \vdots & \\ E_{\mu_l-l+1} \cdots & E_{\mu_l} \end{vmatrix} \tag{24.11}$$

In this section we work out the character formula for the other classical Lie algebras, including analogues of these determinantal formulas. The analogues of the first determinantal formula (24.10) were given by Weyl, but the analogues of (24.11) were found only recently ([D'H], [Ko-Te]). We also pay, at least by way of exercises, the debts to (WCF) that we owe from earlier lectures.

The Symplectic Case

The weights for $\mathfrak{sp}_{2n}\mathbb{C}$ are integral linear combinations of $L_1, \ldots, L_n$. We often write $\mu = (\mu_1, \ldots, \mu_n)$ for the weight $\mu_1 L_1 + \cdots + \mu_n L_n$.

The positive roots are $\{L_i - L_j\}_{i<j}$ and $\{L_i + L_j\}_{i \leq j}$, from which we find

$$\rho = \sum (n+1-i)L_i = L_1 + (L_1 + L_2) + \cdots + (L_1 + \cdots + L_n), \tag{24.12}$$

i.e., $\rho = (n, n-1, \ldots, 1)$.

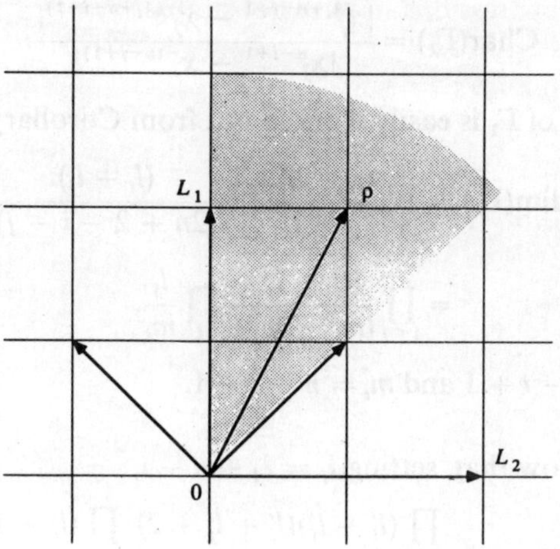

As we saw in Lecture 16, an element in the Weyl group can be written uniquely as a product $\varepsilon\sigma$, where σ is a permutation of $\{L_1, \ldots, L_n\}$, and $\varepsilon = (\varepsilon_1, \ldots, \varepsilon_n)$, with $\varepsilon_i = \pm 1$. Hence

$$A_\mu = \sum_\sigma (-1)^\sigma \sum_\varepsilon (-1)^\varepsilon e\left(\sum_{i=1}^n \varepsilon_i \mu_i L_{\sigma(i)}\right); \tag{24.13}$$

here the sign $(-1)^\varepsilon$ is the product of the ε_i. Now with $x_i = e(L_i)$, this can be written

$$A_\mu = \sum_\sigma (-1)^\sigma \prod_{i=1}^n (x_{\sigma(i)}^{\mu_i} - x_{\sigma(i)}^{-\mu_i})$$

or

$$A_\mu = |x_j^{\mu_i} - x_j^{-\mu_i}|, \tag{24.14}$$

where $|a_{i,j}|$ denotes the determinant of the $n \times n$ matrix $(a_{i,j})$. In particular,

$$A_\rho = |x_j^{n-i+1} - x_j^{-(n-i+1)}|. \tag{24.15}$$

From (24.14) or Exercise A.52 we have

$$A_\rho = \Delta(x_1 + x_1^{-1}, \ldots, x_n + x_n^{-1}) \cdot (x_1 - x_1^{-1}) \cdot \ldots \cdot (x_n - x_n^{-1}), \tag{24.16}$$

where Δ is the discriminant.

Exercise 24.17. Show that

$$A_\rho = \prod_{i<j} (x_i - x_j)(x_i x_j - 1) \cdot \prod_i (x_i^2 - 1)/(x_1 \cdot \ldots \cdot x_n)^n.$$

The character of the irreducible representation Γ_λ with highest weight $\lambda = \sum \lambda_i L_i$, $\lambda_1 \geq \cdots \geq \lambda_n \geq 0$, is therefore:

$$\text{Char}(\Gamma_\lambda) = \frac{|x_j^{\lambda_i+n-i+1} - x_j^{-(\lambda_i+n-i+1)}|}{|x_j^{n-i+1} - x_j^{-(n-i+1)}|}. \tag{24.18}$$

The dimension of Γ_λ is easily worked out from Corollary 24.6:

$$\begin{aligned}\dim(\Gamma_\lambda) &= \prod_{i<j} \frac{(l_i - l_j)}{(j-i)} \cdot \prod_{i\le j} \frac{(l_i + l_j)}{(2n+2-i-j)} \\ &= \prod_{i<j} \frac{(l_i^2 - l_j^2)}{(m_i^2 - m_j^2)} \cdot \prod_i \frac{l_i}{m_i},\end{aligned} \tag{24.19}$$

where $l_i = \lambda_i + n - i + 1$ and $m_i = n - i + 1$.

Exercise 24.20. Show that, setting $l_i' = \lambda_i + n - i$,

$$\dim(\Gamma_\lambda) = \frac{\prod_{i<j} (l_i' - l_j')(l_i' + l_j' + 2) \cdot \prod_i (l_i' + 1)}{(2n-1)! \cdot (2n-3)! \cdot \ldots \cdot 1!}.$$

These formulas give the dimension of the irreducible representation $\Gamma_{a_1,\ldots,a_n}$ with highest weight $a_1\omega_1 + \cdots + a_n\omega_n$, where the ω_i are the fundamental weights, using the relation $\lambda_i = a_i + \cdots + a_n$.

Exercise 24.21. Use Exercise 24.20 to verify that for $\lambda = L_1 + \cdots + L_k$, the dimension of Γ_λ is $2n$ if $k = 1$, and $\binom{2n}{k} - \binom{2n}{k-2}$ if $k \geq 2$. Use this to give another proof that the kernel of the contraction from $\wedge^k V$ to $\wedge^{k-2} V$ is irreducible.

The first determinantal formula for the symplectic group goes as follows. Let

$$J_d(x_1, \ldots, x_n) = H_d(x_1, \ldots, x_n, x_1^{-1}, \ldots, x_n^{-1}),$$

where H_d is the dth complete symmetric polynomial in $2n$ variables. In other words, J_d is the character of the representation $\text{Sym}^d(\mathbb{C}^{2n})$ of $\mathfrak{sp}_{2n}\mathbb{C}$. From Proposition A.50 of Appendix A we have

Proposition 24.22. *If $\lambda = (\lambda_1 \geq \cdots \geq \lambda_r > 0)$, the character of Γ_λ is the determinant of the $r \times r$ matrix whose ith row is*

$$(J_{\lambda_i - i+1} \quad J_{\lambda_i - i+2} + J_{\lambda_i - i} \quad J_{\lambda_i - i+3} + J_{\lambda_i - i-1} \quad \ldots \quad J_{\lambda_i - i+r} + J_{\lambda_i - i-r+2}).$$

For example, for $\lambda = (d)$, i.e., $\lambda = dL_1$, we have $\text{Char}(\Gamma_{(d)}) = J_d$, which is the character of $\text{Sym}^d(\mathbb{C}^{2n})$. In particular, this verifies that the kth *symmetric powers* $\text{Sym}^k(\mathbb{C}^{2n})$ *of the standard representation are all irreducible.* (This, of course, is a special case of the general description given in §17.3, since all the contraction maps vanish on the symmetric powers.)

Exercise 24.23. (i) Find the character of the representation of $\mathfrak{sp}_4\mathbb{C}$ with highest weight $\omega_1 + \omega_2 = 2L_1 + L_2$, verifying that the multiplicities are as we found in §16.2. (ii) Find the character of the representation of $\mathfrak{sp}_6\mathbb{C}$ with highest weight $\omega_1 + \omega_2$, thus verifying the assertion of Exercise 17.4.

The second Giambelli formula in the symplectic case expresses Γ_λ in terms of the basic representations

$$\Gamma_{\omega_k} = \mathrm{Ker}(\wedge^k(\mathbb{C}^{2n}) \to \wedge^{k-2}(\mathbb{C}^{2n}))$$

which are the kernels of the contractions. The character of Γ_{ω_k} is E'_k, where $E'_0 = 1$, $E'_1 = E_1 = x_1 + \cdots + x_n + x_1^{-1} + \cdots + x_n^{-1}$, and

$$E'_k = E_k(x_1, \ldots, x_n, x_1^{-1}, \ldots, x_n^{-1}) - E_{k-2}(x_1, \ldots, x_n, x_1^{-1}, \ldots, x_n^{-1})$$

for $k \geq 2$, where E_k is the kth elementary symmetric polynomial. The formula is

Corollary 24.24. *Let $\mu = (\mu_1, \ldots, \mu_l)$ be the conjugate partition to λ. The character of Γ_λ is equal to the determinant of the $l \times l$ matrix whose ith row is*

$$(E'_{\mu_i-i+1} \quad E'_{\mu_i-i+2} + E'_{\mu_i-i} \quad E'_{\mu_i-i+3} + E'_{\mu_i-i-1} \quad \cdots \quad E'_{\mu_i-i+l} + E'_{\mu_i-i-l+2}).$$

PROOF. This follows from the proposition and Proposition A.44, which equates the two determinants before specializing the variables. □

There is also a simple formula for the character in terms of the characters E_k of $\wedge^k(\mathbb{C}^{2n})$, which also follows from Proposition A.44:

$$\mathrm{Char}(\Gamma_\lambda) = |E_{\mu_i-i+j} - E_{\mu_i-i-j}|. \tag{24.25}$$

Note that $E_{n+k} = E_{n-k}$ (corresponding to the isomorphism $\wedge^{n+k}\mathbb{C}^{2n} \cong \wedge^{n-k}\mathbb{C}^{2n}$) and $E'_{n+k} = -E'_{n-k+2}$. In particular, Corollary 24.24 expresses $\mathrm{Char}(\Gamma_\lambda)$ as a polynomial in the characters of the basic representations Γ_{ω_1}, $\ldots, \Gamma_{\omega_n}$.

The Odd Orthogonal Case

For $\mathfrak{so}_{2n+1}\mathbb{C}$ the weights are $\sum \mu_i L_i$, $\mu = (\mu_1, \ldots, \mu_n)$, with all μ_i integers or all half-integers. The positive roots are $\{L_i - L_j\}_{i<j}$, $\{L_i + L_j\}_{i<j}$, and $\{L_i\}$, so ρ is $\frac{1}{2}(L_1 + \cdots + L_n)$ less than in the case for $\mathfrak{sp}_{2n}$:

$$\rho = \sum (n + \tfrac{1}{2} - i)L_i, \tag{24.26}$$

or

$$\rho = (n - \tfrac{1}{2}, n - \tfrac{3}{2}, \ldots, \tfrac{1}{2}).$$

With $x_i^{\pm 1} = e(\pm L_i)$ and $x_i^{\pm 1/2} = e(\pm L_i/2)$, we have the same formula as before [(24.14)] for A_μ.

Exercise 24.27*. Show that

$$A_\rho = |x_j^{n-i+1/2} - x_j^{-(n-i+1/2)}|$$
$$= \Delta(x_1 + x_1^{-1}, \ldots, x_n + x_n^{-1})\cdot(x_1^{1/2} - x_1^{-1/2})\cdot\ldots\cdot(x_n^{1/2} - x_n^{-1/2}).$$

If Γ_λ is the irreducible representation with highest weight $\lambda = \sum \lambda_i L_i$, $\lambda_1 \geq \cdots \geq \lambda_n \geq 0$, then the character formula can be written

$$\operatorname{Char}(\Gamma_\lambda) = \frac{|x_j^{\lambda_i+n-i+1/2} - x_j^{-(\lambda_i+n-i+1/2)}|}{|x_j^{n-i+1/2} - x_j^{-(n-i+1/2)}|}. \tag{24.28}$$

Similarly,

$$\dim(\Gamma_\lambda) = \prod_{i<j}\frac{(l_i - l_j)}{(j-i)}\cdot\prod_{i\leq j}\frac{(l_i + l_j)}{(2n+1-i-j)}$$
$$= \prod_{i<j}\frac{(l_i^2 - l_j^2)}{(m_i^2 - m_j^2)}\cdot\prod_i \frac{l_i}{m_i}, \tag{24.29}$$

where $l_i = \lambda_i + n - i + \frac{1}{2}$, and $m_i = n - i + \frac{1}{2}$.

Exercise 24.30. Show that, with $l_i' = \lambda_i + n - i$,

$$\dim(\Gamma_\lambda) = \frac{\prod_{i<j}(l_i' - l_j')(l_i' + l_j' + 1)\cdot\prod_i(2l_i' + 1)}{(2n-1)!\cdot(2n-3)!\cdot\ldots\cdot 1!}.$$

These formulas give the dimension of the irreducible representation $\Gamma_{a_1,\ldots,a_n}$ with highest weight $a_1\omega_1 + \cdots + a_n\omega_n$, where the ω_i are the fundamental weights, using the equations

$$\lambda_i = a_i + \cdots + a_{n-1} + \tfrac{1}{2}a_n.$$

Exercise 24.31. Use the dimension formula to verify that for $\lambda = L_1 + \cdots + L_k$, the dimension of Γ_λ is $\binom{2n+1}{k}$. Use this to give another proof that $\bigwedge^k V$ is irreducible for $1 \leq k \leq n$. Verify that the dimension of the spin representation is 2^n, thus reproving that it is irreducible.

Exercise 24.32. Use the dimension formula to verify that the kernel of the contraction

$$\operatorname{Sym}^d(\mathbb{C}^{2n+1}) \to \operatorname{Sym}^{d-2}(\mathbb{C}^{2n+1})$$

is an irreducible representation with highest weight dL_1.

In case the representation is a representation of $\mathrm{SO}_{2n+1}\mathbb{C}$, i.e., the λ_i are all integral, there is a first determinantal formula that expresses Γ_λ in terms of the kernels of the contractions

$$\operatorname{Ker}(\operatorname{Sym}^d(\mathbb{C}^{2n+1}) \to \operatorname{Sym}^{d-2}(\mathbb{C}^{2n+1})).$$

Let K_d denote the character of this kernel, so $K_0 = 1$, $K_1 = x_1 + \cdots + x_n + x_1^{-1} + \cdots + x_n^{-1} + 1$, and

$$K_d = H_d(x_1, \ldots, x_n, x_1^{-1}, \ldots, x_n^{-n}, 1) - H_{d-2}(x_1, \ldots, x_n, x_1^{-1}, \ldots, x_n^{-n}, 1),$$

where H_d is the dth complete symmetric polynomial. From Proposition A.60 we have

Proposition 24.33. *If $\lambda = (\lambda_1 \geq \cdots \geq \lambda_r > 0)$, with the λ_i integral, then the character of Γ_λ is the determinant of the $r \times r$ matrix whose ith row is*

$$(K_{\lambda_i-i+1} \quad K_{\lambda_i-i+2} + K_{\lambda_i-i} \quad K_{\lambda_i-i+3} + K_{\lambda_i-i-1} \quad \cdots \quad K_{\lambda_i-i+r} + K_{\lambda_i-i-r+2}).$$

In particular, for $\lambda = (d)$, the character is K_d, which verifies that the kernel of $\mathrm{Sym}^d(\mathbb{C}^{2n+1}) \to \mathrm{Sym}^{d-2}(\mathbb{C}^{2n+1})$ is irreducible.

Exercise 24.34. Use the character formula to verify that the multiplicities of the representation $\Gamma_{2L_1+L_2}$ of $\mathfrak{so}_5\mathbb{C}$ are as specified in Exercise 18.9.

The second determinantal formula for $SO_{2n+1}\mathbb{C}$ writes Γ_λ in terms of the representations $\bigwedge^k(\mathbb{C}^{2n+1})$, whose characters are

$$E_k = E_k(x_1, \ldots, x_n, x_1^{-1}, \ldots, x_n^{-n}, 1).$$

Applying Proposition 24.33 with Corollary A.46, we have

Corollary 24.35. *Let $\mu = (\mu_1, \ldots, \mu_l)$ be the conjugate partition to λ. The character of Γ_λ is equal to the determinant of the $l \times l$ matrix whose ith row is*

$$(E_{\mu_i-i+1} \quad E_{\mu_i-i+2} + E_{\mu_i-i} \quad \cdots \quad E_{\mu_i-i+l} + E_{\mu_i-i-l+2}).$$

Since $E_{n+k} = E_{n+1-k}$ (corresponding to the isomorphism $\bigwedge^{n+k}\mathbb{C}^{2n+1} \cong \bigwedge^{n+1-k}\mathbb{C}^{2n+1}$), this expresses $\mathrm{Char}(\Gamma_\lambda)$ as a polynomial in $E_1, \ldots, E_n$, with $E_d = \mathrm{Char}(\bigwedge^d\mathbb{C}^{2n+1})$.

The Even Orthogonal Case

For $\mathfrak{so}_{2n}\mathbb{C}$ the weights are the same as in the preceding case. This time the $\{L_i\}$ are not positive roots, however, so ρ is $\frac{1}{2}(L_1 + \cdots + L_n)$ less than in the case of $\mathfrak{so}_{2n+1}\mathbb{C}$, or $L_1 + \cdots + L_n$ less than in the case of $\mathfrak{sp}_{2n}\mathbb{C}$:

$$\rho = \sum (n-i)L_i, \tag{24.36}$$

or

$$\rho = (n-1, n-2, \ldots, 0).$$

The calculation of A_μ is similar, but using only those ε of positive sign. This time

$$\sum_{\varepsilon}(-1)^{\varepsilon} e\left(\sum_{i=1}^{n} \varepsilon_i \mu_i L_{\sigma(i)}\right) = \frac{1}{2}\left[\prod_{i=1}^{n}(x_{\sigma(i)}^{\mu_i} + x_{\sigma(i)}^{-\mu_i}) + \prod_{i=1}^{n}(x_{\sigma(i)}^{\mu_i} - x_{\sigma(i)}^{-\mu_i})\right].$$

This leads to

$$A_\mu = \tfrac{1}{2}(|x_j^{\mu_i} + x_j^{-\mu_i}| + |x_j^{\mu_i} - x_j^{-\mu_i}|). \tag{24.37}$$

Note that the second determinant term vanishes when any μ_i is zero. In particular,

$$A_\rho = \tfrac{1}{2}|x_j^{n-i} + x_j^{-(n-i)}|. \tag{24.38}$$

From (24.14) or Exercise A.66,

$$A_\rho = \Delta(x_1 + x_1^{-1}, \ldots, x_n + x_n^{-1}). \tag{24.39}$$

This gives, with Γ_λ the irreducible representation with highest weight $\lambda = \sum \lambda_i L_i$, $\lambda_1 \geq \cdots \geq |\lambda_n| \geq 0$,

$$\operatorname{Char}(\Gamma_\lambda) = \frac{|x_j^{l_i} + x_j^{-l_i}| + |x_j^{l_i} - x_j^{-l_i}|}{|x_j^{n-i} + x_j^{-(n-i)}|}, \tag{24.40}$$

where $l_i = \lambda_i + n - i$. As before,

$$\begin{aligned}\dim(\Gamma_\lambda) &= \prod_{i<j}\frac{(l_i - l_j)}{(j-i)}\cdot\frac{(l_i + l_j)}{(2n-i-j)} \\ &= \prod_{i<j}\frac{(l_i^2 - l_j^2)}{(m_i^2 - m_j^2)},\end{aligned} \tag{24.41}$$

where $l_i = \lambda_i + n - i$ and $m_i = n - i$. Note that, as expected, the two representations with weights $(\lambda_1, \ldots, \lambda_{n-1}, \pm\lambda_n)$ have the same dimensions.

Exercise 24.42. Show that

$$\dim(\Gamma_\lambda) = 2^{n-1}\frac{\prod_{i<j}(l_i - l_j)(l_i + l_j)}{(2n-2)!\cdot(2n-4)!\cdot\ldots\cdot 2!}.$$

These formulas give the dimension of the irreducible representation $\Gamma_{a_1,\ldots,a_n}$ with highest weight $a_1\omega_1 + \cdots + a_n\omega_n$, where the ω_i are the fundamental weights, using the equations

$$\lambda_i = a_i + \cdots + a_{n-2} + \tfrac{1}{2}(a_{n-1} + a_n), \quad 1 \leq i \leq n-2,$$
$$\lambda_{n-1} = \tfrac{1}{2}(a_{n-1} + a_n), \qquad \lambda_n = \tfrac{1}{2}(-a_{n-1} + a_n).$$

Exercise 24.43. Use the dimension formula to verify that for $\omega = L_1 + \cdots + L_k$, $k < n$, the dimension of Γ_ω is $\binom{2n}{k}$, so $\bigwedge^k(\mathbb{C}^{2n})$ is irreducible. For $\lambda = L_1 + \cdots + L_{n-1} \pm L_n$, the dimension is $\frac{1}{2}\binom{2n}{k}$, so $\bigwedge^n(\mathbb{C}^{2n})$ is the sum of the two corresponding irreducible representations. Verify that the dimension of the two spin representations are 2^{n-1}, proving irreducibility again.

Note that the second term in the numerator in (24.40) changes sign when λ_n is replaced by $-\lambda_n$; in particular, it vanishes when $\lambda_n = 0$. When $\lambda_n = 0$, the representation Γ_λ is a representation of the orthogonal group $O_{2n}\mathbb{C}$. When $\lambda_n \neq 0$, the direct sum of the two representations with highest weights $(\lambda_1, \ldots, \pm\lambda_n)$ is an irreducible representation of $O_{2n}\mathbb{C}$. (See Exercises 23.19 and 23.37.)

Let L_d be the character of $\mathrm{Ker}(\mathrm{Sym}^d(\mathbb{C}^{2n}) \to \mathrm{Sym}^{d-2}(\mathbb{C}^{2n}))$, i.e., $L_d = H_d(x_1, \ldots, x_n, x_1^{-1}, \ldots, x_n^{-n}) - H_{d-2}(x_1, \ldots, x_n, x_1^{-1}, \ldots, x_n^{-n})$. In either case, Proposition A.64 applies to give the first determinantal formula:

Proposition 24.44. *Given integers $\lambda_1 \geq \cdots \geq \lambda_r > 0$, the character of the irreducible representation of $O_{2n}\mathbb{C}$ with highest weight $\lambda = (\lambda_1, \ldots, \lambda_r)$ is the determinant of the $r \times r$ matrix whose ith row is*

$$(L_{\lambda_i-i+1} \quad L_{\lambda_i-i+2} + L_{\lambda_i-i} \quad \cdots \quad L_{\lambda_i-i+r} + L_{\lambda_i-i-r+2}).$$

Again, for $\lambda = (d)$, this verifies that the kernel of the contraction from $\mathrm{Sym}^d(\mathbb{C}^{2n})$ to $\mathrm{Sym}^{d-2}(\mathbb{C}^{2n})$ is irreducible.

The second determinantal formula is the same as in the odd case, but with $E_k = E_k(x_1, \ldots, x_n, x_1^{-1}, \ldots, x_n^{-n})$:

Corollary 24.45. *Let $\mu = (\mu_1, \ldots, \mu_l)$ be the conjugate partition to λ. The character of Γ_λ is equal to the determinant of the $l \times l$ matrix whose ith row is*

$$(E_{\mu_i-i+1} \quad E_{\mu_i-i+2} + E_{\mu_i-i} \quad \cdots \quad E_{\mu_i-i+l} + E_{\mu_i-i-l+2}).$$

Using the fact that $E_{n+k} = E_{n-k}$, this expresses $\mathrm{Char}(\Gamma_\lambda)$ as a polynomial in $E_1, \ldots, E_n$, with $E_d = \mathrm{Char}(\wedge^d \mathbb{C}^{2n})$.

Exercise 24.46*. For each of the orthogonal groups $O_m\mathbb{C}$, show that the character of the irreducible representation with highest weight λ can be written in the form

$$\mathrm{Char}(\Gamma_\lambda) = |h_{\lambda_i-i+j} - h_{\lambda_i-i-j}|,$$

where h_k is the character of $\mathrm{Sym}^k(\mathbb{C}^m)$. Another formula for the dimension of Γ_λ is obtained by substituting $\binom{m}{k}$ for h_k in this determinant.

There are other formulas expressing the characters of general representations in terms of simpler ones. Abramsky, Jahn, and King [A-J-K] give one that can be expressed by the *same* formula for the general linear, symplectic, and orthogonal groups. The general irreducible representations are given by partitions λ or Young diagrams, and in their formula the simpler representations are those corresponding to hooks. To express it, let $(a * b)$ denote the hook with horizontal leg of length $a + 1$ and vertical leg of length $b + 1$, i.e., the partition $(a + 1, 1, \ldots, 1)$, with b 1's. More generally, given $\mathbf{a} = (a_1 > \cdots > a_r \geq 0)$ and $\mathbf{b} = (b_1 > \cdots > b_r \geq 0)$ with a_r or b_r nonzero, let $(\mathbf{a} * \mathbf{b})$ denote the partition whose Young diagram has legs of these lengths to the right of and below the r diagonal boxes (cf. Frobenius's notation, Exercise 4.17). Let $\chi_{(\mathbf{a}*\mathbf{b})}$ denote the character of the corresponding irreducible representation. Their formula is

$$\chi_{(\mathbf{a}*\mathbf{b})} = |\chi_{(a_i*b_j)}|_{1 \le i, j \le r}. \tag{24.47}$$

Taking the degree of both sides gives new formulas for the dimensions of the irreducible representations. These formulas are particularly useful if the rank r of the partition is small.

Exceptional Cases

We will, as a last example, work out the Weyl character formula for the exceptional Lie algebra $\mathfrak{g}_2$, and thereby verify some of the analysis of its representations given in Lecture 22. The remaining four exceptional Lie algebras we will leave as exercises.

To begin with, the value of ρ is easily seen to be $2L_1 + 3L_2$, in terms of the basis L_1, L_2 for the weight lattice introduced in Lecture 22.

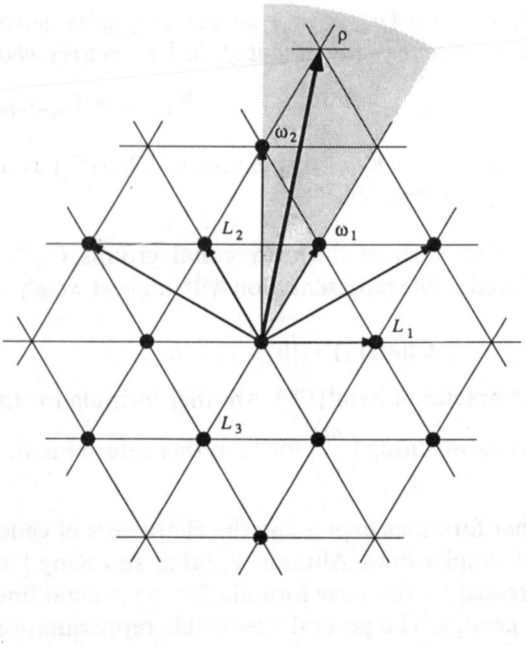

Now, for any weight $\mu = pL_1 + qL_2 + rL_3$, we have

$$\begin{aligned} A_\mu &= \sum_{\sigma \in \mathfrak{S}_3} x_{\sigma(1)}^p \cdot x_{\sigma(2)}^q \cdot x_{\sigma(3)}^r - \sum_{\sigma \in \mathfrak{S}_3} x_{\sigma(1)}^{-p} \cdot x_{\sigma(2)}^{-q} \cdot x_{\sigma(3)}^{-r} \\ &= \Delta(x) \cdot S_{p,q,r}(x) - \Delta(x^{-1}) \cdot S_{p,q,r}(x^{-1}), \end{aligned}$$

where we write x for (x_1, x_2, x_3) and x^{-1} for $(x_1^{-1}, x_2^{-1}, x_3^{-1})$, Δ is the discriminant, and $S_{p,q,r}$ the Schur function. Using the relation $\prod x_i = 1$ we can also write this as

$$= \Delta(x)\cdot(S_{p,q,r}(x) - S_{m-p,m-q,m-r}(x))$$

for any $m \geq \max(p, q, r)$. To make this notation agree with the standard notation for Schur polynomials from Appendix (A.4), note that $S_{p,q,r}$ is the Schur polynomial $S_{(s,t)}$ for the partition (s, t), $s \geq t$, where s is two less than the difference between the largest and smallest of p, q, and r, while t is one less than the difference between the second largest and the smallest; if p, q, and r are not distinct, $S_{p,q,r} = 0$. Thus, for example,

$$\begin{aligned} A_\rho &= \Delta(x)\cdot(S_{(1,1)}(x) - S_{(1)}(x)) \\ &= \Delta(x)\cdot(x_1x_2 + x_1x_3 + x_2x_3 - x_1 - x_2 - x_3). \end{aligned}$$

Now, any irreducible representation Γ_λ of $\mathfrak{g}_2$ has highest weight $\lambda = a\omega_1 + b\omega_2$, where $\omega_1 = L_1 + L_2$ and $\omega_2 = L_1 + 2L_2$ are the two fundamental weights, and a and b are non-negative integers. Then $\lambda + \rho = (a + b + 2)L_1 + (a + 2b + 3)L_2$. The Weyl character formula in this case becomes

Proposition 24.48. *The character of the representation of* $\mathfrak{g}_2$ *with highest weight* $a\omega_1 + b\omega_2$ *is*

$$\operatorname{Char}(\Gamma_{a,b}) = \frac{S_{(a+2b+1,a+b+1)} - S_{(a+2b+1,b)}}{S_{(1,1)} - S_{(1)}}.$$

Exercise 24.49. In the case of the standard representation $\Gamma_{1,0}$, the adjoint representation $\Gamma_{0,1}$, and the representation $\Gamma_{2,0}$, use this formula to verify the multiplicities found in Lecture 22.

We can also work out the dimension formula explicitly in this case. The two fundamental weights ω_1 and ω_2 have inner products

$$(\omega_1, \omega_1) = 1, \qquad (\omega_1, \omega_2) = 3/2, \qquad \text{and} \qquad (\omega_2, \omega_2) = 3;$$

ω_1 and ω_2 are among the positive roots of $\mathfrak{g}_2$, and in terms of these the remaining positive roots are $2\omega_1 - \omega_2$, $3\omega_1 - \omega_2$, $\omega_2 - \omega_1$, and $2\omega_2 - 3\omega_1$. The weight ρ is the sum of the fundamental weights ω_1 and ω_2, so that for an arbitrary weight $\lambda = a\omega_1 + b\omega_2$ we have the following table of inner products:

	$(\cdot, \rho)$	$(\cdot, \lambda)$	$(\cdot, \lambda + \rho)$
$2\omega_1 - \omega_2$	$1/2$	$a/2$	$(a + 1)/2$
$3\omega_1 - \omega_2$	3	$3a/2 + 3b/2$	$3(a + b + 2)/2$
ω_1	$5/2$	$a + 3b/2$	$(2a + 3b + 5)/2$
ω_2	$9/2$	$3a/2 + 3b$	$3(a + 2b + 3)/2$
$-\omega_1 + \omega_2$	2	$a/2 + 3b/2$	$(a + 3b + 4)/2$
$-3\omega_1 + 2\omega_2$	$3/2$	$3b/2$	$3(b + 1)/2$

We conclude that *the dimension of the irreducible representation* $\Gamma_{a,b}$ *of* $\mathfrak{g}_2$ *with highest weight* $\lambda = a\omega_1 + b\omega_2$ *is*

$$\dim(\Gamma_{a,b}) = \frac{(a+1)(a+b+2)(2a+3b+5)(a+2b+3)(a+3b+4)(b+1)}{120}.$$

We can check this in the cases $a = 1$, $b = 0$ and $a = 0$, $b = 1$, getting the dimensions 7 and 14 of the standard and adjoint representations, respectively. In case $a = 2$, $b = 0$ we may verify the result of the explicit calculation in Lecture 22, finding that

$$\dim(\Gamma_{2,0}) = 27$$

and, therefore, deducing that $\wedge^3 V = \Gamma_{2,0} \oplus V \oplus \mathbb{C}$ and $\mathrm{Sym}^2 V = \Gamma_{2,0} \oplus \mathbb{C}$.

Exercise 24.50. Show that $\mathrm{Sym}^a V = \bigoplus_{k=0}^{[a/2]} \Gamma_{a-2k,0}$.

We leave the analogous computations for the remaining four Lie algebras as exercises, using the description of the root systems found in Exercise 21.16. Since we have not said much about the Weyl group in the exceptional cases the formula (WCF) cannot be used directly—not to mention the fact that the orders of these Weyl groups are: $2^7 \cdot 3^2 = 1152$ for $\mathfrak{f}_4$; $2^7 \cdot 3^4 \cdot 5 = 51{,}840$ for $\mathfrak{e}_6$, $2^{10} \cdot 3^4 \cdot 5 \cdot 7 = 2{,}903{,}040$ for $\mathfrak{e}_7$, and $2^{14} \cdot 3^5 \cdot 5^2 \cdot 7 = 696{,}729{,}600$ for $\mathfrak{e}_8$. However, the dimension formula is available.

Exercise 24.51*. For each of the four remaining exceptional Lie algebras, compute ρ = half the sum of the positive roots. For each of the fundamental weights ω, at least for $\mathfrak{f}_4$, compute the dimension of the irreducible representation with highest weight ω. In particular, find the nontrivial representation of minimal dimension. Use this to verify that (E_6) is not isomorphic to (B_6) or (C_6), i.e., that $\mathfrak{e}_6$ is not isomorphic to $\mathfrak{so}_{13}\mathbb{C}$ or $\mathfrak{sp}_{12}\mathbb{C}$.

Exercise 24.52*. List all irreducible representations V of simple Lie algebras $\mathfrak{g}$ such that $\dim V \le \dim \mathfrak{g}$. Note that these include all cases where the corresponding group representation has a Zariski dense orbit, or a finite number of orbits.

LECTURE 25

More Character Formulas

In this lecture we give two more formulas for the multiplicities of an irreducible representation of a semisimple Lie algebra or group. First, Freudenthal's formula (§25.1) gives a straightforward way of calculating the multiplicity of a given weight once we know the multiplicity of all higher ones. This in turn allows us to prove in §25.2 the Weyl character formula, as well as another multiplicity formula due to Kostant. Finally, in §25.3 we give Steinberg's formula for the decomposition of the tensor product of two arbitrary irreducible representations of a semisimple Lie algebra, and also give formulas for some pairs $\mathfrak{h} \subset \mathfrak{g}$ for the decomposition of the restriction to $\mathfrak{h}$ of irreducible representations of $\mathfrak{g}$.

§25.1. Freudenthal's Multiplicity Formula

Freudenthal's formula gives a general way of computing the multiplicities of a representation, i.e., the dimensions of its weight spaces, by working down successively from the highest weight. The result is similar to (but more complicated than) what we did for $\mathfrak{sl}_3\mathbb{C}$ in Lecture 13, where we found the multiplicities along successive concentric hexagons in the weight diagram.

Let Γ_λ be the irreducible representation with highest weight λ, which will be fixed throughout this discussion. Let $n_\mu = n_\mu(\Gamma_\lambda)$ be the dimension of the weight space[1] of weight μ in Γ_λ, i.e., $\text{Char}(\Gamma_\lambda) = \sum n_\mu e(\mu)$. Freudenthal gives a formula for n_μ in terms of multiplicities of weights that are higher than μ.

[1] In the literature, these multiplicities n_μ are often referred to as "inner multiplicities."

Proposition 25.1 (Freudenthal's Multiplicity Formula). *With the above notation,*

$$c(\mu)\cdot n_\mu(\Gamma_\lambda) = 2\sum_{\alpha\in R^+}\sum_{k\geq 1}(\mu+k\alpha,\alpha)n_{\mu+k\alpha},$$

where $c(\mu) = \|\lambda+\rho\|^2 - \|\mu+\rho\|^2$.

Here $\|\beta\|^2 = (\beta, \beta)$, $(\ ,\ )$ is the Killing form, and ρ is half the sum of the positive roots.

Exercise 25.2*. Verify that $c(\mu)$ is positive if $\mu \neq \lambda$ and $n_\mu > 0$.

The proof of Freudenthal's formula uses a *Casimir operator*, denoted C. This is an endomorphism of any representation V of the semisimple Lie algebra $\mathfrak{g}$, and is constructed as follows. Take any basis $U_1, \ldots, U_r$ for $\mathfrak{g}$, and let $U_1', \ldots, U_r'$ be the dual basis with respect to the Killing form on $\mathfrak{g}$. Set

$$C = U_1U_1' + \cdots + U_rU_r',$$

i.e., for any $v \in V$, $C(v) = \sum U_i\cdot(U_i'\cdot v)$.

Exercise 25.3. Verify that C is independent of the choice of basis[2].

The key fact is

Exercise 25.4*. Show that C commutes with every operation in $\mathfrak{g}$, i.e.,

$$C(X\cdot v) = X\cdot C(v) \quad \text{for all } X\in\mathfrak{g},\ v\in V.$$

The idea is to use a special basis for the construction of C, so that each term U_iU_i' will act as multiplication by a constant on any weight space, and this constant can be calculated in terms of multiplicities. Then Schur's lemma can be applied to know that, in case V is irreducible, C itself is multiplication by a scalar. Taking traces will lead to a relation among multiplicities, and a little algebraic manipulation will give Freudenthal's formula.

The basis for $\mathfrak{g}$ to use is a natural one: Choose the basis $H_1, \ldots, H_n$ for the Cartan subalgebra $\mathfrak{h}$, where $H_i = H_{\alpha_i}$ corresponds to the simple root α_i, and let H_i' be the dual basis for the restriction of the Killing form to $\mathfrak{h}$. For each root α, choose a nonzero $X_\alpha \in \mathfrak{g}_\alpha$. The dual basis will then have X_α' in $\mathfrak{g}_{-\alpha}$. In fact, if we let $Y_\alpha \in \mathfrak{g}_{-\alpha}$ be the usual element so that X_α, Y_α, and $H_\alpha = [X_\alpha, Y_\alpha]$ are the canonical basis for the subalgebra $\mathfrak{s}_\alpha \cong \mathfrak{sl}_2\mathbb{C}$ that they span, then

$$X_\alpha' = ((\alpha,\alpha)/2)\,Y_\alpha. \tag{25.5}$$

Exercise 25.6*. Verify (25.5) by showing that $(X_\alpha, Y_\alpha) = 2/(\alpha, \alpha)$.

[2] In fancy language, C is an element of the universal enveloping algebra of $\mathfrak{g}$, but we do not need this.

Now we have the Casimir operator

$$C = \sum H_i H_i' + \sum_{\alpha \in R} X_\alpha X_\alpha',$$

and we analyze the action of C on the weight space V_μ corresponding to weight μ for any representation V. Let $n_\mu = \dim(V_\mu)$. First we have

$$\sum H_i H_i' \text{ acts on } V_\mu \text{ by multiplication by } (\mu, \mu) = \|\mu\|^2. \qquad (25.7)$$

Indeed, $H_i H_i'$ acts by multiplication by $\mu(H_i)\mu(H_i')$. If we write $\mu = \sum r_i \omega_i$, where the ω_i are the fundamental weights, then $\mu(H_i) = r_i$, and if $\mu = \sum r_i' \omega_i'$, with ω_i' the dual basis to ω_i, then similarly $\mu(H_i') = r_i'$. Hence $\sum \mu(H_i)\mu(H_i') = \sum r_i r_i' = (\mu, \mu)$, as asserted.

Now consider the action of $X_\alpha X_\alpha' = ((\alpha, \alpha)/2) X_\alpha Y_\alpha$ on V_μ. Restricting to the subalgebra $\mathfrak{s}_\alpha \cong \mathfrak{sl}_2$ and to the subrepresentation $\bigoplus_i V_{\mu + i\alpha}$ corresponding to the α-string through μ, we are in a situation which we know very well. Suppose this string is

$$V_\beta \oplus V_{\beta - \alpha} \oplus \cdots \oplus V_{\beta - m\alpha},$$

so $m = \beta(H_\alpha)$ [cf. (14.10)], and let k be the integer such that $\mu = \beta - k\alpha$. We assume for now that $k \le m/2$.

On the first term V_β, $X_\alpha Y_\alpha$ acts by multiplication by $m = \beta(H_\alpha) = 2(\beta, \alpha)/(\alpha, \alpha)$, so $X_\alpha X_\alpha'$ acts by multiplication by (β, α). In general, on the part of $V_{\beta - k\alpha}$ which is the image of V_β by multiplication by $(Y_\alpha)^k$, we know [cf. (11.5)] that $X_\alpha Y_\alpha$ acts by multiplication by $(k+1)(m-k)$. This gives us a subspace of V_μ of dimension n_β on which $X_\alpha X_\alpha'$ acts by multiplication by

$$(k+1)((\beta, \alpha) - k(\alpha, \alpha)/2) = (k+1)((\mu, \alpha) + k(\alpha, \alpha)/2).$$

Now peel off the subrepresentation (over $\mathfrak{s}_\alpha$) of V spanned by V_β, and apply the same reasoning to what is left. We have a subspace of $V_{\beta - \alpha}$ of dimension $n_{\beta - \alpha} - n_\beta$ to which the same analysis can be made. From this we get a subspace of V_μ of dimension $n_{\beta - \alpha} - n_\beta$ on which $X_\alpha X_\alpha'$ acts by multiplication by

$$(k)((\mu, \alpha) + (k-1)(\alpha, \alpha)/2).$$

Continuing to peel off subrepresentations, the space V_μ is decomposed into pieces on which $X_\alpha X_\alpha'$ acts by multiplication by a scalar. The trace of $X_\alpha X_\alpha'$ on V_μ is therefore the sum

$$n_\beta \cdot (k+1)((\mu, \alpha) + k(\alpha, \alpha)/2) + (n_{\beta - \alpha} - n_\beta) \cdot (k)((\mu, \alpha) + (k-1)(\alpha, \alpha)/2)$$
$$+ \cdots + ((n_{\beta - k\alpha} - n_{\beta - (k-1)\alpha}) \cdot (1)((\mu, \alpha) + (0)(\alpha, \alpha)/2).$$

Canceling in successive terms, this simplifies to

$$\operatorname{Trace}(X_\alpha X_\alpha'|_{V_\mu}) = \sum_{i=0}^{k} (\mu + i\alpha, \alpha) n_{\mu + i\alpha}. \qquad (25.8)$$

One pleasant fact about this sum is that it may be extended to all $i \ge 0$, since $n_{\mu + i\alpha} = 0$ for $i > k$.

In case $k \geq m/2$, the computation is similar, peeling off representations from the other end, starting with $V_{\beta-m\alpha}$. The only difference is that the action of $X_\alpha Y_\alpha$ on $V_{\beta-m\alpha}$ is zero. The result is

$$\operatorname{Trace}(X_\alpha X'_\alpha|_{V_\mu}) = -\sum_{i=1}^{\infty} (\mu - i\alpha, \alpha) n_{\mu-i\alpha}. \tag{25.9}$$

Exercise 25.10. Show that $X_\alpha X'_\alpha = X_{-\alpha} X'_{-\alpha} + ((\alpha, \alpha)/2)H_\alpha$, and deduce (25.9) directly from (25.8) by replacing α by $-\alpha$.

In fact, (25.8) is valid for all μ and α, as we see from the identity

$$\sum_{i=-\infty}^{\infty} (\mu + i\alpha, \alpha) n_{\mu+i\alpha} = 0. \tag{25.11}$$

Exercise 25.12*. Verify (25.11) by using the symmetry of the α-string through β.

Now we add the assumption that V is irreducible, so C is multiplication by some scalar c. Taking the trace of C on V_μ and adding, we get

$$c n_\mu = (\mu, \mu) n_\mu + \sum_{\alpha \in R} \sum_{i \geq 0} (\mu + i\alpha, \alpha) n_{\mu+i\alpha}. \tag{25.13}$$

Note that when $i = 0$ the two terms for α and $-\alpha$ cancel each other, so the summation can begin at $i = 1$ instead. Rewriting this in terms of the positive weights, and using (25.11) the sums become

$$\sum_{\alpha \in R^+} \sum_{i=1}^{\infty} (\mu + i\alpha, \alpha) n_{\mu+i\alpha} + \sum_{\alpha \in R^+} \sum_{i=1}^{\infty} (\mu - i\alpha, \alpha) n_{\mu-i\alpha}$$

$$= n_\mu \sum_{\alpha \in R^+} (\mu, \alpha) + 2 \sum_{\alpha \in R^+} \sum_{i=1}^{\infty} (\mu + i\alpha, \alpha) n_{\mu+i\alpha}.$$

Summarizing, and observing that $\sum_{\alpha \in R^+} (\mu, \alpha) = (\mu, 2\rho)$, we have

$$c n_\mu = ((\mu, \mu) + (\mu, 2\rho)) n_\mu + 2 \sum_{\alpha \in R^+} \sum_{i=1}^{\infty} (\mu + i\alpha, \alpha) n_{\mu+i\alpha}.$$

Note that $(\mu, \mu) + (\mu, 2\rho) = (\mu + \rho, \mu + \rho) - (\rho, \rho) = \|\mu + \rho\|^2 - \|\rho\|^2$. To evaluate the constant we evaluate on the highest weight space V_λ, where $n_\lambda = 1$ and $n_{\lambda+i\alpha} = 0$ for $i > 0$. Hence,

$$c = (\lambda, \lambda) + (\lambda, 2\rho) = \|\lambda + \rho\|^2 - \|\rho\|^2. \tag{25.14}$$

Combining the preceding two equations yields Freudenthal's formula. □

Exercise 25.15. Apply Freudenthal's formula to the representations of $\mathfrak{sl}_3\mathbb{C}$ considered in §13.2, verifying again that the multiplicities are as prescribed on the hexagons and triangles.

Exercise 25.16. Use Freudenthal's formula to calculate multiplicities for the representations $\Gamma_{1,0}$, $\Gamma_{0,1}$, and $\Gamma_{2,0}$ of $(\mathfrak{g}_2)$.

§25.2. Proof of (WCF); the Kostant Multiplicity Formula

It is not unreasonable to anticipate that Weyl's character formula can be deduced from Freudenthal's inductive formula, but some algebraic manipulation is certainly required. Let

$$\chi_\lambda = \operatorname{Char}(\Gamma_\lambda) = \sum n_\mu e(\mu)$$

be the character of the irreducible representation with highest weight λ. Freudenthal's formula, in form (25.13), reads[3]

$$c \cdot \chi_\lambda = \sum_\mu (\mu, \mu) n_\mu e(\mu) + \sum_\mu \sum_{\alpha \in R} \sum_{i=0}^{\infty} (\mu + i\alpha, \alpha) n_{\mu+i\alpha} e(\mu),$$

where $c = \|\lambda + \rho\|^2 - \|\rho\|^2$. To get this to look anything like Weyl's formula, we must get rid of the inside sums over i. If α is fixed, they will disappear if we multiply by $e(\alpha) - 1$, as successive terms cancel:

$$(e(\alpha) - 1) \cdot \sum_\mu \sum_{i=0}^{\infty} (\mu + i\alpha, \alpha) n_{\mu+i\alpha} e(\mu) = \sum_\mu (\mu, \alpha) n_\mu e(\mu + \alpha).$$

Let $P = \prod_{\alpha \in R} (e(\alpha) - 1) = (e(\alpha) - 1) \cdot P_\alpha$, where $P_\alpha = \prod_{\beta \neq \alpha} (e(\beta) - 1)$. The preceding two formulas give

$$c \cdot P \cdot \chi_\lambda = P \cdot \sum_\mu (\mu, \mu) n_\mu e(\mu) + \sum_{\mu, \alpha} (\mu, \alpha) P_\alpha n_\mu e(\mu + \alpha). \tag{25.17}$$

Note also that

$$P = (-1)^r A_\rho \cdot A_\rho,$$

where r is the number of positive roots, so at least the formula now involves the ingredients that go into (WCF).

We want to prove (WCF): $A_\rho \cdot \chi_\lambda = A_{\lambda+\rho}$. We have seen in §24.1 that both sides of this equation are alternating, and that both have highest weight term $e(\lambda + \rho)$, with coefficient 1. On the right-hand side the only terms that appear are those of the form $\pm e(W(\lambda + \rho))$, for W in the Weyl group. To prove (WCF), it suffices to prove that the only terms appearing with nonzero coefficients in $A_\rho \cdot \chi_\lambda$ are these same $e(W(\lambda + \rho))$, for then the alternating property and the knowledge of the coefficient of $e(\lambda + \rho)$ determine all the coefficients. This can be expressed as:

[3] In this section we work in the ring $\mathbb{C}[\Lambda]$ of finite sums $\sum m_\mu e(\mu)$ with complex coefficients m_μ.

Claim. *The only terms $e(\nu)$ occurring in $A_\rho \cdot \chi_\lambda$ with nonzero coefficient are those with $\|\nu\| = \|\lambda + \rho\|$.*

To see that this is equivalent, note that by definition of A_ρ and χ_λ, the terms in $A_\rho \cdot \chi_\lambda$ are all of the form $\pm e(\nu)$, where $\nu = \mu + W(\rho)$, for μ a weight of Γ_λ and W in the Weyl group. But if $\|\mu + W(\rho)\| = \|\lambda + \rho\|$, since the metric is invariant by the Weyl group, this gives $\|W^{-1}(\mu) + \rho\| = \|\lambda + \rho\|$. But we saw in Exercise 25.2 that this cannot happen unless $\mu = W(\lambda)$, as required.

We are thus reduced to proving the claim. This suggests looking at the "Laplacian" operator that maps $e(\mu)$ to $\|\mu\|^2 e(\mu)$, that is, the map

$$\Delta: \mathbb{C}[\Lambda] \to \mathbb{C}[\Lambda]$$

defined by

$$\Delta\left(\sum m_\mu e(\mu)\right) = \sum (\mu, \mu) m_\mu e(\mu).$$

The claim is equivalent to the assertion that $F = A_\rho \cdot \chi_\lambda$ satisfies the "differential equation"

$$\Delta(F) = \|\lambda + \rho\|^2 F.$$

From the definition $\Delta(\chi_\lambda) = \sum (\mu, \mu) n_\mu e(\mu)$. And $\Delta(A_\rho) = \|\rho\|^2 A_\rho$. In general, since $\|W(\alpha)\| = \|\alpha\|$ for all $W \in \mathfrak{W}$,

$$\Delta(A_\alpha) = \sum (-1)^W \|W(\alpha)\|^2 e(W(\alpha)) = \|\alpha\|^2 A_\alpha.$$

So we would be in good shape if we had a formula for Δ of a product of two functions. One expects such a formula to take the form

$$\Delta(fg) = \Delta(f)g + 2(\nabla f, \nabla g) + f\Delta(g), \tag{25.18}$$

where ∇ is a "gradient," and $(\ ,\ )$ is an "inner product." Taking $f = e(\mu)$, $g = e(\nu)$, we see that we need to have $(\nabla e(\mu), \nabla e(\nu)) = (\mu, \nu) e(\mu + \nu)$. There is indeed such a gradient and inner product. Define a homomorphism

$$\nabla: \mathbb{C}[\Lambda] \to \mathfrak{h}^* \otimes \mathbb{C}[\Lambda] = \operatorname{Hom}(\mathfrak{h}, \mathbb{C}[\Lambda])$$

by the formula $\nabla(e(\mu)) = \mu \cdot e(\mu)$, and define the bilinear form $(\ ,\ )$ on $\mathfrak{h}^* \otimes \mathbb{C}[\Lambda]$ by the formula $(\alpha e(\mu), \beta e(\nu)) = (\alpha, \beta) e(\mu + \nu)$, where (α, β) is the Killing form on $\mathfrak{h}^*$.

Exercise 25.19. With these definitions, verify that (25.18) is satisfied, as well as the Leibnitz rule

$$\nabla(fg) = \nabla(f)g + f\nabla(g).$$

For example, $\nabla(\chi_\lambda) = \sum_\mu n_\mu \mu \cdot e(\mu)$, and, by the Leibnitz rule,

$$\nabla(P) = \sum_{\alpha \in R} P_\alpha \alpha \cdot e(\alpha).$$

But now look at formula (25.17). This reads

$$c \cdot P\chi_\lambda = P\Delta(\chi_\lambda) + (\nabla P, \nabla \chi_\lambda).$$

Since, also by the exercise, $\nabla(P) = 2(-1)^r A_\rho \nabla(A_\rho)$, we may cancel $(-1)^r A_\rho$ from each term in the equation, getting

$$c \cdot A_\rho \chi_\lambda = A_\rho \Delta(\chi_\lambda) + 2(\nabla A_\rho, \nabla \chi_\lambda).$$

By the identity (25.18), the right-hand side of this equation is

$$\Delta(A_\rho \chi_\lambda) - \Delta(A_\rho)\chi_\lambda = \Delta(A_\rho \chi_\lambda) - \|\rho\|^2 A_\rho \chi_\lambda.$$

Since $c = \|\lambda + \rho\|^2 - \|\rho\|^2$, this gives $\|\lambda + \rho\|^2 A_\rho \chi_\lambda = \Delta(A_\rho \chi_\lambda)$, which finishes the proof. □

We conclude this section with a proof of another general multiplicity formula, discovered by Kostant. It gives an elegant closed formula for the multiplicities, but at the expense of summing over the entire Weyl group (although as we will indicate below, there are many interesting cases where all but a few terms of the sum vanish). It also involves a kind of partition counting function. For each weight μ, let $P(\mu)$ be the number of ways to write μ as a sum of positive roots; set $P(0) = 1$. Equivalently,

$$\prod_{\alpha \in R^+} \frac{1}{1 - e(\alpha)} = \sum_\mu P(\mu) e(\mu). \tag{25.20}$$

Proposition 25.21. (Kostant's Multiplicity Formula). *The multiplicity $n_\mu(\Gamma_\lambda)$ of weight μ in the irreducible representation Γ_λ is given by*

$$n_\mu(\Gamma_\lambda) = \sum_{W \in \mathfrak{W}} (-1)^W P(W(\lambda + \rho) - (\mu + \rho)),$$

where ρ is half the sum of the positive roots.

PROOF. Write $(A_\rho)^{-1} = e(-\rho)/\prod(1 - e(-\alpha)) = \sum_\nu P(\nu) e(-\nu - \rho)$. By (WCF),

$$\begin{aligned} \chi_\lambda = A_{\lambda+\rho}(A_\rho)^{-1} &= \sum_{W,\nu} (-1)^W e(W(\lambda + \rho)) P(\nu) e(-\nu - \rho) \\ &= \sum_{W,\nu} (-1)^W P(\nu) e(W(\lambda + \rho) - (\nu + \rho)) \\ &= \sum_{W,\mu} (-1)^W P(W(\lambda + \rho) - (\mu + \rho)) e(\mu), \end{aligned}$$

as seen by writing $\mu = W(\lambda + \rho) - (\nu + \rho)$. □

In fact, the proof shows that Kostant's formula is equivalent to Weyl's formula, cf. [Cart].

One way to interpret Kostant's formula, at least for weights μ close to the highest weight λ of Γ_λ, is as a sort of converse to Proposition 14.13(ii). Recall that this says that Γ_λ will be generated by the images of its highest weight vector v under successive applications of the generators of the negative root spaces; in practice, we used this fact to bound from above the multiplicities of various weights μ close to λ by counting the number of ways of getting from λ to μ by adding negative roots. The problem in making this precise was always that we did not know how many relations there were among these images, if any. Kostant's formula gives an answer: for example, if the difference $\lambda - \mu$ is small relative to λ, we see that the only nonzero term in the sum is the principle term, corresponding to $W = 1$; in this case the answer is that there are no relations other than the trivial ones $X(Y(v)) - Y(X(v)) = [X, Y](v)$. When μ gets somewhat smaller, other terms appear corresponding to single reflections W in the walls of the Weyl chamber for which $W(\lambda + \rho)$ is higher than $\mu + \rho$; we can think of these terms, which all appear with sign -1, as correction terms indicating the presence of relations. As μ gets smaller still, of course, more terms appear of both signs, and this viewpoint breaks down.

To see how this works in practice, the reader can for example carry out the analysis of the example at the end of §13.1.

Exercise 25.22* (Kostant). Prove the following formula for the function P, which can be used to calculate it inductively: $P(0) = 1$, and, for $\mu \neq 0$,

$$P(\mu) = -\sum_{W \neq 1} (-1)^W P(\mu + W(\rho) - \rho).$$

Exercise 25.23* (Racah). Deduce from Kostant's formula and the preceding exercise the following inductive formula for the multiplicities n_μ of μ in Γ_λ: $n_\mu = 1$ if $\mu = \lambda$, and if μ is any other weight of Γ_λ, then

$$n_\mu = -\sum_{W \neq 1} (-1)^W n_{\mu+\rho-W(\rho)}.$$

Show, in fact, that for any weight μ

$$\sum_{W \in \mathfrak{W}} (-1)^W n_{\mu+\rho-W(\rho)} = \sum_{W'} (-1)^{W'},$$

where the second sum is over those $W' \in \mathfrak{W}$ such that $W'(\lambda + \rho) = \mu + \rho$.

Note that Kostant's formula, more than any of the others, shows us directly the pattern of multiplicities in the irreducible representations of $\mathfrak{sl}_3\mathbb{C}$. For one thing, it is easy to represent the function P diagrammatically: in the weight lattice of $\mathfrak{sl}_3\mathbb{C}$, the function $P(\mu)$ will be a constant 1 on the rays $\{aL_2 - aL_1\}_{a\geq 0}$ and $\{aL_3 - aL_2\}_{a\geq 0}$ through the origin in the direction of the two simple positive roots $L_2 - L_1$ and $L_3 - L_2$. It will have value 2 on the translates $\{aL_2 - (a+3)L_1\}_{a\geq -1}$ and $\{aL_3 - (a-3)L_2\}_{a\geq 2}$ of these two rays by the third positive root $L_3 - L_1$: for example, the first of these can be written as

$$\begin{aligned} aL_2 - (a+3)L_1 &= (a+1)\cdot(L_2 - L_1) + L_3 - L_1 \\ &= (a+2)\cdot(L_2 - L_1) + L_3 - L_2; \end{aligned}$$

and correspondingly its value will increase by 1 on each successive translate of these rays by $L_3 - L_1$. The picture is thus

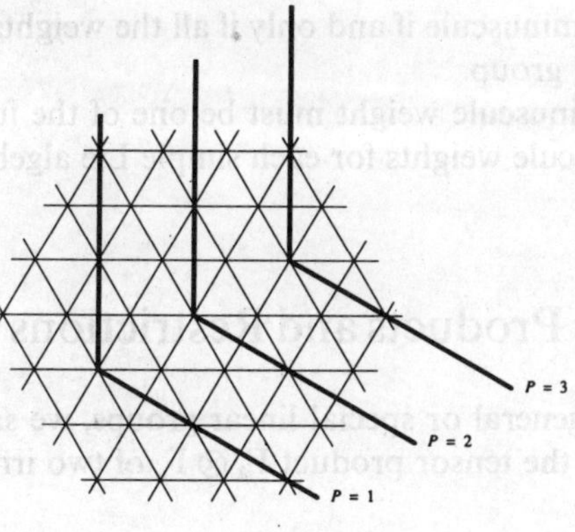

Now, the prescription given in the Kostant formula for the multiplicities is to take six copies of this function flipped about the origin, translated so that the vertex of the outer shell lies at the points $w(\lambda + \rho) - \rho$ and take their alternating sum. Superimposing the six pictures we arrive at

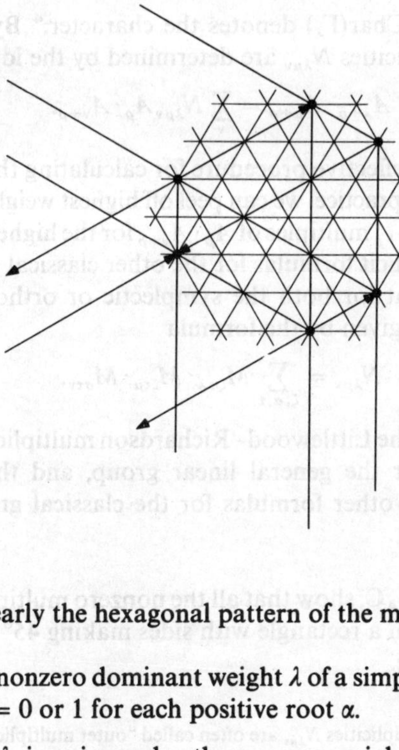

which shows us clearly the hexagonal pattern of the multiplicities.

Exercise 25.24*. A nonzero dominant weight λ of a simple Lie algebra is called *minuscule* if $\lambda(H_\alpha) = 0$ or 1 for each positive root α.

(a) Show that if λ is minuscule, then every weight space of Γ_λ is one dimensional.

(b) Show that λ is minuscule if and only if all the weights of Γ_λ are conjugate under the Weyl group.
(c) Show that a minuscule weight must be one of the fundamental weights. Find the minuscule weights for each simple Lie algebra.

§25.3. Tensor Products and Restrictions To Subgroups

In the case of the general or special linear groups, we saw general formulas for describing how the tensor product $\Gamma_\lambda \otimes \Gamma_\mu$ of two irreducible representations decomposes:

$$\Gamma_\lambda \otimes \Gamma_\mu = \bigoplus_\nu N_{\lambda\mu\nu}\Gamma_\nu.$$

In these cases the multiplicities $N_{\lambda\mu\nu}$ can be described by a combinatorial formula: the Littlewood–Richardson rule. In general, such a decomposition is equivalent to writing

$$\chi_\lambda \chi_\mu = \sum_\nu N_{\lambda\mu\nu}\chi_\nu \tag{25.25}$$

in $\mathbb{Z}[\Lambda]$, where $\chi_\lambda = \mathrm{Char}(\Gamma_\lambda)$ denotes the character.[4] By Weyl's character formula, these multiplicities $N_{\lambda\mu\nu}$ are determined by the identity

$$A_{\lambda+\rho} \cdot A_{\mu+\rho} = \sum_\nu N_{\lambda\mu\nu} A_\rho \cdot A_{\nu+\rho}. \tag{25.26}$$

This formula gives an effective procedure for calculating the coefficients $N_{\lambda\mu\nu}$, if one that is tedious in practice: we can peel off highest weights, i.e., successively subtract from $A_{\lambda+\rho} \cdot A_{\mu+\rho}$ multiples of $A_\rho \cdot A_{\nu+\rho}$ for the highest ν that appears.

There are some explicit formulas for the other classical groups. R. C. King [Ki2] has showed that for both the symplectic or orthogonal groups, the multiplicities $N_{\lambda\mu\nu}$ are given by the formula

$$N_{\lambda\mu\nu} = \sum_{\zeta,\sigma,\tau} M_{\zeta\sigma\lambda} \cdot M_{\zeta\tau\mu} \cdot M_{\sigma\tau\nu}, \tag{25.27}$$

where the M's denote the Littlewood–Richardson multiplicities, i.e., the corresponding numbers for the general linear group, and the sum is over all partitions ζ, σ, τ. For other formulas for the classical groups, see [Mur1], [We1, p. 230].

Exercise 25.28*. For $\mathfrak{so}_4\mathbb{C}$, show that all the nonzero multiplicities $N_{\lambda\mu\nu}$ are 1's, and these occur for ν in a rectangle with sides making 45° angles to the axes. Describe this rectangle.

[4] In the literature these multiplicities $N_{\lambda\mu\nu}$ are often called "outer multiplicities," and the problem of finding them, or decomposing the tensor product, the "Clebsch–Gordan" problem.

Steinberg has also given a general formula for the multiplicities $N_{\lambda\mu\nu}$. Since it involves a double summation over the Weyl group, using it in a concrete situation may be a challenge.

Proposition 25.29 (Steinberg's Formula). *The multiplicity of* Γ_ν *in* $\Gamma_\lambda \otimes \Gamma_\mu$ *is*

$$N_{\lambda\mu\nu} = \sum_{W,W'} (-1)^{WW'} P(W(\lambda + \rho) + W'(\mu + \rho) - \nu - 2\rho),$$

where the sum is over pairs $W, W' \in \mathfrak{W}$, *and* P *is the counting function appearing in Kostant's multiplicity formula.*

Exercise 25.30*. Prove Steinberg's formula by multiplying (25.25) by A_ρ, using (WCF) to get $\chi_\lambda A_{\mu+\rho} = \sum N_{\lambda\mu\nu} A_{\nu+\rho}$. Write out both sides, using Kostant's formula for χ_λ, and compute the coefficient of the term $e(\beta + \rho)$ on each side, for any β. This gives

$$\sum_{W,W'} (-1)^{WW'} P(W(\lambda + \rho) + W'(\mu + \rho) - \beta - 2\rho) = \sum_{W} (-1)^W N_{\lambda,\mu,W(\beta+\rho)-\rho}.$$

Show that for $\beta = \nu$ all the terms on the right are zero but $N_{\lambda\mu\nu}$.

Exercise 25.31 (Racah). Use the Steinberg and Kostant formulas to show that

$$N_{\lambda\mu\nu} = \sum_{W} (-1)^W n_{\nu+\rho-W(\mu+\rho)}(\Gamma_\lambda).$$

The following is the generalization of something we have seen several times:

Exercise 25.32. If λ and μ are dominant weights, and α is a simple root with $\lambda(H_\alpha)$ and $\mu(H_\alpha)$ not zero, show that $\lambda + \mu - \alpha$ is a dominant weight and $\Gamma_\lambda \otimes \Gamma_\mu$ contains the irreducible representation $\Gamma_{\lambda+\mu-\alpha}$ with multiplicity one. So

$$\Gamma_\lambda \otimes \Gamma_\mu = \Gamma_{\lambda+\mu} \oplus \Gamma_{\lambda+\mu-\alpha} \oplus \text{others}.$$

In case $\mu = \lambda$, with $\lambda(H_\alpha) \neq 0$, $\mathrm{Sym}^2(\Gamma_\lambda)$ contains $\Gamma_{\lambda+\mu}$, while $\wedge^2(\Gamma_\lambda)$ contains $\Gamma_{\lambda+\mu-\alpha}$.

Exercise 25.33. If $\lambda + \zeta$ is a dominant weight for each weight ζ of Γ_μ, show that the irreducible representations appearing in $\Gamma_\lambda \otimes \Gamma_\mu$ are exactly the $\Gamma_{\lambda+\zeta}$. In fact, with no assumptions, every component of $\Gamma_\lambda \otimes \Gamma_\mu$ always has this form. One can show that $N_{\lambda\mu\nu}$ is the dimension of

$$\{v \in (\Gamma_\lambda)_{\nu-\mu} : H_i^{l_i+1}(v) = 0, 1 \le i \le n, l_i = \mu(H_i)\}.$$

For this, see [Žel, §131].

For other general formulas for the multiplicities $N_{\lambda\mu\nu}$ see [Kem], [K-N], [Li], and [Kum1], [Kum2].

We have seen in Exercise 6.12 a formula for decomposing the representation Γ_λ of $GL_m\mathbb{C}$ when restricted to the subgroup $GL_{m-1}\mathbb{C}$. In this case the multiplicities of the irreducible components again have a simple combinatorial description. There are similar formulas for other classical groups. In the literature, such formulas are often called "branching formulas," or "modification rules." We will just state the analogues of this formula for the symplectic and orthogonal cases:

For $\mathfrak{so}_{2n}\mathbb{C} \subset \mathfrak{so}_{2n+1}\mathbb{C}$, and Γ_λ the irreducible representation of $\mathfrak{so}_{2n+1}\mathbb{C}$ given by $\lambda = (\lambda_1 \geq \cdots \geq \lambda_n \geq 0)$, the restriction is

$$\mathrm{Res}^{\mathfrak{so}_{2n+1}\mathbb{C}}_{\mathfrak{so}_{2n}\mathbb{C}}(\Gamma_\lambda) = \bigoplus \Gamma_{\bar{\lambda}}, \tag{25.34}$$

the sum over all $\bar{\lambda} = (\bar{\lambda}_1, \ldots, \bar{\lambda}_n)$ with

$$\lambda_1 \geq \bar{\lambda}_1 \geq \lambda_2 \geq \bar{\lambda}_2 \geq \cdots \geq \bar{\lambda}_{n-1} \geq \lambda_n \geq |\bar{\lambda}_n|,$$

with the $\bar{\lambda}_i$ and λ_i simultaneously all integers or all half integers.

For $\mathfrak{so}_{2n-1}\mathbb{C} \subset \mathfrak{so}_{2n}\mathbb{C}$, and Γ_λ the irreducible representation of $\mathfrak{so}_{2n}\mathbb{C}$ given by $\lambda = (\lambda_1 \geq \cdots \geq |\lambda_n|)$,

$$\mathrm{Res}^{\mathfrak{so}_{2n}\mathbb{C}}_{\mathfrak{so}_{2n-1}\mathbb{C}}(\Gamma_\lambda) = \bigoplus \Gamma_{\bar{\lambda}}, \tag{25.35}$$

the sum over all $\bar{\lambda} = (\bar{\lambda}_1, \ldots, \bar{\lambda}_{n-1})$ with

$$\lambda_1 \geq \bar{\lambda}_1 \geq \lambda_2 \geq \bar{\lambda}_2 \geq \cdots \geq \bar{\lambda}_{n-1} \geq |\lambda_n|,$$

with the $\bar{\lambda}_i$ and λ_i simultaneously all integers or all half integers.

For $\mathfrak{sp}_{2n-2}\mathbb{C} \subset \mathfrak{sp}_{2n}\mathbb{C}$, and Γ_λ the irreducible representation of $\mathfrak{sp}_{2n}\mathbb{C}$ given by $\lambda = (\lambda_1 \geq \cdots \geq \lambda_n \geq 0)$, the restriction is

$$\mathrm{Res}^{\mathfrak{sp}_{2n}\mathbb{C}}_{\mathfrak{sp}_{2n-2}\mathbb{C}}(\Gamma_\lambda) = \bigoplus N_{\lambda\bar{\lambda}}\Gamma_{\bar{\lambda}}, \tag{25.36}$$

the sum over all $\bar{\lambda} = (\bar{\lambda}_1, \ldots, \bar{\lambda}_{n-1})$ with $\bar{\lambda}_1 \geq \cdots \geq \bar{\lambda}_{n-1} \geq 0$, and the multiplicity $N_{\lambda\bar{\lambda}}$ is the number of sequences $p_1, \ldots, p_n$ of integers satisfying

$$\lambda_1 \geq p_1 \geq \lambda_2 \geq p_2 \geq \cdots \geq \lambda_n \geq p_n \geq 0$$

and

$$p_1 \geq \bar{\lambda}_1 \geq p_2 \cdots \geq p_{n-1} \geq \bar{\lambda}_{n-1} \geq p_n.$$

As in the case of $GL_n\mathbb{C}$, these formulas are equivalent to identities among symmetric polynomials. The reader may enjoy trying to work them out from this point of view, cf. Exercise 23.43 and [Boe]. A less computational approach is given in [Žel].

As we saw in the case of the general linear group, these branching rules can be used inductively to compute the dimensions of the weight spaces. For example, for $\mathfrak{so}_m\mathbb{C}$ consider the chain

$$\mathfrak{so}_m\mathbb{C} \supset \mathfrak{so}_{m-1}\mathbb{C} \supset \mathfrak{so}_{m-2}\mathbb{C} \supset \cdots \supset \mathfrak{so}_3\mathbb{C}.$$

Decomposing a representation successively from one layer to the next will finally write it as a sum of one-dimensional weight spaces, and the dimension can be read off from the number of "partitions" in chains that start with the given λ. The representations can be constructed from these chains, as described by Gelfand and Zetlin, cf. [Žel, §10].

Similarly, one can ask for formulas for decomposing restrictions for other inclusions, such as the natural embeddings: $\mathrm{Sp}_{2n}\mathbb{C} \subset \mathrm{SL}_{2n}\mathbb{C}$, $\mathrm{SO}_m\mathbb{C} \subset \mathrm{SL}_m\mathbb{C}$, $\mathrm{GL}_m\mathbb{C} \times \mathrm{GL}_n\mathbb{C} \subset \mathrm{GL}_{m+n}\mathbb{C}$, $\mathrm{GL}_m\mathbb{C} \times \mathrm{GL}_n\mathbb{C} \subset \mathrm{GL}_{mn}\mathbb{C}$, $\mathrm{SL}_n\mathbb{C} \subset \mathrm{Sp}_{2n}\mathbb{C}$, $\mathrm{SL}_n\mathbb{C} \subset \mathrm{SO}_{2n+1}\mathbb{C}$, $\mathrm{SL}_n\mathbb{C} \subset \mathrm{SO}_{2n}\mathbb{C}$, to mention just a few. Such formulas are determined in principle by computing what happens to generators of the representation rings, which is not hard: one need only decompose exterior or symmetric products of standard representations, cf. Exercise 23.31. A few closed formulas for decomposing more general representations can also be found in the literature. We state what happens when the irreducible representations of $\mathrm{GL}_m\mathbb{C}$ are restricted to the orthogonal or symplectic subgroups, referring to [Lit3] for the proofs:

For $\mathrm{O}_m\mathbb{C} \subset \mathrm{GL}_m\mathbb{C}$, with $m = 2n$ or $2n+1$, given $\lambda = (\lambda_1 \geq \cdots \geq \lambda_n \geq 0)$,

$$\mathrm{Res}_{\mathrm{O}_m\mathbb{C}}^{\mathrm{GL}_m\mathbb{C}}(\Gamma_\lambda) = \bigoplus N_{\lambda\bar{\lambda}}\Gamma_{\bar{\lambda}}, \tag{25.37}$$

the sum over all $\bar{\lambda} = (\bar{\lambda}_1 \geq \cdots \geq \bar{\lambda}_n \geq 0)$, where

$$N_{\lambda\bar{\lambda}} = \sum_{\delta} N_{\delta\bar{\lambda}\lambda},$$

with $N_{\delta\bar{\lambda}\lambda}$ the Littlewood–Richardson coefficient, and the sum over all $\delta = (\delta_1 \geq \delta_2 \geq \cdots)$ with all δ_i even.

Exercise 23.38. Show that the representation $\Gamma_{(2,2)}$ of $\mathrm{GL}_m\mathbb{C}$ restricts to the direct sum

$$\Gamma_{(2,2)} \oplus \Gamma_{(2)} \oplus \Gamma_{(0)}$$

over $\mathrm{O}_m\mathbb{C}$. (This decomposition is important in differential geometry: the *Riemann–Christoffel* tensor has type (2, 2), and the above three components of its decomposition are the *conformal curvature* tensor, the *Ricci* tensor, and the *scalar* curvature, respectively.)

Similarly for $\mathrm{Sp}_{2n}\mathbb{C} \subset \mathrm{GL}_{2n}\mathbb{C}$,

$$\mathrm{Res}_{\mathrm{Sp}_{2n}\mathbb{C}}^{\mathrm{GL}_{2n}\mathbb{C}}(\Gamma_\lambda) = \bigoplus N_{\lambda\bar{\lambda}}\Gamma_{\bar{\lambda}}, \tag{25.39}$$

the sum over all $\bar{\lambda} = (\bar{\lambda}_1 \geq \cdots \geq \bar{\lambda}_n \geq 0)$, where

$$N_{\lambda\bar{\lambda}} = \sum_{\eta} N_{\eta\bar{\lambda}\lambda},$$

$N_{\eta\bar{\lambda}\lambda}$ is the Littlewood–Richardson coefficient, and the sum is over all $\eta = (\eta_1 = \eta_2 \geq \eta_3 = \eta_4 \geq \cdots)$ with each part occurring an even number of times.

It is perhaps worth pointing out the the decomposition of tensor products is a special case of the decomposition of restrictions: the exterior tensor product $\Gamma_\lambda \boxtimes \Gamma_\mu$ of two irreducible representations of G is an irreducible representation of $G \times G$, and the restriction of this to the diagonal embedding of G in $G \times G$ is the usual tensor product $\Gamma_\lambda \otimes \Gamma_\mu$.

There are also some general formulas, valid whenever $\bar{\mathfrak{g}}$ is a semisimple Lie subalgebra of a semisimple Lie algebra $\mathfrak{g}$. Assume that the Cartan subalgebra $\bar{\mathfrak{h}}$ is a subalgebra of $\mathfrak{h}$, so we have a restriction from $\mathfrak{h}^*$ to $\bar{\mathfrak{h}}^*$, and we assume the half-spaces determining positive roots are compatible. We write $\bar{\mu}$ for weights of $\bar{\mathfrak{g}}$, and we write $\mu \downarrow \bar{\mu}$ to mean that a weight μ of $\mathfrak{g}$ restricts to $\bar{\mu}$. Similarly write $\overline{W}$ for a typical element of the Weyl group of $\bar{\mathfrak{g}}$, and $\bar{\rho}$ for half the sum of its positive weights. If λ (resp. $\bar{\lambda}$) is a dominant weight for $\mathfrak{g}$ (resp. $\bar{\mathfrak{g}}$), let $N_{\lambda\bar{\lambda}}$ denote the multiplicity with which $\Gamma_{\bar{\lambda}}$ appears in the restriction of Γ_λ to $\bar{\mathfrak{g}}$, i.e.,

$$\operatorname{Res}(\Gamma_\lambda) = \bigoplus_{\bar{\lambda}} N_{\lambda\bar{\lambda}} \Gamma_{\bar{\lambda}}.$$

Exercise 25.40*. Show that, for any dominant weight λ of $\mathfrak{g}$ and any weight $\bar{\mu}$ of $\bar{\mathfrak{g}}$,

$$\sum_{\mu \downarrow \bar{\mu}} n_\mu(\Gamma_\lambda) = \sum_{\bar{\lambda}} N_{\lambda\bar{\lambda}} n_{\bar{\mu}}(\Gamma_{\bar{\lambda}}).$$

Exercise 25.41* (Klimyk). Show that

$$N_{\lambda\bar{\lambda}} = \sum_{\overline{W}} (-1)^{\overline{W}} \sum_{\mu \downarrow \bar{\lambda} + \bar{\rho} - \overline{W}(\bar{\rho})} n_\mu(\Gamma_\lambda).$$

Exercise 25.42. Show that if the formula of the preceding exercise is applied to the diagonal embedding of $\mathfrak{g}$ in $\mathfrak{g} \times \mathfrak{g}$, then the Racah formula of Exercise 25.31 results.

For additional formulas of a similar vein, as well as discussions of how they can be implemented on a computer, there are several articles in *SIAM J. Appl. Math.* **25**, 1973.

Finally, we note that it is possible, for any semisimple Lie algebra $\mathfrak{g}$, to make the direct sum of all its irreducible representations into a commutative algebra, generalizing constructions we saw in Lectures 15, §17, and §19. Let $\Gamma_{\omega_1}, \ldots, \Gamma_{\omega_n}$ be the irreducible representations corresponding to the fundamental weights $\omega_1, \ldots, \omega_n$. Let

$$A^\bullet = \operatorname{Sym}^\bullet(\Gamma_{\omega_1} \oplus \cdots \oplus \Gamma_{\omega_n}).$$

This is a commutative graded algebra, the direct sum of pieces

$$A^{\mathbf{a}} = \bigoplus_{a_1, \ldots, a_n} \operatorname{Sym}^{a_1}(\Gamma_{\omega_1}) \otimes \cdots \otimes \operatorname{Sym}^{a_n}(\Gamma_{\omega_n}),$$

where $\mathbf{a} = (a_1, \ldots, a_n)$ is an n-tuple of non-negative integers. Then $A^{\mathbf{a}}$ is the direct sum of the irreducible representation Γ_λ whose highest weight is $\lambda = \sum a_i \omega_i$, and a sum $J^{\mathbf{a}}$ of representations whose highest weight is strictly smaller. As before, weight considerations show that $J^\bullet = \bigoplus_{\mathbf{a}} J^{\mathbf{a}}$ is an ideal in $A^\bullet$, so the quotient

$$A^\bullet / J^\bullet = \bigoplus_\lambda \Gamma_\lambda$$

is the direct sum of all the irreducible representations. The product

$$\Gamma_\lambda \otimes \Gamma_\mu \to \Gamma_{\lambda+\mu}$$

in this ring is often called *Cartan multiplication*; note that the fact that $\Gamma_{\lambda+\mu}$ occurs once in the tensor product determines such a projection, but only up to multiplication by a scalar.

Using ideas of §25.1, it is possible to give generators for the ideal $J^{\bullet}$. If C is the Casimir operator, we know that C acts on all representations and is multiplication by the constant $c_\lambda = (\lambda, \lambda) + (2\lambda, \rho)$ on the irreducible representation with highest weight λ. Therefore, if $\lambda = \sum a_i\omega_i$, the endomorphism $C - c_\lambda I$ of $A^{\mathbf{a}}$ vanishes on the factor Γ_λ, and on each of the representations Γ_μ of lower weight μ it is multiplication by $c_\mu - c_\lambda \neq 0$ [cf. (25.2)]. It follows that

$$J^{\mathbf{a}} = \mathrm{Image}(C - c_\lambda I\colon A^{\mathbf{a}} \to A^{\mathbf{a}}).$$

Exercise 25.43*. Write $C = \sum U_i U_i'$ as in §25.1. Show that for $v_1, \dots, v_m$ vectors in the fundamental weight spaces, with $v_j \in \Gamma_{\alpha_j}$ and $\sum \alpha_j = \sum a_i\omega_i$, the element $(C - c_\lambda I)(v_1 \cdot v_2 \cdot \ldots \cdot v_m)$ is the sum over all pairs j, k, with $1 \leq j < k \leq m$, of the terms

$$\left(\sum_i (U_i(v_j)\cdot U_i'(v_k) + U_i'(v_j)\cdot U_i(v_k)) - 2(\alpha_j, \alpha_k)v_j\cdot v_k\right)\cdot \prod_{l\neq j,k} v_l.$$

From this exercise follows a theorem of Kostant: $J^{\bullet}$ is generated by the elements

$$\sum_i (U_i(v)\cdot U_i'(w) + U_i'(v)\cdot U_i(w)) - 2(\alpha, \beta)v\cdot w$$

for $v \in \Gamma_\alpha$, $w \in \Gamma_\beta$, with α and β fundamental roots. For the classical Lie algebras, this formula can be used to find concrete realizations of the ring. If one wants a similar ring for a semisimple Lie group, one has the same ring, of course, when the group is simply connected; this leads to the ring described in Lectures 15 and 17 for $SL_n\mathbb{C}$ and $Sp_{2n}\mathbb{C}$. For $SO_m\mathbb{C}$, little change is needed when m is odd, but there is more work for m even. Details can be found in [L-T].

LECTURE 26

Real Lie Algebras and Lie Groups

In this lecture we indicate how to complete the last step in the process outlined at the beginning of Part II: to take our knowledge of the classification and representation theory of complex algebras and groups and deduce the corresponding statements in the real case. We do this in the first section, giving a list of the simple classical real Lie algebras and saying a few words about the corresponding groups and their (complex) representations. The existence of a compact group whose Lie algebra has as complexification a given semisimple complex Lie algebra makes it possible to give another (indeed, the original) way to prove the Weyl character formula; we sketch this in §26.2. Finally, we can ask in regard to real Lie groups *G* a question analogous to one asked for the representations of finite groups in §3.5: which of the complex representations *V* of *G* actually come from real ones. We answer this in the most commonly encountered cases in §26.3. In this final lecture, proofs, when we attempt them, are generally only sketched and may require more than the usual fortitude from the reader.

§26.1. Classification of Real Simple Lie Algebras and Groups

Having described the semisimple complex Lie algebras, we now address the analogous problem for real Lie algebras. Since the complexification $\mathfrak{g}_0 \otimes_{\mathbb{R}} \mathbb{C}$ of a semisimple real Lie algebra $\mathfrak{g}_0$ is a semisimple complex Lie algebra and we have classified those, we are reduced to the problem of describing the *real forms* of the complex semisimple Lie algebras: that is, for a given complex Lie algebra $\mathfrak{g}$, finding all real Lie algebras $\mathfrak{g}_0$ with

$$\mathfrak{g}_0 \otimes_{\mathbb{R}} \mathbb{C} \cong \mathfrak{g}.$$

We saw many of the real forms of the classical complex Lie groups and algebras back in Lectures 7 and 8. In this section we will indicate one way to approach the question systematically, but we will only include sketches of proofs.

To get the idea of what to expect, let us work out real forms of $\mathfrak{sl}_2\mathbb{C}$ in detail. To do this, suppose $\mathfrak{g}_0$ is any real Lie subalgebra of $\mathfrak{sl}_2\mathbb{C}$, with $\mathfrak{g}_0 \otimes_{\mathbb{R}} \mathbb{C} = \mathfrak{sl}_2\mathbb{C}$. The natural thing to do is to try to carry out our analysis of semisimple Lie algebras for the real Lie algebra $\mathfrak{g}_0$: that is, find an element $H \in \mathfrak{g}_0$ such that $\mathrm{ad}(H)$ acts semisimply on $\mathfrak{g}_0$, decompose $\mathfrak{g}_0$ into eigenspaces, and so on. The first part of this presents no problem: since the subset of $\mathfrak{sl}_2\mathbb{C}$ of non-semisimple matrices is a proper algebraic subvariety, it cannot contain the real subspace $\mathfrak{g}_0 \subset \mathfrak{sl}_2\mathbb{C}$, so that we can certainly find a semisimple $H \in \mathfrak{g}_0$.

The next thing is to consider the eigenspaces of $\mathrm{ad}(H)$ acting on $\mathfrak{g}$. Of course, $\mathrm{ad}(H)$ has one eigenvalue 0, corresponding to the eigenspace $\mathfrak{h}_0 = \mathbb{R}\cdot H$ spanned by H. The remaining two eigenvalues must then sum to zero, which leaves just two possibilities:

(i) $\mathrm{ad}(H)$ has eigenvalues λ and $-\lambda$, for λ a nonzero real number; multiplying H by a real scalar, we can take $\lambda = 2$. In this case we obtain a decomposition of the vector space $\mathfrak{g}_0$ into one-dimensional eigenspaces

$$\mathfrak{g}_0 = \mathfrak{h}_0 \oplus \mathfrak{g}_2 \oplus \mathfrak{g}_{-2}.$$

We can then choose $X \in \mathfrak{g}_2$ and $Y \in \mathfrak{g}_{-2}$; the standard argument then shows that the bracket $[X, Y]$ is a nonzero multiple of H, which we may take to be 1 by rechoosing X and Y. We thus have the real form $\mathfrak{sl}_2\mathbb{R}$, with the basis

$$H = \begin{pmatrix} 1 & 0 \\ 0 & -1 \end{pmatrix}, \qquad X = \begin{pmatrix} 0 & 1 \\ 0 & 0 \end{pmatrix}, \qquad Y = \begin{pmatrix} 0 & 0 \\ 1 & 0 \end{pmatrix}.$$

(ii) $\mathrm{ad}(H)$ has eigenvalues $i\lambda$ and $-i\lambda$ for λ some nonzero real number; again, adjusting H by a real scalar we may take $\lambda = 1$. In this case, of course, there are no real eigenvectors for the action of $\mathrm{ad}(H)$ on $\mathfrak{g}_0$; but we can decompose $\mathfrak{g}_0$ into the direct sum of $\mathfrak{h}_0$ and the two-dimensional subspace $\mathfrak{g}_{\{i,-i\}}$ corresponding to the pair of eigenvalues i and $-i$. We may then choose a basis B and C for $\mathfrak{g}_{\{i,-i\}}$ with

$$[H, B] = C \quad \text{and} \quad [H, C] = -B.$$

The commutator $[B, C]$ will then be a nonzero multiple of H, which we may take to be either H or $-H$ (we can multiply B and C simultaneously by a scalar μ, which multiplies the commutator $[B, C]$ by μ^2). In the latter case, we see that $\mathfrak{g}_0$ is isomorphic to $\mathfrak{sl}_2\mathbb{R}$ again: these are the relations we get if we take as basis for $\mathfrak{sl}_2\mathbb{C}$ the three vectors

$$H = \begin{pmatrix} 0 & \frac{1}{2} \\ -\frac{1}{2} & 0 \end{pmatrix}, \quad B = \begin{pmatrix} 0 & 1 \\ 1 & 0 \end{pmatrix}, \quad \text{and} \quad C = \begin{pmatrix} 1 & 0 \\ 0 & -1 \end{pmatrix}.$$

Finally, if the commutator $[B, C] = H$, we do get a new example: $\mathfrak{g}_0$ is in this case isomorphic to the algebra

$$\mathfrak{su}_2 = \{A: {}^t\bar{A} = -A \text{ and } \operatorname{trace}(A) = 0\} \subset \mathfrak{sl}_2\mathbb{C},$$

which has as basis

$$H = \begin{pmatrix} i/2 & 0 \\ 0 & -i/2 \end{pmatrix}, \quad B = \begin{pmatrix} 0 & 1/2 \\ -1/2 & 0 \end{pmatrix}, \quad \text{and} \quad C = \begin{pmatrix} 0 & i/2 \\ i/2 & 0 \end{pmatrix}.$$

Exercise 26.1. Carry out this analysis for the real Lie algebras $\mathfrak{so}_3\mathbb{R}$ and $\mathfrak{so}_{2,1}\mathbb{R}$. In particular, give an isomorphism of each with either $\mathfrak{sl}_2\mathbb{R}$ or $\mathfrak{su}_2$.

This completes our analysis of the real forms of $\mathfrak{sl}_2\mathbb{C}$. In the general case, we can try to apply a similar analysis, and indeed at least one aspect generalizes: given a real form $\mathfrak{g}_0 \subset \mathfrak{g}$ of the complex semisimple Lie algebra $\mathfrak{g}$, we can find a real subalgebra $\mathfrak{h}_0 \subset \mathfrak{g}_0$ such that $\mathfrak{h}_0 \otimes \mathbb{C}$ is a Cartan subalgebra of $\mathfrak{g} = \mathfrak{g}_0 \otimes \mathbb{C}$; this is called a *Cartan subalgebra* of $\mathfrak{g}_0$. There is a further complication in the case of Lie algebras of rank 2 or more: the values on $\mathfrak{h}_0$ of a root $\alpha \in R$ of $\mathfrak{g}$ need not be either all real or all purely imaginary. We, thus, need to consider the root spaces $\mathfrak{g}_\alpha$, $\mathfrak{g}_{\bar{\alpha}}$, $\mathfrak{g}_{-\alpha}$, and $\mathfrak{g}_{-\bar{\alpha}}$, and the subalgebra they generate, at the same time. Moreover, as we saw in the above example, whether the values of the roots $\alpha \in R$ of $\mathfrak{g}$ on the real subspace $\mathfrak{h}_0$ are real, purely imaginary, or neither will in general depend on the choice of $\mathfrak{h}_0$.

Exercise 26.2*. In the case of $\mathfrak{g}_0 = \mathfrak{sl}_3\mathbb{R} \subset \mathfrak{g} = \mathfrak{sl}_3\mathbb{C}$, suppose we choose as Cartan subalgebra $\mathfrak{h}_0$ the space spanned over $\mathbb{R}$ by the elements

$$H_1 = \begin{pmatrix} 2 & 0 & 0 \\ 0 & -1 & 0 \\ 0 & 0 & -1 \end{pmatrix} \quad \text{and} \quad H_2 = \begin{pmatrix} 0 & 0 & 0 \\ 0 & 0 & 1 \\ 0 & -1 & 0 \end{pmatrix}.$$

Show that this is indeed a Cartan subalgebra, and find the decomposition of $\mathfrak{g}$ into eigenspaces for the action of $\mathfrak{h} = \mathfrak{h}_0 \otimes \mathbb{C}$. In particular, find the roots of $\mathfrak{g}$ as linear functions on $\mathfrak{h}$, and describe the corresponding decomposition of $\mathfrak{g}_0$.

Judging from these examples, it is probably prudent to resist the temptation to try to carry out an analysis of real semisimple Lie algebras via an analogue of the decomposition $\mathfrak{g} = \mathfrak{h} \oplus (\bigoplus \mathfrak{g}_\alpha)$ in this case. Rather, in the present book, we will do two things. First, we will give the statement of the classification theorem for the real forms of the classical algebras—that is, we will list all the simple real Lie algebras whose complexifications are classical algebras. Second, we will focus on two distinguished real forms possessed by any real semisimple Lie algebra, the *split form* and the *compact form*. These are the two forms that you see most often; and the existence of the latter in particular will be essential in the following section.

For the first, it turns out to be enough to work out the complexifications $\mathfrak{g}_0 \otimes_{\mathbb{R}} \mathbb{C} = \mathfrak{g}_0 \oplus i \cdot \mathfrak{g}_0$ of the real Lie algebras $\mathfrak{g}_0$ we know. The list is:

Real Lie algebra	Complexification
$\mathfrak{sl}_n\mathbb{R}$	$\mathfrak{sl}_n\mathbb{C}$
$\mathfrak{sl}_n\mathbb{C}$	$\mathfrak{sl}_n\mathbb{C} \times \mathfrak{sl}_n\mathbb{C}$
$\mathfrak{sl}_n\mathbb{H} = \mathfrak{gl}_n\mathbb{H}/\mathbb{R}$	$\mathfrak{sl}_{2n}\mathbb{C}$
$\mathfrak{so}_{p,q}\mathbb{R}$	$\mathfrak{so}_{p+q}\mathbb{C}$
$\mathfrak{so}_n\mathbb{C}$	$\mathfrak{so}_n\mathbb{C} \times \mathfrak{so}_n\mathbb{C}$
$\mathfrak{sp}_{2n}\mathbb{R}$	$\mathfrak{sp}_{2n}\mathbb{C}$
$\mathfrak{sp}_{2n}\mathbb{C}$	$\mathfrak{sp}_{2n}\mathbb{C} \times \mathfrak{sp}_{2n}\mathbb{C}$
$\mathfrak{su}_{p,q}$	$\mathfrak{sl}_{p+q}\mathbb{C}$
$\mathfrak{u}_{p,q}\mathbb{H}$	$\mathfrak{sp}_{2(p+q)}\mathbb{C}$
$\mathfrak{u}_n^*\mathbb{H}$	$\mathfrak{so}_{2n}\mathbb{C}$

The last two in the left-hand column are the Lie algebras of the groups $U_{p,q}\mathbb{H}$ and $U_n^*\mathbb{H}$ of automorphisms of a quaternionic vector space preserving a Hermitian form with signature (p, q), and a skew-symmetric Hermitian form, respectively.

We should first verify that the algebras on the right are indeed the complexifications of those on the left. Some are obvious, such as the complexification

$$(\mathfrak{sl}_n\mathbb{R})_{\mathbb{C}} = \mathfrak{sl}_n\mathbb{R} \oplus i \cdot \mathfrak{sl}_n\mathbb{R} = \mathfrak{sl}_n\mathbb{C}.$$

The same goes for $\mathfrak{so}_{p,q}\mathbb{R}$ and $\mathfrak{sp}_{2n}\mathbb{R}$.

Next, consider the complexification of

$$\mathfrak{su}_n = \{A \in \mathfrak{sl}_n\mathbb{C}: {}^t\bar{A} = -A\}.$$

To see that $\mathfrak{sl}_n\mathbb{C} = \mathfrak{su}_n \oplus i \cdot \mathfrak{su}_n$, let $M \in \mathfrak{sl}_n\mathbb{C}$, and write

$$M = \tfrac{1}{2}(M - {}^t\bar{M}) + \tfrac{1}{2}(M + {}^t\bar{M}) = \tfrac{1}{2}A + \tfrac{1}{2}B;$$

then $A \in \mathfrak{su}_n$, $iB \in \mathfrak{su}_n$, and $M = \frac{1}{2}A - i(i/2)B$.

The general case of $\mathfrak{su}_{p,q} \subset \mathfrak{sl}_{p+q}\mathbb{C}$ is similar: if the form is given by $(x, y) = {}^t\bar{x}Qy$, then $\mathfrak{su}_{p,q} = \{A: {}^t\bar{A}Q = -QA\}$. Writing $M \in \mathfrak{sl}_{p+q}\mathbb{C}$ in the form

$$M = \tfrac{1}{2}(M - Q \cdot {}^t\bar{M} \cdot Q) - i \cdot (\tfrac{1}{2}(iM + iQ \cdot {}^t\bar{M} \cdot Q))$$

and using $\bar{Q} = {}^tQ = Q^{-1} = Q$, one sees that $M \in \mathfrak{su}_{p,q} \oplus i \cdot \mathfrak{su}_{p,q}$.

For the complexification of $\mathfrak{sl}_m\mathbb{C}$, embed $\mathfrak{sl}_m\mathbb{C}$ in $\mathfrak{sl}_m\mathbb{C} \times \mathfrak{sl}_m\mathbb{C}$ by $A \mapsto (A, \bar{A})$. Given any pair (B, C), write

$$\begin{aligned}(B, C) &= \tfrac{1}{2}(B + \bar{C}, \bar{B} + C) + \tfrac{1}{2}(B - \bar{C}, -\bar{B} + C)\\ &= \tfrac{1}{2}(B + \bar{C}, \bar{B} + C) - i \cdot (\tfrac{1}{2}(iB + \overline{iC}, \overline{iB} + iC).\end{aligned}$$

For the quaternionic Lie algebra, from the description of $\mathrm{GL}_n\mathbb{H}$ we saw in Lecture 7, we have

$$\mathfrak{gl}_n\mathbb{H} = \{A \in \mathfrak{gl}_{2n}\mathbb{C}: AJ = J\bar{A}\},$$

with $J = \begin{pmatrix} 0 & I \\ -I & 0 \end{pmatrix}$. As before, for $M \in \mathfrak{gl}_{2n}\mathbb{C}$, we can write

$$M = \tfrac{1}{2}(M - J \cdot \overline{M} \cdot J) - i \cdot (\tfrac{1}{2}(iM + iJ \cdot \overline{iM} \cdot J))$$

to see that $\mathfrak{gl}_n\mathbb{H} \otimes_{\mathbb{R}} \mathbb{C} = \mathfrak{gl}_{2n}\mathbb{C}$.

Exercise 26.3. Verify the rest of the list.

The theorem, which also goes back to Cartan, is that *this includes the complete list of simple real Lie algebras associated to the classical complex types* (A_n)–(D_n). In fact, there are an additional 17 simple real Lie algebras associated with the five exceptional Lie algebras. The proof of this theorem is rather long, and we refer to the literature (cf. [H-S], [Hel], [Ar]) for it.

Split Forms and Compact Forms

Rather than try to classify in general the real forms $\mathfrak{g}_0$ of a semisimple Lie algebra $\mathfrak{g}$, we would like to focus here on two particular forms that are possessed by every semisimple Lie algebra and that are by far the most commonly dealt with in practice: the *split form* and the *compact form*.

These represent the two extremes of behavior of the decomposition $\mathfrak{g} = \mathfrak{h} \oplus (\bigoplus \mathfrak{g}_\alpha)$ with respect to the real subalgebra $\mathfrak{g}_0 \subset \mathfrak{g}$. To begin with, the *split form* of $\mathfrak{g}$ is a form $\mathfrak{g}_0$ such that there exists a Cartan subalgebra $\mathfrak{h}_0 \subset \mathfrak{g}_0$ (that is, a subalgebra whose complexification $\mathfrak{h} = \mathfrak{h}_0 \otimes \mathbb{C} \subset \mathfrak{g}_0 \otimes \mathbb{C} = \mathfrak{g}$ is a Cartan subalgebra of $\mathfrak{g}$) whose action on $\mathfrak{g}_0$ has all real eigenvalues—i.e., such that all the roots $\alpha \in R \subset \mathfrak{h}^*$ of $\mathfrak{g}$ (with respect to the Cartan subalgebra $\mathfrak{h} = \mathfrak{h}_0 \otimes \mathbb{C} \subset \mathfrak{g}$) assume all real values on the subspace $\mathfrak{h}_0$. In this case we have a direct sum decomposition

$$\mathfrak{g}_0 = \mathfrak{h}_0 \oplus (\bigoplus \mathfrak{i}_\alpha)$$

of $\mathfrak{g}_0$ into $\mathfrak{h}_0$ and one-dimensional eigenspaces $\mathfrak{i}_\alpha$ for the action of $\mathfrak{h}_0$ (each $\mathfrak{i}_\alpha$ will just be the intersection of the root space $\mathfrak{g}_\alpha \subset \mathfrak{g}$ with $\mathfrak{g}_0$); each pair $\mathfrak{i}_\alpha$ and $\mathfrak{i}_{-\alpha}$ will generate a subalgebra isomorphic to $\mathfrak{sl}_2\mathbb{R}$. As we will see momentarily, this uniquely characterizes the real form $\mathfrak{g}_0$ of $\mathfrak{g}$.

By contrast, in the *compact form* all the roots $\alpha \in R \subset \mathfrak{h}^*$ of $\mathfrak{g}$ (with respect to the Cartan subalgebra $\mathfrak{h} = \mathfrak{h}_0 \otimes \mathbb{C} \subset \mathfrak{g}$) assume all purely imaginary values on the subspace $\mathfrak{h}_0$. We accordingly have a direct sum decomposition

$$\mathfrak{g}_0 = \mathfrak{h}_0 \oplus (\bigoplus \mathfrak{l}_\alpha)$$

of $\mathfrak{g}_0$ into $\mathfrak{h}_0$ and two-dimensional spaces on which $\mathfrak{h}_0$ acts by rotation (each $\mathfrak{l}_\alpha$ will just be the intersection of the root space $\mathfrak{g}_\alpha \oplus \mathfrak{g}_{-\alpha}$ with $\mathfrak{g}_0$); each $\mathfrak{l}_\alpha$ will generate a subalgebra isomorphic to $\mathfrak{su}_2$.

The existence of the split form of a semisimple complex Lie algebra was already established in Lecture 21: one way to construct a real—even rational

—form $\mathfrak{g}_0$ of a semisimple Lie algebra $\mathfrak{g}$ is by starting with any generator X_{α_i} for the root space for each positive simple root α_i, completing it to standard basis X_{α_i}, Y_{α_i}, and $H_i = [X_{\alpha_i}, Y_{\alpha_i}]$ for the corresponding $\mathfrak{s}_{\alpha_i} = \mathfrak{sl}_2\mathbb{C}$, and taking $\mathfrak{g}_0$ to be the real subalgebra generated by these elements. Choosing a way to write each positive root as a sum of simple roots even determined a basis $\{H_i \in \mathfrak{h}, X_\alpha \in \mathfrak{g}_\alpha, Y_\alpha \in \mathfrak{g}_{-\alpha}\}$ for $\mathfrak{g}_0$, as in (21.20). The Cartan subalgebra $\mathfrak{h}_0$ of $\mathfrak{g}_0$ is the real span of these H_i. Note that once $\mathfrak{h}$ is fixed for $\mathfrak{g}$, the real subalgebra $\mathfrak{h}_0$ is uniquely determined as the span of the H_α for all roots α. The algebra $\mathfrak{g}_0$ is determined up to isomorphism; it is sometimes called the *natural* real form of $\mathfrak{g}$. Note that this also demonstrates the uniqueness of the split form: it is the only real form $\mathfrak{g}_0$ of $\mathfrak{g}$ that has a Cartan subalgebra $\mathfrak{h}_0$ acting on $\mathfrak{g}_0$ with all real eigenvalues.

As for the compact form of a semisimple Lie algebra, it owes much of its significance (as well as its name) to the last condition in

Proposition 26.4. *Suppose $\mathfrak{g}$ is any complex semisimple Lie algebra and $\mathfrak{g}_0 \subset \mathfrak{g}$ a real form of $\mathfrak{g}$. Let $\mathfrak{h}_0$ be a Cartan subalgebra of $\mathfrak{g}_0$, $\mathfrak{h} = \mathfrak{h}_0 \otimes \mathbb{C}$ the corresponding Cartan subalgebra of $\mathfrak{g}$. The following are equivalent:*

(i) *Each root $\alpha \in R \subset \mathfrak{h}^*$ of $\mathfrak{g}$ assumes purely imaginary values on $\mathfrak{h}_0$, and for each root α the subalgebra of $\mathfrak{g}_0$ generated by the intersection $\mathfrak{l}_\alpha$ of $(\mathfrak{g}_\alpha \oplus \mathfrak{g}_{-\alpha})$ with $\mathfrak{g}_0$ is isomorphic to $\mathfrak{su}_2$;*
(ii) *The restriction to $\mathfrak{g}_0$ of the Killing form of $\mathfrak{g}$ is negative definite;*
(iii) *The real Lie group G_0 with Lie algebra $\mathfrak{g}_0$ is compact.*

In (iii), G_0 can be taken to be the adjoint form of $\mathfrak{g}_0$. However, a theorem of Weyl ensures that the fundamental group of any such G_0 is finite, so the condition is independent of the choice of G_0. Note also that, by the equivalence with (ii) and (iii), the condition (i) must be independent of the choice of Cartan subalgebra $\mathfrak{h}_0$. This is in contrast with the split case, where we require only that there exist a Cartan subalgebra whose action on $\mathfrak{g}$ has all real eigenvalues; as we saw in the case of $\mathfrak{sl}_2\mathbb{R}$, in the split case a different $\mathfrak{h}_0$ may have imaginary eigenvalues.

PROOF. We start by showing that the first condition implies the second; this will follow from direct observation. To begin with, the value of the Killing form on $H \in \mathfrak{h}_0$ is visibly

$$B(H, H) = \sum (\alpha(H))^2 < 0.$$

Next, the subspaces $\mathfrak{l}_\alpha$ are orthogonal to one another with respect to B, so it remains only to verify $B(Z, Z) < 0$ for a general member $Z \in \mathfrak{l}_\alpha$. To do this, let X and Y be generators of $\mathfrak{g}_\alpha$ and $\mathfrak{g}_{-\alpha} \subset \mathfrak{g}$ respectively, chosen so as to form, together with their commutator $H = [X, Y]$ a standard basis for $\mathfrak{sl}_2\mathbb{C}$. By the analysis of real forms of $\mathfrak{sl}_2\mathbb{C}$ above, we may take as generators of the algebra generated by $\mathfrak{l}_\alpha$ the elements iH, $U = X - Y$ and $V = iX + iY$. If we set

$$Z = aU + bV = (a + ib)\cdot X + (-a + ib)\cdot Y,$$

then we have

$$\begin{aligned}\operatorname{ad}(Z)\circ\operatorname{ad}(Z) = {} & (a+ib)^2\operatorname{ad}(X)\circ\operatorname{ad}(X)\\ & -(a^2+b^2)(\operatorname{ad}(X)\circ\operatorname{ad}(Y)+\operatorname{ad}(Y)\circ\operatorname{ad}(X))\\ & +(a-ib)^2\operatorname{ad}(Y)\circ\operatorname{ad}(Y).\end{aligned}$$

Now, $\operatorname{ad}(X)\circ\operatorname{ad}(X)$ and $\operatorname{ad}(Y)\circ\operatorname{ad}(Y)$ have no trace, so we can write

$$\operatorname{trace}(\operatorname{ad}(Z)\circ\operatorname{ad}(Z)) = -2\cdot(a^2+b^2)\cdot\operatorname{trace}(\operatorname{ad}(X)\circ\operatorname{ad}(Y)). \quad (26.5)$$

By direct examination, in the representation $\operatorname{Sym}^n V$ of $\mathfrak{sl}_2\mathbb{C}$, $\operatorname{ad}(X)\circ\operatorname{ad}(Y)$ acts by multiplication by $(n-\lambda)(n+\lambda-2)/4 \geq 0$ on the λ-eigenspace for H, from which we deduce that the right-hand side of (26.5) is negative.

Next, we show that the second condition implies the third. This is immediate: the adjoint form G_0 is the connected component of the identity of the group $\operatorname{Aut}(\mathfrak{g}_0)$. In particular, it is a closed subgroup of the adjoint group of $\mathfrak{g}$, and it acts faithfully on the real vector space $\mathfrak{g}_0$, preserving the bilinear form B. If B is negative definite it follows that G_0 is a closed subgroup of the orthogonal group $\mathrm{SO}_m\mathbb{R}$, which is compact.

Finally, if we know that G_0 is compact, by averaging we can construct a positive definite inner product on $\mathfrak{g}_0$ invariant under the action of G_0. For any X in $\mathfrak{g}_0$, $\operatorname{ad}(X)$ is represented by a skew-symmetric matrix $A=(a_{i,j})$ with respect to an orthonormal basis of $\mathfrak{g}_0$ (cf. (14.23)), so $B(X,X)=\operatorname{Tr}(A\circ A)=\sum_{i,j}a_{i,j}a_{j,i}=-\sum a_{i,j}^2\leq 0$. In particular, the eigenvalues of $\operatorname{ad}(X)$ must be purely imaginary. Therefore $\alpha(\mathfrak{h}_0)\subset i\mathbb{R}$ and $\bar{\alpha}=-\alpha$ for any root α, from which (i) follows. □

We now claim that *every semisimple complex Lie algebra has a unique compact form.* To see this we need an algebraic notion which is, in fact, crucial to the classification theorem mentioned above: that of *conjugate linear involution.* If $\mathfrak{g}=\mathfrak{g}_0\otimes_{\mathbb{R}}\mathbb{C}$ is the complexification of a real Lie algebra $\mathfrak{g}_0$, there is a map $\sigma\colon\mathfrak{g}\to\mathfrak{g}$ which takes $x\otimes z$ to $x\otimes\bar{z}$ for $x\in\mathfrak{g}_0$ and $z\in\mathbb{C}$; it is conjugate linear, preserves Lie brackets, and σ^2 is the identity. The real algebra $\mathfrak{g}_0$ is the fixed subalgebra of σ, and conversely, given such a conjugate linear involution σ of a complex Lie algebra $\mathfrak{g}$, its fixed algebra $\mathfrak{g}^\sigma$ is a real form of $\mathfrak{g}$. To prove the claim, we start with the split, or natural form, as constructed in Lecture 21 and referred to above. With a basis for $\mathfrak{g}$ chosen as in this construction, it is not hard to show that there is a unique Lie algebra automorphism φ of $\mathfrak{g}$ that takes each element of $\mathfrak{h}$ to its negative and takes each X_α to Y_α (this follows from Claim 21.25). This automorphism φ is a complex linear involution which preserves the real subalgebra $\mathfrak{g}_0$. This automorphism commutes with the associated conjugate linear σ. The composite $\sigma\varphi=\varphi\sigma$ is a conjugate linear involution, from which it follows that its fixed part $\mathfrak{g}_c=\mathfrak{g}^{\sigma\varphi}$ is another real form of $\mathfrak{g}$. This has Cartan subalgebra $\mathfrak{h}_c=\mathfrak{h}^{\sigma\varphi}=i\cdot\mathfrak{h}_0$. We have seen that the restriction of the Killing form to $\mathfrak{h}_0$ is positive definite. It follows that its restriction to $\mathfrak{h}_c$ is negative definite, and hence that $\mathfrak{g}_c$ is a compact form of $\mathfrak{g}$. Finally, this construction of $\mathfrak{g}_c$ from $\mathfrak{g}_0$ is reversible, and from this one can deduce the uniqueness of the compact form.

We may see directly from this construction that

$$\mathfrak{g}_c = \mathfrak{h}_c \oplus \bigoplus_{\alpha \in R^+} \mathfrak{l}_\alpha,$$

where $\mathfrak{l}_\alpha = (\mathfrak{g}_\alpha \oplus \mathfrak{g}_{-\alpha})^{\sigma\varphi}$ is a real plane with $\mathfrak{l}_\alpha \otimes_{\mathbb{R}} \mathbb{C} = \mathfrak{g}_\alpha \oplus \mathfrak{g}_{-\alpha}$ and $[\mathfrak{h}_c, \mathfrak{l}_\alpha] \subset \mathfrak{l}_\alpha$.

Exercise 26.6. Verify that $\{A_j = i \cdot H_j: 1 \le j \le n\}$ is a basis for $\mathfrak{h}_c$, $\{B_\alpha = X_\alpha - Y_\alpha, C_\alpha = i \cdot (X_\alpha + Y_\alpha)\}$ is a basis for $\mathfrak{l}_\alpha$, and the action is given by

$$[A_j, B_\alpha] = p \cdot C_\alpha \quad \text{and} \quad [A_j, C_\alpha] = -p \cdot B_\alpha,$$

where p is the integer $\alpha(H_j)$. In particular, $\mathfrak{h}_c$ acts by rotations on the planes $\mathfrak{l}_\alpha$.

Our classical Lie algebras $\mathfrak{g}$ all came equipped with a natural real form $\mathfrak{g}_0$, and with a basis of the above type. These split forms are:

Complex simple Lie algebra	Split form
$\mathfrak{sl}_{n+1}\mathbb{C}$	$\mathfrak{sl}_{n+1}\mathbb{R}$
$\mathfrak{so}_{2n+1}\mathbb{C}$	$\mathfrak{so}_{n+1,n}$
$\mathfrak{sp}_{2n}\mathbb{C}$	$\mathfrak{sp}_{2n}\mathbb{R}$
$\mathfrak{so}_{2n}\mathbb{C}$	$\mathfrak{so}_{n,n}$

Exercise 26.7. For each of these split forms, find the corresponding compact form $\mathfrak{g}_c$.

Exercise 26.8. Let $\mathfrak{g}_0$ be a real semisimple Lie algebra. Show that a subalgebra $\mathfrak{h}_0$ of $\mathfrak{g}_0$ is a Cartan subalgebra if and only if it is a maximal abelian subalgebra and the adjoint action on $\mathfrak{g}_0$ is semisimple.

Exercise 26.9*. Starting with a real form $\mathfrak{g}_0$ of $\mathfrak{g}$ with associated conjugation σ, show that one can always find a compact form $\mathfrak{g}_c$ of $\mathfrak{g}$ such that $\sigma(\mathfrak{g}_c) = \mathfrak{g}_c$, and such that

$$\mathfrak{g}_0 = \mathfrak{k} \oplus \mathfrak{p},$$

where $\mathfrak{k} = \mathfrak{h}_0 = \mathfrak{g}_0 \cap \mathfrak{g}_c$, and $\mathfrak{p} = \mathfrak{g}_0 \cap (i \cdot \mathfrak{g}_c)$. Such a decomposition is called a *Cartan decomposition* of $\mathfrak{g}_0$. It is unique up to inner automorphism.

Exercise 26.10*. For any real form $\mathfrak{g}_0$ of $\mathfrak{g}$, given by a conjugation σ, show that there is a Cartan subalgebra $\mathfrak{h}$ of $\mathfrak{g}$ that is preserved by σ, so $\mathfrak{g}_0 \cap \mathfrak{h}$ is a Cartan subalgebra of $\mathfrak{g}_0$.

Naturally, the various special isomorphisms between complex Lie algebras ($\mathfrak{sl}_2\mathbb{C} \cong \mathfrak{so}_3\mathbb{C} \cong \mathfrak{sp}_2\mathbb{C}$, etc.) give rise to special isomorphisms among their real forms. For example, we have already seen that

$$\mathfrak{sl}_2\mathbb{R} \cong \mathfrak{su}_{1,1} \cong \mathfrak{so}_{2,1} \cong \mathfrak{sp}_2\mathbb{R},$$

while

$$\mathfrak{su}_2 \cong \mathfrak{so}_3\mathbb{R} \cong \mathfrak{sl}_1\mathbb{H} \cong \mathfrak{u}_1\mathbb{H}$$

(cf. Exercise 26.1). Similarly, each of the remaining three special isomorphisms of complex semisimple Lie algebras gives rise to isomorphisms between their real forms, as follows:

(i) $\mathfrak{so}_4\mathbb{C} \cong \mathfrak{sl}_2\mathbb{C} \times \mathfrak{sl}_2\mathbb{C}$
 compact forms: $\mathfrak{so}_4\mathbb{R} \cong \mathfrak{su}_2 \times \mathfrak{su}_2$
 split forms: $\mathfrak{so}_{2,2} \cong \mathfrak{sl}_2\mathbb{R} \times \mathfrak{sl}_2\mathbb{R}$
 others: $\mathfrak{so}_{3,1} \cong \mathfrak{sl}_2\mathbb{C}$, $\mathfrak{u}_2^*\mathbb{H} \cong \mathfrak{su}_2 \times \mathfrak{sl}_2\mathbb{R}$.

(ii) $\mathfrak{sp}_4\mathbb{C} \cong \mathfrak{so}_5\mathbb{C}$
 compact forms: $\mathfrak{u}_2\mathbb{H} \cong \mathfrak{so}_5\mathbb{R}$
 split forms: $\mathfrak{sp}_4\mathbb{R} \cong \mathfrak{so}_{3,2}$
 other: $\mathfrak{u}_{1,1}\mathbb{H} \cong \mathfrak{so}_{4,1}$.

(iii) $\mathfrak{sl}_4\mathbb{C} \cong \mathfrak{so}_6\mathbb{C}$
 compact forms: $\mathfrak{su}_4 \cong \mathfrak{so}_6\mathbb{R}$
 split forms: $\mathfrak{sl}_4\mathbb{R} \cong \mathfrak{so}_{3,3}$
 others: $\mathfrak{su}_{2,2} \cong \mathfrak{so}_{4,2}$; $\mathfrak{su}_{3,1} \cong \mathfrak{u}_3^*\mathbb{H}$; $\mathfrak{sl}_2\mathbb{H} \cong \mathfrak{so}_{5,1}$.

In addition, the extra automorphism of $\mathfrak{so}_8\mathbb{C}$ coming from triality gives rise to an isomorphism $\mathfrak{u}_4^*\mathbb{H} \cong \mathfrak{so}_{6,2}$.

Exercise 26.11. Verify some of the isomorphisms above. (Of course, in the case of compact and split forms, these are implied by the corresponding isomorphisms of complex Lie algebras, but it is worthwhile to see them directly in any case.)

Real Groups

We turn now to problem of describing the real Lie groups with these Lie algebras. Let G be the adjoint form of the semisimple complex Lie algebra $\mathfrak{g}$. If $\mathfrak{g}_0$ is a real form of $\mathfrak{g}$, the associated conjugate linear involution σ of $\mathfrak{g}$ that fixes $\mathfrak{g}_0$ lifts to an involution $\tilde{\sigma}$ of G. (This follows from the functorial nature of the adjoint form, noting that G is regarded now as a real Lie group.) The fixed points $G^{\tilde{\sigma}}$ of this involution then form a closed subgroup of G; its connected component of the identity G_0 is a real Lie group whose Lie algebra is $\mathfrak{g}_0$. G is called the *complexification* of G_0.

We have seen in §23.1 that if $\Gamma = \Gamma_W$ is the lattice of those elements in $\mathfrak{h}$ on which all roots take integral values, then $2\pi i\Gamma$ is the kernel of the exponential mapping exp: $\mathfrak{h} \to G$ to the adjoint form. If $\mathfrak{h}_0$ is a Cartan subalgebra of $\mathfrak{g}_0$, $T = \exp(\mathfrak{h}_0)$ will be compact precisely when the intersection of $\mathfrak{h}_0$ with the kernel $2\pi i\Gamma$ is a lattice of maximal rank. In this case, T will be a product of n copies of the circle S^1, $n = \dim(\mathfrak{h})$, and, since the Killing form on $\mathfrak{h}_0$ is negative definite, the corresponding real group G_0 will also be compact. Such a G_0 will be a maximal compact subgroup of G.

When $G_0 \subset G$ is a maximal compact subgroup, they have the same irreducible complex representations. Indeed, for any complex group G', each complex homomorphism from G to G' is the extension of a unique real homomorphism from G_0 to G'. This follows from the corresponding fact for Lie algebras and the fact that G_0 and G have the same fundamental group. This is another general fact, which implies the finiteness of the fundamental group of G_0; we omit the proof, noting only that it can be seen directly in the classical cases:

Exercise 26.12*. Prove that $\pi_1(G_0) \to \pi_1(G)$ is an isomorphism for each of the classical adjoint groups.

Exercise 26.13*. The special isomorphisms of real Lie algebras listed above give rise to special isomorphisms of real Lie groups. Can you find these?

It is another general fact that any compact (connected) Lie group is a quotient

$$(G_1 \times G_2 \times \cdots \times G_r \times T)/Z,$$

where the G_i are simple compact Lie groups, $T \cong (S^1)^k$ is a torus, and Z is a discrete subgroup of the center. In particular, its Lie algebra is the direct sum of a semisimple compact Lie algebra and an abelian Lie algebra. This provides another reason why the classification of irreducible representations in the real compact case and the semisimple complex case are essentially the same.

Representations of Real Lie Algebras

Finally, we should say a word here about the irreducible representations (always here in complex vector spaces!) of simple real Lie algebras. In some cases these are easily described in terms of the complex case: for example, the irreducible representations of $\mathfrak{su}_m$ or $\mathfrak{sl}_m\mathbb{R}$ are the same as those for $\mathfrak{sl}_m\mathbb{C}$, i.e., they are the restrictions of the irreducible representations $\Gamma_\lambda = \mathbb{S}_\lambda\mathbb{C}^m$ corresponding to partitions or Young diagrams λ. This is the situation in general whenever the complexification $\mathfrak{g} = \mathfrak{g}_0 \otimes \mathbb{C}$ of the real Lie algebra $\mathfrak{g}_0$ is still simple: the representations of $\mathfrak{g}_0$ on complex vector spaces are exactly the representations of $\mathfrak{g}$. The situation is slightly different when we have a simple real Lie algebra whose complexification is not simple: for example, the irreducible representations of $\mathfrak{sl}_m\mathbb{C}$, regarded as a real Lie algebra, are of the form $\Gamma_\lambda \otimes \overline{\Gamma}_\mu$, where $\overline{\Gamma}_\mu$ is the conjugate representation of Γ_μ. The situation in general is expressed in the following

Exercise 26.14. Show that if $\mathfrak{g}_0$ is a simple real Lie algebra whose complexification $\mathfrak{g}$ is simple, its irreducible representations are the restrictions of (uniquely determined) irreducible representations of $\mathfrak{g}$. If $\mathfrak{g}_0$ is the underlying real algebra of a simple complex Lie algebra, show that the irreducible representations of $\mathfrak{g}_0$ are of the form $V \otimes \overline{W}$, where V and W are (uniquely determined) irreducible representations of the complex Lie algebra.

§26.2. Second Proof of Weyl's Character Formula

The title of this section is perhaps inaccurate: what we will give here is actually a sketch of the first proof of the Weyl character formula. Weyl, in his original proof, used what he called the "unitarian trick," which is to say he introduces the compact form of a given semisimple Lie algebra and uses integration on the corresponding compact group G. (This trick was already described in §9.3, in the context of proving complete reducibility of representations of a semisimple algebra.)

Indeed, the main reason for including this section (which is, after all, logically unnecessary) is to acquaint the reader with the "classical" treatment of Lie groups via their compact forms. This treatment follows very much the same lines as the representation theory of finite groups. To begin with, we replace the average $(1/|G|)\sum_{g \in G} f(g)$ by the integral $\int_G f(g)\, d\mu$, the volume element $d\mu$ chosen to be translation invariant and such that $\int_G d\mu = 1$. If $\rho: G \to \mathrm{Aut}(V)$ is a finite-dimensional representation, with character

$$\chi_V(g) = \mathrm{Trace}(\rho(g)),$$

then $\int_G \rho(g)\, d\mu \in \mathrm{Hom}(V, V)$ is idempotent, and it is the projection onto the invariant subspace V^G. So $\int_G \chi_V(g)\, d\mu = \dim(V^G)$. Applied to $\mathrm{Hom}(V, W)$ as before, since $\chi_{\mathrm{Hom}(V,W)} = \bar{\chi}_V \chi_W$, it follows that

$$\int_G \bar{\chi}_V \chi_W \, d\mu = \dim(\mathrm{Hom}_G(V, W)).$$

So if V and W are irreducible,

$$\int_G \bar{\chi}_V \chi_W \, d\mu = \begin{cases} 1 & \text{if } V \cong W \\ 0 & \text{otherwise.} \end{cases}$$

Up to now, everything is completely analogous to the case of finite groups, and is proved in exactly the same way. The last general fact, analogous to the basic Proposition 2.30, is harder in the compact case:

Peter–Weyl Theorem. The characters of irreducible representations span a dense subspace of the space of continuous class functions.

It is, moreover, the case that the coordinate functions of the irreducible matrix representations span a dense subspace of all continuous (or L^2) functions on G. For the proof of these statements we refer to [Ad] or [B-tD]. Given the fundamental role that (2.30) played in the analysis of representations of finite groups, it is not surprising that the Peter–Weyl theorem is the cornerstone of most treatments of compact groups, even though it has played no role so far in this book.

We now proceed to indicate how the original proof of the Weyl character formula went in this setting. In this section, G will denote a fixed compact group, whose Lie algebra $\mathfrak{g}$ is a real form of the semisimple complex Lie algebra $\mathfrak{g}_{\mathbb{C}} = \mathfrak{g} \otimes_{\mathbb{R}} \mathbb{C}$. We have seen that

$$\mathfrak{g} = \mathfrak{h} \oplus \bigoplus_{\alpha \in R^+} \mathfrak{l}_\alpha,$$

compatible with the usual decomposition $\mathfrak{g}_{\mathbb{C}} = \mathfrak{h}_{\mathbb{C}} \oplus \bigoplus(\mathfrak{g}_\alpha \oplus \mathfrak{g}_{-\alpha})$ when complexified. The real Cartan algebra $\mathfrak{h}$ acts by rotations on the planes $\mathfrak{l}_\alpha$.

Now let $T = \exp(\mathfrak{h}) \subset G$. As before we have chosen $\mathfrak{h}$ so that it contains the lattice $2\pi i\Gamma$ which is the kernel of the exponential map from $\mathfrak{h}_{\mathbb{C}}$ to the simply-connected form of $\mathfrak{g}_{\mathbb{C}}$; so $T \cong (S^1)^n$ is a compact torus.

In this compact case we can realize the Weyl group on the group level again:

Claim 26.15. $N(T)/T \cong \mathfrak{W}$.

Proof. For each pair of roots $\alpha, -\alpha$, we have a subalgebra $\mathfrak{s}_\alpha \cong \mathfrak{sl}_2\mathbb{C} \subset \mathfrak{g}_{\mathbb{C}}$, with a corresponding $\mathfrak{su}_2 \subset \mathfrak{g}$. Exponentiating gives a subgroup $\mathrm{SU}(2) \subset G$. The element $\begin{pmatrix} 0 & 1 \\ -1 & 0 \end{pmatrix}$ acts by Ad, taking H to $-H$, X to Y, and Y to X. It is in $N(T)$, and, with B as in the preceding section, $\begin{pmatrix} 0 & 1 \\ -1 & 0 \end{pmatrix} = \exp\left(\frac{1}{2}\pi i B\right)$. Then $\exp\left(\frac{1}{2}\pi i B\right) \in \mathfrak{g}$ acts by reflection in the hyperplane $\alpha^\perp \subset \mathfrak{h}$. □

Note that $\mathfrak{W}$ acting on $\mathfrak{h}$ takes the lattice $2\pi i\Gamma$ to itself, so $\mathfrak{W}$ acts on $T = \mathfrak{h}/2\pi i\Gamma$ by conjugation.

Theorem 26.16. *Every element of G is conjugate to an element of T. A general element is conjugate to $|\mathfrak{W}|$ such elements of T.*

Sketch of a proof: Note that G acts by left multiplication on the left coset space $X = G/T$. For any $z \in G$, consider the map $f_z\colon X \to X$ which takes yT to zyT. The claim is that f_z must have a fixed point, i.e., there is a y such that $y^{-1}zy \in T$. Since all f_z are homotopic, and X is compact, the Lefschetz number of f_z is the topological Euler characteristic of X. The first statement follows from the claim that this Euler characteristic is not zero. This is a good exercise for the classical groups; see [Bor2] for a general proof. For another proof see Remark 26.20 below.

For the second assertion, check first that any element that commutes with every element of T is in T. Take an "irrational" element x in T so that its multiples are dense in T. Then for any $y \in G$, $yxy^{-1} \in T \Leftrightarrow yTy^{-1} = T$, and $yxy^{-1} = x \Leftrightarrow y \in T$. This gives precisely $|\mathfrak{W}|$ conjugates of x that are in T.

Corollary 26.17. *The class functions on G are the $\mathfrak{W}$-invariant functions on T.*

Suppose G is a real form of the complex semisimple group $G_{\mathbb{C}}$, i.e., G is a real analytic closed subgroup of $G_{\mathbb{C}}$, and the Lie algebra of $G_{\mathbb{C}}$ is $\mathfrak{g}_{\mathbb{C}}$. The characters on $G_{\mathbb{C}}$ can be written $\sum n_\mu e^{2\pi i\mu}$, the sum over μ in the weight lattice Λ; they are invariant under the Weyl group. From what we have seen, they can be identified with $\mathfrak{W}$-invariant functions on the torus T. Let us work this out for the classical groups:

Case ($\mathbf{A}_n$): $G = \mathrm{SU}(n+1)$. The Lie algebra $\mathfrak{su}_{n+1}$ consists of skew-Hermitian matrices,

$$\mathfrak{h} = \mathfrak{su}_{n+1} \cap \mathfrak{sl}_{n+1}\mathbb{R} = \{\text{imaginary diagonal matrices of trace } 0\},$$

and $T = \{\mathrm{diag}(e^{2\pi i\vartheta_1}, \ldots, e^{2\pi i\vartheta_{n+1}}) : \sum \vartheta_j = 0\}$. In this case, the Weyl group $\mathfrak{W}$ is the symmetric group $\mathfrak{S}_{n+1}$, represented by permutation matrices (with one entry ± 1 on each row and column, other entries 0) modulo T. Let $z_i : T \to S^1$ correspond to the ith diagonal entry $e^{2\pi i\vartheta_i}$. So characters on T are symmetric polynomials in $z_1, \ldots, z_{n+1}$ modulo the relation $z_1 \cdot \ldots \cdot z_{n+1} = 1$. Therefore, characters on $\mathrm{SU}(n+1)$ are symmetric polynomials in $z_1, \ldots, z_{n+1}$.

Case ($\mathbf{B}_n$): $G = \mathrm{SO}(2n+1)$. $\mathfrak{h}$ consists of matrices with n 2×2 blocks of the form

$$\begin{pmatrix} \cos(2\pi\vartheta_i) & -\sin(2\pi\vartheta_i) \\ \sin(2\pi\vartheta_i) & \cos(2\pi\vartheta_i) \end{pmatrix}$$

along the diagonal, and one 1 in the lower right corner. Again we see that $T = (S^1)^n$. This time $N(T)$ will have block permutations to interchange the blocks, and also matrices with some blocks $\begin{pmatrix} 0 & 1 \\ 1 & 0 \end{pmatrix}$ in the squares along the diagonal, with the other blocks 2×2 identity matrices, with a ± 1 in the corner to make the determinant positive; these take ϑ_i to $-\vartheta_i$ for each i where a block is $\begin{pmatrix} 0 & 1 \\ 1 & 0 \end{pmatrix}$. This again realizes the Weyl group as a semidirect product of $\mathfrak{S}_n$ and $(\mathbb{Z}/2)^n$. With z_i identified with $e^{2\pi i\vartheta_i}$ again, we see that the characters are the symmetric polynomials in the variables $z_i + z_i^{-1}$, i.e., in $\cos(2\pi\vartheta_1), \ldots, \cos(2\pi\vartheta_n)$.

Case ($\mathbf{D}_n$): $G = \mathrm{SO}(2n)$. $\mathfrak{h}$ is as in the preceding case, but with no lower corner. Since we have no corner to put a -1 in, there can be only an even number of blocks of the form $\begin{pmatrix} 0 & 1 \\ 1 & 0 \end{pmatrix}$, reflecting the fact that $\mathfrak{W}$ is a semidirect product of $(\mathbb{Z}/2)^{n-1}$ and $\mathfrak{S}_n$. This time the invariants are symmetric polynomials in the $z_i + z_i^{-1}$, and one additional $\Pi_i(z_i - z_i^{-1})$.

Case ($\mathbf{C}_n$): $G = \mathrm{Sp}(2n)$. $\mathfrak{h}$ consists of imaginary diagonal matrices, T consists of diagonal matrices with entries $e^{2\pi i\vartheta_i}$. The Weyl group in generated by permutation matrices and diagonal matrices with entries which are 1's and quaternionic j's: $\mathfrak{W}$ is a semidirect product of $(\mathbb{Z}/2)^n$ and $\mathfrak{S}_n$. The invariants are symmetric polynomials in the $z_i + z_i^{-1}$.

The key to Weyl's analysis is to calculate the integral of a class function f on G as a suitable integral over the torus T. For this, consider the map

$$\pi: G/T \times T \to G, \qquad \pi(xT, y) = xyx^{-1}.$$

By what we said earlier, π is a generically finite-sheeted covering, with $|\mathfrak{W}|$ sheets. It follows that

$$\int_G f\,d\mu = \frac{1}{|\mathfrak{W}|}\int_{G/T\times T} \pi^*(f)\pi^* d\mu.$$

Now $\pi^*(f)(xT, y) = f(y)$ since f is a class function. To calculate $\pi^* d\mu$, consider the induced map on tangent spaces

$$\pi_* = d\pi: \mathfrak{g}/\mathfrak{h} \times \mathfrak{h} \to \mathfrak{g}.$$

At the point $(x_0 T, y_0) \in G/T \times T$,

$$(x_0 e^{tx} T, y_0 e^{ty}) \mapsto x_0 e^{tx} y_0 e^{ty} e^{-tx} x_0^{-1}.$$

We want to calculate

$$\frac{d}{dt}(x_0 e^{tx} y_0 e^{ty} e^{-tx} x_0^{-1})|_{t=0}(x_0 y_0 x_0^{-1})^{-1},$$

which is

$$x_0(xy_0 + y_0 y - y_0 x)x_0^{-1}(x_0 y_0^{-1} x_0^{-1}) = x_0(x + y_0 y y_0^{-1} - y_0 x y_0^{-1})x_0^{-1}.$$

Now $y_0 y y_0^{-1} = y$ since $y_0 \in T$ and $y \in \mathfrak{h}$. To calculate the determinant of π_* we can ignore the volume-preserving transformation $x_0(\ )x_0^{-1}$. If we identify $\mathfrak{g}$ with $\mathfrak{g}/\mathfrak{h} \times \mathfrak{h}$, the matrix becomes

$$\begin{pmatrix} I - \mathrm{Ad}(y_0) & 0 \\ 0 & I \end{pmatrix}.$$

So the determinant of π_* is $\det(I - \mathrm{Ad}(y_0))$. Now $(\mathfrak{g}/\mathfrak{h})_{\mathbb{C}} = \bigoplus \mathfrak{g}_\alpha$, and $\mathrm{Ad}(y_0)$ acts as $e^{2\pi i\alpha(y_0)}$ on $\mathfrak{g}_\alpha$. Hence

$$\det(\pi_*) = \prod_{\alpha\in R}(1 - e^{2\pi i\alpha}), \tag{26.18}$$

as a function on T alone, independent of the factor G/T. This gives *Weyl's integration formula*:

$$\int_G f\,d\mu_G = \frac{1}{|\mathfrak{W}|}\int_T f(y)\prod_{\alpha\in R}(1 - e^{2\pi i\alpha(y)})\,d\mu_T. \tag{26.19}$$

Remark 26.20. The same argument gives another proof of the theorem that G is covered by conjugates of T. This amounts to the assertion that the map

$\pi: G/T \times T \to G$ of compact manifolds is surjective. By what we saw above, for a generic point $y_0 \in T$ there are exactly $|\mathfrak{W}|$ points in $\pi^{-1}(y_0)$, and at each of these the Jacobian determinant is the same (nonzero) number. It follows that the topological degree of the map π is $|\mathfrak{W}|$, so the map must be surjective.

Now $(1 - e^{2\pi i\alpha})(1 - e^{-2\pi i\alpha}) = (e^{\pi i\alpha} - e^{-\pi i\alpha})\overline{(e^{\pi i\alpha} - e^{-\pi i\alpha})}$, so if we set

$$\Delta = \prod_{\alpha \in R^+} (e^{\pi i\alpha} - e^{-\pi i\alpha}),$$

then $\det(\pi_*) = \Delta\bar{\Delta}$. As we saw in Lemma 24.3, $\Delta = A_\rho$, where ρ is half the sum of the positive roots and, for any weight μ,

$$A_\mu = \sum_{W \in \mathfrak{W}} (-1)^W e^{2\pi i W(\mu)}.$$

Now we can complete the second proof of Weyl's character formula: the character of the representation with highest weight λ is $A_{\lambda+\rho}/A_\rho$. Since we saw in §24.1 that $A_{\lambda+\rho}/A_\rho$ has highest weight λ and (see Corollary 24.6) its value at the identity is positive, it suffices to show that the integral of $\int_G \chi\bar{\chi} = 1$, where $\chi = A_{\lambda+\rho}/A_\rho$. By Weyl's integration formula,

$$\int_G \chi\bar{\chi} = \frac{1}{|\mathfrak{W}|}\int_T \chi\bar{\chi}\Delta\bar{\Delta} = \frac{1}{|\mathfrak{W}|}\int_T A_{\lambda+\rho}\overline{A_{\lambda+\rho}}$$

$$= \frac{1}{|\mathfrak{W}|}\int_T \sum_{W \in \mathfrak{W}} (-1)^W e^{2\pi i W(\lambda+\rho)} \cdot \sum_{W \in \mathfrak{W}} (-1)^W e^{-2\pi i W(\lambda+\rho)} = 1,$$

which concludes the proof.

§26.3. Real, Complex, and Quaternionic Representations

The final topic we want to take up is the classification of irreducible complex representations of semisimple Lie groups or algebras into those of real, quaternionic, or complex type. To define our terms, given a real semisimple Lie group G_0 or its Lie algebra $\mathfrak{g}_0$ and a representation of G_0 or $\mathfrak{g}_0$ on a complex vector space V we say that the representation V is *real*, or of *real type*, if it comes from a representation of G_0 or $\mathfrak{g}_0$ on a real vector space V_0 by extension of scalars ($V = V_0 \otimes_{\mathbb{R}} \mathbb{C}$); this is equivalent to saying that it has a conjugate linear endomorphism whose square is the identity. It is *quaternionic* if it comes from a quaternionic representation by restriction of scalars, or equivalently if it has a conjugate linear endomorphism whose square is minus the identity. Finally, we say that the representation is *complex* if it is neither of these. (Compare with Theorem 3.37 for finite groups.)

Having completely classified the irreducible representations of the classical complex Lie algebras, and having described all the real forms of these Lie algebras, we have a clear-cut problem: to detemine the type of the restriction of each representation to each real form. Rather than try to answer this in every case, however, we will instead mention some of the ideas that allow us to answer this question, and then focus on the cases of the split forms (where the answer is easy) and the compact forms (where the answer is more interesting, and where we have more tools to play with). We assume the complexification $\mathfrak{g}$ of $\mathfrak{g}_0$ is simple, so irreducible representations of $\mathfrak{g}_0$ are restrictions of unique irreducible representations of $\mathfrak{g}$ (cf. (26.14)); in particular, we have the classification of irreducible representations by dominant weights.

To begin with, the tensor products of two real, or two quaternionic, or of a pair of complex conjugate representations is always real; and exterior powers of real and quaternionic representations are equally easy to analyze, as for finite groups (see Exercise 3.43). Such tensor and exterior powers may not be irreducible, but the following criterion can often be used to describe an irreducible component of highest weight that occurs inside them:

Exercise 26.21*. Suppose W is a representation of a semisimple group G that is real or quaternionic, and suppose W has a highest weight λ that occurs with multiplicity 1. Show that the irreducible representations Γ_λ with highest weight λ has the same type as W.

We may apply this in particular to the tensor product $\Gamma_\lambda \otimes \Gamma_\mu$ of the irreducible representations of $\mathfrak{g}$ with highest weights λ and μ; since the irreducible representation $\Gamma_{\lambda+\mu}$ with highest weight $\lambda + \mu$ appears once in this tensor product, we deduce

Exercise 26.22*. (i) If Γ_λ and Γ_μ are both real or both quaternionic, then $\Gamma_{\lambda+\mu}$ is real. (ii) If Γ_λ is real and Γ_μ is quaternionic, then $\Gamma_{\lambda+\mu}$ is quaternionic. (iii) If Γ_λ and Γ_μ are complex and conjugate, then $\Gamma_{\lambda+\mu}$ is real.

The last two exercises almost completely answer the question of the representations of the split forms of the classical groups: we have

Proposition 26.23. *Every irreducible representation of the split forms* $\mathfrak{sl}_{n+1}\mathbb{R}$, $\mathfrak{so}_{n+1,n}\mathbb{R}$, $\mathfrak{sp}_{2n}\mathbb{R}$, *and* $\mathfrak{so}_{n,n}\mathbb{R}$ *of the classical Lie algebras is real.*

PROOF. In each of these cases, the standard representation V is real, from which it follows that the exterior powers $\wedge^k V$ are real, from which it follows that the symmetric powers $\mathrm{Sym}^{a_k}(\wedge^k V)$ are real. Now, in the cases of $\mathfrak{sl}_{n+1}\mathbb{R}$ and $\mathfrak{sp}_{2n}\mathbb{R}$, we have seen that the highest weights ω_k of the representations $\wedge^k V$ for $k = 1, \ldots, n$ form a set of fundamental weights: that is, every irreducible representation Γ has highest weight $\sum a_k \cdot \omega_k$ for some non-negative integers $a_1, \ldots, a_n$. It follows that Γ appears once in the tensor product

$$\mathrm{Sym}^{a_1} V \otimes \mathrm{Sym}^{a_2}(\wedge^2 V) \otimes \cdots \otimes \mathrm{Sym}^{a_n}(\wedge^n V)$$

and so is real. (Alternatively, Weyl's construction produces real representations when applied to real vector spaces.)

The only difference in the orthogonal case is that some of the exterior powers $\wedge^k V$ of the standard representation must be replaced in this description by the spin representation(s). That the spin representations are real follows from the construction in Lecture 20, cf. Exercise 20.23; the result in this case then follows as before. □

The Compact Case

We turn now to the compact forms of the classical Lie algebras. In this case, the theory behaves very much like that of finite groups, discussed in Lecture 5. Specifically, any action of a compact group G_0 on a complex vector space V preserves a nondegenerate Hermitian inner product (obtained, for example, by choosing one arbitrarily and averaging its translates under the action of G_0). It follows that the dual of V is isomorphic to its conjugate, so that V will be either real or quaternionic exactly when it is isomorphic to its dual V^*. (In terms of characters, this says that the character $\operatorname{Char}(V)$ is invariant under the automorphism of $\mathbb{Z}[\Lambda]$ which takes $e(\mu)$ to $e(-\mu)$; for groups, this says the character is real.) More precisely, an irreducible representation of a compact group/Lie algebra will be real (resp. quaternionic) if and only if it has an invariant nondegenerate symmetric (resp. skew-symmetric) bilinear form. In other words, the classification of an irreducible V is determined by whether

$$V \otimes V = \operatorname{Sym}^2 V \oplus \wedge^2 V$$

contains the trivial representation, and, if so, in which factor. So determining which type a representation belongs to is a very special case of the general plethysm problem of decomposing such representations.

With this said, we consider in turn the algebras $\mathfrak{su}_n$, $\mathfrak{u}_n\mathbb{H}$, and $\mathfrak{so}_m\mathbb{R}$.

Let Γ_λ be the irreducible representation of $\mathfrak{sl}_n\mathbb{C}$ with highest weight $\lambda = \sum a_i \cdot \omega_i$, where $\omega_i = L_1 + \cdots + L_i$, $i = 1, \ldots, n-1$ are the fundamental weights of $\mathfrak{sl}_n\mathbb{C}$. The dual of Γ will have highest weight $\sum a_{n-i} \cdot \omega_i$, so that Γ will be real or quaternionic if and only if $a_i = a_{n-i}$ for all i. We now distinguish three cases:

(i) If n is odd, then the sublattice of weights $\lambda = \sum a_i \cdot \omega_i$ with $a_i = a_{n-i}$ for all i is freely generated by the sums $\omega_i + \omega_{n-i}$ for $i = 1, \ldots, (n-1)/2$. Now, ω_i is the highest weight of the exterior power $\wedge^i V$, so that the irreducible representation with highest weight $\omega_i + \omega_{n-i}$ will appear once in the tensor product

$$\wedge^i V \otimes \wedge^{n-i} V = (\wedge^i V) \otimes (\wedge^i V)^*,$$

which by Exercise 26.21 above is real. It follows that for any weight $\lambda = \sum a_i \cdot \omega_i$ with $a_i = a_{n-i}$ for all i, the irreducible representation Γ_λ is real.

(iia) If $n = 2k$ is even, then the sublattice of weights $\lambda = \sum a_i \cdot \omega_i$ with

$a_i = a_{n-i}$ for all i is freely generated by the sums $\omega_i + \omega_{n-i}$ for $i = 1, \ldots, k-1$, together with the weight ω_k. As before, the irreducible representations with highest weight $\omega_i + \omega_{n-i}$ are all real. Moreover, in case n is divisible by 4 the representation $\wedge^k V$ is real as well, since $\wedge^k V$ admits a symmetric bilinear form

$$\wedge^k V \otimes \wedge^k V \to \wedge^{2k} V = \mathbb{C}$$

given by wedge product. It follows then as before that for any weight $\lambda = \sum a_i \cdot \omega_i$ with $a_i = a_{n-i}$ for all i, the irreducible representation Γ_λ is real.

(iib) In case n is congruent to 2 mod 4, the analysis is similar to the last case except that wedge product gives a skew-symmetric bilinear pairing on $\wedge^k V$. The representation $\wedge^k V$ is thus quaternionic, and it follows that for any weight $\lambda = \sum a_i \cdot \omega_i$ with $a_i = a_{n-i}$ for all i, the irreducible representation Γ_λ is real if a_k is even, quaternionic if a_k is odd. In sum, then, we have

Proposition 26.24. *For any weight* $\lambda = \sum a_i \cdot \omega_i$ of $\mathfrak{su}_n$, *the irreducible representation* Γ_λ *with highest weight* λ *is: complex if* $a_i \neq a_{n-i}$ *for any* i; *real if* $a_i = a_{n-i}$ *for all* i *and* n *is odd, or* $n = 4k$, *or* $n = 4k + 2$ *and* a_{2k+1} *is even; and quaternionic if* $a_i = a_{n-i}$ *for all* i *and* $n = 4k + 2$ *and* a_{2k+1} *is odd.*

Next, we consider the case of the compact form $\mathfrak{u}_n\mathbb{H}$ of $\mathfrak{sp}_{2n}\mathbb{C}$. To begin with, we note that since the restriction to $\mathfrak{u}_n\mathbb{H}$ of the standard representation of $\mathfrak{sp}_{2n}\mathbb{C}$ on $V \cong \mathbb{C}^{2n}$ is quaternionic, the exterior power $\wedge^k V$ is real for k even and quaternionic for k odd. Since the highest weights ω_k of $\wedge^k V$ for $k = 1, \ldots, n$ form a set of fundamental weights, this completely determines the type of the irreducible representations of $\mathfrak{u}_n\mathbb{H}$: we have

Proposition 26.25. *For any weight* $\lambda = \sum a_i \cdot \omega_i$ *of* $\mathfrak{u}_n\mathbb{H}$, *the irreducible representation* Γ_λ *with highest weight* λ *is real if* a_i *is even for all odd* i, *and quaternionic if* a_i *is odd for any odd* i.

Next, we consider the odd orthogonal algebras. Part of this is easy: since the restriction to $\mathfrak{so}_{2n+1}\mathbb{R}$ of the standard representation V of $\mathfrak{so}_{2n+1}\mathbb{C}$ is real, so are all its exterior powers; and it follows that any representation of $\mathfrak{so}_{2n+1}\mathbb{R}$ whose highest weight lies in the sublattice of index two generated by the highest weights of these exterior powers is real. It remains, then, to describe the type of the spin representation; the answer, whose verification we leave as Exercise 26.28 below, is that the spin representation Γ_α of $\mathfrak{so}_{2n+1}\mathbb{C}$ (that is, the irreducible representation whose highest weight is one-half the highest weight of $\wedge^n V$) is real when $n \equiv 0$ or 3 mod 4, and quaternionic if $n \equiv 1$ or 2 mod 4. This yields

Proposition 26.26. *Let* ω_i *be the highest weight of the representation* $\wedge^i V$ *of* $\mathfrak{so}_{2n+1}\mathbb{C}$. *For any weight* $\lambda = a_1\omega_1 + \cdots + a_{n-1}\omega_{n-1} + a_n\omega_n/2$ *of* $\mathfrak{so}_{2n+1}\mathbb{R}$, *the irreducible representation* Γ_λ *with highest weight* λ *is real if* a_n *is even, or if* n *is* *congruent to* 0 *or* 3 mod 4; *if* a_n *is odd and* $n \equiv 1$ *or* 2 mod 4, *then* Γ_λ *is quaternionic.*

(Note that, in each of the last two cases, the fact that every representation is either real or quaternionic follows from the observation that the Weyl group action on the Cartan subalgebra $\mathfrak{h} \subset \mathfrak{g}$ includes multiplication by -1.)

Finally, we have the even orthogonal Lie algebras. As before, the exterior powers of the standard representation V are all real, but we now have two spin representations to deal with, with highest vectors (in the notation of Lecture 19) $\alpha = (L_1 + \cdots + L_n)/2$ and $\beta = (L_1 + \cdots + L_{n-1} - L_n)/2$. The first question is whether these two are self-conjugate or conjugate to each other. In case n is even, as in the case of the symplectic and odd orthogonal algebras, the Weyl group action on the Cartan subalgebra contains multiplication by -1 (the Weyl group contains the automorphism of $\mathfrak{h}^*$ reversing the sign of any even number of the basis elements L_i), so that Γ_α and Γ_β will be isomorphic to their duals; if n is odd, on the other hand, we see that Γ_α will have $-\beta$ as a weight, so that Γ_α and Γ_β will be complex representations dual to each other. We consider these cases in turn.

(i) Suppose first that n is odd, and say λ is any weight, written as

$$\lambda = a_1\omega_1 + \cdots + a_{n-2}\omega_{n-2} + a_{n-1}\beta + a_n\alpha.$$

If $a_{n-1} \neq a_n$, the representation Γ_λ with highest weight λ will not be isomorphic to its dual, and so will be complex. On the other hand, $\Gamma_{\alpha+\beta}$ appears once in $\Gamma_\alpha \otimes \Gamma_\beta = \mathrm{End}(\Gamma_\alpha)$, and so is real; thus, if $a_{n-1} = a_n$, the representation Γ_λ will be real.

(ii) If, by contrast, n is even then all representations of $\mathfrak{so}_{2n}\mathbb{R}$ will be either real or quaternionic. The half-spin representations Γ_α and Γ_β are real if $n \equiv 0$ (mod 4), quaternionic if $n \equiv 2$ (mod 4), a fact that we leave as Exercise 26.28. It follows that, with λ as above, Γ_λ will be real if either n is divisible by 4, or if $a_{n-1} + a_n$ is even; if $n \equiv 2$ mod 4 and $a_{n-1} + a_n$ is odd, Γ_λ will be quaternionic. In sum, then, we have

Proposition 26.27. *The representation* Γ_λ *of* $\mathfrak{so}_{2n}\mathbb{R}$ *with highest weight* $\lambda = a_1\omega_1 + \cdots + a_{n-2}\omega_{n-2} + a_{n-1}\beta + a_n\alpha$ *will be complex if* n *is odd and* $a_{n-1} \neq a_n$; *it will be quaternionic if* $n \equiv 2$ mod 4 *and* $a_{n-1} + a_n$ *is odd; and it will be real otherwise.*

Exercise 26.28*. Verify the statements made above about the types of the spin representation Γ_α of the orthogonal Lie algebras, i.e., that the spin representation Γ_α of $\mathfrak{so}_{2n+1}\mathbb{R}$ is real when $n \equiv 0$ or 3 (mod 4); and quaternionic if $n \equiv 1$ or 2 (mod 4), and that the half-spin representations of $\mathfrak{so}_{2n}\mathbb{R}$ are real if $n \equiv 0$ (mod 4) and quaternionic if $n \equiv 2$ (mod 4). Show, in fact, that the even Clifford algebras $C_m^{\text{even}} \subset C_m = C(0, m)$ are products of one or two copies of matrix algebras over $\mathbb{R}$, $\mathbb{C}$, or $\mathbb{H}$, with $\mathbb{R}$ occurring for $m \equiv 0$ or ± 1 mod 8, $\mathbb{C}$ occurring for $m \equiv \pm 2$ mod 8, and $\mathbb{H}$ for $m \equiv \pm 3$ or 4 mod 8.

Exercise 26.29. Show that for a representation V of a compact group G,

$$\int_G \chi_V(g^2) = \begin{cases} 0 & \text{if } V \text{ is complex} \\ 1 & \text{if } V \text{ is real} \\ -1 & \text{if } V \text{ is quaternionic.} \end{cases}$$

Exercise 26.30*. Show that for a representation V of a compact group, the number of irreducible real components it contains, minus the number of quaternionic representations, is the number of times the trivial representation occurs in $\psi^2 V$ in the representation ring, where ψ^2 is the Adams operation (cf. Exercise 23.39).

APPENDICES

These appendices contain proofs of some of the general Lie algebra facts that were postponed during the course, as well as some results from algebra and invariant theory which were used particularly in the "Weyl construction–Schur functor" descriptions of representations.

The first appendix is a fairly serious excursion in polynomial algebra. It proves some basic facts about symmetric functions, especially the Schur polynomials, which occur as characters of representations of GL_n or SL_n, and gives determinantal formulas for them in terms of other basic symmetric polynomials. The last section of Appendix A includes some new identities among symmetric polynomials, which, when the variables are specialized, express characters of representations of Sp_{2n} and SO_m as determinants in the characters of basic representations.

Appendix B gives a short summary of some basic multilinear facts about exterior and symmetric powers. The first two sections can be used as a reference for the conventions and notations we have followed; the third contains a general discussion of constructions such as contractions, many special cases of which were discussed in the main text.

The next three appendices conclude our discussion of the theory of Lie algebras, which began in Lectures 9, 14, and 21. Proofs are given, by standard methods, of the promised general results on semisimplicity, the theorem on conjugacy of Cartan subalgebras, facts about the Weyl group, Ado's theorem that every Lie algebra has a faithful representation, and Levi's theorem that splits the map from a Lie algebra to its semisimple quotient.

The last appendix develops just enough classical invariant theory to find the polynomial invariants for $SL_n\mathbb{C}$, $Sp_{2n}\mathbb{C}$, and $SO_n\mathbb{C}$. This was the key to our proof that Weyl's construction gives the irreducible representations of the symplectic and orthogonal groups.

APPENDICES

These appendices contain proofs of some of the general Lie algebra facts that were postponed during the course, as well as some results from algebra and invariant theory which were used, particularly in the "Weyl construction–Schur functor" descriptions of representations.

The first appendix is a fairly serious excursion in polynomial algebra. It proves some basic facts about symmetric functions, especially the Schur polynomials, which occur as characters of representations of GL_n or SL_n, and gives determinantal formulas for them in terms of other basic symmetric polynomials. The last section of Appendix A includes some new identities among symmetric polynomials, which, when the variables are specialized, express characters of representations of Sp_{2n} and SO_m as determinants in the characters of basic representations.

Appendix B gives a short summary of some basic multilinear facts about exterior and symmetric powers. The first two sections can be used as a reference for the conventions and notations we have followed; the third contains a general discussion of constructions such as contractions, many special cases of which were discussed in the main text.

The next three appendices conclude our discussion of the theory of Lie algebras, which began in Lectures 9, 14, and 21. Proofs are given, by standard methods, of the promised general results on semisimplicity, the theorem on conjugacy of Cartan subalgebras, facts about the Weyl group, Ado's theorem that every Lie algebra has a faithful representation, and Levi's theorem that splits the map from a Lie algebra to its semisimple quotient.

The last appendix develops just enough classical invariant theory to find the polynomial invariants for SL_nC, $Sp_{2n}C$, and SO_nC. This was the key to our proof that Weyl's construction gives the irreducible representations of the symplectic and orthogonal groups.

APPENDIX A

On Symmetric Functions

§A.1. Basic Symmetric Polynomials and Relations among Them

The vector space of homogeneous symmetric polynomials of degree d in k variables $x_1, \ldots, x_k$ has several important bases, usually indexed by the partitions $\lambda = (\lambda_1 \geq \lambda_2 \geq \cdots \geq \lambda_k \geq 0)$ of d into at most k parts, or by Young diagrams with at most k rows (see §4.1). We list four of these bases, which are all valid for polynomials with integer coefficients, or coefficients in any commutative ring.

First we have the monomials in the complete symmetric polynomials:

$$H_\lambda = H_{\lambda_1} \cdot H_{\lambda_2} \cdot \ldots \cdot H_{\lambda_k}, \tag{A.1}$$

where H_j is the jth *complete symmetric polynomial*, i.e., the sum of all distinct monomials of degree j; equivalently,

$$\prod_{i=1}^{k} \frac{1}{1 - x_i t} = \sum_{j=0}^{\infty} H_j t^j.$$

For example, with three variables,

$$H_{(1,1)} = (x_1 + x_2 + x_3)^2,$$
$$H_{(2,0)} = x_1^2 + x_2^2 + x_3^2 + x_1 x_2 + x_1 x_3 + x_2 x_3.$$

Next are the *monomial symmetric polynomials*:

$$M_\lambda = \sum X^\alpha, \tag{A.2}$$

the sum over all distinct permutations $\alpha = (\alpha_1, \dots, \alpha_k)$ of $(\lambda_1, \dots, \lambda_k)$; here $X^\alpha = x_1^{\alpha_1} \cdot \ldots \cdot x_k^{\alpha_k}$. For example,

$$M_{(1,1)} = x_1x_2 + x_1x_3 + x_2x_3,$$
$$M_{(2,0)} = x_1^2 + x_2^2 + x_3^2.$$

The third are the monomials in the elementary symmetric functions. Unlike the first two, these are parametrized by partitions μ of d in integers no larger than k, i.e., $k \geq \mu_1 \geq \cdots \geq \mu_l \geq 0$. These are exactly the partitions that are conjugate to a partition of d into at most k parts. (The *conjugate* to a partition λ is the partition whose Young diagram is obtained from that of λ by interchanging rows and columns. We denote the conjugate of λ by λ', although the notation $\tilde{\lambda}$ is also common.) For such μ set

$$E_\mu = E_{\mu_1} \cdot E_{\mu_2} \cdot \ldots \cdot E_{\mu_l}, \tag{A.3}$$

where E_j is the jth *elementary symmetric polynomial*, i.e.,

$$E_j = \sum_{i_1 < \cdots < i_j} x_{i_1} \cdot \ldots \cdot x_{i_j}, \qquad \prod_{i=1}^{k} (1 + x_i t) = \sum_{j=0}^{\infty} E_j t^j.$$

For example,

$$E_{(1,1)} = (x_1 + x_2 + x_3)^2,$$
$$E_{(2,0)} = x_1x_2 + x_1x_3 + x_2x_3.$$

The fourth are the *Schur polynomials*, which may be the most important, although they are less often met in modern algebra courses:

$$S_\lambda = \frac{|x_j^{\lambda_i + k - i}|}{|x_j^{k-i}|} = \frac{|x_j^{\lambda_i + k - i}|}{\Delta}, \tag{A.4}$$

where $\Delta = \prod_{i<j}(x_i - x_j)$ is the discriminant, and $|a_{i,j}|$ denotes the determinant of a $k \times k$ matrix. For example,

$$S_{(1,1)} = x_1x_2 + x_1x_3 + x_2x_3,$$
$$S_{(2,0)} = x_1^2 + x_2^2 + x_3^2 + x_1x_2 + x_1x_3 + x_2x_3.$$

The first task of this appendix is to describe some relations among these symmetric polynomials. For example, one sees quickly that

$$S_{(1,1)} = E_{(2,0)} = H_1^2 - H_2,$$
$$S_{(2,0)} = H_{(2,0)} = E_1^2 - E_2,$$
$$S_{(1,0)} \cdot S_{(1,0)} = S_{(1,1)} + S_{(2,0)}.$$

These are special cases of three important formulas involving Schur polynomials, which we state next. The first two are known as *determinantal formulas*. The first is also known as the *Jacobi–Trudy identity*. From geometry, the first two are sometimes called *Giambelli's formulas*, and the third is *Pieri's formula*. The proofs will be given in the next section.

$$S_\lambda = |H_{\lambda_i+j-i}| = \begin{vmatrix} H_{\lambda_1} & H_{\lambda_1+1} \dots H_{\lambda_1+k-1} \\ H_{\lambda_2-1} & H_{\lambda_2} \dots & \\ \vdots & & \\ H_{\lambda_k-k+1} \dots & & H_{\lambda_k} \end{vmatrix}. \tag{A.5}$$

Note that if $\lambda_{p+1} = \dots = \lambda_k = 0$, the determinant on the right is the same as the determinant of the upper left $p \times p$ corner. The second is

$$S_\lambda = |E_{\mu_i+j-i}| = \begin{vmatrix} E_{\mu_1} & E_{\mu_1+1} \dots E_{\mu_1+l-1} \\ E_{\mu_2-1} & E_{\mu_2} \dots & \\ \vdots & & \\ E_{\mu_l-l+1} \dots & & E_{\mu_l} \end{vmatrix}, \tag{A.6}$$

where $\mu = (\mu_1, \dots, \mu_l)$ is the conjugate partition to λ.

The third "Pieri" formula tells how to multiply a Schur polynomial S_λ by a basic Schur polynomial $S_{(m)} = H_m$[1]:

$$S_\lambda S_{(m)} = \sum S_\nu, \tag{A.7}$$

the sum over all ν whose Young diagram can be obtained from that of λ by adding a total of m boxes to the rows, but with no two boxes in the same column, i.e., those $\nu = (\nu_1, \dots, \nu_k)$ with

$$\nu_1 \geq \lambda_1 \geq \nu_2 \geq \lambda_2 \geq \dots \geq \nu_k \geq \lambda_k \geq 0,$$

and $\sum \nu_j = \sum \lambda_j + m = d + m$. For example, the identity

$$S_{(2,1)} \cdot S_{(2)} = S_{(4,1)} + S_{(3,2)} + S_{(3,1,1)} + S_{(2,2,1)}$$

can be seen from the pictures

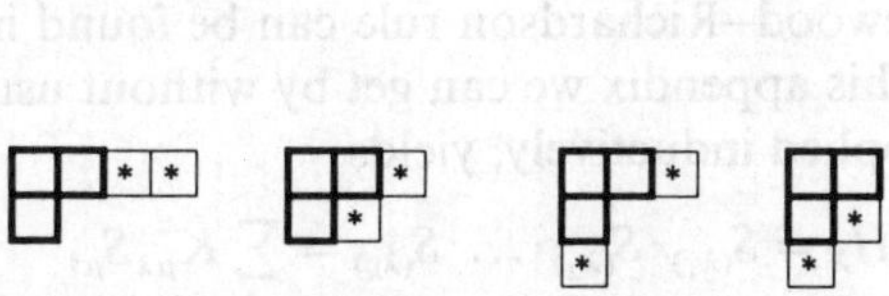

One can use the Pieri and determinantal formulas to multiply any two Schur polynomials, but there is a more direct formula, which generalizes Pieri's formula. This *Littlewood–Richardson rule* gives a combinatorial formula for the coefficients $N_{\lambda\mu\nu}$ in the expansion of a product as a linear combination of Schur polynomials:

[1] When k is fixed, we often omit zeros at the end of partitions, so (m) denotes the partition $(m, 0, \dots, 0)$.

$$S_\lambda \cdot S_\mu = \sum N_{\lambda\mu\nu} S_\nu. \tag{A.8}$$

Here λ is a partition of d, μ a partition of m, and the sum is over all partitions ν of $d + m$ (each with at most k parts). The Littlewood–Richardson rule says that *$N_{\lambda\mu\nu}$ is the number of ways the Young diagram for λ can be expanded to the Young diagram for ν by a strict μ-expansion.* If $\mu = (\mu_1, \ldots, \mu_k)$, a *$\mu$-expansion* of a Young diagram is obtained by first adding μ_1 boxes, according to the above description in Pieri's formula, and putting the integer 1 in each of these μ_1 boxes; then adding similarly μ_2 boxes with a 2, continuing until finally μ_k boxes are added with the integer k. The expansion is called *strict* if, when the integers in the boxes are listed from right to left, starting with the top row and working down, and one looks at the first t entries in this list (for any t between 1 and $\mu_1 + \cdots + \mu_k$), each integer p between 1 and $k - 1$ occurs at least as many times as the next integer $p + 1$.

For example, the equation

$$\begin{aligned} S_{(2,1)} \cdot S_{(2,1)} = S_{(4,2)} + S_{(4,1,1)} + S_{(3,3)} + 2S_{(3,2,1)} \\ + S_{(3,1,1,1)} + S_{(2,2,2)} + S_{(2,2,1,1)} \end{aligned}$$

can be seen by listing the strict (2, 1)-expansions of the Young diagram 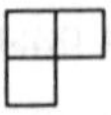:

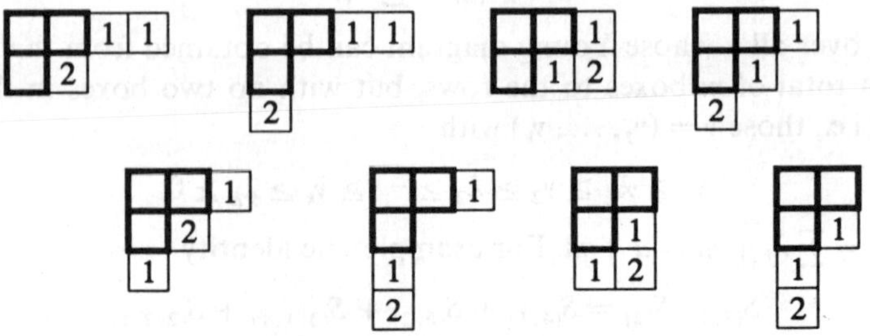

A proof of the Littlewood–Richardson rule can be found in [Mac, §I.9]; for the other results of this appendix we can get by without using it.

Formula (A.7), applied inductively, yields

$$H_\lambda = S_{(\lambda_1)} \cdot S_{(\lambda_2)} \cdot \ldots \cdot S_{(\lambda_k)} = \sum K_{\mu\lambda} S_\mu, \tag{A.9}$$

where *$K_{\mu\lambda}$ is the number of ways one can fill the boxes of the Young diagram of μ with λ_1 1's, λ_2 2's, up to λ_k k's, in such a way that the entries in each row are nondecreasing, and those in each column are strictly increasing.* Such a tableau is called a *semistandard tableau on μ of type λ.* These integers $K_{\mu\lambda}$ are all non-negative, with

$$K_{\lambda\lambda} = 1 \quad \text{and} \quad K_{\mu\lambda} = 0 \quad \text{if } \lambda > \mu, \tag{A.10}$$

i.e., if the first nonvanishing $\lambda_i - \mu_i$ is positive; in addition, $K_{\mu\lambda} = 0$ if λ has more nonzero terms than μ. For example, if $k = 3$, $(K_{\mu\lambda})$ is given by the matrix

	(3)	(2,1)	(1,1,1)
(3)	1	1	1
(2,1)	0	1	2
(1,1,1)	0	0	1

The integers $K_{\mu\lambda}$ are called *Kostka numbers.*

Exercise A.11. Show that $K_{\mu\lambda}$ is nonzero if and only if

$$\lambda_1 + \lambda_2 + \cdots + \lambda_i \leq \mu_1 + \mu_2 + \cdots + \mu_i$$

for all $i \geq 1$.

When $\lambda = (1, 1, \ldots, 1)$, $K_{\mu(1,\ldots,1)}$ *is the number of standard tableaux on the diagram of* μ, where a *standard tableau* is a numbering of the d boxes of a Young diagram by the integers 1 through d, increasing in both rows and columns.

We need one more formula involving Schur polynomials, which comes from an identity of Cauchy. Let $y_1, \ldots, y_k$ be another set of indeterminates, and write $P(x)$ and $P(y)$ for the same polynomial P expressed in terms of variables $x_1, \ldots, x_k$ and $y_1, \ldots, y_k$, respectively. The formula we need is

$$\det\left|\frac{1}{1 - x_i y_j}\right| = \frac{\Delta(x)\Delta(y)}{\prod_{i,j}(1 - x_i y_j)}. \tag{A.12}$$

The proof is by induction on k. To compute the determinant, first subtract the first row from each of the other rows, noting that

$$\frac{1}{1 - x_i y_j} - \frac{1}{1 - x_1 y_j} = \frac{x_i - x_1}{1 - x_1 y_j} \cdot \frac{y_j}{1 - x_i y_j}$$

and factor out common factors. Then subtract the first column from each of the other columns, this time using the equation

$$\frac{y_j}{1 - x_i y_j} - \frac{y_1}{1 - x_i y_1} = \frac{y_j - y_1}{1 - x_i y_1} \cdot \frac{1}{1 - x_i y_j}$$

to factor out common factors. One is left with a matrix whose first row is (1 0 ... 0), and whose lower right square has the original entries. The formula follows by induction (cf. [We1, p. 202]). □

Another form of Cauchy's identity is

$$\frac{1}{\prod_{i,j}(1-x_iy_j)} = \sum_\lambda S_\lambda(x)S_\lambda(y), \tag{A.13}$$

the sum over all partitions λ with at most k terms. To prove this, expand the determinant whose i, j entry is $(1 - x_iy_j)^{-1} = 1 + x_iy_j + x_i^2y_j^2 + \cdots$. One sees that for any $l_1 > \cdots > l_k$ the coefficient of $y_1^{l_1}y_2^{l_2}\cdot \ldots \cdot y_k^{l_k}$ is the determinant $|x_j^{l_i}|$. By symmetry of the x and y variables we have

$$\det\left|\frac{1}{1-x_iy_j}\right| = \sum_l |x_j^{l_i}|\cdot|y_j^{l_i}|. \tag{A.14}$$

Combining (A.12) with (A.4) gives (A.13). □

Expansion of the left-hand side of (A.13) gives

$$\frac{1}{\prod_{i,j}(1-x_iy_j)} = \prod_j\left(\sum_{m=0}^{\infty} H_m(x)y_j^m\right) = \sum_\lambda H_\lambda(x)M_\lambda(y). \tag{A.15}$$

Since the polynomials H_λ as well as the M_μ form a basis for the symmetric polynomials, one can define a bilinear form $\langle\ ,\ \rangle$ on the space of homogeneous symmetric polynomials of degree d in k variables, by requiring that

$$\langle H_\lambda, M_\mu\rangle = \delta_{\lambda,\mu}, \tag{A.16}$$

where $\delta_{\lambda,\mu}$ is 1 if $\lambda = \mu$ and 0 otherwise. The basic fact here is that *the Schur polynomials form an orthonormal basis for this pairing*:

$$\langle S_\lambda, S_\mu\rangle = \delta_{\lambda,\mu}. \tag{A.17}$$

In particular, this implies that the pairing $\langle\ ,\ \rangle$ is *symmetric*. Equation (A.17) is easily deduced from the preceding equations, as follows. Write $S_\lambda = \sum a_{\lambda\gamma}H_\gamma = \sum b_{\gamma\lambda}M_\gamma$, for some integer matrices $a_{\lambda\gamma}$ and $b_{\gamma\lambda}$. Then

$$\langle S_\lambda, S_\mu\rangle = \sum_\gamma a_{\lambda\gamma}b_{\gamma\mu}. \tag{A.18}$$

In order that

$$\sum_\lambda S_\lambda(x)S_\lambda(y) = \sum_{\lambda,\gamma,\rho} a_{\lambda\gamma}H_\gamma(x)b_{\rho\lambda}M_\rho(y)$$

be equal to $\sum_\gamma H_\gamma(x)M_\gamma(y)$, which it must by (A.13) and (A.15), we must have

$$\sum_\lambda b_{\rho\lambda}a_{\lambda\gamma} = \delta_{\rho,\gamma}.$$

This is equivalent to the equation $\sum_\gamma a_{\lambda\gamma}b_{\gamma\mu} = \delta_{\lambda,\mu}$, which by (A.18) implies (A.17).

Because of this duality, formula (A.9) is equivalent to the equation

$$S_\mu = \sum_\lambda K_{\mu\lambda}M_\lambda. \tag{A.19}$$

This gives another formula for these Kostka numbers: $K_{\mu\lambda}$ *is the coefficient of* X^λ *in* S_μ, where $X^\lambda = x_1^{\lambda_1} \cdot \ldots \cdot x_k^{\lambda_k}$.

The identities (A.9) and (A.19) for the basic symmetric polynomials allow us to relate the coefficients of X^λ in any symmetric polynomial P with the coefficients expanding P as a linear combination of the Schur polynomials. If P is any homogeneous symmetric polynomial of degree d in k variables, and λ is any partition of d into at most k parts, define numbers $\psi_\lambda(P)$ and $\omega_\lambda(P)$ by

$$\psi_\lambda(P) = [P]_\lambda, \tag{A.20}$$

where $[P]_\lambda$ denotes the coefficient of $X^\lambda = x_1^{\lambda_1} \cdot \ldots \cdot x_k^{\lambda_k}$ in P, and

$$\omega_\lambda(P) = [\Delta \cdot P]_l, \quad l = (\lambda_1 + k - 1, \lambda_2 + k - 2, \ldots, \lambda_k); \tag{A.21}$$

here $\Delta = \prod_{i<j}(x_i - x_j)$. We want to compare these two collections of numbers, as λ varies over the partitions.

The first numbers $\psi_\lambda(P)$ are the coefficients in the expression

$$P = \sum \psi_\lambda(P) M_\lambda \tag{A.22}$$

for P as a linear combination of the monomial symmetric polynomials M_λ. The integers $\omega_\lambda(P)$ have a similar interpretation in terms of Schur polynomials:

$$P = \sum \omega_\lambda(P) S_\lambda. \tag{A.23}$$

Note from the definition that the coefficient of X^l in $\Delta \cdot S_\lambda$ is 1, and that no other monomial with strictly decreasing exponents appears in $\Delta \cdot S_\lambda$; from this, formula (A.23) is evident. In this terminology we may rewrite (A.19) and (A.9) as

$$K_{\mu\lambda} = \psi_\lambda(S_\mu) = [S_\mu]_\lambda = \text{coefficient of } X^\lambda \text{ in } S_\mu \tag{A.24}$$

and

$$K_{\mu\lambda} = \omega_\mu(H_\lambda) = [\Delta \cdot H_\lambda]_{(\lambda_1 + k - 1, \ldots, \lambda_k)}. \tag{A.25}$$

Lemma A.26. *For any symmetric polynomial P of degree d in k variables,*

$$\psi_\lambda(P) = \sum_\mu K_{\mu\lambda} \cdot \omega_\mu(P).$$

Proof. We have

$$\sum_\lambda \psi_\lambda(P) M_\lambda = P = \sum_\mu \omega_\mu(P) S_\mu = \sum_{\lambda,\mu} \omega_\mu(P) K_{\mu\lambda} M_\lambda$$

$$= \sum_\lambda \left(\sum_\mu K_{\mu\lambda} \omega_\mu(P) \right) M_\lambda,$$

and the result follows, since the M_λ are independent. □

We want to apply the preceding discussion when the polynomial P is a product of sums of powers of the variables. Let $P_j = x_1^j + \cdots + x_k^j$, and for

$\mathbf{i} = (i_1, \ldots, i_d)$, a d-tuple of non-negative integers with $\sum \alpha i_\alpha = d$, set

$$P^{(\mathbf{i})} = P_1^{i_1} \cdot P_2^{i_2} \cdot \ldots \cdot P_d^{i_d}.$$

These *Newton* or *power sum* polynomials form a basis for the symmetric functions with rational coefficients, but not with integer coefficients. Let

$$\omega_\lambda(\mathbf{i}) = \omega_\lambda(P^{(\mathbf{i})}).$$

Equivalently,

$$P^{(\mathbf{i})} = \sum \omega_\lambda(\mathbf{i}) S_\lambda. \tag{A.27}$$

For the proof of Frobenius's formula in Lecture 4 we need a formal lemma about these coefficients $\omega_\lambda(\mathbf{i})$:

Lemma A.28. *For partitions λ and μ of d,*

$$\sum_{\mathbf{i}} \frac{1}{1^{i_1} i_1! \cdot \ldots \cdot d^{i_d} i_d!} \omega_\lambda(\mathbf{i}) \omega_\mu(\mathbf{i}) = \begin{cases} 1 & \text{if } \lambda = \mu \\ 0 & \text{otherwise.} \end{cases}$$

PROOF. We will use Cauchy's formula (A.13). Note that

$$\log\left(\prod_{i,j} (1 - x_i y_j)^{-1}\right) = \sum_{j=1}^{\infty} \frac{1}{j} P_j(x) P_j(y),$$

so

$$\begin{aligned} \frac{1}{\prod (1 - x_i y_j)} &= \prod_j \exp\left(\frac{1}{j} P_j(x) P_j(y)\right) \\ &= \sum_{\mathbf{i}} \frac{1}{1^{i_1} i_1! \cdot \ldots \cdot d^{i_d} i_d!} P^{(\mathbf{i})}(x) P^{(\mathbf{i})}(y) \\ &= \sum_{\mathbf{i}} \frac{1}{1^{i_1} i_1! \cdot \ldots \cdot d^{i_d} i_d!} \sum_\lambda \omega_\lambda(\mathbf{i}) S_\lambda(x) \sum_\mu \omega_\mu(\mathbf{i}) S_\mu(y). \end{aligned}$$

Comparing with (A.13), the conclusion follows. □

Exercise A.29*. Using the pairing $\langle\ ,\ \rangle$ of (A.16), the coefficients $\omega_\lambda(\mathbf{i}) = \omega_\lambda(P^{(\mathbf{i})})$ can be written $\omega_\lambda(\mathbf{i}) = \langle S_\lambda, P^{(\mathbf{i})} \rangle$.

(a) Show that the Newton polynomials are orthogonal for this pairing, and

$$\langle P^{(\mathbf{i})}, P^{(\mathbf{i})} \rangle = 1^{i_1} i_1! 2^{i_2} i_2! \cdot \ldots \cdot d^{i_d} i_d!.$$

Equivalently,

$$S_\lambda = \sum \frac{1}{z(\mathbf{i})} \omega_\lambda(\mathbf{i}) P^{(\mathbf{i})},$$

where the sum is over all partitions $\mathbf{i} = (i_1, \ldots, i_d)$ with $\sum \alpha i_\alpha = d$, and $z(\mathbf{i}) = i_1! 1^{i_1} \cdot i_2! 2^{i_2} \cdot \ldots \cdot i_d! d^{i_d}$.

(b) Show that $\omega_\lambda(\mathbf{i}) = \sum_\nu \langle S_\lambda, M_\nu \rangle \cdot \langle H_\nu, P^{(\mathbf{i})} \rangle$.

We should remark that we have chosen to write our formulas for a fixed number k of variables, since that often simplifies computations when k is small. It is more usual to require the number of variables to be large, at least as large as the numbers being partitioned—or in the limiting ring with an infinite number of variables, cf. Exercise A.32; the formulas for smaller k are then recovered by setting the variables $x_i = 0$ for $i > k$. For example, if $k \geq 2$ we have $S_{(1)}^2 = S_{(2)} + S_{(1,1)}$, which reduces to $S_{(1)}^2 = S_{(2)}$ when $k = 1$.

The next two exercises give formulas for the value of the Schur polynomials when the variables x_i are all set equal to 1; these numbers are the dimensions of the corresponding representations. For a formula for $S_\lambda(1, \ldots, 1)$ involving hook lengths of the Young diagram of λ, see Exercise 6.4.

Exercise A.30*. When $x_i = x^{i-1}$, the numerators in (A.4) are van der Monde determinants, leading to

(i) $$S_\lambda(1, x, x^2, \ldots, x^{k-1}) = x^k \prod_{i<j} \frac{x^{\lambda_i - \lambda_j + j - i} - 1}{x^{j-i} - 1}.$$

Taking the limit as $x \to 1$, one finds

(ii) $$S_\lambda(1, \ldots, 1) = \prod_{i<j} \frac{\lambda_i - \lambda_j + j - i}{j - i}.$$

By (A.5) and (A.6) we have also the following two formulas:

(iii) $$S_\lambda(1, \ldots, 1) = |h_{\lambda_i + j - i}|, \quad \text{where } \sum h_j t^j = \frac{1}{(1-t)^k}.$$

(iv) $$S_\lambda(1, \ldots, 1) = \left| \binom{k}{\mu_i + j - i} \right|, \quad \text{where } (\mu_1, \ldots, \mu_r) = \lambda'.$$

Exercise A.31*. (a) Show that

$$S_\mu = \sum K_{\mu a} X^a,$$

the sum over all monomials $X^a = x_1^{a_1} \cdot \ldots \cdot x_k^{a_k}$, where, for *any* k-tuple a of non-negative integers, $K_{\mu a}$ is the number of ways to number the boxes of the Young diagram of μ with a_1 1's, a_2 2's, ..., a_k k's, with nondecreasing rows and strictly increasing columns. In particular, the right-hand side is a symmetric polynomial, a fact which is not obvious from the definition.

(b) Deduce that $S_\mu(1, \ldots, 1)$ is the number of ways to number the boxes of the Young diagram of μ with integers from 1 to k, with nondecreasing rows and strictly increasing columns (i.e., the number of semistandard tableaux).

Exercise A.32*. The idea of considering symmetric polynomials in an arbitrarily large number of variables can be formalized by working in the ring $\Lambda = \varprojlim \Lambda(k)$, where $\Lambda(k)$ denotes the ring of symmetric polynomials in k variables. Then

$$\Lambda = \mathbb{Z}[H_1, \ldots H_k, \ldots] = \mathbb{Z}[E_1, \ldots E_k, \ldots]$$

is a graded polynomial ring, with H_i and E_i of degree i. A ring homomorphism $\vartheta\colon \Lambda \to \Lambda$ can be defined by requiring

$$\vartheta(E_i) = H_i \qquad \text{for all } i.$$

(i) Show that ϑ is an involution: $\vartheta^2 = \vartheta$. Equivalently,

$$\vartheta(H_i) = E_i.$$

(ii) If λ' is the conjugate partition to λ, show that

$$\vartheta(S_\lambda) = S_{\lambda'}.$$

(iii) If $P_j = x_1^j + \cdots + x_k^j$ is the jth power sum, show that

$$\vartheta(P_j) = (-1)^{j-1}P_j.$$

(iv) Deduce the formula

$$E_\lambda = \sum_\mu K_{\mu\lambda}\Delta_{\mu'}.$$

(v) Deduce a dual form of (A.7):

$$S_\lambda \cdot S_{(1,\ldots,1)} = S_\lambda \cdot E_m = \sum S_\pi,$$

the sum over all partitions π whose Young diagram can be obtained from that of λ by adding m boxes, with no two in any row.

(vi) Show that

$$H_m = \sum \frac{1}{z(\mathbf{i})} P^{(\mathbf{i})}, \qquad E_m = \sum \frac{(-1)^{\sum(i_j-1)}}{z(\mathbf{i})} P^{(\mathbf{i})},$$

where the sums are over all $\mathbf{i} = (i_1, \ldots, i_d)$ with $\sum \alpha i_\alpha = d$, and $z(\mathbf{i}) = i_1!1^{i_1} \cdot i_2!2^{i_2} \cdot \ldots \cdot i_d!d^{i_d}$. Note that

$$\sum (-1)^i H_i t^i = (\sum E_i t^i)^{-1}.$$

§A.2. Proofs of the Determinantal Identities

To prove the Jacobi–Trudi identity (A.5), note the identities

$$x_j^p - E_1 x_j^{p-1} + E_2 x_j^{p-1} - \cdots + (-1)^k E_k x_j^{p-k} = 0, \tag{A.33}$$

for any $1 \le j \le k$, $p \ge k$. And for any $0 \le m < k$ and $p \ge k$,

$$H_{p-m} - E_1 H_{p-m-1} + E_2 H_{p-m-2} + \cdots + (-1)^k E_k H_{p-m-k} = 0. \tag{A.34}$$

Both of these follow immediately from the defining power series for the E_j and H_j. Since these two recursion relations are the same, there are universal polynomials $A(p, q)$ in the variables $E_1, \ldots, E_k$ such that

$$\begin{aligned} x_j^p &= A(p,1)x_j^{k-1} + A(p,2)x_j^{k-2} + \cdots + A(p,k), \\ H_{p-m} &= A(p,1)H_{k-m-1} + A(p,2)H_{k-m-2} + \cdots + A(p,k)H_{-m}. \end{aligned} \tag{A.35}$$

For any integers $\lambda_1, \ldots, \lambda_k$ this leads to matrix identities

$$\begin{aligned} (x_j^{\lambda_i+k-i})_{ij} &= (A(\lambda_i + \mathrm{k} - \mathrm{i}, r))_{ir} \cdot (x_j^{k-r})_{rj}, \\ (H_{\lambda_i+j-i})_{ij} &= (A(\lambda_i + \mathrm{k} - \mathrm{i}, r))_{ir} \cdot (H_{j-r})_{rj}, \end{aligned} \tag{A.36}$$

where $(\ )_{pq}$ denotes the $k \times k$ matrix whose p, q entry is specified between the parentheses. The relations (A.34) also imply:

Lemma (A.37). *The matrices* (H_{q-p}) *and* $((-1)^{q-p}E_{q-p})$ *are lower-triangular matrices with* 1's *along the diagonal, and are inverses of each other.*

The identities (A.36) therefore combine to give

$$(x_j^{\lambda_i+k-i})_{ij} = (H_{\lambda_i+p-i})_{ip} \cdot ((-1)^{q-p}E_{q-p})_{pq} \cdot (x_j^{k-q})_{qj} \tag{A.38}$$

Taking determinants gives (A.5), since the determinant of the matrix in the middle is 1.

Exercise A.39*. Prove the identity

$$|x_j^{l_i}| \cdot \prod_{j=1}^{k} (1 - x_j)^{-1} = \sum |x_j^{m_i}|,$$

the sum over all k-tuples $(m_1, \ldots, m_k)$ of non-negative integers with $m_1 \geq l_1 > m_2 \geq \cdots > m_k \geq l_k$, and deduce Pieri's formula (A.7).

To complete the proofs of the assertions in §A.1, we show that the two determinants appearing in the Giambelli formulas (A.5) and (A.6) are equal, i.e., if $\lambda = (\lambda_1, \ldots, \lambda_k)$ and $\mu = (\mu_1, \ldots, \mu_l)$ are conjugate partitions, then

$$|H_{\lambda_i+j-i}| = |E_{\mu_i+j-i}|. \tag{A.40}$$

Here the H_i and E_i can be any elements (in a commutative ring) satisfying the identity $(\sum H_i t^i) \cdot (\sum (-1)^i E_i t^i) = 1$, with $H_0 = E_0 = 1$ and $H_i = E_i = 0$ for $i < 0$. To prove it, we need a combinatorial characterization of the conjugacy of partitions:

Exercise A.41*. For $\lambda = (\lambda_1, \ldots, \lambda_k)$ and $\mu = (\mu_1, \ldots, \mu_l)$ conjugate partitions, show that the sets

$$\{\lambda_i + n + 1 - i : 1 \leq i \leq k\} \quad \text{and} \quad \{n + j - \mu_j : 1 \leq j \leq l\}$$

form a disjoint union of the set $\{1, \ldots, k + l\}$.

We also need a basic matrix identity which relates minors of a matrix to minors of its inverse (or matrix of cofactors). If $A = (a_{ij})$ is an $r \times r$ matrix, and $S = (s_1, \ldots, s_k)$ and $T = (t_1, \ldots, t_k)$ are two sequences of k distinct integers from $\{1, \ldots, r\}$, let $A_{S,T}$ denote the corresponding minor: $A_{S,T}$ is the determinant of the $k \times k$ matrix whose i, j entry is a_{s_i, t_j}.

Lemma A.42. *Let A and B be $r \times r$ matrices whose product is a scalar matrix $c \cdot I_r$. Let (S, S') and (T, T') be permutations of the sequence $(1, \ldots, r)$, where S and T consists of k integers, S' and T' of $r - k$. Then*

$$c^{r-k} \cdot A_{S,T} = \varepsilon \cdot \det(A) \cdot B_{T',S'},$$

where ε is the product of the signs of the two permutations.

PROOF. By permuting the rows and columns of A, multiplying on the left and right by permutation matrices P and Q corresponding to the two permutations of $(1, \ldots, r)$, we may take the (S, T) minor to the upper left corner:

$$PAQ = \begin{pmatrix} A_1 & A_2 \\ A_3 & A_4 \end{pmatrix}, \qquad A_{S,T} = \det A_1.$$

Then

$$Q^{-1}BP^{-1} = \begin{pmatrix} B_1 & B_2 \\ B_3 & B_4 \end{pmatrix}, \qquad B_{T',S'} = \det B_4.$$

Now taking determinants in the identity

$$\begin{pmatrix} A_1 & A_2 \\ A_3 & A_4 \end{pmatrix} \cdot \begin{pmatrix} I_k & B_2 \\ 0 & B_4 \end{pmatrix} = \begin{pmatrix} A_1 & 0 \\ A_3 & cI_{r-k} \end{pmatrix}$$

gives the equation $\det(PAQ) \cdot \det(B_4) = \det(A_1) \cdot c^{r-k}$. Since ε is the product of the determinants of P and Q, the lemma follows. □

PROOF OF (A.40). Apply the lemma to $A = (H_{q-p})$ and $B = ((-1)^{q-p} E_{q-p})$, with $r = k + l$, and

$$\begin{aligned} S &= (\lambda_1 + k, \lambda_2 + k - 1, \ldots, \lambda_k + 1), \\ S' &= (k + 1 - \mu_1, k + 2 - \mu_2, \ldots, k + l - \mu_l), \\ T &= (k, k - 1, \ldots, 1), \\ T' &= (k + 1, k + 2, \ldots, k + l). \end{aligned}$$

Then

$$A_{S,T} = \det(H_{(\lambda_i + k + 1 - i) - (k + 1 - j)}) = |H_{\lambda_i + j - i}|.$$

Similarly,

$$\begin{aligned} B_{T',S'} &= |(-1)^{\mu_j + i - j} E_{\mu_j + i - j}| = (-1)^{\sum (\mu_j - j)} (-1)^{\sum i} |E_{\mu_j + i - j}| \\ &= (-1)^d |E_{\mu_j + i - j}|, \end{aligned}$$

with $d = \sum \mu_j = \sum \lambda_i$. Since $\varepsilon = (-1)^d$, (A.40) follows. □

§A.3. Other Determinantal Identities

In this final section we prove some variations of these formulas which are useful for calculating characters of symplectic and orthogonal groups. We want to compare minors, not of $H = (H_{i-j})$ and $E = ((-1)^{i-j}E_{i-j})$, but of matrices H^+ and E^- constructed from them by the following procedures:

For an $r \times r$ matrix $H = (H_{i,j})$, and a fixed integer k between 1 and r, H^+ denotes the $r \times r$ matrix obtained from H by folding H along the kth column, and adding each column to the right of the kth column to the column the same distance to the left. That is,

$$H^+_{i,j} = \begin{cases} H_{i,j} + H_{i,2k-j} & \text{if } j < k \\ H_{i,j} & \text{if } j \geq k \end{cases}$$

(with the convention that $H_{p,q} = 0$ if p or q is not between 1 and r). The matrix E^- is obtained by folding E along its kth row, and subtracting rows above this row from those below:

$$E^-_{i,j} = \begin{cases} E_{i,j} - E_{2k-i,j} & \text{if } i > k \\ E_{i,j} & \text{if } i \leq k. \end{cases}$$

Lemma A.43. *If H and E are lower-triangular matrices with 1's along the diagonal, that are inverse to each other, then the same is true for H^+ and E^-.*

PROOF. This is a straightforward calculation: the i, j entry of the matrix $H^+ \cdot E^-$ is

$$\sum_{p=1}^{k-1} (H_{i,p} + H_{i,2k-p})E_{p,j} + H_{i,k}E_{k,j} + \sum_{p=k+1}^{r} H_{i,p}(E_{p,j} - E_{2k-p,j})$$

$$= \sum_{p=1}^{r} H_{i,p}E_{p,j} + \sum_{p=1}^{k-1} H_{i,2k-p}E_{p,j} - \sum_{q=k+1}^{r} H_{i,q}E_{2k-q,j}.$$

The first sum is $\delta_{i,j}$, and the others cancel term by term. □

Proposition A.44. *Let $\lambda = (\lambda_1, \ldots, \lambda_k)$ and $\mu = (\mu_1, \ldots, \mu_l)$ be conjugate partitions. Set*

$$E'_i = E_i \quad \textit{for } i \leq 1, \qquad \textit{and} \qquad E'_i = E_i - E_{i-2} \quad \textit{for } i \geq 2.$$

Then the determinant of the $k \times k$ matrix whose ith *row is*

$$(H_{\lambda_i-i+1} \quad H_{\lambda_i-i+2} + H_{\lambda_i-i} \quad H_{\lambda_i-i+3} + H_{\lambda_i-i-1} \quad \cdots \quad H_{\lambda_i-i+k} + H_{\lambda_i-i-k+2})$$

is equal to the determinant of the $l \times l$ matrix whose ith *row is*

$$(E'_{\mu_i-i+1} \quad E'_{\mu_i-i+2} + E'_{\mu_i-i} \quad E'_{\mu_i-i+3} + E'_{\mu_i-i-1} \quad \cdots \quad E'_{\mu_i-i+l} + E'_{\mu_i-i-l+2}).$$

Each of these determinants is equal to the determinant

$$|E_{\mu_i-i+j} - E_{\mu_i-i-j}|$$

and to the determinant

$$|H''_{\lambda_i-i+j} - H''_{\lambda_i-i-j}|,$$

where $H''_i = H_i$ *for* $i \le 1$, *and for* $i \ge 2$

$$H''_i = H_i + H_{i-2} + H_{i-4} + \cdots + \begin{cases} H_1 & \textit{if } i \textit{ is odd} \\ 1 & \textit{if } i \textit{ is even.} \end{cases}$$

Proof. With $H = (H_{i-j})$ and $E = ((-1)^{q-p}E_{q-p})$ we can apply the basic lemma (A.42) to the new matrices $A = H^+$ and $B = E^-$, and the same permutations (S, S') and (T, T') used in the proof of (A.40). This time

$$A_{S,T} = \det(H^+_{\lambda_i+k+1-i,k-j+1}),$$

and

$$H^+_{\lambda_i+k+1-i,k-j+1} = \begin{cases} H_{\lambda_i-i+j} + H_{\lambda_i-i-j+2} & \text{if } j = 2, \ldots, k \\ H_{\lambda_i-i+1} & \text{if } j = 1. \end{cases}$$

Similarly,

$$B_{T',S'} = \det(E^-_{k+i,k+j-\mu_j}),$$

with

$$E^-_{k+i,k+j-\mu_j} = (-1)^{\mu_j+i-j}(E_{\mu_j+i+j} - E_{\mu_j+i-j}).$$

As before, Lemma A.42 implies that the determinant of the first displayed matrix of the proposition is equal to that of the third. Noting that

$$E_{\mu_j+i+j} - E_{\mu_j+i-j} = E'_{\mu_j+i+j} + E'_{\mu_j+i+j-2} + \cdots + E'_{\mu_j+i-j+2},$$

one can do elementary column operations on the third matrix, subtracting the first column from the third, then the second by the fourth, etc., to see that the second and third determinants are equal. Since $H_i = H''_i - H''_{i-2}$, the same argument shows the equality of the first and fourth determinants. □

Note that in these four formulas, as in the determinantal formulas for Schur polynomials, if a partition has p nonzero terms, only the upper left $p \times p$ subdeterminant needs to be calculated. We denote by $S_{\langle\lambda\rangle}$ the determinant of the proposition:

$$S_{\langle\lambda\rangle} = |H_{\lambda_i-i+1} \quad H_{\lambda_i-i+2} + H_{\lambda_i-i} \quad \ldots \quad H_{\lambda_i-i+k} + H_{\lambda_i-i-k+2}|. \tag{A.45}$$

Dually, set $H'_i = H_i - H_{i-2}$ and $E''_i = E_i + E_{i-2} + E_{i-4} + \cdots$.

Corollary A.46. *The following determinants are equal:*

(i) $|H'_{\lambda_i-i+1} \quad H'_{\lambda_i-i+2} + H'_{\lambda_i-i} \quad \ldots \quad H'_{\lambda_i-i+k} + H'_{\lambda_i-i-k+2}|$,

(ii) $|E_{\mu_i-i+1} \quad E_{\mu_i-i+2} + E_{\mu_i-i} \quad \ldots \quad E_{\mu_i-i+l} + E_{\mu_i-i-l+2}|$,

(iii) $$|E''_{\mu_i-i+j} - E''_{\mu_i-i-j}|,$$

(iv) $$|H_{\lambda_i-i+j} - H_{\lambda_i-i-j}|.$$

Define $S_{[\lambda]}$ to be the determinant of this corollary:

$$S_{[\lambda]} = |H'_{\lambda_i-i+1} \quad H'_{\lambda_i-i+2} + H'_{\lambda_i-i} \quad \dots \quad H'_{\lambda_i-i+k} + H'_{\lambda_i-i-k+2}|. \tag{A.47}$$

Exercise A.48*. Let Λ be the ring of symmetric polynomials, $\vartheta\colon \Lambda \to \Lambda$ the involution of Exercise A.32. Show that

$$\vartheta(S_{\langle\lambda\rangle}) = S_{[\mu]}$$

when λ and μ are conjugate partitions.

For applications to symplectic and orthogonal characters we need to specialize the variables $x_1, \dots, x_k$. First (for the symplectic group Sp_{2n}) take $k = 2n$, let $z_1, \dots, z_n$ be independent variables, and specialize

$$x_1 \mapsto z_1, \ \dots, \ x_n \mapsto z_n, \ x_{n+1} \mapsto z_1^{-1}, \ \dots, \ x_{2n} \mapsto z_n^{-1}.$$

Set

$$J_j = H_j(z_1, \dots, z_n, z_1^{-1}, \dots, z_n^{-1}) \tag{A.49}$$

in the field $\mathbb{Q}(z_1, \dots, z_n)$ of rational functions.

Proposition A.50. *Given integers $\lambda_1 \geq \cdots \geq \lambda_n \geq 0$, we have*

$$\frac{|z_j^{\lambda_i+n-i+1} - z_j^{-(\lambda_i+n-i+1)}|}{|z_j^{n-i+1} - z_j^{-(n-i+1)}|} = |J_\lambda|,$$

where J_λ denotes the $n \times n$ matrix whose ith row is

$$(J_{\lambda_i-i+1} \quad J_{\lambda_i-i+2} + J_{\lambda_i-i} \quad \cdots \quad J_{\lambda_i-i+n} + J_{\lambda_i-i-n+2}).$$

From Proposition A.44 we obtain three other formulas for the right-hand side, e.g.,

$$|J_\lambda| = |e_{\mu_i-i+j} - e_{\mu_i-i-j}|, \tag{A.51}$$

where $e_j = E_j(z_1, \dots, z_n, z_1^{-1}, \dots, z_n^{-1})$, and μ is the conjugate partition to λ.

Exercise A.52. Calculate the denominator of the left-hand side:

$$|z_j^{n-i+1} - z_j^{-(n-i+1)}| = \Delta(\xi_1, \dots, \xi_n)\cdot\zeta_1\cdot\ldots\cdot\zeta_n,$$

where $\xi_j = z_j + z_j^{-1}$ and $\zeta_j = z_j - z_j^{-1}$.

Proof of Proposition A.50. Set

$$\zeta_j(p) = z_j^p - z_j^{-p}, \qquad \xi_j(p) = z_j^p + z_j^{-p}. \tag{A.53}$$

By the same argument that proved the Jacobi–Trudy formula (A.5) via (A.38), the proposition follows from the following lemma:

Lemma A.54. *For $1 \le j \le n$ and any integer $l \ge 0$, $\zeta_j(l)$ is the product of the $1 \times n$, $n \times n$, and $n \times 1$ matrices*

$$(J_{l-n} \quad J_{l-n+1} + J_{l-n-1} \quad \cdots \quad J_{l-1} + J_{l-2n+1}) \cdot ((-1)^{q-p} e_{q-p}) \cdot \begin{pmatrix} \zeta_j(n) \\ \zeta_j(n-1) \\ \vdots \\ \zeta_j(1) \end{pmatrix}$$

PROOF. From (A.37) we can calculate z_j^l and z_j^{-l}, and subtracting gives

$$\zeta_j(l) = \sum_{p=1}^{2n} J_{l-2n+p} s_p, \tag{A.55}$$

where $s_p = \sum_{q=p}^{2n} (-1)^{q-p} e_{q-p} \zeta_j(2n-q)$. Multiplying (A.33) by z_j^{-p} and subtracting we find

$$\begin{aligned} &\zeta_j(p) - e_1 \zeta_j(p-1) + \cdots + (-1)^{p-1} e_p \zeta_j(1) \\ &= (-1)^{p+1} e_{p+1} \zeta_j(1) + (-1)^{p+2} e_{p+2} \zeta_j(2) + \cdots + e_{2n} \zeta_j(2n-p). \end{aligned} \tag{A.56}$$

Note also that

$$(-1)^p e_p = (-1)^{2n-p} e_{2n-p}, \tag{A.57}$$

since $\sum (-1)^p e_p t^p = \prod (1 - z_i t)(1 - z_i^{-1} t) = \prod (1 - \xi_i t + t^2)$. From (A.56) and (A.57) follows

$$s_{2n-p} = s_p = r_{n-p+1}, \tag{A.58}$$

where $r_p = \sum_{q=p}^{n} (-1)^{q-p} e_{q-p} \zeta_j(n+1-q)$. Combining (A.55) and (A.58) concludes the proof. □

Next (for the odd orthogonal groups O_{2n+1}) let $k = 2n+1$, and specialize the variables $x_1, \ldots, x_{2n}$ as above, and $x_{2n+1} \mapsto 1$. We introduce variables $z_j^{1/2}$ and $z_j^{-1/2}$, square roots of the variables just considered, and we work in the field $\mathbb{Q}(z_1^{1/2}, \ldots, z_n^{1/2})$. Set

$$\begin{aligned} K_j &= H_j'(z_1, \ldots, z_n, z_1^{-1}, \ldots, z_n^{-1}, 1) \\ &= H_j(z_1, \ldots, z_n, z_1^{-1}, \ldots, z_n^{-1}, 1) - H_{j-2}(z_1, \ldots, z_n, z_1^{-1}, \ldots, z_n^{-1}, 1), \end{aligned} \tag{A.59}$$

where H_j is the jth complete symmetric polynomial in $2n+1$ variables.

Proposition A.60. *Given integers $\lambda_1 \ge \cdots \ge \lambda_n \ge 0$, we have*

$$\frac{|z_j^{\lambda_i+n-i+1/2} - z_j^{-(\lambda_i+n-i+1/2)}|}{|z_j^{n-i+1/2} - z_j^{-(n-i+1/2)}|} = |K_\lambda|,$$

where K_λ is the $n \times n$ matrix whose ith row is

$$(K_{\lambda_i-i+1} \quad K_{\lambda_i-i+2}+K_{\lambda_i-i} \quad \cdots \quad K_{\lambda_i-i+n}+K_{\lambda_i-i-n+2}).$$

Corollary A.46 gives three alternative expressions for this determinant, e.g.,

$$|K_\lambda| = |h_{\lambda_i-i+j} - h_{\lambda_i-i-j}|, \tag{A.61}$$

where $h_j = H_j(z_1, \ldots, z_n, z_1^{-1}, \ldots, z_n^{-1}, 1)$.

Exercise A.62. Calculate the denominator of the left-hand side:

$$|z_j^{n-i+1/2} - z_j^{-(n-i+1/2)}| = \Delta(\xi_1, \ldots, \xi_n)\cdot\zeta_1(\tfrac{1}{2})\cdot\ldots\cdot\zeta_n(\tfrac{1}{2}).$$

Proof of Proposition A.60. We have $\zeta_j(l) = z_j^l - z_j^{-l}$ and $\xi_j(l) = z_j^l + z_j^{-l}$ in $\mathbb{Q}(z_1^{1/2}, \ldots, z_n^{1/2})$ for l an integer or a half integer. First note that

$$\xi_j(\tfrac{1}{2})\cdot\zeta_j(l) = \zeta_j(l+\tfrac{1}{2}) + \zeta_j(l-\tfrac{1}{2}).$$

Multiplying the numerator and denominator of the left-hand side of the statement of the proposition by $\xi_1(\frac{1}{2})\cdot\ldots\cdot\xi_n(\frac{1}{2})$, the numerator becomes $|\zeta_j(\lambda_i+n-i+1)+\zeta_j(\lambda_i+n-i)|$, and the denominator becomes $|\zeta_j(n-i+1)+\zeta_j(n-i)| = |\zeta_j(n-i+1)|$. We can, therefore, apply Lemma A.54 to calculate the ratio, getting the determinant of a matrix whose entries are sums of certain J_j's. Note that by direct calculation $K_j = J_j + J_{j-1}$, so the terms can be combined, and the ratio is the determinant of the displayed matrix K_λ. □

Finally (for the even orthogonal groups O_{2n}), let $k = 2n$, and specialize the variables $x_1, \ldots, x_{2n}$ as above. Set

$$\begin{aligned} L_j &= H_j'(z_1, \ldots, z_n, z_1^{-1}, \ldots, z_n^{-1}) \\ &= H_j(z_1, \ldots, z_n, z_1^{-1}, \ldots, z_n^{-1}) - H_{j-2}(z_1, \ldots, z_n, z_1^{-1}, \ldots, z_n^{-1}), \end{aligned} \tag{A.63}$$

with H_j the complete symmetric polynomial in $2n$ variables.

Proposition A.64. *Given integers $\lambda_1 \geq \cdots \geq \lambda_n \geq 0$, we have*

$$\frac{|z_j^{\lambda_i+n-i} + z_j^{-(\lambda_i+n-i)}|}{|z_j^{n-i} + z_j^{-(n-i)}|} = \begin{cases} \frac{1}{2}|L_\lambda| & \text{if } \lambda_n > 0 \\ |L_\lambda| & \text{if } \lambda_n = 0, \end{cases}$$

where L_λ is the $n \times n$ matrix whose ith row is

$$(L_{\lambda_i-i+1} \quad L_{\lambda_i-i+2}+L_{\lambda_i-i} \quad \cdots \quad L_{\lambda_i-i+n}+L_{\lambda_i-i-n+2}).$$

As before, there are other expressions for these determinants, e.g.,

$$|L_\lambda| = |h_{\lambda_i-i+j} - h_{\lambda_i-i-j}|, \tag{A.65}$$

where $h_j = H_j(z_1, \ldots, z_n, z_1^{-1}, \ldots, z_n^{-1})$.

Exercise A.66. Calculate the denominator of the left-hand side:

$$|z_j^{n-i} + z_j^{-(n-i)}| = 2 \cdot \Delta(\xi_1, \ldots, \xi_n).$$

PROOF OF PROPOSITION A.64. Note that $\zeta_j \cdot \xi_j(l) = \xi_j(l+1) - \xi_j(l-1)$. Multiplying the numerator and denominator by $\zeta_1 \cdot \ldots \cdot \zeta_n$, the numerator becomes $|\zeta_j(\lambda_i + n - i + 1) - \zeta_j(\lambda_i + n - i - 1)|$ and the denominator becomes

$$|\zeta_j(n - i + 1) - \zeta_j(n - i - 1)| = 2|\zeta_j(n - i + 1)|;$$

this is seen by noting that the bottom row of the matrix on the left is $(\zeta_j(1) - \zeta_j(-1)) = (2\zeta_j(1))$, and performing row reductions starting from the bottom row. The rest of the proof is the same as in the preceding proposition. The only change is when $\lambda_n = 0$, in which case the bottom row in the numerator matrix is the same as that in the denominator. □

Exercise A.67*. Find a similar formula for

$$\frac{|z_j^{\lambda_i+n-i} - z_j^{-(\lambda_i+n-i)}|}{|z_j^{n-i} + z_j^{-(n-i)}|}.$$

APPENDIX B

On Multilinear Algebra

In this appendix we state the basic facts about ténsor products and exterior and symmetric powers that are used in the text. It is hoped that a reader with some linear algebra background can fill in details of the proofs.

§B.1. Tensor Products

The *tensor product* of two vector spaces V and W over a field is a vector space $V \otimes W$ equipped with a bilinear map

$$V \times W \to V \otimes W, \qquad v \times w \mapsto v \otimes w,$$

which is universal: for any bilinear map $\beta: V \times W \to U$ to a vector space U, there is a unique linear map from $V \otimes W$ to U that takes $v \otimes w$ to $\beta(v, w)$. This universal property determines the tensor product up to canonical isomorphism. If the ground field K needs to be mentioned, the tensor product is denoted $V \otimes_K W$.

If $\{e_i\}$ and $\{f_j\}$ are bases for V and W, the elements $\{e_i \otimes f_j\}$ form a basis for $V \otimes W$. This can be used to construct $V \otimes W$. The construction is functorial: linear maps $V \to V'$ and $W \to W'$ determine a linear map from $V \otimes W$ to $V' \otimes W'$.

Similarly one has the tensor product $V_1 \otimes \cdots \otimes V_n$ of n vector spaces, with its universal multilinear map

$$V_1 \times \cdots \times V_n \to V_1 \otimes \cdots \otimes V_n,$$

taking $v_1 \times \cdots \times v_n$ to $v_1 \otimes \cdots \otimes v_n$. (Recall that a map from the Cartesian product to a vector space U is *multilinear* if, when all but one of the factors V_i are fixed, the resulting map from V_i to U is linear.) The construction of tensor products is commutative:

$$V \otimes W \cong W \otimes V, \qquad v \otimes w \mapsto w \otimes v;$$

distributive:

$$(V_1 \oplus V_2) \otimes W \cong (V_1 \otimes W) \oplus (V_2 \otimes W);$$

and associative:

$$(U \otimes V) \otimes W \cong U \otimes (V \otimes W) \cong U \otimes V \otimes W,$$

by $(u \otimes v) \otimes w \mapsto u \otimes (v \otimes w) \mapsto u \otimes v \otimes w$.

In particular, there are *tensor powers* $V^{\otimes n} = V \otimes \cdots \otimes V$ of a fixed space V. By convention, $V^{\otimes 0}$ is the ground field.

If A is an algebra over the ground field, and V is a right A-module, and W a left A-module, there is a tensor product denoted $V \otimes_A W$, which can be constructed as the quotient of $V \otimes W$ by the subspace generated by all $(v \cdot a) \otimes w - v \otimes (a \cdot w)$ for all $v \in V$, $w \in W$, and $a \in A$. The resulting map from $V \times W$ to $V \otimes_A W$ is universal for bilinear maps β from $V \times W$ to vector spaces U that satisfy the property that $\beta(v \cdot a, w) = \beta(v, a \cdot w)$. This tensor product is also distributive.

§B.2. Exterior and Symmetric Powers

The *exterior powers* $\bigwedge^n V$ of a vector space V, sometimes denoted $\mathrm{Alt}^n V$, come equipped with an alternating multilinear map

$$V \times \cdots \times V \to \textstyle\bigwedge^n V, \qquad v_1 \times \cdots \times v_n \mapsto v_1 \wedge \cdots \wedge v_n,$$

that is universal: for $\beta\colon V \times \cdots \times V \to U$ an alternating multilinear map, there is a unique linear map from $\bigwedge^n V$ to U which takes $v_1 \wedge \cdots \wedge v_n$ to $\beta(v_1, \ldots, v_n)$. Recall that a multilinear map β is *alternating* if $\beta(v_1, \ldots, v_n) = 0$ whenever two of the vectors v_i are equal. This implies that $\beta(v_1, \ldots, v_n)$ changes sign when two of the vectors are interchanged.[1] It follows that

$$\beta(v_{\sigma(1)}, \ldots, v_{\sigma(n)}) = \mathrm{sgn}(\sigma)\beta(v_1, \ldots, v_n) \quad \text{for all } \sigma \in \mathfrak{S}_n.$$

The exterior power can be constructed as the quotient space of $V^{\otimes n}$ by the subspace generated by all $v_1 \otimes \cdots \otimes v_n$ with two of the vectors equal. We let

$$\pi\colon V^{\otimes n} \to \textstyle\bigwedge^n V, \qquad \pi(v_1 \otimes \cdots \otimes v_n) = v_1 \wedge \cdots \wedge v_n$$

[1] This follows from the standard polarization.: for two factors, $\beta(v + w, v + w) - \beta(v, v) - \beta(w, w) = \beta(v, w) + \beta(w, v)$.

denote the projection. If $\{e_i\}$ is a basis for V, then

$$\{e_{i_1} \wedge e_{i_2} \wedge \cdots \wedge e_{i_n} : i_1 < i_2 < \cdots < i_n\}$$

is a basis for $\bigwedge^n V$. Define $\bigwedge^0 V$ to be the ground field.

If V and W are vector spaces, there is a canonical linear map from $\bigwedge^a V \otimes \bigwedge^b W$ to $\bigwedge^{a+b}(V \oplus W)$, which takes $(v_1 \wedge \cdots \wedge v_a) \otimes (w_1 \wedge \cdots \wedge w_b)$ to $v_1 \wedge \cdots \wedge v_a \wedge w_1 \wedge \cdots \wedge w_b$. This determines an isomorphism

$$\bigwedge^n(V \oplus W) \cong \bigoplus_{a=0}^{n} \bigwedge^a V \otimes \bigwedge^{n-a} W. \tag{B.1}$$

(From this isomorphism the assertion about bases of $\bigwedge^n V$ follows by induction on the dimension.)

The *symmetric powers* $\mathrm{Sym}^n V$, sometimes denoted $S^n V$, comes with a universal symmetric multilinear map

$$V \times \cdots \times V \to \mathrm{Sym}^n V, \qquad v_1 \times \cdots \times v_n \mapsto v_1 \cdot \ldots \cdot v_n.$$

Recall that a multilinear map $\beta: V \times \cdots \times V \to U$ is *symmetric* if it is unchanged when any two factors are interchanged, or

$$\beta(v_{\sigma(1)}, \ldots, v_{\sigma(n)}) = \beta(v_1, \ldots, v_n) \quad \text{for all } \sigma \in \mathfrak{S}_n.$$

The symmetric power can be constructed as the quotient space of $V^{\otimes n}$ by the subspace generated by all $v_1 \otimes \cdots \otimes v_n - v_{\sigma(1)} \otimes \cdots \otimes v_{\sigma(n)}$, or by those in which σ permutes two successive factors. Again we let

$$\pi: V^{\otimes n} \to \mathrm{Sym}^n V, \qquad \pi(v_1 \otimes \cdots \otimes v_n) = v_1 \cdot \ldots \cdot v_n,$$

denote the projection. If $\{e_i\}$ is a basis for V, then

$$\{e_{i_1} \cdot e_{i_2} \cdot \ldots \cdot e_{i_n} : i_1 \le i_2 \le \cdots \le i_n\}$$

is a basis for $\mathrm{Sym}^n V$. So $\mathrm{Sym}^n V$ can be regarded as the space of homogeneous polynomials of degree n in the variables e_i. Define $\mathrm{Sym}^0 V$ to be the ground field. As before, there are canonical isomorphisms

$$\mathrm{Sym}^n(V \oplus W) \cong \bigoplus_{a=0}^{n} \mathrm{Sym}^a V \otimes \mathrm{Sym}^{n-a} W. \tag{B.2}$$

The exterior powers $\bigwedge^n V$ and symmetric powers $\mathrm{Sym}^n V$ can also be realized as subspaces of $V^{\otimes n}$, assuming, as we have throughout, that the ground field has characteristic 0. We will denote the inclusions by ι, so we have

$$V^{\otimes n} \overset{\pi}{\twoheadrightarrow} \bigwedge^n V \overset{\iota}{\rightarrowtail} V^{\otimes n}, \qquad V^{\otimes n} \overset{\pi}{\twoheadrightarrow} \mathrm{Sym}^n V \overset{\iota}{\rightarrowtail} V^{\otimes n}.$$

The imbedding $\iota: \bigwedge^n V \to V^{\otimes n}$ is defined by

$$\iota(v_1 \wedge \cdots \wedge v_n) = \sum_{\sigma \in \mathfrak{S}_n} \mathrm{sgn}(\sigma) v_{\sigma(1)} \otimes \cdots \otimes v_{\sigma(n)}. \tag{B.3}$$

(This is well defined since the right-hand side is alternating.) The image of ι is the space of anti-invariants of the right action of $\mathfrak{S}_n$ on $V^{\otimes n}$:

$$(v_1 \otimes \cdots \otimes v_n)\cdot \sigma = v_{\sigma(1)} \otimes \cdots \otimes v_{\sigma(n)}, \quad v_i \in V, \sigma \in \mathfrak{S}_n. \tag{B.4}$$

(The anti-invariants are the vectors $z \in V^{\otimes n}$ such that $z \cdot \sigma = \operatorname{sgn}(\sigma) z$ for all $\sigma \in \mathfrak{S}_n$.) Moreover, if $A = \iota \circ \pi$, then $(1/n!)A$ is the projection onto this anti-invariant subspace.[2] (Often the coefficient $1/n!$ is put in front of the formula for ι; this makes no essential difference, but leads to awkward formulas for contractions.)

Similarly we have $\iota\colon \operatorname{Sym}^n V \to V^{\otimes n}$ by

$$\iota(v_1 \cdot \ldots \cdot v_n) = \sum_{\sigma \in \mathfrak{S}_n} v_{\sigma(1)} \otimes \cdots \otimes v_{\sigma(n)}. \tag{B.5}$$

The image of ι is the space of invariants of the right action of $\mathfrak{S}_n$ on $V^{\otimes n}$. If $A = \iota \circ \pi$, then $(1/n!)A$ is the projection onto this invariant subspace.

The wedge product $\wedge$ determines a product

$$\textstyle\bigwedge^m V \otimes \bigwedge^n V \xrightarrow{\wedge} \bigwedge^{m+n} V, \tag{B.6}$$

$$(v_1 \wedge \cdots \wedge v_m) \otimes (v_{m+1} \wedge \cdots \wedge v_{m+n}) \mapsto v_1 \wedge \cdots \wedge v_m \wedge v_{m+1} \wedge \cdots \wedge v_{m+n},$$

which is associative and skew-commutative. This product is compatible with the projection from the tensor powers onto the exterior powers, but care must be taken for the inclusion of exterior in tensor powers, since for example $v \wedge w$ is sent to $v \otimes w - w \otimes v$ [not to $\frac{1}{2}(v \otimes w - w \otimes v)$] by ι. In general, the diagram

$$\begin{array}{ccc} \bigwedge^m V \otimes \bigwedge^n V & \xrightarrow{\ \wedge\ } & \bigwedge^{m+n} V \\ {\scriptstyle \iota\otimes\iota}\big\downarrow & & \big\downarrow{\scriptstyle \iota} \\ V^{\otimes m} \otimes V^{\otimes n} & \longrightarrow & V^{\otimes(m+n)} \end{array} \tag{B.7}$$

commutes when the bottom horizontal map is defined by the formula

$$\begin{aligned} &(v_1 \otimes \cdots \otimes v_m) \otimes (v_{m+1} \otimes \cdots \otimes v_{m+n}) \\ &\quad \mapsto \sum \operatorname{sgn}(\sigma) v_{\sigma(1)} \otimes \cdots \otimes v_{\sigma(m)} \otimes v_{\sigma(m+1)} \otimes \cdots \otimes v_{\sigma(m+n)}, \end{aligned} \tag{B.8}$$

the sum over all "shuffles," i.e., permutations σ of $\{1, \ldots, m+n\}$ that preserve the order of the subsets $\{1, \ldots, m\}$ and $\{m+1, \ldots, m+n\}$.

Similarly the symmetric powers have a commutative product $(v_1 \cdot \ldots \cdot v_m) \otimes (v_{m+1} \cdot \ldots \cdot v_{m+n}) \mapsto v_1 \cdot \ldots \cdot v_m \cdot v_{m+1} \cdot \ldots \cdot v_{m+n}$, with a similar compatibility. Note that $v^2 \in \operatorname{Sym}^2 V$ is sent to $2v \otimes v$ in $V \otimes V$, $v^n \in \operatorname{Sym}^n V$ to $n!(v \otimes \cdots \otimes v)$ in $V^{\otimes n}$, and generally one has the analogue of (B.7), changing each "sgn(σ)" to "1" in formula (B.8).

All these mappings are compatible with linear maps of vector spaces $V \to W$, and in particular commute with the left actions of the general linear group $GL(V) = \operatorname{Aut}(V)$ of automorphisms, or the algebra $\operatorname{End}(V) = \operatorname{Hom}(V, V)$ of endomorphisms, on $V^{\otimes n}$, $\bigwedge^n V$, and $\operatorname{Sym}^n V$.

[2] It is this factor which limits our present discussion to vector spaces over fields of characteristic 0.

It is sometimes convenient to make algebras out of the direct sum of all of the tensor, exterior, or symmetric powers. The *tensor algebra* $T^{\cdot}V$ is the sum $\bigoplus_{n\geq 0} V^{\otimes n}$, with product determined by the canonical isomorphism $V^{\otimes n} \otimes V^{\otimes m} \to V^{\otimes(n+m)}$. The *exterior algebra* $\wedge^{\cdot}V$ is the sum $\bigoplus_{n\geq 0} \wedge^n V$, which is the quotient of $T^{\cdot}V$ by the two-sided ideal generated by all $v \otimes v$ in $V^{\otimes 2}$. The *symmetric algebra* $\mathrm{Sym}^{\cdot}V$ is the sum $\bigoplus_{n\geq 0} \mathrm{Sym}^n V$, which is the quotient of $T^{\cdot}V$ by the two-sided ideal generated by all $v \otimes w - w \otimes v$ in $V^{\otimes 2}$.

Exercise B.9. The algebra $\mathrm{Sym}^{\cdot}V$ is a commutative, graded algebra, which satisfies the universal property that any linear map from V to the first graded piece C^1 of a commutative graded algebra $C^{\cdot}$ determines a homomorphism $\mathrm{Sym}^{\cdot}V \to C^{\cdot}$ of graded algebras. Use this to show that $\mathrm{Sym}^{\cdot}(V \oplus W) \cong \mathrm{Sym}^{\cdot}V \otimes \mathrm{Sym}^{\cdot}W$, and deduce the isomorphism (B.2). Prove the analogous assertions for $\wedge^{\cdot}V$, in the category of skew-commutative graded algebras. In particular, construct an isomorphism $\wedge^{\cdot}(V \oplus W) \cong \wedge^{\cdot}V \hat{\otimes} \wedge^{\cdot}W$, where $\hat{\otimes}$ denotes the skew-commutative tensor product: it is the usual tensor product additively, but the product has $(a \otimes b)\cdot(c \otimes d) = (-1)^{\deg(b)\deg(c)}(a \cdot b) \otimes (c \cdot d)$ for homogeneous elements a and c in the first algebra, and b and d in the second. In particular, this proves (B.1).

§B.3. Duals and Contractions

Although only a few simple contractions are used in the lectures, and most of these are written out by hand where needed, it may be useful to see the general picture.

If V^* denotes the dual space to V, there are contraction maps

$$c^i_j\colon V^{\otimes p} \otimes (V^*)^{\otimes q} \to V^{\otimes(p-1)} \otimes (V^*)^{\otimes(q-1)},$$

for any $1 \leq i \leq p$ and $1 \leq j \leq q$, determined by evaluating the jth coordinate of $(V^*)^{\otimes q}$ on the ith coordinate of $V^{\otimes p}$:

$$\begin{aligned} &c^i_j(v_1 \otimes \cdots \otimes v_p \otimes \varphi_1 \otimes \cdots \otimes \varphi_q) \\ &\quad = \varphi_j(v_i)v_1 \otimes \cdots \otimes \hat{v}_i \otimes \cdots \otimes v_p \otimes \varphi_1 \otimes \cdots \otimes \hat{\varphi}_j \otimes \cdots \otimes \varphi_q. \end{aligned} \tag{B.10}$$

More generally if $I = (i_1, \ldots, i_n)$ and $J = (j_1, \ldots, j_n)$ are two sequences of n distinct indices from $\{1, \ldots, p\}$ and $\{1, \ldots, q\}$, respectively, there is a contraction

$$c^I_J\colon V^{\otimes p} \otimes (V^*)^{\otimes q} \to V^{\otimes(p-n)} \otimes (V^*)^{\otimes(q-n)} \tag{B.11}$$

which takes $v_1 \wedge \cdots \wedge v_p \otimes \varphi_1 \otimes \cdots \otimes \varphi_q$ to

$$\prod_{\alpha=1}^{n} \varphi_{j_\alpha}(v_{i_\alpha})v_1 \otimes \cdots \otimes \hat{v}_{i_1} \otimes \cdots \otimes \hat{v}_{i_2} \otimes \cdots \otimes v_p \otimes \varphi_1 \otimes \cdots \otimes \hat{\varphi}_{j_1} \otimes \cdots \otimes \varphi_q.$$

For example, if $p = q = n$ and $I = J = (1, \ldots, n)$, this contraction $V^{\otimes n} \otimes (V^*)^{\otimes n} \to \mathbb{C}$ identifies $(V^*)^{\otimes n}$ with the dual space of $V^{\otimes n}$.

Now $(V^{\otimes n})^*$ consists of n-multilinear forms on V, and $(\bigwedge^n V)^*$ consists of alternating n multilinear forms on V; in particular, $(\bigwedge^n V)^*$ is a subspace of $(V^{\otimes n})^*$; this is the inclusion via π^*. The composite

$$\bigwedge^n(V^*) \to (V^*)^{\otimes n} \to (V^{\otimes n})^*,$$

where the first map is the inclusion $\imath$ and the second is the isomorphism of the preceding paragraph, maps $\bigwedge^n(V^*)$ isomorphically onto the subspace $(\bigwedge^n V)^*$. Explicitly,

$$\bigwedge^n(V^*) \xrightarrow{\cong} (\bigwedge^n V)^*,$$

$$\varphi_1 \wedge \cdots \wedge \varphi_n \mapsto [v_1 \wedge \cdots \wedge v_n \mapsto \sum \operatorname{sgn}(\sigma)\varphi_{\sigma(1)}(v_1) \cdot \ldots \cdot \varphi_{\sigma(n)}(v_n) = \det(\varphi_j(v_i))].$$

This dual pairing $\bigwedge^n V \otimes \bigwedge^n(V^*) \to K$ is often denoted $\langle\ ,\ \rangle$.

There is a similar isomorphism of $\operatorname{Sym}^n(V^*)$ with $\operatorname{Sym}^n(V)^*$, but without the signs "sgn(σ)."

Exercise B.12. If $e_1, \ldots, e_m$ is a basis for V, with e_i^* the dual basis for V^*, then $\{e_{i_1} \wedge \cdots \wedge e_{i_n}: 1 \le i_1 < \cdots < i_n \le m\}$ is a basis for $\bigwedge^n V$, and $\{e_1^{i_1} \cdot \ldots \cdot e_m^{i_m}: i_\alpha \ge 0, \sum i_\alpha = n\}$ is a basis for $\operatorname{Sym}^n V$. Show that, via the above isomorphisms, the dual bases for $\bigwedge^n(V^*)$ and $\operatorname{Sym}^n(V^*)$ are

$$\{e_{i_1}^* \wedge \cdots \wedge e_{i_n}^*\} \quad \text{and} \quad \left\{\frac{1}{\prod_\alpha (i_\alpha!)}(e_1^*)^{i_1} \cdot \ldots \cdot (e_m^*)^{i_m}\right\}.$$

There are related contractions, sometimes called internal products, and denoted $\lrcorner$ and $\llcorner$, on exterior and symmetric powers. For the exterior powers they are maps:

$$\begin{aligned} &\bigwedge^p V \otimes \bigwedge^{p+q}(V^*) \to \bigwedge^q(V^*), \qquad x \otimes \alpha \mapsto x \lrcorner \alpha; \\ &\bigwedge^{p+q} V \otimes \bigwedge^p(V^*) \to \bigwedge^q(V), \qquad x \otimes \alpha \mapsto x \llcorner \alpha. \end{aligned} \tag{B.13}$$

These can be defined most simply as transposes of wedge products, i.e., they are determined by the identities

$$\langle z, x \lrcorner \alpha\rangle = \langle z \wedge x, \alpha\rangle \quad \text{for } z \in \bigwedge^q V$$

and

$$\langle x \llcorner \alpha, \beta\rangle = \langle x, \alpha \wedge \beta\rangle \quad \text{for } \beta \in \bigwedge^q(V^*).$$

(The relation of this definition to the contraction maps c_J^I above is expressed in Exercise. B.16.) Note that when $q = 0$, these contractions reduce to the previous duality pairing between $\bigwedge^p V$ and $\bigwedge^p(V^*)$.

For symmetric powers, the internal products are defined similarly:

$$\begin{aligned} &\operatorname{Sym}^p V \otimes \operatorname{Sym}^{p+q}(V^*) \to \operatorname{Sym}^q(V^*), \qquad x \otimes \alpha \mapsto x \lrcorner \alpha; \\ &\operatorname{Sym}^{p+q} V \otimes \operatorname{Sym}^p(V^*) \to \operatorname{Sym}^q(V), \qquad x \otimes \alpha \mapsto x \llcorner \alpha. \end{aligned} \tag{B.14}$$

Exercise B.15. For $v, w \in V$, and $\varphi, \psi \in V^*$, show that

$$v \lrcorner (\varphi \wedge \psi) = \psi(v)\varphi - \varphi(v)\psi \quad \text{and} \quad (v \wedge w) \llcorner \varphi = \varphi(v)w - \varphi(w)v.$$

More generally, for if $x = v_1 \wedge \cdots \wedge v_p$ and $\alpha = \varphi_1 \wedge \cdots \wedge \varphi_{p+q}$, with $v_i \in V$ and $\varphi_j \in V^*$, then

(i) $$x \lrcorner \alpha = \sum \operatorname{sgn}(\sigma)\varphi_{\sigma(q+1)}(v_1)\cdot \ldots \cdot \varphi_{\sigma(q+p)}(v_p)\cdot \varphi_{\sigma(1)} \wedge \cdots \wedge \varphi_{\sigma(q)},$$

the sum over all permutations σ of $\{1, \ldots, p+q\}$ that preserve the order of $\{1, \ldots, q\}$. If $x = v_1 \wedge \cdots \wedge v_{p+q}$ and $\alpha = \varphi_1 \wedge \cdots \wedge \varphi_p$, then

(ii) $$x \llcorner \alpha = \sum \operatorname{sgn}(\sigma)\varphi_1(v_{\sigma(1)})\cdot \ldots \cdot \varphi_p(v_{\sigma(p)})\cdot v_{\sigma(p+1)} \wedge \cdots \wedge v_{\sigma(p+q)},$$

the sum over all permutations that preserve the order of $\{p+1, \ldots, p+q\}$. Verify these formulas and use them to give formulas for these internal products in terms of standard bases. State and verify analogous formulas for symmetric powers. In particular, for $v, w \in V$, $\varphi, \psi \in V^*$,

$$v \lrcorner (\varphi \cdot \psi) = \psi(v)\varphi + \varphi(v)\psi \quad \text{and} \quad (v \cdot w) \llcorner \varphi = \varphi(v)w + \varphi(w)v.$$

For example, $v \lrcorner (\varphi^2) = 2\varphi(v)\varphi$ and $(v^2) \llcorner \varphi = 2\varphi(v)v$.

Exercise B.16. Using formula (ii) of the preceding exercise, show that the contraction map $\llcorner$ may be given as $1/p!q!$ times the composition of the maps

$$\textstyle\bigwedge^{p+q} V \otimes \bigwedge^p (V^*) \to V^{\otimes(p+q)} \otimes (V^*)^{\otimes p} \to V^{\otimes q} \to \bigwedge^q V,$$

where the middle map is the contraction map c_J^I of (B.11), with $I = J = \{1, \ldots, p\}$, and the other maps come from ι and π. Prove the same formulas (with the same scalar factor) for the other internal products.

Exercise B.17. In the situation of formula (ii), suppose the v_i are independent, and let W be the $(p+q)$-dimensional subspace of V that they span; suppose the φ_i are independent, and let Z be the p-codimensional subspace of V of the common zeros of the φ_i. Show that $x \llcorner \alpha = 0$ if $\dim(W \cap Z) > q$, and otherwise $x \llcorner \alpha = u_1 \wedge \cdots \wedge u_q$ for some vectors u_i that span $W \cap Z$.

Exercise B.18. Prove the formulas

$$(x \wedge y) \lrcorner \alpha = x \lrcorner (y \lrcorner \alpha) \quad \text{and} \quad x \llcorner (\alpha \wedge \beta) = (x \llcorner \alpha) \llcorner \beta.$$

State and verify the analogous formulas for symmetric powers.

For a detailed development of these ideas, see [Bour, *Algebra*, Chap. 3].

APPENDIX C
On Semisimplicity

§C.1. The Killing Form and Cartan's Criterion

We recall first the Jordan decomposition of a linear transformation X of a finite-dimensional complex vector space V as a sum of its semisimple and nilpotent parts: $X = X_s + X_n$, where X_s is the semisimple part of X, and X_n the nilpotent part. It is uniquely characterized by the fact that X_s is semisimple (diagonalizable), X_n is nilpotent, and X_s and X_n commute with each other. In fact, X_s and X_n can be written as polynomials in X, so any endomorphism that commutes with X automatically commutes with X_s and X_n. One case of the invariance of Jordan decomposition is an easy calculation:

Exercise C.1*. For any $X \in \mathfrak{gl}(V)$, the endomorphism $\operatorname{ad}(X)$ of $\mathfrak{gl}(V)$ satisfies

$$\operatorname{ad}(X)_s = \operatorname{ad}(X_s) \quad \text{and} \quad \operatorname{ad}(X)_n = \operatorname{ad}(X_n).$$

There is a Killing form B_V defined on $\mathfrak{gl}(V)$ by the formula

$$B_V(X, Y) = \operatorname{Tr}(X \circ Y), \tag{C.2}$$

where Tr is the trace and $\circ$ denotes composition of transformations. As in (14.23), the identity

$$B_V(X, [Y, Z]) = B_V([X, Y], Z) \tag{C.3}$$

holds for all X, Y, Z in $\mathfrak{gl}(V)$.

The Killing form B on a Lie algebra $\mathfrak{g}$ is that of Exercise C.1 for the adjoint representation: $B(X, Y) = B_{\mathfrak{g}}(\mathrm{ad}(X), \mathrm{ad}(Y))$. This was introduced in Lecture 14, where a few of its properties were proved. Here we use the Killing form to characterize solvability and semisimplicity of the Lie algebra.

If $\mathfrak{g}$ is solvable, by Lie's theorem its adjoint representation can be put in upper-triangular form. It follows that $\mathscr{D}\mathfrak{g} = [\mathfrak{g}, \mathfrak{g}]$ acts by strictly upper-triangular matrices. So if X is in $\mathscr{D}\mathfrak{g}$ and Y in $\mathfrak{g}$, then $\mathrm{ad}(X) \circ \mathrm{ad}(Y)$ is strictly upper triangular; in particular its trace $B(X, Y)$ is zero. Cartan's criterion is that this characterizes solvability:

Proposition C.4. *The Lie algebra* $\mathfrak{g}$ *is solvable if and only if* $B(\mathfrak{g}, \mathscr{D}\mathfrak{g}) = 0$.

We will prove something that looks a little weaker, but will turn out to be a little stronger. We prove:

Theorem C.5 (Cartan's criterion). *If* $\mathfrak{g}$ *is a subalgebra of* $\mathfrak{gl}(V)$ *and* $B_V(X, Y) = 0$ *for all* X *and* Y *in* $\mathfrak{g}$, *then* $\mathfrak{g}$ *is solvable.*

For this, it suffices to show that every element of $\mathscr{D}\mathfrak{g}$ is nilpotent, for then by Engel's theorem $\mathscr{D}\mathfrak{g}$ must be a nilpotent ideal, and therefore $\mathfrak{g}$ is solvable.

So take $X \in \mathscr{D}\mathfrak{g}$, and let $\lambda_1, \ldots, \lambda_r$ be its eigenvalues (counted with multiplicity) for X as an endomorphism of V. We must show the λ_i are all zero. These eigenvalues satisfy some obvious relations; for example, $\sum \lambda_i \lambda_i = \mathrm{Tr}(X \circ X)) = B_V(X, X) = 0$. What we need to show is

$$\bar{\lambda}_1 \lambda_1 + \cdots + \bar{\lambda}_r \lambda_r = 0. \tag{C.6}$$

To prove this, take a basis for V so that X is in Jordan canonical form, with $\lambda_1, \ldots, \lambda_r$ down the diagonal; the semisimple part $D = X_s$ of X is this diagonal transformation. Let $\bar{D}$ be the endomorphism of V given by the diagonal matrix with $\bar{\lambda}_1, \ldots, \bar{\lambda}_r$ down the diagonal. Since $\mathrm{Tr}(\bar{D} \circ X) = \sum \bar{\lambda}_i \lambda_i$, it suffices to prove

$$\mathrm{Tr}(\bar{D} \circ X) = 0. \tag{C.7}$$

Since X is a sum of commutators $[Y, Z]$, with Y and Z in $\mathfrak{g}$, $\mathrm{Tr}(\bar{D} \circ X)$ is a sum of terms of the form $\mathrm{Tr}(\bar{D} \circ [Y, Z]) = \mathrm{Tr}([\bar{D}, Y] \circ Z)$. So we will be done if we know that $[\bar{D}, Y]$ belongs to $\mathfrak{g}$, for our hypothesis is that $\mathrm{Tr}(\mathfrak{g} \circ \mathfrak{g}) \equiv 0$. That is, we are reduced to showing

$$\mathrm{ad}(\bar{D})(\mathfrak{g}) \subset \mathfrak{g}. \tag{C.8}$$

For this it suffices to prove that $\mathrm{ad}(\bar{D})$ can be written as a polynomial in $\mathrm{ad}(X)$, for we know that $\mathrm{ad}(X)^k(Y)$ is in $\mathfrak{g}$ if X and Y are in $\mathfrak{g}$. Since $\mathrm{ad}(D) = \mathrm{ad}(X_s) = \mathrm{ad}(X)_s$ is a polynomial in $\mathrm{ad}(X)$, it suffices to show that $\mathrm{ad}(\bar{D})$ can be written as a polynomial in $\mathrm{ad}(D)$. This is a simple computation: using the usual basis $\{E_{ij}\}$ for $\mathfrak{gl}(V)$, $\mathrm{ad}(D)$ and $\mathrm{ad}(\bar{D})$ are complex conjugate diagonal matrices, and any such are polynomials in each other. □

We can prove now that if $\mathfrak{g}$ is a Lie algebra for which $B(\mathscr{D}\mathfrak{g}, \mathscr{D}\mathfrak{g}) \equiv 0$, then $\mathfrak{g}$ is solvable, which certainly implies Proposition C.4. By what we just proved, the image of $\mathscr{D}\mathfrak{g}$ by the adjoint representation in $\mathfrak{gl}(\mathfrak{g})$ is solvable. Since the kernel of the adjoint map is abelian, this makes $\mathscr{D}\mathfrak{g}$ solvable (cf. Exercise 9.8), and by definition this makes $\mathfrak{g}$ solvable. □

Exercise C.9. Show that a Lie algebra $\mathfrak{g}$ is solvable if and only if $B(\mathrm{ad}(X), \mathrm{ad}(X)) = 0$ for all X in $\mathfrak{g}$.

It is easy to deduce from Cartan's criterion a criterion for semisimplicity—part of which we saw in Lecture 14, but there assuming some facts we had not proved yet:

Proposition C.10. *A Lie algebra $\mathfrak{g}$ is semisimple if and only if its Killing form B is nondegenerate.*

PROOF. By (C.3) the null-space $\mathfrak{s} = \{X \in \mathfrak{g}: B(X, Y) = 0 \text{ for all } Y \in \mathfrak{g}\}$ is an ideal. Suppose $\mathfrak{g}$ is semisimple. By Cartan's criterion, the image $\mathrm{ad}(\mathfrak{s}) \subset \mathfrak{gl}(\mathfrak{g})$ is solvable; as in the preceding proof, $\mathfrak{s}$ is then solvable, so $\mathfrak{s} = 0$ by the definition of semisimple. Conversely, if B is nondegenerate, we must show that any abelian ideal $\mathfrak{a}$ in $\mathfrak{g}$ must be zero. If $X \in \mathfrak{a}$ and $Y \in \mathfrak{g}$, then $A = \mathrm{ad}(X) \circ \mathrm{ad}(Y)$ maps $\mathfrak{g}$ into $\mathfrak{a}$ and $\mathfrak{a}$ to 0, so $\mathrm{Tr}(A) = 0$. This implies that $\mathfrak{a} \subset \mathfrak{s} = 0$, as required. □

Corollary C.11. *A semisimple Lie algebra is a direct product of simple Lie algebras.*

PROOF. For any ideal $\mathfrak{h}$ of $\mathfrak{g}$, the annihilator

$$\mathfrak{h}^\perp = \{X \in \mathfrak{g}: B(X, Y) = 0 \text{ for all } Y \in \mathfrak{h}\}$$

is an ideal, by (C.3) again. By Cartan's criterion, $\mathfrak{h} \cap \mathfrak{h}^\perp$ is solvable, hence zero, so $\mathfrak{g} = \mathfrak{h} \oplus \mathfrak{h}^\perp$. The decomposition follows by a simple induction. □

It follows that $\mathfrak{g} = \mathscr{D}\mathfrak{g}$, and that all ideals and images of $\mathfrak{g}$ are semisimple. In fact:

Exercise C.12*. Show that if $\mathfrak{g}$ is a direct product of simple Lie algebras, the only ideals in $\mathfrak{g}$ are sums of some of the factors. In particular, the decomposition into simple factors is unique (not just up to isomorphism).

Exercise C.13*. Show that if $\mathfrak{g}$ is semisimple, the adjoint map $\mathrm{ad}: \mathfrak{g} \to \mathfrak{gl}(\mathfrak{g})$ is an isomorphism of $\mathfrak{g}$ onto the algebra $\mathrm{Der}(\mathfrak{g})$ of derivations of $\mathfrak{g}$.

Exercise C.14. Show that if $\mathfrak{g}$ is nilpotent then its Killing form is identically zero, and find a counterexample to the converse.

§C.2. Complete Reducibility and the Jordan Decomposition

We repeat that this section is optional, since the results can be deduced from the existence of a compact group such that the complexification of its Lie algebra is a given semisimple Lie algebra. We include here the standard algebraic approach. A finite-dimensional representation of a Lie algebra $\mathfrak{g}$ will be called a $\mathfrak{g}$-module, and a $\mathfrak{g}$-invariant subspace a submodule.

Proposition C.15. *Let V be a representation of the semisimple Lie algebra $\mathfrak{g}$ and $W \subset V$ a submodule. Then there exists a submodule $W' \subset V$ complementary to W.*

PROOF. Since the image of $\mathfrak{g}$ by the representation is semisimple, we may assume $\mathfrak{g} \subset \mathfrak{gl}(V)$. We will require a slight generalization of the Casimir operator $C_V \in \operatorname{End}(V)$ which was used in §25.1 in the proof of Freudenthal's formula. We take a basis $U_1, \ldots, U_r$ for $\mathfrak{g}$, and a dual basis $U_1', \ldots, U_r'$, but this time with respect to the Killing form B_V defined in Exercise C.1: $B_V(X, Y) = \operatorname{Tr}(X \circ Y)$. (Note by Cartan's criterion that B_V is nondegenerate.) Then C_V is defined by the formula $C_V(v) = \sum U_i \cdot (U_i' \cdot v)$.

As before, a simple calculation shows that C_V is an endomorphism of V that commutes with the action of $\mathfrak{g}$. Its trace is

$$\operatorname{Tr}(C_V) = \sum \operatorname{Tr}(U_i \circ U_i') = \sum B_V(U_i, U_i') = \dim(\mathfrak{g}). \tag{C.16}$$

We note also that since C_V maps any submodule W to itself, and since it commutes with $\mathfrak{g}$, its kernel $\operatorname{Ker}(C_V)$ and image are submodules.

Note first that all one-dimensional representations of a semisimple $\mathfrak{g}$ are trivial, since $\mathscr{D}\mathfrak{g}$ must act trivially on a one-dimensional representation, and $\mathfrak{g} = \mathscr{D}\mathfrak{g}$.

We proceed to the proof itself. As should be familiar from Lecture 9, the basic case to prove is when $W \subset V$ is an irreducible invariant subspace of codimension one. Then C_V maps W into itself, and C_V acts trivially on V/W. But now by Schur's lemma, since W is irreducible, C_V is multiplication by a scalar on W. This scalar is not zero, or (C.16) would be contradicted. Hence $V = W \oplus \operatorname{Ker}(C_V)$, which finishes this special case.

It follows easily by induction on the dimension that the same is true whenever $W \subset V$ has codimension one. For if W is not irreducible, let Z be a nonzero submodule, and find a complement to $W/Z \subset V/Z$ (by induction), say Y/Z. Since Y/Z is one dimensional, find (by induction) U so that $Y = Z \oplus U$. Then $V = W \oplus U$.

By the same argument, it suffices to prove the statement of the theorem when W is irreducible. Consider the restriction map

$$\rho\colon \operatorname{Hom}(V, W) \to \operatorname{Hom}(W, W),$$

a homomorphism of $\mathfrak{g}$-modules. The second contains the one-dimensional submodule $\mathrm{Hom}_{\mathfrak{g}}(W, W)$. By the preceding case, there is a one-dimensional submodule of $\rho^{-1}(\mathrm{Hom}_{\mathfrak{g}}(W, W)) \subset \mathrm{Hom}(V, W)$ which maps onto $\mathrm{Hom}_{\mathfrak{g}}(W, W)$ by ρ. Since one-dimensional modules are trivial, this means there is a $\mathfrak{g}$-invariant ψ in $\mathrm{Hom}(V, W)$ such that $\rho(\psi) = 1$. But this means that ψ is a $\mathfrak{g}$-invariant projection of V onto W, so $V = W \oplus \mathrm{Ker}(\psi)$, as required. □

We will apply this to prove the invariance of Jordan decomposition (Theorem 9.20). The essential point is:

Proposition C.17. *Let $\mathfrak{g}$ be a semisimple Lie subalgebra of $\mathfrak{gl}(V)$. Then for any element $X \in \mathfrak{g}$, the semisimple part X_s and the nilpotent part X_n are also in $\mathfrak{g}$.*

PROOF. The idea is to write $\mathfrak{g}$ as an intersection of Lie subalgebras of $\mathfrak{gl}(V)$ for which the conclusion of the theorem is easy to prove. For example, we know $\mathfrak{g} \subset \mathfrak{sl}(V)$ since $\mathfrak{g} = \mathscr{D}\mathfrak{g}$, and clearly X_s and X_n are traceless if X is. Similarly, if V is not irreducible, for any submodule W of V, let

$$\mathfrak{s}_W = \{Y \in \mathfrak{gl}(V)\colon Y(W) \subset W \text{ and } \mathrm{Tr}(Y|_W) = 0\}.$$

Then $\mathfrak{g}$ is also a subalgebra of $\mathfrak{s}_W$, and X_s and X_n are also in $\mathfrak{s}_W$.

Since $[X, \mathfrak{g}] \subset \mathfrak{g}$, it follows that $[p(X), \mathfrak{g}] \subset \mathfrak{g}$ for any polynomial $p(T)$. Hence $[X_s, \mathfrak{g}] \subset \mathfrak{g}$ and $[X_n, \mathfrak{g}] \subset \mathfrak{g}$. In other words, X_s and X_n belong to the Lie subalgebra $\mathfrak{n}$ of $\mathfrak{gl}(V)$ consisting of those endomorphisms A such that $[A, \mathfrak{g}] \subset \mathfrak{g}$. So $\mathfrak{n}$ gives us another subalgebra to work with. Now we claim that $\mathfrak{g}$ is the intersection of $\mathfrak{n}$ and all the algebras $\mathfrak{s}_W$ for all submodules W of V. This claim, as we saw, will finish the proof. Let $\mathfrak{g}'$ be the intersection of all these Lie algebras. Then $\mathfrak{g}$ is an ideal in $\mathfrak{g}'$ since $\mathfrak{g}' \subset \mathfrak{n}$.

By the complete reducibility theorem we can find a submodule U of $\mathfrak{g}'$ so that $\mathfrak{g}' = \mathfrak{g} \oplus U$. Since $[\mathfrak{g}, \mathfrak{g}'] \subset \mathfrak{g}$, we must have $[\mathfrak{g}, U] = 0$. To show that U is 0, it suffices to show that for any $Y \in U$ its restriction to any irreducible submodule W of V is zero (noting that Y preserves W since $Y \in \mathfrak{s}_W$, and that V is a sum of irreducible submodules). But since Y commutes with $\mathfrak{g}$, Schur's lemma implies that the restriction of Y to W is multiplication by a scalar, and the assumption that $Y \in \mathfrak{s}_W$ means that $\mathrm{Tr}(Y|_W) = 0$, so $Y|_W = 0$, as required. □

Now if $\mathfrak{g}$ is a semisimple algebra, the adjoint representation ad embeds $\mathfrak{g}$ in $\mathfrak{gl}(\mathfrak{g})$. For any X in $\mathfrak{g}$ the theorem implies that the semisimple and nilpotent parts of $\mathrm{ad}(X)$ are in $\mathfrak{g}$. We write these X_s and X_n. The decomposition $X = X_s + X_n$ may be called the *absolute* Jordan decomposition. Note that $[X_s, X_n] = 0$. It follows easily from the definition that if $\rho\colon \mathfrak{g} \to \mathfrak{g}'$ is a homomorphism from one semisimple Lie algebra onto another, then $\rho(X_s) = \rho(X)_s$ and $\rho(X_n) = \rho(X)_n$. (This follows for example from the fact that $\mathfrak{g}'$ is obtained from $\mathfrak{g}$ by factoring out some of its simple ideals.) In fact, the absolute decomposition determines all others:

Corollary C.18. *If $\rho: \mathfrak{g} \to \mathfrak{gl}(V)$ is any representation of a semisimple Lie algebra $\mathfrak{g}$, then $\rho(X_s)$ is the semisimple part of $\rho(X)$ and $\rho(X_n)$ is the nilpotent part of $\rho(X)$.*

PROOF. We just saw that $\rho(X_s)$ and $\rho(X_n)$ are the semisimple and nilpotent parts of $\rho(X)$ as regarded in the semisimple Lie algebra $\mathfrak{g}' = \rho(\mathfrak{g})$. Apply the theorem to $\mathfrak{g}' \subset \mathfrak{gl}(V)$. □

It follows that an element X in a semisimple Lie algebra that is semisimple in one faithful representation is semisimple in all representations.

§C.3. On Derivations

In this final section we collect a few facts relating the Killing form, solvability, and nilpotency with derivations of Lie algebras, mainly for use in Appendix E. We first prove a couple of lemmas related to the Lie–Engel theory of Lecture 9. For these $\mathfrak{g}$ is any Lie algebra, $\mathfrak{r} = \mathrm{Rad}(\mathfrak{g})$ denotes its radical, and $\mathscr{D}\mathfrak{g} = [\mathfrak{g}, \mathfrak{g}]$.

Lemma C.19. *For any representation $\rho: \mathfrak{g} \to \mathfrak{gl}(V)$, every element of $\rho(\mathscr{D}\mathfrak{g} \cap \mathfrak{r})$ is a nilpotent endomorphism.*

PROOF. It suffices to treat the case where the representation V is irreducible, for if W were a proper subrepresentation, we would know the result by induction on the dimension for W and V/W, which implies it for V. We may replace $\mathfrak{g}$ by its image, so we may assume ρ is injective. In this case we show that $\mathscr{D}\mathfrak{g} \cap \mathfrak{r} = 0$. We may assume $\mathfrak{r} \neq 0$. Consider the largest integer k such that $\mathfrak{a} = \mathscr{D}^k\mathfrak{r}$ is not zero. This $\mathfrak{a}$ is an abelian ideal of $\mathfrak{g}$. It suffices to show that $\mathscr{D}\mathfrak{g} \cap \mathfrak{a} = 0$, for if $k > 0$, then $\mathfrak{a} \subset \mathscr{D}\mathfrak{g}$.

We need three facts:

(i) If $\mathfrak{g} \subset \mathfrak{gl}(V)$ is an irreducible representation and $\mathfrak{b}$ is any ideal of $\mathfrak{g}$ that consists of nilpotent transformations of V, then $\mathfrak{b} = 0$. (Indeed, by Engel's theorem,

$$W = \{v \in V: X(v) = 0 \text{ for all } X \in \mathfrak{b}\}$$

is nonzero, and by Lemma 9.13, W is preserved by $\mathfrak{g}$. Since V is irreducible, $W = V$, which says that $\mathfrak{b} = 0$.)

(ii) A transformation X is nilpotent exactly when $\mathrm{Tr}(X^n) = 0$ for all positive integers n. (This is seen by writing X in Jordan canonical form.)

(iii) $\mathrm{Tr}([X, Y] \cdot Z) = 0$ whenever $[Y, Z] = 0$. (This follows from the identity (C.3): $\mathrm{Tr}([X, Y] \cdot Z) = \mathrm{Tr}(X \cdot [Y, Z])$.)

Next we can see that $[\mathfrak{g}, \mathfrak{a}] = 0$. For if $X \in \mathfrak{g}$ and $Y \in \mathfrak{a}$, then $[X, Y] \in \mathfrak{a}$; since $\mathfrak{a}$ is abelian, Y commutes with $[X, Y]$ and hence with powers of $[X, Y]$.

Applying (iii) with $Z = [X, Y]^{n-1}$ gives $\mathrm{Tr}([X, Y]^n) = 0$ for $n > 0$, and (ii) and (i) imply that $[\mathfrak{g}, \mathfrak{a}] = 0$.

Finally we show that $\mathscr{D}\mathfrak{g} \cap \mathfrak{a} = 0$. If $X, Y \in \mathfrak{g}$ and $[X, Y] \in \mathfrak{a}$, then $[Y, [X, Y]] = 0$ by the preceding step, so again Y commutes with powers of $[X, Y]$, and the same argument shows that $\mathrm{Tr}([X, Y]^n) = 0$, and (ii) and (i) again show that $\mathscr{D}\mathfrak{g} \cap \mathfrak{a} = 0$. □

Lemma C.20. *For any Lie algebra* $\mathfrak{g}$, $[\mathfrak{g}, \mathfrak{r}]$ *is nilpotent.*

Proof. Look at the images $\bar{\mathfrak{g}}$ and $\bar{\mathfrak{r}}$ of $\mathfrak{g}$ and $\mathfrak{r}$ by the adjoint representation ad: $\mathfrak{g} \to \mathfrak{gl}(\mathfrak{g})$. By Lemma C.19 and Engel's theorem, $[\bar{\mathfrak{g}}, \bar{\mathfrak{r}}]$ is a nilpotent ideal of $\bar{\mathfrak{g}}$. Since the kernel of the adjoint representation is the center of $\mathfrak{g}$, it follows that the quotient of $[\mathfrak{g}, \mathfrak{r}]$ by a central ideal is nilpotent, which implies that $[\mathfrak{g}, \mathfrak{r}]$ itself is nilpotent. □

An ideal $\mathfrak{a}$ of a Lie algebra $\mathfrak{g}$ is called *characteristic* if any derivation of $\mathfrak{g}$ maps $\mathfrak{a}$ into itself. Note that an ideal is just a subspace that is preserved by all inner derivations $D_X = \mathrm{ad}(X)$. It follows from the definitions that if $\mathfrak{a}$ is any ideal in $\mathfrak{g}$, then any characteristic ideal in $\mathfrak{a}$ is automatically an ideal in $\mathfrak{g}$.

The following simple construction is useful for turning questions about general derivations into questions about inner derivations. Given any Lie algebra $\mathfrak{g}$ and a derivation D of $\mathfrak{g}$, let $\mathfrak{g}' = \mathfrak{g} \oplus \mathbb{C}$, and define a bracket on $\mathfrak{g}'$ by

$$[(X, \lambda), (Y, \mu)] = ([X, Y] + \lambda D(Y) - \mu D(X), 0).$$

It is easy to verify that $\mathfrak{g}'$ is a Lie algebra containing $\mathfrak{g} = \mathfrak{g} \oplus 0$ as an ideal, and that, setting $\xi = (0, 1)$, the restriction of $D_\xi = \mathrm{ad}(\xi)$ to $\mathfrak{g}$ is the given derivation D.

As a simple application of this construction, if B is the Killing form on $\mathfrak{g}$, we have the identity

$$B(D(X), Y) + B(X, D(Y)) = 0 \tag{C.21}$$

for any derivation D of $\mathfrak{g}$, and any X and Y in $\mathfrak{g}$. Indeed, if B' is the Killing form on $\mathfrak{g}'$, (C.3) gives $B'([\xi, X], Y) + B'(X, [\xi, Y]) = 0$; since $\mathfrak{g}$ is an ideal in $\mathfrak{g}'$, B is the restriction of B' to $\mathfrak{g}$, and (C.21) follows.

From (C.21) it follows that if $\mathfrak{a}$ is a characteristic ideal of $\mathfrak{g}$, then its orthogonal complement with respect to the Killing form is also a characteristic ideal of $\mathfrak{g}$.

Proposition C.22. *For any Lie algebra* $\mathfrak{g}$, $\mathrm{Rad}(\mathfrak{g})$ *is the orthogonal complement to* $\mathscr{D}\mathfrak{g}$ *with respect to the Killing form.*

Proof. To see that $\mathfrak{r} = \mathrm{Rad}(\mathfrak{g})$ is contained in $\mathscr{D}\mathfrak{g}^\perp$, i.e., that $\mathscr{D}\mathfrak{g}$ is perpendicular to $\mathfrak{r}$, let $X, Y \in \mathfrak{g}$ and $Z \in \mathfrak{r}$. Recalling that $B([X, Y], Z) = B(X, [Y, Z])$, it suffices to show that $B(X, [Y, Z]) = 0$. Let $\mathfrak{h}$ be the subalgebra of $\mathfrak{g}$ generated by $\mathfrak{r}$ and X. Then $[\mathfrak{h}, \mathfrak{h}] \subset \mathfrak{r}$, so $\mathfrak{h}$ is solvable, so by Lie's theorem, under the adjoint action, $\mathfrak{h}$ acts on $\mathfrak{g}$ by upper-triangular matrices. By Lemma C.19, $[Y, Z]$ acts on $\mathfrak{g}$ by nilpotent transformations. It follows that $X \circ [Y, Z]$ also acts nilpotently on $\mathfrak{g}$, from which it follows that $B(X, [Y, Z]) = \mathrm{Tr}(X \circ [Y, Z]) = 0$, as required.

Since $\mathscr{D}\mathfrak{g}$ is a characteristic ideal, $(\mathscr{D}\mathfrak{g})^\perp$ is an ideal. It is solvable by Cartan's criterion (Proposition C.4), since

$$B(\mathscr{D}\mathfrak{g}^\perp, \mathscr{D}(\mathscr{D}\mathfrak{g}^\perp)) \subset B(\mathscr{D}\mathfrak{g}^\perp, \mathscr{D}\mathfrak{g}) = 0.$$

It follows that $\mathscr{D}\mathfrak{g}^\perp \subset \mathfrak{r}$, which concludes the proof. □

Corollary C.23. *If* $\mathfrak{a}$ *is an ideal in a Lie algebra* $\mathfrak{g}$, *then*

$$\mathrm{Rad}(\mathfrak{a}) = \mathrm{Rad}(\mathfrak{g}) \cap \mathfrak{a}.$$

PROOF. Since $\mathrm{Rad}(\mathfrak{a})$ is a characteristic ideal of an ideal, it is an ideal of $\mathfrak{g}$. Since it is solvable, it must be contained in the radical of $\mathfrak{g}$. This shows the inclusion $\subset$; the opposition inclusion is clear since $\mathrm{Rad}(\mathfrak{g}) \cap \mathfrak{a}$ is a solvable ideal in $\mathfrak{a}$. □

Proposition C.24. *If* D *is a derivation of a Lie algebra* $\mathfrak{g}$, *then* $D(\mathrm{Rad}(\mathfrak{g}))$ *is contained in a nilpotent ideal of* $\mathfrak{g}$.

PROOF. Construct $\mathfrak{g}' = \mathfrak{g} \oplus \mathbb{C}$ as before, with $\xi = (0, 1)$. Since $\mathrm{Rad}(\mathfrak{g}) \subset \mathrm{Rad}(\mathfrak{g}')$, we have

$$D(\mathrm{Rad}(\mathfrak{g})) = [\xi, \mathrm{Rad}(\mathfrak{g})] \subset [\mathfrak{g}', \mathrm{Rad}(\mathfrak{g}')] \cap \mathfrak{g}.$$

By Lemma C.20, $[\mathfrak{g}', \mathrm{Rad}(\mathfrak{g}')]$ is a nilpotent ideal in $\mathfrak{g}'$, so its intersection with $\mathfrak{g}$ is also nilpotent. □

Just as with the notion of solvability, any Lie algebra $\mathfrak{g}$ contains a largest nilpotent ideal, usually called the *nil radical* of $\mathfrak{g}$, and denoted $\mathrm{Nil}(\mathfrak{g})$ or $\mathfrak{n}$. Proposition C.24 says that any derivation maps $\mathfrak{r}$ into $\mathfrak{n}$, which includes the result of Lemma C.20 that $[\mathfrak{g}, \mathfrak{r}] \subset \mathfrak{n}$. The existence of this ideal follows from:

Lemma C.25. *If* $\mathfrak{a}$ *and* $\mathfrak{b}$ *are nilpotent ideals in a Lie algebra* $\mathfrak{g}$, *then* $\mathfrak{a} + \mathfrak{b}$ *is also a nilpotent ideal.*

PROOF. An ideal $\mathfrak{a}$ is nilpotent iff there is a positive integer k so that all k-fold brackets $[X_1, [X_2, [\cdots, [X_{k-1}, X_k] \cdots]]]$ are zero when each X_i is in $\mathfrak{a}$. Equivalently, all m-fold brackets of m elements of $\mathfrak{g}$ are zero if at least k of them are in $\mathfrak{a}$. If k is chosen to work for $\mathfrak{a}$ and for $\mathfrak{b}$, it is easy to verify that $2k$ works for the sum $\mathfrak{a} + \mathfrak{b}$, since any bracket of $2k$ elements, each from $\mathfrak{a}$ or from $\mathfrak{b}$, contains at least k elements from $\mathfrak{a}$ or from $\mathfrak{b}$. □

Since $\mathrm{Nil}(\mathfrak{g}) \subset \mathrm{Rad}(\mathfrak{g})$, it follows from Proposition C.24 that $\mathrm{Nil}(\mathfrak{g})$ is a characteristic ideal of $\mathfrak{g}$. The same reasoning as in Corollary C.23 gives:

Corollary C.26. *If* $\mathfrak{a}$ *is an ideal in a Lie algebra* $\mathfrak{g}$, *then*

$$\mathrm{Nil}(\mathfrak{a}) = \mathrm{Nil}(\mathfrak{g}) \cap \mathfrak{a}.$$

If $\mathfrak{g}$ is a Lie algebra, its *universal enveloping algebra* $U = U(\mathfrak{g})$ is the quotient of the tensor algebra of $\mathfrak{g}$ modulo the two-sided ideal generated by all $X \otimes Y - Y \otimes X - [X, Y]$ for all X, Y in $\mathfrak{g}$. It is an associative algebra, with a map $\iota\colon \mathfrak{g} \to U$ such that

$$\iota([X, Y]) = [\iota(X), \iota(Y)] = \iota(X)\iota(Y) - \iota(Y)\iota(X),$$

and satisfying the universal property: for any linear map φ from $\mathfrak{g}$ to an associative algebra A such that $\varphi([X, Y]) = [\varphi(X), \varphi(Y)]$ for all X, Y, there is a unique homomorphism of algebras $\tilde{\varphi}\colon U \to A$ such that $\varphi = \tilde{\varphi} \circ \iota$. For example, a representation $\rho\colon \mathfrak{g} \to \mathfrak{gl}(V)$ determines an algebra homomorphism $\tilde{\rho}\colon U(\mathfrak{g}) \to \mathrm{End}(V)$. Conversely, any representation arises in this way.

We will need the following easy lemma:

Lemma C.27. *For any derivation* D *of a Lie algebra* $\mathfrak{g}$, *there is a unique derivation* $\tilde{D}$ *of the associative algebra* $U(\mathfrak{g})$ *such that* $\tilde{D} \circ \iota = \iota \circ D$.

PROOF. Define an endomorphism of the tensor algebra of $\mathfrak{g}$ which is zero on the zeroth tensor power, and on the nth tensor power is

$$X_1 \otimes \cdots \otimes X_n \mapsto DX_1 \otimes X_2 \otimes \cdots \otimes X_n + X_1 \otimes DX_2 \otimes \cdots \otimes X_n + \cdots + X_1 \otimes X_2 \otimes \cdots \otimes DX_n.$$

This is well defined, since it is multilinear in each factor, and it is easily checked to be a derivation of the tensor algebra; denote it by D'. To see that D' passes to the quotient $U(\mathfrak{g})$ one checks routinely that it vanishes on generators for the ideal of relations. □

Exercise C.28. If D is an inner derivation by an element X in $\mathfrak{g}$, verify that $\tilde{D}$ is the inner derivation by the element $\iota(X)$.

It is a fact that the canonical map ι embeds $\mathfrak{g}$ in $U(\mathfrak{g})$. The *Poincaré–Birkhoff–Witt* theorem asserts that, in fact, if $U(\mathfrak{g})$ is filtered with the nth piece generated by all products of at most n products of elements of $\iota(\mathfrak{g})$, then the associated graded ring is the symmetric algebra on $\mathfrak{g}$. Equivalently, if $X_1, \ldots, X_r$ is a basis for $\mathfrak{g}$, then the monomials $X_1^{i_1} \cdot \ldots \cdot X_r^{i_r}$ form a basis for $U(\mathfrak{g})$. We do not need this theorem, but we will use the fact that these monomials generate $U(\mathfrak{g})$; this follows by a simple induction, using the equations $X_i \cdot X_j - X_j \cdot X_i = [X_i, X_j]$ to rearrange the order in products.

APPENDIX D

Cartan Subalgebras

Our task here is to prove the basic general facts that were stated in Lecture 14 about the decomposition of a semisimple Lie algebra $\mathfrak{g}$ into a Cartan algebra $\mathfrak{h}$ and a sum of root spaces $\mathfrak{g}_\alpha$, including the existence of such $\mathfrak{h}$ and its uniqueness up to conjugation.

§D.1. The Existence of Cartan Subalgebras

Note that if we have a decomposition as in Lecture 14, and H is any element of $\mathfrak{h}$ such that $\alpha(H) \neq 0$ for all roots α, then $\mathfrak{h}$ is determined by H: $\mathfrak{h} = \mathfrak{c}(H)$, where

$$\mathfrak{c}(H) = \{X \in \mathfrak{g}: [H, X] = 0\}. \tag{D.1}$$

The elements of $\mathfrak{h}$ with this property are called *regular*. They form a Zariski open subset of $\mathfrak{h}$: the complement of the union of the hyperplanes defined by the equations $\alpha = 0$. In particular, regular elements are dense in $\mathfrak{h}$. If $H \in \mathfrak{h}$ is not regular, then $\mathfrak{c}(H)$ is larger than $\mathfrak{h}$, since it contains other root spaces. Note that all elements of $\mathfrak{h}$ are also semisimple, i.e., they are equal to their semisimple parts.

Of course, this discussion depends on knowing the decomposition which we are trying to prove. But it suggests one way to construct and characterize

Cartan subalgebras: they should be subalgebras of the form $\mathfrak{c}(H)$ for some semisimple element H, that are minimal in some sense. We can measure this minimality simply by dimension.

Definition D.2. The *rank* n of a semisimple Lie algebra $\mathfrak{g}$ is the minimum of the dimension of $\mathfrak{c}(H)$ as H varies over all semisimple elements of $\mathfrak{g}$. A semisimple element H is called *regular* if $\mathfrak{c}(H)$ has dimension n. A *Cartan subalgebra* of $\mathfrak{g}$ is an abelian subalgebra all of whose elements are semisimple, and that is not contained in any larger such subalgebra. Our first main goal is

Proposition D.3. *If H is regular, then $\mathfrak{c}(H)$ is a Cartan subalgebra.*

For any semisimple element H, $\mathfrak{g}$ decomposes into eigenspaces for the adjoint action of H:

$$\mathfrak{g} = \bigoplus_{\lambda} \mathfrak{g}_\lambda(H) = \mathfrak{c}(H) \oplus \bigoplus_{\lambda \neq 0} \mathfrak{g}_\lambda(H), \tag{D.4}$$

where $\mathfrak{g}_\lambda(H) = \{X \in \mathfrak{g}: [H, X] = \lambda X\}$, and $\mathfrak{c}(H) = \mathfrak{g}_0(H)$. There is a similar decomposition even if H (or $\mathfrak{g}$) is not semisimple, but replacing the eigenspace by $\mathfrak{g}_\lambda(H) = \{X \in \mathfrak{g}: (\mathrm{ad}(H) - \lambda I)^k(X) = 0 \text{ for large } k\}$.

Exercise D.5. Without assuming that H is semisimple, show that $[\mathfrak{g}_\lambda(H), \mathfrak{g}_\mu(H)] \subset \mathfrak{g}_{\lambda+\mu}(H)$, by proving the identity

$$\begin{aligned}(\mathrm{ad}(H) - (\lambda + \mu)I)^k([X, Y]) \\ = \sum_{j=0}^{k} \binom{k}{j} [(\mathrm{ad}(H) - \lambda I)^j(X), (\mathrm{ad}(H) - \mu I)^{k-j}(Y)]\end{aligned}$$

Let us (temporarily) call an arbitrary element $H \in \mathfrak{g}$ regular if $\dim(\mathfrak{g}_0(H)) \leq \dim(\mathfrak{g}_0(X))$ for all $X \in \mathfrak{g}$.

Lemma D.6. *If H is regular, then $\mathfrak{g}_0(H)$ is abelian.*

PROOF. Consider how the Killing form B respects the decomposition (D.4)—again knowing what to expect from Lecture 14. If Y is in $\mathfrak{g}_\lambda(H)$ with $\lambda \neq 0$, then $\mathrm{ad}(Y)$ maps each eigenspace to a different eigenspace (by Exercise D.5), as does $\mathrm{ad}(Y) \circ \mathrm{ad}(X)$ for $X \in \mathfrak{g}_0(H)$. The trace of such an endomorphism is zero, i.e., $B(X, Y) = 0$ for such X and Y.

Because $\mathfrak{g}$ is semisimple, B is nondegenerate. Since we have shown that $\mathfrak{g}_0(H)$ is perpendicular to the other weight spaces, it follows that the restriction of B to $\mathfrak{g}_0(H)$ is nondegenerate.

Consider the Jordan decomposition $X = X_s + X_n$ of an element X in $\mathfrak{g}_0(H)$. Since $\mathrm{ad}(X_n) = \mathrm{ad}(X)_n$ is nilpotent, X_n belongs to $\mathfrak{g}_0(H)$, so $X_s = X - X_n$ does also. Then $\mathrm{ad}(X_s) = \mathrm{ad}(X)_s$ is nilpotent and semisimple on $\mathfrak{g}_0(H)$, so it vanishes there. But this already shows that $\mathrm{ad}(X) = \mathrm{ad}(X_s) + \mathrm{ad}(X_n)$ is a nilpotent endomorphism of $\mathfrak{g}_0(H)$ for any $X \in \mathfrak{g}_0(H)$. Hence, by Engel's theorem, $\mathfrak{g}_0(H)$ is nilpotent, so by Lie's theorem $\mathfrak{g}$ has a basis in which the endomorphisms $\operatorname{ad}(X)$ are upper-triangular for all $X \in \mathfrak{g}_0(H)$. It follows that for any elements in $\mathfrak{g}_0(H)$, the trace of products of their adjoint actions on $\mathfrak{g}$ is independent of the order of composition. In particular, for $X, Y, Z \in \mathfrak{g}_0(H)$, the trace of $\operatorname{ad}([X, Y]) \circ \operatorname{ad}(Z)$ on $\mathfrak{g}$ is zero, i.e., $B([X, Y], Z]) \equiv 0$. But since B is nondegenerate on $\mathfrak{g}_0(H)$, $[X, Y] = 0$, so $\mathfrak{g}_0(H)$ is abelian. □

It follows immediately that $\mathfrak{g}_0(H)$ is not contained in any larger abelian subalgebra, since any element that commutes with H is in $\mathfrak{g}_0(H)$ by definition. To finish the proof of the proposition we must prove the following lemma, which also shows that the temporary definition of regular agrees with the first one:

Lemma D.7. *If H is regular, then any element of $\mathfrak{g}_0(H)$ is semisimple.*

Proof. We saw that if X is in $\mathfrak{g}_0(H)$ then X_n is also. Using the same basis as in the preceding proof, we see that $\operatorname{ad}(X_n)$ has a strictly upper-triangular matrix. Hence, $B(X_n, Y) = \operatorname{Tr}(\operatorname{ad}(X_n) \circ \operatorname{ad}(Y)) = 0$ for all Y in $\mathfrak{g}_0(H)$. By the nondegeneracy again, $X_n = 0$, as required. □

It follows from Lemma D.6 that if H is regular, and X is in $\mathfrak{g}_0(H)$, then $\mathfrak{g}_0(X)$ contains $\mathfrak{g}_0(H)$, and they are equal exactly when X is also regular.

Problem D.8*. Prove that if H is regular in any Lie algebra, then $\mathfrak{g}_0(H)$ is a nilpotent Lie algebra.

Exercise D.9. Show that a subalgebra is a Cartan subalgebra if and only if it consists entirely of semisimple elements and is contained in no larger subalgebra with this property.

§D.2. On the Structure of Semisimple Lie Algebras

Let $\mathfrak{h}$ be a Cartan subalgebra of a semisimple Lie algebra $\mathfrak{g}$. Under the adjoint representation it consists of commuting semisimple endomorphisms. It is then a standard linear algebra fact that this action is simultaneously diagonalizable:

$$\mathfrak{g} = \bigoplus \mathfrak{g}_\alpha, \tag{D.10}$$

where the eigenspaces are parametrized by some set of linear forms $\alpha \in \mathfrak{h}^*$, including $\alpha = 0$, and where

$$\mathfrak{g}_\alpha = \{X \in \mathfrak{g}: [H, X] = \alpha(H) \cdot X \text{ for all } H \in \mathfrak{h}\}.$$

In particular, $\mathfrak{g}_0$ is the centralizer of $\mathfrak{h}$ in $\mathfrak{g}$. The nonzero α are called *roots*.

Lemma D.11. $\mathfrak{h} = \mathfrak{g}_0$.

PROOF. Since $\mathfrak{h}$ is abelian, $\mathfrak{h}$ is contained in $\mathfrak{g}_0$. If $\mathfrak{h}$ corresponds to a regular element H, i.e., $\mathfrak{h} = \mathfrak{g}_0(H)$, anything that commutes with H must be in $\mathfrak{h}$, so $\mathfrak{g}_0$ is contained in $\mathfrak{h}$. □

If $\mathfrak{h}$ is constructed from the regular element H, then by definition $\mathfrak{g}_\lambda(H)$ is the direct sum of those $\mathfrak{g}_\alpha$ for which $\alpha(H) = \lambda$. Note that the decomposition (D.10) may be finer than (D.4), but that if H is chosen to be an element of $\mathfrak{h}$ such that the $\alpha(H)$ are distinct for distinct roots α, then the decompositions coincide.

Our next task is to study the other eigenspaces $\mathfrak{g}_\alpha$. As before, we have $[\mathfrak{g}_\alpha, \mathfrak{g}_\beta] \subset \mathfrak{g}_{\alpha+\beta}$. It follows that if $\alpha + \beta \neq 0$, and if $X \in \mathfrak{g}_\alpha$ and $Y \in \mathfrak{g}_\beta$, then $\mathrm{ad}(X) \circ \mathrm{ad}(Y)$ is nilpotent, so its trace is zero, i.e.,

$$\textit{If } \alpha + \beta \neq 0, \textit{ then } B(\mathfrak{g}_\alpha, \mathfrak{g}_\beta) = 0. \tag{D.12}$$

Now for any root α, if $-\alpha$ were not a root, this implies $\mathfrak{g}_\alpha$ is perpendicular to all $\mathfrak{g}_\beta$ (including $\beta = 0$), which would contradict the nondegeneracy of B. So we get one of the facts asserted in Lecture 14:

$$\textit{If } \alpha \textit{ is a root, then } -\alpha \textit{ is also a root.} \tag{D.13}$$

Moreover, the pairing $B\colon \mathfrak{g}_\alpha \times \mathfrak{g}_{-\alpha} \to \mathbb{C}$ is nondegenerate. Another fact also follows easily:

$$\textit{The roots } \alpha \textit{ span } \mathfrak{h}^*. \tag{D.14}$$

For if not there would be a nonzero $X \in \mathfrak{h}$ with $\alpha(X) = 0$ for all roots α, which means that $[X, Y] = 0$ for all Y in all $\mathfrak{g}_\alpha$. But then X is in the center of $\mathfrak{g}$, which is zero by semisimplicity of $\mathfrak{g}$.

Now let α be a root, let $X \in \mathfrak{g}_\alpha$, $Y \in \mathfrak{g}_{-\alpha}$, and take any $H \in \mathfrak{h}$. Then

$$B(H, [X, Y]) = B([H, X], Y) = \alpha(H)B(X, Y). \tag{D.15}$$

This cannot be zero for all H, X, and Y without contradicting what we have just proved. In particular,

$$\textit{For any root } \alpha, [\mathfrak{g}_\alpha, \mathfrak{g}_{-\alpha}] \neq 0. \tag{D.16}$$

Let $T_\alpha \in \mathfrak{h}$ be the element dual to α via the pairing B on $\mathfrak{h}$, i.e., characterized by the identity $B(T_\alpha, H) = \alpha(H)$ for all H in $\mathfrak{h}$. We claim next that

$$[X, Y] = B(X, Y)T_\alpha \quad \text{for all } X \in \mathfrak{g}_\alpha, Y \in \mathfrak{g}_{-\alpha}. \tag{D.17}$$

To see it, pair both sides with an arbitrary element H of $\mathfrak{h}$. Using (D.15), we have

$$B(H, B(X, Y)T_\alpha) = B(H, T_\alpha)B(X, Y) = \alpha(H)B(X, Y) = B(H, [X, Y]),$$

as required. Next we show that

$$\alpha(T_\alpha) \neq 0. \tag{D.18}$$

Suppose this were false. Choose $X \in \mathfrak{g}_\alpha$, $Y \in \mathfrak{g}_{-\alpha}$ such that $B(X, Y) = c \neq 0$. Then $[X, Y] = cT_\alpha$, so X, Y, and T_α span a Lie subalgebra $\mathfrak{s}$ of $\mathfrak{g}$. If $\alpha(T_\alpha) = 0$, $\mathfrak{s}$ is solvable. Since $[X, Y] \in \mathscr{D}\mathfrak{s}$, it follows that $\operatorname{ad}([X, Y])$ is a nilpotent endomorphism of $\mathfrak{g}$. But then T_α is nilpotent; but all elements of $\mathfrak{h}$ are semisimple, so $T_\alpha = 0$, a contradiction. This gives another claim from Lecture 14:

$$\textit{For any root } \alpha,\ [[\mathfrak{g}_\alpha, \mathfrak{g}_{-\alpha}], \mathfrak{g}_\alpha] \neq 0. \tag{D.19}$$

For with X and Y as above, $[[X, Y], X] = c\cdot[T_\alpha, X] = c\cdot\alpha(T_\alpha)X \neq 0$.

The last remaining fact about root spaces left unproved from Lecture 14 is

$$\textit{For any root } \alpha,\ \mathfrak{g}_\alpha \textit{ is one-dimensional.} \tag{D.20}$$

By what we have seen, we can find $X \in \mathfrak{g}_\alpha$, $Y \in \mathfrak{g}_{-\alpha}$, so that $H = [X, Y] \neq 0$, and $\alpha(H) \neq 0$. Adjusting by scalars, they generate a subalgebra $\mathfrak{s}$ isomorphic to $\mathfrak{sl}_2\mathbb{C}$, with standard basis H, X, Y, so in particular $\alpha(H) = 2$. Consider the adjoint action of $\mathfrak{s}$ on the sum $V = \mathfrak{h} \oplus \bigoplus \mathfrak{g}_{k\alpha}$, the sum over all nonzero complex multiples $k\alpha$ of α. From what we know about the weights of representations of $\mathfrak{s}$, the only k that can occur are integral multiples of $\frac{1}{2}$.

Now $\mathfrak{s}$ acts trivially on $\operatorname{Ker}(\alpha) \subset \mathfrak{h} \subset V$, and it acts irreducibly on $\mathfrak{s} \subset V$. Together these cover the zero weight space $\mathfrak{h}$, since H is not in $\operatorname{Ker}(\alpha)$. So the only even weights occurring can be 0 and ± 2. In particular,

$$2\alpha \textit{ cannot be a root.} \tag{D.21}$$

But this implies that $\frac{1}{2}\alpha$ cannot be a root, which says that 1 is not a weight occurring in V. But then there can be no other representations occurring in V, i.e., $V = \operatorname{Ker}(\alpha) \oplus \mathfrak{s}$, which proves (D.20). □

§D.3. The Conjugacy of Cartan Subalgebras

We show that any two Cartan subalgebras are conjugate by an inner automorphism of the adjoint subgroup of $\operatorname{Aut}(\mathfrak{g})$. Fix one Cartan subalgebra $\mathfrak{h}$, and consider the decomposition (D.10). For any element X in a root space $\mathfrak{g}_\alpha$, $\operatorname{ad}(X) \in \mathfrak{gl}(\mathfrak{g})$ is nilpotent, as we have seen, so its exponential $\exp(\operatorname{ad}(X)) \in GL(\mathfrak{g})$ is just a finite polynomial in $\operatorname{ad}(X)$. Set

$$e(X) = \exp(\operatorname{ad}(X)).$$

Let $E(\mathfrak{h})$ be the subgroup of $\operatorname{Aut}(\mathfrak{g})$ generated by all such $e(X)$. We want to prove now that this group is independent of the choice of $\mathfrak{h}$, and that all Cartan subalgebras are conjugate by elements in this group. (We will see in the next section that $E(\mathfrak{h})$ is the connected component of $\operatorname{Aut}(\mathfrak{g})$, i.e., that it is the adjoint group.) The proof will be a kind of complex algebraic analogue of the corresponding argument for compact tori that was sketched in Lecture 26.

Theorem D.22. *Let* $\mathfrak{h}$ *and* $\mathfrak{h}'$ *be two Cartan subalgebras of* $\mathfrak{g}$. *Then* (i) $E(\mathfrak{h}) = E(\mathfrak{h}')$, *and* (ii) *there is an element* $g \in E = E(\mathfrak{h})$ *so that* $g(\mathfrak{h}) = \mathfrak{h}'$.

Proof. Fix a Cartan subalgebra $\mathfrak{h}$. Let $\alpha_1, \ldots, \alpha_r$ be its roots. Consider the mapping

$$F: \mathfrak{g}_{\alpha_1} \times \cdots \times \mathfrak{g}_{\alpha_r} \times \mathfrak{h} \to \mathfrak{g}$$

defined by $F(X_1, \ldots, X_r, H) = e(X_1) \circ \cdots \circ e(X_r)(H)$. Note that F is a polynomial mapping from one complex vector space to another of the same dimension. We want to show that not only is the image of F dense, but that, if $\mathfrak{h}_{\text{reg}}$ denotes the set of regular elements in $\mathfrak{h}$, then

$$F(\mathfrak{g}_{\alpha_1} \times \cdots \times \mathfrak{g}_{\alpha_r} \times \mathfrak{h}_{\text{reg}}) \textit{ contains a Zariski open set,} \tag{D.23}$$

i.e., it contains the complement of a hypersurface defined by a polynomial equation.

Suppose that this claim is proved. It follows that for any other Cartan subalgebra $\mathfrak{h}'$, the corresponding image also contains a Zariski open set. But two nonempty Zariski open sets always meet. In this case this means $E(\mathfrak{h}) \cdot \mathfrak{h}_{\text{reg}}$ meets $E(\mathfrak{h}') \cdot \mathfrak{h}'_{\text{reg}}$. That is, there are $g \in E(\mathfrak{h})$, $H \in \mathfrak{h}_{\text{reg}}$, $g' \in E(\mathfrak{h}')$, $H' \in \mathfrak{h}'_{\text{reg}}$ such that $g(H) = g'(H')$. But then since H and H' are regular,

$$g(\mathfrak{h}) = g(\mathfrak{g}_0(H)) = \mathfrak{g}_0(g(H)) = \mathfrak{g}_0(g'(H')) = g'(\mathfrak{g}_0(H')) = g'(\mathfrak{h}').$$

This proves the conjugacy of $\mathfrak{h}$ and $\mathfrak{h}'$. And since

$$E(\mathfrak{h}) = gE(\mathfrak{h})g^{-1} = E(g(\mathfrak{h})) = E(g'(\mathfrak{h}')) = g'E(\mathfrak{h}')(g')^{-1} = E(\mathfrak{h}'),$$

both statements of the theorem are proved. □

To prove (D.23), we use a special case of a very general fact from basic algebraic geometry: if $F: \mathbb{C}^N \to \mathbb{C}^N$ is a polynomial mapping whose derivative $dF_*|_P$ is invertible at some point P, then for any nonempty Zariski open set $U \subset \mathbb{C}^N$, $F(U)$ contains a nonempty Zariski open set. For the proof we refer to any basic algebraic geometry text, e.g., [Ha], or to [Bour, VI, App. A]. So it suffices to show that $dF_*|_P$ is surjective at a point $P = (0, \ldots, 0, H)$, where $H \in \mathfrak{h}_{\text{reg}}$. This is a simple calculation:

Exercise D.24*. Show that $dF_*|_P(0, \ldots, 0, Z) = Z$ for $Z \in \mathfrak{h}$, and that $dF_*|_P(0, \ldots, 0, Y, 0, \ldots, 0, 0) = \operatorname{ad}(Y)(H) = -\operatorname{ad}(H)(Y)$ for $Y \in \mathfrak{g}_{\alpha_i}$. Conclude that the image of $dF_*|_P$ contains $\mathfrak{h}$ and each root space, so $dF_*|_P$ is surjective. □

We remark that although this section, like the preceding appendix, was written for complex Lie algebras, a simple "base change" argument shows that the results extend to Lie algebras over any algebraically closed field of characteristic zero. Some, such as Cartan's criterion, then follow over any field of characteristic zero, by extending to an algebraic closure.

§D.4. On the Weyl Group

In this section we complete the proofs of some of the general facts about the Weyl group that were stated in Lectures 14 and 21. The notation will be as in those sections: $\mathbb{E}$ is the real space generated by the roots R; $\mathfrak{W}$ is the Weyl group, generated by the involutions W_α of $\mathbb{E}$ determined by

$$W_\alpha(\beta) = \beta - \beta(H_\alpha)\alpha = \beta - 2\frac{(\beta, \alpha)}{(\alpha, \alpha)}\alpha,$$

where $(\ ,\ )$ denotes the Killing form (or any inner product invariant for the Weyl group). We consider a decomposition

$$R = R^+ \cup R^-$$

into positive and negative roots, given by some $l\colon \mathbb{E} \to \mathbb{R}$ as in Lecture 14, and we let $S \subset R^+$ be the set of simple roots for this decomposition. Note that for any W in the Weyl group,

$$R = W(R^+) \cup W(R^-)$$

is the decomposition into positive and negative roots for the linear map $l \circ W^{-1}$. We want to show that every decomposition arises this way. To prove this we need some simple variations of the ideas in §21.1.

Lemma D.25. *If α is a simple root, then W_α permutes all the other positive roots, i.e., W_α maps $R^+ \setminus \{\alpha\}$ to itself.*

Proof. This follows from the expression of positive roots as sums $\beta = \sum m_i \alpha_i$, with the α_i simple, and the m_i non-negative integers. If $\alpha = \alpha_i$, $W_\alpha(\beta)$ differs from β only by an integral multiple of α_i. If $\beta \neq \alpha_i$, $W_\alpha(\beta)$ still has some positive coefficients, so it must be a positive root. □

Let $\mathfrak{W}_0$ be the subgroup of $\mathfrak{W}$ generated by the W_α, as α varies over the simple roots. (We will soon see that $\mathfrak{W}_0 = \mathfrak{W}$.)

Lemma D.26. *Any root β can be written in the form $\beta = W(\alpha)$ for some $\alpha \in S$ and $W \in \mathfrak{W}_0$. In particular, $R = \mathfrak{W}(S)$.*

Proof. It suffices to do this for positive roots, since $\mathfrak{W}_0(\alpha) = \mathfrak{W}_0 W_\alpha(\alpha) = -\mathfrak{W}_0(\alpha)$ for any $\alpha \in S$. If β is positive but not simple, write $\beta = \sum m_i \alpha_i$ as above, and induct on the level $\sum m_i$. As in the previous lemma, there is a simple root γ so that $W_\gamma(\beta)$ is a positive root of lower level. By induction, $W_\gamma(\beta) = W(\alpha)$ for $\alpha \in S$ and $W \in \mathfrak{W}_0$, so $\beta = W_\gamma W(\alpha)$, as required. □

Lemma D.27. *The Weyl group is generated by the reflections in the simple roots, i.e., $\mathfrak{W} = \mathfrak{W}_0$.*

Proof. Given a root β, we must show that W_β is in $\mathfrak{W}_0$. By the preceding lemma, write $\beta = U(\alpha)$ for some $U \in \mathfrak{W}_0$, $\alpha \in S$. Then

$$W_\beta = W_{U(\alpha)} = U \cdot W_\alpha \cdot U^{-1}, \tag{D.28}$$

since both sides act the same on β and $\beta^\perp$. □

Proposition D.29. *The Weyl group acts simply transitively on the set of decompositions of R into positive and negative roots.*

Proof. For the transitivity, suppose $R = Q^+ \cup Q^-$ is another decomposition. We induct on the number of roots that are in R^+ but not in Q^+. If this number is zero, then $R^+ = Q^+$. Otherwise there must be some simple root α that is not in Q^+. It suffices to prove that $W_\alpha(Q^+)$ has more roots in common with R^+ than Q^+ does, for then by induction we can write $W_\alpha(Q^+) = W(R^+)$ for some $W \in \mathfrak{W}$, so $Q^+ = W_\alpha W(R^+)$, as required. In fact, we have by Lemma D.25,

$$W_\alpha(Q^+) \cap R^+ \supset W_\alpha(Q^+ \cap R^+) \cup \{\alpha\} = W_\alpha(Q^+ \cap R^+ \cup \{-\alpha\}),$$

and this proves the assertion.

For simple transitivity, we must show that if an element W in the Weyl group takes R^+ to itself, then it must be the identity. If not, write W as a product of reflections in simple roots,

$$W = W_1 \cdot \ldots \cdot W_r,$$

with r minimal, with W_i the reflection in the simple root β_i. Let $\alpha = \beta_r$. It suffices to show that

$$W_1 \cdot \ldots \cdot W_r = W_1 \cdot \ldots \cdot W_{s-1} W_{s+1} \cdot \ldots \cdot W_{r-1}$$

for some s, $1 \le s \le r-2$. Let $U_s = W_{s+1} \cdot \ldots \cdot W_{r-1}$. This equation is equivalent to the equation $W_s U_s W_r = U_s$, or $U_s W_r U_s^{-1} = W_s$, or $U_s(\alpha) = \beta_s$ (since by (D.28), $W_{U(\alpha)} = U W_\alpha U^{-1}$).

To finish the proof we must find an s so that $U_s(\alpha) = \beta_s$. Note that $U_{r-2}(\alpha) = W_{r-1}(\alpha)$ is a positive root (by Lemma D.25, since $\beta_{r-1} \neq \alpha$). On the other hand, the hypothesis implies that

$$U_0(\alpha) = W_1 \cdot \ldots \cdot W_{r-1}(\alpha) = W_1 \cdot \ldots \cdot W_r(-\alpha) = -W(\alpha)$$

is a negative root. So there must be some s with $1 \le s \le r-2$ such that $U_s(\alpha)$ is positive and $U_{s-1}(\alpha)$ is negative. This means that W_s takes the positive root $U_s(\alpha)$ to the negative root $U_{s-1}(\alpha)$. But by Lemma D.25 again, this can happen only if W_s is the reflection in the root $U_s(\alpha)$, i.e., $\beta_s = U_s(\alpha)$. □

The simple roots S for a decomposition $R = R^+ \cup R^-$ are called a *basis* for the roots. Since S and R^+ determine each other, the proposition is equivalent to the assertion that *the Weyl group acts simply transitively on the set of bases.*

Exercise D.30. For $W \in \mathfrak{W}$, set $l(W) = \#(R^+ \cap W(R^-))$. Show that W can be written as a product of $l(W)$ reflections in simple roots, but no fewer.

If Ω_α denotes the hyperplane in $\mathbb{E}$ perpendicular to the root α, the (closed) *Weyl chambers* are the closures of the connected components of the complement $\mathbb{E}\backslash\bigcup\Omega_\alpha$ of these hyperplanes. For a decomposition $R = R^+ \cup R^-$ with simple roots S, the set

$$\mathscr{W} = \{\beta \in \mathbb{E}: (\beta, \alpha) \geq 0, \forall\alpha \in R^+\} = \{\beta \in \mathbb{E}: (\beta, \alpha) \geq 0, \forall\alpha \in S\}$$

is one of these Weyl chambers. The fact that every Weyl chamber arises this way follows from

Lemma D.31. *For any β in $\mathbb{E}$ there is some $W \in \mathfrak{W}$ such that $(W(\beta), \alpha) \geq 0$ for all $\alpha \in S$.*

PROOF. Let ρ be half the sum of the positive roots. It follows from Lemma D.25 that $W_\alpha(\rho) = \rho - \alpha$ for any simple root α. Take W in $\mathfrak{W}$ to maximize the inner product $(W(\beta), \rho)$. Then for all $\alpha \in S$,

$$(W_\alpha W(\beta), \rho) = (W(\beta), W_\alpha \rho) = (W(\beta), \rho - \alpha) = (W(\beta), \rho) - (W(\beta), \alpha)$$

cannot be larger than $(W(\beta), \rho)$, so $(W(\beta), \alpha) \leq 0$. □

Thus, the orbit of one Weyl chamber by the Weyl group covers $\mathbb{E}$, so all Weyl chambers are conjugate to each other by the action of the Weyl group. So all arise by partitioning R into positive and negative roots. This partitioning is uniquely determined by the Weyl chamber. In fact, the walls of a Weyl chamber are the hyperplanes Ω_α as α varies over the n corresponding simple roots, $n = \dim(\mathbb{E})$. From the proposition we have:

Corollary D.32. *The Weyl group acts simply transitively on Weyl chambers.*

Exercise D.33*. Let $\mathfrak{G}$ be the group of automorphisms of $\mathbb{E}$ that map R to itself.

(i) Show that $\mathfrak{W}$ is a normal subgroup of $\mathfrak{G}$.
(ii) Let $\mathfrak{R}$ be the automorphisms in $\mathfrak{G}$ which map a given set of simple roots S to itself. Show that $\mathfrak{G}$ is a semidirect product of $\mathfrak{W}$ and $\mathfrak{R}$.
(iii) Show that $\mathfrak{R}$ is isomorphic to the group of automorphisms of the Dynkin diagram.
(iv) Compute $\mathfrak{R}$ for each of the simple groups.

Our next goal is to show that the lattice $\mathbb{Z}\{H_\alpha: \alpha \in R\} \subset \mathfrak{h}$ has a basis of elements H_α where α varies over the simple roots. This is analogous to the statement we have proved that the root lattice Λ_R in $\mathfrak{h}^*$ is generated by simple roots. The first statement can be deduced from the second, using the Killing form to map $\mathfrak{h}$ to $\mathfrak{h}^*$, $H \mapsto (H, -)$, where $(\ , \ )$ is the Killing form. We saw in Lecture 14 that this map takes H_α to $\alpha' = (2/(\alpha, \alpha))\alpha$. Given a root system R in a Euclidean space $\mathbb{E}$, to each root α one can define its *coroot* α' in $\mathbb{E}$ by the formula

$$\alpha' = \frac{2}{(\alpha, \alpha)}\alpha.$$

Let $R' = \{\alpha' : \alpha \in R\}$ be the set of coroots. For any $0 \neq \alpha \in \mathfrak{h}$, set $\alpha' = (2/(\alpha, \alpha))\alpha$, and for any $\alpha, \beta \in \mathfrak{h}^*$, set $n_{\beta\alpha} = 2(\beta, \alpha)/(\alpha, \alpha)$. Let $R = R^+ \cup R^-$ be a decomposition of R into positive and negative roots, and let S be the corresponding set of simple positive roots.

Lemma D.34. (i) *The set R' of coroots forms a root system in $\mathbb{E}$.*

(ii) *The set $S' = \{\alpha' : \alpha \in S\}$ is a set of simple roots for R'.*

(iii) *For $\alpha, \beta \in S$, $n_{\beta'\alpha'} = n_{\alpha\beta}$.*

PROOF. It is a straightforward calculation that $n_{\beta'\alpha'} = n_{\alpha\beta}$. It follows by another short calculation that if W_α denotes the reflection in the hyperplane perpendicular to α, then $W_{\alpha'}(\beta') = (W_\alpha(\beta))'$. The four defining properties of a root system specified in §21.1 follow immediately from this. It is clear that if R^+ is the set of roots in R that are positive for a functional l on $\mathbb{E}$, then $(R^+)' = \{\alpha' : \alpha \in R^+\}$ is the corresponding set of positive roots for R'. Roots in R^+ are those that can be written as a nonnegative linear combinations of roots in S, and this property characterizes S. Since α' is a positive multiple of α for any α, it follows that roots in $(R^+)'$ are those that can be written as non-negative linear combinations of roots in S', which proves (ii). □

The root system R' is called the *dual* of R.

Exercise D.35. Find the dual of each type of simple root system.

Proposition D.36. (i) *The elements H_α for $\alpha \in S$ generate the lattice $\mathbb{Z}\{H_\alpha : \alpha \in R\}$.*

(ii) *If $\omega_\alpha \in \mathfrak{h}$ are defined by the property that $\omega_\alpha(H_\beta) = \delta_{\alpha,\beta}$, then the elements ω_α generate the weight lattice Λ_W.*

(iii) *The nonnegative integral linear combinations of the fundamental weights ω_α are precisely the weights in $\mathscr{W} \cap \Lambda_W$, where $\mathscr{W}$ is the closed Weyl chamber corresponding to R^+.*

PROOF. The isomorphism $\mathfrak{h} \to \mathfrak{h}^*$ given by the Killing form takes H_α to the coroot α'. By the lemma and the fact that all positive roots are sums of simple roots, the set $\{\alpha' : \alpha \in S\}$ spans the same lattice as $\{\alpha' : \alpha \in R\}$. This proves (i), and it follows that the weights are precisely those elements in $\mathfrak{h}$ that take integral values on the set $\{H_\alpha : \alpha \in S\}$. The rest of the proposition follows, noting that

$$\begin{aligned}\mathscr{W} &= \{\beta \in \mathbb{E} : \beta(H_\alpha) \geq 0 \text{ for all } \alpha \in R^+\} \\ &= \{\beta \in \mathbb{E} : \beta(H_\alpha) \geq 0 \text{ for all } \alpha \in S\}.\end{aligned}$$ □

If we identify $\mathfrak{h}$ with $\mathfrak{h}^*$ by means of the Killing form, we can regard $\mathfrak{W}$ as a group of automorphisms of $\mathfrak{h}$. By means of this, the reflection W_α corresponding to a root α becomes the automorphism of $\mathfrak{h}$ which takes an element H to $H - \alpha(H) \cdot H_\alpha$. We have a last debt (Fact 14.11) to pay about the Weyl group:

Proposition D.37. *Every element of the Weyl group is induced by an automorphism of* $\mathfrak{g}$ *which maps* $\mathfrak{h}$ *to itself.*

PROOF. It suffices to produce the generating involutions W_α in this way. The claim is that if X_α and Y_α are generators of $\mathfrak{g}_\alpha$ and $\mathfrak{g}_{-\alpha}$ as usual, then $\vartheta_\alpha = e(X_\alpha)e(-Y_\alpha)e(X_\alpha)$ is such an automorphism, where, as in the preceding section, we write $e(X)$ for $\exp(\mathrm{ad}(X))$. We must show that $\vartheta_\alpha(H) = H - \alpha(H) \cdot H_\alpha$ for all H in $\mathfrak{h}$. It suffices to do this for H with $\alpha(H) = 0$, and for $H = H_\alpha$, since such together span $\mathfrak{h}$. If $\alpha(H) = 0$, then $[X_\alpha, H] = [Y_\alpha, H] = 0$, so $\vartheta_\alpha(H) = H$, which takes care of this case. For $H = H_\alpha$, it suffices to calculate on the subalgebra $\mathfrak{s}_\alpha = \mathbb{C}\{H_\alpha, X_\alpha, Y_\alpha\} \cong \mathfrak{sl}_2\mathbb{C}$, and this is a simple calculation:

Exercise D.38. (a) For $\mathfrak{sl}_2\mathbb{C}$ with its standard basis, show that $\vartheta = e(X)e(Y)e(X)$ maps H to $-H$, X to $-Y$, and Y to $-X$.

(b) Show that if G is a Lie group with Lie algebra $\mathfrak{g}$, then ϑ_α is induced by the element $\exp(\frac{1}{2}\pi(X_\alpha - Y_\alpha))$ of G.

We need a refinement of the preceding calculation. For a root α and a nonzero complex number t, define two automorphisms of $\mathfrak{g}$:

$$\vartheta_\alpha(t) = e(t \cdot X_\alpha) \circ e(-(t)^{-1} \cdot Y_\alpha) \circ e(t \cdot X_\alpha)$$

and

$$\Phi_\alpha(t) = \vartheta_\alpha(t) \circ \vartheta_\alpha(-1).$$

Lemma D.39. *The automorphism* $\Phi_\alpha(t)$ *is the identity on* $\mathfrak{h}$, *and for any root* β, *it is multiplication by* $t^{\beta(H_\alpha)}$ *on* $\mathfrak{g}_\beta$.

PROOF. Look first in $\mathfrak{sl}_2$, with $X = X_\alpha$, $Y = Y_\alpha$. It is simplest to calculate in the covering $\mathrm{SL}_2\mathbb{C}$ of the adjoint group. Here $\vartheta_\alpha(t)$ lifts to

$$\begin{aligned}\exp(tX) \cdot \exp(-t^{-1}Y) \cdot \exp(tX) &= \begin{pmatrix} 1 & t \\ 0 & 1 \end{pmatrix} \cdot \begin{pmatrix} 1 & 0 \\ -t^{-1} & 1 \end{pmatrix} \cdot \begin{pmatrix} 1 & t \\ 0 & 1 \end{pmatrix} \\ &= \begin{pmatrix} 0 & t \\ -t^{-1} & 0 \end{pmatrix},\end{aligned}$$

so $\Phi_\alpha(t)$ lifts to

$$\begin{pmatrix} 0 & t \\ -t^{-1} & 0 \end{pmatrix} \cdot \begin{pmatrix} 0 & -1 \\ 1 & 0 \end{pmatrix} = \begin{pmatrix} t & 0 \\ 0 & t^{-1} \end{pmatrix}.$$

To see how $\Phi_\alpha(t)$ acts on $\mathfrak{g}_\beta$, for $\beta \neq \pm\alpha$, it suffices to consider the action of the $\mathrm{SL}_2\mathbb{C}$ corresponding to $\mathfrak{s}_\alpha = \mathbb{C}\{H_\alpha, X_\alpha, Y_\alpha\}$ on the α-string through β, i.e., on $\bigoplus \mathfrak{g}_{\beta+k\alpha}$. We know that this is an irreducible representation of $SL_2\mathbb{C}$, and the weight of $\mathfrak{g}_\alpha$ is $\beta(H_\alpha)$. It follows that $\begin{pmatrix} t & 0 \\ 0 & t^{-1} \end{pmatrix}$ acts by multiplication by $t^{\beta(H_\alpha)}$. Similarly on $\mathfrak{h}$ it acts by multiplication by $t^0 = 1$. □

Putting the preceding results together, we can give a description of the automorphism group $\mathrm{Aut}(\mathfrak{g})$ of $\mathfrak{g}$. Let $E = E(\mathfrak{h})$ be the subgroup generated by elements $\exp(\mathrm{ad}(Z))$, as Z varies over root spaces $\mathfrak{g}_\alpha$, $\alpha \neq 0$, as in §D.3. Let G be the adjoint form of $\mathfrak{g}$, so we have

$$E \subset G \subset \mathrm{Aut}^0(\mathfrak{g}) \subset \mathrm{Aut}(\mathfrak{g}),$$

where $\mathrm{Aut}^0(\mathfrak{g})$ is the connected component of the identity.

Proposition D.40. *We have* $E = G = \mathrm{Aut}^0(\mathfrak{g})$, *and* $\mathrm{Aut}(\mathfrak{g})/\mathrm{Aut}^0(\mathfrak{g})$ *is isomorphic to the automorphism group of the Dynkin diagram.*

PROOF. Fix the Cartan algebra $\mathfrak{h}$ and positive roots R^+. Let $\mathrm{Aut}(\mathfrak{g})'$ be the group of automorphisms of $\mathfrak{g}$ that map $\mathfrak{h}$ to itself, and similarly denote by primes the intersections of subgroups with $\mathrm{Aut}(\mathfrak{g})'$. We leave it to the reader to construct a finite subgroup K of $\mathrm{Aut}(\mathfrak{g})'$ which maps isomorphically onto the automorphism group of the Dynkin diagram, and which meets G only in the identity element (see Exercise 22.25 for a direct case-by-case approach, or use (21.25)). It then suffices to prove that $\mathrm{Aut}(\mathfrak{g})$ is a semidirect product of E and K, i.e., that $\mathrm{Aut}(\mathfrak{g}) = E \cdot K$.

To see this, start with any element σ in $\mathrm{Aut}(\mathfrak{g})$. By Theorem D.22, there is a $\tau_1 \in E$ with $\sigma(\mathfrak{h}) = \tau_1(\mathfrak{h})$. Then $\sigma_1 = \tau_1^{-1} \cdot \sigma$ is in $\mathrm{Aut}(\mathfrak{g})'$. By Proposition D.29 and the proof of Proposition D.37 there is a $\tau_2 \in E'$ so that $\sigma_2 = \tau_2^{-1} \cdot \sigma_1$ maps R^+ to R^+. This element may permute the simple roots, but there is some $k \in K$ so that $\sigma_3 = \sigma_2 \cdot k^{-1}$ is the identity on the set of simple roots. Now σ_3 is the identity on $\mathfrak{h}$ and it is multiplication by some nonzero scalar c_β on each $\mathfrak{g}_\beta$. By the nonsingularity of the Cartan matrix there is some nonzero complex number t and some $\lambda \in \Lambda_R$ so that $c_\beta = t^{\lambda(H_\beta)}$ for every simple root β. From Lemma D.39 it follows that there is a τ in E' so that τ and σ_3 agree on each $\mathfrak{g}_\beta$ for each simple root β, and both are the identity on $\mathfrak{h}$. But it then follows from the uniqueness theorem (Claim 21.25) that $\sigma_3 = \tau$. Hence

$$\sigma = \tau_1 \cdot \tau_2 \cdot \sigma_3 \cdot k \in E \cdot K,$$

as required. □

Exercise D.41. Show that any two Borel subalgebras of a semisimple Lie algebra are conjugate.

APPENDIX E
Ado's and Levi's Theorems

§E.1. Levi's Theorem

The object of this section is to prove Levi's theorem:

Theorem E.1. *Let* $\mathfrak{g}$ *be a Lie algebra with radical* $\mathfrak{r}$. *Then there is a subalgebra* $\mathfrak{l}$ *of* $\mathfrak{g}$ *such that* $\mathfrak{g} = \mathfrak{r} \oplus \mathfrak{l}$.

PROOF. There are several simple reductions. First, we may assume there is no nonzero ideal of $\mathfrak{g}$ that is properly contained in $\mathfrak{r}$. For if $\mathfrak{a}$ were such an ideal, by induction on the dimension of $\mathfrak{g}$, $\mathfrak{g}/\mathfrak{a}$ would have a subalgebra complementary to $\mathfrak{r}/\mathfrak{a}$, and this subalgebra has the form $\mathfrak{l}/\mathfrak{a}$, with $\mathfrak{l}$ as required. In particular, we may assume $\mathfrak{r}$ is abelian, since otherwise $\mathscr{D}\mathfrak{r}$ is a proper ideal in $\mathfrak{r}$ which is an ideal in $\mathfrak{g}$ by Corollary C.23. We may also assume that $[\mathfrak{g}, \mathfrak{r}] = \mathfrak{r}$, for if $[\mathfrak{g}, \mathfrak{r}] = 0$ then the adjoint representation factors through $\mathfrak{g}/\mathfrak{r}$, and since $\mathfrak{g}/\mathfrak{r}$ is semisimple, the submodule $\mathfrak{r} \subset \mathfrak{g}$ has a complement, which is the required $\mathfrak{l}$.

Now $V = \mathfrak{gl}(\mathfrak{g})$ is a $\mathfrak{g}$-module via the adjoint representation: for $X \in \mathfrak{g}$ and $\varphi \in V$,

$$X \cdot \varphi = [\mathrm{ad}(X), \varphi] = \mathrm{ad}(X) \circ \varphi - \varphi \circ \mathrm{ad}(X).$$

In other words, for $X, Y \in \mathfrak{g}$ and $\varphi \in V$,

$$(X \cdot \varphi)(Y) = [X, \varphi(Y)] - \varphi([X, Y]). \tag{E.2}$$

The trick is to consider the following subspaces of V:

$$C = \{\varphi \in V: \varphi(\mathfrak{g}) \subset \mathfrak{r} \text{ and } \varphi|_{\mathfrak{r}} \text{ is multiplication by a scalar}\}$$
$$\cup$$
$$B = \{\varphi \in V: \varphi(\mathfrak{g}) \subset \mathfrak{r} \text{ and } \varphi(\mathfrak{r}) = 0\}$$
$$\cup$$
$$A = \{\text{ad}(X): X \in \mathfrak{r}\}.$$

These are easily checked to be $\mathfrak{g}$-submodules of V, included in each other as indicated. And C/B is a trivial $\mathfrak{g}$-module of rank 1, i.e. $C/B = \mathbb{C}$, by taking φ in C to the scalar λ such that $\varphi|_{\mathfrak{r}} = \lambda \cdot I$. (Note that $C/B \neq 0$ since one can find an endomorphism of the vector space $\mathfrak{g}$ which is the identity on $\mathfrak{r}$ and zero on a vector space complement to $\mathfrak{r}$.) We claim also that

$$\mathfrak{g} \cdot C \subset B \quad \text{and} \quad \mathfrak{r} \cdot C \subset A. \tag{E.3}$$

To prove these let $\varphi \in C$, and assume the restriction of φ to $\mathfrak{r}$ is multiplication by the scalar c. If $X \in \mathfrak{g}$ and $Y \in \mathfrak{r}$, then by (E.2),

$$(X \cdot \varphi)(Y) = [X, cY] - c[X, Y] = 0,$$

so $X \cdot \varphi \in B$; this proves the first inclusion. If $X \in \mathfrak{r}$, and $Y \in \mathfrak{g}$, then $[X, \varphi(Y)] \in [\mathfrak{r}, \mathfrak{r}] = 0$, so

$$(X \cdot \varphi)(Y) = -\varphi([X, Y]) = [-cX, Y],$$

and $X \cdot \varphi = \text{ad}(-cX)$ is in A, which proves the second inclusion.

This means that the map $C/A \to C/B = \mathbb{C}$ is a surjection of $\mathfrak{g}/\mathfrak{r}$-modules, which must split since $\mathfrak{g}/\mathfrak{r}$ is semisimple. In other words, there is an element φ in C such that $\varphi|_{\mathfrak{r}} = \text{id}_{\mathfrak{r}}$ and $\mathfrak{g} \cdot \varphi$ is contained in A. Now let

$$\mathfrak{l} = \{X \in \mathfrak{g}: X \cdot \varphi = 0\}.$$

It is easy to check that $\mathfrak{l}$ is a subalgebra of $\mathfrak{g}$. We must verify: (i) $\mathfrak{l} \cap \mathfrak{r} = 0$; and (ii) $\mathfrak{g} = \mathfrak{l} + \mathfrak{r}$. For the first, if X is a nonzero element of the intersection, then, as we saw above, $X \cdot \varphi = \text{ad}(-X)$, so $\text{ad}(X) = 0$. Hence $[\mathfrak{g}, X] = 0$, so $\mathbb{C} \cdot X$ is a nonzero ideal in $\mathfrak{r}$, contradicting our assumptions. For (ii), let $X \in \mathfrak{g}$. Then $X \cdot \varphi$ is in A, so $X \cdot \varphi = \text{ad}(Y)$ for some Y in $\mathfrak{r}$. We saw that $\text{ad}(Y) = -Y \cdot \varphi$, so $(X + Y) \cdot \varphi = 0$, i.e., $X + Y$ belongs to $\mathfrak{l}$. Hence $X = (X + Y) - Y$ is in the sum of $\mathfrak{l}$ and $\mathfrak{r}$. □

This proves the existence of Levi subalgebras $\mathfrak{l}$ of any Lie algebra. We have no need to prove the companion fact that any two Levi subalgebras are conjugate, cf. [Bour, I, §6.8].

§E.2. Ado's Theorem

The goal is Ado's theorem that every Lie algebra is linear, i.e., is a subalgebra of $\mathfrak{gl}(V)$ for some vector space V, which is the same as saying it has a finite-dimensional faithful representation. As in the previous section, there are some easy steps, and then a clever argument is needed to create an appropriate representation.

We start, of course, with the adjoint representation, which is about the only representation we have for an abstract Lie algebra $\mathfrak{g}$. Since the kernel of the adjoint representation is the center $\mathfrak{c}$ of $\mathfrak{g}$, it suffices to find a representation of $\mathfrak{g}$ which is faithful on $\mathfrak{c}$. For then the sum of this representation and the adjoint representation is a faithful representation of $\mathfrak{g}$.

The abelian Lie algebra $\mathfrak{c}$ has a faithful representation by nilpotent matrices. For example, when $\mathfrak{c} = \mathbb{C}$ is one dimensional, one can take the representation $\lambda \mapsto \left(\begin{smallmatrix} 0 & \lambda \\ 0 & 0 \end{smallmatrix}\right)$; in general a direct sum of such representations will suffice.

We can choose a sequence of subalgebras

$$\mathfrak{c} = \mathfrak{g}_0 \subset \mathfrak{g}_1 \subset \cdots \subset \mathfrak{g}_p = \mathfrak{n} \subset \mathfrak{g}_{p+1} \subset \cdots \subset \mathfrak{g}_q = \mathfrak{r} \subset \mathfrak{g}_{q+1} = \mathfrak{g},$$

each an ideal in the next, with $\mathfrak{n} = \mathrm{Nil}(\mathfrak{g})$ the largest nilpotent ideal of $\mathfrak{g}$, and $\mathfrak{r} = \mathrm{Rad}(\mathfrak{g})$ the largest solvable ideal; as in §9.1 we may assume $\dim(\mathfrak{g}_i/\mathfrak{g}_{i-1}) = 1$ for $i \leq q$. The plan is to start with a faithful representation of $\mathfrak{g}_0$, and construct successively representations of each $\mathfrak{g}_i$ which are faithful on $\mathfrak{c}$. The conditions we will need to make this step are that $\mathfrak{g}_i = \mathfrak{g}_{i-1} \oplus \mathfrak{h}_i$ with $\mathfrak{g}_{i-1}$ a solvable ideal in $\mathfrak{g}_i$ and $\mathfrak{h}_i$ a subalgebra of $\mathfrak{g}_i$. We can achieve this by taking $\mathfrak{h}_i$ to be any one-dimensional vector space complementary to $\mathfrak{g}_{i-1}$ for $i \leq q$. Similarly to go from $\mathfrak{r}$ to $\mathfrak{g}$, use Levi's theorem to write $\mathfrak{g} = \mathfrak{r} \oplus \mathfrak{h}$ for a subalgebra $\mathfrak{h}$.

Call a representation ρ of a Lie algebra $\mathfrak{g}$ a *nilrepresentation* if $\rho(X)$ is a nilpotent endomorphism for every X in $\mathrm{Nil}(\mathfrak{g})$. A stronger version of Ado's theorem is:

Theorem E.4. *Every Lie algebra has a faithful finite-dimensional nilrepresentation.*

The crucial step is:

Proposition E.5. *Let $\mathfrak{g}$ be a Lie algebra which is a direct sum of a solvable ideal $\mathfrak{a}$ and a subalgebra $\mathfrak{h}$. Let σ be a nilrepresentation of $\mathfrak{a}$. Then there is a representation ρ of $\mathfrak{g}$ such that*

$$\mathfrak{h} \cap \mathrm{Ker}(\rho) \subset \mathrm{Ker}(\sigma).$$

If $\mathrm{Nil}(\mathfrak{g}) = \mathrm{Nil}(\mathfrak{a})$ *or* $\mathrm{Nil}(\mathfrak{g}) = \mathfrak{g}$, *then ρ may be taken to be a nilrepresentation.*

Ado's theorem follows readily from this proposition. Starting with a faithful representation ρ_0 of $\mathfrak{c} = \mathfrak{g}_0$ by nilpotent matrices, one uses the proposition to construct successively nilrepresentations ρ_i of $\mathfrak{g}_i$. The displayed condition assures that they are all faithful on $\mathfrak{c}$. Note that if $i \leq p$, $\mathrm{Nil}(\mathfrak{g}_i) = \mathfrak{g}_i$, while if $i > p$ we have $\mathrm{Nil}(\mathfrak{g}_i) = \mathrm{Nil}(\mathfrak{g}_{i-1}) = \mathfrak{n}$ by Corollary C.26, so the hypotheses assure that all representations can be taken to be nilrepresentations. □

Suppose $\mathfrak{g} = \mathfrak{a} \oplus \mathfrak{h}$ is a Lie algebra which is a direct sum of an ideal $\mathfrak{a}$ and a subalgebra $\mathfrak{h}$. Let $U = U(\mathfrak{a})$ be the universal enveloping algebra of $\mathfrak{a}$. Any

Y in $\mathfrak{a}$ determines a linear endomorphism L_Y of U, which is simply left multiplication by the image of Y in U. Any X in $\mathfrak{g}$ determines an inner derivation $Y \mapsto [X, Y]$ of $\mathfrak{a}$; let D_X be the corresponding derivation of U, cf. Lemma C.27. For each X in $\mathfrak{g}$ we define a linear mapping $T_X: U \to U$ by writing $X = Y + Z$ with Y in $\mathfrak{a}$ and Z in $\mathfrak{h}$, and setting

$$T_X = L_Y + D_Z.$$

A straightforward calculation shows that

$$T_{[X_1, X_2]} = T_{X_1} \circ T_{X_2} - T_{X_2} \circ T_{X_1}. \tag{E.6}$$

If $\mathfrak{gl}(U)$ denotes the infinite-dimensional Lie algebra of endomorphisms of U, with the usual bracket $[A, B] = A \circ B - B \circ A$, this means that the mapping $\mathfrak{a} \to \mathfrak{gl}(U)$, $X \mapsto T_X$, is a homomorphism of Lie algebras.

Suppose $\sigma: \mathfrak{a} \to \mathfrak{gl}(V)$ is a finite-dimensional representation of $\mathfrak{a}$. Let $\tilde{\sigma}: U \to \text{End}(V)$ be the corresponding homomorphism of algebras, as in §C.3, and let I be the kernel of $\tilde{\sigma}$. The basic step is:

Lemma E.7. *Assume that $\mathfrak{a}$ is solvable. Suppose I is an ideal of $U = U(\mathfrak{a})$ satisfying the following two properties*: (i) *U/I is finite dimensional*; (ii) *the image of every element in* Nil($\mathfrak{a}$) *in U/I is nilpotent. Then there is an ideal $J \subset I$ of U satisfying properties* (i) *and* (ii), *and also* (iii) *for every derivation D of $\mathfrak{a}$, the corresponding derivation of U maps J into itself.*

Granting this lemma, we prove Proposition E.5 as follows. From the representation σ we constructed an ideal I in $U = U(\mathfrak{a})$, with $U/I \subset \text{End}(V)$, so condition (i) is satisfied; the fact that σ is a nilrepresentation implies that condition (ii) also holds. Let J be an ideal whose existence is asserted in the lemma. Because of (iii), each of the endomorphisms T_X of U maps J into itself, and so determines an endomorphism $\overline{T_X}$ of U/J. By (E.6), the mapping $X \mapsto \overline{T_X}$ is a homomorphism of Lie algebras from $\mathfrak{g}$ to $\mathfrak{gl}(U/J)$. This is the representation ρ required in the proposition.

We first verify that $\text{Ker}(\rho) \cap \mathfrak{a} \subset \text{Ker}(\sigma)$. Note that if X is in $\mathfrak{a}$, then $\overline{T_X}$ is just left multiplication by X on U/J, so if $\rho(X)$ vanishes, the image of X in U must be in J; since $J \subset I$, X maps to zero in $U/I \subset \text{End}(V)$, so $\sigma(X) = 0$, as required.

It remains to show that, under either of the additional hypotheses, ρ is a nilrepresentation. Note first that each X in $\mathfrak{a}$ acts on U/J by left multiplication, and if X is in Nil($\mathfrak{a}$), by (ii) its image in U/J is nilpotent. Thus $\rho(X)$ is nilpotent for every X in Nil($\mathfrak{a}$). In particular, this shows that ρ is a nilrepresentation when Nil($\mathfrak{g}$) = Nil($\mathfrak{a}$).

In the other case, $\mathfrak{g}$ is nilpotent, so $\mathfrak{a}$ is also nilpotent, and the preceding shows that $\rho(Y)$ is nilpotent for every Y in $\mathfrak{a}$. We need a slightly stronger assertion than this. Let $A \subset \text{End}(U/J)$ be the associative algebra (with unit) generated by $\rho(\mathfrak{g})$, and let $P \subset A$ be the two-sided ideal generated by $\rho(\mathfrak{a})$. The claim is that P is a nilpotent ideal, i.e., that $P^k = P \cdot \ldots \cdot P = 0$ for some k. To see this, note that there is a k such that every product of k elements of $\rho(\mathfrak{a})$ is zero; this follows from Engel's theorem, putting the action in strictly upper-triangular form. To show that $P^k = 0$, we must show that any product of elements in $\rho(\mathfrak{g})$ which contains at least k members from $\rho(\mathfrak{a})$ is zero. But if x is in $\rho(\mathfrak{g})$ and y is in $\rho(\mathfrak{a})$, we have

$$x \cdot y = y \cdot x + [x, y],$$

and $[x, y]$ is in $\rho(\mathfrak{a})$, so terms from $\rho(\mathfrak{a})$ can be successively moved to the left until the product is a sum of products each beginning with k terms from $\rho(\mathfrak{a})$.

Now if $\mathfrak{g}$ is nilpotent, for any Z in $\mathfrak{h}$ (or in $\mathfrak{g}$), $\mathrm{ad}(Z)$ is a nilpotent endomorphism of $\mathfrak{g}$, and hence of $\mathfrak{a}$. By the Leibnitz rule for derivations, it follows that the corresponding derivation D_Z of U is nilpotent on any element, although the power required to annihilate an element may be unbounded. However, since U/J is finite dimensional, it follows readily that the induced derivation of U/J is nilpotent. In other words, $\rho(Z)$ is nilpotent for every Z in $\mathfrak{h}$. Given X in $\mathfrak{g}$, write $X = Y + Z$ with $Y \in \mathfrak{a}$ and $Z \in \mathfrak{h}$. Choose k as in the preceding paragraph, and choose l so that $\rho(Z)^l = 0$. It follows that $\rho(X)^{kl} = (\rho(Y) + \rho(Z))^{kl}$ vanishes, since, when the latter is expanded, each summand either has $\rho(Y)$ occurring at least k times, or else $\rho(Z)^l$ occurs somewhere in the product. □

To finish, we must prove Lemma E.7. Let Q be the two-sided ideal in the algebra U/I generated by the image of $\mathrm{Nil}(\mathfrak{a})$. Since U/I is generated by the image of $\mathfrak{a}$, the same argument as in the paragraph before last shows that $Q^k = 0$ for some k. Write $Q = K/I$ for an ideal K of U, and set $J = K^k$. Clearly $J \subset I$, and we claim that J satisfies the conditions (i)–(iii) of the lemma.

To see that J has finite codimension, let $x_1, \ldots, x_n$ be a basis for the image of $\mathfrak{a}$ in U, and choose monic polynomials p_i such that $p_i(x_i)$ is in K; this is possible since U/K is finite dimensional. Therefore, $p_i(x_i)^k$ is in J, so the images of the x_i satisfy monic equations in U/J. Since U is generated by the monomials $x_1^{i_1} \cdot \ldots \cdot x_r^{i_r}$, it follows readily that U/J is spanned by a finite number of these elements.

Property (ii) is clear from the construction, for if $x \in U$ is the image of an element of $\mathrm{Nil}(\mathfrak{a})$, some power x^p is in I by assumption, so x^{pk} is in $I^k \subset K^k = J$.

For (iii), if D is a derivation of $\mathfrak{a}$, since $\mathfrak{a}$ is solvable, it follows from Proposition C.24 that D maps $\mathfrak{a}$ into $\mathrm{Nil}(\mathfrak{a})$. The corresponding derivation of U therefore maps U into K, from which it follows that it maps $J = K^k$ to itself. □

As before, the results of this section also apply to real Lie algebras: if $\mathfrak{g}$ is real, a faithful representation (complex) representation of $\mathfrak{g} \otimes \mathbb{C}$ is automatically a faithful real representation, and embeds $\mathfrak{g}$ is some $\mathfrak{gl}_n\mathbb{R}$.

APPENDIX F

Invariant Theory for the Classical Groups

The object is to derive just enough invariant theory for the classical groups to verify the claims made in the text. We follow a classical, constructive approach, using an identity of Capelli.

§F.1. The Polynomial Invariants

Let $V = \mathbb{C}^n$, regarded as the standard representation of $\mathrm{GL}_n\mathbb{C}$, so of any of the subgroups $G = \mathrm{SL}_n\mathbb{C}$, $\mathrm{O}_n\mathbb{C}$, $\mathrm{SO}_n\mathbb{C}$, or $\mathrm{Sp}_n\mathbb{C}$ (for n even); $e_1, \ldots, e_n$ denotes a standard basis for V, compatible with one of the standard realizations of G. The goal is to find those polynomials $F(x^{(1)}, \ldots, x^{(m)})$ of m variables on V which are invariant by G. For example, if $Q\colon V \otimes V \to \mathbb{C}$ is the bilinear form determining the orthogonal or symplectic group, the polynomials $Q(x^{(i)}, x^{(j)})$ are invariants. In addition, if G is a subgroup of $\mathrm{SL}(V)$, the *bracket* $[x^{(1)}\, x^{(2)} \ldots x^{(n)}]$, given by the determinant,

$$[x^{(1)} \quad x^{(2)} \quad \ldots \quad x^{(n)}] = \det(x_j^{(i)}), \tag{F.1}$$

is an invariant of G. The *first fundamental theorem* of invariant theory for these groups asserts that any invariant is a polynomial function of these basic invariants. This is the goal of this appendix.

We denote by S^d the homogeneous polynomial functions of degree d on V, i.e., $S^d = \mathrm{Sym}^d(V^*)$. For an m-tuple $\mathbf{d} = (d_1, \ldots, d_m)$ of non-negative integers, let $S^{\mathbf{d}} = S^{d_1} \otimes \cdots \otimes S^{d_m}$ be the polynomials on $V^{\oplus m}$ which are homogeneous of degree d_i in the ith variable. Note that

$$\mathrm{Sym}^k(V^{\oplus m})^* = \bigoplus S^{\mathbf{d}},$$

the sum over all $\mathbf{d}$ with $d_1 + d_2 + \cdots + d_m = k$, which identifies elements of $S^{\mathbf{d}}$ with functions of m-tuples in V. We write $F(x^{(1)}, \ldots, x^{(m)})$ for such a polynomial, with usual abbreviations to $F(x)$ for $m = 1$, $F(x, y)$ for $m = 2$, $F(x, y, z)$ for $m = 3$.

When $m = 1$ we have already found the invariants: for $\mathrm{SL}_n\mathbb{C}$ and $\mathrm{Sp}_n\mathbb{C}$ all symmetric powers S^d are irreducible, so there are no invariants unless $d = 0$; for $\mathrm{SO}_n\mathbb{C}$ the kernel of the map $S^d \to S^{d-2}$ (contracting with the given quadratic form Q) is irreducible, so by induction one sees that there are no invariants if d is odd, whereas if d is even, the invariants are scalar multiples of the polynomial $Q(x, x)^{d/2}$. (These results will be proved again below.)

In theory one could follow procedures outlined in the text to decompose the tensor products of the known representations S^{d_i} to find out how the trivial representation occurs in $S^{\mathbf{d}}$. Except in small degrees and dimensions, however, this is rather impractical.

To describe the G-invariant polynomials in $S^{\mathbf{d}}$, we will carry out an induction, first with respect to the total degree $\sum d_i$, then with respect to the individual multidegrees ordered *antilexicographically*: $\mathbf{d}' < \mathbf{d}$ means that either $\sum d_i' < \sum d_i$ or $\sum d_i' = \sum d_i$ and the largest i for which d_i' and d_i differ has $d_i' < d_i$.

For integers i and j between 1 and m there is a canonical "polarization" map D_{ij} which takes a polynomial F of m variables to the polynomial

$$D_{ij}(F) = \sum_{k=1}^{n} x_k^{(i)} \frac{\partial F}{\partial x_k^{(j)}}. \tag{F.2}$$

This operator lowers the jth degree by 1, while it increases the ith degree by 1, i.e., it maps $S^{\mathbf{d}}$ to $S^{\mathbf{d}'}$, where $\mathbf{d}'$ is the same sequence of multi-indices as $\mathbf{d}$, but with $d_j' = d_j - 1$ and $d_i' = d_i + 1$; if $d_j = 0$ set $S^{\mathbf{d}'} = 0$. When $j = i$, note that by *Euler's formula*, D_{ii} is multiplication by d_i. Note also that these D_{ij} are derivations:

$$D_{ij}(F_1 \cdot F_2) = D_{ij}(F_1) \cdot F_2 + F_1 \cdot D_{ij}(F_2). \tag{F.3}$$

These maps may be described intrinsically in terms of the multilinear algebra of Appendix B, as follows. Since only two factors are involved, it suffices to look at the map D_{12} when there are only two factors. In this case the map

$$D_{12} \colon S^d \otimes S^e \to S^{d+1} \otimes S^{e-1}$$

is defined by

$$u_1 \cdot \ldots \cdot u_d \otimes w_1 \cdot \ldots \cdot w_e \mapsto \sum_{i=1}^{e} u_1 \cdot \ldots \cdot u_d \cdot w_i \otimes w_1 \cdot \ldots \cdot \hat{w}_i \cdot \ldots \cdot w_e.$$

Equivalently, D_{12} is the composite

$$S^d \otimes S^e \to S^d \otimes (S^1 \otimes S^{e-1}) = (S^d \otimes S^1) \otimes S^{e-1} \to S^{d+1} \otimes S^{e-1},$$

where the second is determined by the product $S^d \otimes S^1 \to S^{d+1}$ of symmetric powers, and the first by the dual map $S^e \to S^1 \otimes S^{e-1}$ (which takes $F(x)$ to $\sum_k x_k \otimes \partial F/\partial x_k$). This shows, if there were any doubt, that the D_{ij} are maps of $\mathrm{GL}(V)$-modules, i.e., that they are independent of choice of coordinates.

Note that $D_{ji} \circ D_{ij}$ maps $S^{\mathbf{d}}$ to itself. Explicitly, for $\mathbf{d} = (d, e)$,

$$\begin{aligned} D_{21} \circ D_{12}(F) &= \sum_k y_k \frac{\partial}{\partial x_k}\left(\sum_l x_l \frac{\partial F}{\partial y_l}\right) \\ &= \sum_k y_k \frac{\partial F}{\partial y_k} + \sum_{k,l} y_k x_l \frac{\partial^2 F}{\partial y_l \partial x_k} \\ &= e \cdot F + \sum_{k,l} y_k x_l \frac{\partial^2 F}{\partial y_l \partial x_k}. \end{aligned}$$

A first idea is that, if F is an invariant by a group $G \subset \mathrm{GL}(V)$, then $D_{ij}(F)$ will also be an invariant, and these invariants will be known by induction if $i < j$, so one can describe the possible $D_{ji} \circ D_{ij}(F)$ that arise. If one also knew the second term in the above expression for this, one could determine $e \cdot F$, which suffices to determine F, provided e is not zero.

In general, it is not evident how to proceed, but in case $\dim V = 2$, and $\mathbf{d} = (d, e)$, this can idea can be carried through as follows. Some of the terms in the second term also occur in the expression

$$[xy] \cdot \Omega(F) = (x_1 y_2 - x_2 y_1) \cdot \left(\frac{\partial^2 F}{\partial x_1 \partial y_2} - \frac{\partial^2 F}{\partial x_2 \partial y_1}\right).$$

The rest occur in

$$de \cdot F = d \cdot \left(\sum y_l \frac{\partial F}{\partial y_l}\right) = \sum x_k y_l \frac{\partial^2 F}{\partial x_k \partial y_l}.$$

Comparing the preceding three formulas gives the identity

$$(d+1)e \cdot F = D_{21} \circ D_{12}(F) + [xy] \cdot \Omega(F). \tag{F.4}$$

From this identity it is easy to find all invariants for one of our subgroups of $\mathrm{GL}_2\mathbb{C}$ and for functions of two variables. We will do it for $G = \mathrm{SO}_2\mathbb{C}$, as it illustrates the ideas of the general case—even though G is not semisimple, and the results can be seen directly by identifying G with $\mathbb{C}^*$. We assume the simple case of functions of one variable has been checked: only multiples of $Q(x, x)^{d/2}$ are invariant. Suppose $F \in S^d \otimes S^e$ is an invariant of $G = \mathrm{SO}_2\mathbb{C}$, with $e > 0$. We claim that F is a polynomial in the bracket function $[xy]$ and the polynomials $Q(x, y)$, $Q(x, x)$, and $Q(y, y)$. Either directly or from the above identity one sees that $\Omega(F)$ is also an $\mathrm{SO}_2\mathbb{C}$-invariant, and by induction it is a polynomial in these basic polynomials. Similarly by the antilexicographic induction we know that $D_{12}(F)$ is a polynomial in the the basic invariants. It therefore suffices to verify that D_{21} preserves polynomials in the four basic invariants. By the derivation property (F.3) it is enough to compute the effect of D_{21} on the basic invariants, and this is easy:

$$D_{21}[xy] = 0, \qquad D_{21}Q(x, y) = Q(y, y),$$

$$D_{21}Q(x, x) = 2Q(x, y), \quad D_{21}Q(y, y) = 0.$$

By (F.4) we conclude that $(d + 1)e \cdot F$ is a polynomial in the basic invariants, which concludes the proof.

This plan of attack, in fact, extends to find all polynomial invariants of all the classical subgroups of GL(V). What is needed is an appropriate generalization of the identity (F.4). About a century ago Capelli found such an identity. The clue is to write (F.4) in the more suggestive form

$$\begin{vmatrix} D_{11} + 1 & D_{12} \\ D_{21} & D_{22} \end{vmatrix} (F) = [xy] \cdot \Omega(F),$$

where the determinant on the left is evaluated by expanding as usual, but being careful to read the composition of operators from left to right, since they do not commute.

This is the formula which generalizes. If F is a function of m variables from V, and dim $V = m$, define, following Cayley,

$$\Omega(F) = \sum_{\sigma \in \mathfrak{S}_m} \operatorname{sgn}(\sigma) \frac{\partial^m F}{\partial x_{\sigma(1)}^{(1)} \cdot \ldots \cdot \partial x_{\sigma(m)}^{(m)}}; \tag{F.5}$$

in symbols, Ω is given by the determinant

$$\begin{vmatrix} \dfrac{\partial}{\partial x_1^{(1)}} & \dfrac{\partial}{\partial x_1^{(2)}} & \cdots & \dfrac{\partial}{\partial x_1^{(m)}} \\ \dfrac{\partial}{\partial x_2^{(1)}} & \cdots & & \dfrac{\partial}{\partial x_2^{(m)}} \\ \vdots & & & \vdots \\ \dfrac{\partial}{\partial x_m^{(1)}} & \cdots & & \dfrac{\partial}{\partial x_m^{(m)}} \end{vmatrix}$$

The *Capelli identity* is the formula:

$$\begin{vmatrix} D_{11} + m - 1 & D_{12} & \cdots & D_{1m} \\ D_{21} & D_{22} + m - 2 & \cdots & D_{2m} \\ \vdots & & & \vdots \\ D_{m1} & D_{m2} & \cdots & D_{mm} \end{vmatrix} = [x^{(1)} x^{(2)} \ldots x^{(m)}] \cdot \Omega. \tag{F.6}$$

This is an identity of operators acting on functions $F = F(x^{(1)}, \ldots, x^{(m)})$ of m variables, with $m = n = \dim V$, and as always the determinant is expanded with compositions of operators reading from left to right. Note the important corollary: if the number of variables is greater than the dimension, $m > n$, then

$$\begin{vmatrix} D_{11}+m-1 & D_{12} & \cdots & D_{1m} \\ D_{21} & D_{22}+m-2 & \cdots & D_{2m} \\ \vdots & & & \vdots \\ D_{m1} & D_{m2} & \cdots & D_{mm} \end{vmatrix}(F) = 0. \tag{F.7}$$

This follows by regarding F as a function on $\mathbb{C}^m$ which is independent of the last $m - n$ coordinates. Since $\Omega(F) = 0$ for such a function, (F.7) follows from (F.6).

We will prove Capelli's identity in §F.3. Now we use it to compute invariants. Let K denote the operator on the left-hand side of these Capelli identities. The expansion of K has a main diagonal term, the product of the diagonal entries $D_{ii} + m - i$, which are scalars on multihomogeneous functions. Note that in any other product of the expansion, the last nondiagonal term which occurs is one of the D_{ij} with $i < j$. Since the diagonal terms commute with the others, we can group the products that precede a given D_{ij} into one operator, so we can write, for $F \in S^{\mathbf{d}}$,

$$K(F) = \rho \cdot F - \sum_{i<j} P_{ij} D_{ij}(F),$$

where $\rho = (d_1 + m - 1) \cdot (d_2 + m - 2) \cdot \ldots \cdot (d_m)$, and each P_{ij} is a linear combination of compositions of various D_{ab}. Capelli's identities say that

$$\rho \cdot F = \sum_{i<j} P_{ij} D_{ij}(F) \qquad \text{if } m > n; \tag{F.8}$$

$$\rho \cdot F = \sum_{i<j} P_{ij} D_{ij}(F) + [x^{(1)} \ldots x^{(m)}] \cdot \Omega(F) \quad \text{if } m = n. \tag{F.9}$$

Just as in the above special case, if F is an invariant of a group G, each $D_{ij}(F)$ is also an invariant in a $S^{\mathbf{d}'}$ where we will know all such invariants by induction. If G is a subgroup of $\mathrm{SL}(V)$, and $m = n$, then $\Omega(F)$ is also an invariant, as follows from the definition or Capelli's identity.

Invariants for $\mathrm{SL}_n\mathbb{C}$.

Let $F \in S^{\mathbf{d}}$ be an invariant of the group $\mathrm{SL}_n\mathbb{C}$. We must show that F can be written as a polynomial in the basic bracket polynomials. In particular, if $m < n$, we must verify that there are no invariants except the constants in $S^0 = \mathbb{C}$. This is a simple consequence of the fact that for a dense open set of m-tuples of vectors—namely, those which are linearly independent—there is an automorphism of $\mathrm{SL}_n\mathbb{C}$ taking them to a fixed m-tuple of independent vectors, say $e_1, \ldots, e_m$. So an invariant function must take the same value on all such m-tuples. By the density, it must be constant.

For $m \geq n$, we proceed by induction as indicated above. All $D_{ij}F$ are known to be invariants (for $i < j$), as is $\Omega(F)$, so these are polynomials in the brackets.

To complete the proof, by Capelli's identities (F.8) and (F.9), it suffices to see that the operators D_{ab} all take brackets to scalar multiples of brackets. This is an obvious calculation: D_{ab} takes a bracket $[x^{(i_1)} x^{(i_2)} \dots x^{(i_n)}]$ to zero if b does not appear as one of the superscripts, or to the bracket with the variable $x^{(b)}$ replaced by $x^{(a)}$ if $x^{(b)}$ does occur; the latter is zero if $x^{(a)}$ also occurs and is a bracket otherwise. To avoid repeats, one needs only consider brackets where the superscripts are increasing. This completes the proof of

Proposition F.10. *Polynomial invariants $F(x^{(1)}, \dots, x^{(m)})$ of $\mathrm{SL}_n\mathbb{C}$ can be written as polynomials in the brackets*

$$[x^{(i_1)} x^{(i_2)} \dots x^{(i_n)}], \quad 1 \le i_1 < i_2 < \dots < i_n \le m.$$

Exercise F.11. Show that the only polynomial invariants of $\mathrm{GL}_n\mathbb{C}$ are the constants.

Invariants for $\mathrm{Sp}_n\mathbb{C}$

Let $r = n/2$, and let Q be the skew form defining the symplectic group $\mathrm{Sp}_n\mathbb{C}$, e.g. $Q(x, y) = \sum_{i=1}^r x_i y_{r+i} - x_{r+i} y_i$ in standard coordinates. Note first that the brackets are not needed:

Exercise F.12*. Show that the bracket $[x^{(1)} x^{(2)} \dots x^{(n)}]$ is equal to

$$\sum \mathrm{sgn}(\sigma) Q(x^{(\sigma(1))}, x^{\sigma(2)}) \cdot Q(x^{(\sigma(3))}, x^{\sigma(4)}) \cdot \ldots \cdot Q(x^{(\sigma(n-1))}, x^{\sigma(n)}),$$

where the sum is over all permutations σ of $\{1, \dots, n\}$ such that $\sigma(2i-1) < \sigma(2i)$ for $1 \le i \le r$ and $\sigma(i-1) < \sigma(i)$ for $2 \le i \le r$.

Let T_n^m be the assertion that any $\mathrm{Sp}_n\mathbb{C}$-invariant polynomial in m variables from $\mathbb{C}^n$ can be written as a polynomial in the basic polynomials $Q(x^{(i)}, x^{(j)})$. The antilexicographic induction using the Capelli identities is the same as before, and gives the implications

$$T_n^{n-1} \Rightarrow T_n^n \Rightarrow T_n^m \quad \text{for all } m > n.$$

The only variation here is to verify that the operators D_{ab} preserve polynomials in the basic invariants, and $D_{ab}Q(x^{(i)}, x^{(j)})$ is again zero or another basic invariant.

The situation where $m < n$ is a little more complicated than that for the special linear group, however—which is hardly surprising since there are nontrivial invariants for $\mathrm{Sp}_n\mathbb{C}$ in this range. Note that T_n^m implies $T_n^{m'}$ for $m' < m$, so it suffices to prove T_n^{n-1}. This is done by induction on $r = n/2$, i.e., by proving the implication $T_{n-2}^{n-1} \Rightarrow T_n^{n-1}$. To prove this, consider the restriction F' of an invariant polynomial F on $V = \mathbb{C}^n$ to the subspace $V' = \mathbb{C}^{n-2}$ perpendicular to the plane spanned by e_r and e_n. This restriction is an invariant of the group $\mathrm{Sp}_{n-2}\mathbb{C}$. By induction, F' is a polynomial in the basic invariants. Since $Q(x^{(i)}, x^{(j)})$ restricts to the corresponding invariant on V', there is a polynomial in these $Q(x^{(i)}, x^{(j)})$ such that F and this polynomial have the same restriction to V'. Subtracting, it suffices to prove that if an invariant F restricts to zero on V', then F is zero.

We show first that the restriction of F to the larger subspace $W = V' \oplus \mathbb{C}e_r$ must be zero. Fix $y^{(1)}, \ldots, y^{(m)}$ in V', and consider the function of m complex variables.

$$h(t_1, \ldots, t_m) = F(y^{(1)} + t_1 e_r, \ldots, y^{(m)} + t_m e_r).$$

The fact that F is invariant by automorphisms in $\mathrm{Sp}_n\mathbb{C}$ which fix V' and send e_r to $\alpha \cdot e_r$ and e_n to $\alpha^{-1} \cdot e_n$ shows that

$$h(\lambda t_1, \ldots, \lambda t_m) = h(t_1, \ldots, t_m) \quad \text{for all } \lambda \neq 0.$$

Since h is a polynomial, it must be constant, so $h(t_1, \ldots, t_m) = h(0, \ldots, 0) = 0$, as required.

Since F is invariant, it follows that the restriction of F to any hyperplane of the form $g \cdot W$, for any $g \in \mathrm{Sp}_n\mathbb{C}$ is zero. It is not hard to verify that every hyperplane in $\mathbb{C}^n$ has this form. So any $n - 1$ vectors lie in such an hyperplane, and so F is identically zero. This finishes the proof for the symplectic group:

Proposition F.13. *Polynomial invariants $F(x^{(1)}, \ldots, x^{(m)})$ of $\mathrm{Sp}_n\mathbb{C}$ can be written as polynomials in functions*

$$Q(x^{(i)}, x^{(j)}), \quad 1 \leq i < j \leq m.$$

Invariants for $\mathrm{SO}_n\mathbb{C}$

This time brackets may be needed, as well as the functions given by the symmetric form Q, but products of brackets are not required:

Exercise F.14. Prove the identity

$$[x^{(1)}\, x^{(2)} \ldots x^{(n)}] \cdot [y^{(1)}\, y^{(2)} \ldots y^{(n)}] = |Q(x^{(i)}, y^{(j)})|_{1 \leq i,j \leq n}$$

for any variables $x^{(1)}, \ldots, x^{(n)}, y^{(1)}, \ldots, y^{(n)}$.

Let T_n^m be the assertion that any $\mathrm{SO}_n\mathbb{C}$-invariant polynomial in m variables can be written as a polynomial in the brackets and the invariants $Q(x^{(i)}, x^{(j)})$, where we take $Q(x, y) = \sum_{i=1}^{n} x_i y_i$ to be the form determining the orthogonal group. The proofs of the implications $T_n^{n-1} \Rightarrow T_n^n \Rightarrow T_n^m$ for $m > n$ are exactly as in the preceding cases, and require no further comment. As before, it remains to prove T_n^{n-1}, and, by induction on n, it suffices to prove the implication $T_{n-1}^{n-1} \Rightarrow T_n^{n-1}$.

Let $V' = \mathbb{C}^{n-1}$ be the orthogonal complement to e_n. The restriction F' to V' of an $\mathrm{SO}_n\mathbb{C}$-invariant polynomial F is $\mathrm{SO}_{n-1}\mathbb{C}$-invariant, and by induction we know it is a polynomial in the restrictions of the basic polynomials $Q(x^{(i)}, x^{(j)})$ and in the bracket $[x^{(1)} \ldots x^{(n-1)}]$. An apparent snag is met here, however, since this bracket is not the restriction of an invariant on V. By Exercise F.14, we can write

$$F' = A + B \cdot [x^{(1)} \ldots x^{(n-1)}],$$

where A and B are polynomials in the Q's alone. In particular, A and B are *even*, i.e., they are invariants of the full orthogonal group $\mathrm{O}_{n-1}\mathbb{C}$. But F' is also even, since any element of $\mathrm{O}_{n-1}\mathbb{C}$ is the restriction of some element in $\mathrm{SO}_n\mathbb{C}$ (mapping e_n to $\pm e_n$). Since the bracket is taken to minus itself by automorphisms of determinant -1, we must have $F' = A$. This means that we can subtract a polynomial in the invariants $Q(x^{(i)}, x^{(j)})$ from F, so we can assume $F' = 0$. Therefore, the restriction of F to any hyperplane of the form $g \cdot V'$, $g \in \mathrm{SO}_n\mathbb{C}$, is zero. But it is easy to verify that $(n-1)$-tuples in such hyperplanes form an open dense subset of all $(n-1)$-tuples in $\mathbb{C}^n$ (the condition is that there be an orthogonal vector e with $Q(e \cdot e) \neq 0$). This proves:

Proposition F.15. *Polynomial invariants $F(x^{(1)}, \ldots, x^{(m)})$ of $\mathrm{SO}_n\mathbb{C}$ can be written as polynomials in functions*

$$Q(x^{(i)}, x^{(j)}) \quad \textit{and} \quad [x^{(i_1)} x^{(i_2)} \ldots x^{(i_n)}],$$

with $1 \leq i \leq j \leq m$, $1 \leq i_1 < i_2 < \cdots < i_n \leq m$.

Exercise F.16*. Show that the polynomial invariants of $\mathrm{O}_n\mathbb{C}$ can be written as polynomials in the functions $Q(x^{(i)}, x^{(j)})$, $1 \leq i < j \leq m$. Show that *odd* polynomial invariants of $\mathrm{O}_n\mathbb{C}$, i.e., polynomials F which are taken to $\det(g) \cdot F$ by g in $\mathrm{O}_n\mathbb{C}$, can be written as linear combinations of even invariants times brackets.

§F.2. Applications to Symplectic and Orthogonal Groups

We consider the symplectic group $\mathrm{Sp}_n\mathbb{C}$ and the orthogonal group $\mathrm{O}_n\mathbb{C}$ together, letting Q denote the corresponding skew or symmetric form. The results in the first section, applied to the case $\mathbf{d} = (1, \ldots, 1)$, say that the invariants in $(V^*)^{\otimes m}$ are all polynomials in the polynomials $Q(x^{(i)}, x^{(j)})$, and by degree considerations m must be even, and they are all linear combinations of products

$$Q(x^{(\sigma(1))}, x^{(\sigma(2))}) \cdot Q(x^{(\sigma(3))}, x^{(\sigma(4))}) \cdot \ldots \cdot Q(x^{(\sigma(m-1))}, x^{(\sigma(m))}) \qquad \text{(F.17)}$$

for permutations σ of $\{1, \ldots, m\}$ such that $\sigma(2i-1) < \sigma(2i)$ for $1 \leq i \leq m/2$. Regarding $Q \in V^* \otimes V^*$, these are obtained from the invariant $Q \otimes \cdots \otimes Q$ ($m/2$ times) by permuting the factors. In other words, one pairs off the m components, and inserts Q in the place indicated by each pair.

The form Q gives an isomorphism of V with V^*, which takes v to $Q(v, -)$. Using this we can find all invariants of tensor products $(V^*)^{\otimes k} \otimes (V)^{\otimes l}$, via the isomorphism

$$(V^*)^{\otimes(k+l)} = (V^*)^{\otimes k} \otimes (V^*)^{\otimes l} \cong (V^*)^{\otimes k} \otimes (V)^{\otimes l}.$$

They are linear combinations of the images of the above invariants under this identification. To see what they are, we just need to see what happens to Q under the isomorphisms $V^* \otimes V^* \cong V^* \otimes V$ and $V^* \otimes V^* \cong V \otimes V$:

Exercise F.18. (i) Verify that under the canonical isomorphism

$$V^* \otimes V^* \cong V^* \otimes V = \operatorname{Hom}(V, V) = \operatorname{End}(V)$$

Q maps to the identity endomorphism. (ii) Let ψ be the image of Q under the canonical isomorphism $V^* \otimes V^* \cong V \otimes V$. Verify that

$$\psi = \sum_{i=1}^{r} e_i \otimes e_{r+i} - e_{r+i} \otimes e_i \quad \text{for } G = \mathrm{Sp}_n\mathbb{C},\ n = 2r;$$

$$\psi = \sum_{i=1}^{n} e_i \otimes e_i \quad \text{for } G = \mathrm{O}_n\mathbb{C}.$$

For the applications in Lectures 17 and 19, we need only the case $l = k$, but we want to reinterpret these invariants by way of the canonical isomorphism

$$(V^*)^{\otimes 2d} \cong (V^*)^{\otimes d} \otimes (V)^{\otimes d} \cong \operatorname{Hom}(V^{\otimes d}, V^{\otimes d}) = \operatorname{End}(V^{\otimes d}). \quad \text{(F.19)}$$

In §§17.3 and 19.5 we defined endomorphisms $\vartheta_I \in \operatorname{End}(V^{\otimes d})$ for each pair I of integers from $\{1, \ldots, d\}$; for I the first pair,

$$\vartheta_I(v_1 \otimes v_2 \otimes v_3 \otimes \cdots \otimes v_d) = Q(v_1, v_2) \cdot \psi \otimes v_3 \otimes \cdots \otimes v_d;$$

the case for general I is a permutation of this. We claim that an invariant in $(V^*)^{\otimes 2d}$ of the form (F.17) is taken by the isomorphism (F.19) to a composition of operators ϑ_I and permutations σ in $\mathfrak{S}_d$. This is simply a matter of unraveling the definitions, which may be simpler to follow pictorially than notationally. The invariant in (F.17) is described by pairing the integers from 1 to $2d$. These pairs are either from the first d, the last d, or one of each. For example, if $d = 5$ the pairings could be as indicated:

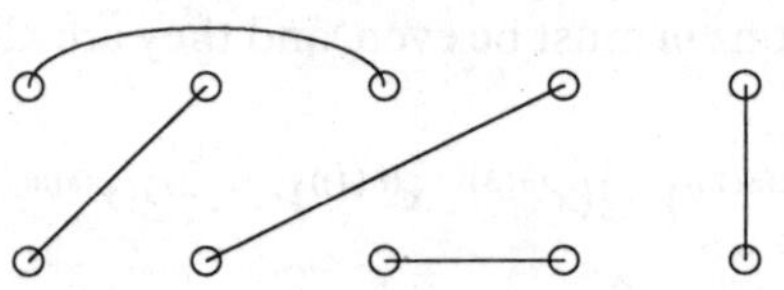

for the pairs {1, 3}, {8, 9}, {2, 6}, {4, 7}, {5, 10}. Composing before and after with permutations, this can be changed to

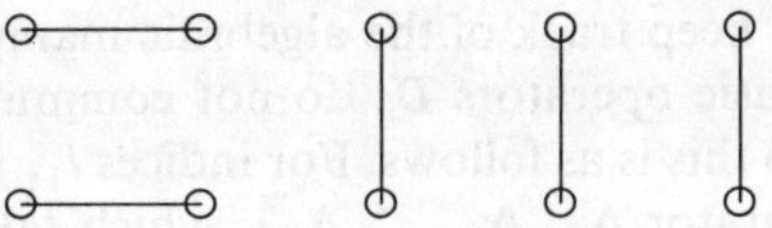

The corresponding endomorphism of $V^{\otimes 5}$ becomes ϑ_I, $I = \{1, 2\}$. The general invariant one gets can be expressed in the form

$$\sigma \circ \vartheta_{I_1} \circ \vartheta_{I_2} \circ \cdots \circ \vartheta_{I_p} \circ \tau,$$

where σ and τ permute the d factors, and the pairs I_j are the first p pairs: $I_j = \{2j - 1, 2j\}$.

Now let A be the subalgebra of the ring $\mathrm{End}(V^{\otimes d})$ generated by all $g \otimes \cdots \otimes g$ for g in the group $G = \mathrm{Sp}_n\mathbb{C}$ (or $\mathrm{O}_n\mathbb{C}$). By the simplicity of the group, we know that A is a semisimple algebra of endomorphisms. We have just computed that the ring B of commutators of A is the ring generated by all permutations in $\mathfrak{S}_d$ and the operators ϑ_I. By the general theory of semisimple algebras, cf. §6.2, A must be the commutator algebra of B. In English, *any endomorphism of $V^{\otimes d}$ which commutes with permutations and with the operators ϑ_I must be a finite linear combination of operators of the form $g \otimes \cdots \otimes g$ for g in G.* This is precisely the fact from invariant theory that was used in the text.

We remark that a similar procedure can be used for $\mathrm{SL}_n\mathbb{C}$, but since in this case V and V^* are not isomorphic, to do this one must first do some more work to compute invariants in tensor products of covariant and contravariant factors. The idea is simple enough: use the canonical isomorphism $V \cong \wedge^{n-1}(V^*)$ to turn each V factor into several V^* factors. Tracing through the invariants by this procedure is rather complicated, however, and we refer to [We1, II.8] for details. We did not need this analysis, because it was easy to work the commutator story the other way around, showing that the commutator of $\mathbb{C}[\mathfrak{S}_d]$ is the algebra generated by all $g \otimes \cdots \otimes g$ for g in $\mathrm{SL}_n\mathbb{C}$ (or $\mathrm{GL}_n\mathbb{C}$). This can, in turn, be run backwards:

Exercise F.20*. Use the fact that the the $\mathrm{GL}_n\mathbb{C}$-invariants of $\mathrm{End}(V^{\otimes d})$ are generated by permutations to show that the $\mathrm{GL}_n\mathbb{C}$-invariants of $(V^*)^{\otimes d} \otimes V^{\otimes d}$ are obtained by pairing off the factors and contracting. There are no $\mathrm{GL}_n\mathbb{C}$-invariants in $(V^*)^{\otimes k} \otimes V^{\otimes l}$ if $k \neq l$. For $\mathrm{SL}_n\mathbb{C}$-invariants, one also has determinant factors when $k - l$ is a multiple of the dimension.

We also omit any discussion of the *second fundamental theorems*, which describe the relations among the generators of the rings of invariants (but see the discussions at the ends of Lectures 17 and 19). These results can also be found in [We1].

§F.3. Proof of Capelli's Identity

The proof is not essentially different from the case $m = 2$, once one has a good notational scheme to keep track of the algebraic manipulations which come about because the basic operators D_{ij} do not commute with each other. A convenient way to do this is as follows. For indices $i_1, j_1, \dots, i_p, j_p$ between 1 and m, define an operator $\Delta_{i_1j_1}\Delta_{i_2j_2}\dots\Delta_{i_pj_p}$ which takes a function F of m variables $x^{(1)}, \dots, x^{(m)}$ to the function

$$\Delta_{i_1j_1}\cdots\Delta_{i_pj_p}(F) = \sum_{k_1,\dots,k_p=1}^{n} x_{k_1}^{(i_1)}\cdot\ldots\cdot x_{k_p}^{(i_p)}\cdot\frac{\partial^p F}{\partial x_{k_1}^{(j_1)}\cdot\ldots\cdot\partial x_{k_p}^{(j_p)}}.$$

For $p = 1$, Δ_{ij} is just the operator D_{ij}, but for $p > 1$, this is *not* the composition of the operators $\Delta_{i_kj_k}$. Note that the order of the terms in the expression $\Delta_{i_1j_1}\cdots\Delta_{i_pj_p}$ is unimportant.

We can form determinants of $p \times p$ matrices with entries these Δ_{ij}, which act on functions by expanding the determinant as usual, with each of the $p!$ products operating as above. For example, for the $m \times m$ matrix (Δ_{ij}),

$$|\Delta_{ij}|(F) = \sum_{\sigma\in\mathfrak{S}_m}\operatorname{sgn}(\sigma)\cdot\Delta_{1\sigma(1)}\Delta_{2\sigma(2)}\cdot\ldots\cdot\Delta_{m\sigma(m)}(F).$$

The matrix (Δ_{ij}) is a product of matrices $(x_k^{(i)})\cdot(\partial/\partial x_k^{(j)})$, and taking determinants gives the

Lemma F.21. *For $m = n$, $|\Delta_{ij}|(F) = [x^{(1)}\dots x^{(m)}]\cdot\Omega(F)$.*

To prove Capelli's identity (F.6), then, we must prove the following identity of operators on functions $F(x^{(1)}, \dots, x^{(m)})$:

$$\begin{vmatrix} D_{11}+m-1 & D_{12} & \cdots & D_{1m} \\ D_{21} & D_{22}+m-2 & \cdots & D_{2m} \\ \vdots & & & \vdots \\ D_{m1} & D_{m2} & \cdots & D_{mm}\end{vmatrix} = \begin{vmatrix} \Delta_{11} & \Delta_{12} & \cdots & \Delta_{1m} \\ \Delta_{21} & \Delta_{22} & \cdots & \Delta_{2m} \\ \vdots & & & \vdots \\ \Delta_{m1} & \Delta_{m2} & \cdots & \Delta_{mm}\end{vmatrix} \tag{F.22}$$

This is a formal identity, based on the simple identities:

$$D_{qp}\circ D_{ab} = D_{qp}\Delta_{ab} = \Delta_{qp}\Delta_{ab} \quad \text{if } p \neq a;$$

$$D_{qp}\circ D_{ab} = \Delta_{qp}\Delta_{ab} + D_{qb} \quad \text{if } p = a.$$

Similarly, if $p \neq a_k$ for all k, then

$$D_{qp}\circ\Delta_{a_1b_1}\cdots\Delta_{a_rb_r} = \Delta_{qp}\Delta_{a_1b_1}\cdots\Delta_{a_rb_r}; \tag{F.23}$$

while if there is just one k with $p = a_k$, then

$$D_{qp}\circ\Delta_{a_1b_1}\cdots\Delta_{a_rb_r} = \Delta_{qp}\Delta_{a_1b_1}\cdots\Delta_{a_rb_r} + \Delta_{a_1b_1}\cdots\Delta_{qb_k}\cdots\Delta_{a_rb_r} \tag{F.24}$$

where in the last term the Δ_{qb_k} replaces $\Delta_{a_kb_k}$.

We prove (F.22) by showing inductively that all $r \times r$ minors of the two matrices of (F.22) which are taken from the last r columns are equal (as operators on functions F as always). This is obvious when $r = 1$. We suppose it has been proved for $r = m - p$, and show it for $r + 1$. By induction, we may replace the last r columns of the matrix on the left by the last r columns of the matrix on the right. The difference of a minor on the left and the corresponding minor on the right will then be a maximal minor of the matrix

$$\begin{vmatrix} D_{1p} - \Delta_{1p} & \Delta_{1p+1} & \cdots & \Delta_{1m} \\ D_{2p} - \Delta_{2p} & \Delta_{2p+1} & \cdots & \Delta_{2m} \\ \vdots & & & \vdots \\ D_{pp} - \Delta_{pp} + r & \Delta_{pp+1} & \cdots & \Delta_{pm} \\ \vdots & & & \vdots \\ D_{mp} - \Delta_{mp} & \Delta_{mp+1} & \cdots & \Delta_{mm} \end{vmatrix},$$

so we must show that all maximal minors of this matrix are zero. Suppose the minor chosen is that using the q_ith rows, for $1 \le q_0 < q_1 < \cdots < q_r \le m$. Expanding along the left column, this determinant is

$$E_0 M_0 - E_1 M_1 + E_2 M_2 - \cdots + (-1)^r E_r M_r, \tag{F.25}$$

where $E_k = D_{q_k p} - \Delta_{q_k p}$ if $q_k \ne p$, and $E_k = D_{pp} - \Delta_{pp} + r$ if $q_k = p$, and M_k is the corresponding cofactor ($r \times r$) determinant:

$$\sum_{\sigma \in \mathfrak{S}_r} \operatorname{sgn}(\sigma) \Delta_{q_0 p+\sigma(1)} \cdots \Delta_{q_{k-1} p+\sigma(k)} \Delta_{q_{k+1} p+\sigma(k+1)} \cdots \Delta_{q_r p+\sigma(r)}. \tag{F.26}$$

To show that (F.25) is zero, there are two cases. In the first case, the pth row is not included in the minor, i.e., $q_i \ne p$ for all i. In this case each term $E_i M_i$ is zero, since $E_i = D_{q_i p} - \Delta_{q_i p}$, and all the products in the expansion of M_i are of the form $\Delta_{a_1 b_1} \ldots \Delta_{a_r b_r}$ with all $a_i \ne p$, and the assertion follows from (F.23).

In the second case, the pth row is included, i.e., $q_k = p$ for some k. As in the first case, $(D_{pp} - \Delta_{pp}) M_k = 0$, and since $E_k = D_{pp} - \Delta_{pp} + r$, we have

$$E_k M_k = r \cdot M_k.$$

We claim that each of the other terms $E_i M_i$, for $i \ne k$, is equal to $(-1)^{k-i+1} M_k$, from which it follows that the alternating sum in (F.25) is zero. When M_i is written out as in (F.26), and it is multiplied by $E_i = D_{q_i p} - \Delta_{q_i p}$, an application of (F.24) shows that one gets the same determinant as (F.26), but expanded with the q_ith row moved between the q_{k-1}th and the q_{k+1}th rows. This transposition of rows accounts for the sign $(-1)^{k-i+1}$, yielding $E_i M_i = (-1)^{k-i+1} M_k$, as required. □

Exercise F.27. Find a GL(V)-linear surjection from $S^{d_1} \otimes \cdots \otimes S^{d_n}$ onto $\bigwedge^n V^* \otimes S^{d_1 - 1} \otimes \cdots \otimes S^{d_n - 1}$ that realizes the map $F \mapsto [x^{(1)} \ldots x^{(n)}] \cdot \Omega(F)$.

Hints, Answers, and References

Note: Usually answers or references are given only for more theoretical exercises, or those which may be referred to elsewhere.

Lecture 1

(1.3) The hypotheses ensure that $\wedge^n V$ is trivial, and the bilinear map $\wedge^k V \otimes \wedge^{n-k} V \to \wedge^n V = \mathbb{C}$ is a perfect pairing, i.e., it makes each space the dual of the other, cf. §B.3.

(1.4) For (b), take the function α to the function α', where $\alpha'(g) = \alpha(g^{-1})$.

(1.13) Yes. See Exercise 6.18.

(1.14) If H is a Hermitian inner product on V, let $\tilde{H}: V \to V^*$ be the conjugate linear map given by $v \mapsto H(v, \cdot)$. If H' is another, the composite $(\tilde{H}')^{-1} \circ \tilde{H}$ is linear, and a G-homomorphism if H and H' are G-invariant. Apply Schur's lemma.

Lecture 2

(2.3) For a general formula expressing complete symmetric polynomials and elementary symmetric polynomials in terms of sums of powers, see Exercise A.32(vi).

(2.4) Look at the induced action on $\wedge^k V$.

(2.7) $V^{\otimes n} = U^{\oplus a} \oplus U'^{\oplus b} \oplus V^{\oplus c}$, with $a = b = \frac{1}{3}(2^{n-1} + (-1)^n)$, and $c = \frac{1}{3}(2^n + (-1)^{n-1})$.

(2.25) Answers: (i) $U \oplus V \oplus U' \oplus V'$; (ii) $U \oplus V^{\oplus 2} \oplus V' \oplus W$.

(2.29) The regular representation will do.

(2.33) For (c) use characters or the isomorphism

$$\mathrm{Hom}_G(V \otimes W, U) \cong \mathrm{Hom}_G(W, V^* \otimes U).$$

(2.34) Schur's lemma applies to L.

(2.35) Apply the preceding exercise, with L_0 given by a matrix of indeterminates. For details, see [Se2, §2.2].

(2.36) Show that $(\chi, \chi) = 1$, and compute the sum of the squares of these representations. Reference: [Se2, §3.2].

(2.37) If φ is the character of an irreducible representation, and χ is the character of V, let $a_n = (\varphi, \chi^n)$, and consider the power series

$$\sum_{n=0}^{\infty} a_n t^n = \frac{1}{|G|} \sum_{n=0}^{\infty} \sum_{C} |C| \overline{\varphi(C)} \chi(C)^n t^n = \frac{1}{|G|} \sum_{C} \frac{|C| \overline{\varphi(C)}}{(1 - \chi(C)t)}.$$

Here C runs over conjugacy classes. Since $\chi(C) = \dim(V)$ only for $C = [e]$, the right-hand side is a nontrivial rational function; in particular a_n cannot be zero for all positive n.

(2.38) This is another theorem of Burnside. If C is a conjugacy class in G, $\varphi = \sum_{g \in C} g\colon V \to V$ is a G-map, so multiplication by a scalar λ_C, and $\lambda_C \cdot \dim V = \mathrm{Trace}(\varphi) = |C| \cdot \chi_V(C)$. The λ_C are algebraic integers, since the elements $\sum_{g \in C} e_g$, as C varies over the conjugacy classes, generate the center of the group ring $\mathbb{Z}[G]$, which is a finitely generated abelian group. Now

$$\sum_{C} |C| \cdot \overline{\chi_V(C)} \chi_V(C) = |G|,$$

so $|G|/\dim V = \sum_C \lambda_C \cdot \overline{\chi_V(C)}$ is an algebraic integer. In fact, the dimension of V divides the index of the center of G, cf. [Se2, p. 53].

(2.39) In case the character χ is $\mathbb{Z}$-valued, the equation $\sum |\chi(g)|^2 = |G|$ shows that $|G|$ is the sum of $|G|$ non-negative integers, one of which, $|\chi(e)|^2$, is greater than 1, so at least one must be 0. In general, the values of χ are algebraic integers, since they are sums of roots of unity. Let $\chi_1, \ldots \chi_m$ be the characters obtained from χ by the action of the Galois group $\mathrm{Gal}(\overline{\mathbb{Q}}/\mathbb{Q})$ (or $\mathrm{Gal}(\mathbb{C}/\mathbb{Q})$) on χ; these characters are also characters of irreducible representations of G. Now if $\chi(g) \neq 0$, then $\prod_i \chi_i(g)$ is a nonzero integer, so $|\prod_i \chi_i(g)|^2 \geq 1$. Since the arithmetic mean is at least the geometric mean, $\sum_i |\chi_i(g)|^2 \geq m$. Therefore,

$$m|G| = \sum_{i=1}^{m} \sum_{g \in G} |\chi_i(g)|^2 \geq m|G|,$$

and we must have equality for every $g \in G$. In particular, if d is the degree of the representation, $md^2 = \sum_i |\chi_i(e)|^2 = m$, so $d = 1$.

Lecture 3

(3.5) Use the fact that $g = (12345)$ is conjugate to its inverse, so $\chi(g) = \chi(g^{-1}) = \overline{\chi(g)}$ is real.

(3.25) See §5.1.

(3.26) If $H \subset G$ is the subgroup of order 7, there are three one-dimensional representations from G/H, and two three-dimensional representations induced from H. For generalizations, see [Se2, §8.2].

(3.30) W is embedded in the space of W-valued functions on G by sending $w \in W$ to the function which takes $h \in H$ to $h \cdot w$ and all other cosets to zero. Note that if $\{g_\sigma\}$ is a set of coset representatives, the map $f \mapsto \sum g_\sigma \otimes f(g_\sigma^{-1})$ gives an isomorphism from $\mathrm{Hom}_H(\mathbb{C}G, W)$ to $\mathbb{C}G \otimes_{\mathbb{C}H} W$.

(3.32) For (b), identify the right-hand side with the trace of an endomorphism of $\mathbb{C}G$. For (c), take φ to be the characteristic function of an element g and apply (b).

(3.33) F is the determinant of left multiplication by the element $a = \sum x_g e_g \in \mathbb{C}G$ on the regular representation, and F_ρ is the determinant of left multiplication by a on the irreducible $\mathbb{C}G$-module V_ρ corresponding to ρ. The factorization of F follows from the decomposition of the regular representation. The irreducibility of F_ρ follows from the irreducibility of a matrix whose entries are indeterminates, using Proposition 3.29. Fixing g in G, set the variables $x_e = 1$ and $x_h = 0$ for $h \neq g$; the coefficient of x_g in the determinant of left multiplication by $1 + x_g e_g$ on V_ρ is $\chi_\rho(g)$.

(3.34) See Exercises 3.8 and 3.9.

(3.38) V can be replaced by V^*; $V \otimes V = \mathrm{Sym}^2 V \oplus \wedge^2 V$ contains at most one copy of the trivial representation. If $\mathrm{Sym}^2 V$ contains the trivial representation, then

$$|G| = \sum_{g \in G} \chi_{\mathrm{Sym}^2 V}(g) = \tfrac{1}{2}\left(\sum \chi_V(g)^2 + \sum \chi_V(g^2)\right).$$

Otherwise, the right-hand side is zero; similarly for $\wedge^2 V$. Note that if χ_V is real, then $\sum \chi_V(g)^2 = |G|$.

(3.41) Reference: [Se2, §13.2].

(3.42) Reference: [Ja-Ke, p. 12].

(3.43) Consider the endomorphism $J \otimes J$ of $V \otimes W$.

(3.44) For $G = \mathbb{Z}/3$, the rank of $R_{\mathbb{R}}(G)$ is 2, whereas that of $R(G)$ is 3.

(3.45) See [Se2, §12] for details.

Lecture 4

(4.4) Right multiplication by a gives a map $Aab \to Aba$, and right multiplication by b gives a map back. The composites are multiplications by nonzero scalars. More generally, if $A = \mathbb{C}G$ is a group algebra, call an element $a = \sum a_g e_g$ *Hermitian* if $\hat{a} = \bar{a}$, i.e., $a_{g^{-1}} = \overline{a_g}$. If a and b are idempotents which are Hermitian, then $Aab \cong Aba$.

(4.6) A basis for $V_{(d-1,1)} = \mathbb{C}\mathfrak{S}_d \cdot c_\lambda$ is $v_2, \ldots, v_d$, where

$$v_j = \sum_{g(d)=j} e_g - \sum_{h(1)=j} e_h.$$

Note that $v_d = c_\lambda$, $v_1 + \cdots + v_d = 0$, and $g \cdot v_j = v_i$ if $g(j) = i$. A basis for $V \subset \mathbb{C}^d$ is v_2, $\ldots$, v_d, where $v_j = e_j - e_{j-1}$. For the case $s > 1$, use (4.10) or see (4.43).

(4.13) Note that the hook lengths of the boxes in the first column are the numbers l_1, $\ldots$, l_k. Induct from the diagram obtained by omitting the first column.

(4.14) Induct as in the preceding exercise by removing the first column, considering separately the cases when the remaining diagram is one of the exceptions.

(4.15) Frobenius [Fro1] gives these and analogous formulas for $\lambda = (d-3, 3)$, $(d-3, 1, 1, 1)$, $(d-4, 4), \ldots$.

(4.16) Using Frobenius's formula, the coefficient of $x_1^{l_1} \cdot \ldots \cdot x_k^{l_k}$ in $\Delta \cdot (x_1^d + \cdots + x_k^d)$ can be nonzero only if $l_1 = d$, so λ has the prescribed form; the coefficient of $x_2^{k-1} x_3^{k-2} \cdot \ldots \cdot x_k$ in $\Delta(0, x_2, \ldots, x_k)$ is $(-1)^{k-1}$.

(4.19) See Exercise 4.51 for a general procedure for decomposing tensor products.

(4.20) Use Frobenius's formula as in Exercise 4.16 to show that $\chi_\lambda(g) = (-1)^{k-1}\chi_\mu(h)$, where $\mu = (\lambda_2 - 1, \lambda_3 - 1, \ldots, \lambda_k - 1)$ and $h \in \mathfrak{S}_{d-q_1}$ is the product of cycles of lengths $q_2, \ldots, q_r$.

(4.24) If $\lambda < \mu$ use the anti-involution $\hat{}$ of A induced by the map $g \mapsto g^{-1}$, $g \in \mathfrak{S}_d$, noting that $\hat{c}_\lambda = (a_\lambda b_\lambda)\hat{} = \hat{b}_\lambda \hat{a}_\lambda = b_\lambda a_\lambda$, so $(c_\lambda \cdot x \cdot c_\mu)\hat{} = \hat{c}_\mu \cdot \hat{x} \cdot \hat{c}_\lambda = b_\mu \cdot (a_\mu \cdot \hat{x} \cdot b_\lambda) \cdot a_\lambda = 0$.

(4.40) Note that the ψ_λ's are related to the χ_λ's by the same equations as the symmetric polynomials H_λ's to the Schur polynomials S_λ's, cf. (A.9) in the appendix. The equation (A.5) for the S_λ's in terms of the H_λ's therefore implies the determinantal formula.

(4.43) Use Frobenius reciprocity and (4.42) to prove the general formula. To prove that $V_{(d-s,1,\ldots,1)} \cong \wedge^s V$, argue by induction on d. Note that the restriction of $\wedge^s V$ splits into a sum of two exterior powers of the standard representation, and from anything but a hook one can remove at least three boxes.

(4.44) The induced representation of V_λ by the inclusion of $\mathfrak{S}_d$ in $\mathfrak{S}_{d+m}$ is $V_\lambda \circ V_{(m)}$. Use the transitivity of induction, Exercise 3.16(b).

(4.45) For (a), see [Jam, pp. 79–83]. For (b), using (4.33), the coefficient of X^a in $(x_1^m + \cdots + x_k^m) \cdot P^{(\mathrm{i})}$ is the sum of the coefficients of $X^a x_i^{-m}$ in $P^{(\mathrm{i})}$, summing over those i for which $a_i \geq m$. Use the determinantal formula to write $\chi_\lambda(g)$ as a sum $\Sigma \pm \chi_\mu(h)$, and show that the μ which occur are those obtained by removing skew hooks. Reference: [Boe, pp. 192–196].

(4.46) See Exercise A.11. In fact, this condition is equivalent to the condition that $K_{\rho\lambda} \leq K_{\rho\mu}$ for all ρ, or to the condition that U_λ is isomorphic to $U_\mu \oplus W$, for some representation W, cf. [L-V].

(4.47) References: For the first construction see [Jam], [Ja-Ke]; for the second, see [Pe2].

(4.48) There are several ways to do this: (i) Use the methods of this lecture to show that the value of the character of U'_λ on the class C_{i} is $[\vartheta(P^{(\mathrm{i})})]_{\lambda'}$, where ϑ is the involution defined in Exercise A.32. Then apply Lemma A.26. (ii) Show that $U'_\lambda \otimes U'$ is isomorphic to $U_{\lambda'}$ and use Corollary 4.39. (iii) Use Exercise 4.40 or 4.44.

(4.49) Use Exercise A.32(v).

(4.51) (a) Note that $\chi_\lambda = \sum_{\mathrm{i}} \omega_\lambda(\mathrm{i})\xi_{(\mathrm{i})}$, and $\xi_{(\mathrm{i})} = (1/z(\mathrm{i}))\sum_\nu \omega_\nu(\mathrm{i})\chi_\nu$, where $\xi_{(\mathrm{i})}$ is the characteristic function of the conjugacy class $C_{(\mathrm{i})}$. Therefore,

$$\chi_\lambda \chi_\mu = \sum_{\mathrm{i}} \omega_\lambda(\mathrm{i})\omega_\mu(\mathrm{i})\xi_{(\mathrm{i})},$$

from which the required formula follows. For other procedures and tables for small d see [Ja-Ke], [Co], and [Ham].

(b) $V_\lambda \otimes V_{(d)} = V_\lambda$, and $V_\lambda \otimes V_{(1,\ldots,1)} = V_{\lambda'}$, which prove the corresponding results for $C_{\lambda(d)\mu}$ and $C_{\lambda(1,\ldots,1)\mu}$. Use (a) to permute the subscripts.

(4.52) For (a), the described map from Λ to R is surjective by the determinantal formula of Exercise 4.40; it is an isomorphism since R_n and Λ_n are free of the same rank. For (f), note that $P^{(\mathbf{i})}$ corresponds to the character $\sum_\lambda \chi_\lambda(C_{(\mathbf{i})})\chi_\lambda$, which by Exercise 2.21 is the class function which is zero outside the conjugacy class $C_{(\mathbf{i})}$, and whose value on $C_{(\mathbf{i})}$ is $z(\mathbf{i})$.

For more on this correspondence, see [Bu], [Di2], [Mac]. In [Kn] a λ-ring structure on this ring is related to representation theory. In [Liu] this Hopf algebra is used to derive many of the facts about representations of $\mathfrak{S}_d$ from scratch. In [Ze] a similar approach is also used for representations of $\mathrm{GL}_n(\mathbb{F}_q)$.

More about representations of the symmetric groups can also be found in [Foa] and [J-L].

Lecture 5

(5.2) Consider the class functions on H which are invariant by conjugation by an element not in H.

(5.4) *Step* 1. (i) Inverses of elements of c' are conjugate to elements of c' if m is even, and to elements of c'' if m is odd; $\chi(g^{-1}) = \overline{\chi(g)}$. (ii) (ϑ, ϑ) is

$$\frac{2}{d!}(\#c' \cdot |u - v|^2 + \#c'' \cdot |v - u|^2) = \frac{2}{d!}\frac{d!}{q_1 \cdot \ldots \cdot q_r}|u - v|^2.$$

(iii) If λ corresponded to $p \neq q$, the values of χ'_λ and χ''_λ on the corresponding conjugacy classes $c'(p)$ and $c''(p)$ would be the same number, say w, and Exercise 4.20 implies that $2w = \pm 1$. Since w is an algebraic integer, this is impossible. Therefore, λ corresponds to q, and now from Exercise 4.20 we get the additional equation $u + v = (-1)^m$.

Step 2. (ii) Information about the characters χ' and χ'' of X' and X'' is easily determined from Exercise 3.19, and the fact that the characters of the factors are known by induction. In particular, since $c'(q)$ and $c''(q)$ each decomposes into two conjugacy classes in H, we have

$$\chi'(c'(q)) = \frac{\varepsilon_1 + \sqrt{\varepsilon_1 q_1}}{2} \cdot \frac{\varepsilon' + \sqrt{\varepsilon' q'}}{2} + \frac{\varepsilon_1 - \sqrt{\varepsilon_1 q_1}}{2} \cdot \frac{\varepsilon' - \sqrt{\varepsilon' q'}}{2}$$

$$= \frac{\varepsilon + \sqrt{\varepsilon q_1 \cdot \ldots \cdot q_r}}{2},$$

where $\varepsilon_1 = (-1)^{(q_1-1)/2}$, $\varepsilon' = (-1)^{(d-q_1-r+1)/2}$, $\varepsilon = \varepsilon_1 \cdot \varepsilon'$, and $q' = q_2 \cdot \ldots \cdot q_r$; and similarly for the other values. (iv) The character of Y takes equal values on each pair of conjugate classes. (Reference: [Fro2], [Boe]).

(5.5) Reference: [Ja-Ke].

(5.9) If N is a normal subgroup properly between $\{\pm 1\}$ and $\mathrm{SL}_2(\mathbb{F}_q)$, one of the nontrivial characters χ must take the value $\chi(1)$ identically on N.

(5.11) Reference: [Ste1].

Lecture 6

(6.4) Compare (1) of the theorem with formulas (4.11) and (4.12). For a procedure to construct a basis of $\mathbb{S}_\lambda V$, see Exercise 6.28.

(6.10) By (4.41), there is an isomorphism of $\mathbb{C}\mathfrak{S}_{d+m}$-modules:

$$\mathbb{C}\mathfrak{S}_{d+m} \otimes_{\mathbb{C}(\mathfrak{S}_d \times \mathfrak{S}_m)} (V_\lambda \boxtimes V_\mu) \cong \bigoplus_\nu N_{\lambda\mu\nu} V_\nu.$$

Tensoring on the left with the right $\mathbb{C}\mathfrak{S}_{d+m}$-module $V^{\otimes(d+m)} = V^{\otimes d} \otimes_{\mathbb{C}} V^{\otimes m}$, and noting that $\mathbb{C}(\mathfrak{S}_d \times \mathfrak{S}_m) = \mathbb{C}\mathfrak{S}_d \otimes \mathbb{C}\mathfrak{S}_m$,

$$(V^{\otimes d} \otimes_{\mathbb{C}} V^{\otimes m}) \otimes_{\mathbb{C}\mathfrak{S}_d \otimes \mathbb{C}\mathfrak{S}_m} (V_\lambda \otimes V_\mu) \cong \bigoplus_\nu N_{\lambda\mu\nu} \mathbb{S}_\nu V.$$

(This also uses the general fact: if $A \to B$ is a ring homomorphism, N a left A-module, and M a right B-module, then $M \otimes_B (B \otimes_A N) \cong M \otimes_A N$.) The left-hand side of the displayed equation is

$$(V^{\otimes d} \otimes_{\mathbb{C}\mathfrak{S}_d} V_\lambda) \otimes_{\mathbb{C}} (V^{\otimes m} \otimes_{\mathbb{C}\mathfrak{S}_m} V_\mu) \cong \mathbb{S}_\lambda(V) \otimes \mathbb{S}_\mu(V),$$

which concludes the proof.

(6.11) (a) The key observation is that

$$(V \oplus W)^{\otimes d} = \bigoplus (V^{\otimes a} \otimes W^{\otimes b}) \otimes_{\mathbb{C}(\mathfrak{S}_a \times \mathfrak{S}_b)} \mathbb{C}(\mathfrak{S}_d),$$

the sum over all a, b with $a + b = d$. Tensoring this on the right with the $\mathbb{C}(\mathfrak{S}_d)$-module V_ν one gets

$$(V \oplus W)^{\otimes d} = \bigoplus (V^{\otimes a} \otimes W^{\otimes b}) \otimes_{\mathbb{C}(\mathfrak{S}_a \times \mathfrak{S}_b)} \operatorname{Res}_{a,b} V_\nu,$$

where $\operatorname{Res}_{a,b}$ denotes the restriction to $\mathfrak{S}_a \times \mathfrak{S}_b$. Then use Exercise 4.43 to decompose this restriction.

(b) By Frobenius reciprocity, the representation induced by V_ν via the diagonal embedding of $\mathfrak{S}_d$ in $\mathfrak{S}_d \times \mathfrak{S}_d$ is $\bigoplus C_{\lambda\mu\nu} V_\lambda \boxtimes V_\mu$. With $A = \mathbb{C}\mathfrak{S}_d$, this says

$$(A \otimes A) \otimes_A Ac_\nu = \bigoplus C_{\lambda\mu\nu}(Ac_\lambda \otimes Ac_\mu).$$

Tensor this with the right $(A \otimes A)$-module $(V \otimes W)^{\otimes d} = V^{\otimes d} \otimes W^{\otimes d}$. The special case follow from Exercise 4.51(b).

(6.13) Use Exercise A.32(iv), or write the left side as $V^{\otimes d} \otimes A \cdot b_\lambda$ and use Exercise 4.48.

(6.14) These come from the realizations of the representation $V_\lambda = Ac_\lambda$ as the image of the maps $Ab_\lambda \to Aa_\lambda$ given by right multiplication by a_λ, and similarly $Aa_\lambda \to Ab_\lambda$ by right multiplication by b_λ.

(6.15) It is clear that if one allows T to vary over all tableaux with strictly increasing columns but no conditions on the rows, then the corresponding v_T span the first space $\bigotimes_i (\wedge^{\mu_i} V)$; to show that the v_T for T semistandard span the image the key point is to show how to interchange elements in successive rows. Once it is checked that the elements span, the independence can be deduced from the fact that the number of semistandard tableaux is the same as the dimension. For a direct proof of both spanning and independence, see [A-B-W]—but note that their partitions are all the conjugates of ours. See also Proposition 15.55.

(6.16) Use Exercise 6.14 to realize each $\mathbb{S}_\lambda V$ which occurs as the image in $V^{\otimes d} \otimes V^{\otimes d}$ of a symmetrizing map, and check whether this image is invariant or anti-invariant by the map which permutes the two factors.

(6.17) (a) Identifying the dm elements on which $\mathfrak{S}_{dm}$ acts with the set of pairs $\{(i, j) | 1 \le i \le d, 1 \le j \le m\}$ determines embeddings of the groups $\mathfrak{S}_d \times \cdots \times \mathfrak{S}_d$ (m factors) and $\mathfrak{S}_m$ in $\mathfrak{S}_{dm}$. Let

$$c' = c_\lambda \otimes \cdots \otimes c_\lambda \in \mathbb{C}\mathfrak{S}_d \otimes \cdots \otimes \mathbb{C}\mathfrak{S}_d = \mathbb{C}(\mathfrak{S}_d \times \cdots \times \mathfrak{S}_d) \subset \mathbb{C}\mathfrak{S}_{dm},$$
$$c'' = c_\mu \in \mathbb{C}\mathfrak{S}_m \subset \mathbb{C}\mathfrak{S}_{dm}.$$

Then $c = c' \cdot c''$ is the required element of $\mathbb{C}\mathfrak{S}_{dm}$. For a combinatorial description of plethysm see [Mac, §I.8].

(b) The answers are

$$\mathrm{Sym}^2(\mathbb{S}_{(2,2)} V) = \mathbb{S}_{(4,4)} V \oplus \mathbb{S}_{(4,2,2)} V \oplus \mathbb{S}_{(3,3,1,1)} V \oplus \mathbb{S}_{(2,2,2,2)} V;$$
$$\Lambda^2(\mathbb{S}_{(2,2)} V) = \mathbb{S}_{(4,3,2)} V \oplus \mathbb{S}_{(3,2,2,1)} V.$$

Reference: [Lit2, p. 278].

(6.18) Their characters are the same. In fact, if x and y are eigenvalues of an endomorphism of V, the trace on the left-hand side is $\sum f(k) x^k y^{pq-k}$, where $f(k)$ is the number of partitions of k into at most p integers each at most q. This number is symmetric in p and q, by conjugating partitions.

(6.19) The facts about skew Schur polynomials are straightforward generalizations of corresponding facts for regular Schur polynomials given in Appendix A; proofs of (i)–(iv) can be found in [Mac]. To see that the two descriptions of $V_{\lambda/\mu}$ agree see the hint for Exercise 4.4(a). Skew Schur functors are discussed in [A-B-W], where the construction of a basis is given; from this the character formula (viii) follows. Then (iv) implies (v) and (ix).

(6.20) References, with proofs of similar statements in arbitrary characteristic (where the results, however, are weaker), are [Pe1] and [Jam].

(6.21) References: [A-B-W] and [P-W].

(6.29) A reference for the general theory of semisimple algebras and its applications to group theory is [C-R, §26].

Lecture 7

(7.1) One way to show that a symplectic transformation has determinant 1, cf. [Di1], is to show that the group $\mathrm{Sp}_{2n}\mathbb{C}$ is generated by those which fix a hyperplane, i.e., transformations of the form $v \mapsto v + \lambda Q(v, u)u$ for some vector u and scalar λ. Another, cf. Exercise F.12, is to write the determinant as a polynomial expression in terms of the form Q.

(7.2) Consider the action on the quadric $Q(v, v) = 1$.

(7.11) For any y, the image of the map $x \mapsto xyx^{-1}y^{-1}$ is discrete only if y is central.

(7.13) $\mathrm{PGL}_n\mathbb{C}$ acts by conjugation on $n \times n$ matrices.

Lecture 8

(8.10) (b) $\operatorname{ad}[X, Y](Z) = [[X, Y], Z]$, and $[\operatorname{ad} X, \operatorname{ad} Y](Z) = (\operatorname{ad} X \circ \operatorname{ad} Y - \operatorname{ad} Y \circ \operatorname{ad} X)(Z) = [X, [Y, Z]] - [Y, [X, Z]]$.

(8.16) The kernel of Ad is the center $Z(G)$, cf. Exercise 7.11.

(8.17) Use statement (ii), noting that W is G-invariant if it is $\tilde{G}$-invariant, $\tilde{G}$ the universal covering of G.

(8.24) With A, B, C, D $n \times n$ matrices,

$$\mathrm{Sp}_{2n}(\mathbb{R}) = \left\{\begin{pmatrix} A & B \\ C & D \end{pmatrix} : {}^tAC = {}^tCA, {}^tBD = {}^tDB, {}^tAD - {}^tCB = I\right\}.$$

$$\mathfrak{sp}_{2n}\mathbb{R} = \left\{\begin{pmatrix} A & B \\ C & D \end{pmatrix} \middle| {}^tB = B, {}^tC = C, {}^tA = -D\right\}.$$

(8.28) The automorphisms of $G = \tilde{G}/C$ are the automorphisms of $\tilde{G}$ which preserve C.

(8.29) The point is that the commutator of two vector fields is again a vector field, which can be checked in local coordinates.

(8.35) Both signs are plus.

(8.38) Reference: [Ho1].

(8.42) For $h \in H$, $H_0 \cdot h$ gives a coordinate neighborhood of h. For another approach to Proposition 8.41, with more details, see [Hel, §II.2].

(8.43) For an example, take any simply connected group which contains a torus of dimension greater than one, say SU(3), and take an irrational line in the torus.

Lecture 9

(9.7) If H is an abelian subgroup of G, and the claim holds for G/H, show that it holds for G. Or, if G is realized as a group of nilpotent matrices, apply Campbell–Hausdorff.

(9.10) If each $\operatorname{ad}(X)$ is nilpotent, the theorem gives a flag $\mathfrak{g} = V_0 \supset V_1 \supset \cdots \supset V_k = 0$, with $[\mathfrak{g}, V_i] \subset V_{i+1}$, from which it follows that $\mathscr{D}_i\mathfrak{g} \subset V_i$.

(9.21) If $\mathfrak{g}$ had an abelian ideal $\mathfrak{a}$, semisimplicity of the adjoint representation would mean that there is a surjection $\mathfrak{g} \to \mathfrak{a}$ of Lie algebras. But an abelian Lie algebra has lots of representations that are not semisimple.

(9.24) For the last statement, note that the adjoint representation is semisimple. Or see Corollary C.11.

(9.25) Reference: [Bour, I] for this (as well as for details for many other statements in Lecture 9).

(9.27) the adjoint representation is semisimple.

Lecture 10

(10.1) Any holomorphic map from E to $\mathbb{C}$ must be constant.

(10.2) An isomorphism $G_n \cong G_m$ would lift to a map $G \to G$; show that this map would have to be an isomorphism.

(10.4) By hypothesis, the Lie algebra $\mathfrak{g}$ of G has an ideal $\mathfrak{h}$ with abelian quotient; use the corresponding exact sequence of groups, with the corresponding long exact homotopy sequence (cf. §23.1), and an induction on the dimension of G.

Lecture 11

(11.11) Verify the combinatorial formula

$$\left(\sum_{i=0}^{a} x^{a-2i}\right)\left(\sum_{j=0}^{b} x^{b-2j}\right) = \sum_{k=0}^{a}\left(\sum_{l=0}^{a+b-2k} x^{a+b-2k-2l}\right).$$

Reference: [B-tD, p. 87]

(11.19) Given two points on C there is a 2-dimensional vector space of quadrics containing C and the chord between the points.

(11.20) Answer: it is the subspace of the space of quadrics spanned by the squares of the osculating planes to the twisted cubic curve.

(11.23) Answer: the cones over the curve, with vertex a varying point in $\mathbb{P}^3$.

(11.25) Look at the chordal variety of the rational normal curve in $\mathbb{P}^4$.

(11.32) The sum for $\alpha \geq k$ corresponds to the quadrics containing the osculating $(k-1)$-planes to the curve.

(11.34) See Exercise 6.18.

(11.35) Reference: [Mur1, §15].

Lecture 13

(13.3) For V standard, $\mathbb{S}_{(a+b,b)}V \cong \Gamma_{a,b}$. See §15.3 for details.

(13.8) If $a, b > 0$, $V \otimes \Gamma_{a,b} = \Gamma_{a+1,b} \oplus \Gamma_{a-1,b+1} \oplus \Gamma_{a,b-1}$, cf. §15.3.

(13.20) Warning: writing out the eigenvalue diagram and performing the algorithm above is probably not the way to do this.

(13.22) The tangent planes to the Veronese surface should span a subrepresentation.

(13.24) See §23.3 for a general description of these closed orbits.

More applications of representation theory to geometry can be found in [Don] and [Gre].

Lecture 14

(14.15) The fact that $[\mathfrak{g}_\alpha, \mathfrak{g}_\beta] = \mathfrak{g}_{\alpha+\beta}$ is proved in Claim 21.19.

(14.33) See the proof of Proposition 14.31.

(14.34) If $\mathrm{Rad}(\mathfrak{g}) \cap \mathfrak{g}_\alpha \neq 0$, then $\mathrm{Rad}(\mathfrak{g}) \supset \mathfrak{s}_\alpha \cong \mathfrak{sl}_2$, which is not solvable. If $\mathrm{Rad}(\mathfrak{g}) \cap \mathfrak{h} \ni H$, and $\alpha(H) \neq 0$, then $\mathfrak{g}_\alpha = [H, \mathfrak{g}_\alpha] \subset \mathrm{Rad}(\mathfrak{g})$. Use the fact that $[\mathfrak{h}, \mathrm{Rad}(\mathfrak{g})] \subset \mathrm{Rad}(\mathfrak{g})$ to conclude that $\mathrm{Rad}(\mathfrak{g}) = \mathrm{Rad}(\mathfrak{g}) \cap \mathfrak{h} + \sum \mathrm{Rad}(\mathfrak{g}) \cap \mathfrak{g}_\alpha = 0$. For a stronger theorem, see [Va, §4.4].

(14.35) If $\mathfrak{b}' \supset \mathfrak{b}$, then $\mathfrak{b}' \supset \mathfrak{h}$, so $\mathfrak{b}'$ is a direct sum of $\mathfrak{h}$ and some root spaces $\mathfrak{g}_\alpha$ for $\alpha \in T$, $T \supsetneqq R^+$. Then T contains some $-\alpha$ together with α, so $\mathfrak{b}' \supset \mathfrak{s}_\alpha \cong \mathfrak{sl}_2$, which is not solvable.

(14.36) For $\mathfrak{sl}_m\mathbb{C}$, $B(X, Y) = 2m\,\mathrm{Tr}(X \circ Y)$. For $\mathfrak{so}_m\mathbb{C}$, the coefficient is $(m - 2)$, and for $\mathfrak{sp}_m\mathbb{C}$, the coefficient is $(m + 2)$.

Lecture 15

(15.19) See also Exercise 6.20.

(15.20) See Pieri's formulas (6.9), (6.8).

(15.21) Use the dimension formula (15.17).

(15.31) See Exercise 6.20.

(15.32) This is Exercise 6.16 in another notation (and restricted to the special linear group).

(15.33) See Exercise 6.16.

(15.51) Use Weyl's unitary trick with the group $\mathrm{U}(n)$.

(15.52) See Exercise 6.18.

(15.54) Show by induction on r that $r!$ times the difference is an integral linear combination of generators for $I^{\bullet}$. For details see [Tow2].

(15.57) The analogue of (15.53) is valid for these products of minors, and that can be used as in Proposition 15.55 to show that the e_T for semistandard T generate D_λ. The same e_T as in Proposition 15.55 is a highest weight vector. For more on this construction, see [vdW]; we learned it from J. Towber.

For other realizations of the representations of $\mathrm{GL}_n\mathbb{C}$, see [N-S].

Lecture 16

(16.7) With $v = (e_1 \wedge e_2)^2$, calculate as in §13.1; the two vectors $X_{2,1}V_2X_{2,1}V_2v$ and $X_{2,1}X_{2,1}V_2V_2v$ are proportional, and $V_2X_{2,1}X_{2,1}V_2v$ is independent of them.

Lecture 17

(17.18) (i) Note that $\Psi_{\{1,2\}}\colon \wedge^{s-2}V \to \wedge^s V$ is surjective if $s > n$. See Exercise 6.14 for the second statement. (ii) This can be done by direct calculation, as in [Wel, p. 155] for the harder case of the orthogonal group. Or, show that $\mathbb{S}_\lambda(V)$ has a highest weight vector with weight λ, and this cannot occur in any $\Psi_I(V^{\langle d-2\rangle})$.

(17.22) This follows from the theorem and the corresponding result for the general linear groups. Or see Exercise 6.30.

Lecture 19

(19.3)

$$V_{p,q}(v_I) = \begin{cases} 0 & \text{if } \{p,q\} \cap I = \varnothing \text{ or } \{p,q\} \subset I \\ \pm v_{I\setminus\{q\}\cup\{p\}} & \text{if } p \notin I \text{ and } q \in I \\ \pm v_{I\setminus\{p\}\cup\{q\}} & \text{if } q \notin I \text{ and } p \in I. \end{cases}$$

The first assertion follows readily. If $w = \sum a_I v_I$, with the fewest number of nonzero coefficients, and a_J and a_K are nonzero, choose $q \in J \setminus K$, $p \notin J \cup K$ (possible since $2k < m$); then $V_{p,q}(v_J) \neq 0$, $V_{p,q}(v_K) = 0$, and so $V_{p,q}(w)$ is a nonzero vector with fewer nonzero coefficients.

(19.4) The multiplicity of $L_1 + \cdots + L_a - L_{n-b} - \cdots - L_n$ in $\wedge^k V$ is $\binom{2r}{r}$ if $k - a - b = 2r$. For $\Gamma_{2\alpha}$ or $\Gamma_{2\beta}$ the multiplicity is $\frac{1}{2}\binom{2r}{r}$ if r is positive, by symmetry under replacing any Γ_p by $-\Gamma_p$. For Γ_α the weights are $\frac{1}{2}(\varepsilon_1 L_1 + \cdots + \varepsilon_n L_n)$, with $\varepsilon_i = \pm 1$, and $\prod \varepsilon_i = 1$; the multiplicities are all one since these are conjugate under the Weyl group; similarly for Γ_β but with $\prod \varepsilon_i = -1$.

(19.21) For generalizations, see §23.2.

Lecture 20

(20.17) If f spans $\wedge^n W'$, and u_0 spans U with $Q(u_0, u_0) = 1$, then $f \cdot (1 + (-1)^n u_0)$ is such a generator. See Exercise 20.12.

(20.21) If x is in the center, take an orthogonal basis $\{v_i\}$, write out $x = \sum a_I v_I$ in terms of the basis, and look at the equations $x \cdot v_j = v_j \cdot x$ for all j. Note that $v_I \cdot v_j = (-1)^{|I|} v_j \cdot v_I$ if $j \notin I$, whereas $v_I \cdot v_j = (-1)^{|I|-1} v_j \cdot v_I$ if $j \in I$. Conclude that $a_I = 0$ if $|I|$ is odd and there is some $j \notin I$ or if $|I|$ is even and there is some $j \in I$. A similar argument works if x is odd. Reference [A-B-S, p. 7].

(20.22) If $X = a \wedge b$, $[X, v] = \frac{1}{2}(a \cdot b \cdot v - b \cdot a \cdot v - v \cdot a \cdot b + v \cdot b \cdot a)$, which is

$$\tfrac{1}{2}(2Q(b,v)a - a \cdot v \cdot b - 2Q(a,v)b + b \cdot v \cdot a - 2Q(a,v)b + a \cdot v \cdot b + 2Q(b,v)a - b \cdot v \cdot a)$$
$$= 2Q(b,v)a - 2Q(a,v)b = \varphi_{a \wedge b}(v).$$

(20.23) Reference: [Por], but note that his $C(p, q)$ is our $C(q, p)$. See also [A-B-S].

(20.32) If $Q(v-w, v-w) \neq 0$, then $R_{v-w}(v) = w$. Otherwise, $R_{v+w}(v) = -w$, and $R_w(-w) = w$. For (b) compose a given element of O(Q) with an element constructed by (a) to get one fixed on a line, and write, by induction on the dimension, the restriction to the perpendicular hyperplane as a product of reflections.

(20.33) By Exercise 20.22, $X \cdot v = [X, v]$. See also Exercise 8.24.

(20.36) If v_i are a basis for V with $Q(v_i, v_j) = -\delta_{i,j}$, then $\omega = v_1 \cdot \ldots \cdot v_m$. If $m \equiv 2$ (4), the center is cyclic of order four, while if $m \equiv 0$ (4), it is the Klein four group.

(20.37) Show that $\mathfrak{so}(Q)$ acts by traceless endomorphisms. For example, the trace of H_i on S^+ is the number of $I \subset \{1, \ldots, n\}$ such that $|I|$ is even and $i \in I$, minus the number with $i \notin I$.

(20.38) For the first statement of (a), choose f spanning $\bigwedge^n W'$ so that, for the chosen generator of $\bigwedge^n W$, $\tau(f) \cdot e \cdot f = f$. For the second, when m is even, $x(s)f = x \cdot s \cdot f$ by Exercise 20.12, so $\beta(x(s), x(t))f = \tau(x \cdot s \cdot f) \cdot (x \cdot t \cdot f) = \tau(s \cdot f) \cdot \tau(x) \cdot x \cdot (t \cdot f) = \tau(s \cdot f) \cdot (t \cdot f) = \beta(s, t)$. The odd case can be reduced to the even case by imbedding $C(Q)$ into a larger Clifford algebra as in Exercise 20.40.

(20.43) Reference: [Por].

(20.44) For example, the transposition of α_1 and α_4 is achieved by the matrix

$$\begin{bmatrix} \frac{1}{2} & \frac{1}{2} & \frac{1}{2} & \frac{1}{2} \\ \frac{1}{2} & \frac{1}{2} & -\frac{1}{2} & -\frac{1}{2} \\ \frac{1}{2} & -\frac{1}{2} & \frac{1}{2} & -\frac{1}{2} \\ \frac{1}{2} & -\frac{1}{2} & -\frac{1}{2} & \frac{1}{2} \end{bmatrix}$$

(20.50) Reference: [Ch2, §4.3].

(20.51) Reference: [Ch2, §4.2–4.5], [Jac1].

Other references include [L-M], [Ca1], [B-tD], [Hus], [P-S].

Lecture 21

(21.9) If $\alpha_1, \ldots, \alpha_r$ are the vectors, and we have a nontrivial relation

$$v = \sum_{i \le k} n_i \alpha_i = \sum_{j > k} n_j \alpha_j,$$

with non-negative coefficients, then $(v, v) = \sum_{i,j} n_i n_j (\alpha_i, \alpha_j) \le 0$, so $v = 0$. But v lies on the same side of the hyperplane.

(21.15) The first is ruled out by considering

$$u = e_2, \qquad v = (3e_3 + 2e_4 + e_5)/\sqrt{6}, \qquad w = (3e_6 + 2e_7 + e_8)/\sqrt{6},$$

with $1 > (e_1, u)^2 + (e_1, v)^2 + (e_1, w)^2 = 1/4 + 3/8 + 3/8 = 1$. For the second, use

$$u = e_2, \qquad v = (2e_3 + e_4)/\sqrt{3}, \qquad w = (5e_5 + 4e_6 + 3e_7 + 2e_8 + e_9)/\sqrt{15},$$

with $(e_1, u)^2 + (e_1, v)^2 + (e_1, w)^2 = 1/4 + 1/3 + 5/12 = 1$.

(21.16) Using the characterization that $\omega_i(H_{\alpha j}) = \delta_{i,j}$, one can write the fundamental weights ω_i in terms of the basis L_i. The tables in [Bour, Ch. 6] also express them in terms of the simple roots.

(E6): $\omega_1 = 2\frac{\sqrt{3}}{3}L_6,$

$\omega_2 = \frac{1}{2}(L_1 + L_2 + L_3 + L_4 + L_5) + \frac{\sqrt{3}}{2}L_6,$

$\omega_3 = \frac{1}{2}(-L_1 + L_2 + L_3 + L_4 + L_5) + 5\frac{\sqrt{3}}{6}L_6,$

$\omega_4 = L_3 + L_4 + L_5 + \sqrt{3}L_6,$

$\omega_5 = L_4 + L_5 + 2\frac{\sqrt{3}}{3}L_6,$

$\omega_6 = L_5 + \frac{\sqrt{3}}{3}L_6;$

(E_7): $\omega_1 = \sqrt{2}L_7,$

$\omega_2 = \frac{1}{2}(L_1 + L_2 + L_3 + L_4 + L_5 + L_6) + \sqrt{2}L_7,$

$\omega_3 = \frac{1}{2}(-L_1 + L_2 + L_3 + L_4 + L_5 + L_6) + 3\frac{\sqrt{2}}{2}L_7,$

$\omega_4 = L_3 + L_4 + L_5 + L_6 + 2\sqrt{2}L_7,$

$\omega_5 = L_4 + L_5 + L_6 + 3\frac{\sqrt{2}}{2}L_7,$

$\omega_6 = L_5 + L_6 + \sqrt{2}L_7,$

$\omega_7 = L_6 + \frac{\sqrt{2}}{2}L_7;$

(E_8) $\omega_1 = 2L_8,$

$\omega_2 = \frac{1}{2}(L_1 + L_2 + L_3 + L_4 + L_5 + L_6 + L_7 + 5L_8),$

$\omega_3 = \frac{1}{2}(-L_1 + L_2 + L_3 + L_4 + L_5 + L_6 + L_7 + 7L_8),$

$\omega_4 = L_3 + L_4 + L_5 + L_6 + L_7 + 5L_8,$

$\omega_5 = L_4 + L_5 + L_6 + L_7 + 4L_8,$

$\omega_6 = L_5 + L_6 + L_7 + 3L_8,$

$\omega_7 = L_6 + L_7 + 2L_8,$

$\omega_8 = L_7 + L_8;$

(F_4): $\omega_1 = L_1 + L_2,$

$\omega_2 = 2L_1 + L_2 + L_3,$

$\omega_3 = \frac{1}{2}(3L_1 + L_2 + L_3 + L_4),$

$\omega_4 = L_1;$

(G_2) $\omega_1 = \frac{1}{2}(L_1 + \sqrt{3}L_2) = 2\alpha_1 + \alpha_2,$

$\omega_2 = \sqrt{3}L_2 = 3\alpha_1 + 2\alpha_2.$

(21.17) The only cases of the same rank that have the same number of roots are (B_n) and (C_n) for all n, and (B_6), (C_6), and (E_6); (B_n) has n roots shorter than the others, (C_n) n roots longer, and in (E_6) all the roots are the same length.

(21.18) For the matrices see [Bour, Ch. 6] or [Hu1, p. 59]. The determinants are: $n+1$ for (A_n); 2 for (B_n), (C_n) and (E_7); 4 for (D_n); 3 for (E_6); and 1 for (G_2), (F_4), and (E_8).

(21.23) See Lecture 22.

The proof of Lemma 21.20 is from [Jac1, p. 124], where details can be found.

For more on Dynkin diagrams and classification, see [Ch3], [Dem], [Dy-O], [LIE], and [Ti1].

Lecture 22

(22.5) Use the fact that $B(Y, Z) = 6\,\mathrm{Tr}(Y \circ Z)$ on $\mathfrak{sl}_3\mathbb{C}$, and the formula $[e_i, e_j^*] = 3 \cdot E_{i,j} - \delta_{i,j} \cdot I$, giving

$$B([e_i, e_j^*], Z) = 6 \cdot \mathrm{Tr}((3 \cdot E_{i,j} - \delta_{i,j} \cdot I) \circ Z) = 18 \cdot \mathrm{Tr}(E_{i,j} \circ Z) = 18 \cdot e_j^*(Z \cdot e_i).$$

(22.13) Hint: use the dihedral group symmetry.

(22.15) Answer: $\mathfrak{sl}_3\mathbb{C} \times \mathfrak{sl}_3\mathbb{C}$.

(22.20) For (b), apply ψ to both sides of (22.17), and evaluate both sides of (22.18) on w. Note that $\psi((v \wedge w) \wedge \varphi) = (v \wedge w)(\varphi \wedge \psi) = ((\varphi \wedge \psi) \wedge v)(w)$.

(22.21) For a triple $J = \{p < q < r\} \subset \{1, \dots, 9\}$, let $e_J = e_p \wedge e_q \wedge e_r$ and similarly for φ_J. For triples J and K the essential calculation (see Exercise 22.5) is to verify that $e_J * \varphi_K$ is 1/18 times

$$\begin{array}{ll} 0 & \text{if } \# J \cap K \le 1; \\ \pm E_{m,n} & \text{if } K = \{p, q, n\},\ J = \{p, q, m\},\ m \ne n; \\ E_{p,p} + E_{q,q} + E_{r,r} - \tfrac{1}{3} I & \text{if } K = J = \{p, q, r\}; \end{array}$$

the sign in front of $E_{m,n}$ is the product of the signs of the permutations that put the two sets in order. Verify that $(v \wedge w) \wedge \varphi = 18((w * \varphi) \cdot v - (v * \varphi) \cdot w)$. For Freudenthal's construction, see [Fr2], [H-S].

(22.24) For $\mathfrak{sl}_{n+1}\mathbb{C}$, such an involution takes $E_{i,j}$ to $(-1)^{j-i+1} E_{n+2-j,n+2-i}$; the fixed algebra is $\{X : {}^t X M = -MX\}$, where $M = (m_{ij})$, with $m_{ij} = 0$ if $i + j \ne n + 2$, and otherwise $m_{ij} = (-1)^i$. This M is symmetric if n is even, skew if n is odd, so the fixed subalgebra for (A_{2m}) is the Lie algebra $\mathfrak{so}_{2m+1}\mathbb{C}$ of (B_m), and that for (A_{2m-1}) is the Lie algebra $\mathfrak{sp}_{2m}\mathbb{C}$ of (C_m). For (D_n), the fixed algebra is $\mathfrak{so}_{2n-1}\mathbb{C}$, corresponding to (B_{n-1}), while for the rotation of (D_4), the fixed algebra is $\mathfrak{g}_2$. For a description of possible automorphisms of simple Lie algebras, see [Jac1, §IX].

(22.25) Answer: For $\mathfrak{sl}_{n+1}\mathbb{C}$, $X \mapsto -X^t$. For $\mathfrak{so}_{2n}\mathbb{C}$, $n \ge 5$, $X \mapsto PXP^{-1}$, where P is the automorphism of $\mathbb{C}^{2n}$ that interchanges e_n and e_{2n} and preserves the other basic vectors. For the other automorphisms of $\mathfrak{so}_8\mathbb{C}$, see Exercise 20.44.

(22.27) References: [Her], [Jac3, p. 777], [Pos], [Hu1, §19.3].

(22.38) Reference [Ch2, §4.5], [Jac4, p. 131], [Jac1], [Lo, p. 104].

Lecture 23

(23.3) The map takes $z = x + iy$ to (u, v) with $u = x/\|x\|$, $v = y$.

(23.10) Since $\rho(\exp(\sum a_j H_j)) = (e^{a_1}, \ldots, e^{a_n}, e^{-a_1}, \ldots)$, to be in the kernel we must have $a_j = 2\pi i \cdot n_j$, and then $\exp(\sum a_j H_j) = (-1)^{\sum n_j}$.

(23.11) Note that the surjectivity of the fundamental groups is equivalent to the connectedness of $\pi^{-1}(H)$ when $\pi: \tilde{G} \to G$ is the universal covering, which is equivalent to the Cartan subgroup of $\tilde{G}$ containing the center of $\tilde{G}$.

(23.17) Note that $\Gamma(G) = \pi_1(H)$ surjects onto $\pi_1(G)$, and there is an exact sequence

$$0 \to \pi_1(G) \to \mathrm{Center}(\tilde{G}) \to \mathrm{Center}(G) \to 0.$$

(23.19) When m is odd, the representations are the representations of $\mathrm{SO}_m\mathbb{C}$, and the products of those by the one-dimensional alternating (determinant) representation. When $m = 2n$, the representations of $\mathrm{SO}_m\mathbb{C}$ with highest weights $(\lambda_1, \ldots, \lambda_n)$ and $(\lambda_1, \ldots, -\lambda_n)$ are conjugate, so that, if $\lambda_n \neq 0$, they correspond to one irreducible representation of $\mathrm{O}_{2n}\mathbb{C}$, whose underlying space can be identified with $\Gamma_{(\lambda_1, \ldots, \lambda_n)} \oplus \Gamma_{(\lambda_1, \ldots, -\lambda_n)}$. If $\lambda_n = 0$, then Γ_λ is an irreducible representation of $\mathrm{O}_m\mathbb{C}$. In either case, the representations correspond to partitions $\lambda = (\lambda_1 \geq \cdots \geq \lambda_n \geq 0)$. See §19.5 for an argument.

(23.31) See Exercises 19.6, 19.7, and 19.16.

(23.36) For (b), consider $(D^+)^2 \cdot (D^-)^2 = (D^+ \cdot D^-)^2$.

(23.37) Reference: [B-tD, VI §7].

(23.38) For $\mathfrak{sl}_{n+1}\mathbb{C}$, $\Gamma_\lambda^* = \Gamma_{(\lambda_1, \lambda_1 - \lambda_n, \ldots, \lambda_1 - \lambda_2)}$; for $\mathfrak{so}_{2n}\mathbb{C}$, n odd, $\Gamma_\lambda^* = \Gamma_{(\lambda_1, \lambda_2, \ldots, \lambda_{n-1}, -\lambda_n)}$.

(23.39) Reference: [Bour, VIII §7, Exer. 11].

(23.42) Compute highest weight vectors in the (external) tensor product of two irreducible representations, to verify that it is irreducible with highest weight the sum of the two weights.

(23.43) See Exercise 20.40 and Theorems 17.5 and 19.2.

(23.51) An isotropic $(n-1)$-plane is automatically contained in an isotropic n-plane. These are two-step flag varieties, corresponding to omitting two nodes.

(23.62) For (b), use the fact that $B \cdot n' \cdot B$ is open in G. For (c), if μ is a weight, $f(x^{-1}wy) = \mu(x)\lambda(y)f(w)$ for x and y in B, so with $x \in H$ and $w = n'$,

$$\mu(x)f(w) = f(x^{-1}w) = f(wx) = \lambda(x)f(w).$$

Other references on homogeneous spaces include [B-G-G], [Hel], and [Hi].

Lecture 24

(24.4) (a) is proved in Lemma D.25, and (b) follows. For (c), note that by the definition of ρ as half the sum of the positive roots, $\rho - W(\rho)$ is the sum of those positive β such that $W(\beta)$ is negative.

(24.27) This is Exercise A.62.

(24.46) This follows from formulas (A.61) and (A.65).

(24.51) In the following the fundamental weights are numbered as in the answer to Exercise 21.16:

(F_4): $$\rho = 8\alpha_1 + 15\alpha_2 + 21\alpha_3 + 11\alpha_4;$$

$$\dim(\Gamma_{\omega_i}),\ (i = 1, 2, 3, 4):\quad 52,\ 1274,\ 273,\ 26.$$

(E_6): $$\rho = L_2 + 2L_3 + 3L_4 + 4L_5 + 4\sqrt{3}L_6$$
$$= 8\alpha_1 + 11\alpha_2 + 15\alpha_3 + 21\alpha_4 + 15\alpha_5 + 8\alpha_6;$$

$$\dim(\Gamma_{\omega_i}),\ (i = 1, \ldots, 6):\quad 27,\ 78,\ 351,\ 2925,\ 351,\ 27.$$

(E_7): $$\rho = L_2 + 2L_3 + 3L_4 + 4L_5 + 5L_6 + 17\sqrt{2}/2L_7$$
$$= \tfrac{1}{2}(34\alpha_1 + 49\alpha_2 + 66\alpha_3 + 96\alpha_4 + 75\alpha_5 + 52\alpha_6 + 27\alpha_7);$$

$$\dim(\Gamma_{\omega_i}),\ (i = 1, \ldots, 7):\quad 133,\ 912,\ 8645,\ 365750,\ 27664,\ 1539,\ 56.$$

(E_8): $$\rho = L_2 + 2L_3 + 3L_4 + 4L_5 + 5L_6 + 6L_7 + 23L_8$$
$$= 46\alpha_1 + 68\alpha_2 + 91\alpha_3 + 135\alpha_4 + 110\alpha_5 + 84\alpha_6 + 57\alpha_7 + 29\alpha_8;$$

$$\dim(\Gamma_{\omega_i}),\ (i = 1, \ldots, 8):\quad 3875,\ 147250,\ 6696000,\ 6899079264,\ 146325270,$$
$$2450240,\ 30380,\ 248.$$

(24.52) Using the dimension formula as in Exercise 24.9, it suffices to check which fundamental weights correspond to small representations, and then which sums of these are still small. The results are:

(A) $n \geq 1$; $\dim G = n^2 + 2n$; $\dim \Gamma_{\omega_k} = \binom{n+1}{k}$;

the dominant weights whose representations have dimension at most $\dim G$ are:

$\omega_1, \omega_2, \omega_{n-1}, \omega_n$;

$2\omega_1, 2\omega_n$, of dimension $\binom{n+2}{2}$;

$\omega_1 + \omega_n$, of dimension $n^2 + 2n$;

ω_3 for $n = 5$; ω_3, ω_4 for $n = 6$; ω_3, ω_5 for $n = 7$.

(B_n) $n \geq 2$; $\dim G = 2n^2 + n$; $\dim \Gamma_{\omega_k} = \binom{2n+1}{k}$ for $k < n$, and $\dim \Gamma_{\omega_n} = 2^n$, giving:

ω_1, ω_2;

ω_n for $n = 3, 4, 5, 6$;

$2\omega_2$, of dimension 10, for $n = 2$.

(C_n) $n \geq 3$; $\dim G = 2n^2 + n$; $\dim \Gamma_{\omega_k} = \binom{2n}{k} - \binom{2n}{k-2}$, giving:

ω_1, ω_2;

$2\omega_1$, of dimension $2n^2 + n$;

ω_3 for $n = 3$.

(D_n) $n \geq 4$; $\dim G = 2n^2 - n$; $\dim \Gamma_{\omega_k} = \binom{2n}{k}$ for $k \leq n-2$, and

$\dim \Gamma_{\omega_{n-1}} = \dim \Gamma_{\omega_n} = 2^{n-1}$, giving:

ω_1, ω_2;

ω_{n-1}, ω_n for $n = 4, 5, 6, 7$.

(E_6) $\dim G = 78$; $\omega_1, \omega_2, \omega_6$.

(E_7) $\dim G = 133$; ω_1, ω_7.

(E_8) $\dim G = 248$; ω_8.

(F_4) $\dim G = 52$; ω_1, ω_4.

(G_2) $\dim G = 14$; ω_1, ω_2.

For irreducible representations of general Lie groups with this property, see [S-K].

Other references with character formulas include [ES-K], [Ki1], [Ki2], [Kl], [Mur2], and [Ra].

Lecture 25

(25.2) Changing μ by an element of the Weyl group, one can assume μ is also dominant and $\lambda - \mu$ is a sum of positive roots. Then $\|\lambda\| > \|\mu\|$, and $c(\mu) = (\lambda, \lambda) - (\mu, \mu) + (\lambda - \mu, 2\rho) > 0$.

(25.4) A direct calculation gives

$$C(X \cdot v) - X \cdot C(v) = \sum U_i \cdot [U_i', X] \cdot v + \sum [U_i, X] \cdot U_i' \cdot v.$$

To see that this is zero, write $[U_i, X] = \sum \alpha_{ij} U_j$; then by (14.23), $\alpha_{ij} = ([U_i, X], U_j') = -([U_j', X], U_i)$, so $[U_j', X] = -\sum \alpha_{ij} U_i'$. The terms in the above sums then cancel in pairs.

(25.6) By (14.25), $(H_\alpha, H_\alpha) = \alpha(H_\alpha)(X_\alpha, Y_\alpha) = 2(X_\alpha, Y_\alpha)$. Use Exercise 14.28.

(25.12) The symmetry gives

$$(\beta - i\alpha, \alpha)n_{\beta - i\alpha} + (\beta - (m-i)\alpha, \alpha)n_{\beta-(m-i)\alpha} = (2\beta - m\alpha, \alpha)n_{\beta - i\alpha} = 0$$

since $2(\beta, \alpha) = m(\alpha, \alpha)$, so the terms cancel in pairs.

(25.22) We have

$$\sum_{W,\mu} (-1)^W P(\mu + W(\rho) - \rho)e(-\mu) = \sum_W (-1)^W (e(W(\rho) - \rho)) / \prod_{\alpha \in R^+} (1 - e(-\alpha)),$$

and the right-hand side is 1 by Lemma 24.3.

(25.23) We have

$$\begin{aligned}\sum_W (-1)^W n_{\mu+\rho-W(\rho)} &= \sum_{W,W'} (-1)^{WW'} P(W'(\lambda + \rho) - ((\mu + \rho - W(\rho)) + \rho)) \\ &= \sum_{W'} (-1)^{W'} \sum_W (-1)^W P((W'(\lambda + \rho) - \mu - \rho) + W(\rho) - \rho),\end{aligned}$$

and the inner sum is zero unless $W'(\lambda+\rho)=\mu+\rho$. Note that if μ is a root of Γ_λ, this happens only if $\mu=\lambda$ by Exercise 25.2.

(25.24) The minuscule weights are:

(A_n): $\omega_1, \ldots, \omega_n$,

(B_n): ω_1,

(C_n): ω_n,

(D_n): $\omega_1, \omega_{n-1}, \omega_n$,

(E_6): ω_1, ω_6,

(E_7): ω_7.

Reference: [Bour, VIII, §7.3].

(25.28) One easy way is to use the isomorphism $\mathfrak{so}_4\mathbb{C} \cong \mathfrak{sl}_2\mathbb{C} \times \mathfrak{sl}_2\mathbb{C}$.

(25.30) $N_{\lambda\mu\gamma}$ is zero by definition when γ is not in the closed positive Weyl chamber $\mathscr{W}$, and $W(\nu+\rho)-\rho$ is not in $\mathscr{W}$ if $W \neq 1$. Reference: [Hu1].

(25.40) The weight space of the restriction of Γ_λ corresponding to $\bar{\mu}$ is the direct sum of the weight spaces of Γ_λ corresponding to those μ which restrict to $\bar{\mu}$.

(25.41) Use the preceding exercise and Exercise 25.23.

(25.43) Using the action of a Lie algebra on a tensor product, the action of C on $v_1 \cdot \ldots \cdot v_m$ is a sum over terms where U_i and U_i' act on different elements or the same element. Grouping the terms accordingly leads to the displayed formula. See [L-T, I, pp. 19–20].

Lecture 26

(26.2) In terms of the basis L_1, L_2 of $\mathfrak{h}^*$ dual to $\{H_1, H_2\}$, eigenvalues are $\pm iL_2$ and $\pm 3L_1 \pm iL_2$.

(26.9) Reference: [Hel, §III.7].

(26.10) Constructing $\mathfrak{h} = \mathfrak{g}_0(H)$ as in Appendix D, take H so that $\sigma(H) = H$.

(26.12) See Exercise 23.6.

(26.13) Reference: [Hel, §X.6.4].

(26.21) If a conjugate linear endomorphism $\varphi: W \to W$ did not map Γ_λ to itself, there would be another factor U of W and an isomorphism of Γ_λ with U^*; the highest weight of $(\Gamma_\lambda)^*$ cannot be lower than λ.

(26.22) See Exercise 3.43 and Exercise 26.21.

(26.28) References: [A-B-S], [Hus], [Por]. See also Exercise 20.38.

(26.30) Use the identity $\psi^2[V] = [V \otimes V] - 2[\wedge^2 V]$.

Other references on real forms are [Gi1], [B-tD], [Va].

Appendix A

(A.29) (b) Use $P^{(i)} = \sum_\nu \langle H_\nu, P^{(i)} \rangle M_\nu$.

(A.30) Some of these formulas also follow from Weyl's character formula.

(A.31) For part (a), when $a_1 \geq a_2 \geq \cdots \geq a_k$, this is (A.19). The proof of (A.9) shows that for any $a = (a_1, \ldots, a_k)$,

$$H_{a_1} \cdot H_{a_2} \cdot \ldots \cdot H_{a_k} = \sum K_{\mu a} S_\mu,$$

which shows that the $K_{\mu a}$ are unchanged when the a_i's are reordered. For a purely combinatorial proof see [Sta, §10].

(A.32) For (i) compare the generating functions $E(t) = \sum E_i t^i = \prod(1 + x_i t)$ and $H(t) = \sum H_i t^i = 1/E(-t)$; (ii) follows from (A.5) and (A.6). For (iii), note that $P(t) = \sum P_j t^j = \sum x_i t/(1 - x_i t) = tH'(t)/H(t)$. Exponentiate this to get (vi). For details and more on this involution, see [Mac] or [Sta], where it is used to derive basic identities among symmetric polynomials.

(A.39) References: [Mac], [Sta], [Fu, §A.9.4].

(A.41) See [Mac, p. 33] or [Fu, p. 420].

(A.48) Since $\vartheta(E_i') = H_i'$ and $\vartheta(E_i'') = H_i''$,

$$\vartheta(S_{\langle\lambda\rangle}) = \vartheta(|H''_{\lambda_i - i + j} - H''_{\lambda_i - i - j}|) = |E''_{\mu_i - i + j} - E''_{\mu_i - i - j}| = S_{[\mu]}.$$

(A.67) Answer: $\frac{1}{2}\zeta_1 \cdot \ldots \cdot \zeta_n$ times the determinant of the matrix whose ith row is

$$(J_{\lambda_i - i} \quad J_{\lambda_i - i + 1} + J_{\lambda_i - i - 1} \quad \ldots \quad J_{\lambda_i - i + n - 1} + J_{\lambda_i - i - n + 1}).$$

More on symmetric polynomials can be found in [Mac], [Sta], [L-S], and references listed in these sources. Some of the identities in §A.3 are new, although results along these lines can be found in [We1], [Lit1], [Lit2] and [Ko-Te]; other identities involving the determinants discussed in §A.3 can be found in [Mac, §I.5]. Discussions of Schur functions and representation theory can be found in [Di2] and [Lit2].

Appendix C

(C.1) Take a basis in which X has Jordan canonical form, and compute using the corresponding basis E_{ij} for $\mathfrak{gl}(V)$.

(C.12) If $\mathfrak{g} = \bigoplus \mathfrak{g}_i$, and $\mathfrak{h}$ is a simple ideal, $\mathfrak{h} = [\mathfrak{g}, \mathfrak{h}] = \bigoplus [\mathfrak{g}_i, \mathfrak{h}]$, so $\mathfrak{h}$ is contained in some $\mathfrak{g}_i$.

(C.13) Since for $\delta \in \mathrm{Der}(\mathfrak{g})$ and $X \in \mathfrak{g}$, $\mathrm{ad}(\delta(X)) = [\delta, \mathrm{ad}(X)]$, $\mathrm{ad}(\mathfrak{g})$ is an ideal in the Lie algebra $\mathrm{Der}(\mathfrak{g})$. Therefore, $[\mathrm{ad}(\mathfrak{g})^\perp, \mathrm{ad}(\mathfrak{g})] = 0$; in particular, if $\delta \in \mathrm{ad}(\mathfrak{g})^\perp$ and $X \in \mathfrak{g}$, then $\mathrm{ad}(\delta(X)) = [\delta, \mathrm{ad}(X)] = 0$. So $\mathrm{ad}(\mathfrak{g})^\perp = 0$ and $\mathrm{ad}(\mathfrak{g}) = \mathrm{Der}(\mathfrak{g})$.

Appendix D

(D.8) To show $\mathrm{ad}(X)$ is nilpotent on $\mathfrak{g}_0(H)$ for X in $\mathfrak{g}_0(H)$, consider the complex line from H to X: set $H(z) = (1 - z)H + zX$. Then $\mathrm{ad}(H(z))$ preserves each eigenspace $\mathfrak{g}_\lambda(H)$. By continuity, for z sufficiently near 0, $\mathrm{ad}(H(z))$ is a nonsingular transformation of $\mathfrak{g}_\lambda(H)$ for $\lambda \neq 0$, which implies that $\mathfrak{g}_0(H(z))$ is contained in $\mathfrak{g}_0(H)$, and by the regularity of H, $\mathfrak{g}_0(H(z)) = \mathfrak{g}_0(H)$ for small z.

This means that there is an integer k so that $\mathrm{ad}(H(z))^k(Y) = 0$ for all $Y \in \mathfrak{g}_0(H)$ and all small z. But $\mathrm{ad}(H(z))^k(Y)$ is a polynomial function of z, so it must vanish identically. Hence, setting $z = 1$, $\mathrm{ad}(X)^k$ vanishes on $\mathfrak{g}_0(H)$, as asserted.

(D.24) See [Bour, VII, §3] for details.

(D.33) References: [Se3, §V.11], [Hu1, §12.2].

Appendix E

Proofs of both of these theorems can be found, together with many other related results, in [Bour I]. See also [Se3], [Pos], [Va], [Jac1].

Appendix F

(F.12) Check that the right-hand side is multilinear, alternating, and takes the value 1 on a standard basis. Or see [We1, §VI.1].

(F.16) $\mathrm{SO}_n\mathbb{C}$-invariants can be written in the form $A + \sum A_i B_i$ where A and the A_i are polynomials in the $Q(x^{(i)}, x^{(j)})$ and the B_i are brackets. Such is taken to $A + \det(g) \sum A_i B_i$ by g in $\mathrm{O}_n\mathbb{C}$. For an odd (resp. even) invariant the first (resp. the second) term must vanish.

(F.20) Reference: [We1, II.6], or [Br, p. 866].

There are many elementary references for invariant theory, such as [D-C], [Pr], [Sp1], and [Ho2]; the last contains a proof of Capelli's formula. There are also many modern approaches to invariant theory, some which can be found in [DC-P], [Sch] and [Vu] and references described therein; some of these also contain some invariant theory for exceptional groups. For a more conceptual and representation-theoretic approach to Capelli's identity, see [Ho3]. Weyl's book [We1] remains an excellent reference for invariant theory of the orthogonal and symplectic groups together with the related [Br], [We2].

Bibliography

[A-B] M. F. Atiyah and R. Bott, A Lefschetz fixed point formula for elliptic complexes: II. Applications, *Ann. Math.* **9** (1968), 451–491.

[A-B-S] M. F. Atiyah, R. Bott, and A. Shapiro, Clifford modules, *Topology* **3**, Supp. 1 (1964), 3–38.

[A-B-W] K. Akin, D. A. Buchsbaum, and J. Weyman, Schur functors and Schur complexes, *Adv. Math.* **44** (1982), 207–278.

[Ad] J. F. Adams, *Lectures on Lie Groups*, W. A. Benjamin, Inc., New York, 1969.

[Ahl] L. V. Ahlfors, *Complex Analysis, Second Edition*, McGraw-Hill, New York 1966.

[A-J-K] Y. J. Abramsky, H. A. Jahn, and R. C. King, Frobenius symbols and the groups S_s, GL(n), O(n), and Sp(n), *Can. J. Math.* **25** (1973), 941–959.

[And] G. E. Andrews, The Theory of Partitions, *Encyclopedia of Mathematics and Its Applications*, vol. 2, Addison-Wesley, Reading, MA, 1976.

[Ar] S. K. Araki, On root systems and an infinitesimal classification of irreducible symmetric spaces, *J. Math. Osaka City Univ.* **13** (1963), 1–34.

[A-T] M. Atiyah and D. O. Tall, Group representations, λ-rings, and the J-homomorphism, *Topology* **8** (1969), 253–297.

[B-G-G] I. N. Bernstein, I. M. Gelfand, and S. I. Gelfand, Schubert cells and cohomology of the spaces G/P, *Russ. Math. Surv.* **28** (1973), 1–26.

[Boe] H. Boerner, *Representations of Groups*, Elsevier North-Holland, Amsterdam, 1970.

[Bor1] A. Borel, *Linear Algebraic Groups*, W. A. Benjamin, 1969 and (GTM 126), Springer-Verlag, New York, 1991.

[Bor2] A. Borel, Topology of Lie groups and characteristic classes, *Bull. Amer. Math. Soc.* **61** (1955), 397–432.

[Bot] R. Bott, On induced representations, in *The Mathematical Heritage of Hermann Weyl*, Proc. Symp. Pure Math Vol. 48, American Mathematical Society, Providence, RI 1988, pp. 1–13.

[Bour] N. Bourbaki, *Lie Groups and Lie Algebras, Chapters 1–3*, Springer-Verlag, New York, 1989; *Groupes et algèbres de Lie, Chapitres 4, 5 et 6*, Masson, Paris, 1981; *Groupes et algèbres de Lie, Chapitres 7 et 8*, Diffusion C.C.L.S., Paris, 1975; *Algebra 1*, Chapter 3, Springer-Verlag, New York, 1989.

[Br] R. Brauer, On algebras which are connected with the semisimple continuous groups, *Ann. Math.* **38** (1937), 857–872.

[B-tD] T. Bröcker and T. tom Dieck, *Representations of Compact Lie Groups*, Springer-Verlag, New York, 1985.

[Bu] J. Burroughs, Operations in Grothendieck rings and the symmetric group, *Can. J. Math.* **26** (1974), 543–550.

[Ca1] E. Cartan, *The Theory of Spinors*, Hermann, Paris, 1966, and Dover Publications, 1981.

[Ca2] E. Cartan, Le principe de dualité et la théorie des groupes simples et semi-simples, Bull. Sci. Math. **49** (1925), 361–374.

[Cart] P. Cartier, On H. Weyl's character formula, *Bull. Amer. Math. Soc.* **67** (1961), 228–230.

[Ch1] C. Chevalley, *Theory of Lie Groups*, Princeton University Press, Princeton, NJ, 1946.

[Ch2] C. Chevalley, *The Algebraic Theory of Spinors*, Columbia University Press, New York, 1954.

[Ch3] Séminaire C. Chevalley 1956–1958, *Classification des Groupes de Lie Algébriques, Secrétariat mathématique*, Paris, 1958.

[Ch-S] C. Chevalley and R. D. Schafer, The exceptional simple Lie algebras F_4 and E_6, *Proc. Natl. Acad. Sci. USA.* **36** (1950), 137–141.

[Co] A. J. Coleman, *Induced Representations with Applications to S_n and* GL(n), *Queens Papers Pure Appl. Math.* **4** (1966).

[C-R] C. W. Curtis and I. Reiner, *Representation Theory of Finite Groups and Associative Algebras*, Interscience Publishers, New York, 1962.

[D-C] J. Dieudonné and J. Carrell, *Invariant Theory, Old and New*, Academic Press, New York, 1971.

[DC-P] C. De Concini and C. Procesi, A characteristic free approach to invariant theory, *Adv. Math.* **21** (1976), 330–354.

[Dem] M. Demazure, *A, B, C, D, E, F*, etc., *Springer Lecture Notes* 777, Springer-Verlag, Heidelberg, 1980, pp. 221–227.

[D'H] E. D'Hoker, Decompositions of representations into basis representations for the classical groups, J. Math. Physics **25** (1984), 1–12.

[Di1] J. Dieudonné, *Sur les Groupes Classiques*, Hermann, Paris, 1967.

[Di2] J. Dieudonné, Schur functions and group representations, in *Young tableaux and Schur functors in algebra and geometry*, *Astérisque* **87–88** (1981), 7–19.

[Dia] P. Diaconis, *Group Representations in Probability and Statistics*, Institute of Mathematical Statistics, Hayward, CA, 1988.

[Don] R. Donagi, On the geometry of Grassmannians, *Duke Math. J.* **44** (1977), 795–837.

[Dor] L. Dornhoff, *Group Representation Theory*, Parts *A* and *B*, Marcel Dekker, New York, 1971, 1972.

[Dr] D. Drucker, Exceptional Lie algebras and the structure of hermitian symmetric spaces, *Mem. Amer. Math. Soc.* **208** (1978).

[Dy-O] E. B. Dynkin and A. L. Oniščik, Compact global Lie groups, *Amer. Math. Soc. Transl., Series 2* **21** (1962), 119–192.

[ES-K] N. El Samra and R. C. King, Reduced determinantal forms for characters of the classical Lie groups, *J. Phys. A: Math. Gen.* **12** (1979), 2305–2315.

[Foa] D. Foata (ed.), *Combinatoire et Représentation du Groupe Symétrique, Strasbourg 1976*, Springer Lecture Notes 579, Springer-Verlag, Heidelberg, 1977.

[Fr1] H. Freudenthal, *Oktaven, Ausnahmegruppen und Oktavengeometrie*, Mathematisch Instituut der Rijksuniversiteit te Utrecht, Utrecht, 1951, 1960.

[Fr2] H. Freudenthal, Lie groups in the foundations of geometry, *Adv. Math.* **1** (1964), 145–190.

[Fr-dV] H. Freudenthal and H. de Vries, *Linear Lie Groups*, Academic Press, New York, 1969.

[Fro1] F. G. Frobenius, Über die Charaktere der symmetrischen Gruppe, *Sitz. König. Preuss. Akad. Wissen.* (1900), 516–534; Gesammelte Abhandlungen III, Springer-Verlag, Heidelberg, 1968, pp. 148–166.

[Fro2] F. G. Frobenius, Über die Charaktere der alternirenden Gruppe, *Sitz. König. Preuss. Akad. Wissen.* (1901), 303–315; Gesammelte Abhandlungen III, Springer-Verlag, Heidelberg, 1968, pp. 167–179.

[Fu] W. Fulton, *Intersection Theory*, Springer-Verlag, New York, 1984.

[G-H] P. Griffiths and J. Harris, *Principles of Algebraic Geometry*, Wiley-Interscience, New York, 1978.

[Gil] R. Gilmore, *Lie Groups, Lie Algebras, and Some of Their Applications*, Wiley, New York, 1974.

[G-N-W] C. Greene, A. Nijenhuis, and H. S. Wilf, A probabilistic proof of a formula for the number of Young tableaux of a given shape, *Adv. Math.* **31** (1979), 104–109.

[Gr] J. A. Green, The characters of the finite general linear group, *Trans. Amer. Math. Soc.* **80** (1955), 402–447.

[Gre] M. L. Green, Koszul cohomology and the geometry of projective varieties, I, II, *J. Diff. Geom.* **19** (1984), 125–171; **20** (1984), 279–289.

[Grie] R. L. Griess, Automorphisms of extra special groups and nonvanishing degree 2 cohomology, *Pacific J. Math.* **48** (1973), 403–422.

[Ha] J. Harris, *Algebraic Geometry*, Springer-Verlag, New York, to appear.

[Ham] M. Hamermesh, *Group Theory and its Application to Physical Problems*, Addison-Wesley, Reading, MA, 1962 and Dover, 1989.

[Har] G. H. Hardy, *Ramanujan*, Cambridge University Press, Cambridge, MA, 1940.

[Hel] S. Helgason, *Differential Geometry, Lie Groups, and Symmetric Spaces*, Academic Press, New York, 1978.

[Her] R. Hermann, *Spinors, Clifford and Cayley Algebras*, Interdisciplinary Mathematics Volume VII, 1974.

[Hi] H. Hiller, *Geometry of Coxeter Groups*, Pitman, London, 1982.

[Ho1] R. Howe, Very basic Lie theory, *Amer. Math. Monthly* **90** (1983), 600–623; **91** (1984), 247.

[Ho2] R. Howe, The classical groups and invariants of binary forms, *Proc. Symp. Pure Math.* **48** (1988), 133–166.

[Ho3] R. Howe, Remarks on classical invariant theory, *Trans. Amer. Math. Soc.* **313** (1989), 539–569.

[H-P] W. V. D. Hodge and D. Pedoe, *Methods of Algebraic Geometry*, vols. 1 and 2, Cambridge University Press, Cambridge, MA, 1947, 1952; 1968.

[H-S] M. Hausner and J. T. Schwarz, *Lie groups*; *Lie algebras*, Gordon and Breach, New York, 1968.

[Hu1] J. E. Humphreys, *Introduction to Lie Algebras and Representation Theory*, Springer-Verlag, New York, 1972, 1980.

[Hu2] J. E. Humphreys, *Linear Algebraic Groups*, Springer-Verlag, New York, 1975, 1981.

[Hu3] J. E. Humphreys, Representations of SL(2, *p*), *Amer. Math. Monthly* **82** (1975), 21–39.

[Hur] A. Hurwitz, Über die Anzahl der Riemann'schen Flächen mit gegebenen Verzweigungspunkten, *Math. Ann.* **55** (1902), 53–66.

[Hus] D. Husemoller, *Fibre Bundles*, second edition, Springer-Verlag, New York, 1975.

[In] R. E. Ingram, Some characters of the symmetric group, *Proc. Amer. Math. Soc.* **1** (1950), 358–369.

[Iv] B. Iversen, The geometry of algebraic groups, *Adv. Math.* **20** (1976), 57–85.

[Jac1] N. Jacobson, *Lie Algebras*, Wiley, New York 1962, and Dover, 1979.

[Jac2] N. Jacobson, *Exceptional Lie Algebras*, Marcel Dekker, New York 1971.

[Jac3] N. Jacobson, Cayley numbers and simple Lie algebras of type G, *Duke Math. J.* **5** (1939), 775–783.

[Jac4] N. Jacobson, Triality and Lie algebras of type D_4, *Rend. Circ. Mat. Palermo* (2) **13** (1964), 129–153.

[Ja-Ke] G. James and A. Kerber, *The Representation Theory of the Symmetric Group*, Encyclopedia of Mathematics and Its Applications, vol. 16, Addison-Wesley, Reading, MA, 1981.

[Jam] G. D. James, *The Representation Theory of the Symmetric Groups*, Springer Lecture Notes 682, Springer-Verlag, Heidelberg, 1978.

[J-L] T. Józefiak and A. Lascoux (eds.), *Young Tableaux and Schur Functors in Algebra and Geometry, Toruń, Poland* 1980, *Astérisque* 87–88, 1981.

[Ke] A. Kerber, *Representations of Symmetric Groups I*, Springer Lecture Notes 240, Springer-Verlag, Heidelberg, 1971.

[Kem] G. Kempf, Tensor products of representations, *Amer. J. Math.* **109** (1987), 401–415.

[Ki1] R. C. King, The dimensions of irreducible tensor representations of the orthogonal and symplectic groups, *Can. J. Math.* **23** (1971), 176–188.

[Ki2] R. C. King, Modification rules and products of irreducible representations of the unitary, orthogonal and symplectic groups, *J. Math. Phys.* **12** (1971), 1588–1598.

[Kir] A. A. Kirillov, *Elements of the Theory of Representations*, Springer-Verlag, New York, 1976.

[Kl] A. U. Klymyk, Multiplicities of weights of representations and multiplicities of representations of semisimple Lie algebras, *Sov. Math. Dokl.* **8** (1967), 1531–1534.

[K-N] G. Kempf and L. Ness, Tensor products of fundamental representations, *Can. J. Math.* **40** (1988), 633–648.

[Kn] D. Knutson, *λ-Rings and the Representation Theory of the Symmetric Group*, Springer Lecture Notes 308, Springer-Verlag, Heidelberg 1973.

[Kos] B. Kostant, A formula for the multiplicity of a weight, *Trans. Amer. Math. Soc.* **93** (1959), 53–73.

[Ko-Te] K. Koike and I. Terada, Young-diagrammatic methods for the representation theory of the classical groups of type B_n, C_n, and D_n, *J. Algebra* **107** (1987), 466–511.

[Ku] J. P. S. Kung (ed.), *Young Tableaux in Combinatorics, Invariant Theory, and Algebra*, Academic Press, New York, 1982.

[Kum1] S. Kumar, Proof of the Parthasarathy–Ranga Rao–Varadarajan conjecture, *Invent. Math.* **93** (1988), 117–130.

[Kum2] S. Kumar, A refinement of the PRV conjecture, *Invent. Math.* **97** (1989), 305–311.

[Le] W. Ledermann, *Introduction to Group Characters*, Cambridge University Press, Cambridge, MA, 1977.

[Li] P. Littelmann, A Littlewood–Richardson rule for classical groups, *C. R. Acad. Sci. Paris* **306** (1988), 299–303.

[LIE] Séminaire Sophus LIE 1954/1955, *Théorie des Algèbres de Lie, Topologie des groupes de Lie*, École Normale Supérieure, Paris, 1955.

[Lit1] D. E. Littlewood, *The Theory of Group Characters and Matrix Representations of Groups*, second ed., Oxford University Press, Oxford, 1950.

[Lit2] D. E. Littlewood, *A University Algebra*, William Heinemann Ltd, London, 1950; second ed., 1958, and Dover, 1970.

[Lit3] D. E. Littlewood, On invariants under restricted groups, *Philos. Trans. Roy. Soc. A* **239** (1944), 387–417.

[Liu] A. Liulevicius, Arrows, symmetries and representation rings, *J. Pure Appl. Algebra* **19** (1980), 259–273.

[L-M] H. B. Lawson and M.-L. Michelson, *Spin Geometry*, Princeton University Press, Princeton, NJ, 1989.

[L-M-S] V. Lakshmibai, C. Musili, and C. S. Seshadri, Geometry of *G/P*, *Bull. Amer. Math. Soc.* **1** (1979), 432–435.

[Lo] Loos, *Symmetric Spaces*, W. A. Benjamin, New York, 1969.

[L-S] A. Lascoux and M. P. Schützenberger, *Formulaire raisonné de fonctions symmetriques*, U. E. Maths, Paris VII, L.A. 248, 1985.

[L-T] G. Lancaster and J. Towber, Representation-functors and flag-algebras for the classical groups I, II, *J. Algebra* **59** (1979), 16–38; **94** (1985), 265–316.

[L-VdV] R. Lazarsfeld and A. Van de Ven, *Topics in the Geometry of Projective Space*, DMV Seminar Band 4, Birkhäuser, Boston, MA, 1984.

[L-V] R. A. Liebler and M. R. Vitale, Ordering the partition characters of the symmetric group, *J. Algebra* **25** (1973), 487–489.

[Mac] I. G. Macdonald, *Symmetric Functions and Hall Polynomials*, Clarendon Press, Oxford, 1979.

[Mack] G. W. Mackey, *Introduction to The Racah–Wigner Algebra in Quantum Theory*, by L. C. Biedenharn and J. D. Louck, Encyclopedia of Mathematics and Its Applications, vol. 9, Addison-Wesley, Reading, MA, 1981.

[M-S] G. Musili and C. S. Seshadri, *Standard monomial theory*, Springer Lecture Notes **867** (1981), 441–476.

[Mur1] F. D. Murnaghan, *The Theory of Group Representations*, The Johns Hopkins Press, Baltimore, 1938.

[Mur2] F. D. Murnaghan, *The Unitary and Rotation Groups*, Spartan Books, Washington, DC, 1962.

[No] K. Nomizu, *Lie Groups and Differential Geometry*, Mathematics Society of Japan, Tokyo, 1956.

[N-S] M. A. Naimark and A. I. Stern, *Theory of Group Representations*, Springer-Verlag, New York, 1982.

[Pe1] M. H. Peel, Hook representations of symmetric groups, *Glasgow Math. J.* **12** (1971), 136–149.

[Pe2] M. H. Peel, Specht modules and the symmetric groups, *J. Algebra* **36** (1975), 88–97.

[Por] I. R. Porteous, *Topological Geometry*, second edition, Cambridge University Press, Cambridge, MA, 1981.

[Pos] M. Postnikov, *Lie Groups and Lie Algebras*, MIR, Moscow 1986.

[Pr] C. Procesi, *A Primer of Invariant Theory*, Brandeis Lecture Notes 1, 1982.

[P-S] A. Pressley and G. Segal, *Loop Groups*, Clarendon Press, Oxford, 1986.

[P-W] P. Pragacz and J. Weyman, On the construction of resolutions of determinantal ideals: a survey, Springer Lecture Notes 1220 (1986), 73–92.

[Qu] D. Quillen, The mod 2 cohomology rings of extra-special 2-groups and the spinor groups, *Math. Ann.* **194** (1971), 197–212.

[Ra] G. Racah, Lectures on Lie groups, in *Group Theoretical Concepts and Methods in Elementary Particle Physics*, Gordon and Breach, New York, 1964, 1–36.

[Sc] R. D. Schafer, *An Introduction to Nonassociative Algebras*, Academic Press, New York, 1966.

[Sch] G. W. Schwarz, On classical invariant theory and binary cubics, *Ann. Inst. Fourier* **37** (1987), 191–216.

[Se1] J-P. Serre, *Lie Algebras and Lie Groups*, W. A. Benjamin, New York, 1965.

[Se2] J-P. Serre, *Linear Representations of Finite Groups*, Springer-Verlag, New York, 1977.

[Se3] J-P. Serre, *Complex Semi-simple Lie Algebras*, Springer-Verlag, New York, 1987.

[S-K] M. Sato and T. Kimura, A classification of irreducible prehomogeneous vector spaces and their relative invariants, *Nagoya Math. J.* **65** (1977), 1–155.

[Sp] T. A. Springer, *Invariant Theory*, Springer Lecture Notes 565, Springer-Verlag, Heidelberg, 1977.

[Sta] R. P. Stanley, Theory and Application of Plane Partitions, Parts 1 and 2, *Studies Appl. Math.* **1** (1971), 167–188, 259–279.

[Ste1] R. Steinberg, The representations of GL(3, q), GL(4, q), PGL(3, q), and PGL(4, q), *Can. J. Math.* **3** (1951), 225–235.

[Ste2] R. Steinberg, *Conjugacy classes in Algebraic Groups*, Springer Lecture Notes 366, Springer-Verlag, Heidelberg, 1974.

[S-W] D. H. Sattinger and O. L. Weaver, *Lie Groups and Algebras with Applications to Physics, Geometry, and Mechanics*, Springer-Verlag, New York, 1986.

[Ti1] J. Tits, Groupes simples et géométries associées, *Proc. Intern. Cong. Math. Stockholm* (1962), 197–221.

[Ti2] J. Tits, Sur les constantes de structure et le théorème d'existence des algèbres de Lie semi-simples, *Publ. Math.* I.H.E.S. **31** (1965), 21–58.

[To] M. L. Tomber, Lie algebras of type F, *Proc. Amer. Math. Soc.* **4** (1953), 759–768.

[Tow1] J. Towber, Two new functors from modules to algebras, *J. Algebra* **47** (1977), 80–104.

[Tow2] J. Towber, Young symmetry, the flag manifold, and representations of GL(n), *J. Algebra* **61** (1979), 414–462.

[Va] V. S. Varadarajan, *Lie Groups, Lie Algebras, and Their Representations*, Springer-Verlag, New York, 1974, 1984.

[vdW] B. L. van der Waerden, Reihenentwicklungen und Überschiebungen in der Invariantentheorie, insbesondere im quarternären Giebiet, *Math. Ann.* **113** (1936), 14–35.

[Vu] T. Vust, Sur la théorie des invariants des groupes classiques, *Ann. Inst. Fourier* **26** (1976), 1–31.

[Wa] Z-X. Wan, *Lie Algebras*, Pergamon Press, New York, 1975.

[We1] H. Weyl, *Classical Groups*, Princeton University Press, Princeton, NJ, 1939; second edition, 1946.

[We2] H. Weyl, Über Algebren, die mit der Komplexgruppe in Zusammenhang stehen, und ihre Darstellungen, *Math. Zeit.* **35** (1932), 300–320.

[Ze] A. V. Zelevinsky, *Representations of Finite Classical Groups*, Springer Lecture Notes 869, Springer-Verlag, Heidelberg 1981.

[Žel] D. P. Želobenko, *Compact Lie Groups and Their Representations*, Translations of Mathematical Monographs, vol. 40, American Mathematical Society, Providence, RI, 1973.

Index of Symbols

Index